PHYSICOCHEMICAL
AND ENVIRONMENTAL
PLANT PHYSIOLOGY

PHYSICOCHEMICAL AND ENVIRONMENTAL PLANT PHYSIOLOGY

Park S. Nobel

Department of Biology
University of California, Los Angeles
Los Angeles, California

ACADEMIC PRESS, INC.

 Harcourt Brace Jovanovich, Publishers

San Diego   New York   Boston

London   Sydney   Tokyo   Toronto

Copyright © 1991, 1983, 1974, 1970 by Academic Press, Inc.
All Rights Reserved.
No part of this publication may be reproduced or transmitted in any form or
by any means, electronic or mechanical, including photocopy, recording, or
any information storage and retrieval system, without permission in writing
from the publisher.

Academic Press, Inc.
San Diego, California 92101

United Kingdom Edition published by
Academic Press Limited
24–28 Oval Road, London NW1 7DX

Library of Congress Cataloging-in-Publication Data

Nobel, Park S.
    Physicochemical and environmental plant physiology / Park S.
Nobel.
        p.   cm.
    Includes bibliographical references.
    ISBN   0-12-520020-X   (alk. paper). -- ISBN   0-12-520021-8   (pbk.:
alk. paper)
    1. Plant physiology. 2. Botanical chemistry. 3. Plant
physiological ecology. 4. Plant cells and tissues.   I. Title.
QK711.2.N623   1991
581.1--dc20
                                                        90-580
                                                        CIP

Printed in the United States of America
91 92 93 94   9 8 7 6 5 4 3 2 1

Contents

Preface

Plant physiology is the study of cellular, organ, and organism function—we will specifically consider water relations, solute transport, photosynthesis, transpiration, respiration, and environmental interactions. A physiologist endeavors to understand such topics in physical and chemical terms; accurate models can then be constructed and responses to the internal and the external environment predicted. A primary objective of this book is to use elementary chemistry, physics, and mathematics to explain and develop concepts that are key to an understanding of various areas of plant physiology in particular and physiology in general. The intent is to provide a rigorous development, not a compendium of facts. Numerous references provide further details, although in some cases the enunciated principles carry the reader to the forefront of current research. Calculations are used to indicate the physiological consequences of the various equations, and problems at the end of chapters provide further such exercises.

Chapters 1 through 3 describe water relations and ion transport for plant cells. In Chapter 1, after discussing the concept of diffusion, we consider the physical barriers to diffusion imposed by cellular and organelle membranes. Another physical barrier associated with plant cells is the cell wall, which limits the size of the cells. In the treatment of the movement of water through cells in response to specific forces in Chapter 2, we employ the thermodynamic argument of chemical potential gradients. Chapter 3 considers solute movement into and out of plant cells, leading to an explanation of electrical potential differences across membranes and establishing the formal criteria for distinguishing diffusion from active transport. Based on concepts from irreversible thermodynamics, an important parameter called the reflection coefficient is derived, which permits a precise evaluation of the influence of osmotic pressures on flow. The thermodynamic arguments used to describe ion and water movements are equally applicable to animal cells.

The next three chapters deal primarily with the interconversion of various forms of energy. In Chapter 4 we consider the properties of light and its absorption. After light is absorbed, its radiant energy usually is rapidly converted to heat. However, the arrangement of photosynthetic pigments and their special molecular structures allow some radiant energy from the sun to be converted by plants into chemical energy. In Chapter 5 we discuss the particular features of chlorophyll and the accessory pigments for photosynthesis that allow this energy conversion. Light energy absorbed by chloroplasts leads to the formation of ATP and NADPH. These compounds represent currencies for carrying chemical and electrical (redox potential) energy, respectively. How much energy they actually carry is discussed in Chapter 6.

In the last three chapters we consider the various forms in which energy and matter enter and leave a plant as it interacts with its environment. The physical quantities involved in an energy budget analysis are presented in Chapter 7 so that the relative importance of the various factors affecting the temperature of leaves or other plant parts can be evaluated quantitatively. The resistances (or their reciprocals, conductances) affecting the movement of both water vapor during transpiration and carbon dioxide during photosynthesis are discussed in detail for leaves in Chapter 8. The movement of water from the soil through the plant to the atmosphere is discussed quantitatively in Chapter 9. Because these and other topics depend on material introduced elsewhere in the book, the text is extensively cross-referenced.

This text has evolved from *Biophysical Plant Physiology and Ecology* (W. H. Freeman, 1983), which evolved from *Introduction to Biophysical Plant Physiology* (W. H. Freeman, 1974), which in turn evolved from *Plant Cell Physiology: A Physicochemical Approach* (W. H. Freeman, 1970). Some of the changes in the most recent version involve the deletion of certain nonstandard terminology and approaches. The text has been updated based on the ever increasing quality of plant research, such as on cell growth and water relations, membrane channels, mechanisms of active transport, the bioenergetics of chloroplasts and mitochondria, stomatal control, and CO_2 diffusion within leaves. One-third of the figures are new or modified, nearly 200 new references are incorporated, the appendices on constants and conversion factors are doubled in the number of entries, and the solutions to problems are given for the first time. Many other changes have emanated from the best laboratory for any book, the classroom.

The author gratefully acknowledges stimulating discussions with many colleagues, including Julian Schroeder and Philip Thornber, as well as the researchers and publishers who granted permission to use their figures. The text was read and substantially improved by the comments of David Alm, Ove Hoegh-Guldberg, Michael Loik, Francis Lopez, Gretchen North, and David Tissue. Marjorie Macdonald and Carol Lee did an outstanding job on the manuscript preparation.

Park S. Nobel
September 25, 1990

Symbols and Abbreviations

Where appropriate, typical units are indicated in parentheses.

Quantity	Description
a	absorptance or absorptivity (dimensionless)
a^{st}	mean area of stomata (m^{-2})
a_{IR}	absorptance or absorptivity in infrared region (dimensionless)
a_j	activity of species j (same as concentration)*
a.t.	subscript indicating active transport
Å	angström (10^{-10} m)
A	electron acceptor
A	area (m^2)
A	absorbance (also called "optical density") (dimensionless)
ABA	abscisic acid
ADP	adenosine diphosphate
ATP	adenosine triphosphate
b	nonosmotic volume (m^3)
b	optical path length (m)
bl	superscript for boundary layer
c	centi (as a prefix), 10^{-2}
c	superscript for cuticle
c_d	drag coefficient (dimensionless)
c_j	concentration of species j (mol m^{-3})†
$\bar{c}_s$	a mean concentration of solute s
cal	calorie
chl	superscript for chloroplast
clm	superscript for chloroplast limiting membranes
cw	superscript for cell wall
cyt	superscript for cytosol

---

*The activity, a_j, is often considered to be dimensionless, in which case the activity coefficient, γ_j, has the units of reciprocal concentration ($a_j = \gamma_j c_j$, Eq. 2.5).

†We note that mol litre^{-1}, or molarity (M), is a concentration unit of widespread use, although it is not an SI unit.

C	superscript for conduction
C	capacitance, electrical (F)
C^j	capacitance for water storage in component j (m^3 MPa^{-1})
C'	capacitance/unit area (F m^{-2})
Chl	chlorophyll
Cl	subscript for chloride ion
C_P	volumetric heat capacity (J m^{-3} °C^{-1})
Cyt	cytochrome
d	deci (as a prefix), 10^{-1}
d	depth or distance (m)
d	diameter (m)
dyn	dyne
D	electron donor
D	dielectric constant (dimensionless)
D_j	diffusion coefficient of species j (m^2 s^{-1})
e	electron
e	superscript for water evaporation site
e_{IR}	emissivity or emittance in infrared region (dimensionless)
eV	electron volt
E	light energy (J)
E	kinetic energy (J)
E	electrical potential (mV)
E_j	redox potential of species j (mV)
$E_j^{*,H}$	midpoint redox potential of species j referred to standard hydrogen electrode (mV)
E_M	electrical potential difference across a membrane (mV)
E_{N_j}	Nernst potential of species j (mV)
f	femto (as a prefix), 10^{-15}
F	farad
F	Faraday's constant (coulomb mol^{-1})
F	average cumulative leaf area/ground area (dimensionless)
FAD	flavin adenine dinucleotide (oxidized form)
FADH$_2$	reduced form of flavin adenine dinucleotide
FMN	flavin mononucleotide
g	gram
g_j	conductance of species j (mm s^{-1} with Δc_j and mmol m^{-2} s^{-1} with ΔN_j)

G	giga (as a prefix), 10^9
G	Gibbs free energy (J)
Gr	Grashof number (dimensionless)
G/n_j	Gibbs free energy/mole of some product or reactant (J mol^{-1})
h	hour
ha	hectare
h	height (m)
h_c	heat convection coefficient (W m^{-2} °C^{-1})
hv	a quantum of light energy
H	subscript for heat
i	superscript for inside
i	electrical current (ampere)
ias	superscript for intercellular air spaces
in	superscript for inward
in vitro	in a test tube, beaker, flask (literally, in glass)
in vivo	in a living organism (literally, in the living)
I	electrical current (ampere)
IR	infrared
j	subscript for species j
J	joule
J_j	flux density of species j (mol m^{-2} s^{-1})
J_j^{in}	inward flux density (influx) of species j (mol m^{-2} s^{-1})
J_j^{out}	outward flux density (efflux) of species j (mol m^{-2} s^{-1})
J_{V_j}	volume flux density of species j (m^3 m^{-2} s^{-1}, i.e., m s^{-1})
J_V	total volume flux density (m s^{-1})
k	kilo (as a prefix), 10^3
k	foliar absorption coefficient (dimensionless)
k_j	first-order rate constant for the jth process (s^{-1})
K	temperature on kelvin scale
K	subscript for potassium ion
K	equilibrium constant (concentration raised to some power)
K_h	hydraulic conductance per unit length (m^4 MPa^{-1} s^{-1})
K^j	thermal conductivity coefficient of region j (W m^{-1} °C^{-1})
K_j	partition coefficient of species j (dimensionless)
K_j	concentration for half-maximal uptake rate of species j (Michaelis constant) (mol m^{-3}, or M)
K_j	eddy diffusion coefficient of gaseous species j (m^2 s^{-1})

$K_{\text{pH 7}}$	equilibrium constant at pH 7
l	litre
l	superscript for lower
l	length (m), e.g., mean distance across leaf in wind direction
ln	natural or Napierian logarithm (to the base e, where e is 2.71828 . . .)
log	common or Briggsian logarithm (to the base 10)
L^{soil}	hydraulic conductivity coefficient of the soil (m^2 Pa^{-1} s^{-1})
L_{jk}	Onsager or phenomenological coefficient (flux density per unit force)
L_P	hydraulic conductivity coefficient (in irreversible thermodynamics) (m Pa^{-1} s^{-1})
L_w	water conductivity coefficient (m Pa^{-1} s^{-1})
m	milli (as a prefix), 10^{-3}
m	metre
m_j	mass per mole of species j (molar mass)(kg mol^{-1})
max	subscript for maximum
memb	superscript for membrane
mes	superscript for mesophyll
min	subscript for minimum
min	minute
mol	mole, a mass equal to the molecular weight of the species in grams; contains Avogadro's number of molecules
M	mega (as a prefix), 10^6
M	molar (mol litre^{-1})
M_j	amount of species j per unit area (mol m^{-2})
n	nano (as a prefix), 10^{-9}
n	number of stomata per unit area
$n(E)$	number of moles with energy of E or greater
n_j	amount of species j (mol)
N	newton
Na	subscript for sodium ion
NAD$^+$	nicotinamide adenine dinucleotide (oxidized form)
NADH	reduced form of nicotinamide adenine dinucleotide
NADP$^+$	nicotinamide adenine dinucleotide phosphate (oxidized form)
NADPH	reduced form of nicotinamide adenine dinucleotide phosphate
N_j	mole fraction of species j (dimensionless)
Nu	Nusselt number (dimensionless)

o	superscript for outside
0	subscript for initial value (at $t = 0$)
out	superscript for outward
p	pico (as a prefix), 10^{-12}
p	period (s)
pH	$-\log(a_{H^+})$
pl	superscript for plasmalemma
ps	superscript for photosynthesis
P	pigment
P	hydrostatic pressure (MPa)
Pa	pascal
P_j	permeability coefficient of species j (m s^{-1})
P_j	partial pressure of gaseous species j (kPa)
PPFD	photosynthetic photon flux density (400 to 700 nm)
q	number of electrons transferred per molecule (dimensionless)
Q	charge (coulomb)
Q_{10}	temperature coefficient (dimensionless)
r	superscript for respiration
r	radius (m)
r	reflectivity (dimensionless)
r + pr	superscript for respiration plus photorespiration
r_j	resistance for gaseous species j (s m^{-1})
R	electrical resistance (ohm)
R	gas constant (J mol^{-1} K^{-1})
R^j	resistance of component j across which water moves as a liquid (MPa s m^{-3})
Re	Reynolds number (dimensionless)
RH	relative humidity (%)
s	subscript for solute
s	second
s_j	amount of species j (mol)
st	superscript for stoma(ta)
surf	superscript for surface
surr	superscript for surroundings
S	singlet
$S_{(\pi,\pi)}$	singlet ground state
$S_{(\pi,\pi*)}$	singlet excited state where a π electron has been promoted to a $\pi*$ orbital

S	magnitude of net spin (dimensionless)
S	total flux density of solar irradiation, i.e., global irradiation (W m^{-2})
t	time (s)
ta	superscript for turbulent air
T	superscript for transpiration
T	triplet
$T_{(\pi,\pi*)}$	excited triplet state
T	temperature (K, °C)
u	superscript for upper
u_j	mobility of species j (velocity per unit force)
u_+	mobility of monovalent cation
u_-	mobility of monovalent anion
U	kinetic energy (J mol^{-1})
U_B	minimum kinetic energy to cross barrier (J mol^{-1})
UV	ultraviolet
v	magnitude of velocity (m s^{-1})
v	wind speed (m s^{-1})
v^{wind}	wind speed (m s^{-1})
v_j	magnitude of velocity of species j (m s^{-1})
v_{CO_2}	rate of photosynthesis per unit volume (mol m^{-3} s^{-1})
vs.	versus
V	volt
V	subscript for volume
V	volume (m^3)
$\bar{V}_j$	partial molal volume of species j (m^3 mol^{-1})
V_{max}	maximum rate of CO$_2$ fixation (mol m^{-3} s^{-1})
w	subscript for water
wv	subscript for water vapor
W	watt (J s^{-1})
x	distance (m)
z	altitude (m)
z_j	charge number of ionic species j (dimensionless)
α	contact angle (°)
γ_j	activity coefficient of species j (dimensionless, but see a_j)
$\gamma_\pm$	mean activity coefficient of cation–anion pair (dimensionless)
δ	delta, a small quantity of something, e.g., δ^- refers to a small fraction of an electronic charge

δ	distance (m)
δ^{bl}	thickness of air boundary layer (mm)
Δ	delta, the difference or change in the quantity that follows it
ε	volumetric elastic modulus (MPa)
ε_λ	absorption coefficient at wavelength λ (m^2 mol^{-1})
η	viscosity (N s m^{-2}, Pa s)
λ	wavelength of light (nm)
λ_{max}	wavelength position for the maximum absorption coefficient in an absorption band or for the maximum photon (or energy) emission in an emission spectrum
μ	micro (as a prefix), 10^{-6}
μ_j	chemical potential of species j (J mol^{-1})
ν	frequency of electromagnetic radiation (s^{-1}, hertz)
ν	kinematic viscosity (m^2 s^{-1})
π	ratio of circumference to diameter of a circle (3.14159)
π	an electron orbital in a molecule, or an electron in such an orbital
π^*	an excited or antibonding electron orbital in a molecule, or an electron in such an orbital
Π	total osmotic pressure (MPa)
Π_j	osmotic pressure of species j (MPa)
Π_s	osmotic pressure due to solutes (MPa)
ρ	density (kg m^{-3})
ρ	resistivity, electrical (ohm m)
ρ^j	hydraulic resistivity of component j (MPa s m^{-2})
σ	surface tension (N m^{-1})
σ	reflection coefficient (dimensionless)
σ_j	reflection coefficient of species j (dimensionless)
σ_L	longitudinal stress (MPa)
σ_T	tangential stress (MPa)
τ	matric pressure (MPa)
τ	lifetime (s)
τ_j	lifetime for the jth de-excitation process (s)
ϕ_j	osmotic coefficient of species j (dimensionless)
Φ_i	quantum yield or efficiency for ith de-excitation pathway (dimensionless)
Ψ	water potential (MPa)
Ψ_Π	osmotic potential (MPa)
$^\circ C$	degree Celsius (see p. 17)

°	angular degree
*	superscript for a standard or reference state
*	superscript for a molecule in an excited electronic state
*	superscript for saturation of air with water vapor
∞	infinity

CHAPTER 1

Cells and Diffusion

CELL STRUCTURE

Before formally considering diffusion and other such topics, we will outline the structure of certain plant cells and tissues, thus introducing most of the anatomical terms used throughout the book.

Generalized Plant Cell

Figure 1.1 depicts a representative leaf cell from a higher plant and illustrates the larger subcellular structures observed. Surrounding the *protoplast* is the *cell wall,* composed of cellulose and other polysaccharides; these polymers help provide rigidity to individual cells as well as to the whole plant. Because the cell wall contains numerous relatively large interstices, it does not serve as the main permeability barrier to the entry of water or small solutes into plant cells. The main barrier, a cell membrane known as the plasma membrane or *plasmalemma,* occurs inside the cell wall and surrounds the *cytoplasm.* The permeability of this membrane varies with the particular solute, so the plasmalemma can regulate what enters and leaves a plant cell. The cytoplasm contains organelles like *chloroplasts* and *mitochondria,* distinct membrane-surrounded compartments in which energy can be converted from one form to another. Also in the cytoplasm are microbodies such as peroxisomes, numerous ribosomes, and proteins, as well as many other macromolecules and structures that influence the

1

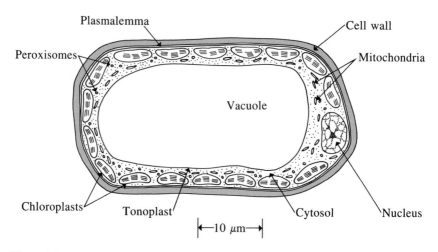

Plasmalemma Cell wall Peroxisomes Mitochondria Vacuole Chloroplasts Tonoplast Cytosol Nucleus

|←—10 μm—→|

Figure 1.1

Schematic representation of a mature mesophyll cell from the leaf of a higher plant, suggesting some of the complexity resulting from the presence of many membrane-surrounded subcellular bodies.

thermodynamic properties of water. The term *cytoplasm* includes the organelles (but generally not the nucleus), whereas the increasingly used term *cytosol* refers to the cytoplasmic solution delimited by the plasmalemma and the tonoplast (to be discussed next) but exterior to all the organelles.

In mature cells of higher plants and many lower (less advanced) plants, there is a large central aqueous phase, the *vacuole,* which is surrounded by the *tonoplast*. This membrane is usually quite large in area, because the central vacuole can occupy up to about 90% of the volume of a mature cell. The aqueous solution in the central vacuole contains mainly inorganic ions or organic acids as solutes, although considerable amounts of sugars and amino acids may be present in some species. Water uptake into this vacuole occurs during cell growth (for methods to isolate various cell components, see Linskens and Jackson, 1985; for comments on membranes, see Kotyk *et al.*, 1988).

One immediate impression of plant cells is the great prevalence of membranes. In addition to surrounding the cytoplasm, membranes also separate various compartments in the cytoplasm. Diffusion of substances across these membranes is much more difficult than is diffusion within the compartments. Thus organelle and vacuolar membranes can control the contents and consequently the reactions occurring in the particular compartments that they surround. Diffusion can also impose limitations on the overall size of a cell, because the time for diffusion increases with the square of the distance, as we will quantitatively consider in the next section.

Although many plant cells share most of the features indicated in Figure 1.1, they are remarkably diverse in size. The nearly spherical cells of the green alga *Chlorella* are approximately 4×10^{-6} metre (4 μm) in diameter. On the other hand, some species of the intertidal green alga *Valonia* have multinucleated cells as large as 20 mm in diameter. The genera *Chara* and *Nitella* include fresh and brackish water green algae having large internodal cells (Fig. 3.9) that may be 100 mm long and 1 mm in diameter. Such giant algal cells have proved extremely useful for studying ion fluxes, as we will consider in Chapter 3.

Leaf Cells

A transverse section of a leaf can illustrate various cell types and anatomical features that are important for photosynthesis and transpiration. Leaves are generally 4 to 10 cells thick, which corresponds to a few hundred μm (Fig. 1.2). An *epidermis* occurs on both the upper and the lower sides of a leaf and is usually one layer of cells thick. Except for the *guard cells* (Fig. 1.2), the

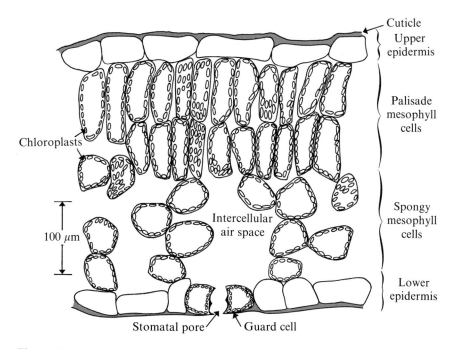

Figure 1.2

Schematic transverse section through a leaf, indicating the arrangement of various cell types. There are often about 30 to 40 mesophyll cells per stoma.

cytoplasm of epidermal cells generally is colorless because very few if any chloroplasts are present there (depending on the plant species). Epidermal cells have a relatively thick waterproof *cuticle* on the atmospheric side (Fig. 1.2). The cuticle contains *cutin,* a diverse group of complex polymers composed principally of esters of 16- and 18-carbon monocarboxylic acids that have two or three hydroxyl groups. Cutin is relatively inert and also resists enzymatic degradation by microorganisms; thus it is often well preserved in fossil material. We will be mainly concerned with its role in preventing water loss from a leaf.

Between the two epidermal layers is the *mesophyll* tissue, which is usually differentiated into chloroplast-containing "palisade" and "spongy" cells. The palisade cells are often elongated at right angles to the upper epidermis and are found immediately beneath it (Fig. 1.2). The spongy mesophyll cells, between the palisade mesophyll cells and the lower epidermis, are loosely packed, and intercellular air spaces are conspicuous. In fact, most of the surface area of both spongy and palisade mesophyll cells is exposed to the air in the intercellular spaces. A spongy mesophyll cell is often rather spherical, is about 40 μm in diameter, and contains approximately 40 chloroplasts. (As Fig. 1.2 illustrates, the cells are by no means geometrically regular, so dimensions here indicate only approximate size.) A neighboring palisade cell is generally more oblong; it can be 80 μm long, can contain 60 chloroplasts, and might be represented by a cylinder 30 μm in diameter with hemispherical ends. In many leaves about 70% of the chloroplasts are in the palisade cells, which generally outnumber the spongy mesophyll cells nearly two to one.

The pathway of least resistance for gases to cross an epidermis—and thus to enter or to exit from a leaf—is through the adjustable space between a pair of guard cells (Fig. 1.2). This pore is called a *stoma* or *stomate* (plural: stomata and stomates, respectively), and it can open and close. The stomatal pores allow for the entry into the leaf of CO_2 and for the exit of photosynthetically produced O_2. The inevitable loss of water vapor by transpiration also occurs mainly through the stomatal pores, as we will discuss in Chapter 8. Stomata thus serve as a control, helping to strike a balance between freely admitting the CO_2 needed for photosynthesis and at the same time preventing excessive loss of the water vapor from the plant. Air pollutants such as ozone (O_3) and sulfur dioxide (SO_2) also enter plants primarily through the open stomata.

Cells of Vascular Tissue

The *xylem* and the *phloem* make up the vascular systems found contiguously in the roots, stems (Fig. 1.3), and leaves of plants. In a tree trunk the phloem

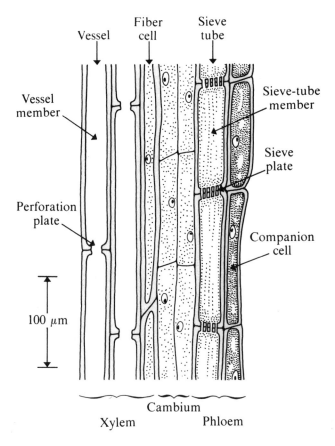

Figure 1.3

Highly idealized longitudinal section through part of a vascular bundle in a stem, illustrating various anatomical aspects of the xylem and the phloem. The new cells forming in the xylem initially contain cytoplasm. Fiber cells, which occur in the xylem, are generally quite tapered and provide structural support; the nucleated companion cells are metabolically involved with the sieve-tube members of the phloem.

constitutes a layer of the bark and the xylem constitutes almost all of the wood. The xylem commonly provides structural support for plants. Water conduction in the xylem of a tree often occurs only in the outermost annual ring,[1] which

---

1. The rings in trees are not always annual. In many desert species a ring forms when large xylem cells are produced after a suitable rainy period followed by smaller cells, and this can occur more than once or sometimes not at all in a given year. Moreover, trees from the wet tropics may have no annual rings.

lies just inside the *vascular cambium* (region of meristematic activity from which xylem and phloem cells differentiate). Outside the functioning phloem are other phloem cells that can be shed as pieces of bark slough off. Phloem external to the xylem, as in a tree, is the general pattern for the stems of plants. As we follow the vascular tissue from the stem along a petiole and into a leaf, we observe that the xylem and the phloem often form a vein, which sometimes conspicuously protrudes from the lower surface of the leaf. Reflecting the orientation in the stem or the trunk, the phloem is found abaxial to the xylem in the vascular tissue of a leaf; i.e., the phloem is located on the side of the lower epidermis. The vascular system branches and rebranches as it crosses a dicotyledonous leaf, becoming smaller (in cross section) at each step. In contrast to the reticulate venation in dicotyledons, monocotyledons characteristically have parallel-veined leaves. Individual mesophyll cells in the leaf are never further than a few cells from the vascular tissue.

The movement of water and nutrients from the soil to the upper portions of a plant occurs primarily in the xylem. The xylem sap generally contains about 10 mol m^{-3} (10 mM)^2 inorganic nutrients plus organic forms of nitrogen that are metabolically produced in the root. The xylem is a tissue of various cell types that we will consider in more detail in the final chapter when water movement in plants is discussed in terms of water potential. The conducting cells in the xylem are the narrow, elongated *tracheids* and the *vessel members,* which tend to be shorter and wider than the tracheids. Vessel members are joined end-to-end in long linear files, their adjoining end walls or *perforation plates* having from one large to many small holes. The conducting cells lose their protoplasts, and the remaining cell walls thus form a low-resistance channel for the passage of solutions. Xylem sap moves from the root, up the stem, through the petiole, and then to the leaves in these hollow xylem "cells," motion occurring in the direction of decreasing hydrostatic pressure. Some solutes leave the xylem along the stem on the way to a leaf, and others diffuse or are actively transported across the plasmalemmas of various leaf cells adjacent to the conducting cells of the xylem.

The movement of most organic compounds throughout the plant takes place in the other vascular tissue, the phloem. A portion of the photosynthetic products made in the mesophyll cells of the leaf diffuses or is actively transported across cellular membranes until it reaches the conducting cells of the leaf phloem. By means of the phloem, the photosynthetic products—which are then often mainly

2. Molarity (moles of solute per litre of solution, symbolized by M) is a useful unit for concentration, but it is not recommended by the international unit convention, Le Système International d'Unités (SI) (Goldman and Bell, 1986). Nevertheless, we will use molarity in addition to the SI unit of mol m^{-3}. We also note that SI recommends the spellings "litre" and "metre."

in the form of sucrose—are distributed throughout the plant. The carbohydrates produced by photosynthesis and certain other substances generally move in the phloem toward regions of lower concentration, although diffusion is not the mechanism for the movement, as we will indicate in Chapter 9. The phloem is a tissue consisting of several types of cells. In contrast to the xylem, however, the conducting cells of the phloem contain cytoplasm. They are *sieve cells* and *sieve-tube members* and are joined end-to-end, thus forming a transport system throughout the plant. Although these phloem cells often contain no nuclei at maturity, they remain metabolically active.

Root Cells

Roots anchor plants in the ground as well as absorb water and nutrients from the soil and then conduct these substances upward to the stem. To help understand uptake into a plant, we will examine the functional zones that occur along the length of a root.

At the extreme tip of a root is the *root cap* (Fig. 1.4a). It consists of relatively

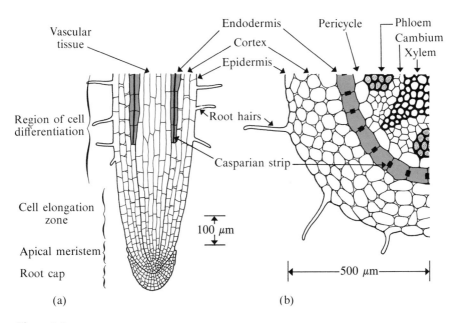

Figure 1.4

Schematic diagrams of a root: (a) longitudinal section, indicating the zones that can occur near the root tip, and (b) cross-sectional view approximately 10 mm back from the tip, indicating the arrangement of the various cell types.

undifferentiated cells that are scraped off as the root grows into new regions of the soil. Cell walls in the root cap are often mucilaginous, which can reduce friction with soil particles. Proximal to the root cap is a meristematic region where the cells rapidly divide. Cells in this *apical meristem* tend to be isodiametric and have thin cell walls. Next is a region of *cell elongation* in the direction of the root axis. Such elongation mechanically pushes the root tip through the soil, causing cells of the root cap to slough off by abrasion with soil particles. Sometimes the region of dividing cells is not spatially distinct from the elongation zone. Also, cell size and the extent of the zones vary with both plant species and physiological status (Esau, 1977).

The next region indicated in Figure 1.4a is that of *cell differentiation*, where the cells begin to assume more highly specialized functions. The cell walls become thicker and elongation is greatly diminished. The epidermal cells develop fine projections, radially outward from the root, called *root hairs*. These root hairs greatly increase the surface area across which water and nutrients can enter a plant. As we follow a root toward the stem, the root surface generally becomes less permeable to water and the root interior becomes more involved with conducting water toward the stem. Water movement into the root is discussed in Chapter 9, so the discussion here will be restricted to some morphological features.

The region of the root where water absorption most readily occurs generally has little or no waxy cuticle. Figure 1.4b shows a cross section of a root at the level where root hairs are found. Starting from the outside, we observe first the root epidermis and then a number of layers of cells known as the *cortex*. There are abundant intercellular air spaces in the cortex, facilitating the diffusion of O_2 and CO_2 within this tissue (such air spaces generally are lacking in vascular tissue). Inside the cortex is a single layer of cells, the *endodermis*. The radial and transverse walls of the endodermal cells are impregnated with waxy material, including *suberin*, forming a band around the cells known as the *casparian strip*, which prevents passage of water and solutes across that part of the cell wall. Because there are no air spaces between endodermal cells, and the radial walls are blocked by the waterproof casparian strip, water must pass through the lateral walls and enter the cytoplasm of endodermal cells to continue across the root. The endodermal cells can represent the only place in the entire pathway for water movement from the soil, through the plant, to the air where it is mandatory that the water enter a cell's cytoplasm.[3] In the rest of the pathway, water can move

---

3. In the roots of many species a subepidermal layer or layers of *hypodermis* occurs. Radial walls of hypodermal cells can also be blocked with a waxy material analogous to the casparian strip in the endodermis, in which case the layers are often termed an *exodermis*.

in cell walls or in the hollow lumens of xylem vessels, a region referred to as the *apoplast*.

Immediately inside the endodermis is the *pericycle,* which is typically one cell thick in angiosperms. The cells of the pericycle can divide and form a meristematic region that can produce lateral or branch roots in the region just above the root hairs. Radially inside the pericycle is the vascular tissue. The phloem generally occurs in two to eight or more strands located around the root axis. The xylem generally radiates out between the phloem strands, so water does not have to cross the phloem to reach the root xylem. As in stems, the tissue between the xylem and the phloem is the vascular cambium, which through cell division and differentiation produces xylem to the inside and phloem to the outside.

Our rather elementary discussion of leaves, vascular tissues, and roots leads to the following oversimplified but useful picture. The roots take up water from the soil along with nutrients required for growth. These are conducted in the xylem to the leaves. Leaves of a photosynthesizing plant lose the water to the atmosphere along with a release of O_2 and an uptake of CO_2. Carbon from the latter ends up in photosynthate translocated in the phloem back to the root. Thus the xylem and the phloem serve as the "plumbing" that connects the two types of plant organs functionally interacting with the environment. To understand the details of such physiological processes we must turn to fields like calculus, thermodynamics, and photochemistry for the analytical methods. Our next step is to bring the abstract ideas of these fields into the realm of cells and plants, which means we need to make calculations using appropriate assumptions and approximations.

We will begin by describing diffusion (Ch. 1). To discuss water (Ch. 2) and solutes (Ch. 3), we will introduce the thermodynamic concept of chemical potential. This leads to a quantitative description of fluxes, electrical potentials across membranes, and the energy requirements for active transport of solutes. Some important energy conversion processes take place in the organelles. For instance, light energy is absorbed (Ch. 4) by photosynthetic pigments located in the internal membranes of chloroplasts (Ch. 5) and then converted into other forms of energy useful to a plant (Ch. 6) or dissipated as heat (Ch. 7). Leaves (Ch. 8) as well as groups of plants (Ch. 9) also interact with the environment through exchanges of water vapor and CO_2. In our problem-solving approach to these topics, we will pay particular attention to dimensions and ranges for the parameters as well as to the insights that can be gained by developing the relevant formulae and then making calculations.

DIFFUSION

Diffusion is a spontaneous process leading to the net movement of a substance from some region to an adjacent one where that substance has a lower concentration. It takes place in both the liquid and the gas phases associated with plants. Diffusion results from the random thermal motions of the molecules either of the solute(s) and the solvent in the case of a solution or of gases in the case of air. The net movement caused by diffusion is a statistical phenomenon: there is a greater probability of molecules moving from the concentrated to the dilute region than vice versa. In other words, more molecules per unit volume are present in the concentrated region than in the dilute one, so more are available for diffusing toward the dilute region than are available for movement in the opposite direction. If isolated from external influences, diffusion of a neutral species tends to even out concentration differences originally present in adjoining regions of a liquid or a gas. In fact, the randomizing tendency of such molecular Brownian movement is a good example of the increase in entropy, or decrease in order, that accompanies all spontaneous processes.

Diffusion is involved in many plant processes, such as gas exchange and the movement of nutrients toward root surfaces. For instance, diffusion is the mechanism for most, if not all, steps by which CO_2 from the air reaches the sites of photosynthesis in chloroplasts. CO_2 diffuses from the atmosphere up to the leaf surface and then diffuses through the stomatal pores. After entering the leaf, CO_2 diffuses within intercellular air spaces (Fig. 1.2). Next, CO_2 diffuses across the wet cell wall, crosses the plasmalemma of a leaf mesophyll cell, and then diffuses through the cytosol to reach the chloroplasts. Finally, CO_2 enters a chloroplast and diffuses up to the enzymes that are involved in carbohydrate formation. If the enzymes were to fix all the CO_2 in their vicinity, and no other CO_2 were to diffuse in from the atmosphere surrounding the plant, photosynthetic processes would stop.[4] In this chapter we will develop the mathematical formulation necessary for understanding both diffusion across a membrane and diffusion in a solution.

Fick's First Law

In 1855, Fick was one of the first to examine diffusion quantitatively (Fick, 1855). For such an analysis we need to consider the concentration (c_j) of some

---

4. We note that in solution "CO_2" can occur in the form of bicarbonate (HCO_3^-) and also that the crossing of membranes does not have to be by diffusion, refinements that we will return to in Chapter 8.

species j in a solution; the subscript j indicates that we are considering only one species out of the many that could be present. We will assume that the concentration of species j in some region is less than in a neighboring one. A net migration of molecules occurs by diffusion from the concentrated region to the dilute region. Such a molecular flow down a concentration gradient is analogous to the flow of heat from a warmer region to a cooler one. The analogy is actually rather good (especially for gases), because both processes depend on the random thermal motion of the molecules. In fact, the differential equations and their solutions that are used to describe diffusion are those that had previously been developed to describe heat flow (see Crank, 1975; Starzak, 1984).

To express diffusion quantitatively, we will consider a diffusive flux or flow of species j. We will restrict our attention to diffusion involving planar fronts of uniform concentration, a relatively simple situation that fortunately has widespread application to situations of interest in biology. We will let J_j be the amount of some species j crossing a certain area per unit time, e.g., moles of particles per m^2 in a second, which is termed the *flux density*.[5] Reasoning by analogy with heat flow, Fick deduced that the "force," or causative agent, leading to the net molecular movement is the concentration gradient. A gradient indicates how a certain parameter changes with distance; the gradient in concentration of species j in the x-direction is represented by $\partial c_j / \partial x$.[6] In general, *the flux density of some substance is proportional to the appropriate force*, a relation that we will use repeatedly in this text. In the present instance the driving force is the negative of the concentration gradient of species j, which we will represent by $-\partial c_j / \partial x$ for diffusion in one dimension. To help appreciate why a minus sign occurs, recall that the direction of net diffusion is toward regions of lower concentration. We can now write the following relation showing the dependence of the flux density on the driving force:

$$J_j = -D_j \frac{\partial c_j}{\partial x} \tag{1.1}$$

Equation 1.1 is commonly known as *Fick's first law* of diffusion, where D_j is the *diffusion coefficient* of species j. For J_j in mol m^{-2} s^{-1} and c_j in mol m^{-3}

---

5. Although the SI convention recommends the term flux density, much of the diffusion literature refers to J_j as a flux. Moreover, many symbols have been used for flux density (e.g., A, D, E, F, I, J, M, Q, U, and V), some of which (such as A for assimilation and E for evaporation) conflict with those used for other common variables (A for area and E for electric field or potential). We have chosen J because of its lack of conflict and the long precedent for its use (e.g., Onsager used J for the flux densities of heat and mass in the early 1930s).

6. The partial derivative $\partial c_j / \partial x$ indicates how much c_j changes as we move along the x-axis when other variables, such as time and position along the y-axis, are held constant.

(hence, $\partial c_j/\partial x$ in mol m^{-4}), D_j has units of m^2 s^{-1}. Because D_j varies with concentration and temperature, it is properly called a *coefficient* in the general case. In certain applications, however, we can obtain sufficient accuracy by treating D_j as if it were a constant. The partial derivative is used in Equation 1.1 to indicate the change in concentration in the x-direction of Cartesian coordinates at some moment in time (constant t) and for specified values of y and z. For most of the cases that we will consider, the flux density in the x-direction has the same magnitude at any value of y and z, meaning that we are dealing with one-dimensional, planar fluxes. By convention, a net flow in the direction of increasing x is positive (from left to right in Fig. 1.5). Because a net flow occurs toward regions of lower concentration, we again note that the minus sign is needed in Equation 1.1. Fick's first law—which has been amply demonstrated experimentally—is the starting point for our discussion of diffusion.

Continuity Equation and Fick's Second Law

As we indicated earlier, diffusion in a solution is important for the movement of solutes across plant cells and tissues. How rapid are such processes? For

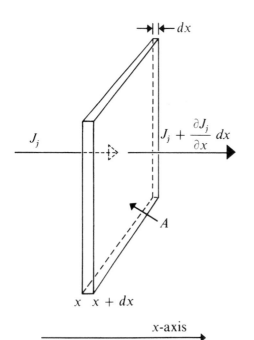

Figure 1.5

Diagram showing the dimensions and the flux densities that form the geometrical basis for the continuity equation. The same general figure is used to discuss water flow in Chapter 2 and solute flow in Chapter 3.

example, if we release a certain amount of material in one location, how long will it take before we can detect that substance at various distances? To discuss such phenomena adequately, we must determine the dependence of the concentration on both time and distance. We can readily derive such a time-distance relationship if we first consider the concept of conservation of mass, which is necessary if we are to transform Equation 1.1 into an expression that is convenient for describing the actual solute distributions caused by diffusion. In particular, we want to eliminate J_j from Equation 1.1 so that we can see how c_j depends on x and t.

The amount of species j per unit time crossing a given area, here considered to be a planar area perpendicular to the x-axis (Fig. 1.5), can change with position along the x-axis. Let us imagine a volume element of thickness dx in the direction of flow and of cross-sectional area A (Fig. 1.5). At x, we will let the flux density across the surface of area A be J_j. At $x + dx$, the flux density has changed to $J_j + (\partial J_j/\partial x)dx$, where $\partial J_j/\partial x$ is the gradient of the flux density of j in the x-direction; i.e., the rate of change of J_j with position, $\partial J_j/\partial x$, times the distance, dx, gives the overall change in the flux density, $(\partial J_j/\partial x)dx$. The change in the amount of species j in the volume $A\ dx$ in unit time for this one-dimensional problem is simply the amount flowing into the volume element per unit time, $J_j A$, minus that flowing out, $[J_j + (\partial J_j/\partial x)dx]A$. The change in the amount of species j in the volume element in unit time can also be expressed as the change in the concentration of species j with time, $\partial c_j/\partial t$, multiplied by the volume in which the change in concentration occurs, $A\ dx$. Equating these two different expressions that describe the change in the amount of species j in the volume $A\ dx$, we obtain the following relation:

$$J_j A - \left(J_j + \frac{\partial J_j}{\partial x}\,dx \right)A = \frac{\partial c_j}{\partial t}\,A\ dx \tag{1.2}$$

The two $J_j A$ terms on the left side of Equation 1.2 cancel each other. After division through by $A\ dx$, Equation 1.2 leads to the very useful expression known as the *continuity equation:*

$$-\frac{\partial J_j}{\partial x} = \frac{\partial c_j}{\partial t} \tag{1.3}$$

The continuity equation is a mathematical way of stating that matter cannot be created or destroyed under ordinary conditions. Thus, if the flux density of some species decreases as we move in the x-direction ($\partial J_j/\partial x < 0$), Equation 1.3 indicates that its concentration must consequently be increasing with time, as the material is then accumulating locally. When we substitute Fick's first law (Eq. 1.1) into the continuity equation (Eq. 1.3), we obtain *Fick's second law*. For the important special case of constant D_j, this general equation for diffusion becomes

$$\frac{\partial c_j}{\partial t} = -\frac{\partial}{\partial x}\left(-D_j \frac{\partial c_j}{\partial x}\right) = D_j \frac{\partial^2 c_j}{\partial x^2} \qquad (1.4)$$

Solution of Equation 1.4, which is the most difficult differential equation to be encountered in this book, describes how the concentration of some solute changes with position and time as a result of diffusion. To determine the particular function that satisfies this important differential equation, we need to know the specific conditions for the situation under consideration. Nevertheless, a representative solution useful for the consideration of diffusion under simple initial and boundary conditions will be sufficient for the present purpose of describing the characteristics of solute diffusion in general terms. For example, we will assume that there are no obstructions in the x-direction and that species j is initially placed in a plane at the origin ($x = 0$). In this case, the following expression for the concentration of species j satisfies the differential form of Fick's second law when D_j is constant, and also satisfies our rather simple conditions[7]:

$$c_j = \frac{M_j}{2(\pi D_j t)^{1/2}} e^{-x^2/4D_j t} \qquad (1.5)$$

In Equation 1.5, M_j is the total amount of solute j per unit area initially ($t = 0$) placed in a plane located at the origin of the x-direction (i.e., at $x = 0$, while y and z can have any value, which defines the plane considered here), and c_j is its concentration at position x at any later time t. For M_j to have this useful meaning, the factor $1/[2(\pi D_j)^{1/2}]$ is necessary in Equation 1.5.[8] Moreover, the

---

7. To show that Equation 1.5 is a possible solution of Fick's second law, it can be substituted into Equation 1.4 and the differentiations performed (M_j and D_j are constant; $\partial ax^n/\partial x = anx^{n-1}$, $\partial e^{ax^n}/\partial x = anx^{n-1} e^{ax^n}$, and $\partial uv/\partial x = u\,\partial v/\partial x + v\,\partial u/\partial x$; see Lauffer, 1989). The solution of Equation 1.4 becomes progressively more difficult when more complex initial and boundary conditions or molecular interactions (which cause variations in D_j) are considered. See Crank (1975) and Jacobs (1967) for analytical solutions in such cases.

8. Note that $\int_{-\infty}^{\infty} c_j(x, t)\,dx = M_j$, where $c_j(x, t)$ is the concentration function that depends on position and time as given by Equation 1.5 and the probability integral, $\int_{-\infty}^{\infty} e^{-a^2 u^2}\,du$, equals $\sqrt{\pi}/a$.

solute in this case is allowed to diffuse for an unlimited distance in either the plus or the minus x-direction and no additional solute is added at times $t > 0$. Often this idealized situation can be achieved by inserting a radioactive tracer in a plane at the origin of the x-direction. Equation 1.5 is only one of the possible solutions to the second-order partial differential equation representing Fick's second law. The form is relatively simple compared with other solutions, and, more important, the initial condition of having a finite amount of material released at a particular location is realistic for certain applications to biological problems.

Time–Distance Relation for Diffusion

Although the functional form of c_j given by Equation 1.5 is only one particular solution to Fick's second law (Eq. 1.4) and is restricted to the case of constant D_j, it nevertheless proves to be an extremely useful expression for understanding diffusion. It can be used to relate the distance a substance diffuses to the time necessary to reach that distance. The expression involves the diffusion coefficient of species j, D_j, which can be determined experimentally. In fact, Equation 1.5 itself is often employed to determine a particular D_j.

In the plane at the origin of the x-direction ($x = 0$), Equation 1.5 indicates that the concentration is $M_j/[2(\pi D_j t)^{1/2}]$, which becomes infinitely large as t is turned back to 0, the starting time. This infinite value for c_j at $x = 0$ corresponds to having all the solute initially placed in a plane at the origin. For t greater than 0, the material begins to diffuse away from the origin. The distribution of molecules along the x-axis at two successive times is indicated in Figures 1.6a and 1.6b, while 1.6c explicitly shows the movement of the concentration profiles along the time axis. Because the total amount of species j does not change (it remains at M_j per unit area of the y-z plane, i.e., in a volume element parallel to the x-axis and extending from x values of $-\infty$ to $+\infty$), the area under each of the concentration profiles is the same.

Comparing Figure 1.6a with 1.6b, we see that the average distance of the diffusing molecules from the origin increases with time. Also, Figure 1.6c shows how the concentration profiles flatten out as time increases, as the diffusing species is then distributed over a greater region of space. In estimating how far molecules diffuse in time t, a useful parameter is the distance x_e at which the concentration has dropped to $1/e$ or 37% of its value in the plane at the origin. Although somewhat arbitrary, this parameter describes the shift of the statistical distribution of the molecules with time. From Equation 1.5, the concentration at the origin is $M_j/[2(\pi D_j t)^{1/2}]$. (See also Fig. 1.6.) The concentration therefore drops to $1/e$ ($= e^{-1}$) of the value at the origin when the exponent of e in Equation

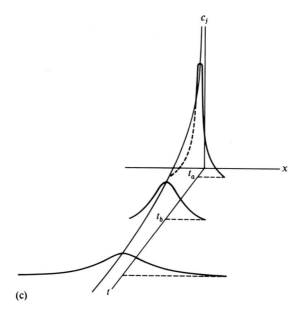

Figure 1.6

Concentration of species j, c_j, as a function of position x for molecules diffusing according to Fick's second law. The molecules were initially placed in a plane at the origin of the x-direction, i.e., at $x = 0$. For a given value of x, c_j is the same throughout a plane in the y- and z-directions. (a) Distribution of concentrations along the x-axis occurring at a time t_a, (b) distribution occurring at a subsequent time t_b, and (c) portrayal of the concentration profiles as a function of time. Note that x_e is the location where the concentration of species j has dropped to $1/e$ of its value at the origin.

1.5 is -1. From Equation 1.5, -1 then equals $-x_e^2/4D_jt_e$, so the distance x_e satisfies

$$x_e^2 = 4D_jt_e \tag{1.6}$$

The distance x_e along the x-axis is also indicated in Figures 1.6a and 1.6b.

Equation 1.6 is an extremely important relationship that indicates a fundamental characteristic of diffusion processes: the distance a population of molecules of a particular species diffuses—for the one-dimensional case where the molecules are released in a plane at the origin—is proportional to the square root of both the diffusion coefficient of the species and the time for diffusion. In other words, the time to diffuse a given distance increases with the square of that distance. An individual molecule may diffuse a greater or lesser distance in time t_e than is indicated by Equation 1.6 (see Fig. 1.6), because the latter refers to the time required for the concentration of species j at position x_e to become $1/e$ of the value at the origin; i.e., we are dealing with the characteristics of a whole population of molecules, not the details of an individual molecule. Furthermore, the factor 4 is rather arbitrary, because some criterion other than $1/e$ would cause this numerical factor to be somewhat different, although the basic form of Equation 1.6 would be preserved. For example, the numerical factor is 2.8 if the criterion is to drop to half of the value at the origin.

Table 1.1 lists the magnitudes of diffusion coefficients for various solutes in water at 25°C.[9] For ions and other small molecules, D_j's in aqueous solutions are approximately 10^{-9} m^2 s^{-1}. Because proteins have higher relative molecular masses (i.e., higher molecular weights[10]) than the small solutes, their diffusion coefficients are lower (Table 1.1). Also, because of the greater frictional interaction between water molecules and fibrous proteins than with the more compact globular ones, fibrous proteins often have approximately twofold lower diffusion coefficients than do globular proteins of the same molecular weight.

---

9. The symbol °C in this text represents degrees on the Celsius temperature scale. By the SI system (Goldman and Bell, 1986), the Celsius degree as well as the kelvin unit or a kelvin (abbreviated K) is $1/273.16$ of the thermodynamic temperature of the triple point of water (0.01000°C) and absolute zero is at -273.15°C. Although it is still in use, the term "centigrade" is no longer recommended.

10. Relative molecular mass, an expression that is preferred over the commonly used "molecular weight," is a dimensionless number indicating the molecular mass of a substance relative to that of a neutral carbon atom with six protons and six neutrons (^{12}C) taken as 12.00000. For proteins, the molecular mass is often expressed in kilodaltons (kDa; sometimes kD and even kd is used), where 1 dalton is $^1/_{12}$ the mass of ^{12}C. For instance, sucrose has a relative molecular mass, or molecular weight, of 342 and a molecular mass of 342 daltons.

Table 1.1

Diffusion coefficients in aqueous solutions and air. Values are for dilute solutions at 25°C or air under standard atmospheric pressure at 20°C (see Fasman, 1976; Weast and Lide, 1989).

Small solutes in water		Globular proteins in water	
Substance	D_j (m² s⁻¹)	Molecular mass (kilodaltons)	D_j (m² s⁻¹)
Alanine	0.92×10^{-9}	15	10^{-10}
Citrate	0.66×10^{-9}	1000	10^{-11}
Glucose	0.67×10^{-9}		
Glycine	1.1×10^{-9}	Gases in air	
Sucrose	0.52×10^{-9}		
		Gas	D_j (m² s⁻¹)
Ca^{2+} (with Cl^-)	1.2×10^{-9}		
K^+ (with Cl^-)	1.9×10^{-9}	CO_2	1.51×10^{-5}
Na^+ (with Cl^-)	1.5×10^{-9}	H_2O	2.42×10^{-5}
CO_2	1.7×10^{-9}	O_2	1.95×10^{-5}

To illustrate the time–distance consequences of Equation 1.6, we will quantitatively consider the diffusion of small molecules in an aqueous solution. How long, on the average, would it take a small solute with a D_j of 10^{-9} m² s⁻¹ to diffuse 50 μm, the distance across a typical leaf cell? From Equation 1.6, the time required for the population of molecules to shift so that the concentration at this distance is $1/e$ of the value at the origin is

$$t_e = \frac{(50 \times 10^{-6} \text{ m})^2}{(4)(10^{-9} \text{ m}^2 \text{ s}^{-1})} = 0.6 \text{ s}$$

Thus, diffusion is a fairly rapid process over subcellular distances.

Next, let us consider the diffusion of the same substance over a distance of 1 m. The time needed is

$$t_e = \frac{(1 \text{ m})^2}{(4)(10^{-9} \text{ m}^2 \text{ s}^{-1})} = 2.5 \times 10^8 \text{ s} \cong 8 \text{ years}$$

Diffusion is indeed not rapid over long distances. Thus, inorganic nutrients in xylary sap would not ascend a tree by diffusion at a rate sufficient to sustain life. On the other hand, diffusion is often adequate for the movement of solutes within leaf cells and especially inside organelles such as chloroplasts and mitochondria. In summary, diffusion in a solution is fairly fast over short distances (less than about 100 μm) but extremely slow for very long distances.

In the cytoplasm of living cells there is a mechanical mixing due to *cytoplasmic streaming,* which can lead to much more rapid movement than by diffusion. This cytoplasmic streaming, whose cessation is often a good indicator that cellular damage has occurred, requires energy. Energy is usually supplied in the form of adenosine triphosphate (ATP), and movement may involve a similar mechanism to that in muscle (Kamiya, 1981).

Diffusion of gases in the air surrounding and within leaves is necessary for both photosynthesis and transpiration. For instance, water vapor evaporating from the cell walls of mesophyll cells diffuses across the intercellular air spaces (Fig. 1.2) to reach the stomata and from there diffuses across an air boundary layer into the atmosphere (considered in detail in Ch. 8). CO_2 diffuses from the atmosphere through the open stomata to the surfaces of mesophyll cells, and the photosynthetically evolved O_2 traverses the same pathway in the reverse direction, also primarily by diffusion. The experimentally determined diffusion coefficients of these three gases in air at sea level (standard atmospheric pressure) and 20°C are about 10^{-5} m^2 s^{-1} (see Table 1.1). Such diffusion coefficients in air are approximately 10^4 times greater than the D_j describing the diffusion of a small solute in a liquid, indicating that diffusion coefficients depend markedly on the medium. In particular, many more intermolecular collisions occur per unit time in a liquid phase than in a less dense gas phase. Thus, a molecule can move further in air than in an aqueous solution before being influenced by other molecules. Most cells in animals are bathed by fluids, so for larger animals a circulatory system is necessary to transport O_2 to their cells and to remove CO_2. Plants, on the other hand, often have conspicuous intercellular air spaces where the large values of D_{O_2} and D_{CO_2} in a gas phase facilitate diffusion.

Although the relation between diffusion coefficients and molecular weight can be complex, especially in solution, molecules with higher molecular weights tend to have lower diffusion coefficients. For instance, in a gas phase CO_2 has a lower diffusion coefficient than O_2, which has a lower diffusion coefficient than H_2O (Table 1.1), as would be expected based on their relative molecular weights (44, 32, and 18, respectively). Diffusion coefficients depend inversely on the viscosity of the medium (discussed on p. 124, where the temperature dependence of D_j is also considered). Because diffusion coefficients of gases in air are inversely proportional to ambient pressure (Equation 8.9), they become larger at higher altitudes.

The pathways for the movement of gas molecules in the intercellular air spaces of a leaf can be quite tortuous. Nevertheless, calculations using Equation 1.6

can give useful estimates of the diffusion times for the many processes involving gaseous movements in a leaf. An upper limit for the diffusion pathway in such intercellular air spaces of a leaf might be 1 000 μm. Using Equation 1.6 and the diffusion coefficients given in Table 1.1, we can calculate that the times needed for water vapor, CO_2, and O_2 to diffuse 1 000 μm in air are from 10 to 16 ms. Diffusion of molecules in a gas is therefore a fairly rapid process. However, when illumination and temperature are not limiting, the rate of photosynthesis in plants is generally limited by the amount of CO_2 diffusing to the chloroplasts, and the rate of transpiration is determined by diffusion of water vapor from the cell walls within the leaf to the outside air. In the latter case, the limitation posed by diffusion helps prevent excessive water loss from the plant and therefore is physiologically useful.

MEMBRANE STRUCTURE

The plasmalemma presents a major barrier to the diffusion of solutes into and out of plant cells, the organelle membranes play an analogous role for the various subcellular compartments, and the tonoplast performs this function for the vacuole. For instance, although H_2O and CO_2 readily penetrate the plasmalemma, ATP and metabolic intermediates usually do not diffuse across it at an appreciable rate. Before we mathematically describe the penetration of membranes by solutes, we will briefly review certain features of the structure of membranes.

Membrane Models

Gorter and Grendel in 1925 estimated that the lipids extracted from erythrocytes (red blood cells), when spread as a monomolecular layer on water, cover an area about twice the surface area of the cells (see Kotyk *et al.*, 1988; Silver, 1985; Starzak, 1984). The amount of lipid present is apparently sufficient to form a double layer in the membrane. Moreover, the penetration of a series of species across membranes often depends primarily on the relative lipid solubility of the molecules (Davson and Danielli, 1952; Diamond and Wright, 1969; Stein, 1986). This circumstantial evidence led to the concept of a biological membrane composed primarily of a lipid bilayer.

To help understand the properties of lipid bilayers, we must consider the charge distribution within lipid molecules. The arrangement of atoms in the hydrocarbon (containing only C and H) region of lipid molecules leads to bonding in which no local charge imbalance develops. This hydrocarbon part of the molecule is thus nonpolar (no substantial, local, intramolecular separation of positive and negative charge), tends to avoid water, and is therefore called *hydrophobic*. Most lipid molecules in membranes also have a phosphate or an amine group or both. This region of the molecule becomes charged in aqueous solutions. Such charged regions interact electrostatically with the polar parts of other molecules. Because they interact attractively with water, the polar regions are termed *hydrophilic*.

Many of the phospholipids in membranes are glycerol derivatives that have two esterified fatty acids plus a charged side chain joined by a phosphate ester linkage.[11] A typical example is phosphatidyl choline (also called lecithin), a major component of most membranes:

$$
\begin{array}{l}
\quad\quad\quad\quad\; O \\
\quad\quad\quad\quad\; \| \\
CH_2\!-\!O\!-\!C\!-\!R \\
\\
\quad\quad\quad\quad\; O \\
\quad\quad\quad\quad\; \| \\
CH\!-\!O\!-\!C\!-\!R' \\
\\
\quad\quad\quad\quad\; O \\
\quad\quad\quad\quad\; \| \\
CH_2\!-\!O\!-\!P\!-\!O\!-\!CH_2CH_2\overset{+}{N}(CH_3)_3 \\
\quad\quad\quad\quad\; | \\
\quad\quad\quad\quad\; O^-
\end{array}
$$

where R and R' are the hydrocarbon parts of the fatty acids. Various fatty acids commonly esterified to this phospholipid include palmitic (16C and no double bonds, represented as 16:0), palmitoleic (16:1), stearic (18:0), oleic (18:1), linoleic (18:2), and linolenic (18:3). (For example, the major fatty acid combined in various ways in the membranes of higher plant chloroplasts is linolenic.) The hydrocarbon side chains of the esterified fatty acids affect the packing of the lipid molecules in a membrane. As the number of double bonds in the fatty acid side chain increases, the side chain tends to be less straight, so that the area per lipid molecule in a monolayer becomes greater. This change in intermolecular distances affects the permeability of such lipid layers and, presumably, that of biological membranes.

To form a bilayer in a membrane, the lipid molecules have their nonpolar portions adjacent to each other (Fig. 1.7), facilitating hydrophobic interactions. The polar or hydrophilic regions are then on the outside. These outer charged

---

11. Esterification refers to the chemical joining of an acid and an alcohol with the removal of water. For an ether, two alcohols are joined with the elimination of water.

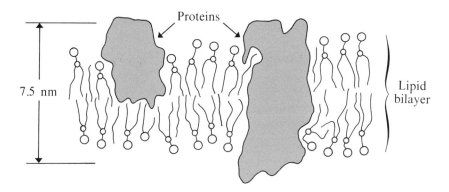

Figure 1.7

Membrane model where globular proteins are interspersed within a lipid bilayer. The ionic "head" of phospholipids is represented by ○—○ and the fatty acid side chains leading to the nonpolar "tail" are indicated by the two wavy lines emanating from the head.

regions of the lipids attract water molecules and the charged parts of proteins, which are both present in membranes. For instance, up to 70% of the dry weight of mitochondrial membranes can be protein, whereas chloroplast membranes are generally less than 50% protein. Membranes also contain a small amount of steroids and oligosaccharides (composed of 2 to 10 monosaccharide residues, often bound to proteins). Membranes can be about half water by weight. When the water content of membranes is reduced below about 20%, as can occur during prolonged desiccation, the lipid bilayer configuration and membrane integrity are lost (Simon, 1974). Thus, a minimal water content is necessary to create the hydrophilic environment that stabilizes the lipid bilayer.

Thermodynamic predictions and experimental results indicate that membrane proteins occur as globular forms embedded within the membranes, and indeed some proteins extend all the way across a membrane (Fig. 1.7). Globular proteins apparently have their hydrophobic portions buried within the membrane, and a hydrophilic portion sticks out into the aqueous solution next to the membrane. In such proteins those amino acids whose side chains dissociate—aspartate, glutamate, arginine, lysine—tend to be exposed to water, whereas amino acids with hydrophobic side chains—leucine, isoleucine, valine—tend to the interior of the membrane where they interact with the fatty acid side chains of the phospholipids, which are also hydrophobic. In this model the lipid occurs as a bilayer, and the hydrophilic portions of the lipids interact directly with water on either side of the membrane (Fig. 1.7). The lipid bilayer arrangement between the globular proteins helps account for the good correlation between lipid solubility and membrane permeation.

Organelle Membranes

Both mitochondria and chloroplasts are surrounded by two membranes and have extensive internal membrane systems, and both are highly involved with cellular metabolism. Specific proteins implementing electron transfer in respiration and in photosynthesis occur in the interior membranes of mitochondria and chloroplasts, respectively. Such membrane subunits are able to move or vibrate thermally because of their own kinetic energy while remaining in the membrane. In fact, diffusion of globular proteins *within* the plane of the membrane has been observed. Diffusion coefficients for such movements are generally about 10^{-14} to 10^{-13} m^2 s^{-1}, compared with nearly 10^{-10} m^2 s^{-1} for the same proteins in solution, suggesting that membranes have rather high viscosities (Singer and Nicolson, 1972). Such diffusion of membrane proteins could allow successive interactions of some bound substrate with various enzymes located in the membrane. Furthermore, the side-by-side location of various components involved with electron transfer in a semisolid part of the membrane system can ensure an orderly, rapid, directed passage of electrons from enzyme to enzyme.

The proteins involved with electron transfer vary in size and shape, and thus the internal membranes of chloroplasts and mitochondria are not uniform and regular. Proteins taking part in ion transport and possibly cell wall synthesis may be embedded in the plasmalemma, and other proteins involved with transport occur in the tonoplast. On the other hand, many globular proteins in membranes can serve a structural role.

Of the two mitochondrial membranes (Fig. 1.8), the outer one is much more permeable than the inner one to sucrose, various small anions and cations (e.g., H^+), adenine nucleotides, and many other compounds. The inner membrane invaginates to form the mitochondrial *cristae,* in which are embedded the enzymes responsible for electron transfer and the accompanying ATP formation. For instance, the inner membrane system has various dehydrogenases, an ATPase, and cytochromes a, a_3, b, and c associated with it. (Cytochromes are discussed in Chs. 5 and 6.) These proteins with enzymatic activity occur in a globular form, and they can be an integral part of the membrane (Fig. 1.7) or loosely bound to its periphery (as appears to be the case for cytochrome c). Electron microscopy has revealed small particles attached by stalks to the cristae; these particles are proteins involved in the phosphorylation accompanying respiration.

Inside the inner membrane of a mitochondrion is a viscous region known as the *matrix* (Fig. 1.8). The citric acid cycle (Krebs cycle) enzymes, as well as others, are located there. For substrates to be catabolized via the citric acid cycle, they must cross two membranes to pass from the cytosol to the inside of a mitochondrion. Often the slowest or rate-limiting step in the oxidation of such

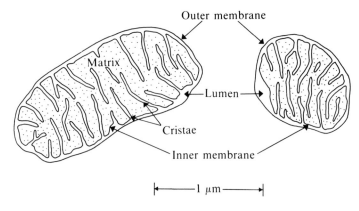

Outer membrane

Matrix

Lumen

Cristae

Inner membrane

|← 1 μm →|

Figure 1.8

Representative mitochondria in cross section, as seen in an electron micrograph of a 50-nm-thick section of plant tissue.

substrates is their entry into the mitochondrial matrix. Because the inner mitochondrial membrane is highly impermeable to most molecules, transport across the membrane using a "carrier" or "transporter" (see Ch. 3) is generally invoked to explain how various substances get into the matrix. These carriers, situated in the inner membrane, might shuttle important substrates from the lumen between the outer and the inner mitochondrial membranes to the matrix. Because of the inner membrane, important ions and substrates in the mitochondrial matrix do not leak out. Such permeability barriers between various subcellular compartments serve to improve the overall efficiency of a cell.

Chloroplasts (Fig. 1.9) are also surrounded by two *limiting* membranes, the outer one of which is more permeable than the inner one to small solutes. These limiting membranes are relatively high in galactolipid and low in protein (Douce and Joyard, 1979; Heber and Heldt, 1981). The internal *lamellar* membranes of chloroplasts are about half lipid and half protein by dry weight. The chlorophyll and most other photosynthetic pigments are bound to proteins. These proteins, as well as other components involved with photosynthetic electron transfer (see Chs. 5 and 6), are anchored in the lamellar membranes, apparently by hydrophobic forces. Each lamella consists of a pair of closely apposed membranes 6 to 8 nm thick. In many regions of a chloroplast the lamellae form flattened sacs called *thylakoids* (Fig. 1.9). When seen in an electron micrograph, a transverse section of a thylakoid shows a pair of apposed membranes joined at the ends.

The organization of lamellar membranes within chloroplasts varies greatly with environmental conditions during development and among plant groups. The

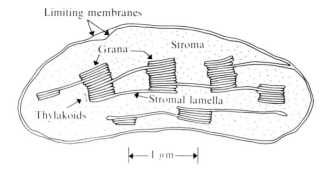

Figure 1.9
Generalized chloroplast in cross section from a leaf mesophyll cell.

chloroplasts of red algae appear to have the simplest internal structure, because the lamellae are in the form of single large thylakoids separated by appreciable distances from each other. For most higher plant chloroplasts, the characteristic feature is stacks of about 10 or more thylakoids known as *grana* (Fig. 1.9), which are typically 0.4 to 0.5 μm in diameter, 10 to 50 of these grana occurring in a single chloroplast. The lamellar extensions between grana are called inter-granal or stromal lamellae. The remainder of the chloroplast volume is known as the *stroma*, which contains the enzymes involved with the fixation of CO_2 into the various products of photosynthesis (Ch. 8). Cyanobacteria and other photosynthetic bacteria do not contain chloroplasts, but their photosynthetic pigments are also generally located in membranes, often in lamellae immediately underlying the cell membrane. In some photosynthetic bacteria the lamellae can pinch off and form discrete subcellular bodies often referred to as *chromato-phores*, which are about 0.1 μm in diameter.

"Microbodies" are also quite numerous in plant cells. They are generally divided into two classes, *glyoxysomes* and *peroxisomes*, the latter being about three times as prevalent as mitochondria in many leaf cells. Microbodies are usually spherical and from 0.4 to 1.5 μm in diameter. They differ from mito-chondria and chloroplasts in that they are surrounded by a single membrane. Because both glyoxysomes and peroxisomes carry out only a portion of an overall metabolic pathway, these subcellular compartments depend on reactions in the cytosol and in other organelles. In Chapter 8 we will briefly consider the role of peroxisomes in photorespiration, where O_2 is consumed and CO_2 is released in the light in mesophyll cells. Glyoxysomes contain the enzymes necessary for the breakdown of fatty acids and are prevalent in fatty tissues of germinating seeds. Other membrane-bound structures recognized in plant cells include the endoplasmic reticulum, the Golgi apparatus, and small vesicles associated with them that are involved in the transport of various cellular components (see Tolbert, 1980).

MEMBRANE PERMEABILITY

With this knowledge of the general structure of membranes, let us now turn to a quantitative analysis of the interactions between membranes and diffusing solutes. In Chapter 3 we will discuss active transport, which is important for moving specific solutes across membranes, thereby overcoming limitations posed by diffusion.

The rate-limiting step for the movement of many molecules into and out of plant cells is diffusion through the plasmalemma. Because of the close packing and the interactions between the component molecules of a membrane, such diffusion is greatly restricted compared with the relatively free movement in an aqueous phase like the cytosol. In other words, the solute has great difficulty in threading its way between the molecules composing the membrane phase. The average diffusion coefficient of a small solute in a membrane is often about 10^6 times lower than in the adjacent aqueous solutions. Also, a membrane represents a different type of molecular environment than does an aqueous solution, so the relative solubility of a species in the two phases must be taken into account to describe membrane permeability. As we will show, Fick's first law can be modified to describe the diffusion of molecules across a membrane. Once past the barrier presented by the membrane, the molecules may be distributed throughout the cell relatively rapidly by diffusion as well as by cytoplasmic streaming.

Concentration Difference across a Membrane

The driving force for diffusion of uncharged molecules into or out of plant cells can be regarded as the negative of the concentration gradient of that species across the plasmalemma. Because the actual concentration gradient of some species j is not known in the plasmalemma (or in any other membrane for that matter), the driving force is generally approximated by the negative of the average gradient of that species across the membrane:

$$-\frac{\partial c_j}{\partial x} \cong -\frac{\Delta c_j}{\Delta x} = -\frac{c_j^i - c_j^o}{\Delta x} = \frac{c_j^o - c_j^i}{\Delta x} \tag{1.7}$$

where c_j^o is the concentration of species j outside the cell, c_j^i is its concentration in the cytosol, and Δx is the thickness of the plasmalemma, which acts as the barrier restricting the penetration of the molecules into the cell (see Fig. 1.10). The concentrations c_j^o and c_j^i in Equation 1.7 can also represent values in the phases separated by any other membrane, such as the tonoplast. If c_j^i is less than c_j^o as we move in the positive x-direction (Fig. 1.10), $\partial c_j / \partial x$ is then negative, so $-\partial c_j / \partial x$ is positive; the net flux density would then occur in the positive x-direction (see Eq. 1.1, $J_j = -D_j \, \partial c_j / \partial x$). Alternatively, if we designate a net flow into a cell as positive—which is the usual convention—then the minus sign in $-\partial c_j / \partial x$ is incorporated into the concentration difference $(c_j^o - c_j^i)$ that we must use to describe correctly the flux density, namely, $J_j = D_j (c_j^o - c_j^i)/\Delta x$.

The magnitude of what may be called the "effective" concentration difference across a membrane is made somewhat uncertain by the existence of unstirred boundary layers (Fig. 1.10) (Dainty, 1976). Boundary layers in which there is no turbulent mixing occur at the interface between any fluid (liquid or gas) and

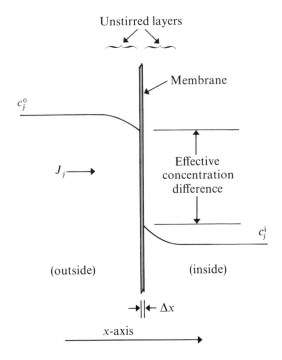

Figure 1.10

The effect of unstirred layers on the concentration of species j near a membrane across which the solute is diffusing.

a solid, such as the air boundary layers on either side of a leaf (discussed in Ch. 7). A substance moves across these relatively unstirred layers next to the membrane or other solid not by mechanical mixing but by diffusion, indicating that a concentration gradient must also exist in the boundary layers (see Fig. 1.10). When the mixing in the solutions adjacent to a membrane is increased—for example, either by the turbulence resulting from cytoplasmic streaming on the inside of the cell or by rapid stirring on the outside—the thicknesses of the unstirred layers are reduced. But they are not eliminated. Under actual experimental conditions with vigorous stirring, the external unstirred layer may be from 10 to 100 μm thick, which is much thicker than is the membrane itself (Green and Otori, 1970). Because of cytoplasmic streaming in plant cells, the internal unstirred layer is generally thinner than the external one. Although diffusion is more rapid in aqueous solutions than in membranes, an unstirred layer can represent an appreciable distance for molecules to diffuse across. In fact, diffusion through the unstirred layer in some cases can become the rate-limiting factor for the entry into cells or organelles of those molecules that rapidly penetrate the membrane itself. For convenience, the difference in concentration across a membrane will be represented by $c_j^i - c_j^o$, but this is an overestimate of the effective concentration difference, as is indicated in Figure 1.10.

The difference in concentration determining the diffusion of molecules across a membrane is the concentration just inside one side of the membrane minus that just within the other side. In Equation 1.7 the concentrations are those in the aqueous phases outside on either side of the membrane. Because membranes are quite different as solvents compared with aqueous solutions, the concentrations just inside the membrane can differ appreciably from those just outside in the aqueous solution. A correction factor must therefore be applied to Equation 1.7 to give the actual concentration gradient existing in the membrane. This factor is known as the *partition coefficient*, K_j. It is defined as the ratio of the concentration of a solute in the material of the membrane to that in equilibrium outside it in the aqueous phase (c_j^o or c_j^i), and so it is dimensionless.

As a guide to the situation in a membrane, the partition coefficient is generally determined by measuring the ratio of the equilibrium concentration of some solute in a lipid phase—such as olive oil, which mimics the membrane lipids—to the concentration in an adjacent and immiscible aqueous phase (which mimics the solutions on either side of the membrane). This rather simple convention for obtaining partition coefficients is based on the high lipid content of membranes and the numerous experimental results showing that the relative ease of penetration of a membrane depends mainly on the lipid solubility of the molecules. Partition coefficients have a wide range of values, most being between 10^{-6} and 10 (see Wright and Diamond, 1969, where the effect on K_j of various lipids as solvents is described). For example, an inorganic ion might have a K_j near 10^{-5} for membranes, and a nonpolar hydrophobic substance might have a value near 1. Usually, K_j is assumed to be the same coming from either side of the membrane. These considerations lead us to write $K_j(c_j^o - c_j^i)$ for the concentration difference leading to diffusion across a membrane, where K_j is characteristic of a particular species j.

Permeability Coefficient

We will next use Fick's first law ($J_j = - D_j \, \partial c_j / \partial x$, Eq. 1.1) to obtain an expression convenient for describing the movement of a substance across a membrane. The negative concentration gradient will be replaced by $(c_j^o - c_j^i)/\Delta x$ (Eq. 1.7). Moreover, because D_j is the diffusion coefficient of species j within the substance of the membrane, we must use the actual concentration drop within the membrane, $K_j(c_j^o - c_j^i)$. These various considerations lead us to the following expression describing the diffusion of species j across a membrane or other barrier:

$$J_j = D_j \frac{K_j(c_j^o - c_j^i)}{\Delta x}$$

$$= P_j(c_j^o - c_j^i) \tag{1.8}$$

where P_j is called the *permeability coefficient* of species j.

Because permeability coefficients have so many applications in physiology, we will define P_j explicitly:

$$P_j = \frac{D_j K_j}{\Delta x} \tag{1.9}$$

which follows directly from Equation 1.8. The permeability coefficient conveniently replaces three quantities that describe the diffusion of some solute across a membrane or other barrier. In this regard, the partition coefficients of membranes are generally determined by using a lipid phase such as olive oil or ether, not the actual lipids occurring in membranes, and thus K_j will have some uncertainty. Also, measurements of Δx and D_j for membranes tend to be rather indirect. On the other hand, P_j is a single, readily measured quantity characterizing the diffusion of some solute across a membrane or other barrier. The units of P_j are length per time, e.g., m s^{-1}.

A typical permeability coefficient for a small nonelectrolyte or uncharged molecule (e.g., isopropanol or phenol) is 10^{-6} m s^{-1} for the plasmalemma (Altman and Dittmer, 1966; Stein, 1986), whereas P_j for a small ion (e.g., K$^+$ or Na$^+$) might be about 10^{-9} m s^{-1}. The lower permeability coefficients for charged particles are mainly due to the much lower partition coefficients of these solutes compared with nonelectrolytes, which tend to be hydrophobic (recall that $P_j = D_j K_j / \Delta x$, Eq. 1.9). In other words, because of its charge, the electrolyte tends to be much less soluble in a membrane than is a neutral molecule, so the effective concentration gradient driving the charged species across the membrane is generally smaller for given concentrations in the aqueous phases on either side of the barrier. By comparison, water can rapidly enter or leave cells and can have a permeablity coefficient of about 10^{-4} m s^{-1} for Characean algae, although lower values can occur for other plant cells (Stadelmann, 1977). (Actually, P_{water} is difficult to measure because of the relative importance of the unstirred layers.) Although adequate measurements on plant membranes have so far not been made for O_2 and CO_2, their P_j's for the plasmalemma are probably quite high. For instance, O_2, which is lipophilic and has a fairly high partition coefficient of 4.4, has a very high permeability coefficient of about 0.3 m s^{-1} for erythrocyte membranes at 25°C (Fischkoff and Vanderkooi, 1975).

Diffusion and Cellular Concentration

Instead of calculating the flux density of some species diffusing into a cell—the amount entering per unit area per unit time—we often focus our attention on the total amount of that species diffusing in over a certain time interval. Let s_j be the amount of species j inside the cell, where s_j can be expressed in moles. If that substance is not involved in any other reactions, ds_j/dt represents the rate of entry of species j into the cell. The flux density J_j is the rate of entry of the substance per unit area or $(1/A)$ (ds_j/dt), where A is the area of the cellular membrane across which the substance is diffusing. Using this expression for J_j, we can express Equation 1.8 as follows:

$$ds_j/dt = J_j A = P_j A(c_j^o - c_j^i) \qquad (1.10)$$

When the external concentration of species $j (c_j^o)$ is greater than the internal one (c_j^i), species j will enter the cell and ds_j/dt is positive, as Equation 1.10 indicates.

A question that often arises in cellular physiology is how the internal concentration of a penetrating solute changes with time. How long, for example, would it take the internal concentration of an initially absent species to build up to 70% of the external concentration? To solve the general case, we must first express s_j in Equation 1.10 in a suitable fashion and then integrate. It is useful to introduce the approximation—particularly appropriate to plant cells—that the cell volume does not change during the time interval of interest. In other words, because of the rigid cell wall, we will assume that the plant cell has a nearly constant volume (V) during the entry or the exit of the solute being considered.

The average internal concentration of species $j(c_j^i)$ is equal to the amount of that particular solute inside the cell (s_j) divided by the cellular volume (V), or s_j/V. Therefore, s_j can conveniently be replaced by Vc_j^i, and thus ds_j/dt in Equation 1.10 equals $V\,dc_j^i/dt$ if the cell volume does not change with time. We are also assuming that species j is not produced or consumed by any reaction within the cell, so the diffusion of species j into or out of the cell is the only process affecting c_j^i. For simplicity, we are treating the concentration of species j as if it were uniform within the cell. (In a more complex case we might need a relation of the form of Equation 1.10 for each membrane-surrounded compartment.) Moreover, we are presupposing either mechanical mixing or rapid diffusion inside the cell, so that the drop in concentration ($c_j^o - c_j^i$) occurs essentially only across the membrane. Finally, let us assume that P_j is independent of concentration, a condition that is often satisfactorily met over a limited range of concentration. We can now proceed with the integration of Equation 1.10. Upon replacement of ds_j/dt by $V\,dc_j^i/dt$, rearrangement to separate variables, and the insertion of integral signs, Equation 1.10 becomes

$$\int_{c_j^i(0)}^{c_j^i(t)} \frac{dc_j^i}{c_j^o - c_j^i} = \frac{P_j A}{V} \int_0^t dt \tag{1.11}$$

where $c_j^i(0)$ is the initial internal concentration of species j, i.e., when $t = 0$, and $c_j^i(t)$ is the concentration at a later time t.

The volume outside certain unicellular algae and other membrane-surrounded entities can be quite large compared with V. In such cases the external concentration (c_j^o) does not change appreciably. Such an approximation can also be appropriate for experiments of short duration or when special arrangements are made to maintain c_j^o at some fixed value. For those cases in which c_j^o is constant, Equation 1.11 can be integrated to give the following expression[12]:

$$\frac{P_j A t}{V} = \ln \frac{c_j^o - c_j^i(0)}{c_j^o - c_j^i(t)} \tag{1.12}$$

---

12. Note that $\int dx/(a - x) = -\int d(a - x)/(a - x) = -\ln(a - x) + b$, where ln is the natural logarithm to the base e. See Appendix III for certain logarithmic identities.

Starting with a known c_j^o outside a cell and determining the internal concentration both initially, i.e., for $t = 0$, and at some subsequent time t, we can calculate P_j from Equation 1.12 if A/V is known. Even when A/V is not known, the relative permeability coefficients for two substances can be determined from the time dependencies of the respective c_j's. The above derivation can easily be extended to the case where c_j^o is zero. In that case, P_jAt/V equals $\ln[c_j^i(0)/c_j^i(t)]$, an expression that can be used to describe the diffusion of a photosynthetic product out of a chloroplast or some substance from a cell into a large external solution initially devoid of that species. Such studies can be facilitated by using radioactive tracers, which generally are initially present only in one region.

When diffusion occurs from all directions across a membrane surrounding a cell (or an organelle), the time to reach a given internal concentration can be directly proportional to the dimensions of the cell. For convenience in illustrating this observation, let us consider a spherical cell of radius r. The volume V then equals $(4/3)\pi r^3$ and the surface area A is $4\pi r^2$. Hence, V/A is $r/3$. Equation 1.12 then becomes

$$t^{\text{sphere}} = \frac{r}{3P_j} \ln \frac{c_j^o - c_j^i(0)}{c_j^o - c_j^i(t)} \tag{1.13}$$

For the same P_j, a given level inside a small cell is reached sooner than it would be inside a large cell. When diffusion across the membrane into a spherical cell occurs from all directions, the time to reach a given concentration inside is *linearly* proportional to the radius (assuming a uniform c_j^i). This contrasts with the time it takes for a planar front to diffuse in one direction, which is proportional to the square of the distance traveled ($x_e^2 = 4D_jt_e$, Eq. 1.6).

Using Equation 1.13, we can calculate the time required for a substance initially absent from a cell [$c_j^i(0) = 0$] to achieve an internal concentration equal to half the external concentration [$c_j^i(t) = \frac{1}{2}c_j^o$]. The conditions on the concentrations mean that [$c_j^o - c_j^i(0)]/[c_j^o - c_j^i(t)$] equals $(c_j^o - 0)/(c_j^o - \frac{1}{2}c_j^o)$, which is 2. Substituting these relations into Equation 1.13 indicates that the time needed is $(r/3P_j) \ln 2$, where $\ln 2$ equals 0.693. For a cell 50 μm in diameter—which is a reasonable dimension for spongy mesophyll cells in many leaves (see Fig. 1.2)—the time for an initially absent nonelectrolyte with a P_j of 10^{-6} m s^{-1} to reach half of the external concentration therefore is

$$t_{1/2}^{\text{sphere}} = \frac{(25 \times 10^{-6} \text{ m})}{(3)(10^{-6} \text{ m s}^{-1})} (0.693) = 6 \text{ s}$$

and the time is 1 000-fold greater (about 2 hours) for an electrolyte with a P_j of 10^{-9} m s^{-1}. Hence, the external membrane is an extremely good barrier for electrolytes, markedly hindering their entry into or exit from the cell. On the other hand, small nonelectrolytes can fairly readily diffuse in and out of plant cells, depending to a large extent on their relative lipid solubility.

CELL WALLS

Cell walls play many roles in plants. Their rigidity helps determine the size and the shape of a cell and ultimately the morphology of a plant. This supportive role is performed in conjunction with the internal hydrostatic pressure, which causes a distension of the cell walls. The cell wall is also intrinsically involved in many aspects of the ion and water relations of plant cells. Because it surrounds the plasmalemma of each cell, all fluxes of water and solutes into or out of a plant cell protoplast must cross the cell wall, usually by diffusion. The cell wall generally has a large negative charge, so it can interact differently with cations than with anions. Water evaporating from a plant during transpiration comes directly from cell walls. (Some aspects of this process will be considered quantitatively in Chs. 2 and 8.) The space surrounded by the cell wall in certain specialized cells can act as a channel through which solutions move. In the xylem, for example, the conducting cells have lost their protoplasts, with the result that the pathway for the conduction of solutions in the xylem consists essentially of hollow pipes, or conduits, made exclusively of cell walls.

Cell walls vary from about 0.1 to 10 μm in thickness, and they are generally divided into three regions: primary wall, secondary wall, and middle lamella. The *primary cell wall* surrounds dividing meristematic cells as well as the elongating cells during the period of cell enlargement. The cell wall often becomes thickened by the elaboration of a *secondary cell wall* inside the primary one (Fig. 1.11), which makes the cell mechanically much less flexible. Hence, cells whose walls have undergone secondary thickening, such as xylem vessel members, are generally incapable of subsequent elongation. The cell wall also includes an amorphous region between contiguous cells called the *middle lamella*. Although it contains some cellulose, the middle lamella may be composed mainly of the calcium salts of pectic acids, which causes adjacent cells to adhere together.

Chemistry and Morphology

Cellulose is the most abundant organic component in living organisms. It is the characteristic substance of the plant cell wall, constituting from 25% to 50% of

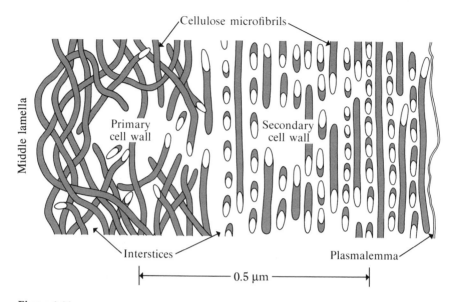

Cellulose microfibrils

Middle lamella

Primary cell wall

Secondary cell wall

Interstices

Plasmalemma

0.5 μm

Figure 1.11

Hypothetical thin section through a cell wall, indicating the cellulose microfibrils in the primary and secondary cell walls. The interstices are filled with noncellulosic material including appreciable water, the solvent for solute diffusion across the cell wall.

the cell wall organic material. Cellulose is a linear (unbranched) polysaccharide consisting of 1,4-linked β-D-glucopyranose units:

The polymer is about 0.8 nm in its maximum width, 0.33 nm^2 in cross-sectional area, and can contain about 10 000 glucose residues with their rings in the same plane. In the cell wall these polymers are organized into *microfibrils* that can be 5 nm by 9 nm in cross section. These microfibrils apparently consist of an inner core of about 50 parallel chains of cellulose arranged in a crystalline array surrounded by a similar number of cellulose and other polymers in a paracrystalline array (Preston, 1974). Microfibrils are the basic unit of the cell wall and are readily observed in electron micrographs. Although great variation exists, they tend to be interwoven in the primary cell wall and parallel to each other in the secondary cell wall (see Fig. 1.11).

Interstices between the cellulose microfibrils are usually 5 to 30 nm across. In these interstices is a matrix of amorphous components that can occupy a larger volume of the cell wall than do the microfibrils themselves. In fact, by weight, the main constituent of the cell wall is actually water, some consequences of which we will consider here and in Chapter 2.

The cell wall matrix contains noncellulosic polysaccharides such as pectin, lignins (secondary cell wall only), a small amount of protein, bound and free water, appreciable calcium, other cations, and sometimes silicates. Lignins are complex polymers based on phenylpropanoid subunits (a 6-carbon ring, to which is attached a 3-carbon chain as in phenylalanine, a lignin precursor) plus certain other residues. Lignins constitute the second most abundant class of organic molecules in living organisms and are about half as prevalent as cellulose. They are quite resistant to enzymatic degradation, so lignins are important in peat and coal formation. They also make the cell wall more rigid. Because the plant cannot break down the lignin polymers, cells are unable to expand after extensive lignification of their cell walls, as occurs in the secondary walls of xylem vessel members. Pectin consists primarily of 1,4-linked α-D-galacturonic acid residues, the carboxyl groups of which are normally dissociated and have a negative charge (galacturonic acid differs from galactose by having —COOH instead of —CH$_2$OH in the 6-carbon position). The negative charge of the dissociated carboxyl groups leads to the tremendous cation-binding capacity of cell walls. In particular, much of the divalent cation calcium (Ca^{2+}) is bound, which may help link the various polymers together. Polymers based on 1,4-linked β-D-xylopyranose units (xylans), as well as on many other residues, can also be extracted from cell walls. They are loosely referred to as hemicelluloses, e.g., xylans, mannans, galactans, and glucans. In general, hemicelluloses tend to have a low molecular weight (in the ten thousands) compared with pectin (about five to ten times larger) or cellulose (for further details on cell wall chemistry, see Brett and Hillman, 1985; Fry, 1988; Preston, 1974; Tolbert, 1980). The presence of negatively charged pectins hinders the entry of anions into plant cells. As we will consider next, however, ions and other solutes generally pass through the cell wall much more easily than through the plasmalemma.

Diffusion across Cell Walls

How would we calculate the ease with which molecules might diffuse across a cell wall? A good place to start is Equation 1.9, which indicates that the permeability coefficient of species j equals $D_j K_j / \Delta x$. We must first consider what we mean by D_j and K_j in a cell wall. The diffusion of solutes across a cell wall occurs mainly in the water located in the numerous interstices, which are often about 10 nm in diameter. Thus, the movement from the external solution up to the plasmalemma can be in aqueous channels through the cell wall. By the definition of a partition coefficient (solubility in barrier/solubility in water), we recognize that K_j would be unity for the water-filled interstices of the cell wall, it would be very low for the solid phases (cellulose, lignin, and the other polymers present in the wall), and it would have some intermediate value for the cell wall as a whole. Because diffusion takes place primarily in the cell wall water, we will let K_j be unity. If we use this convenient definition for K_j in cell walls, we must be careful how we define D_j. In particular, diffusion coefficients must then be expressed on the basis of the whole cell wall and not on the much smaller area presented by the interstices. The value of D_j in the water of the interstices could be very similar to that in a free solution, whereas its value averaged over the whole cell wall would be considerably less, because the interstices are not straight channels through the cell wall and because they do not occupy the entire cell wall volume. Such effective diffusion coefficients of small solutes are at least three times lower when averaged over the cell wall than are the D_j's of the same species in an extended aqueous solution.

Next, let us calculate the permeability coefficient for a solute that has a diffusion coefficient of 2×10^{-10} m^2 s^{-1} for a cell wall. We will assume a representative value of 1 μm for the cell wall thickness. Using Equation 1.9 ($P_j = D_j K_j / \Delta x$), we can thus estimate that P_j for the cell wall is

$$P_j = \frac{(2 \times 10^{-10} \text{ m}^2 \text{ s}^{-1})(1)}{(10^{-6} \text{ m})} = 2 \times 10^{-4} \text{ m s}^{-1}$$

Most of the permeability coefficients for small solutes crossing the plasmalemma range from 10^{-10} to 10^{-6} m s^{-1}. Hence, a cell wall generally has a higher permeability coefficient than does a membrane, which means that the wall is usually more permeable for small solutes than is the plasmalemma. For comparison, let us consider a permeability coefficient that might be appropriate for an unstirred liquid layer adjacent to a cell wall or membrane. Specifically, D_j for a small solute may be 10^{-9} m^2 s^{-1} in water, K_j is unity in the aqueous solution, and let us assume that Δx is 30 μm for the unstirred layer. In such a case we have

$$P_j = \frac{(10^{-9} \text{ m}^2 \text{ s}^{-1})(1)}{(30 \times 10^{-6} \text{ m})} = 3 \times 10^{-5} \text{ m s}^{-1}$$

which is less than the above value for a cell wall.

Molecules diffuse less readily across a given distance in a plasmalemma than in a cell wall or the adjacent unstirred layer. For the above numerical values, D_jK_j is 10^{-9} m² s⁻¹ in the aqueous solution and 2×10^{-10} m² s⁻¹ in the cell wall, yet for a plasmalemma about 8 nm thick, D_jK_j is only 10^{-18} to 10^{-14} m² s⁻¹. Membranes do indeed provide very effective barriers for the diffusion of solutes.

Secondary cell walls are often interrupted by localized pits. A pit in the wall of a given cell usually occurs opposite a complementary pit in an adjacent cell, the cytoplasm of two adjacent cells being brought into close proximity at such a pit-pair. The local absence of extensive cell wall substance facilitates the diffusion of molecules from one cell into the other. An easier way for molecules to move between plant cells is by means of the *plasmodesmata* (singular: *plasmodesma*). These are fine, membrane-flanked, cytoplasmic threads that pass from a protoplast, through a pore in the cell wall, directly into the protoplast of a second cell (Fig. 1.12). The pores generally occur in locally thin regions of the primary cell wall, referred to as primary pit-fields, which can contain many plasmodesmata. (If a secondary cell wall is deposited, openings in that wall occur over the primary pit-fields.) Plasmodesmata can be 50 nm in diameter

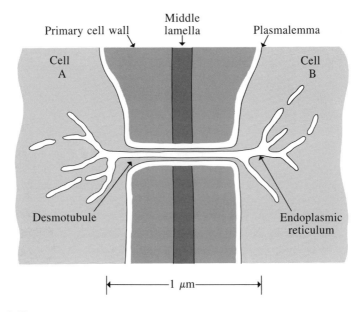

Figure 1.12

Longitudinal section through a plasmodesma, showing that the primary cell walls are locally thinner (which is generally the case) and that the plasmalemmas of the two adjacent cells are continuous.

(range, 20 to 200 nm) and typically have a frequency of 2 to 10 per μm^2 of cell surface (Robards, 1975). They usually occupy about 0.1% to 0.5% of the surface area of a cell. The continuum of communicating cytoplasm created by such intercellular connections is commonly referred to as the *symplasm*.[13] The plasmodesmata are not simple aqueous channels between cells, because they contain desmotubules and have other ultrastructural characteristics including special constrictions and associations of the desmotubules with the membranes of the endoplasmic reticulum. Nevertheless, the presence of the plasmodesmata can provide a particularly effective pathway for solute movement between adjacent cells where permeability barriers in the form of either cell walls or membranes can be avoided.

Because solute movement from cell to cell through the symplasm may be quite important physiologically, let us make a rather oversimplified calculation so that we can compare it with the concomitant diffusion across the plasmalemma. We will use Fick's first law as presented in Equation 1.8, namely, $J_j = D_j K_j \Delta c_j / \Delta x = P_j \Delta c_j$. Let us consider two adjacent cells in which the glucose concentration differs by 1 mol m^{-3} (1 mM). Glucose is rather insoluble in membrane lipids and might have a permeability coefficient of about 10^{-9} m s^{-1} for a plasmalemma. For movement from cell to cell across the two plasmalemmas in series, the effective P_j would be 0.5×10^{-9} m s^{-1} (inclusion of the cell wall as another series barrier for diffusion would only slightly reduce the effective P_j in this case; see Problem 1.4 for handling the permeability coefficient of barriers in series). Thus, the flux density of glucose across the plasmalemmas toward the cell with the lower concentration is

$$J_{glucose}^{plasmalemma} = (0.5 \times 10^{-9} \text{ m s}^{-1})(1 \text{ mol m}^{-3}) = 5 \times 10^{-10} \text{ mol m}^{-2} \text{ s}^{-1}$$

Let us next consider the flux for the symplasm. In the aqueous part of the plasmodesmata, $D_{glucose}$ should be similar to its value in water, 0.7×10^{-9} m^2 s^{-1} (see Table 1.1), and $K_{glucose}$ would be unity. We will let the plasmodesmata be 1 μm long. The flux density in the pores thus is

$$J_{glucose}^{pores} = \frac{(0.7 \times 10^{-9} \text{ m}^2 \text{ s}^{-1})(1)}{(10^{-6} \text{ m})} (1 \text{ mol m}^{-3}) = 7 \times 10^{-4} \text{ mol m}^{-2} \text{ s}^{-1}$$

If the aqueous channels or pores occupy 10% of the area of the plasmodesmata, which in turn occupy 0.2% of the surface area of the cells, then the rate of

---

13. The term *symplast* is defined as the continuum of proto*plasts* together with the plasmodesmata which interconnect them, and hence means the same as symplasm (compare *apoplast*).

glucose diffusion through the plasmodesmata per unit area of the *cells* would be (0.1) (0.002) $J_{glucose}^{pores}$, or 14×10^{-8} mol m^{-2} s^{-1}. This is more than a hundredfold greater than the simultaneously occurring diffusion across the plasmalemmas.

The permeability coefficients of the plasmalemmas for phosphorylated (charged) sugars like ribose-5-phosphate or ribulose-1,5-diphosphate would be less than that for glucose, and the diffusion coefficients in the plasmodesmata would be about the same as for glucose. Thus, the discrepancy between flux densities of charged species through the plasmodesmata and those across the plasmalemmas would be even greater than the difference calculated here for glucose. The joining of cytoplasms into a symplasm indeed facilitates the diffusion of solutes from one cell to another.

Stress–Strain Relations of Cell Wall

Cell walls of mature plant cells are generally quite resistant to mechanical stretching, especially when there is appreciable thickening of the secondary cell walls. Nevertheless, cell walls will stretch when a stress is applied. Reversible elastic properties are generally described by a measure of elasticity known as *Young's modulus,* which is the ratio of applied stress (force per unit area) to the resulting strain (fractional change in length):

$$\text{Young's modulus} = \frac{\text{stress}}{\text{strain}} = \frac{\text{force/area}}{\Delta l / l} \qquad (1.14)$$

Because $\Delta l / l$ is dimensionless, Young's modulus has the dimensions of force per unit area, or pressure. A high value of this modulus of elasticity means that a large stress must be applied to produce a noticeable strain.

Young's modulus for dry cotton fibers, which are nearly pure cellulose, is quite large—about 10^{10} N m^{-2}, or 10 000 MPa, which is 5% of that for steel (Preston, 1974).[14] Because of both the complicated three-dimensional array of microfibrils in the cell wall and the presence of many other components, Young's modulus for a cell wall is considerably less than for pure cellulose. For example, the modulus of elasticity for the cell wall of *Nitella* is about 700 MPa (Kamiya

---

14. The newton, abreviated N, and the pascal, abbreviated Pa, are the SI units for force and pressure, respectively (1 N m^{-2} = 1 Pa). Pressures in plant studies are often expressed in bars, where 1 bar equals 10^6 dynes cm^{-2}, 0.987 atmosphere, or 0.1 MPa. See Appendix II for further conversion factors for pressure.

et al., 1963). We will use this value when we indicate the fractional stretching that can occur for plant cells.

The hydrostatic (turgor) pressure (force per unit area in a liquid), P, acts uniformly in all directions in a cell. This internal pressure pushes against the plasmalemma, which itself is closely appressed to the cell wall. This tends to cause the cell to expand and also leads to tensions (stresses) in the cell wall. The magnitude of these stresses varies with the physiological state of the plant as well as with direction, an aspect that we will consider next.

A cylinder, which closely approximates the shape of the large internodal cells of *Chara* or *Nitella* (Fig. 3.9), is a useful geometrical model for evaluating the various cell wall stresses. We will let the radius of the cylinder be r (Fig. 1.13). The force on an end wall is the pressure, P, times the area of the end wall, πr^2, so it equals $P\pi r^2$. This force is balanced by a force arising from the *longitudinal stress*, σ_L, that occurs in the cell wall. The area over which σ_L acts is shown by the cut portion of the cell wall in the left part of Figure 1.13. The longitudinal stress occurs in an annulus of circumference $2\pi r$ (approximately) and a width equal to the thickness of the cell wall, t. The longitudinal force in the cell wall is thus $(\sigma_L)(2\pi r)(t)$. This force is an equal and opposite reaction to $P\pi r^2$, so $\sigma_L 2\pi r t$ equals $P\pi r^2$, or

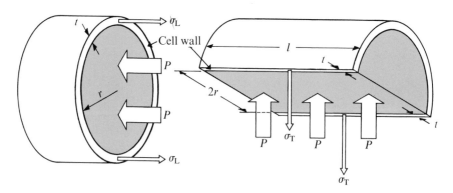

Figure 1.13

Schematic sections of a hypothetical cylindrical cell resembling the internodal cells of *Nitella* or *Chara*, illustrating various dimensions and the stresses (σ_L and σ_T) existing in the cell wall. One way to calculate the stresses is to imagine that the cellular contents are removed, leaving only the cell wall, which has a uniform hydrostatic pressure P acting perpendicular to its inside surface. The projection of this P over the appropriate area gives the force acting in a certain direction, and the reaction to this force is an equal force in the opposite direction in the cell wall. By dividing the force in the cell wall by the area over which it occurs, we can determine the cell wall stress.

The longitudinal stress acts parallel to the major axis of the cylinder and resists the lengthwise deformation of the cell.

A *tangential stress,* σ_T, also exists in the cell wall in response to the internal pressure; it limits the radial expansion of the cell. To determine the magnitude of this stress, we will consider a cell split in half along its axis (right side of Fig. 1.13). The split part of the cell has an area of $2rl$ in the plane of the cut; this area is acted on by the pressure P, leading to a force of $P2rl$. This force is resisted by the tangential stress in the cell wall. As shown in Figure 1.13, σ_T acts along two cell wall surfaces, each of width t and length l. The total tangential force in the cut part of the cell wall is thus the area $(2tl)$ times the tangential stress. Equating this force $(\sigma_T 2tl)$ with that due to the internal pressure in the cell $(P2rl)$, we obtain the following relationship for the tangential stress:

$$\sigma_T = \frac{rP}{t} \tag{1.16}$$

The tangential stress in the cell wall given by Equation 1.16 is twice as large as the longitudinal stress (Eq. 1.15). Thus, this simple cylindrical model illustrates the important point that the stresses in a cell wall can vary with direction (Lockhart, 1965). Young's modulus also varies with direction, reflecting the anisotropic orientation of the microfibrils in the cell wall.

To estimate the magnitudes of the stresses and the resulting strains in the cell wall, let us consider a *Nitella* or *Chara* cell 1 mm in diameter with a cell wall 5 μm thick. In this case, r/t is

$$\frac{r}{t} = \frac{(0.5 \times 10^{-3} \text{ m})}{(5 \times 10^{-6} \text{ m})} = 100$$

A reasonable estimate of P is 0.5 MPa (5 bars). Using Equation 1.15, we can calculate the longitudinal stress in the cell wall of such a cell:

$$\sigma_L = \frac{(100)(0.5 \text{ MPa})}{(2)} = 25 \text{ MPa}$$

which is an appreciable tension. By Equation 1.16, the tangential stress would be twice as great, or 50 MPa. Ignoring changes in the radial direction, we can calculate the strain produced by the longitudinal stress using the definition of Young's modulus (Eq. 1.14) and its particular value in the longitudinal direction for the cell wall of *Nitella,* 700 MPa. From Equation 1.14, the fractional change in length is then

$$\frac{\Delta l}{l} = \frac{(25 \text{ MPa})}{(700 \text{ MPa})} = 0.036$$

i.e., only about 4%. [If changes in the radial direction were included, the length change would be about 3% in the present case (Vinters et al., 1977).] Hence, even with an internal pressure of 0.5 MPa, the cell wall (and consequently the whole cell) is not extended very much. The cell wall is indeed rigid and therefore well suited both for delimiting individual cells and for providing the structural support of a plant.

Elastic properties of plant cell walls must be considered when analyzing the cell expansion accompanying growth as well as other features of water movement. For this, we are interested in what volume change (ΔV) is caused by a given pressure change (ΔP); this relation can be quantified using the volumetric elastic modulus (ε):

$$\varepsilon = \frac{\Delta P}{\Delta V/V} \tag{1.17}$$

Equation 1.17 indicates that $\Delta V/V = \Delta P/\varepsilon$, so cells with a higher ε will have a smaller fractional change in volume for a given ΔP, i.e., they are more rigid. Values of ε generally range from 1 to 50 MPa (Dainty, 1976; Zimmermann, 1978), meaning that cells change in volume by 0.2% to 10% for each 0.1 MPa change in pressure, but only a few plant species have been examined so far. Also, ε depends both on P (it is smaller at lower P), V, and the developmental stage of the cells. We will return to a consideration of the elastic modulus when discussing the water relations of cells in Chapter 2. We note here that Young's modulus (Eq. 1.14) and the volumetric elastic modulus (Eq. 1.17) both represent an applied pressure divided by a fractional change in size.

Before concluding this discussion of cell walls, we note that the case of elasticity or reversible deformability is only one extreme of stress-strain behavior. At the opposite extreme is plastic (irreversible) extension. If the amount of strain is directly proportional to the time that a certain stress is applied, and if the strain persists when the stress is removed, we have viscous flow. The cell wall exhibits intermediate properties and is said to be *viscoelastic*. When a given stress is applied to a viscoelastic material, the resulting strain is approximately proportional to the logarithm of time. Such extension is partly elastic (reversible) and partly plastic (irreversible). Underlying the viscoelastic behavior of the cell wall are the crosslinks between the various polymers. For example, if a bond from one cellulose polymer to another is broken while the cell wall is under tension, the opportunity exists for a new bond to form in a less strained configuration, leading to an irreversible or plastic extension of the cell wall. The quantity responsible for the tension in the cell wall—which in turn leads to such viscoelastic extension—is the hydrostatic pressure within the cell.

Problems

1.1 A thin layer of some solution is inserted into a long column of water. One hour later the concentration of the solute is 100 mol m^{-3} (0.1 M) at the plane of insertion and 37 mol m^{-3} (0.037 M) at a distance 3 mm away. (a) What is its diffusion coefficient? (b) When the concentration 90 mm away is 37% of the value at the plane of insertion, how much time has elapsed? (c) How many moles of solute per unit area were initially inserted into the column of water? (d) Suppose that a trace amount of a substance having a diffusion coefficient 100 times smaller than that of the main solute was also initially introduced. For the time in (b), where would its concentration drop to 1/e of the value at the plane of insertion?

1.2 Let us suppose that mitochondria with a volume of 0.30 μm^3 each and a density of 1 110 kg m^{-3} (= 1.10 g cm^{-3}) diffuse like a chemical species. (a) What is the "molecular weight" of mitochondria? (b) Suppose that a chemically similar species of molecular weight 200 has a diffusion coefficient of 0.5 × 10^{-9} m^2 s^{-1}. If diffusion coefficients are inversely proportional to the cube root of molecular weights for this series of similar species, what is $D_{\text{mitochondria}}$? (c) If we assume that Equation 1.6 can adequately describe such motion, how long would it take on the average for a mitochondrion to diffuse 0.2 μm (a distance just discernible using a light microscope)? How long would it take for the mitochondrion to diffuse 50 μm (the distance across a typical leaf cell)? (d) If D_{ATP} is 0.3 × 10^{-9} m^2 s^{-1}, how long would it take ATP to diffuse 50 μm? Is it more expedient for mitochondria or ATP to diffuse across a cell?

1.3 Suppose that an unstirred air layer 1 mm thick is adjacent to a guard cell with a cell wall 2 μm thick. (a) Assume that an (infinitely) thin layer of $^{14}CO_2$ is introduced at the surface of the guard cell. If D_{CO_2} is 10^6 times larger in air than in the cell wall, what are the relative times for $^{14}CO_2$ to diffuse across the two barriers? (b) If it takes $^{14}CO_2$ just as long to cross an 8 nm plasmalemma as it does to cross the cell wall, what are the relative sizes of the two diffusion coefficients (assume that the $^{14}CO_2$ was introduced in a plane between the two barriers)? (c) Assuming that the partition coefficient for CO_2 is 100 times greater in the cell wall than in the plasmalemma, in which barrier is the permeability coefficient larger, and by how much?

1.4 Without correcting for the effect of an unstirred layer 20 μm thick outside a membrane 7.5 nm in thickness, the apparent (total) permeability coefficients were measured to be 1.0 × 10^{-4} m s^{-1} for D$_2$O, 2.0 × 10^{-5} m s^{-1} for methanol, and 3.0 × 10^{-8} m s^{-1} for L-leucine. For barriers in series, the overall permeability coefficient for species j (P_j^{total}) is related to those of the individual barriers (P_j^i) as follows: $1/P_j^{\text{total}} = \Sigma_i \, 1/P_j^i$. For purposes of calculation, we will assume that in the present case the unstirred layer on the inner side of the membrane is negligibly thin. (a) What is P_j for the external unstirred layer for each of the compounds? Assume that D_{D_2O} is 2.6

$\times$ 10^{-9} m^2 s^{-1}, D_{methanol} is 0.80×10^{-9} m^2 s^{-1}, and D_{leucine} is 0.20×10^{-9} m^2 s^{-1} in water at 25°C. (b) What are the permeability coefficients of the three compounds for the membrane? (c) From the results in (a) and (b), what are the main barriers for the diffusion of the three compounds in this case? (d) What are the highest possible values at 25°C of P_j^{total} for each of the three compounds moving across an unstirred layer of 20 μm and an extremely permeable membrane in series?

1.5 Consider a solute having a permeability coefficient of 10^{-6} m s^{-1} for the plasmalemma of a cylindrical *Chara* cell that is 100 mm long and 1 mm in diameter. Assume that its concentration remains essentially uniform within the cell. (a) How much time would it take for 90% of the solute to diffuse out into a large external solution initially devoid of that substance? (b) How much time would it take if diffusion occurred only at the two ends of the cell? (c) How would the times calculated in (a) and (b) change for 99% of the solute to diffuse out? (d) How would the times change if P_j were 10^{-8} m s^{-1}?

1.6 A cylindrical *Nitella* cell is 100 mm long and 1 mm in diameter, a spherical *Valonia* cell is 10 mm in diameter, and a spherical *Chlorella* cell is 4 μm in diameter. (a) What is the area/volume in each case? (b) Which cell has the largest amount of surface area per unit volume? (c) If it takes 1 s for the internal concentration of ethanol, which is initially absent from the cells, to reach half the external concentration for *Chlorella*, how long would it take for *Nitella* and *Valonia*? Assume that P_{ethanol} is the same for all the cells. (d) Assume that the cell walls are equal in thickness. For a given internal pressure, which cell would have the highest cell wall stress (consider only the lateral wall for *Nitella*)?

References

Altman, P. L., and D. S. Dittmer. 1966. *Environmental Biology*. Federation of American Societies for Experimental Biology, Bethesda, Maryland.

Brett, C. T., and J. R. Hillman, eds. 1985. *Biochemistry of Plant Cell Walls*. Cambridge University Press, Cambridge, England.

Crank, J. 1975. *The Mathematics of Diffusion*, 2nd ed. Clarendon Press (Oxford University Press), Oxford, England.

Dainty, J. 1976. Water relations of plant cells. In *Transport in Plants II, Part A, Cells*, U. Lüttge and M. G. Pitman, eds. *Encyclopedia of Plant Physiology, New Series*, Vol. 2. Springer-Verlag, Berlin. Pp. 12–35.

Davson, H., and J. F. Danielli. 1952. *The Permeability of Natural Membranes*, 2nd ed. Cambridge University Press, Cambridge, England.

Diamond, J. M., and E. M. Wright. 1969. Biological membranes: The physical basis of ion and nonelectrolyte selectivity. *Annual Review of Physiology 31*:581–646.

Douce, R., and J. Joyard. 1979. Structure and function of the plastid envelope. *Advances in Botanical Research 7*:1–117.

Esau, K. 1977. *Anatomy of Seed Plants*, 2nd ed. Wiley, New York.

Fasman, G. D., ed. 1976. *Handbook of Biochemistry and Molecular Biology*, 3rd ed., *Proteins*, Vol. II. CRC Press, Cleveland, Ohio.

Fick, A. (1855). Über Diffusion. *Poggendorffs Annalen 94*:59–86.

Fischkoff, S., and J. M. Vanderkooi, 1975. Oxygen diffusion in biological and artificial membranes determined by the fluorochrome pyrene. *Journal of General Physiology 65*:663–676.

Fry, S. C. 1988. *The Growing Plant Cell Wall: Chemical and Metabolic Analysis*. Wiley, New York.

Goldman, D. T., and R. J. Bell, eds. 1986. *The International System of Units* (SI), 5th ed. National Bureau of Standards, Washington, D.C.

Green, K., and T. Otori. 1970. Direct measurements of membrane unstirred layers. *Journal of Physiology 207*:93–102.

Heber, U., and H. W. Heldt. 1981. The chloroplast envelope: Structure, function, and role in leaf metabolism. *Annual Review of Plant Physiology 32*:139–168.

Jacobs, M. H. 1967. *Diffusion Processes*. Springer-Verlag, New York.

Kamiya, N. 1981. Physical and chemical basis of cytoplasmic streaming. *Annual Review of Plant Physiology 32*:205–236.

Kamiya, N., M. Tazawa, and T. Takata. 1963. The relation of turgor pressure to cell volume in *Nitella* with special reference to mechanical properties of the cell wall. *Protoplasma 57*:501–521.

Kotyk, A., K. Janáček, and J. Koryta. 1988. *Biophysical Chemistry of Membrane Functions*. Wiley, Chichester, England.

Lauffer, M. A. 1989. *Motion in Biological Systems*. Liss, New York.

Linskens, H. F., and J. F. Jackson, eds. 1985. *Cell Components. Modern Methods of Plant Analysis, New Series*, Vol. 1. Springer-Verlag, Berlin.

Lockhart, J. A. 1965. Cell extension. In *Plant Biochemistry*, J. Bonner and J. E. Varner, eds. Academic Press, New York. Pp. 826–849.

Preston, R. B. 1974. *The Physical Biology of Plant Cell Walls*. Chapman and Hall, London.

Robards, A. W. 1975. Plasmodesmata. *Annual Review of Plant Physiology 26*:13–29.

Silver, B. L. 1985. *The Physical Chemistry of Membranes*. Allen & Unwin, Boston, Massachusetts.

Simon, E. W. 1974. Phospholipids and plant membrane permeability. *New Phytologist 73*:377–420.

Singer, S. J., and G. L. Nicolson. 1972. The fluid mosaic model of the structure of cell membranes. *Science 175*:720–731.

Stadelmann, E. J. 1977. Passive transport parameters of plant cell membranes. In *Regulation of Cell Membrane Activities in Plants*, E. Marré and O. Cifferi, eds. North-Holland, Amsterdam. Pp. 3–18.

Starzak, M. E. 1984. *The Physical Chemistry of Membranes*. Academic Press, Orlando, Florida.

Stein, W. D. 1986. *Transport and Diffusion across Cell Membranes*. Academic Press, Orlando, Florida.

Tolbert, N. E., ed. 1980. *The Plant Cell*. In *The Biochemistry of Plants: a Comprehensive Treatise*, P. K. Stumpf and E. E. Conn, eds., Vol. 1. Academic Press, New York.

Vinters, H., J. Dainty, and M. T. Tyree. 1977. Cell wall elastic properties of *Chara corallina*. *Canadian Journal of Botany 55*:1933–1939.

Weast, R. C., and D. R. Lide, eds. 1989. *CRC Handbook of Chemistry and Physics*, 70th ed. CRC Press, Boca Raton, Florida.

Wright, E. M., and J. M. Diamond. 1969. Patterns of non-electrolyte permeability. *Proceedings of the Royal Society of London, Series B 172*:227–271.

Zimmermann, U. 1978. Physics of turgor- and osmoregulation. *Annual Review of Plant Physiology 29*:121–148.

CHAPTER 2

Water

Water is the main constituent of most plant cells, as our discussion of both vacuoles and cell walls in the previous chapter has already suggested. The actual cellular water content varies with cell type and physiological condition. For example, the carrot root is about 85% water by weight, and the young inner leaves of lettuce contain up to 95% water. Water makes up only 5% of certain dry seeds and spores, but when these become metabolically active, an increase in water content is essential for the transformation.

The physical and the chemical properties of water make it suitable for a variety of purposes in plants. It is the medium in which diffusion of solutes takes place in plant cells. It is a solvent and often a participant in biochemical reactions. Its incompressibility means that water uptake can lead to cell expansion and that intracellular hydrostatic pressures can help support plants. It is well suited to temperature regulation, because it has a high heat of vaporization, a high thermal capacity per unit mass, and a high thermal conductivity for a liquid. Water is also an extremely good general solvent, in part owing to the small size of its molecules. Its polar character makes water suitable for dissolving other polar substances. Its high dielectric constant makes it a particularly appropriate solvent for ions. This latter property has far-reaching consequences for life, because most biologically important solutes are electrically charged. The mineral nutrients needed for growth and the organic products of photosynthesis are transported throughout the plant in aqueous solutions. In actively growing land plants, water continuity exists from the soil, through the plant, to the evaporation sites in the leaves. Water is relatively transparent to visible irradiation, enabling sunlight to reach chloroplasts within the cells in leaves and to reach plants submerged at appreciable depths in lakes and oceans. Water is also intrinsically involved with metabolism. It is the source of the O_2 evolved in photosynthesis and the hydrogen used for CO_2 reduction. The generation of the important energy currency, ATP, involves the extraction of the components of water from ADP plus phosphate; in other words, such a phosphorylation is a *dehydration* process taking place in an *aqueous* solution under biological conditions. At the ecological level, the facts that ice floats and that water is densest near 4°C are important for organisms living in streams or rivers subject to freezing. Water existing as a solid at temperatures within the biological realm also has many other important consequences for water exchange, metabolism, and ultimately cell survival. Thus, an understanding of the physical and chemical properties of water is crucial for an understanding of physiology and ecology.

A number of isotopically different forms of water can be prepared, greatly facilitating experimental studies. If deuterium (^2H) replaces both of the usual hydrogens, the result is "heavy water," or deuterium oxide, with a molecular weight of 20. The entry of water into chemical reactions in the plant can then be studied by analyzing the deuterium content of the various substances involved as reactants or products. Tritium (^3H), a radioactive isotope with a half-life of 12.4 years, can also be incorporated into water. Such tritiated water has been used to measure the rate of water diffusion in plant tissue. Another alternative for tracing the pathway of water is to replace the usual ^{16}O isotope with ^{18}O. This "labeling" of water with ^{18}O was used to determine that the O_2 evolved in photosynthesis comes from water and not from CO_2 (Ch. 5).

PHYSICAL PROPERTIES

Water differs markedly from substances having closely related electronic structures, which can be illustrated by considering the series CH_4, NH_3, H_2O, HF, and Ne (Fig. 2.1). Each molecule contains 10 protons and 10 electrons, but the number of hydrogen atoms decreases along the series from methane to neon. The relatively high melting point and boiling point for water, compared with substances having a similar electronic structure, are consequences of its strong intermolecular forces. In other words, thermal agitation does not easily disrupt the water-water bonding. This strong attraction between molecules is responsible for many characteristic properties of water.

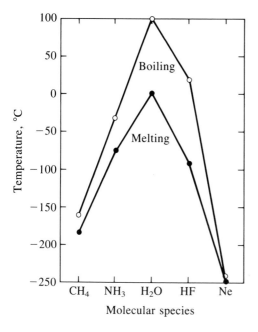

Figure 2.1

Boiling points and melting points for molecules with 10 protons and 10 electrons, showing the strikingly high values for water. Most biological processes take place from 0°C to 50°C, where water can be in the liquid state but the other substances cannot.

Hydrogen Bonding—Thermal Relations

The strong intermolecular forces in water result from the structure of the H_2O molecule (Fig. 2.2). The internuclear distance between the oxygen and each of the two hydrogens is approximately 0.099 nm; the H—O—H bond angle is about 105°. The oxygen atom is strongly electronegative and tends to draw electrons away from the hydrogen atoms. The oxygen thus has a partial negative charge (δ^- in Fig. 2.2), and the two hydrogens each have a partial positive charge (δ^+). These positively charged hydrogens are electrostatically attracted to the negatively charged oxygens of two neighboring water molecules. This leads to *hydrogen bonding* between water molecules, with an energy of about 20 kilo-joules (kJ) mol^{-1} of hydrogen bonds. Such bonding of water molecules to each other leads to increased order in aqueous solutions. In fact, liquid water becomes nearly crystalline in local regions, which affects the molecular interactions and orientations that occur in aqueous solutions.

Ice is a coordinated crystalline structure in which essentially all the water molecules are joined by hydrogen bonds. When enough energy is added to melt the ice, some of these intermolecular hydrogen bonds are broken. The heat of fusion of ice at 0°C is 6.0 kJ mol^{-1}. Total rupture of the intermolecular hydrogen

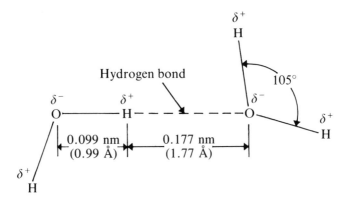

Figure 2.2

Schematic structure of water molecules, indicating the hydrogen bonding resulting from the electrostatic attraction between the net positive charge on a hydrogen (δ^+) in one molecule and the net negative charge on an oxygen (δ^-) in a neighboring water molecule. Depending on the model, the δ^+ is about 0.3 of a protonic charge and the δ^- is about -0.6 (see Hermans *et al.*, 1984; Israelachvili, 1985).

bonds involving both of its hydrogens would require 40 kJ mol^{-1} of water.[1] The heat of fusion thus indicates that (6.0 kJ mol^{-1})/(40 kJ mol^{-1}), or at most 0.15 (15%), of the hydrogen bonds are broken when ice melts. (Some energy is needed to overcome van der Waals attractions,[2] so less than 15% of the hydrogen bonds are actually broken upon melting.) Conversely, over 85% of the hydrogen bonds remain intact for liquid water at 0°C. Because 75.4 J is needed to heat 1 mol of water by 1°C (the thermal capacity of water on a mole basis), it takes (0.0754 kJ mol^{-1} °C^{-1})(25°C), or 1.9 kJ mol^{-1}, to heat water from 0°C to 25°C. Even if all this energy were used to break hydrogen bonds (1.9 kJ mol^{-1} could break about 5% of the possible hydrogen bonds in water), over 80% of the bonds

---

1. The actual magnitude of the hydrogen bond energy assigned to ice depends somewhat on the particular operational definition used in the measurement of the various bonding energies. Therefore, the values in the literature vary somewhat (see Eisenberg and Kauzmann, 1969; Schuster *et al.*, 1976).

2. Van der Waals forces are the electrostatic attractions between electrons in one molecule and the nucleus of an adjacent molecule minus the molecules' interelectronic and internuclear repulsive forces (see Castellan, 1983). In about 1930, London showed these forces to be caused by the attraction between an electric dipole in some molecule and the electric dipole induced in an adjacent one (London, 1937). Therefore, van der Waals forces result from random fluctuations of charge and are important only for molecules that are very close together—in particular, for neighboring molecules.

would still remain intact at 25°C. Such bonding leads to the semicrystalline order observed in aqueous solutions. The extensive intermolecular hydrogen bonding present in the liquid state contributes to the unique and biologically important properties of water that we will discuss throughout this chapter.

The energy required to separate molecules from a liquid and move them into an adjacent vapor phase without a change of temperature is called the *heat of vaporization*. For water, the heat of vaporization at 100°C is 2.26 MJ kg^{-1}, or 40.7 kJ mol^{-1}. Per unit mass, this is the highest heat of vaporization of any known liquid, and it reflects the large amount of energy required to disrupt the extensive hydrogen bonding in aqueous solutions. More pertinent to plants is the heat of vaporization of water at temperatures encountered by them. At 25°C each mole of water evaporated requires 44.0 kJ (see App. I), meaning that substantial heat loss by a plant accompanies the evaporation of water in transpiration. Most of this vaporization energy is needed to break hydrogen bonds so that the water molecules can become separated in the gaseous phase. For example, if 80% of the hydrogen bonds remained at 25°C, then (0.80) (40 kJ mol^{-1}), or 32 kJ mol^{-1}, would be needed to rupture them. Additional energy is needed to overcome the attractive van der Waals forces and for the expansion involved in going from a liquid to a gas. The heat loss accompanying the evaporation of water is one of the primary means of temperature regulation in land plants, dissipating much of the energy gained from the absorption of solar irradiation. (Energy budgets will be considered in detail in Ch. 7.)

Surface Tension

Water has an extremely high surface tension, evident at an interface between water and air. Surface tension can be defined as the force per unit length that can pull perpendicular to a line in the plane of the surface. Because of its high surface tension, water can support a steel needle carefully placed on its surface. The surface tension at such an air-water interface is 0.0728 N m^{-1} at 20°C (see App. I for values at other temperatures). Another way of understanding surface tension is to regard it as the amount of energy required to expand a surface by unit area. In particular, we note that surface tension has the dimensions not only of force per unit length, but also of energy per unit area, which we turn to next.

To see why energy is required to expand the water surface, let us consider water molecules brought from the interior of an aqueous phase to an air-water interface. If this involves only an expansion of the surface area—i.e., if there is no accompanying movement of other water molecules from the surface to the interior—then a loss in the water-water attraction from some of the intermolecular hydrogen bonds occurs with no compensating air-water attraction. Energy is thus needed to break the hydrogen bonds that are lost in moving water molecules from the interior of the solution to the air-water interface (0.0728 J to increase the area by 1 m^2). In fact, the term "surface free energy" is more appropriate from a thermodynamic point of view than the conventional term "surface tension."

The surface tension of an aqueous solution usually is only slightly influenced by the composition of the adjacent gas phase, but it can be markedly affected by certain solutes. Molecules are relatively far apart in a gas—the density of dry air at 0°C and one standard atmosphere (0.1013 MPa, 1.013 bar, or 760 mm Hg) is 1.293 kg m^{-3}—so the frequency of interactions between molecules in the gas phase and those in the liquid phase is relatively low. Certain solutes, such as sucrose or KCl, do not preferentially collect at the air-liquid interface and consequently have little effect on the surface tension of an aqueous solution. But because of their hydrophobic regions, fatty acids and certain lipids may become concentrated at interfaces, which can greatly reduce surface tensions. For example, 10 mol m^{-3} (10 mM) caproic acid (a 6-carbon fatty acid) lowers the surface tension by about 21% below the value for pure water, and only 0.05 mol m^{-3} capric acid (a 10C fatty acid) lowers it by about 34% (see Bull, 1964). Substances such as soaps (salts of fatty acids), other emulsifiers, and denatured proteins with large hydrophobic side chains may collect nearly exclusively at the interface and reduce the surface tension of aqueous solutions by over 70%. A common feature of such "surfactant" molecules is that they have both polar and nonpolar regions.

Capillary Rise

The intermolecular attraction occurring between like molecules in the liquid state, such as the water-water attraction based on hydrogen bonds, is called *cohesion*. The attractive interaction between a liquid and a solid phase, such as between water and the walls of a small capillary (a cylindrical tube with a small internal diameter), is called *adhesion*. When the water-wall attraction is appreciable compared with the water-water cohesion, the walls are said to be *wettable,* and water then rises in such a vertical capillary. At the opposite extreme, where the intermolecular cohesive forces within the liquid are substantially greater than is the adhesion between the liquid and the wall material, the upper level of the liquid in such a capillary is lower than the surface of the solution. Capillary depression occurs for liquid mercury in glass capillaries. For water in glass capillaries or in xylem vessels, the attraction between the water molecules and the walls is great, so the liquid rises. Because capillary rise has important implications in plant physiology, we will discuss its characteristics quantitatively.

As an example appropriate to the evaluation of water ascent in a plant, let us consider a capillary of inner radius r with wettable walls dipping into some aqueous solution (Fig. 2.3a). The strong adhesion of water molecules to the wettable wall causes the fluid to rise up the inner wall of the capillary. Because a strong water-water cohesion occurs in the bulk solution, water is concomitantly pulled up into the lumen of the capillary as water rises along the wall. Viewed another way, the air-water surface greatly resists being stretched, a property reflected in the high surface tension of water at air-water interfaces. This resistance minimizes the area of the air-water interface, a condition achieved if water also moves up in the lumen as well as along the inner wall of the capillary. The rise of water along the wall of the capillary is thus transmitted to a volume of fluid. We will call the height that the liquid rises in the capillary h and the contact angle that the liquid makes with the inner capillary wall α (Fig. 2.3). The extent of the rise depends on α, so the properties of the contact angle will now be examined a little more closely.

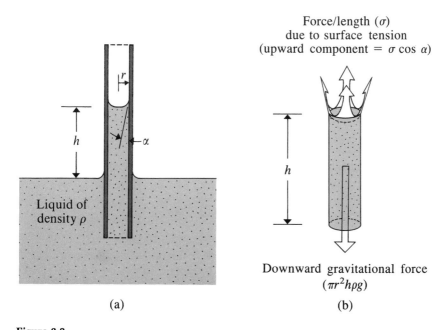

(a)　　　　　　　　　　　　(b)

Figure 2.3

Capillary rise of a liquid: (a) variables involved, and (b) force diagram indicating that surface tension projected in the upward direction is balanced by gravity acting downward.

The size of the contact angle depends on the magnitude of the liquid-solid adhesive force compared with that of the liquid-liquid cohesive force. Specifically, Young's equation indicates that

$$\text{adhesion} = \frac{1 + \cos \alpha}{2} \text{ cohesion} \qquad (2.1)$$

[Equation 2.1 is sometimes called the Young and Dupré equation; for an authoritative treatment, see Adamson (1982), Israelachvili (1985), or Rosen (1989).] When the adhesive force equals (or exceeds) the cohesive force in the liquid, $\cos \alpha$ is unity; the contact angle α then equals zero ($\cos 0° = 1$). This is the case for water in capillaries made of clear, smooth glass or having walls with polar groups on the exposed surface. When the adhesive force equals half the cohesive force, $\cos \alpha$ is zero and the contact angle is then 90° ($\cos 90° = 0$) by Equation 2.1. In this case, the level of the fluid in the capillary is the same as that in the bulk of the solution. This latter condition is closely approached for water-polyethylene adhesion, where α equals 94°. As the liquid-solid adhesive force becomes relatively less compared with the intermolecular cohesion in the liquid phase, the contact angle (indicated in Fig. 2.3a) increases toward 180°, and capillary depression occurs. For instance, water has an α of about 110° with paraffin, and the contact angle for mercury interacting with a glass surface is about 150°. In such cases, the level of the liquid is lower in the capillary than in the bulk solution (see Fig. 2.3a, but imagine that $\alpha > 90°$, so the liquid curves downward where it intersects the surface of the capillary).

We can calculate the extent of capillary rise by considering the balance of two forces: (1) gravity acting downward and (2) surface tension, which leads to an upward force in the case of a wettable wall in a vertical tube (see Fig. 2.3b). The force pulling upward acts along the inside perimeter of the capillary, a distance of $2\pi r$, with a force per unit length equal to σ, the surface tension. The component of this force acting vertically upward is $2\pi r \sigma \cos \alpha$, where α is the contact angle illustrated in Figure 2.3a. This upward force is balanced by the gravitational force acting on the liquid of density ρ and volume of approximately $\pi r^2 h$.[3] The gravitational force is the mass involved times the gravitational acceleration g ($F = ma$, Newton's second law of motion). In the present case, gravity acts on a mass of fluid of approximately $\pi r^2 h \rho$, so the gravitational force is $\pi r^2 h \rho g$, acting downward. This force is balanced by $2\pi r \sigma \cos \alpha$ pulling in

---

3. Some fluid is also held in the "rim" of the meniscus, as indicated in Figure 2.3a. For narrow capillaries, the fluid in the meniscus increases the effective height of the column by about $r/3$.

the opposite direction, so the extent of rise, h, is given by equating the two forces ($\pi r^2 h \rho g = 2 \pi r \sigma \cos \alpha$), which leads to

$$h = \frac{2\sigma \cos \alpha}{r \rho g} \tag{2.2a}$$

Equation 2.2a indicates the readily demonstrated property that the extent of liquid rise in a capillary is inversely proportional to the radius of the tube. For water in glass capillaries as well as in many of the fine channels encountered in plants where the cell walls have a large number of exposed polar groups, the contact angle can be near zero, in which case $\cos \alpha$ in Equation 2.2a can be set equal to 1. The density of water at 20°C is 998 kg m^{-3} (actually, ρ in Eq. 2.2a is the difference between the liquid density and the density of the displaced air, the latter being about 1 kg m^{-3}), and the acceleration due to gravity, g, is about 9.80 m s^{-2} (App. I). Using a surface tension for water at 20°C of 0.0728 N m^{-1} (App. I), we obtain the following relation between the height of the rise and the capillary radius when the contact angle is zero:

$$h_{(m)} = \frac{1.49 \times 10^{-5} \text{ m}^2}{r_{(m)}} \tag{2.2b}$$

where the subscripts (m) in Equation 2.2b mean that the dimensions involved are expressed in metres.

Capillary Rise in the Xylem

Although Equation 2.2 refers to the height of capillary rise only in a static sense, it still has important implications concerning the movement of water in plants. To be specific, let us consider a xylem vessel having a lumen radius of 20 μm. From Equation 2.2b, we calculate that water will rise in it to the following height:

$$h_{(m)} = \frac{(1.49 \times 10^{-5} \text{ m}^2)}{(20 \times 10^{-6} \text{ m})} = 0.75 \text{ m}$$

Such a capillary rise would be sufficient to account for the extent of the ascent of water in small plants, although it says nothing about the rate of such movement (which we consider in Ch. 9). For water to reach the top of a 30-m tree by capillary action, however, the vessel would have to be 0.5 μm in radius. This is much smaller than that observed for xylem vessels, indicating that capillary rise in channels of the size of the xylem cells cannot account for the extent of the water rise in tall trees. Furthermore, the lumens of the xylem vessels are not open to the air at the upper end, and thus they are not really analogous to the capillary depicted in Figure 2.3.

The numerous interstices in the cell walls of xylem vessels form a meshwork of many small, tortuous capillaries, which can lead to an extensive capillary rise of water in a tree. A representative "radius" for these channels in the cell wall might be 5 nm. According to Equation 2.2b, a capillary of 5 nm radius could support a water column of 3 km—far in excess of the needs of any plant. In other words, the cell wall could act as a very effective wick for water rise in its numerous small interstices, although the actual rate of such movement up a tree is generally far from sufficient to replace the water lost by transpiration (Ch. 9).

Because of the appreciable water-wall attraction that can develop both at the top of a xylem vessel and in the numerous interstices of its cell walls, water already present in the lumen of a xylem vessel can be sustained or supported at great heights. The upward force, transmitted to the rest of the solution in the xylem vessel by water-water cohesion, overcomes the gravitational pull downward. Thus, the key to sustaining water already present in the xylem vessel against the pull of gravity is the very potent attractive interaction (adhesion) between water and the cell wall surfaces in the vessel. Two aspects of this warrant further consideration. First, what happens if the lumen of the xylem vessel becomes filled with air? Will water then refill it? The capillary rise of water is not sufficient to refill most air-filled xylem vessels greater than about 1 m in length, so that, in general, air-filled vessels are lost for conduction, such as for the inner annual rings of most trees.[4] The second aspect is more subtle, and we will consider it more extensively.

Tensile Strength, Viscosity

The pulling on water columns that occurs in capillary rise and in the sustaining of water in the xylem requires that water be put under tension, or negative pressure. In other words, water must have a definite tensile strength—the max-

---

4. A positive hydrostatic pressure can occur in the root xylem (termed *root pressure*, see Ch. 9), which can help refill vessels in herbaceous species as well as in certain woody species in the spring.

imum tension (force per unit area) that it can withstand before breaking. The intermolecular hydrogen bonds lead to this tensile strength by resisting the pulling apart of water in a column. According to experiment, water withstands negative pressures (tensions) of up to approximately 30 MPa (300 bars) at 20°C without breaking. [Tensile strengths depend on the wall material, the diameter of the vessel in which the determinations are made, and any solutes present in the water (Hayward, 1971; Oertli, 1971).] This tensile strength of water is nearly 10% of that of copper or aluminum and is high enough to meet the demands encountered for water movement in plants. The great cohesive forces between water molecules thus allow an appreciable tension to exist in an uninterrupted water column in a wettable capillary or tube such as a xylem vessel. This property is important for the continuous movement of water from the root through the plant to the surrounding atmosphere during transpiration (discussed in Ch. 9).

In contrast to metals, which are in a stable state, water under tension is actually in a *metastable state* (a situation in which a change is ready to occur but does not without some outside impulse). If gas bubbles form in the water under tension in the xylem vessels, the water column can be ruptured (Tyree and Sperry, 1989). Thus, the introduction of another phase can disrupt the metastable state describing water under tension. Minute air bubbles sometimes spontaneously form in the xylem sap. These usually adhere to the walls of the xylem vessel, and the gas in them slowly redissolves. If, however, they grow large enough or a number of small bubbles coalesce, forming an *embolism,* the continuity of water can be interrupted and so that xylem vessel could cease functioning. Moreover, freezing of the solution in xylem vessels can lead to bubbles in the ice (the solubility of gases like CO_2, O_2, and N_2 is quite low in ice), and these bubbles could then interrupt the water columns upon thawing. Most plants are thus damaged by freezing and thawing of xylem sap and the consequent loss of water continuity in the xylem vessels.

Hydrogen bonding also influences the *viscosity* of water. The viscosity of a fluid indicates the resistance to flow, reflecting the cohesion within the fluid as well as transfer of molecular momentum between layers of the fluid. Thus, it is a measure of the difficulty for one layer to slide past an adjacent layer when a shearing force is applied. Because hydrogen bonding can restrict the slipping of adjacent liquid layers, the viscosity of water is relatively high compared with solvents that have little or no hydrogen bonding, e.g., acetone, benzene, chloroform, and other organic solvents with small molecules. The decrease in the viscosity of water as the temperature rises reflects the breaking of hydrogen bonds and also the lessening of other attractive forces (e.g., van der Waals forces) accompanying the greater thermal motion of the molecules.

Electrical Properties

Another important physical characteristic of water is its extremely high dielectric constant, which is also a consequence of its molecular structure. A high dielectric constant for a solvent lowers the electrical forces between charged solutes. To quantify the magnitude of electrical effects occurring in a fluid, let us consider two ions having charges Q_1 and Q_2 and separated from each other by a distance r. The electrical force exerted by one ion on the other is expressed by *Coulomb's law*:

$$\text{electrical force} = \frac{Q_1 Q_2}{4\pi\varepsilon_0 D r^2} \qquad (2.3)$$

where ε_0 is a proportionality constant known as the permittivity of a vacuum and D is a dimensionless quantity called the dielectric constant. D equals unity in a vacuum, and it is 1.0006 in air at 0°C at a pressure of one standard atmosphere.

Any substance composed of highly polar molecules, such as water, generally has a high dielectric constant. Specifically, D for water is 80.2 at 20°C and 78.4 at 25°C—extremely high values for a liquid (the decrease in the dielectric constant of water with increasing temperature primarily reflects increasing thermal motion and decreased hydrogen bonding). By contrast, the dielectric constant of the nonpolar liquid hexane is only 1.87, a low value typical of many organic solvents. Based on these two vastly different dielectric constants and using Equation 2.3, we can calculate that the attractive electrical force between ions such as Na^+ and Cl^- is (80.2/1.87), or 43 times greater in hexane than in water. The much stronger attraction between Na^+ and Cl^- in hexane reduces the amount of NaCl that will dissociate, compared with the dissociation of this salt in aqueous solutions. Stated another way, the much weaker electrical forces between ions in aqueous solutions, compared with those in organic solvents, allows a larger concentration of ions to be in solution. Water is thus a good solvent for charged particles.

The electrostatic interaction between ions and water partially cancels or screens out the local electrical fields of the ions (Fig. 2.4). Cations attract the negatively charged oxygens of the water molecule, and anions attract its positively charged hydrogen atoms. The water molecules orient around the charged particles and produce local electrical fields opposing the fields of the ions. The resultant screening or shielding diminishes the electrical interaction between the ions and

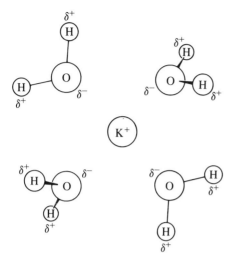

Figure 2.4

Orientation of water molecules around a potassium ion, indicating the charge arrangement that leads to screening of the local electrical field.

allows more of them to remain in solution, which is the molecular basis for water's high dielectric constant (see Fig. 2.4).

The energy of attraction between water and nonpolar molecules is usually less than the energy required to break water-water hydrogen bonds. Nonpolar compounds are therefore not very soluble in water. Certain substances, such as detergents, phospholipids, and proteins, can have both a nonpolar and also a polar region in the same molecule. In aqueous solutions these compounds can form aggregates, called *micelles,* in which the nonpolar groups of the molecules are in the center, while the charged or polar groups are external and interact with water. The lack of appreciable attractive electrostatic interaction between polar or charged species and the nonpolar (hydrophobic) regions of membrane lipids underlies the ability of biological membranes to limit the passage of such solutes into and out of cells and organelles.

CHEMICAL POTENTIAL

The chemical potential of species j indicates the free energy associated with it and available for performing work. Because there are many concepts to be mastered before understanding such a statement, we will first briefly explore the concept of free energy. (Further details are presented later in this chapter, in Ch. 6, and in App. IV.) We will introduce specific terms in the chemical potential of species j and then consider the extremely important case of water.

Free Energy and Chemical Potential

Plants and animals require a continual input of free energy. If we were to remove the sources of free energy, organisms would drift toward equilibrium and consequent cessation of life. The ultimate source of free energy is the sun. Photosynthesis converts its plentiful radiant energy into free energy, which is stored first in intermediate energy "currencies" like ATP and then in the altered chemical bonds that result when CO_2 and H_2O react to form a carbohydrate and O_2. When the overall photosynthetic reactions are reversed during respiration, the available free energy is reconverted into suitable energy currencies like ATP. In turn, such free energy is used to perform biological work, such as transporting amino acids into cells, pumping blood, powering electrical reactions in the brain, or lifting weights. For every such process on both a molecular and a global scale, the free energy decreases. In fact, the structure of cells, as well as that of ecosystems, is governed by the initial supply of free energy and by the inexorable laws of thermodynamics that describe the expenditure of free energy.

Instead of viewing the whole universe, a thermodynamicist focuses on some small part of it referred to as a "system." Such a system might be K^+ dissolved in water, a plant cell, a leaf, or even a whole plant. Something happens to the system; the K^+ concentration is increased, some protein is made in a cell, a leaf abscises, or water moves up a tree. We then say that the system changes from state A to state B. The minimum amount of work needed to cause the change from A to B is the free energy increase associated with it. Alternatively, the free energy decrease in going from state B back to state A represents the maximum amount of work that can be derived from the transition. (If there were no frictional or other irreversible losses, we could dispense with the words "minimum" and "maximum.") Thus, the limits to the work done on or by a system when it undergoes a transition from one state to another involve changes in free energy. Knowledge of the free energy under one condition compared with another allows us to predict the direction of spontaneous change or movement: a spontaneous change in a system at constant temperature and pressure proceeds in the direction of decreasing free energy.

Most systems of interest in biology are subject to a constant pressure (atmospheric) and remain at a constant temperature, at least for short periods. In discussing the energetics of processes for systems at constant temperature and pressure, the appropriate quantity is known as the *Gibbs free energy*. The Gibbs free energy has a very useful property: it decreases for a spontaneous process at constant temperature and pressure. Under such conditions, the decrease in Gibbs free energy equals the maximum amount of energy available for work, whereas if it increases for some transition, the change in Gibbs free energy represents the minimum amount of work required. We will hence restrict our attention to changes that occur when the overall system is at a constant temperature and subjected to a constant pressure. (An equation for the Gibbs free energy is presented in Ch. 6 and App. IV deals with a number of formal matters associated with its use.) For present purposes, we note that the Gibbs free energy of an entire system is made up of additive contributions from each of the species present. We will therefore shift our emphasis to the individual components of the system and the manner in which their free energy can be described.

To every chemical component in a system we can assign a free energy per mole of that species. This quantity is called the *chemical potential* of species j and is given the symbol μ_j. We can view μ_j as a property of species j, indicating how that species will react to a given change, e.g., a transition of the system from state A to state B. During the transition, the chemical potential of species j changes from μ_j^A to μ_j^B. If μ_j^B is less than μ_j^A, then the free energy per mole of species j decreases. Such a process can take place spontaneously as far as species j is concerned—water flowing downhill is an example of such a spontaneous process. In principle, we can harness the spontaneous change to do work. The maximum amount of work that can be done per mole of species j is $\mu_j^A - \mu_j^B$ (Fig. 2.5). Suppose that μ_j^B equals μ_j^A. As far as species j is concerned, no work is involved in the transition from state A to state B; species j is then at equilibrium (Fig. 2.5). We will show later not only that living systems as a whole are far from equilibrium, but also that many chemical species are not in equilibrium across cellular membranes. Finally, let us consider the case where μ_j^B is greater than μ_j^A. A change increasing the chemical potential of a substance can occur only if some other change in the system supplies the free energy required. The pumping of blood along arteries, using ATP to cause contraction of the heart muscles, is an example where the increase in chemical potential of water in the blood is accompanied by an even larger decrease in the free energy associated with ATP (we will consider ATP in detail in Ch. 6). The minimum amount of work that must be done per mole of species j to cause the energetically uphill transition is $\mu_j^B - \mu_j^A$ (Fig. 2.5).

The condition for a spontaneous change—a decrease in chemical potential— has important implications for discussing fluxes from one region to another. In particular, we can use the chemical potential difference between two locations as a measure of the "driving force" for the movement of that component. Thus, the larger is the chemical potential difference, $\mu_j^A - \mu_j^B$, the faster that the spontaneous change takes place—in this case, the larger is the flux density of species j from region A to region B.

The chemical potential of a substance clearly depends on its chemical composition, but it is also influenced by other factors. Here we shall discuss in an

Initial chemical potential	Final chemical potential	Comment
μ_j	μ_j^B	Transition can occur spontaneously as far as species j is concerned; maximum work available per mole of species j equals $\mu_j^{A} - \mu_j^{B}$
μ_j^A	μ_j^B	Equilibrium for species j
μ_j^A	μ_j^B	Change will not occur unless free energy equal to at least $\mu_j^{B} - \mu_j^{A}$ is supplied per mole of species j

Figure 2.5

Possible changes in the chemical potential of species j that can accompany a transition from initial state A to final state B, as might occur in a chemical reaction or in crossing a membrane. The heights of the bars representing μ_j correspond to the relative values of the chemical potential.

intuitive way the environmental factors that might influence μ_j. To begin with, we note that chemical potential depends on the "randomness" (entropy) of the system, and that concentration is a quantitative description of this randomness. For example, the passive process of diffusion discussed in Chapter 1 describes neutral molecules spontaneously moving from some region to another where their concentration is lower. In more formal terms, the net diffusion of species j proceeds in the direction of decreasing μ_j, which in this case is the same as that of decreasing concentration. The concept of chemical potential was thus implicit in the development of Fick's equations in Chapter 1. The importance of concentration gradients—or differences—to the movement of a substance was also established in the discussion of diffusion. Here we will present the effect of concentration on chemical potential in a somewhat more sophisticated manner, using that part of the concentration that is thermodynamically active.

Chemical potential also depends on pressure,[5] which for situations of interest in biology usually means hydrostatic pressure. The existence of pressure gradients can cause movements of fluids—the flow of crude oil in long-distance pipelines, blood in arteries, and sap in xylem vessels—which is another way of saying that pressure differences can affect the direction for spontaneous changes.

Because electrical potential also affects the chemical potential of charged particles, it must be considered when predicting the direction of their movement. By definition, work must be done to move a positively charged particle to a higher electrical potential. Accordingly, if an electrical potential difference is imposed across an electrolyte solution initially containing a uniform concentration of ions throughout, then cations will spontaneously move in one direction (toward the cathode, i.e., to regions of lower electrical potential), and anions will move in the opposite direction (toward the anode). It is clear, then, that to describe the chemical potential of a charged species we must include an electrical term in μ_j.

Another contributor to chemical potential is gravity. We can readily appreciate that position in a gravitational field affects μ_j because work must be done to move a substance vertically upward. Although the gravitational term can be neglected for ion and water movements across plant cells and membranes, it is important for water movement in a tall tree and in the soil.

In summary, the chemical potential of a species depends on its concentration, the pressure, the electrical potential, and gravity. We can compare the chemical potentials of a species on two sides of a barrier to decide whether or not the species is in equilibrium. If μ_j is the same on the two sides, we would not expect a net movement of species j to occur spontaneously across the barrier. We will use a comparison of the relative values of the chemical potential of species j at various locations to predict the directions for spontaneous movement of that chemical substance (toward lower μ_j), just as we can use a comparison of temperatures to predict the direction for heat flow (toward lower T). We will also find that $\Delta\mu_j$ from one region to another gives a convenient measure of the driving force on species j.

---

5. Although local pressure effects can be readily incorporated into μ_j, the system as a whole must experience a constant external pressure for the Gibbs free energy to have the useful properties discussed above.

Analysis of Chemical Potential

Because it has proved experimentally valid, we will represent the chemical potential of any species j by the following sum:

$$\mu_j = \mu_j^* + RT \ln a_j + \bar{V}_j P + z_j FE + m_j gh \qquad (2.4)$$

One measure of the elegance of a mathematical expression is the amount of information it contains. Based on this criterion, Equation 2.4 represents one of the most elegant relations in all of biology. After defining and describing the various contributions to μ_j indicated in Equation 2.4, we will consider the various terms in greater detail for the important special case where species j is water.

Chemical potential, like electrical or gravitational potential, must be expressed relative to some arbitrary energy level. An unknown additive constant, or reference level, μ_j^*, is therefore included in Equation 2.4. Because it contains an unknown constant, the actual value of the chemical potential is not determinable. But for our applications of chemical potential we are interested in the difference in the chemical potential between two particular locations, so only relative values are important anyway. Specifically, because μ_j^* is added to each of the chemical potentials being compared, it cancels out when the chemical potential in one location is subtracted from that in another to obtain the chemical potential difference between the two locations. Figure 2.5 further illustrates the usefulness of considering differences in chemical potential instead of actual values. Note that the units of μ_j and of μ_j^* are energy per mole of the substance, e.g., J mol^{-1}.

In certain cases we can adequately describe the chemical properties of species j by using the concentration of that solute, c_j. Owing to molecular interactions, however, this usually requires that the total solute concentration be low. Molecules of solute species j interact with each other as well as with other solutes in the solution, and this influences the behavior of species j. Such intermolecular interactions increase as the solution becomes more concentrated. The use of concentrations for describing the thermodynamic properties of some solute thus indicates an approximation, except in the limiting case of infinite dilution where interactions between solute molecules are negligible. Where exactness is required, *activities*—which may be regarded as "corrected" concentrations—are employed. Consequently, for general thermodynamic considerations, as in Equation 2.4, the influence of the amount of a particular species j on its chemical potential is handled not by its concentration but by its activity, a_j. The activity of solute j is related to its concentration by means of an activity coefficient, γ_j:

$$a_j = \gamma_j c_j \qquad (2.5)$$

The activity coefficient is usually less than unity, because the thermodynamically *effective* concentration of a species—its *activity*—is generally less than is its actual concentration.[6]

For an ideal solute, γ_j is unity, and the activity of species j equals its concentration. This condition can be approached for real solutes in certain dilute aqueous solutions, especially for neutral species. Activity coefficients for charged species can be appreciably less than unity because of the importance of electrical interactions (discussed in Ch. 3).

The activity of a solvent is defined differently from that of a solute (the solvent is the species having the highest mole fraction in a solution; mole fraction is defined on p. 71). For a solvent, a_j is $\gamma_j N_j$, where N_j is its mole fraction. An ideal solvent has $\gamma_{solvent}$ equal to 1, meaning that the interactions of solvent molecules with the surrounding molecules are indistinguishable from their interactions in the pure solvent. An ideal solution has all activity coefficients equal to unity.

In the expression for chemical potential (Eq. 2.4), the term involving the activity is $RT \ln a_j$. Therefore, the greater is the activity of species j—or, loosely speaking, the higher is its concentration—the larger will be its chemical potential. The logarithmic form can be "justified" in a number of different ways, all based on agreement with empirical observations. (Justifications are given in Ch. 3 and App. IV—we also note that a logarithmic term appears in the expression for work done in compressing a gas.) The factor RT multiplying $\ln a_j$ in Equation 2.4, where R is the gas constant (see App. I) and T is temperature on the absolute scale, results in units of energy per mole for the activity term.

The term $\overline{V}_j P$ in Equation 2.4 represents the effect of pressure on chemical potential. Because essentially all measurements in biology are made on systems subjected to atmospheric pressure, it is convenient to define P as the pressure in excess of this, and we will adopt such a convention here. $\overline{V}_j$ is the differential increase in volume of a system when a differential amount of species j is added, with other species, temperature, pressure, electrical potential, and gravitational position remaining constant:

$$\overline{V}_j = \left(\frac{\partial V}{\partial n_j} \right)_{n_i, T, P, E, h} \tag{2.6}$$

The subscript n_i on the partial derivative in Equation 2.6 means that the number of moles of each species present, other than species j, is held constant when the

---

6. Usually a_j is considered to be dimensionless, in which case γ_j has units of reciprocal concentration, although sometimes a_j is expressed in molality (and even molarity).

derivative is taken; the other four subscripts are included to remind us of the additional variables that must be held constant.

$\overline{V}_j$ is called the *partial molal volume* of species j. (μ_j is actually the partial molal Gibbs free energy, $(\partial G/\partial n_j)_{n_i, T, P, E, h}$, as discussed in Ch. 6 and App. IV.) The partial molal volume of a substance is often nearly equal to the volume of a mole of that species, but because there is in general a slight change in total volume when substances are mixed, the two are not exactly equal.[7]

Because work is often expressed as pressure times change in volume, we note that $\overline{V}_j P$ has the correct units for μ_j—energy per mole ($\overline{V}_j$ represents the volume per mole). To help justify the form of the pressure term, let us imagine that the solution containing species j is built up by adding small volumes of species j while the system is maintained at a constant pressure P. The work done to add a mole of species j is then the existing pressure times some volume change of the system characterizing a mole of that species, namely, $\overline{V}_j$. (In App. IV the inclusion of the $\overline{V}_j P$ term in the chemical potential of species j is justified more formally.)

The influence of electrical potential on the chemical potential of an ion is expressed by the term $z_j F E$ in Equation 2.4, where z_j is an integer representing the charge number of species j, F is a constant known as Faraday's constant (Ch. 3), and E is the electrical potential. Because water is uncharged ($z_w = 0$), the electrical term does not contribute to its chemical potential. But electrical potential is of central importance when discussing ions and the origin of membrane potentials; both of these are examined in detail in Chapter 3, where we explicitly consider the $z_j F E$ term.

Equation 2.4 also includes a gravitational term $m_j g h$ expressing the amount of work required to raise an object of mass per mole m_j to a vertical height h, where g is the gravitational acceleration (close to 9.8 m s^{-2}; see App. I for details). In the case of water, which is the primary concern of this chapter, the mass per mole m_w is 0.018016 kg mol^{-1} (18.016 g mol^{-1}). The gravitational term in the chemical potential of water not only is important for the fall of rain, snow, or hail, but also affects the percolation of water downward through porous soil and the upward movement of water in a tree. The gravitational term $m_j g h$ can have units of (kg mol^{-1}) (m s^{-2}) (m), or kg m^2 s^{-2} mol^{-1}, which is J mol^{-1}. (Conversion factors among energy units are given in App. II.)

---

7. The addition of small concentrations of certain salts can actually cause an aqueous solution to contract—a phenomenon known as electrostriction—which in this special case would result in a negative value for $\overline{V}_j$.

Standard State

The additive constant term μ_j^* in Equation 2.4 is the chemical potential of species j for a specific reference state. From the preceding definitions of the various quantities involved, this reference state is attained when the following conditions hold: the activity of species j is unity ($RT \ln a_j = 0$); the hydrostatic pressure equals atmospheric pressure ($\overline{V}_j P = 0$); the species is uncharged or the electrical potential is zero ($z_j FE = 0$); we are at the zero level for the gravitational term ($m_j gh = 0$); and the temperature equals the temperature of the system under consideration. Under these conditions, μ_j equals μ_j^* (Eq. 2.4).

As just indicated, an activity of unity is defined in different ways for the solute and the solvent. To describe liquid properties, such as the dielectric constant, the heat of vaporization, and the boiling point, the most convenient standard state is that of the pure solvent. For a solvent, $a_{solvent}$ equals $\gamma_{solvent}$ times $N_{solvent}$, so the activity is unity when the mole fraction $N_{solvent}$ is unity ($\gamma_{solvent} = 1$ for pure solvent). Specifically, the properties of a solvent (mole fraction equal to unity when pure) are fully expressed when no solute is present. Thus, the standard reference state for water is pure water at atmospheric pressure and at the temperature and gravitational level of the system under consideration.

Water has an activity of unity when N_w (see Eq. 2.8) is one. The concentration of water on a molality basis (number of moles of a substance per kg of water for aqueous solutions) is then $1/(0.018016 \text{ kg mol}^{-1})$, or 55.5 molal. The accepted convention for a solute, on the other hand, is that a_j is unity when $\gamma_j c_j$ equals 1 molal. For example, if γ_j equals one, a solution with a 1 molal concentration of solute j has an activity of 1 molal for that solute. Thus, the standard state for an ideal solute is when its concentration is 1 molal, in which case $RT \ln a_j$ is zero.[8] A special convention is used for the standard state of a gas such as CO_2 or O_2 in an aqueous solution—namely, the activity is unity when the solution is in equilibrium with a gas phase containing that gas at a pressure of one atmosphere. (At other pressures, the activity is proportional to the partial pressure of that gas in the gas phase.)

The three conventions just introduced for the standard state (solvent, solute, and gases) are generally all inappropriate for biological situations. Essentially never is a solute at a concentration of 1 molal, an important gas at a pressure of 1 atmosphere, or a pure solvent present (except sometimes for water). Hence,

---

8. Because molality involves the moles of a substance per kg of solvent, both SI quantities, it is a legitimate SI unit, whereas moles per litre (M) is not recommended (the analogous SI concentration unit is moles m^{-3} = moles per 1000 litres = mM).

care must be exercised when accepting chemical data based on standard states and interpreting the consequences in a biological context.

The value of the chemical potential for a solute in the standard state, μ_j^*, depends on the solvent. To be explicit, we will next argue that μ_j^* for a polar molecule is smaller for an aqueous solution (where the species readily dissolves) than it is for an organic phase such as olive oil (where it is not as soluble). Let us consider a two-phase system consisting of water with an overlying layer of olive oil, similar to that discussed in Chapter 1 in relation to measuring partition coefficients. If we put the polar solute into our system, shake the water and the olive oil together, and wait for the two phases to separate, the solute concentration will be higher in the bottom or water phase. Let us next analyze this event using symbols, where phase A represents water and phase B is olive oil. Waiting for the phase separation is equivalent to waiting for the attainment of equilibrium, so μ_j^A is then equal to μ_j^B. In the present case $\mu_j^{*,A} + RT \ln a_j^A$ then equals $\mu_j^{*,B} + RT \ln a_j^B$ by Equation 2.4. But the polar solute is more soluble in water than it is in olive oil; hence $RT \ln a_j^A$ is greater than $RT \ln a_j^B$. Because we are at equilibrium, $\mu_j^{*,A}$ must be less than $\mu_j^{*,B}$, as we already indicated.

Hydrostatic Pressure

Because of the rigid cell walls, fairly large hydrostatic pressures can exist in plant cells, whereas hydrostatic pressures in animal cells generally are relatively small. In addition to being involved in the support of a plant, hydrostatic pressures are important for the movement of water and solutes in the xylem and in the phloem. The term expressing the effect of pressure on the chemical potential of water is $\overline{V}_w P$ (see Eq. 2.4), where $\overline{V}_w$ is the partial molal volume of water and P is the hydrostatic pressure in the aqueous solution in excess of the ambient atmospheric pressure. The approximate value of $\overline{V}_w$ can be estimated using the following argument. The density of water is about $1\ 000$ kg m^{-3}; therefore, when 1 mol or 18.0×10^{-3} kg of water is added to water, the volume increases by 18.0×10^{-6} m^3. Actually, using the definition of $\overline{V}_w$ as a partial derivative (see Eq. 2.6), we need to add only an infinitesimally small amount of water (dn_w) and to observe the infinitesimal change in volume of the system (dV). We thus find that $\overline{V}_w$ for pure water is 18.0×10^{-6} m^3 mol^{-1} (18.0 cm^3 mol^{-1}). Although $\overline{V}_w$ can be influenced by the various solutes present, it is generally close to 18.0×10^{-6} m^3 mol^{-1} for a dilute solution, a value that we will use for calculations in this book.

Various units are used for expressing pressures (see footnote, p. 39). A pressure of one standard atmosphere, or 0.1013 MPa, can support a column of mercury

760 mm high or a column of water 10.35 m high. Research concerning the water relations of plants often used pressures in bars. However, as indicated in Chapter 1, the SI unit for pressure is the pascal (Pa), which is 1 N m^{-2}; a quantity of convenient size for hydrostatic pressures is often the MPa (1 MPa = 10 bars = 9.87 atmospheres). (An extensive list of conversion factors for pressure units is given in App. II, which also includes values for related quantities such as RT.) Because pressure is force per unit area, it is dimensionally the same as energy per unit volume, e.g., 1 Pa = 1 N m^{-2} = 1 J m^{-3}. $\bar{V}_w$ has the units of m^3 mol^{-1}, so $\bar{V}_w P$ and hence μ_w can be expressed in J mol^{-1}.

Water Activity and Osmotic Pressure

The presence of solutes in an aqueous solution tends to decrease the activity of water (a_w). As a first approximation, we can view the decrease in a_w when more and more solutes are added as a dilution effect. In other words, the mole fraction of water becomes lower when the solutes are added. As its activity thus decreases, the chemical potential of water is lowered (consider the $RT \ln a_j$ term in μ_j). We also know that the presence of solutes can lead to an *osmotic pressure* (Π) in the solution. An increase in the concentration of solutes raises the osmotic pressure, indicating that Π and a_w change in opposite directions. In fact, the osmotic pressure and water activity are related, a fundamental definition of Π being

$$RT \ln a_w = -\bar{V}_w \Pi \qquad (2.7)$$

where the subscript w refers to water. As solutes are added, a_w decreases from its value of unity for pure water, $\ln a_w$ is therefore negative, and Π (defined by Eq. 2.7) is positive, which is consistent with the arguments presented above. Using Equation 2.7, $RT \ln a_w$ in the chemical potential of water (see Eq. 2.4) can be replaced by $-\bar{V}_w \Pi$.

Unfortunately, there is much variation in the use of the terms "osmotic pressure" and "osmotic potential," as well as the algebraic sign of the quantity. Osmotic pressures were originally measured using an osmometer (Fig. 2.6), a device employing a membrane that in the ideal case is permeable to water but not to any of the solutes present. When pure water is placed on one side of the membrane, and some solution on the other, a net diffusion of water occurs toward the side with the solutes. To counteract this tendency and establish equilibrium, a hydrostatic pressure is necessary on the solution side. This pressure

Figure 2.6

Schematic diagram indicating the principle behind an osmometer in which a semipermeable membrane (permeable to water, but not to solutes) separates pure water (region A) from water containing solutes (region B). Water tends to diffuse to where it has a lower mole fraction, in this case into region B. This causes the solution to rise in the open central column until at equilibrium the hydrostatic pressure (P) at the horizontal dashed line is equal to the osmotic pressure (Π) of the solution. Alternatively, we can apply a hydrostatic pressure P to the right-hand column to prevent a net diffusion of water into region B, this P again being equal to Π.

is often called the *osmotic pressure*.[9] It depends on the presence of the solutes. Does an isolated solution have an osmotic pressure? In the sense of requiring an applied hydrostatic pressure to maintain equilibrium, the answer is no. Yet the same solution when placed in an osmometer can manifest an osmotic pressure. Thus, some physical chemistry texts say that a solution has an osmotic *potential*— i.e., it could show an osmotic pressure if placed in an osmometer—and that osmotic potentials are a property characteristic of solutions. Many plant physiology texts (e.g., Salisbury and Ross, 1991) have defined the *negative* of the same quantity as the osmotic potential, so that the water potential (a quantity we will introduce shortly) will depend directly on the osmotic potential, rather

---

9. If we represent this applied hydrostatic pressure by Π, then the chemical potential of water on the right side of the semipermeable membrane in Figure 2.6 is $\mu_w^* + RT \ln a_w + \overline{V}_w \Pi$. This solution is in equilibrium with pure water on the left side of the membrane, where μ_w equals μ_w^*. Hence, $\mu_w^* + RT \ln a_w + \overline{V}_w \Pi$ equals μ_w^*, or $RT \ln a_w + \overline{V}_w \Pi$ equals zero, which is the same as Equation 2.7.

than on its negative. We will refer to Π, defined by Equation 2.7, as the osmotic pressure (see Van Holde, 1985) and to its negative as the osmotic potential. (The osmotic potential will be symbolized by Ψ_Π, a component of the water potential Ψ.)

A direct and convenient way of evaluating osmotic pressure and water activity as defined by Equation 2.7 is to use another *colligative* property. The four colligative properties of a solution—those that depend on the concentration of solutes regardless of their nature, at least in a dilute enough solution—are the freezing point depression, the boiling point elevation, the lowering of the vapor pressure of the solvent, and the osmotic pressure. Thus, if the freezing point of cell sap is measured, $\Pi^{cell\ sap}$ can be calculated using the freezing point depression of $1.86°C$ for a 1 molal solution together with the *Van't Hoff relation*.

The Van't Hoff Relation

For many purposes in biology osmotic pressures are directly related to the concentration of solutes instead of expressing Π in terms of the water activity, a_w, as is done in Equation 2.7. In general, the greater the concentration of solutes, the more negative is $\ln a_w$ and the larger is the osmotic pressure. Thus, some way of expressing a_w in terms of the properties of the solutes can lead to an expression for Π as a function of the various solute concentrations. The ensuing derivation not only will show how a_w can be so expressed, but will also indicate the many approximations necessary to achieve a rather simple and common expression for Π.

The activity of water equals $\gamma_w N_w$, where γ_w is the activity coefficient of water and N_w is its mole fraction. The *mole fraction* of a species is the ratio of the number of moles of that species to the total number of moles of all species in the system. Thus, N_w is given by

$$N_w = \frac{n_w}{n_w + \sum_j n_j} = \frac{n_w + \sum_j n_j - \sum_j n_j}{n_w + \sum_j n_j} = 1 - \frac{\sum_j n_j}{n_w + \sum_j n_j} \quad (2.8)$$

where n_w is the number of moles of water, n_j is the number of moles of solute j, and the summation Σ_j is over all solutes in the system being considered. Equation 2.8 also expresses the familiar relation that N_w equals 1 minus the mole fraction of solutes.

For an ideal solution, γ_w equals 1. An activity coefficient of unity is approached for dilute solutions, in which case n_w is much greater than $\Sigma_j n_j$. (The expression $n_w \gg \Sigma_j n_j$ defines a dilute solution, where the symbol $\gg$ means "is much greater than.") Using Equation 2.8 and assuming a dilute ideal solution, we obtain the following relations for $\ln a_w$:

$$\ln a_w = \ln N_w = \ln \left(1 - \frac{\displaystyle\sum_j n_j}{n_w + \displaystyle\sum_j n_j} \right) \cong - \frac{\displaystyle\sum_j n_j}{n_w + \displaystyle\sum_j n_j} \cong - \frac{\displaystyle\sum_j n_j}{n_w}$$

(2.9)

The penultimate step in Equation 2.9 is based on the series expansion of a logarithm: $\ln (1 - x) = -x - x^2/2 - x^3/3 - \cdots$, a series that converges rapidly for $|x| \ll 1$. The last two steps employ the approximation $(n_w \gg \Sigma_j n_j)$ relevant to a dilute solution. Equation 2.9 is thus restricted to dilute ideal solutions. Nevertheless, it is a useful expression, clearly indicating that, in the absence of solutes $(\Sigma_j n_j = 0)$, $\ln a_w$ is zero and a_w is unity, and that the presence of solutes decreases the activity of water from the value of one for pure water.

To obtain a rather familiar form for expressing the osmotic pressure, we can incorporate the approximation for $\ln a_w$ given by Equation 2.9 into the definition of osmotic pressure given by Equation 2.7 $(RT \ln a_w = -\overline{V}_w \Pi)$, which yields

$$\Pi_s \cong - \frac{RT}{\overline{V}_w} \left(- \frac{\displaystyle\sum_j n_j}{n_w} \right) = RT \sum_j \frac{n_j}{\overline{V}_w n_w} = RT \sum_j c_j \qquad (2.10)$$

where $\overline{V}_w n_w$ is the total volume of water in the system (essentially the total volume of the system for a dilute aqueous solution), $n_j/\overline{V}_w n_w$ is the number of moles of species j per volume of water and so is the concentration of species j (c_j), and the summations are over all solutes.

Osmotic pressure is often expressed by Equation 2.10, known as the *Van't Hoff relation*, but this is justified only in the limit of *dilute ideal* solutions. As we have already indicated, an ideal solution has ideal solutes dissolved in an ideal solvent. The string of equalities in Equation 2.9 assumes that γ_w is unity, so the subsequently derived expression (Eq. 2.10) strictly applies only when water acts as an ideal solvent $(\gamma_w = 1.00)$. To emphasize that we are neglecting any factors that cause γ_w to deviate from unity and thereby affect the measured osmotic pressure (such as the interaction between water and colloids, which we will discuss shortly), Π_s instead of Π has been used in Equation 2.10, and we will follow this convention throughout the book. The effect of solute concentration on osmotic pressure as described by Equation 2.10 is portrayed in Figure 2.7 and was first clearly recognized by the botanist Pfeffer in 1877 (Pfeffer, 1877). Both the measurement of osmotic pressure and a recognition of its effects have been and still are crucial for an understanding of water relations in biology.

Relatively dilute solutions have fairly large osmotic pressures (Fig. 2.7). The cellular fluid expressed from young leaves of plants like pea or spinach often contains about 0.3 mol osmotically active particles per kg of water (about 300 mol m^{-3}). This fluid is referred to as 0.3 osmolal by analogy with molality, which refers to the total concentration. For example, 0.1 molal CaCl$_2$ is about 0.26 osmolal because most of the CaCl$_2$ is dissociated (Fig. 2.7). We note that molality (moles of solute/kg solvent) is a concentration unit that is independent of temperature and suitable for colligative properties; e.g., the freezing point depression is 1.86°C for a 1 osmolal aqueous solution. But mol m^{-3} and molarity

Figure 2.7

Relationship between concentration and osmotic pressure at 20°C for a nonelectrolyte (sucrose) and two readily dissociating salts (NaCl and CaCl$_2$). The different slopes indicate the different degrees of dissociation for the three substances. Data for osmotic pressure are based on the freezing point depression (see Weast and Lide, 1989).

(moles of solute/litre of solution, M) are often more convenient units. Below about 0.2 M, molarity and molality are nearly the same for a low molecular weight solute in an aqueous solution, but at high concentrations the molarity can be considerably less than the molality. In fact, the deviations from linearity in Figure 2.7 indicate the greater values for osmolality, to which osmotic pressure is proportional, compared with osmolarity. Such deviations are apparent at lower concentrations for sucrose, which has a relatively high molecular weight of 342, than for $CaCl_2$ or NaCl (molecular weights of 111 and 58, respectively). For instance, at 300 mmolar, sucrose is 321 mmolal, $CaCl_2$ is 303 mmolal, and NaCl is 302 mmolal (at these concentrations, $CaCl_2$ is about 78% dissociated and NaCl about 83% dissociated, leading to the higher osmotic pressures for these dissociable salts than for sucrose, Fig. 2.7).

Using Equation 2.10 ($\Pi_s \cong RT \Sigma_j c_j$), we can calculate the osmotic pressure for the cell sap in a leaf. At 20°C, RT is 0.002437 m^3 MPa mol^{-1} (App. II), so Π_s for a 0.3 osmolar solution is

$$\Pi_s = (0.002437 \text{ m}^3 \text{ MPa mol}^{-1})(300 \text{ mol m}^{-3}) = 0.73 \text{ MPa}$$

The osmotic pressure of cell sap pressed out of mature leaves of most plants is 0.6 to 3 MPa (Altman and Dittmer, 1966); for comparison, sea water has a Π_s of 2.5 MPa (Ψ_{Π_s} of -2.5 MPa).

Matric Pressure

For certain applications in plant physiology, another term is frequently included in the chemical potential of water, namely $\overline{V}_w \tau$, where τ is the *matric pressure* (also called the *matric potential*). Although the matric pressure does not represent any new force (all energetic considerations are already fully described by the terms in μ_w for hydrostatic pressure or water activity), use of this term is sometimes convenient for dealing explicitly with interactions occurring at interfaces, although these interfacial forces can also be adequately represented by their contributions to Π or P. In other words, the matric pressure does not represent a new or different force or a new or different contribution to μ_w—but it can be used as a bookkeeping device for handling interfacial interactions. To help make this statement more meaningful, we will briefly consider the influence of liquid-solid interfaces on the chemical potential of water at the surfaces of colloids.[10]

---

10. "Colloid" is a generic term for solid particles approximately 0.002 to 1 μm in diameter suspended in a liquid, e.g., proteins, ribosomes, and even some membrane-bounded organelles.

When water molecules are associated with interfaces such as those provided by membranes or the surfaces of colloidal particles suspended in an aqueous solution, they have less tendency either to react chemically in the bulk solution or to escape into a surrounding vapor phase. The presence of interfaces thus lowers the thermodynamic activity of the water (a_w), especially near their surfaces. The presence of solutes also lowers water activity (Eq. 2.10). As a useful first approximation, we can consider that these two different effects lowering water activity are additive in a solution containing both ordinary solutes and colloids or other interfaces. Actual osmotic pressure (Π) depends on the actual activity of water regardless of the reason for the departure of a_w from unity; i.e., Π still equals $-(RT/\overline{V}_w)\ln a_w$ (Eq. 2.7). Incorporating these various concepts (and recalling that $a_w = \gamma_w N_w$), we can write the following relation:

$$\Pi = -\frac{RT}{\overline{V}_w} \ln \gamma_w N_w = -\frac{RT}{\overline{V}_w} \ln \gamma_w - \frac{RT}{\overline{V}_w} \ln N_w = \tau + \Pi_s \quad (2.11)$$

where τ represents a matric pressure resulting from the water-solid interactions at the surfaces of the colloids and other interfaces that lower γ_w from unity. (Note that if we let $\tau = -(RT/\overline{V}_w) \ln \gamma_w$, Eq. 2.11, then $\gamma_w < 1$ means $\tau > 0$.)

We note that Π_s in Equation 2.11 is the osmotic pressure of all solutes, including the *concentration* of the colloids, as represented by Equation 2.10 ($\Pi_s \cong RT \Sigma_j c_j$). In other words, in the derivation of Π_s we were actually dealing only with $-(RT/\overline{V}_w) \ln N_w$, because γ_w was set equal to unity. For an exact treatment when many interfaces are present—e.g., in the cytosol of a typical cell—we cannot set γ_w equal to unity, because the activity of water, and hence the osmotic pressure (Π), is strongly affected by proteins, other colloids, and other interfaces. In such a case, Equation 2.11 suggests a simple way in which a matric pressure may be related to the reduction of the activity coefficient of water caused by the interactions at interfaces. Equation 2.11 should not be viewed as a relation defining matric pressures for all situations, but rather where it might be useful to represent interfacial interactions by a separate term that can be added to Π_s, the effect of the solutes on Π.

Although Π and a_w may be the same throughout some system, both Π_s and τ in Equation 2.11 may vary. For example, water activity in the bulk of the solution may be predominantly lowered by the presence of the solutes, whereas at or near the surface of colloids the main factor decreasing a_w from unity could be the interfacial attraction and binding of water. Such interfacial interactions do not change the mole fraction of water, but they do reduce its activity coefficient, γ_w.

Other areas of plant physiology where matric potentials and pressures have been invoked are descriptions of the chemical potential of water in soil and in cell walls. (Cell walls will be further considered at the end of this chapter, and soil matric potentials will be mentioned in Ch. 9.) Surface tension effects at the numerous air-water interfaces for the interstices of a cell wall or among soil particles lead to a tension in the cell wall water. Such a tension is a negative hydrostatic pressure; i.e., P is less than zero. Although it is not necessary to do so, P is sometimes used to refer only to positive pressures, and a negative P in cell wall water has been called a positive matric pressure (or potential). Some books define this same negative hydrostatic pressure in the water in cell wall interstices or between soil particles as a negative matric potential. It is far more straightforward and consistent to refer to the quantity that reduces μ_w in such pores as a negative P.

Water Potential

From the definition of chemical potential (Eq. 2.4) and the formal expression for osmotic pressure (Eq. 2.7), we can write the chemical potential of water (μ_w) as

$$\mu_w = \mu_w^* - \overline{V}_w \Pi + \overline{V}_w P + m_w g h \qquad (2.12)$$

where the electrical term $(z_j FE)$ is not included because water carries no net charge $(z_w = 0)$. The quantity $\mu_w - \mu_w^*$ has proved to be of considerable importance for discussing the water relations of plants. It represents the work involved in moving one mole of water from a pool of pure water at atmospheric pressure, at the same temperature as the system under consideration, and at the zero level for the gravitational term to some arbitrary point in a system (at constant pressure and temperature). A difference between two locations in the value of $\mu_w - \mu_w^*$ indicates that water is not in equilibrium, so there will be a tendency for water to flow toward the region where $\mu_w - \mu_w^*$ is lower.

A quantity proportional to $\mu_w - \mu_w^*$ commonly used in studies of plant water relations is the *water potential*, Ψ. This is defined as

$$\Psi = \frac{\mu_w - \mu_w^*}{\overline{V}_w} = P - \Pi + \rho_w g h \qquad (2.13a)$$

which follows directly from Equation 2.12 plus the identification of $m_w/\overline{V}_w$, which is the mass per mole of water/volume per mole of water, or mass/volume for water, as the density of water, ρ_w. Equation 2.13a indicates that an increase in hydrostatic pressure raises the water potential, and an increase in osmotic pressure lowers it. We will consider the values of Ψ on both sides of cellular and subcellular membranes in the next two sections.

We have already mentioned that work must be performed to raise an object in the gravitational field of the earth. Thus, vertical position also affects the chemical potential, and consequently the term $\rho_w gh$ should be included in the water potential given by Equation 2.13a, at least when water moves an appreciable distance vertically in the gravitational field. The magnitude of $\rho_w g$ is 0.0098 MPa m^{-1}; if water moves 10 m vertically upward in a tree, the gravitational contribution to the water potential is increased by about 0.1 MPa. For certain applications in this text, such as considerations of chemical reactions or the crossing of membranes, there is very little or no change in vertical position, and the gravitational term can then be omitted from μ_j and Ψ.

As we mentioned when introducing osmotic pressure, a number of conventions are used to describe the osmotic and other water potential terms. One such convention is to define Ψ as follows:

$$\Psi = \Psi_P + \Psi_\Pi + \Psi_h \qquad (2.13b)$$

where Ψ_P ($= P$) is called the hydrostatic potential or pressure potential, Ψ_Π ($= -\Pi$) is the osmotic potential, and Ψ_h ($= \rho_w gh$) is the gravitational potential. Although uniformity of expression is a cherished ideal, one should not be too alarmed that various conventions are used, because persons from many fields and backgrounds have contributed to our understanding of plants. Because Ψ is so important in understanding plant water relations, a method for measuring it is illustrated in Figure 2.8.

CENTRAL VACUOLE AND CHLOROPLASTS

The water relations of both the large central vacuole and the chloroplasts will be considered next, using the water potential defined above. We will focus our attention on the situation where there is no net flow of water across the limiting membranes surrounding these subcellular compartments and for the special case of nonpenetrating solutes. Hence, we will be considering the idealization of semipermeable membranes (see Fig. 2.6). The more general and biologically realistic case of penetrating solutes will be discussed in Chapter 3 after a consideration of the properties of solutes and the introduction of concepts from irreversible thermodynamics.

The central vacuole occupies up to 90% of a mature plant cell, which means

Figure 2.8

Schematic diagram of a "pressure bomb," which can be used to measure the xylem pressure, P^{xylem}, averaged over the material placed in the chamber. To make a measurement, a severed part of a plant is placed in the chamber with its freshly cut end protruding through a rubber seal. The air pressure (P^{air}) in the chamber is then gradually increased until it just causes the exudation of xylem sap at the cut end (generally viewed with a magnifying glass or a dissecting microscope). At this stage, the resulting pressure of the sap, which equals $P^{xylem} + P^{air}$, is zero, so P^{xylem} equals $-P^{air}$. If the xylem osmotic pressure can be ignored, P^{xylem} is approximately equal to Ψ^{xylem}, which can be the same as the water potential of the other tissue in the chamber (if water equilibration has been achieved) (for details, see Scholander *et al.*, 1964; Turner, 1981, 1988).

that under such circumstances most of the cellular water is in the vacuole. The vacuolar volume is generally about 10^4 μm^3 for the mesophyll cells in a leaf, but can be much larger in certain algal cells. Because the vacuole is nearly as large as the cell containing it, special procedures are required to remove the vacuole from a plant cell without rupturing its surrounding membrane, the tonoplast (Boller and Wiemken, 1986; Marin, 1987). Chloroplasts are much smaller than the central vacuole, often having volumes near 30 μm^3 *in vivo* (although the sizes vary considerably with the plant species). When chloroplasts are carefully isolated, suitable precautions will ensure that their limiting membranes remain intact. Such intact chloroplasts can be placed in solutions having various osmotic pressures, and the resulting movement of water into or out of the plastids can be precisely measured.

Water Relations of the Central Vacuole

To predict whether and in what direction water will move, we need to know the value of the water potential in the various compartments under consideration. At equilibrium, the water potential is the same in all communicating phases, such as those separated by membranes. For example, when water is in equilibrium across the tonoplast, the value of the water potential is the same in the vacuole as it is in the cytosol. No force then drives water across this membrane, and no net flow of water occurs into or out of the vacuole.

The tonoplast most likely does not have an appreciable difference in hydrostatic pressure across it. A higher internal hydrostatic pressure would cause an otherwise slack (folded) tonoplast to be mechanically pushed outward. The observed lack of such motion thus indicates that ΔP is close to zero across a typical tonoplast. If the tonoplast were taut, ΔP would cause a stress in the membrane, analogous to the cell wall stresses discussed in the last chapter; namely, the stress would be $r \, \Delta P/2t$ for a spherical vacuole. But the tensile strength of biological membranes is low—membranes can rupture when a stress of 0.2 to 1.0 MPa develops in them (Kotyk et al., 1988; Wolfe and Steponkus, 1981). For a tonoplast 7 nm thick with a maximum stress of 0.5 MPa surrounding a spherical vacuole 14 μm in radius, the maximum hydrostatic pressure difference across the tonoplast is

$$\Delta P = \frac{2t\sigma}{r} = \frac{(2)(7 \times 10^{-9} \text{ m})(0.5 \text{ MPa})}{(14 \times 10^{-6} \text{ m})} = 0.5 \times 10^{-3} \text{ MPa}$$

which is a very small pressure difference. Thus, P may have essentially the same value in the cytosol and in the vacuole. With this simplifying assumption, and for the equilibrium situation, where Ψ is the same in the two phases, Equation 2.13a ($\Psi = P - \Pi + \rho_w gh$, where $\Delta h = 0$ across a membrane) gives

$$\Pi^{\text{cytosol}} = \Pi^{\text{vacuole}} \tag{2.14}$$

where Π^{cytosol} is the osmotic pressure in the cytosol and Π^{vacuole} is that in the central vacuole.

The vacuole appears to be a relatively homogeneous aqueous phase, whereas the cytoplasm is a more complex phase containing many colloids and membrane-bounded organelles. Because the vacuole contains few colloidal or other interfaces, any matric pressure in it is probably negligible compared with the osmotic pressure resulting from the vacuolar solutes. Expressing osmotic pressure by Equation 2.11 ($\Pi = \Pi_s + \tau$) and assuming that $\tau^{vacuole}$ is negligible, we can replace $\Pi^{vacuole}$ in Equation 2.14 by $\Pi_s^{vacuole}$; i.e., the decrease in vacuolar water activity is due almost solely to the presence of solutes. This also means that water is acting like an ideal solvent ($\gamma_w = 1$). On the other hand, water activity in the cytosol presumably is considerably lowered both by solutes and by the presence of interfaces. Unlike the case for the vacuole, individual water molecules in the cytosol are never very far from proteins, organelles, or other colloids, so interfacial interactions can appreciably affect $a_w^{cytosol}$. Thus, both $\Pi_s^{cytosol}$ and $\tau^{cytosol}$ contribute to $\Pi^{cytosol}$ ($\Pi = \tau + \Pi_s$, Eq. 2.11). Hence, Equation 2.14 indicates that $\Pi_s^{cytosol} + \tau^{cytosol}$ equals $\Pi_s^{vacuole}$. Because $\tau^{cytosol}$ is positive, we conclude that at equilibrium the osmotic pressure in the vacuole due to solutes, $\Pi_s^{vacuole}$, must be larger than $\Pi_s^{cytosol}$. Equation 2.14 thus leads to the prediction that the vacuole has a higher concentration of osmotically active solutes than does the cytosol.

There are various possible roles for the large central vacuole in plant cells (see Raven, 1987). Because of the large central vacuole, the cytoplasm occupies a thin layer around the periphery of the cell (see Fig. 1.1); therefore, for its volume, the cytoplasm has a relatively large surface area across which diffusion can readily occur. Vacuoles also provide large, relatively simple compartments in which hydrostatic pressures lead to the cellular turgidity necessary for the growth and support of a plant. The vacuole can also act as a storage reservoir for toxic products or metabolites; e.g., the nocturnal storage of organic acids such as malic acid takes place in the vacuoles of Crassulacean acid metabolism plants (to be discussed in Ch. 8). Also, certain secondary chemical products such as phenolics, alkaloids, tannins, glucosides, and flavonoids such as the anthocyanins can accumulate in vacuoles.

Boyle–Van't Hoff Relation

The volume of a chloroplast or other membrane-bounded body changes in response to variations in the osmotic pressure of the external solution, Π^o. This is a consequence of the properties of membranes, which generally allow water to move readily across them but restrict the passage of certain solutes, such as sucrose. For example, if Π^o outside a membrane-bounded aqueous compartment were raised, we would expect that water would flow out across the membrane to the region of lower water potential and thereby decrease the compartment volume, whereas in general there would be very little movement of solutes in such a case. This differential permeability leads to the "osmometric behavior" characteristic of many cells and organelles. The conventional expression quantifying this volume response to changes in the external osmotic pressure is the *Boyle–Van't Hoff relation*:

$$\Pi^\circ(V - b) = RT \sum_j \varphi_j n_j = RTn \qquad (2.15)$$

where Π° is the osmotic pressure of the external solution; V is the volume of the cell or organelle; b is the so-called nonosmotic volume (osmotically inactive volume), which is frequently considered to be the volume of a solid phase within volume V that is not penetrated by water; n_j is the number of moles of species j within $V - b$; φ_j is the osmotic coefficient, generally construed as a correction factor indicating the relative osmotic effect of species j; and $n = \Sigma_j \varphi_j n_j$ is the apparent number of osmotically active moles in $V - b$ (Dick, 1966; Nobel, 1969).

In this chapter we will derive the Boyle–Van't Hoff relation using the chemical potential of water, and in Chapter 3 we will extend the treatment to penetrating solutes by employing irreversible thermodynamics. Although the Boyle–Van't Hoff expression will be used to interpret the osmotic responses only of chloroplasts, the equations that will be developed are quite general and can be applied equally well to mitochondria, whole cells, or other membrane-surrounded bodies.

The Boyle–Van't Hoff relation applies to the equilibrium situation where the water potential is the same on either side of the membranes surrounding chloroplasts. When Ψ^i equals Ψ°, net water movement across the membrane ceases, and the volume of a chloroplast is constant. (The superscript i refers to the inside of the cell or organelle, the superscript o to the outside.) If we were to measure the chloroplast volume under such conditions, the external solution would generally be at atmospheric pressure ($P^\circ = 0$). By Equation 2.13a ($\Psi = P - \Pi$, when the gravitational term is ignored), the water potential in the external solution is then

$$\Psi^\circ = -\Pi^\circ \qquad (2.16)$$

Like Ψ° (Eq. 2.16), the water potential inside the chloroplasts, Ψ^i, also depends on osmotic pressure. But the internal hydrostatic pressure (P^i) may be different from atmospheric pressure and should be included in the expression for Ψ^i. In addition, macromolecules and solid-liquid interfaces inside the chloroplasts can lead to a lowering of the activity coefficient of water, which we can represent by a matric pressure τ^i (see Eq. 2.11). To allow for this possibility, the internal osmotic pressure (Π^i) will be considered to be the sum of the solute and the interfacial contributions, in the manner expressed by Equation 2.11. Using Equation 2.13a, Ψ^i is therefore

$$\Psi^i = P^i - \Pi^i = P^i - \Pi^i_s - \tau^i \tag{2.17}$$

We will now re-express Π^i_s and then equate Ψ^o to Ψ^i. For a dilute solution inside the chloroplasts, Π^i_s equals $RT \sum_j n^i_j/(\overline{V}_w n^i_w)$ by Equation 2.10. Equating the water potential outside the chloroplast (Eq. 2.16) to that inside (Eq. 2.17), the condition for water equilibrium across the limiting membranes of the chloroplast becomes

$$\Pi^o = RT \frac{\sum_j n^i_j}{\overline{V}_w n^i_w} + \tau^i - P^i \tag{2.18}$$

Although this expression may look a little frightening, it concisely states that at equilibrium the water potential is the same on both sides of the membranes and that we can explicitly recognize various possible contributors to the two Ψ's involved.

To appreciate the refinements that this thermodynamic treatment introduces into the customary expression describing the osmotic responses of cells and organelles, let us compare Equation 2.18 with Equation 2.15, the conventional Boyle–Van't Hoff relation. The volume of water inside the chloroplast is $\overline{V}_w n^i_w$, because n^i_w is the number of moles of internal water and $\overline{V}_w$ is the volume per mole of water. This factor in Equation 2.18 can be identified with $V - b$ in Equation 2.15. Instead of being designated the "nonosmotic volume," b is more appropriately called the "nonwater volume," for it includes the volume of the internal solutes, colloids, and membranes. In other words, the total volume (V) minus the nonwater volume (b) equals the volume of internal water ($\overline{V}_w n^i_w$). We also note that the possible hydrostatic and matric contributions included in Equation 2.18 are neglected in the usual Boyle–Van't Hoff relation. In summary, although certain approximations and assumptions are incorporated into Equation 2.18 (e.g., that solutes do not cross the limiting membranes, and that we have dilute solutions), it is still a more satisfactory expression for describing osmotic responses of cells and organelles than is the conventional Boyle–Van't Hoff relation.

Osmotic Responses of Chloroplasts

To illustrate the use of Equation 2.18 in interpreting osmotic data, we will consider osmotic responses of pea chloroplasts suspended in external solutions of various osmotic pressures. It is customary to plot the volume V versus the reciprocal of the external osmotic pressure, $1/\Pi^o$, so certain algebraic manipulations are needed to put Equation 2.18 into a more convenient form. By transferring $\tau^i - P^i$ to the left side of Equation 2.18 and then multiplying both sides by $\overline{V}_w n_w^i/(\Pi^o - \tau^i + P^i)$, $\overline{V}_w n_w^i$ can be shown to equal $RT \Sigma_j n_j^i/(\Pi^o - \tau^i + P^i)$. The measured chloroplast volume V can be represented by $\overline{V}_w n_w^i + b$, i.e., as the sum of the aqueous and nonaqueous contributions. Hence, V should equal $RT \Sigma_j n_j^i/(\Pi^o - \tau^i + P^i) + b$, if the modified form of the Boyle–Van't Hoff relation is obeyed. Figure 2.9 indicates that the volume of pea chloroplasts varies linearly with $1/\Pi^o$ over a considerable range of external osmotic pressures. Consequently, $\tau^i - P^i$ in both Equation 2.18 and the above expression must be either negligibly small for pea chloroplasts, compared with the values of Π^o employed, or perhaps proportional to Π^o. For simplicity, we will consider only the observed proportionality between V and $1/\Pi^o$. In other words, we will assume that $V - b$ (or $\overline{V}_w n_w^i$) equals $RT \Sigma_j n_j^i/\Pi^o$ for pea chloroplasts. We will return to such considerations in Chapter 3, where we will further refine the Boyle–

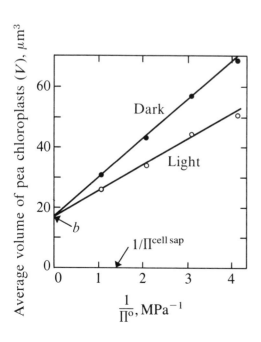

Figure 2.9

Volumes of chloroplasts at various external osmotic pressures, Π^o. Pea chloroplasts were isolated from plants in the light or the dark as indicated. [Source: P. S. Nobel, *Biochimica et Biophysica Acta 172*:134–143 (1969). Data reprinted by permission.]

Van't Hoff relation to include the more realistic case in which solutes can cross the surrounding membranes.

The relatively simple measurement of the volumes of pea chloroplasts for various external osmotic pressures can yield a considerable amount of information about the organelles. If we measure the volume of the isolated chloroplasts at the same osmotic pressure as in the cytosol, we can determine the chloroplast volume in the plant cell. Cell "sap" expressed from young pea leaves can have an osmotic pressure of 0.70 MPa; such sap comes mainly from the central vacuole, but because we expect $\Pi^{cytosol}$ to be essentially equal to $\Pi^{vacuole}$ (Eq. 2.14), $\Pi^{cell\ sap}$ is about the same as $\Pi^{cytosol}$ (some uncertainty exists, because during extraction the cell sap can come into contact with water in the cell walls). At an external osmotic pressure of 0.70 MPa (indicated by an arrow in Fig. 2.9), pea chloroplasts have a volume of 29 μm^3 when isolated from illuminated plants and 35 μm^3 when isolated from plants in the dark. Because these volumes occur at approximately the same osmotic pressure as found in the cell, they are presumably reliable estimates of pea chloroplast volumes *in vivo*.

The data in Figure 2.9 indicate that light affects the chloroplast volume *in vivo*. Although the basis for the size changes is not fully understood, chloroplasts in many plants do have a larger volume in the dark than in the light. The decrease in volume upon illumination of the plants is observed by both phase contrast and electron microscopy as a flattening or decrease in thickness of the chloroplasts. This flattening, which amounts to about 20% of the thickness for pea chloroplasts *in vivo,* is in the vertical direction for the chloroplast depicted in Figure 1.9. We also note that the slopes of the osmotic response curves in Figure 2.9 are equal to $RT \sum_j n_j^i$. The slope is 50% greater for pea chloroplasts from plants in the dark than from those in the light, suggesting that chloroplasts in the dark contain more osmotically active particles. In fact, illumination causes an efflux of K^+ and Cl^- from the chloroplasts. Also, the 20% decrease in chloroplast volume in the light *in vivo* reflects an exodus of nearly 40% of the internal water, leading to a concentrating effect on the remaining solutes such as Mg^{2+}, an ion that is important for enzymes involved in photosynthesis.

The intercept on the ordinate in Figure 2.9 is the chloroplast volume theoretically attained in an external solution of infinite osmotic pressure—a $1/\Pi^o$ of zero is the same as a Π^o of infinity. For such an infinite Π^o, all of the internal water would be removed ($n_w^i = 0$), and the volume, which is obtained by extrapolation, would be that of the nonaqueous components of the chloroplasts.[11]

---

11. Some uncertainty exists for water that is tightly bound to proteins and other substances. Some of this water presumably remains bound even at the hypothetical infinite osmotic pressures considered here. Such water should not be counted as part of the internal water ($\overline{V}_w n_w^i$).

Thus, the intercept on the ordinate of a V-versus-$1/\Pi^\circ$ plot corresponds to b in the conventional Boyle–Van't Hoff relation (Eq. 2.15). This intercept (indicated by an arrow in Fig. 2.9) equals 17 μm^3 for chloroplasts both in the light and in the dark. (The extra solutes in a chloroplast in the dark correspond to less than 0.1 μm^3 of solids.) Using the chloroplast volumes obtained for a Π° of 0.70 MPa and 17 μm^3 for the nonwater volume b, we find that the fractional water content of pea chloroplasts in the dark is

$$\frac{V - b}{V} = \frac{(35~\mu m^3 - 17~\mu m^3)}{(35~\mu m^3)} = 0.51$$

or 51%. In the light it is only 41%. These relatively low water contents in the organelles are consistent with the extensive internal lamellar system (see Fig. 1.9) and the abundance of CO_2 fixation enzymes in chloroplasts. Thus, osmometric responses of cells and organelles can be used to provide information on their fractional water content.

WATER POTENTIAL AND PLANT CELLS

In this section we will shift our emphasis from a consideration of the water relations of subcellular bodies to those of whole cells, and extend the development to include the case of water fluxes. Whether water enters or leaves a plant cell, how much, and the rate of movement, all depend on the water potential outside compared with that inside. The external water potential Ψ° can often be varied experimentally, and the direction as well as the magnitude of the resulting water movement will give information about Ψ^i. Moreover, the equilibrium value of Ψ° can be used to estimate the internal osmotic pressure Π^i. We will also consider various ways of examining the relationships between Ψ^i, Π^i, and P^i.

A loss of water from plant shoots—indeed, sometimes even an uptake—occurs at cell-air interfaces. As we would expect, the chemical potential of water in cells compared with that in the adjacent air determines the direction for net water movement at such locations. Thus, we must obtain an expression for the water potential in a vapor phase, and then relate this Ψ_{wv} to Ψ for the liquid phases in a cell. We will specifically consider the factors influencing the water potential at the plant cell-air interface, namely, in the cell wall. We will find that $\Psi^{cell~wall}$ is dominated by a negative hydrostatic pressure resulting from surface tension effects in the cell wall pores.

Incipient Plasmolysis

For usual physiological conditions, a positive hydrostatic pressure exists inside a plant cell; i.e., it is under turgor. By suitably adjusting the solutes in an external solution bathing such a turgid cell, it is possible to reduce P^i to zero and thereby to obtain an estimate of Π^i, as the following argument will indicate (see Fig. 2.10).

Let us place the cell in pure water ($\Pi^o = 0$) at atmospheric pressure ($P^o = 0$). Ψ^o is then zero (ignoring the gravitational term, $\Psi = P - \Pi$, Eq. 2.13a), and Ψ^i is $P^i - \Pi^i$ (the inner phase can be the cytosol). If the cell can be assumed to be in equilibrium with this external solution ($\Psi^o = \Psi^i$), P^i must then equal Π^i (Fig. 2.10). Suppose that Π^o is now gradually raised from its initial zero value—for example, by adding solute to the external solution. If the cell remains in equilibrium, then Π^o equals $\Pi^i - P^i$ ($P^o = 0$, because the external solution is at atmospheric pressure). As Π^o is increased, P^i will tend to decrease, while Π^i usually does not change very much. More precisely, because the cell wall is quite rigid, the cell will not appreciably change its volume in response to small changes in P^i. (Because the cell wall has elastic properties—see Ch. 1—some water flows out as P^i decreases and the cell shrinks, a matter to which we will return shortly.) If the cell volume does not change in response to changes in P^i and no internal solutes leak out, Π^i will remain constant. As the external osmotic pressure is raised, Π^o will eventually become equal to Π^i. In such a plant cell at equilibrium P^i is 0, which means that no internal hydrostatic pressure is exerted

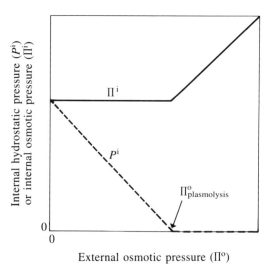

Internal hydrostatic pressure (P^i) or internal osmotic pressure (Π^i)

Π^i

P^i

$\Pi^o_{plasmolysis}$

0

0

External osmotic pressure (Π^o)

Figure 2.10

Responses of an initially turgid cell ($P^i > 0$) placed in pure water ($\Pi^o = 0$) to changes in the external osmotic pressure (Π^o). At the point of incipient plasmolysis (plasmalemma just beginning to pull away from the cell wall), P^i is reduced to zero, which is hence also called the turgor loss point. Constancy of cell volume and water equilibrium are assumed at each step as Π^o is raised (actually, Π^i increases a few percent as Π^o is raised from 0 to $\Pi^o_{plasmolysis}$, because the cell shrinks slightly accompanying the decrease of P^i to zero).

against the cell wall when Π^o equals Π^i (Fig. 2.10). The cell will thus lose its turgidity. If Π^o is increased even further, an appreciable amount of water will flow out of the cell, plasmolysis will occur as the plasmalemma pulls away from the rigid cell wall, and the internal osmotic pressure will increase (the same amount of solutes in less water; see Fig. 3.13). Ignoring for the moment any overall volume changes of the cell, we find that the condition under which water just begins to move out of the cell—referred to as the point of *incipient plasmolysis* and illustrated in Figure 2.10—is

$$\Pi^o_{plasmolysis} = \Pi^i \tag{2.19}$$

Equation 2.19 suggests that only a relatively simple measurement ($\Pi^o_{plasmolysis}$) is needed to estimate the osmotic pressure (Π^i) occurring inside an individual plant cell (see Stadelmann, 1966).

The existence of an internal hydrostatic pressure within a plant cell leads to stresses in its cell wall and a resulting elastic deformation or strain, as indicated in Chapter 1. The decrease of P^i to 0, known as the turgor loss point—which occurs in taking a turgid plant cell to the point of incipient plasmolysis—must therefore be accompanied by a contraction of the cell as the wall stresses are relieved. This decrease in volume means that some water will actually flow out of the cell before the point of incipient plasmolysis is reached. If no internal solutes enter or leave as the cell shrinks, then the osmotic pressure inside will increase (the same amount of solutes in a smaller volume). As a useful first approximation, we can assume that the change in osmotic pressure reciprocally follows the change in volume; i.e., the product of Π^i and the cellular volume is approximately constant. With this assumption, the osmotic pressure in the cell—described by Equation 2.19 and determined by using the technique of measuring the point of incipient plasmolysis—can be corrected to its original value by using the ratio of the initial volume to the final volume of the cell. The change in volume is only a few percent for most plant cells, in which case fairly accurate estimates of Π^i can be obtained from plasmolytic studies alone.

Measurements of $\Pi^o_{plasmolysis}$ often give values near 0.7 MPa for cells in storage tissues like onion bulbs or carrot roots and in young leaves of pea or spinach. These values of the external osmotic pressure provide information on various contributors to the water potential inside the cell, as Equation 2.19 indicates. For the case in which water is in equilibrium within the plant cell at the point of incipient plasmolysis, Π^i in Equation 2.19 is the osmotic pressure both in the cytosol and in the vacuole. Under certain circumstances Π^i can be replaced by $\Pi^i_s + \tau^i$ (Eq. 2.11), where Π^i_s is the osmotic pressure contributed by the internal solutes and τ^i is the matric pressure. This relation was invoked when the implications of Equation 2.14 ($\Pi^{\text{cytosol}} = \Pi^{\text{vacuole}}$) were discussed, and the various arguments presented at that time extend to the present case. Specifically, because the possible matric pressure in the vacuole is probably negligible, $\Pi^o_{\text{plasmolysis}}$ should be a rather good estimate of Π_s^{vacuole}. Moreover, τ^{cytosol} is most probably significant, as the water activity in the cytosol can be lowered by the many interfaces present there. Thus, $\Pi^o_{\text{plasmolysis}}$ is an upper limit for Π_s^{cytosol}. We see that determination of the external osmotic pressure at the point of incipient plasmolysis provides information on the osmotic pressure existing in different compartments within the plant cell.

Höfler Diagrams and Pressure–Volume Curves

Let us next examine the relationship between cellular water potential components and protoplast volume in a way introduced by Höfler in 1920 and now referred to as a *Höfler diagram* (Fig. 2.11) (see McClendon, 1982; Salisbury and Ross, 1991; Turner, 1988). The protoplast volume is maximal when the cell is in equilibrium with pure water, in which case the internal water potential Ψ^i is zero (right-hand side of Fig. 2.11 and left-hand side of Fig. 2.10). As Ψ^i decreases, P^i decreases until the point of incipient plasmolysis and zero turgor is reached. The decrease in P^i causes a slight shrinkage of the cell and hence a slight concentrating of the solutes, so Π^i increases slightly as P^i decreases. Indeed, the product of protoplast volume and Π^i is essentially constant if the amount of solutes in the cell does not change, as we indicated above. Figure 2.11 also shows that Ψ^i is equal to $-\Pi^i$ when the protoplast volume decreases below the value for incipient plasmolysis, as P^i is then zero (based on Eq. 2.13, $\Psi^i = P^i - \Pi^i = \Psi^i_P + \Psi^i_{\Pi}$ when the gravitational term is ignored). As the cell goes from incipient plasmolysis to full turgor, Figure 2.11 indicates that P^i increases slowly at first as the cell expands and then more rapidly as the cell wall approaches the limit of its expansion. Thus, a Höfler diagram directly demonstrates how Ψ^i, P^i, and Π^i change as cell water content and hence volume change, such as during the drying and wetting cycles experienced by plants.

We can readily extend our discussion to include *pressure–volume,* or *P–V, curves,* which have proved useful for analyzing the water relations of plant organs such as leaves. To obtain such curves, we can place an excised leaf in a pressure chamber (Fig. 2.8) and increase the air pressure in the chamber until liquid just becomes visible at the cut end of the xylem, which is viewed with a dissecting microscope or a hand-held magnifying lens so that water in individual conducting cells in the xylem can be observed. When the leaf is excised, the tension in the xylem causes the water contained therein to recede into the leaf,

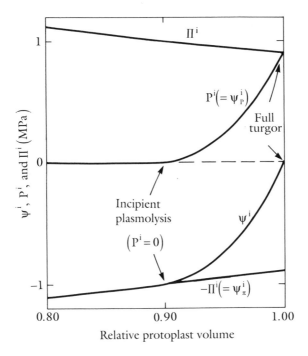

Figure 2.11

A modified Höfler diagram showing the relationship between the water potential (Ψ^i), hydrostatic pressure (P^i), and osmotic pressure (Π^i) in a plant cell for various protoplast volumes. For a group of cells, relative water content is often used in such diagrams instead of relative protoplast volume. The indicated 10% decrease in volume from full turgor to incipient plasmolysis is characteristic of plant cells (see McClendon, 1982; Turner, 1988).

so this applied air pressure is needed to force water back to the cut surface. We next increase the air pressure in the pressure chamber to force out some liquid, which is carefully collected and its volume determined, and the new balancing air pressure is recorded. In this way, we obtain data step by step for a plot of the reciprocal of the balancing air pressure in the chamber, which corresponds approximately to the negative of the reciprocal of the tissue water potential, versus the volume of extruded xylem sap, a pressure–volume curve (Fig. 2.12). After forcing out considerable liquid (the maximum air pressure used in pressure chambers rarely exceeds 10 MPa), we can extrapolate the straight-line portion of the curve to the right to obtain the extrudable liquid at an infinite applied air pressure (1/pressure = 0 means that pressure = ∞). This intercept, which occurs when $-1/\Psi^i$ equals zero, is generally considered to represent the water volume of the symplasm. The apoplastic water volume is obtained as the difference

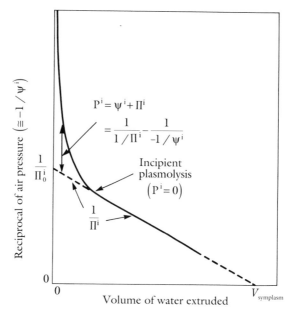

Figure 2.12

Relation between the reciprocal of leaf water potential determined with a pressure chamber (Fig. 2.8) and the volume of xylem sap extruded as the air pressure in the chamber is increased. The solid line indicates a typical range for data points (Turner, 1988). The reciprocal of the internal osmotic pressure ($1/\Pi^i$) including the value at full turgor ($1/\Pi_0^i$), the internal hydrostatic pressure (P^i), the point of incipient plasmolysis and turgor loss, and the volume of symplastic water ($V_{symplasm}$) can all be readily determined from such a P–V curve.

between the total water volume of the tissue (determined from the fresh weight of the fully turgid tissue minus the dry weight obtained after drying the tissue in an oven at about 80°C until no further weight change occurs) minus this symplastic water volume, although some movement of water from the apoplasm to the symplasm may occur as the air pressure is increased in the pressure chamber.

The straight-line portion on the right-hand side of the pressure–volume curve in Figure 2.12 corresponds to the inverse relationship between internal osmotic pressure Π^i and protoplasm volume V^i discussed above [$\Pi^i V^i$ = constant = Π^i ($V_{symplasm} - V_{extruded}$), where $V_{symplasm}$ is the symplastic water volume at full turgor for all the cells in the chamber and $V_{extruded}$ is the volume of water extruded]. When the cells are at full turgor ($\Psi^i = 0$) and no sap has been extruded, the internal osmotic pressure has its minimum value, Π_0^i, which can be obtained by extrapolating the straight-line portion to the left in Figure 2.12.

The hydrostatic pressure inside the cell (P^i) can also be deduced from such a pressure volume curve using the difference between $-1/\Psi^i$ and the extrapolated line for $1/\Pi^i$. Moreover, because the change in volume (ΔV) for a specific change in pressure (ΔP) can be determined from the left-hand part of the curve in Figure 2.12, we can also determine the volumetric elastic modulus ε averaged over the cells in the pressure chamber [$\varepsilon = \Delta P/(\Delta V/V)$, Eq. 1.17]. Because pressure-volume curves can be used to estimate symplastic water volume, apoplastic water volume, internal osmotic pressure, internal hydrostatic pressure, the point of incipient plasmolysis, and the volumetric elastic modulus, they have played a prominent role in the study of plant water relations (see Scholander *et al.*, 1964; Turner, 1988).

Chemical Potential and Water Potential of Water Vapor

Water molecules in an aqueous solution continually escape into a surrounding gas phase. At the same time, water molecules also condense back into the liquid. The rates of escape and condensation depend on the chemical activities of water in the gas phase and in the liquid phase, the two rates becoming equal at equilibrium. The gas phase adjacent to the solution then contains as much water as it can hold at that temperature and still remain in equilibrium with the liquid. It is thus *saturated* with water vapor for the particular solution under consideration. The partial pressure exerted by the water vapor in equilibrium with pure water is known as the *saturation vapor pressure*.

The vapor pressure at equilibrium depends on the temperature and the solution, but it is independent of the relative or absolute amounts of liquid and vapor. When air adjacent to pure water is saturated with water vapor (100% relative humidity), the gas phase has the maximum water vapor pressure possible at that temperature—unless it is supersaturated, a metastable, nonequilibrium situation. This saturation vapor pressure of pure water increases markedly with temperature; e.g., it increases from 0.61 kPa at 0°C to 7.38 kPa at 40°C (see App. I and Fig. 8.5). Thus, heating air at constant pressure causes the relative humidity to drop dramatically (which can lead to dry environments inside buildings that during cold weather draw air in from the outside and then heat it).

As solutes are added to the liquid phase and the activity of its water is thereby lowered, water molecules have less tendency to leave the solution. Hence, the water vapor pressure in the gas phase at equilibrium becomes less—this is one of the colligative properties of solutions that we mentioned earlier. In fact, for dilute solutions the actual partial pressure of water vapor (P_{wv}) at equilibrium depends linearly on the mole fraction of water (N_w) in the liquid phase. This is

Raoult's law, which is also mentioned in Appendix IV. For pure water, N_w equals one and P_{wv} has its maximum value, namely P^*_{wv}, the saturation vapor pressure.

The chemical potential of water vapor μ_{wv}, which depends on the partial pressure of water vapor P_{wv}, can be represented as follows:

$$\mu_{wv} = \mu^*_{wv} + RT \ln \frac{P_{wv}}{P^*_{wv}} + m_{wv}gh \qquad (2.20)$$

where P^*_{wv} is the saturation vapor pressure in equilibrium with pure liquid water at atmospheric pressure and at the same temperature as the system under consideration, and m_{wv} is the mass per mole of water vapor, which is the same as the mass per mole of water, m_w. To handle deviations from ideality, P_{wv} is replaced by $\gamma_{wv}P_{wv}$, where γ_{wv} is the activity coefficient (more properly, the "fugacity" coefficient) of water vapor. If water vapor obeyed the ideal gas law—$P_jV = n_jRT$, where P_j is the partial pressure of species j in a volume V containing n_j moles of that gas—γ_{wv} would equal unity. This is a good approximation for situations of interest in biology.

Let us next consider the various parameters at the surface of pure water ($a_w = 1$) at atmospheric pressure ($P = 0$) in equilibrium with its vapor. In such a case, P_{wv} equals P^*_{wv}, and we can take the zero level of the gravitational term as the surface of the liquid. Hence, μ_w equals μ^*_w and μ_{wv} equals μ^*_{wv}. Because there is no tendency for a net gain or loss of water molecules from the liquid phase at equilibrium ($\mu_w = \mu_{wv}$), we conclude that the constant μ^*_w must equal the constant μ^*_{wv} when the two phases are at the same temperature. Because the arguments are unchanged at other elevations, the same gravitational term must be included in μ_w and μ_{wv} ($m_wgh = m_{wv}gh$ because $m_w = m_{wv}$). The form for the pressure effects in the chemical potential for water vapor is more subtle and is discussed next and in Appendix IV.

Equations 2.4 and 2.12 indicate that μ_w contains a term dependent on the pressure P applied to the system ($\mu_w = \mu^*_w + RT \ln a_w + \overline{V}_wP + m_wgh$), whereas μ_{wv} has no such term apparent ($\mu_{wv} = \mu^*_{wv} + RT \ln P_{wv}/P^*_{wv} + m_wgh$, Eq. 2.20). Because μ_w equals μ_{wv} at equilibrium and a pressure can be applied to such a system, P_{wv} must depend on P. This dependency is embodied in the *Gibbs equation* (see Castellan, 1983; Noggle, 1985; App. IV), one form of which is $RT \ln P_{wv}/P^0_{wv} = \overline{V}_wP$, where P^0_{wv} is the partial pressure of water vapor at atmospheric pressure. By adding and subtracting $RT \ln P^0_{wv}$ from the right-hand side of Equation 2.20 and substituting $\overline{V}_wP$ for $RT \ln P_{wv}/P^0_{wv}$, we obtain $\mu_{wv} = \mu^*_{wv} + RT \ln P^0_{wv}/P^*_{wv} + RT \ln P_{wv}/P^0_{wv} + m_wgh = \mu^*_{wv} + RT \ln P^0_{wv}/P^*_{wv} + \overline{V}_wP + m_wgh$. In practice the $\overline{V}_wP$ term is generally ignored, because most air phases considered in plant physiology are at atmospheric pressure, so P equals 0 (P is conventionally defined as the pressure in excess of atmospheric). In any case, the difference between P_{wv} and P_{wv}^0 is generally very small, as an increase of 0.01 MPa (0.1 atmosphere) in the gas phase surrounding a plant increases P_{wv} by only 0.007%. To appreciate the motion of water vapor that can be induced by changes in P, we need only consider the phenomenon of wind (recall that changes in any term of a chemical potential, which represent changes in energy, can lead to motion). We also note that different air pressures can occur in certain plants—ranging from water lilies to rice—which can lead to internal "winds" leading to mass flow of gases (briefly considered in Ch. 8).

Let us now consider the water potential of water vapor in a gas phase such as air, Ψ_{wv}. Using a definition of Ψ analogous to that in Equation 2.13a [$\Psi = (\mu_w - \mu_w^*)/\overline{V}_w$] and defining μ_{wv} by Equation 2.20, we have

$$\Psi_{wv} = \frac{\mu_{wv} - \mu_w^*}{\overline{V}_w} = \frac{\mu_{wv} - \mu_{wv}^*}{\overline{V}_w}$$

$$= \frac{RT}{\overline{V}_w} \ln \frac{P_{wv}}{P_{wv}^*} + \rho_w g h$$

$$= \frac{RT}{\overline{V}_w} \ln \frac{\% \text{ relative humidity}}{100} + \rho_w g h \qquad (2.21)$$

where $m_{wv}/\overline{V}_w$ (the same as $m_w/\overline{V}_w$) has been replaced by the density of water, ρ_w. We note that $\overline{V}_w$, not $\overline{V}_{wv}$, is used in the definition of Ψ_{wv} in Equation 2.21. This is necessary because the fundamental term representing free energy per mole is the chemical potential—we want to compare $\mu_j - \mu_j^*$ for the two phases when predicting changes at an interface—and thus the proportionality factor between $\mu_w - \mu_w^*$ or $\mu_{wv} - \mu_{wv}^*$ ($= \mu_w - \mu_w^*$) and the more convenient terms, Ψ or Ψ_{wv}, must be the same in each case, namely, $\overline{V}_w$. Equation 2.21 also indicates that P_{wv}/P_{wv}^* equals the % relative humidity/100; relative humidity is a readily measured quantity and has been extensively used in studying the water relations of plants and animals.

What happens to P_{wv} as we move vertically upward from pure water at atmospheric pressure in equilibrium with water vapor in the air? If we let h be zero at the surface of the water, Ψ equals zero ($\Psi = P - \Pi + \rho_w g h$, Eq. 2.13a), and because the water vapor is by supposition in equilibrium with the liquid phase, Ψ_{wv} is also zero. Equation 2.21 indicates that, as we go vertically upward in the gas phase, $\rho_w g h$ must make an increasingly positive contribution to Ψ_{wv}. Because we are at equilibrium, the other term in Ψ_{wv}, $(RT/\overline{V}_w) \ln (P_{wv}/P_{wv}^*)$, must make a compensating negative contribution. At equilibrium P_{wv} must therefore decrease with altitude. We can further appreciate this conclusion simply by noting that gravity attracts the molecules of water vapor and other gases toward the earth, so air pressure is higher at sea level than it is on the top of a mountain.

Rather great heights are needed before the partial pressure of water vapor is reduced significantly. In particular, let us consider a 7% reduction in P_{wv} from P_{wv}^* (its value at the water surface at equilibrium at atmospheric pressure) to $0.93 P_{wv}^*$. Using Equation 2.21 and values at 20°C in Appendix I, we find that the height necessary is

$$h = -\frac{1}{\rho_w g} \frac{RT}{\overline{V}_w} \ln \frac{P_{wv}}{P_{wv}^*}$$

$$= -\frac{1}{(0.0098 \text{ MPa m}^{-1})} (135 \text{ MPa}) \ln \left(\frac{0.93 P_{wv}^*}{P_{wv}^*} \right) = 1\ 000 \text{ m}$$

In other words, not until we reach a height of 1 km does the partial pressure of water vapor at equilibrium decrease by 7% from its value at sea level, indicating that the gravitational term generally has relatively little influence on Ψ_{wv} over the distances involved for an individual plant.

Plant–Air Interface

Water equilibrium across the plant–air interface occurs when the water potential in the leaf cells equals that of the surrounding atmosphere. (This presupposes that the leaf and the air are at the same temperature, an aspect that we will reconsider in Ch. 8.) To measure Ψ^{leaf}, the leaf can be placed in a closed chamber and the relative humidity adjusted until the leaf does not gain or lose water. Such a determination is experimentally demanding, because small changes in relative humidity have large effects on Ψ_{wv}, as we will next show.

Extremely large negative values are possible for Ψ_{wv}. In particular, $RT/\overline{V}_w$ at 20°C is 135 MPa (1 350 bars). In the expression for Ψ_{wv} (Eq. 2.21), this relatively large factor multiplies ln (% relative humidity/100), and a wide range of relative humidities can occur in the air. By Equation 2.21 with $\rho_w g h$ ignored, a relative humidity of 100% corresponds to a water potential in the vapor phase of 0 (ln 1 = 0). This Ψ_{wv} is in equilibrium with pure water at atmospheric pressure, which also has a water potential of 0. For a relative humidity of 99%, Ψ_{wv} given by Equation 2.21 is

$$\Psi_{wv} = (135 \text{ MPa}) \ln \left(\frac{99}{100} \right) = -1.36 \text{ MPa}$$

Table 2.1

Magnitudes of certain water vapor parameters useful for understanding water movement at the plant-air interface. Data are for 20°C. Ψ_{wv} is calculated from Equation 2.21 ignoring the gravitational term. We note that $P_{wv}/P_{wv}^* \times 100 =$ relative humidity in %.

Relative humidity (%)	P_{wv} (kPa)	Ψ_{wv} (MPa)
100.0	2.34	0.00
99.6	2.33	−0.54
99.0	2.32	−1.36
96.0	2.25	−5.51
90.0	2.11	−14.2
50.0	1.17	−93.6
0.0	0.00	−∞

Hence, going from 100% to 99% relative humidity corresponds to a decrease in water potential of 1.36 MPa (Table 2.1). Small changes in relative humidity do indeed reflect large differences in the water potential in air! Finally, we note that a relative humidity of 50% at 20°C leads to a Ψ_{wv} of −94 MPa.

Because of the large negative values of Ψ_{wv}^{air}, water tends to diffuse out of leaves. The actual value of Ψ^{leaf} depends on the ambient conditions as well as the plant type and its physiological status. The range of values for most mesophytes is −0.3 to −3 MPa, with −0.5 MPa being typical for leaves of a garden vegetable such as lettuce. From Equation 2.21, the relative humidity corresponding to a water potential of −0.5 MPa is 99.6%.[12] Even during a rainstorm the relative humidity of the air rarely exceeds 99%. Because relative humidity is lower than 99.6% under most natural conditions, water is continually being lost from a leaf having an internal water potential of −0.5 MPa. A few desert plants have a Ψ^{leaf} as low as −5.5 MPa. Even in this case of adaptation to arid climates, however, water still tends to leave the plant unless the relative humidity is about 96% (Table 2.1).[13]

---

12. Such an extremely high value for the relative humidity in equilibrium with a Ψ^{leaf} of −0.5 MPa indicates why it is difficult to determine Ψ^{leaf} by measuring the Ψ_{wv}^{air} for which no water is gained or lost by the leaf. See Figure 2.8, Boyer (1969), and Turner (1981) for ways to measure Ψ and Ψ_{wv}.

13. Structural modifications and physiological responses are generally more important than a low Ψ^{leaf} for adapting to xerophytic conditions. We will discuss this topic in Chapters 7 and 8.

Pressure in the Cell Wall Water

Plant cells come into contact with air at the cell walls bounding the intercellular air spaces (see Fig. 1.2). Thus, the water potential in the cell walls must be considered with respect to Ψ_{wv} in the adjacent gas phase. The main contributor to Ψ in cell wall water is often the negative hydrostatic pressure arising from surface tension effects at the numerous air-liquid interfaces of the cell wall interstices. In turn, $P^{\text{cell wall}}$ can be related to the geometry of the cell wall pores and the contact angles that occur.

The magnitude of the negative hydrostatic pressure that develops in cell wall water can be analyzed by considering the pressure that occurs near the surface in a liquid within a cylindrical pore—the argument is basically the same as the one presented earlier in this chapter in discussing capillary rise. One of the forces acting on the fluid in a narrow pore of radius r is the result of surface tension, σ. This force equals $2\pi r\sigma \cos \alpha$ (see Fig. 2.3), where α is the contact angle, a quantity that can be essentially zero for wettable walls. Because the adhesive forces at the wall are transmitted to the rest of the fluid by means of cohesion, a tension or negative hydrostatic pressure develops in the fluid. The total force resisting the surface adhesion can be regarded as this tension times the area over which it acts, πr^2, and hence the force is $\pi r^2 \times (-P)$. Equating the two forces gives the following expression for the pressure developed in the fluid contained within a cylindrical pore:

$$P = \frac{2\pi r\sigma \cos \alpha}{-\pi r^2} = -\frac{2\sigma \cos \alpha}{r} \qquad (2.22)$$

Adhesion of water at interfaces generally creates negative hydrostatic pressures in the rest of the fluid (Eq. 2.22 describes this P near the air-water interface, where the gravitational term can be ignored). Such negative hydrostatic pressures arising from interfacial interactions have sometimes been treated in plant physiology as positive matric pressures, a convention that we mentioned earlier.

The strong water-wall adhesive forces, which are transmitted throughout the cell wall interstices by water-water hydrogen bonding, can greatly reduce the water potential in the cell wall, as the following calculation indicates. At 20°C the surface tension of water is 7.28×10^{-8} MPa m (App. I), the voids between the fibers in the cell wall are often about 10 nm across ($r = 5$ nm), and $\cos \alpha$ can equal one for wettable walls. For water in such cylindrical pores, Equation 2.22 indicates that P could be

$$P = -\frac{(2)(7.28 \times 10^{-8} \text{ MPa m})(1)}{(5 \times 10^{-9} \text{ m})} = -29 \text{ MPa}$$

(Fig. 2.13a). This is an estimate of the negative hydrostatic pressure or tension that *could* develop in the aqueous solution within cell wall interstices of typical dimensions, supporting the contention that $\Psi^{cell\ wall}$ can be markedly less than zero. Moreover, the previous discussion of the tensile strength of water indicates that the hydrogen bonding in water can withstand such a tension. We will next consider the actual pressures that develop in the cell wall water.

Let us begin by considering a Ψ_{wv} of -5.5 MPa, which corresponds to a relative humidity of 96% (Table 2.1). According to Equation 2.22, water in cylindrical cell wall pores with radii of 26 nm has a P of -5.5 MPa (see Fig. 2.13b), so such water is in equilibrium with air of 96% relative humidity in the intercellular spaces of the leaf. If the relative humidity near the cell wall surface were decreased, water in such interstices could be lost, but it would remain in the finer pores, where a more negative P or larger tension can be created (Fig. 2.13a and c). Hence, if the plant material is exposed to air of moderate to low relative humidity—as can happen for herbarium specimens—then larger and larger tensions resulting from interfacial interactions develop in the cell walls as the plant material dries. Water is then retained only in finer and finer pores of the wall, as is consistent with Equation 2.22.

We will now show that the availability of water adjacent to that in wettable cell walls affects $P^{cell\ wall}$ and the actual contact angle in the interstices. Suppose that pure water in mesophyll cell wall interstices 10 nm across is in equilibrium with water vapor in the intercellular air spaces where the relative humidity is 99%. As we calculated above, Ψ_{wv} for this relative humidity is -1.36 MPa at 20°C. Hence, at equilibrium the (pure) water in the cell wall interstices would be under a hydrostatic pressure of -1.36 MPa ($\Psi = P - \Pi + \rho_w gh$, Eq.

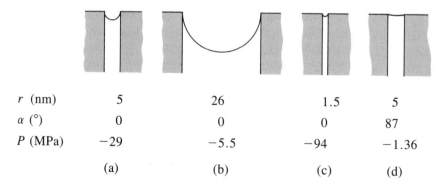

r (nm)	5	26	1.5	5
α (°)	0	0	0	87
P (MPa)	-29	-5.5	-94	-1.36
	(a)	(b)	(c)	(d)

Figure 2.13

Influence of radius r and contact angle α on the hydrostatic pressure P near the surface in cylindrical pores open to air.

2.13a). From Equation 2.22 we can calculate the contact angle where P can be -1.36 MPa for pores 5 nm in radius:

$$\cos \alpha = -\frac{rP}{2\sigma} = -\frac{(5 \times 10^{-9} \text{ m})(-1.36 \text{ MPa})}{(2)(7.28 \times 10^{-8} \text{ MPa m})} = 0.0467$$

and so α is 87°. For the wettable cell walls, α *can* be 0°, in which case cos α is unity and the maximum negative pressure develops, namely -29 MPa for the pores 5 nm in radius. On the other hand, when water is available, it can be pulled into the interstices by such possible tensions and thereby cause α for wettable walls to increase toward 90°, which leads to a decrease in cos α and consequently a decrease in the tension developed. In the present example, α is 87°, P is -1.36 MPa, and the pore 5 nm in radius is nearly filled (Fig. 2.13d). Thus, the water in the interstices can be nearly flush with the cell wall surface. In fact, the large tensions that could be present in the cell wall generally do not occur in living cells, because water is usually available within the plant and is "pulled" into the interstices, thus nearly filling them, depending of course on their dimensions (see Eq. 2.22).

The only contribution to the water potential of the cell wall that we have been considering is the negative pressure that develops because of air-liquid interfaces, as can be the case for pure water. When the gravitational term in Equation 2.13a is ignored, $\Psi^{\text{cell wall}}$ equals $P^{\text{cell wall}} - \Pi^{\text{cell wall}}$. The bulk solution in the cell wall pores generally has an average osmotic pressure of 0.3 to 1.5 MPa. The local $\Pi^{\text{cell wall}}$ next to the Donnan phase (Ch. 3) is considerably greater because of the many ions present there; to help compensate, $P^{\text{cell wall}}$ is also higher near the Donnan phase than it is in the bulk of the cell wall water (activity coefficients of both ions and water are also lower near the Donnan phase; see Richter and Dainty, 1989).

Water Flux

When the water potential inside a cell differs from that outside, the water is no longer in equilibrium and we can expect a net water movement toward the region of lower water potential. This volume flux density of water, J_{V_w}, is assumed to be proportional to the difference in water potential ($\Delta\Psi$) across the membrane or membranes restricting the flow. The proportionality factor indicating the permeability to water flow at the cellular level is expressed by a *water conductivity coefficient*, L_w:

$$J_{V_w} = L_w \, \Delta\Psi = L_w(\Psi^{\circ} - \Psi^{i}) \tag{2.23}$$

In Equation 2.23, J_{V_w} is the volume flow of water per unit area of the barrier per unit time. It can have units of $m^3\ m^{-2}\ s^{-1}$, or $m\ s^{-1}$, which is a velocity. In fact, J_{V_w} is the average velocity of water moving across the barrier being considered.[14] L_w can have units of $m\ s^{-1}\ Pa^{-1}$, in which case the water potentials would be expressed in Pa. (Experimentally, L_w is usually the same as L_P, a coefficient also describing water conductivity that we will introduce in Ch. 3 in a rather different manner.)

When Equation 2.23 is applied to cells, Ψ^o is the water potential in the external solution, and Ψ^i usually refers to the water potential in the vacuole. L_w then indicates the conductivity for water flow across the cell wall, the plasmalemma, and the tonoplast, all in series. For a group of barriers in series the overall water conductivity coefficient of the pathway, L_w, is related to the conductivities of the individual barriers by

$$1/L_w = \sum_j 1/L_{w_j} \qquad (2.24)$$

where L_{w_j} is the water conductivity coefficient of barrier j (see Problem 1.4 for the analogous case of the permeability coefficient for barriers in series). To see why this is so, let us go back to Equation 2.23, $J_{V_w} = L_w\ \Delta\Psi$. When the barriers are in series, J_{V_w} is the same across each one, so J_{V_w} equals $L_{w_j}\ \Delta\Psi_j$, where $\Delta\Psi_j$ is the drop in water potential across barrier j, and $\Sigma_j\ \Delta\Psi_j = \Delta\Psi$. We therefore obtain the following equalities: $J_{V_w}/L_w = \Delta\Psi = \Sigma_j\ \Delta\Psi_j = \Sigma_j\ J_{V_w}/L_{w_j} = J_{V_w}\ \Sigma_j\ 1/L_{w_j}$; hence, $1/L_w$ must equal $\Sigma_j\ 1/L_{w_j}$, as indicated in Equation 2.24. (We also note that $1/L_w$ corresponds to a resistance, and the resistance of a group of resistors in series is the sum of the resistances, $\Sigma_j\ 1/L_{w_j}$.)

We shall now consider a membrane separating two solutions that differ only in osmotic pressure. This will help us to relate L_w to the permeability coefficient of water, P_w, and also to view Fick's first law [$J_j = P_j(c_j^o - c_j^i)$, Eq. 1.8]. in a slightly different way. The appropriate form for Fick's first law when describing the diffusion of the solvent water is $J_{V_w} = P_w(N_w^o - N_w^i)$, where N_w is the mole fraction of water. By Equation 2.23, J_{V_w} also equals $L_w\ (\Psi^o - \Psi^i)$, which becomes $L_w(\Pi^i - \Pi^o)$ in our case (recall that $\Psi = P - \Pi + \rho_w gh$, Eq. 2.13a). Thus, a volume flux density of water toward a region of higher osmotic pressure (Eq. 2.23) is equivalent to J_{V_w} toward a region where water has a lower mole

---

14. To help see why this is so, consider a volume element of cross-sectional area A, extending back from the barrier for a length equal to the average water velocity $\bar{v}_w$ multiplied by dt (e.g., let dx in Fig. 1.5 equal $\bar{v}_w\ dt$). In time dt this volume element, $A\bar{v}_w\ dt$, crosses area A of the barrier, and for such a period the volume flow of water per unit area and per unit time, J_{V_w}, equals $(A\bar{v}_w\ dt)/(A\ dt)$, which is $\bar{v}_w$. Also see Figure 3.3.

fraction (Fick's first law applied to water). Let us pursue this one step further. Using Equations 2.8 through 2.10, we can obtain the following: $N_w \cong 1 - \Sigma_j \ n_j/n_w \cong 1 - \bar{V}_w \Pi_s/RT$. Therefore, $N_w^o - N_w^i$ approximately equals $(1 - \bar{V}_w \Pi_s^o/RT) - (1 - \bar{V}_w \Pi_s^i/RT)$, or $\bar{V}_w (\Pi_s^i - \Pi_s^o)/RT$. When we incorporate this last relation into our two expressions for the volume flux density of water, we obtain $J_{V_w} = L_w(\Pi^i - \Pi^o) = P_w(N_w^o - N_w^i) = P_w \bar{V}_w (\Pi_s^i - \Pi_s^o)/RT$ (Π^i is here the same as Π_s^i). Hence, we derive the following relationship:

$$L_w = \frac{P_w \bar{V}_w}{RT} \qquad (2.25)$$

which states that the water conductivity coefficient (L_w) is proportional to the water permeability coefficient (P_w).

Next, we will estimate a possible value for L_w. P_w can be 10^{-4} m s^{-1} for plasmalemmas (Ch. 1), RT is 2.437×10^3 m^3 Pa mol^{-1} at 20°C (App. I), and $\bar{V}_w$ is 1.8×10^{-5} m^3 mol^{-1}. Thus, L_w can be

$$L_w = \frac{(10^{-4} \text{ m s}^{-1})(1.8 \times 10^{-5} \text{ m}^3 \text{ mol}^{-1})}{(2.437 \times 10^3 \text{ m}^3 \text{ Pa mol}^{-1})} = 7 \times 10^{-13} \text{ m s}^{-1} \text{ Pa}^{-1}$$

L_w for plant cells ranges from about 10^{-13} to 2×10^{-12} m s^{-1} Pa^{-1} (Boyer, 1985; Smith and Nobel, 1986; Tomos, 1988).

Cell Growth

When the value of the water conductivity coefficient is known, the water potential difference necessary to give an observed water flux can be calculated using Equation 2.23. For the internodal cells of *Nitella* and *Chara*, L_w for water entry is about 10^{-12} m s^{-1} Pa^{-1} (see Dainty, 1963). For convenience of calculation, we will consider cylindrical cells 100 mm long and 1 mm in diameter as an approximate model for such algal cells (see Fig. 3.9). The surface area across which the water flux occurs is $(2\pi r)(l)$, where r is the cell radius and l is the cell length. Thus, the area equals $(2\pi)(0.5 \times 10^{-3} \text{ m})(100 \times 10^{-3} \text{ m})$, or 3.14×10^{-4} m^2. [The area of each end of the cylinder (πr^2) is much less than $2\pi r l$; in any case, a water flux from the external solution across them would not be expected, because they are in contact with other cells, not the bathing solution; see Fig. 3.9.] The volume of a cylinder is $\pi r^2 l$, which equals $(\pi)(0.5 \times 10^{-3}$ m)$^2(100 \times 10^{-3}$ m), or 7.9×10^{-8} m^3, in the present case. Internodal cells of *Nitella* and *Chara* grow relatively slowly—a change in volume of about 1% per day is a possible growth rate for the fairly mature cells that were used in determining L_w. This growth rate means that the water content increases by about 1% of the volume per day (1 day $= 8.64 \times 10^4$ s). The volume flux density for water is the rate of volume increase divided by the surface area across which water enters. Using the indicated value of L_w and Equation 2.23, we can then calculate the drop in water potential involved:

$$\Delta\Psi = \frac{J_{V_w}}{L_w} = \frac{\dfrac{1}{A}\dfrac{dV}{dt}}{L_w} = \frac{\dfrac{1}{(3.14 \times 10^{-4} \text{ m}^2)}\dfrac{(0.01)(7.9 \times 10^{-8} \text{ m}^3)}{(8.64 \times 10^4 \text{ s})}}{(10^{-12} \text{ m s}^{-1} \text{ Pa}^{-1})} = 29 \text{ Pa}$$

In other words, the internal water potential (Ψ^i) needed to sustain the water influx accompanying a growth of about 1% per day is only 29 Pa less than the outside water potential (Ψ^o). *Nitella* and *Chara* can grow in pond water, which is a dilute aqueous solution often having a water potential near -7 kPa (-0.07 bar). Thus, Ψ^i need be only slightly more negative than -7 kPa to account for the influx of water accompanying a growth of 1% per day.

The enlargement of cells during growth depends on two simultaneously occurring processes—the uptake of water and the increase in area of the cell wall, as was clearly recognized by Lockhart in 1965 (Boyer, 1985; Cosgrove, 1986; Lockhart, 1965a,b). Because water uptake involves spontaneous water movement toward lower water potential, the irreversible aspect of growth depends on the yielding of the cell wall material. To describe growth of cells, we often use the relative rate of volume increase, $(1/V)\,dV/dt$. For water entering a single cell, $(1/V)\,dV/dt$ can be represented by $(1/V)J_{V_w}A$, which using Equation 2.23 we can express as follows:

$$\frac{1}{V}\frac{dV}{dt} = \left(\frac{A}{V}\right)L_w(\Psi^o - \Psi^i) \tag{2.26}$$

Equation 2.26 is often called a *growth equation* (to put this into the form generally found in the literature, L_w should be replaced by L_P and Ψ by $P - \sigma\Pi$, changes based on irreversible thermodynamics that we will introduce in Ch. 3). At the end of Chapter 1 we indicated that cell enlargement requires the yielding of the cell wall, a plastic or irreversible, viscoelastic process that is proposed to occur when the internal hydrostatic pressure exceeds some critical or threshold value, $P^i_{\text{threshold}}$. This leads to another growth equation of the following form:

$$\frac{1}{V}\frac{dV}{dt} = \varphi(P^i - P^i_{\text{threshold}}) \tag{2.27}$$

where φ is a coefficient describing irreversible cell wall yielding ($P^i \geqslant P^i_{threshold}$). Our present calculation for *Nitella* and *Chara* growing at a rate of 1% per day indicates that Ψ^o is nearly the same as Ψ^i. In this case, cell extension is controlled primarily by yielding of the cell wall, without which P^i and Ψ^i would increase slightly and thereby decrease the rate of water entry. For other cases, the uptake or internal production of osmotically active solutes can exert greater control on cell growth, including effects that involve the reversible cell extension described by the volumetric elastic modulus ε (see Boyer, 1985; Cosgrove, 1986; Ortega *et al.*, 1988).

Kinetics of Volume Changes

We will next indicate how cell wall and membrane properties influence the kinetics of reversible swelling or shrinking for plant cells. When the water potential outside a cell or group of cells is changed, water movement will be induced. A useful expression describing the time constant for the resulting volume change is

$$t_e = \frac{V}{AL_w(\varepsilon + \Pi^i)} \tag{2.28}$$

where V is the initial volume of a cell, A is the area across which water enters or leaves, ε is the volumetric elastic modulus (Eq. 1.17), and Π^i is the initial internal osmotic pressure (see Cosgrove, 1986; Dainty, 1976; Zimmermann and Steudle, 1978). The time constant t_e represents the time to complete all but $1/e$ or 37% of the volume shift. (The halftime, which is the time required to shift halfway from the initial to final volume, equals ln 2 times t_e.) Equation 2.28 indicates that t_e for the swelling or shrinking is small for a cell with a low V/A (a large available surface area per unit volume), a high L_w (which usually means a high plasmalemma permeability to water), or a high ε (a very rigid cell wall). Figure 2.14 shows a device that can be used to measure ΔV and ΔP directly, thus allowing a determination of ε [consider Eq. 1.17, $\varepsilon = \Delta P/(\Delta V/V)$] as well as other water-relations parameters (a *P–V* curve, Fig. 2.12, gives an indirect way of determining P and ε).

Let us next estimate the time constant for volume changes that might occur for a mesophyll cell (Fig. 1.2). We will assume that it has a volume of 10^4 μm^3

Figure 2.14

Schematic diagram of "pressure probe," an apparatus that can be used to measure P, L_w (or L_P, see Ch. 3), and ε for individual plant cells. The intracellular hydrostatic pressure is transmitted to the pressure transducer via an oil-filled microcapillary introduced into the cell. Volume can be changed by adjusting the micrometer and observing the motion of the interface between the cell sap and the oil (in the drawing the cell region is greatly enlarged relative to the rest of the apparatus). See Hüsken *et al.* (1978) for details.

and that water enters from a neighboring cell over an area of 100 μm^2, so that V/A is 100 μm. We will let L_w be 10^{-13} m s^{-1} Pa^{-1}, ε be 20 MPa, and Π^i be 1 MPa (usually Π^i is much less than ε, and so Π^i is often ignored when calculating t_e). From Equation 2.28, the time constant is

$$t_e = \frac{(100 \times 10^{-6} \text{ m})}{(10^{-13} \text{ m s}^{-1} \text{ Pa}^{-1})(21 \times 10^6 \text{ Pa})} = 50 \text{ s}$$

The volume changes caused by water transport between adjacent cells would be more rapid if their contacting area were larger and if L_w and ε were higher. Time constants for volume changes have important implications for the water exchanges underlying stomatal opening (discussed in Ch. 8) and capacitance effects (discussed in Ch. 9).

Problems

2.1 The rise of water in a long vertical capillary 2 mm in diameter with a wettable inner wall is 15 mm. (a) If the capillary is tilted 45° from the horizontal, what is the vertical height of the rise? Assume that the contact angle does not change upon tilting. (b) If sucrose is added to the solution so that the density becomes 1 200 kg m^{-3} (= 1.2 g cm^{-3}), what is the height of the rise? (c) If the capillary extends only 7.5 mm above the main surface of the water, what is the contact angle in the capillary? (d) If the wall of the capillary is so treated that the contact angle becomes 60°, what is the height of the rise? (e) What is the rise of water in a capillary similar to the original one, but with a 1 μm radius? (f) In which of the 5 cases is the greatest weight supported by capillary (surface tension) forces?

2.2 A neutral solute at equilibrium has a concentration of 0.1 molal on one side of a barrier permeable only to water and a c_{solute} of 1 molal on the other side. Let the partial molal volume of the solute be 40×10^{-6} m^3 mol^{-1}, the temperature be 20°C, and the activity coefficients of the solute and the water be unity, except in (c). (a) What is the hydrostatic pressure difference across the barrier? (b) What would it be if the barrier were permeable to solute only? (c) If γ_{solute} is 0.5 on the more concentrated side and if other conditions are unchanged, what is ΔP across the barrier permeable to the solute only? (d) What would happen at equilibrium if the barrier were permeable to both water and solute? (e) If P on the 0.1 molal side is the same as atmospheric pressure, what is the chemical potential of the solute there?

2.3 A solution assumed to be ideal contains 80 g of sorbitol (molar mass of 0.182 kg mol^{-1}) in 1 kg of water at 20°C. (a) What are N_w and a_w in the solution? (b) What is the osmotic pressure of the sorbitol solution? (c) By what percent is the activity of water reduced for an osmotic pressure of 1 MPa at 20°C compared with the value of a_w for pure water? (d) If we assume that activity coefficients are unity, what concentration of a solute corresponds to a Π of 1 MPa at 20°C? (e) A 0.25 mol m^{-3} solution of a particular polymer has a measured osmotic pressure at 20°C of 0.01 MPa. What is the osmotic pressure predicted by the Van't Hoff relation (Eq. 2.10)? Explain any discrepancies. (f) In the vacuole of a certain cell, the mole fraction of water is 0.98, the hydrostatic pressure is 0.8 MPa, and the temperature is 20°C. Assuming activity coefficients are unity, what is the water potential in the vacuole?

2.4 A tank 10 m tall and open at the top is filled with an ideal solution at 20°C. We will assume that the system is in equilibrium and that the zero level for the gravitational term is at the top of the tank, where Ψ is -0.600 MPa. (a) What is the water potential 0.1 m below the surface and at the bottom of the tank? (b) What is Π at the upper surface, 0.1 m below the surface, and at the bottom of the tank? (c) What are P and the gravitational term at the three levels in (b)? (d) What relative humidity would be in equilibrium with the water in the tank?

2.5 Chloroplasts are isolated from a plant cell whose cytosol has an osmotic pressure of 0.4 MPa at 20°C. When the chloroplasts are placed in solutions at 20°C containing an impermeant solute, the volumes are 36 μm^3 at an external osmotic pressure of

0.33 MPa, 28 μm^3 at 0.5 MPa, and 20 μm^3 at 1.0 MPa. Assume that the activity coefficients are unity. (a) What is the volume of the chloroplasts in the plant cell? (b) What is the nonaqueous volume per chloroplast? (c) What volume fraction of the chloroplast is occupied by water *in vivo*? (d) What is the amount of osmotically active particles per chloroplast?

2.6 A spherical algal cell 1 mm in diameter has a water conductivity coefficient L_w equal to 10^{-12} m s^{-1} Pa^{-1}. Let the internal osmotic pressure be 1.0 MPa, the internal hydrostatic pressure be 0.6 MPa at 20°C, and the volumetric elastic modulus be 5 MPa. (a) What is the initial net volume flux density of water into or out of the cell when it is placed in pure water at atmospheric pressure? (b) What is the time constant for the change in (a)? (c) What is the water flux density at the point of incipient plasmolysis? (d) What is the water flux density when the external solution is in equilibrium with a gas phase at 97% relative humidity? (e) Assume that the water in the cell walls is in equilibrium with the internal cellular water. What are cos α and the contact angle at the air-water interface for cylindrical cell wall pores 20 nm in diameter? Assume that $\Pi^{cell\ wall}$ is negligible.

References

Adamson, A. W. 1982. *Physical Chemistry of Surfaces*, 4th ed. Wiley, New York.

Altman, P. L., and D. S. Dittmer. 1966. *Environmental Biology*. Federation of American Societies for Experimental Biology, Bethesda, Maryland.

Boller, T., and A. Wiemken. 1986. Dynamics of vacuolar compartmentation. *Annual Review of Plant Physiology 37*:137–164.

Boyer, J. S. 1969. Measurement of the water status of plants. *Annual Review of Plant Physiology 20*:351–364.

Boyer, J. S. 1985. Water transport. *Annual Review of Plant Physiology 36*:473–516.

Bull, H. B. 1964. *An Introduction to Physical Biochemistry*. Davis, Philadelphia, Pennsylvania.

Castellan, G. W. 1983. *Physical Chemistry*, 3rd ed. Addison-Wesley, Reading, Massachusetts.

Cosgrove, D. 1986. Biophysical control of plant growth. *Annual Review of Plant Physiology 37*:377–405.

Dainty, J. 1963. Water relations of plant cells. *Advances in Botanical Research 1*:279–326.

Dainty, J. 1976. Water relations of plant cells. In *Transport in Plants II, Part A, Cells*, U. Lüttge and M. G. Pitman, eds. *Encyclopedia of Plant Physiology, New Series*, Vol. 2. Springer-Verlag, Berlin. Pp. 12–35.

Dick, D. A. T. 1966. *Cell Water*. Butterworths, London.

Eisenberg, D., and W. Kauzmann. 1969. *The Structure and Properties of Water*. Oxford University Press, Oxford, England.

Hayward, A. T. J. 1971. Negative pressure in liquids: Can it be harnessed to serve man? *American Scientist 59*:434–443.

Hermans, J., H. J. C. Berendsen, W. F. Van Gunsteren, and J. P. M. Postma. 1984. A consistent empirical potential for water–protein interactions. *Biopolymers 23*:1513–1518.

Hüsken, D., E. Steudle, and U. Zimmermann. 1978. Pressure probe technique for measuring water relations of cells in higher plants. *Plant Physiology 61*:158–163.

Israelachvili, J. N. 1985. *Intermolecular and Surface Forces with Applications to Colloidal and Biological Systems.* Academic Press, New York.

Kotyk, A., K. Janáček, and J. Koryta. 1988. *Biophysical Chemistry of Membrane Functions.* Wiley, Chichester, England.

Lockhart, J. A. 1965a. An analysis of irreversible plant cell elongation. *Journal of Theoretical Biology 8*:264–275.

Lockhart, J. A. 1965b. Cell extensions. In *Plant Biochemistry*, J. Bonner and J. E. Varner, eds. Academic Press, New York. Pp. 826–849.

London, F. 1937. The general theory of molecular forces. *Transactions of the Faraday Society 33*:8–26.

Marin, B., ed. 1987. *Plant Vacuoles: Their Importance in Solute Compartmentation in Cells and Their Applications in Plant Biotechnology.* Plenum, New York.

McClendon, J. H. 1982. Water relations curves for plant cells: Toward a realistic Höfler diagram for textbooks. *What's New in Plant Physiology 13*:17–20.

Nobel, P. S. 1969. The Boyle–Van't Hoff relation. *Journal of Theoretical Biology 23*:375–379.

Noggle, J. H. 1985. *Physical Chemistry.* Little, Brown, Boston, Massachusetts.

Oertli, J. J. 1971. The stability of water under tension in the xylem. *Zeitschrift für Pflanzenphysiologie 65*:195–209.

Ortega, J. K. E., R. G. Keanini, and K. J. Manica. 1988. Pressure probe technique to study transpiration in *Phycomyces* sporangiophores. *Plant Physiology 87*:11–14.

Pfeffer, W. F. P. 1877. *Osmotiche Untersuchungen. Studien zur Zell Mechanik.* W. Engelmann, Leipzig.

Raven, J. A. 1987. The role of vacuoles. *New Phytologist 106*:357–422.

Richter, C., and J. Dainty. 1989. Ion behavior in plant cell walls. I. Characterization of the *Sphagnum russowii* cell wall ion exchanger. *Canadian Journal of Botany 67*:451–459.

Rosen, M. J. 1989. *Surfactants and Interfacial Phenomena*, 2nd ed. Wiley, New York.

Salisbury, F. B., and C. W. Ross. 1991. *Plant Physiology*, 4th ed. Wadsworth, Belmont, California. In press.

Scholander, P. F., H. T. Hammel, E. A. Hemmingsen, and E. D. Bradstreet. 1964. Hydrostatic pressure and osmotic potential in leaves of mangroves and some other plants. *Proceedings of the National Academy of Sciences of the United States of America 52*:119–125.

Schuster, P., G. Zundel, and C. Sandorfy, eds. 1976. *The Hydrogen Bond*, 3 vols. North-Holland, Amsterdam.

Smith, J. A. C., and P. S. Nobel. 1986. Water movement and storage in a desert succulent: Anatomy and rehydration kinetics for leaves of *Agave deserti. Journal of Experimental Botany 37*:1044–1053.

Stadelmann, E. J. 1966. Evaluation of turgidity, plasmolysis, and deplasmolysis of plant cells. In *Methods in Cell Physiology*, Vol. II, D. M. Prescott, ed. Academic Press, New York. Pp. 143–216.

Tomos, A. D. 1988. Cellular water relations of plants. In *Water Science Reviews 3*, F. Franks, ed. Cambridge University Press, New York. Pp. 186–277.

Turner, N. C. 1981. Techniques and experimental approaches for the measurement of plant water status. *Plant and Soil 58*:339–366.

Turner, N. C. 1988. Measurement of plant water status by the pressure chamber techniques. *Irrigation Science 9*:289–308.

Tyree, M. T., and J. S. Sperry. 1989. Vulnerability of xylem to cavitation and embolism. *Annual Review of Plant Physiology and Plant Molecular Biology 40*:19–38.

Van Holde, K. E. 1985. *Physical Biochemistry*, 2nd ed. Prentice-Hall, Englewood Cliffs, New Jersey.

Weast, R. C., and D. R. Lide, eds. 1989. *CRC Handbook of Chemistry and Physics*, 70th ed. CRC Press, Boca Raton, Florida.

Wolfe, J., and P. L. Steponkus. 1981. The stress–strain relation of the plasma membrane of isolated plant protoplasts. *Biochimica et Biophysica Acta 643*:663–668.

Zimmermann, U., and E. Steudle. 1978. Physical aspects of water relations of plant cells. *Advances in Botanical Research 6*:45–117.

CHAPTER 3

Solutes

In this chapter we turn our attention to the properties of solutes. We will compare chemical potentials in the aqueous phases on the two sides of a membrane or across some other region to predict the direction of passive solute fluxes as well as the driving forces leading to such motion. We will also show how the fluxes of charged species can account for the electrical potential differences observed across biological membranes.

Many solute properties are intertwined with those of the ubiquitous solvent, water. For example, the osmotic pressure term in the chemical potential of water is mainly due to the decrease of the water activity caused by the solutes ($RT \ln a_w = -\bar{V}_w \Pi$, Eq. 2.7). The movement of water through the soil to the root and then to the xylem can influence the entry of dissolved nutrients, and the subsequent distribution of these nutrients throughout the plant depends on water movement in the xylem (and the phloem in some cases). In contrast with water, however, solute molecules can carry a net positive or negative electrical charge. For such charged particles, the electrical term must be included in their chemical potential, which leads to a consideration of electrical phenomena in general and an interpretation of the electrical potential differences across membranes in particular. Whether an observed ionic flux of some species into or out of a cell can be accounted for by the passive process of diffusion depends on the differences in both the concentration of that species and the electrical potential between the inside and the outside. Ions can also be actively transported across membranes, in which case metabolic energy is involved in the moving of solutes.

When both solutes and water traverse the same barrier, we should replace the classical thermodynamic approach with one based on irreversible thermodynamics. The various forces and fluxes are then viewed as interacting with each other, which means that the movement of water across a membrane influences the movement of solutes, and vice versa. Using this more general approach, we will show that the osmotic pressure difference *effective* in causing a volume flux across a membrane permeable to both water and solutes is less than the *actual* osmotic pressure difference across that membrane.

CHEMICAL POTENTIAL OF IONS

We introduced the concept of chemical potential in Chapter 2. Because a substance spontaneously tends to move to regions where its chemical potential is lower, this quantity is useful for analyzing passive movements of solutes. Using a linear combination of the various contributors, we expressed the chemical potential of any species j by Equation 2.4 as follows: $\mu_j = \mu_j^* + RT \ln a_j + \overline{V}_j P + z_j FE + m_j gh$. Because water is uncharged ($z_w = 0$), the electrical term does not enter into its chemical potential. For ions, however, $z_j FE$ becomes important. In fact, for charged solutes under most conditions of biological interest, differences in the electrical term are usually far larger than are changes in the terms for hydrostatic pressure or gravity. In particular, for movements across membranes, Δh is zero, so we will omit the gravitational term in this chapter. Changes in the $\overline{V}_j P$ term can also usually be ignored because they are relatively small, as we will indicate below. Therefore, the chemical potential generally used when dealing with ions is $\mu_j^* + RT \ln a_j + z_j FE$, commonly referred to as the *electrochemical* potential. This emphasizes the role played by electrical potentials, but the expression "chemical potential" represents the sum of *all* the different contributors affecting a particular species, so the special term "electrochemical" potential is not really necessary when dealing with ions.

Electrical Potential

The difference in electrical potential E between two locations is a measure of the amount of electrical work involved in moving a charge from one location to the other. Specifically, the work in joules (J) equals the charge in coulombs (C) times the potential difference in volts (V). The zero level for electrical potential is arbitrary. Because both the initial electrical potential and the final electrical potential must be expressed relative to the same zero level, the arbitrariness is of no consequence when the electrical potential difference (final potential minus initial potential) is determined.

The charge carried by an ion of species j is a positive or negative integer z_j (the *charge number*) times the charge of a proton. For instance, z_j is $+1$ for K^+ and -2 for sulfate (SO_4^{2-}). The electrical charge carried by a proton is commonly called the *electronic charge*, because it is the same in magnitude as the charge on an electron, although opposite in sign. A single proton has a single charge of 1.602×10^{-19} coulomb, abbreviated C; thus, a mole (Avogadro's number) of protons has a charge equal to (6.022×10^{23} protons mol^{-1}) (1.602×10^{-19} C/proton), or 9.65×10^4 C mol^{-1}. Such a unit, consisting of Avogadro's number of electronic charges (i.e., one mole of single, positive charges), is called Faraday's constant, F. This quantity, which appears in the electrical term of the chemical potential (Eq. 2.4), equals 9.65×10^4 C mol^{-1} or 9.65×10^4 J mol^{-1} V^{-1}.

To illustrate the rather small contribution of the pressure term $\bar{V}_j P$ to differences in the chemical potential of a charged substance, we will compare it with the contribution of the electrical term $z_j FE$. Using a fairly typical electrical potential difference across a biological membrane of 100 mV, we ask what hydrostatic pressure difference (ΔP) for a small monovalent cation [$z_j = +1$, $\bar{V}_j = 3.0 \times 10^{-5}$ m^3 mol^{-1} (30 cm^3 mol^{-1})] makes the same contribution to the chemical potential. Equating the pressure term to the electrical term, ΔP is

$$\Delta P = \frac{z_j F\ \Delta E}{\bar{V}_j} = \frac{(1)(9.65 \times 10^4 \text{ J mol}^{-1} \text{ V}^{-1})(0.100 \text{ V})}{(3.0 \times 10^{-5} \text{ m}^3 \text{ mol}^{-1})}$$

$$= 3.2 \times 10^8 \text{ J m}^{-3} = 320 \text{ MPa}$$

Because most hydrostatic pressures encountered in cells are less than 1 MPa, contributions of the pressure term to the chemical potential differences of ions across biological membranes are usually negligible compared with electrical contributions and hence can generally be ignored.

Electroneutrality and Membrane Capacitance

Another consequence of the relatively large magnitude of electrical effects is the general occurrence of *electroneutrality*. Thus, in most aqueous regions that are large compared with atomic dimensions, the total electrical charge carried by the cations is essentially equal in magnitude to that carried by the anions. What is the situation for plant cells? Do we ever have an excess of negative or positive

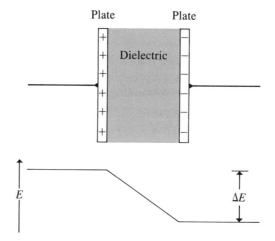

Figure 3.1

A parallel plate capacitor. Each plate represents a conductor in which charges can freely move. Thus, each plate has a particular electrical potential, reflecting its uncompensated positive or negative charges. Between the plates is a region (often called the dielectric) that charges cannot cross; as a result, an electrical potential difference ΔE occurs across this region. The higher the capacitance, the greater is the charge that can be on the plates for a given ΔE across the dielectric. Capacitance is proportional to the dielectric constant D (introduced in Ch. 2, see Eq. 2.3 and the following discussion). Membranes act as dielectrics (D near 3 for the lipid phase) separating the aqueous conducting phases on either side.

charges in cells or organelles? If there were a net charge in some region, such as near a membrane, an electrical potential difference would exist from one part of the region to another. Can we relate the size of the electrical potential difference to the net charge?

To relate charges and ΔE's, we need to introduce a new term, *capacitance*. Electrical capacitance is the coefficient of proportionality between a net charge and the resulting electrical potential difference. A high capacitance means that the region has the *capacity* to have many uncompensated charges separated across it without at the same time developing a large electrical potential difference across that region (see Fig. 3.1). The magnitude of such an electrical potential difference, ΔE, is related to the capacitance of the region (C) as follows[1]:

$$Q = C \, \Delta E \qquad (3.1)$$

---

1. In keeping with most biological literature, E in electrical equations throughout this book refers to electrical potentials, not electric field intensities, as it does in many physics texts.

where Q is the net charge. The unit for capacitance is the *farad* (F), which equals 1 C V^{-1}. The capacitances of most biological membranes are approximately the same per unit area—about 10 mF m^{-2} (1 μF cm^{-2}; Hope and Walker, 1975; Lüttge and Pitman, 1976).

For convenience in estimating the net charge inside a cell that would lead to a typical membrane potential, we will consider a spherical cell of radius r. Suppose that the uncanceled or net charges have a concentration c. For a conductor—a body in which electrical charges can freely move, such as an aqueous solution—the uncanceled or net charges do not remain uniformly distributed. Rather, they repel each other and hence collect at the inner surface of the sphere. The quantity c is therefore the hypothetical concentration of the net charge if it were uniformly distributed through the interior of the sphere. The charge Q within the sphere of radius r is then $(4/3)\pi r^3 cF$. (The concentration c can be expressed in mol m^{-3} and here refers to the average net concentration of uncompensated singly charged particles; F, which is expressed in C mol^{-1}, converts amount to electrical charge.) The capacitance C of the sphere is $4\pi r^2 C'$, where $4\pi r^2$ is the surface area of a sphere and C' is the capacitance per unit area. After substituting these values of Q and C into Equation 3.1, we obtain $(4/3)\pi r^3 cF = 4\pi r^2 C'\ \Delta E$, which leads to the following expression for the electrical potential difference in the case of a spherical capacitor:

$$\Delta E = \frac{rcF}{3C'} \tag{3.2}$$

Equation 3.2 gives the electrical potential difference from the center of the sphere to just outside its surface. For a conductor, the internal uncompensated charges are found near the surface. In that case, ΔE actually occurs close to the bounding surface (such as a membrane) surrounding the spherical body under consideration. Equation 3.2 indicates that the electrical potential difference developed is directly proportional to the average concentration of net charge enclosed and inversely proportional to the capacitance per unit area of the sphere.

To apply Equation 3.2 to a specific situation, let us consider a spherical cell with a radius of 30 μm and an electrical potential difference across the membrane (inside negative) of -100 mV, a value close to that occurring for many cells. If the membrane capacitance per unit area has a typical value of 10 mF m^{-2} (10^{-2} C V^{-1} m^{-2}), to what net charge concentration in the cell does this potential drop correspond? Using Equation 3.2, we obtain

$$c = \frac{3C'\ \Delta E}{rF} = \frac{(3)(10^{-2}\ \text{C V}^{-1}\ \text{m}^2)(-0.100\ \text{V})}{(30 \times 10^{-6}\ \text{m})(9.65 \times 10^4\ \text{C mol}^{-1})}$$

$$= -1.0 \times 10^{-3}\ \text{mol m}^{-3}$$

The sign of the charge is negative, indicating more internal anions than cations. We see that the average concentration of the net (uncanceled) charge leading to a considerable electrical potential difference is rather small, an important point that we will consider next.

It is instructive to compare the net charge concentration averaged over the volume of the cell (c in Eq. 3.2) with the total concentration of anions and cations in a plant cell. Specifically, because the positive and the negative ions in plant cells can each have a total concentration near 100 mol m^{-3} (0.1 M), the above-calculated excess of 10^{-3} mol m^{-3} (1 μM) is only about one extra negative charge per 10^5 anions. Expressed another way, the total charge of the cations inside such a cell equals or compensates that of the anions to within about one part in 100 000. When cations are taken up by a cell to any appreciable extent, either anions must accompany them or cations must be released from inside the cell, or both. Otherwise, marked departures from electrical neutrality would occur in some region, and sizable electrical potential differences would build up.

Activity Coefficients of Ions

We will now turn our attention from the electrical term in the chemical potential to the activity term—specifically, to the chemical activity itself. As indicated in Chapter 2, the activity of species j, a_j, is its thermodynamically effective concentration. For charged particles in an aqueous solution, this activity can differ appreciably from the actual concentration c_j—a fact that has not always been adequately recognized in dealing with ions. By Equation 2.5, a_j equals $\gamma_j c_j$, where γ_j is the activity coefficient of species j.

A quantitative description of the dependence of the activity coefficients of ions on the concentration of the various species in a solution was developed by Debye and Hückel in the 1920s (for further details on the Debye–Hückel theory, see Bull, 1964; Castellan, 1983; Edsall and Wyman, 1958; Lauffer, 1989; Noggle, 1985). In a local region around a particular ion, the electrostatic forces describable by relations such as Equation 2.3 constrain the movement of other ions. As the concentration increases, the average distance between the ions decreases, thereby facilitating ion-ion interactions. Equation 2.3 [electrical force $= Q_1 Q_2/(4\pi\varepsilon_0 D r^2)$] indicates, for example, that the electrostatic interaction between two charged particles varies inversely as the square of the distance between them, so electrical forces greatly increase as the ions get closer together. When ions of opposite sign attract each other, the various other interactions of both ions are restricted, thus lowering their thermodynamically effective concentration or activity.

A simple and approximate form of the Debye–Hückel equation appropriate for estimating the values of activity coefficients of ions in relatively dilute aqueous solutions at 25°C is

$$\ln \gamma_{\pm} \cong \frac{1.17 \, z_{+} z_{-} \sqrt{\frac{1}{2} \sum_{j} c_j z_j^2}}{32 + \sqrt{\frac{1}{2} \sum_{j} c_j z_j^2}} \tag{3.3}$$

where z_{+} is the charge number of the cation, z_{-} that of the anion, concentrations are expressed in mol m^{-3} (numerically equal to mM), and the summations are over all charged species.[2] Because we cannot have a solution of one type of ion by itself in which to measure or to calculate γ_{+} or γ_{-}, activity coefficients of ions occur as the products of those of cations and anions. Hence $\gamma_{\pm}$ in Equation 3.3 represents the mean activity coefficient of some cation-anion pair with charge numbers z_{+} and z_{-}. Because z_{-} is negative, Equation 3.3 indicates that $\ln \gamma_{\pm}$ is also negative, and $\gamma_{\pm}$ is less than one. The activities of ions in aqueous solutions are thus less than their concentrations, as we would indeed expect.

To estimate $\gamma_{\pm}$ under conditions approximating those that might occur in a plant or animal cell, let us consider an aqueous solution containing 100 mol m^{-3} (100 mM) of both monovalent cations and anions and 25 mol m^{-3} of both divalent cations and anions. For this solution, $\frac{1}{2} \sum_{j} c_j z_j^2$, known as the *ionic strength*, is as follows:

$$\frac{1}{2} \sum_{j} c_j z_j^2 = \frac{1}{2}[(100 \text{ mol m}^{-3})(1)^2 + (100 \text{ mol m}^{-3})(-1)^2$$

$$+ (25 \text{ mol m}^{-3})(2)^2 + (25 \text{ mol m}^{-3})(-2)^2]$$

$$= 200 \text{ mol m}^{-3}$$

From Equation 3.3, we can calculate $\ln \gamma_{\pm}$ for the monovalent ions as

$$\ln \gamma_{\pm} \cong \frac{(1.17)(1)(-1)\sqrt{200}}{32 + \sqrt{200}} = -0.156$$

---

2. For very dilute aqueous solutions (much more dilute than typically occur in plant or animal cells), $\sqrt{\frac{1}{2}\sum_{j} c_j z_j^2}$ is neglected compared with 32 in the denominator of Equation 3.3, leading to $\ln \gamma_{\pm} \cong 0.037 z_{+} z_{-} \sqrt{\frac{1}{2}\sum_{j} c_j z_j^2}$, which is known as the Debye–Hückel limiting law (Castellan, 1983; Noggle, 1985).

This corresponds to a mean activity coefficient of only 0.70, a value considerably less than one.

The activity coefficient of a particular ionic species depends on all the ions in the solution, as indicated by the ionic strength terms in Equation 3.3. Therefore, even when some particular ionic species is itself dilute, its activity coefficient can nevertheless be appreciably less than one because of the many electrostatic interactions with other ions. The departure from unity for activity coefficients is even greater for divalent and trivalent ions than for monovalent ions, as the $z_+ z_-$ factor in Equation 3.3 indicates. Thus, although activity coefficients of ions are often set equal to unity for convenience, this is not always justified. A practical difficulty arising under most experimental situations is that c_j is much easier to determine than is a_j, especially for compartments like the cytosol or the interior of a chloroplast. In those circumstances where a_j has been replaced by c_j, caution must be exercised in the interpretations or conclusions. The activity coefficients for nonelectrolytes and water are generally much closer to unity than are those for ions; hence the assumption involved in replacing a_j by c_j for such neutral species is not as severe as it is for the charged substances. For instance, $\gamma_{sucrose}$ is about 0.96 for 300 mol m^{-3} (0.3 M) sucrose. An all-inclusive theory for activity coefficients is quite complicated and beyond the scope of this text. Equation 3.3 is presented to illustrate quantitatively that activity coefficients of ions can be appreciably less than 1.00 under conditions that can occur in a plant cell (see Pytkowicz, 1979).

Nernst Potential

Having considered the electrical and the activity terms in some detail, we now turn to the role of these quantities in the chemical potential of ions. Specifically, we will consider the deceptively simple yet extremely important relationship between the electrical potential difference across a membrane and the accompanying distribution of ions across it at equilibrium.

When ions of some species j are in equilibrium across a membrane, its chemical potential outside (o) is the same as inside (i); i.e., μ_j^o equals μ_j^i. A difference in the hydrostatic pressure term generally makes a negligible contribution to the chemical potential differences of ions across membranes, so $\bar{V}_j P$ can be omitted from μ_j in the present case. With this approximation and the definition of chemical potential (Eq. 2.4 without the pressure and the gravitational terms), the condition for equilibrium of ionic species j across the membrane is

$$\mu_j^* + RT \ln a_j^o + z_j FE^o = \mu_j^* + RT \ln a_j^i + z_j FE^i \qquad (3.4)$$

The term μ_j^* in Equation 3.4 is a constant referring to the same standard state of species j in the aqueous solutions on both sides of the membrane, so it can be canceled from the equation.

Upon solving Equation 3.4 for the electrical potential difference $E^i - E^o$ across the membrane *at equilibrium,* we obtain the following important relationship:

$$E_{N_j} = E^i - E^o = \frac{RT}{z_j F} \ln \frac{a_j^o}{a_j^i} = 2.303 \frac{RT}{z_j F} \log \frac{a_j^o}{a_j^i} \qquad (3.5a)$$

which at 25°C becomes

$$E_{N_j} = \frac{25.7}{z_j} \ln \frac{a_j^o}{a_j^i} \text{ mV} = \frac{59.2}{z_j} \log \frac{a_j^o}{a_j^i} \text{ mV} \qquad (3.5b)$$

The electrical potential difference E_{N_j} in Equation 3.5 is called the *Nernst potential* of species j; i.e., $E^i - E^o = \Delta E = E_{N_j}$ in this case, where the subscript N stands for Nernst, who first derived this relation about 1900.[3] We derived it by assuming equality of the chemical potentials of some charged species on two sides of a membrane. The natural logarithm (ln) is often replaced by 2.303 log, where log is the common logarithm to the base 10; and 2.303 RT/F is replaced by 58.2 mV at 20°C, 59.2 mV at 25°C (as above), and 60.2 mV at 30°C (Apps. I, II, and III).

Equation 3.5, the *Nernst equation,* is an equilibrium statement showing how the internal and the external activities of ionic species j are related to the electrical potential difference across a membrane (see Fig. 3.2). At equilibrium, a tenfold difference in activity of a monovalent ion across some membrane is energetically equivalent to a 59 mV difference in electrical potential (at 25°C). Hence, a relatively small electrical potential drop can balance a large difference in activity across a membrane. For some calculations, γ_j^o/γ_j^i is set equal to one (a less stringent assumption than setting both γ_j^o and γ_j^i equal to one). Under this condition, a_j^o/a_j^i in Equation 3.5 becomes the ratio of the concentrations, c_j^o/c_j^i ($a_j = \gamma_j c_j$, Eq. 2.5). Such a substitution may be justified when the ionic strengths on the two sides of a membrane are approximately the same, but it can lead to errors when the outside solution is much more dilute than the internal one, as would occur for *Chara* or *Nitella* in pond water.

---

3. Nernst made many important contributions to the understanding of the physical chemistry of solutions; he was awarded the Nobel Prize in chemistry in 1920.

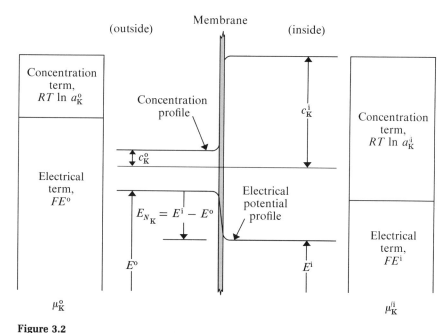

Figure 3.2

Equilibrium of K^+ across a membrane. When K^+ is in equilibrium across some membrane, the side with the higher concentration must be at a lower electrical potential for its chemical potential to be unchanged ($\mu_K^o = \mu_K^i$) in crossing the membrane (μ_K^* is the same on the two sides of the membrane). The electrical potential difference across the membrane is then the Nernst potential for K^+, E_{N_K}. Note that E^o and E^i must be expressed relative to some arbitrary baseline for electrical potentials.

Throughout the rest of this book we will represent the *actual* electrical potential drop existing across a membrane, $E^i - E^o$, by E_M, where the subscript M refers to membrane. Hence, for both equilibrium and nonequilibrium situations, we have $E^i - E^o = \Delta E = E_M$. When a particular ionic species j is in equilibrium across some membrane, E_M equals E_{N_j}, the Nernst potential for that species. However, regardless of the actual electrical potential difference existing across a membrane (E_M), a Nernst potential for each individual ionic species j can always be *calculated* from Equation 3.5 by using the ratio of the outside to the inside activity (a_j/a_j) or concentration (c_j/c_j), e.g., $E_{N_j} = (RT/z_jF) \ln (c_j/c_j)$.

If some ionic species cannot penetrate a membrane or is actively transported across it, E_{N_j} can differ markedly from E_M. In fact, the minimum amount of energy needed to transport ions across a membrane is proportional to the difference between E_{N_j} and E_M. The aqueous phases designated as inside and outside can have more than one membrane intervening between them. For instance, the vacuolar sap and an external solution—the regions often considered experimentally—have two membranes separating them, the plasmalemma and the tonoplast. The thermodynamic arguments remain the same in the case of multiple membranes, with E_M and E_{N_j} referring to the electrical potential differences between the two regions under consideration, regardless of how many membranes occur between them.

Example of E_{N_K}

Data obtained on K^+ for the large internodal cells of the green freshwater alga *Chara corallina* are convenient for illustrating the use of the Nernst equation (see Vorobiev, 1967). The K^+ activity in the medium bathing the cells, a_K^o (where the element symbol K as a subscript is the conventional notation for K^+), is 0.096 mol m^{-3} (0.096 mM), and a_K^i in the vacuole is 48 mol m^{-3}. By Equation 3.5 (with a factor of 58.2, because the measurement was made at 20°C), the Nernst potential for K^+ is

$$E_{N_K} = \frac{(58.2 \text{ mV})}{(1)} \log \left(\frac{0.096 \text{ mol m}^{-3}}{48 \text{ mol m}^{-3}} \right) = -157 \text{ mV}$$

The measured electrical potential of the vacuole relative to the external solution, E_M, is -155 mV—very close to the calculated Nernst potential for K^+. Thus, K^+ in this case may be in equilibrium between the external solution and the vacuole. The K^+ concentration in the vacuole of these *Chara* cells is 60 mol m^{-3} (60 mM). The activity coefficient for K^+ in the vacuole (γ_K^i) equals a_K^i/c_K^i ($a_j = \gamma_j c_j$, Eq. 2.5), so γ_K^i is (48 mol m^{-3})/(60 mol m^{-3}), or 0.80, and the K^+ activity coefficient in the bathing solution (γ_K^o) is about 0.96. For this example, where there are large differences in the internal and the external concentrations, the ratio γ_K^o/γ_K^i is (0.96)/(0.80), or 1.20, which differs appreciably from 1.00. If concentrations instead of activities had been used in Equation 3.5, E_{N_K} would be -162 mV, which is somewhat lower than the measured potential of -155 mV. Calculating from the concentration ratio, the suggestion that K^+ is in equilibrium from the bathing solution to the vacuole could not be made with much confidence, if at all.

Equilibrium does not require that the various forces acting on a substance are zero—rather, it requires that they cancel each other. In the above example for *Chara,* the factors that tend to cause K^+ to move are the differences both in its activity (or concentration) and the electrical potential across the membranes. The activity of K^+ is much higher in the vacuole than in the external solution (see

Fig. 3.2). The activity term in the chemical potential therefore represents a driving force on K^+ directed from the inside of the cell to the bathing solution. The electrical potential is lower inside the cell, as is indicated in Figure 3.2. Hence, the electrical driving force on the positively charged K^+ tends to cause its entry into the cell. At equilibrium these two tendencies for movement in opposite directions are balanced, and no net K^+ flux occurs. As indicated above, the electrical potential difference existing across a membrane when K^+ is in equilibrium is the Nernst potential for K^+, E_{N_K}. Generally, c_K^i and a_K^i for both plant and animal cells are much higher than c_K^o and a_K^o, and K^+ is often close to equilibrium across the cellular membranes. From these observations, we can expect that the interiors of cells are usually at negative electrical potentials compared with the outside solutions, as is indeed the case. The chemical potentials of ions are generally not equal in all regions of interest, so passive movements toward lower μ_j's occur, the topic we turn to next.

FLUXES AND DIFFUSION POTENTIALS

Fluxes of many different solutes are continually occurring across biological membranes. Inward fluxes move mineral nutrients into cells, and certain products of metabolism flow out of cells. The primary concern of this section is the passive fluxes of ions toward lower chemical potentials. First, we will indicate that the passive flux density of some species is directly proportional to the driving force causing the movement. Next, the driving force is expressed in terms of the relevant components of the chemical potential. We then examine the consequences of electroneutrality when there are simultaneous passive fluxes of more than one type of ion. This leads us to an expression describing the electrical potential difference across a membrane in terms of the properties of the ions penetrating it.

   Before discussing the relation between fluxes and chemical potentials, we will briefly consider fluxes already mentioned or which may be familiar from other contexts. In Chapter 1 we discussed Fick's first law of diffusion (Eq. 1.1), which says that the flux density of (neutral) species j equals $-D_j \, \partial c_j / \partial x$, where we can consider that the force is the negative gradient of the concentration. In Chapter 9 we will use Darcy's law (Eq. 9.7) and Poiseuille's law (Eq. 9.11), both of which indicate that the volume flux density of a solution is proportional to $-\partial P / \partial x$. Ohm's law, which describes electrical effects, can be written $\Delta E = IR$, where I is the current (charge moving per unit time) and R is the resistance across which the electrical potential difference is ΔE. The current per unit area A is the charge flux density, J_c, which equals $-(1/\rho) \, \partial E / \partial x$, where ρ is the electrical resistivity and the negative gradient of the electrical potential represents the driving force. We usually replace $-\partial E/\partial x$ by $\Delta E/\Delta x$, which leads to $I/A = (1/\rho)\,\Delta E/\Delta x$. For this to conform with the usual expression of Ohm's law ($\Delta E = IR$), $\rho\,\Delta x/A$ must equal R. The gravitational force, $-m_j g$, is the negative gradient of the potential energy in a gravitational field; i.e., $-\partial m_j gh/\partial h = -m_j g$, where the minus sign indicates that the force is directed toward decreasing altitudes, namely, toward the center of the earth. The gravitational force leads to the various forms of precipitation as well as to the percolation of water down through the soil.

We have just considered examples of fluxes depending on each of the four variable terms in the chemical potential ($\mu_j = \mu_j^* + RT \ln a_j + \overline{V}_j P + z_j FE + m_j gh$, Eq. 2.4). Specifically, the activity term ($RT \ln a_j$) leads to Fick's first law, the pressure term ($\overline{V}_j P$) accounts for Darcy's law and Poiseuille's law, the electrical term ($z_j FE$) yields Ohm's law, and the gravitation term ($m_j gh$) is responsible for fluxes caused by gravity. In each case, flux density is found by experiment to be directly proportional to the appropriate driving force. We can generalize such relationships because nearly all transport phenomena can be represented by the following statement: flux density is proportional to the appropriate force. Force is represented by the negative gradient of a suitable potential, which for convenience we often take as the change in potential over some finite distance. But we have already shown that the chemical potential is an elegant way of summarizing all the factors that can contribute to the motion of a substance. It should not be surprising, therefore, that *in general* the flux density of species j is proportional to the negative gradient of its chemical potential, $-\partial \mu_j/\partial x$.

Flux and Mobility

Let us now consider a charged substance capable of crossing a particular membrane. When its μ_j depends on position, a net passive movement or flux of that species tends to occur toward the region where its chemical potential is lower. The negative gradient of the chemical potential of species j, $-\partial \mu_j/\partial x$, acts as the driving force for this flux. (As in Ch. 1, our discussion will apply to the one-dimensional case, so we will let $\partial \mu_j/\partial y = \partial \mu_j/\partial z = 0$.) The greater is $-\partial \mu_j/\partial x$, the larger will be the flux of species j in the x-direction. As a useful approximation, we will assume that the flux density is proportional to $-\partial \mu_j/\partial x$, where the minus sign means that a net positive flux density occurs in the direction of decreasing chemical potential. The magnitude of a flux density across some plane is also proportional to the local concentration of that species, c_j. That is, for a given driving force on species j, the amount of the substance that moves is proportional to how much of it is actually present—the greater the concentration, the greater is the flux density. Thus, for the one-dimensional case of crossing a plane perpendicular to the x-axis, the flux density J_j can be expressed as

$$J_j = u_j c_j \left(-\frac{\partial \mu_j}{\partial x} \right)$$

$$= \bar{v}_j c_j \tag{3.6}$$

where u_j is a coefficient called the *mobility* of species j. A mobility is generally the ratio of some velocity to the force causing the motion. We will return to a consideration of u_j shortly.

As we have already indicated, the top line of Equation 3.6 is a representative example from the large class of expressions relating various flows to their causative forces. In this particular case, J_j is the rate of flow of moles of species j across unit area of a plane and can be expressed in mol m^{-2} s^{-1}. Such a molar flux density of a species j divided by its local concentration c_j gives the mean velocity $\bar{v}_j$ with which this species moves across the plane; when c_j is in mol m^{-3}, J_j/c_j can have units of (mol m^{-2} s^{-1})/(mol m^{-3}), or m s^{-1}. Perhaps this important point can be better appreciated by considering it in the following way. Suppose that the average velocity of species j moving perpendicularly toward area A of the plane of interest is $\bar{v}_j$, and consider a volume element of cross-sectional area A extending back from the plane for a distance $\bar{v}_j \, dt$ (see Fig. 3.3). In time dt, all molecules in the volume element $(\bar{v}_j \, dt) \times (A)$ will cross area A, so the number of moles of species j crossing the plane in this interval is $[(\bar{v}_j \, dt)A] \times (c_j)$, where c_j is the number of moles per unit volume. Hence, the molar flux density J_j (which is the number of moles crossing unit area per unit time) is $(\bar{v}_j \, dt \, A \, c_j)/(A \, dt)$, or $\bar{v}_j c_j$ (Fig. 3.3). In other words, the mean velocity of species j moving across the plane, $\bar{v}_j$, times the number of those molecules per unit volume that can move, c_j, equals the flux density of that species, J_j, as the bottom line of Equation 3.6 indicates. The upper line of Equation 3.6 states that this average velocity, J_j/c_j, equals the mobility of species j, u_j, times $-\partial \mu_j/\partial x$, the latter being the force on j that causes it to move. Thus mobility is the proportionality factor between the mean velocity of motion ($\bar{v}_j = J_j/c_j$) and the causative force ($-\partial \mu_j/\partial x$). The greater is the mobility of some species, the larger is its velocity in response to a given force.

The particular form of the chemical potential to be substituted into Equation 3.6 depends on the application. For charged particles moving across biological membranes, the appropriate μ_j is $\mu_j^* + RT \ln a_j + z_j FE$. (As we mentioned

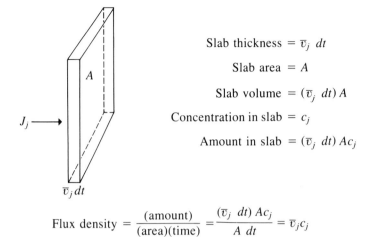

$$\text{Slab thickness} = \bar{v}_j \; dt$$

$$\text{Slab area} = A$$

$$\text{Slab volume} = (\bar{v}_j \; dt) \, A$$

$$\text{Concentration in slab} = c_j$$

$$\text{Amount in slab} = (\bar{v}_j \; dt) \, A c_j$$

$$\text{Flux density} = \frac{(\text{amount})}{(\text{area})(\text{time})} = \frac{(\bar{v}_j \; dt) \, A c_j}{A \; dt} = \bar{v}_j c_j$$

Figure 3.3

Geometrical argument showing that the flux density J_j across surface area A equals $\bar{v}_j c_j$, where $\bar{v}_j$ is the mean velocity of species j. Note that $\bar{v}_j \; dt$ is an infinitesimal distance representing the slab thickness. By the definition of $\bar{v}_j$, all the molecules of species j in the slab cross A in time dt.

before, the $\overline{V}_j P$ term makes only a relatively small contribution to the $\Delta\mu_j$ of an ion, so it is not included, and the $m_j g h$ term in Eq. 2.4 can also be ignored here.) For treating the one-dimensional case described by Equation 3.6, μ_j must be differentiated with respect to x, $\partial\mu_j/\partial x$, which leads to $RT \, \partial \ln a_j/\partial x +$ $z_j F \, \partial E/\partial x$ when T is constant.[4] The quantity $\partial \ln a_j/\partial x$ equals $(1/a_j) \, \partial a_j/\partial x$, which is $(1/\gamma_j c_j) \, \partial\gamma_j c_j/\partial x$ $(a_j = \gamma_j c_j$, Eq. 2.5). Using the above form of μ_j appropriate for charged solutes and this expansion of $\partial \ln a_j/\partial x$, the net flux density of species j in Equation 3.6 can be written

$$J_j = -\frac{u_j R T}{\gamma_j} \frac{\partial \gamma_j c_j}{\partial x} - u_j c_j z_j F \frac{\partial E}{\partial x} \qquad (3.7)$$

Equation 3.7, which is often called the Nernst–Planck equation, is a general expression for the one-dimensional flux density of species j either across a membrane or in a solution in terms of two components of the driving force: the gradients in activity and in electrical potential (see Starzak, 1984; Stein, 1986).

Before proceeding, let us examine the first term on the right-hand side of Equation 3.7. When γ_j varies across a membrane, $\partial\gamma_j/\partial x$ can be considered to represent a driving force on species j. In keeping with common practice, we

---

4. Actually, we must assume isothermal conditions for $-\partial\mu_j/\partial x$ to represent the force precisely; i.e., T is a constant here, and hence $\partial(RT \ln a_j)/\partial x = RT \, \partial \ln a_j/\partial x$.

will ignore this possible force. That is, we will assume that γ_j can be treated as a constant—in any case $\partial\gamma_j/\partial x$ would be extremely difficult to measure. For constant γ_j, the first term on the right-hand side of Equation 3.7 becomes $-u_jRT(\partial c_j/\partial x)$; i.e., it is proportional to the concentration gradient. In the absence of electrical potential gradients ($\partial E/\partial x = 0$), or for neutral solutes ($z_j = 0$), Equation 3.7 indicates that J_j equals $-u_jRT(\partial c_j/\partial x)$. But this is just the type of flux described by Fick's first law ($J_j = -D_j\partial c_j/\partial x$, Eq. 1.1), with u_jRT taking the place of the diffusion coefficient, D_j. In other words, we can replace u_jRT in Equation 3.7 by D_j.

Because D_j equals u_jRT, we can appreciate that diffusion coefficients depend on temperature. Moreover, u_j is generally inversely proportional to the viscosity, which decreases as T increases. Thus, the dependence of D_j on T can be pronounced. For instance, for gases in air, D_j depends approximately on $T^{1.8}$ (see Fuller et al., 1966). Consequently, the temperature should be specified when the value of a diffusion coefficient is given. In the absence of electrical effects and for constant γ_j, the net flux of species j given by Equation 3.7 equals $-D_j\,\partial c_j/\partial x$ when u_jRT is replaced by D_j, as we have just indicated. Thus, Fick's first law ($J_j = -D_j\,\partial c_j/\partial x$, Eq. 1.1) is a special case of our general flux relation, the Nernst–Planck equation (Eq. 3.7), where we first ignored the pressure and gravitational effects and then omitted the electrical effects. But this eventual reduction of Equation 3.7 to Fick's first law is what we should expect, because the only driving force considered when we presented Fick's first law (Ch. 1) was the concentration gradient. Such agreement between our present thermodynamic approach and the *seemingly* more empirical Fick's first law is quite important. It serves to justify the logarithmic term for activity in the chemical potential ($\mu_j = \mu_j^* + RT \ln a_j + \overline{V}_jP + z_jFE + m_jgh$, Eq. 2.4). In other words, if the activity of species j appeared in Equation 2.4 in a form other than $\ln a_j$, Equation 3.7 would not reduce to Fick's first law under the appropriate conditions. Because Fick's first law has been amply demonstrated experimentally, such a disagreement between theory and practice would necessitate some modification in the expression used to define chemical potential.[5]

---

5. As indicated in Chapter 2, the terms in the chemical potential can be justified or "derived" by different methods. The forms of some terms in μ_j can be readily appreciated because they follow from familiar definitions of work, e.g., the gravitational term and the electrical term. The above comparison with Fick's first law indicates that $RT \ln a_j$ is the appropriate form for the activity term. Another derivation of the $RT \ln a_j$ term will be found in Appendix IV, together with a discussion of the pressure term for both liquids and gases. Some of these derivations incorporate conclusions from empirical observations. Moreover, the fact that the chemical potential can be expressed as a series of terms that can be added together agrees with experiment. Thus, a thermodynamic expression for the chemical potential such as Equation 2.4 can (1) summarize the results of previous observations, (2) stand the test of experiments, and (3) lead to new and useful predictions.

In contrast to the case for a neutral solute, the flux of an ion also depends on an electrical driving force, represented in Equation 3.7 by $-\partial E/\partial x$. A charged solute spontaneously tends to move in an electrical potential gradient, a cation moving in the direction of lower or decreasing electrical potential. Of course, the concentration gradient also affects charged particles. If a certain type of ion were present in some region but absent in an adjacent one, the ions would tend to diffuse into the latter region. As such charged particles diffuse toward regions of lower concentration, an electrical potential is created. This electrical potential difference is referred to as a "diffusion potential," which we will discuss in detail below. The interrelationship between concentration and electrical effects is extremely important in biology. Most membrane potentials can be treated as diffusion potentials resulting from different rates of movement of the various ions across a membrane.

Diffusion Potential in a Solution

We will now use Equation 3.7 to derive the electrical potential difference created by ions diffusing down a concentration gradient in a solution containing one type of cation and its accompanying anion. This case is simpler than the biologically more realistic one that follows and thus may more clearly illustrate the relationship between concentration gradients and the accompanying electrical potential differences. We will assume that the cations and the anions are initially placed at one side of the solution. In time, they will *diffuse* across the solution toward regions of lower concentration. In general, one ionic species will have a higher mobility, u_j, than the other. The more mobile ions will tend to diffuse faster than their oppositely charged partners, resulting in a microscopic charge separation. This slight charge separation sets up an electrical potential gradient leading to a *diffusion potential*. Using certain simplifying assumptions, we will calculate the magnitude of the electrical potential difference so created.

For convenience of analysis, let us consider a solution containing only a monovalent cation $(+)$ and its accompanying monovalent anion $(-)$. We will assume that their activity coefficients are constant. As our previous calculations of electrical effects have indicated, solutions are essentially neutral in regions that are large compared with atomic dimensions. Thus, c_+ equals c_-, so the concentration of either species can be designated c. Furthermore, no charge imbalance will develop in time, which means that the flux density of the cation across some plane in the solution, J_+, equals that of the anion across the same plane, J_-. (A very small charge imbalance does initially develop, which sets up the electrical potential gradient, but this uncompensated flux density is transitory and in any case negligible compared with J_+ or J_-.) Both flux densities, J_+ and J_-, can be expressed by Equation 3.7 and then equated to each other, which gives

$$-u_+RT\frac{\partial c}{\partial x} - u_+cF\frac{\partial E}{\partial x} = -u_-RT\frac{\partial c}{\partial x} + u_-cF\frac{\partial E}{\partial x} \tag{3.8}$$

where the plus sign on the right-hand side occurs because the monovalent anion carries a negative charge ($z_- = -1$). Rearrangement of Equation 3.8 yields the following expression for the electrical potential gradient:

$$\frac{\partial E}{\partial x} = \frac{u_- - u_+}{u_+ + u_-}\frac{RT}{Fc}\frac{\partial c}{\partial x} \tag{3.9}$$

Equation 3.9 indicates that a nonzero $\partial E/\partial x$ occurs when the mobility of the cation differs from that of the anion and a concentration gradient exists. If u_- is greater than u_+, the anions move (diffuse) faster than the cations toward regions of lower concentration. As some individual anion moves ahead of its "partner" cation, an electric field is set up in such a direction as to speed up the cation and slow down the anion until they both move at the same speed, thus preserving electrical neutrality.

To obtain the difference in electrical potential produced by diffusion between planes of differing concentration, we must integrate Equation 3.9. At this point we will restrict our consideration to the *steady state* case in which neither E nor c changes with time.[6] In going along the x-axis from region I to region II, the change in the electrical potential term in Equation 3.9 is $\int_I^{II} (\partial E/\partial x)\, dx$, which becomes $\int_I^{II} dE$ for the steady-state condition[7] and so equals $E^{II} - E^I$. The integral of the concentration term is $\int_I^{II} (1/c)(\partial c/\partial x)\, dx$, which becomes $\int_I^{II} dc/c$, or $\ln(c^{II}/c^I)$. Using these two relations, integration of Equation 3.9 leads to

$$E^{II} - E^I = \frac{u_- - u_+}{u_+ + u_-}\frac{RT}{F}\ln\frac{c^{II}}{c^I} \tag{3.10a}$$

---

6. At equilibrium, μ_j does not change with time or position for communicating regions and no net flux of that solute occurs ($J_j = 0$). In a steady state μ_j changes with position but not with time, a condition often used to approximate problems of biological interest; J_j is then constant. Actually, it is a matter of judgment whether μ_j is constant enough in time to warrant describing the system as being in a steady state. Similarly, constancy of μ_j for appropriate time and distance intervals is necessary before indicating that a system is in equilibrium.

7. A total differential like dE equals $(\partial E/\partial x)\, dx + (\partial E/\partial t)\, dt$, and dc is $(\partial c/\partial x)\, dx + (\partial c/\partial t)dt$. For a steady-state condition, both $\partial E/\partial t$ and $\partial c/\partial t$ are zero. Consequently, dE then equals $(\partial E/\partial x)\, dx$ and dc equals $(\partial c/\partial x)\, dx$.

or, at 25°C,

$$E^{\mathrm{II}} - E^{\mathrm{I}} = 59.2 \frac{u_- - u_+}{u_+ + u_-} \log \frac{c^{\mathrm{II}}}{c^{\mathrm{I}}} \qquad \mathrm{mV} \qquad (3.10b)$$

where ln has been replaced by 2.303 log, and the value 59.2 mV has been substituted for $2.303\,RT/F$ at 25°C (App. I). In the general case, the anions have different mobilities than do the cations. As the ions diffuse to regions of lower chemical potential in the solution, an electrical potential difference—given by Equation 3.10 and called a diffusion potential—is set up across the section where the concentration changes from c^{I} to c^{II}.

An example of a diffusion potential describable by Equation 3.10 occurs at the open end of the special micropipettes used for measuring electrical potential differences across membranes (Fig. 3.4). The fine tip of the micropipette is open and provides an electrically conducting pathway into the cell or tissue. Ions diffusing from this fine tip give rise to a diffusion potential between the interior of the micropipette and the aqueous compartment into which the tip is inserted. To estimate the magnitude of this potential for KCl as the electrolyte, we will assume that there is 3 000 mol m^{-3} (3 M) KCl in the micropipette (Fig. 3.4) and 100 mol m^{-3} in the cell. The chloride mobility u_{Cl} is about 1.04 times u_{K} (Briggs et al., 1961), so the diffusion potential calculated from Equation 3.10b at 25°C is

$$E^{\mathrm{cell}} - E^{\mathrm{micropipette}} = (59\ \mathrm{mV}) \left(\frac{1.04\,u_{\mathrm{K}} - u_{\mathrm{K}}}{1.04\,u_{\mathrm{K}} + u_{\mathrm{K}}} \right) \log \left(\frac{100\ \mathrm{mol\ m}^{-3}}{3\ 000\ \mathrm{mol\ m}^{-3}} \right)$$

$$= -2\ \mathrm{mV}$$

For NaCl as the electrolyte, u_{Cl} is $1.52\,u_{\mathrm{Na}}$, and the diffusion potential for the same concentration ratio is -18 mV. Thus, from the point of view of minimizing the diffusion potential across the fine tip, KCl is a much more suitable electrolyte for micropipettes than is NaCl. In fact, KCl is employed in nearly all micropipettes used for measuring membrane potentials (generally, c_{K} inside cells is also higher than c_{Na}, and it does not vary as much from cell to cell as does c_{Na}). In any case, the closer u_- is to u_+, the smaller will be the diffusion potential for a given concentration ratio from one region to another.

Membrane Fluxes

As is the case for diffusion potentials in open solution, membrane potentials also depend on the different mobilities of the various ions and on their concentration

Via saltbridge
to half-cell

Via saltbridge
to half-cell

3 M
KCl

Micropipette

External solution

Cell wall

Plasmalemma

Tonoplast

Vacuole

1 μm

Figure 3.4

Glass micropipette filled with 3 M KCl (a highly conducting solution) and inserted into the vacuole of a plant cell so that the electrical potential difference across the plasmalemma and tonoplast in series can be measured. The micropipette is carefully inserted into the cell with a micromanipulator while observing with a light microscope. The micropipette tip must be strong enough to penetrate the cell wall and yet fine enough not to disturb the cell mechanically or electrically. The finer is the tip, the higher is its resistance—a 1-μm-diameter tip usually has a resistance of about 10^6 ohms, which is generally acceptable. The micropipette is connected via a saltbridge containing an electrically conducting solution to a half-cell (discussed in Ch. 6). This half-cell plus another one in electrical contact with the external solution are connected to a high-input resistance device (generally at least 10^{10} ohms) so that the membrane potentials can be measured without drawing much current (see G. P. Findlay and A. B. Hope in Lüttge and Pitman, 1976; Nobel, 1974).

gradients. In this case, however, the "solution" in which the diffusion takes place toward regions of lower chemical potential is the membrane itself. We noted in Chapter 1 that a membrane is often the main barrier, and thus the rate-limiting step, for the diffusion of molecules into and out of cells or organelles. We would therefore expect it to be the phase across which the diffusion potential is expressed. Under biological conditions, a number of different types of ions are present, so the situation is more complex than for the single cation-anion pair analyzed above. Furthermore, the quantities of interest, such as $\partial E/\partial x$ and γ_j, are those within the membrane, where they are not accessible to experimental measurement. We will consider ways of dealing with these difficulties in this section.

To calculate membrane diffusion potentials, we must make certain assumptions before an integration of fluxes or gradients can be made. As a start, we will assume that the electrical potential (E) varies linearly with distance across a membrane. This means that $\partial E/\partial x$ is a constant equal to $E_M/\Delta x$, where E_M is the electrical potential difference across the membrane—i.e., $E^i - E^o = \Delta E = E_M$—and Δx is the membrane thickness. This assumption of a constant electric field across the membrane—the electric field equals $-\partial E/\partial x$ in our one-dimensional case—was originally suggested by Goldman in 1943 (see Goldman, 1943; Hodgkin and Katz, 1949). It appreciably simplifies the integration conditions leading to an expression describing the electrical potential difference across membranes. As another very useful approximation, we will assume that the activity coefficient of species j, γ_j, is constant across the membrane (a less severe restriction than setting it equal to one). As noted in Chapter 1, a partition coefficient is needed to describe concentrations within a membrane because the solvent properties of a membrane differ from those of the adjoining aqueous solutions where the concentrations are actually determined. Thus, c_j in Equation 3.7 should be replaced by $K_j c_j$, where K_j is the partition coefficient for species j. Incorporating these various simplifications and conditions, we can rewrite Equation 3.7 as follows:

$$J_j = -u_j RT K_j \frac{\partial c_j}{\partial x} - u_j K_j c_j z_j F \frac{E_M}{\Delta x} \qquad (3.11)$$

We next transfer the electrical term to the left-hand side of Equation 3.11, factor out $u_j z_j F E_M/\Delta x$, and divide both sides by $K_j c_j + J_j \, \Delta x/(u_j z_j F E_M)$, which puts all terms containing concentration on the same side of the equation. Because the two most convenient variables are c_j and x, their differentials should appear on opposite sides of the equation. After next multiplying each term by $dx/(u_j RT)$, we transform Equation 3.11 to the following expression:

$$\frac{z_j F E_M}{RT \, \Delta x} dx = -\frac{K_j \, dc_j}{K_j c_j + \dfrac{J_j \, \Delta x}{u_j z_j F E_M}} \qquad (3.12)$$

where $(\partial c_j/\partial x) \, dx$ has been replaced by dc_j in anticipation of the restriction to steady-state conditions.

When the term $J_j\Delta x/(u_j z_j FE_M)$ is constant, Equation 3.12 can be readily integrated from one side of the membrane to the other. The factors Δx, z_j, F, and E_M are all fixed. For convenience, the mobility (u_j) of each species is also assumed to be constant within the membrane. When the flux of species j does not change with time or position, then J_j across any plane parallel to and within the membrane is the same, and species j is neither accumulating nor being depleted in any of the regions of interest. Consequently, $\partial c_j/\partial t$ is zero, while $\partial c_j/\partial x$ is nonzero—which is the steady-state condition (see footnote on p. 126). Our restriction to a steady state therefore means that J_j is constant and $(\partial c_j/\partial x)$ dx equals dc_j, a relation that we have already incorporated into Equation 3.12. With these restrictions, the quantity $J_j\Delta x/(u_j z_j FE_M)$ is constant and we can integrate Equation 3.12 from one side of the membrane (the outside, o) to the other (the inside, i). Because $\int K_j dc_j/(K_j c_j + b)$ equals $\ln(K_j c_j + b)$ and $\int_{x^o}^{x^i} dx$ equals Δx, integration of Equation 3.12 gives

$$\frac{z_j FE_M}{RT} = \ln \frac{\left(K_j c_j^o + \dfrac{J_j\,\Delta x}{u_j z_j FE_M} \right)}{\left(K_j c_j^i + \dfrac{J_j\,\Delta x}{u_j z_j FE_M} \right)} \tag{3.13}$$

After taking exponentials of both sides of Equation 3.13 to put it into a more convenient form, and then multiplying by $K_j c_j^i + J_j\,\Delta x/(u_j z_j FE_M)$, it becomes

$$K_j c_j^i e^{z_j FE_M/RT} + \frac{J_j\,\Delta x}{u_j z_j FE_M}\, e^{z_j FE_M/RT} = K_j c_j^o + \frac{J_j\,\Delta x}{u_j z_j FE_M} \tag{3.14}$$

A quantity of considerable interest in Equation 3.14 is J_j, the net flux density of species j. This equation can be readily solved for J_j, giving

$$J_j = J_j^{in} - J_j^{out} \tag{3.15}$$
$$= \left(\frac{K_j u_j z_j FE_M}{\Delta x} \right) \left(\frac{1}{e^{z_j FE_M/RT} - 1} \right) (c_j^o - c_j^i e^{z_j FE_M/RT})$$

where J_j^{in} is the influx or inward flux density of species j, J_j^{out} is its efflux, and their difference is the net flux density. (The net flux density can represent either a net influx or a net efflux, depending on which of the unidirectional components, J_j^{in} or J_j^{out}, is larger.) We will use Equation 3.15 to derive the Goldman equation, describing the diffusion potential across membranes, and later to derive the Ussing–Teorell equation, a relation obeyed by certain passive fluxes.[8]

Equation 3.15 shows how the passive flux of some charged species j depends on its internal and external concentrations and on the electrical potential difference across the membrane. For most cell membranes, E_M is negative; i.e., the inside of the cell is at a lower electrical potential than is the outside. For a cation (z_j a positive integer) and a negative E_M, the terms in the first two parentheses on the right-hand side of Equation 3.15 are both negative; hence their product is positive. For an anion (z_j a negative integer) and a negative E_M, both parentheses are positive; hence their product is still positive. (For $E_M \geq 0$, the product of the first two parentheses is also positive for both anions and cations.) Thus, the sign of J_j depends on the value of c_j^o relative to that of $c_j^i e^{z_j FE_M/RT}$. When c_j^o is greater than $c_j^i e^{z_j FE_M/RT}$, the expression in the last parenthesis of Equation 3.15 is positive, and a net inward flux density of species j occurs ($J_j > 0$). Such a condition should be contrasted with Equation 1.8 [$J_j = P_j(c_j^o - c_j^i)$], where the net flux density in the absence of electrical effects was inward when c_j^o was simply larger than c_j^i, as would adequately describe the situation for neutral solutes. But knowledge of the concentration difference alone is not sufficient to predict the magnitude or even the direction of the flux of ions—we must also consider the electrical potential difference between the two regions.[9]

Membrane Diffusion Potential—Goldman Equation

Passive fluxes of ions, which can be described by Equation 3.15 and are caused by gradients in the chemical potentials of the various species, lead to an electrical

---

8. Although the mathematical manipulations necessary to get from Equation 3.11 to Equation 3.15 are lengthy and cumbersome, the resulting expression is extremely important for our understanding of both membrane potentials and the passive fluxes of ions. Moreover, throughout this text we have generally presented the actual steps involved in a particular derivation to avoid statements such as "it can *easily* be shown" that such and such follows from so and so—expressions that can be most frustrating and often are untrue.

9. We might ask what happens to the flux density expressed by Equation 3.15 as E_M approaches zero? In particular, the quantity within the first parentheses goes to zero and that within the second parentheses goes to infinity. To resolve this limit situation, we can use the series expansion for an exponential ($e^x = 1 + x + \dfrac{x^2}{2!} + \cdots$, App. III) and keep only the first two terms. The denominator within the second parentheses then becomes $1 + z_j FE_M/RT - 1$, or $z_j FE_M/RT$. Hence, the product of the quantities within the first two parentheses becomes $K_j u_j RT/\Delta x$, which is P_j. Now setting E_M to zero in the last term, Equation 3.15 becomes $J_j = P_j(c_j^o - c_j^i)$, which is Equation 1.8, the appropriate form for the flux density in the absence of electrical effects ($E_M = 0$ and/or $z_j = 0$).

potential difference (diffusion potential) across a membrane. We can determine the magnitude of this electrical potential difference by considering the contributions from all ionic fluxes across the membrane and the condition of electroneutrality. Certain assumptions are needed, however, to keep the equations manageable. Under usual biological situations not all anions and cations can easily move through the membranes. Many divalent cations do not readily enter or leave plant cells passively, meaning that their mobility in the membranes is small. Such ions usually do not make a large enough contribution to the fluxes into or out of plant cells to influence markedly the diffusion potentials across the membranes. Thus, we will omit them in the present analysis, which nevertheless is rather complicated.

For many plant cells, the total ionic flux consists mainly of movements of K^+, Na^+, and Cl^- (Dainty 1962; Gutknecht and Dainty, 1968; Hope, 1971). These three ions generally have fairly high concentrations in and near plant cells, and we would therefore expect them to make substantial contributions to the total ionic flux density. More specifically, a flux density of species j depends on the product of its concentration and its mobility ($J_j = -u_j c_j \, \partial\mu_j/\partial x$, Eq. 3.6); ions having relatively high local concentrations or moving in the membrane fairly easily (high u_j) will therefore tend to be the major contributors to the total ionic flux density. In some cases there may be a sizable flux density of H^+ or OH^- (which can have high u_j's) as well as of other ions, the restriction here to three ions being partially in the interest of algebraic simplicity. However, the real justification for considering only K^+, Na^+, and Cl^- is that the diffusion potentials calculated for the passive fluxes of these three ions across certain membranes are in good agreement with the measured electrical potential differences.

Our previous electrical calculations indicated that aqueous solutions are essentially neutral, a condition that we will again invoke here. In other words, because of the large effects on electrical potentials resulting from small amounts of uncanceled charge, the net charge needed to cause the electrical potential difference across a membrane (E_M) is negligible compared with the ambient concentrations of ions. Furthermore, the steady-state fluxes of the ions across the membrane will not change this condition of electrical neutrality, so no net charge is transported by the algebraic sum of the various charge movements across the membrane, i.e., $\Sigma_j z_j J_j = 0$. When the bulk of the ionic flux density consists of K^+, Na^+, and Cl^- movements, this important condition of electroneutrality can be described by equating the cationic flux densities ($J_K + J_{Na}$) to the anionic one (J_{Cl}), which leads to the following relation (as mentioned above, the various ions are indicated by using only their element symbols as subscripts):

$$J_K + J_{Na} - J_{Cl} = 0 \qquad (3.16)$$

Equation 3.16 describes the net ionic flux densities leading to the electrical potential differences across a membrane. After substituting the expressions for the various J_j's into Equation 3.16, we will solve the resulting equation for the diffusion potential across a membrane, E_M.

To obtain a useful expression for E_M in terms of measurable parameters, it is convenient to introduce a permeability coefficient for species j, P_j. In Chapter 1, such a permeability coefficient was defined as $D_j K_j / \Delta x$, where D_j is the diffusion coefficient of species j, K_j is its partition coefficient, and Δx is the membrane thickness (Eq. 1.9). Upon comparing Equation 3.7 with Equation 1.1 ($J_j = -D_j \, \partial c_j / \partial x$), we see that $u_j RT$ takes the place of the diffusion coefficient of species j, D_j, as already pointed out. The quantity $K_j u_j RT / \Delta x$ can thus be equated to $K_j D_j / \Delta x$, which is the permeability coefficient P_j of species j for some particular membrane. In this way, the unknown mobility of species j in a given membrane, the unknown thickness of the membrane, and the unknown partition coefficient for the solute can all be replaced by one parameter, which describes the permeability properties of that solute crossing the particular membrane under consideration.

With all the preliminaries out of the way, let us now derive the expression for the diffusion potential across a membrane for the case where most of the ionic flux density is due to K^+, Na^+, and Cl^- movements. Using the permeability coefficients of the three ions and substituting in the net flux density of each species as defined by Equation 3.15, Equation 3.16 becomes

$$P_K \left(\frac{1}{e^{FE_M/RT} - 1} \right) (c_K^o - c_K^i e^{FE_M/RT})$$

$$+ \, P_{Na} \left(\frac{1}{e^{FE_M/RT} - 1} \right) (c_{Na}^o - c_{Na}^i e^{FE_M/RT}) \qquad (3.17)$$

$$+ \, P_{Cl} \left(\frac{1}{e^{-FE_M/RT} - 1} \right) (c_{Cl}^o - c_{Cl}^i e^{-FE_M/RT}) = 0$$

where z_K and z_{Na} have been replaced by 1, z_{Cl} has been replaced by -1, and FE_M/RT has been canceled from each of the terms for the three net flux densities. To simplify this unwieldy expression, the quantity $1/(e^{FE_M/RT} - 1)$ can be canceled from each of the three terms in Equation 3.17—note that $1/(e^{FE_M/RT} - 1)$ in the last term is the same as $-e^{FE_M/RT}/(e^{FE_M/RT} - 1)$. Equation 3.17 then assumes a much more manageable form:

$$P_K c_K^o - P_K c_K^i e^{FE_M/RT} + P_{Na} c_{Na}^o - P_{Na} c_{Na}^i e^{FE_M/RT}$$

$$- P_{Cl} c_{Cl}^o \, e^{FE_M/RT} + P_{Cl} c_{Cl}^i = 0 \qquad (3.18)$$

After solving Equation 3.18 for $e^{FE_M/RT}$ and taking logarithms, we obtain the following expression for the electrical potential difference across a membrane:

$$E_M = \frac{RT}{F} \ln \frac{(P_K c_K^o + P_{Na} c_{Na}^o + P_{Cl} c_{Cl}^i)}{(P_K c_K^i + P_{Na} c_{Na}^i + P_{Cl} c_{Cl}^o)} \tag{3.19}$$

Equation 3.19 is generally known as the *Goldman*, or *constant field*, *equation*. As mentioned above, the electrical field equals $-\partial E/\partial x$, which Goldman in 1943 set equal to a constant (here $-E_M/\Delta x$) to facilitate the integration across a membrane.[10] In 1949 Hodgkin and Katz applied the general equation derived by Goldman to the specific case of K^+, Na^+, and Cl^- diffusing across a membrane (Hodgkin and Katz, 1949), and Equation 3.19 is sometimes referred to as the *Goldman–Hodgkin–Katz equation* (Starzak, 1984).

Equation 3.19 gives the diffusion potential across a membrane. We derived it by assuming independent passive movements of K^+, Na^+, and Cl^- across a membrane in which $\partial E/\partial x$, γ_j, J_j, and u_j are all constant. We used the negative gradient of its chemical potential as the driving force for the net flux density of each ion. Thus, Equation 3.19 gives the electrical potential difference arising from the different tendencies of K^+, Na^+, and Cl^- to *diffuse* across a membrane to regions of lower chemical potential. When other ions cross some particular membrane in appreciable amounts, they will also make a contribution to its membrane potential. However, the inclusion of divalent and trivalent ions in the derivation of an expression for E_M complicates the algebra considerably (e.g., if Ca^{2+} is also considered, Eq. 3.18 has fourteen terms on the left-hand side instead of six, and the equation becomes a quadratic in powers of $e^{FE_M/RT}$). But the flux densities of such ions are often rather small, in which case Equation 3.19 can be adequate for describing the membrane potential.

Application of the Goldman Equation

In certain cases, the quantities in Equation 3.19—namely, the permeabilities and the internal and the external concentrations of K^+, Na^+, and Cl^-—have all been measured. The validity of the Goldman equation can then be checked. In other words, we can compare the predicted diffusion potential with the actual electrical potential difference measured across the membrane.

---

10. The assumption of a constant electric field in the membrane is actually not essential for obtaining Equation 3.19; we could invoke Gauss's law and perform a more difficult integration. See Goldman (1943) for a consideration of the constant field situation in a general case.

As a specific example, we will use the Goldman equation to evaluate the membrane potential across the plasmalemma of *Nitella translucens*. The concentrations of K^+, Na^+, and Cl^- in the external bathing solution and in the cytosol of *Nitella translucens* are given in Table 3.1. The ratio of the permeability of Na^+ to that of K^+, P_{Na}/P_K, is about 0.18 for *Nitella translucens* (MacRobbie, 1962). The plasmalemma of *Nitella* is much less permeable to Cl^- than it is to K^+, probably only 0.1% to 1% as much. Thus, for purposes of calculation, we will let P_{Cl}/P_K be 0.003. Using these relative permeability coefficients and the concentrations given in Table 3.1, the Goldman equation (Eq. 3.19) predicts the following membrane potential ($RT/F = 25.3$ mV at 20°C, App. I):

$$E_M =$$

$$(25.3 \text{ mV}) \ln \frac{(P_K)(0.1 \text{ mM}) + (0.18P_K)(1.0 \text{ mM}) + (0.003P_K)(65 \text{mM})}{(P_K)(119 \text{ mM}) + (0.18P_K)(14 \text{ mM}) + (0.003P_K)(1.3 \text{ mM})}$$

$$= -140 \text{ mV}$$

Thus, we expect the cytosol to be electrically negative with respect to the external bathing solution, as is indeed the case. In fact, the measured value of the electrical potential difference across the plasmalemma of *Nitella translucens* is -138 mV at 20°C (Table 3.1). This excellent agreement between the observed electrical

Table 3.1
Concentrations, potentials, and fluxes of various ions for *Nitella translucens* in the light and the dark. The superscript o refers to concentrations in the external bathing solution and superscript i refers to the cytosol. The Nernst potentials (E_{N_j}) were calculated from Equation 3.5 using concentration ratios and a numerical factor of 58.2 mV because the measurement temperature was 20°C. The potential across the plasmalemma (E_M) was -138 mV. The fluxes indicated for the dark refer to values soon after cessation of illumination. (See MacRobbie, 1962; Spanswick and Williams, 1964.)

Ion	c_j^o (mol m^{-3} = mM)	c_j^i	E_{N_j} (mV)	$\dfrac{c_j^o}{c_j^i e^{z_j FE_M/RT}}$	Light J_j^{in}	J_j^{out} (nmol m^{-2} s^{-1})	Dark J_j^{in}	J_j^{out}
Na^+	1.0	14	-67	17	5.5	5.5	5.5	1.0
K^+	0.1	119	-179	0.20	8.5	8.5	2.0	(8.5)
Cl^-	1.3	65	99	0.000085	8.5	8.5	0.5	—

potential difference and that calculated from the Goldman equation supports the contention that the membrane potential is a diffusion potential. This can be fairly easily checked by varying the external concentration of K^+, Na^+, and/or Cl^- and then seeing whether the membrane potential changes in accordance with Equation 3.19

As discussed above, the different ionic concentrations on the two sides of a membrane help set up the passive ionic fluxes creating the diffusion potential, but the actual contribution of a particular ionic species to E_M also depends on the ease with which that type of ion can cross the membrane. Based on the relative permeabilities and concentrations, the major contribution to the electrical potential difference across the plasmalemma of *Nitella* comes from the K^+ flux, with Na^+ and Cl^- playing secondary roles. If the Cl^- terms are omitted from Equation 3.19 (i.e., if P_{Cl} is set equal to zero), the calculated membrane potential is -153 mV, compared with -140 mV when Cl^- is included. This relatively small difference between the two potentials is a reflection of the relatively low permeability coefficient for chloride crossing the plasmalemma of *Nitella*, so the Cl^- flux has less effect on E_M than does the K^+ flux. In fact, the relatively high permeability and the high concentration of K^+ ensure that it will have a major influence on the membrane potential. However, Cl^- must be included in the Goldman equation to get an accurate prediction of the membrane potential for many plant cells (Cl^- is less important for E_M of most animal cells).

Restriction to the three ions indicated has proved to be adequate for treating the diffusion potential across certain membranes, and the Goldman or constant field expression in the form of Equation 3.19 has found widespread application. But changes in the amount of Ca^{2+} or H^+ in the external medium do cause some deviations from the predictions of Equation 3.19 for the electrical potential difference across the plasmalemma of *Nitella*. Thus, the diffusion potential is influenced by the properties of the particular membrane being considered, and ions other than K^+, Na^+, and Cl^- may have to be included in specific cases. To allow for the influence of the passive flux of H^+ on E_M, for example, we could include $P_H c_H^o$ in the numerator of the logarithm in the Goldman equation (Eq. 3.19) and $P_H c_H^i$ in the denominator (note that a movement of H^+ in one direction has the same effect on E_M as a movement of OH^- in the other). In fact, H^+ can be the most important ion influencing the electrical potential difference across certain membranes (Lüttge and Higinbotham, 1979; Lüttge and Pitman, 1976; Smith and Raven, 1979). But the main mechanism for the H^+ effect on E_M may be via an electrogenic pump (discussed below), not the passive diffusion described by the Goldman equation. Such a pump actively transports H^+ from the cytosol out across the plasmalemma.

Donnan Potential

Another type of electrical potential difference encountered in biological systems is associated with immobile or fixed charges in some region adjacent to an aqueous phase containing small mobile ions—a *Donnan potential*. When a plant cell is placed in a KCl solution, for example, a Donnan potential arises between the interior of the cell wall and the bulk of the bathing fluid. The electrical potential difference arising from electrostatic interactions at such a solid–liquid interface can be regarded as a special type of diffusion potential, as we will show shortly.

Pectin and other macromolecules in the cell wall have a large number of carboxyl groups (—COOH) from which hydrogen ions dissociate. This gives the cell wall a net negative charge, as indicated in Chapter 1. Cations such as Ca^{2+} are electrostatically attracted to the negatively charged cell wall, and the overall effect is an exchange of H^+ for Ca^{2+} and other cations.[11] The region containing the immobile charges—such as dissociated carboxyl groups in the case of the cell wall—is generally referred to as the *Donnan phase* (Fig. 3.5). At equilibrium, a distribution of oppositely charged ions electrostatically attracted to these immobile charges occurs between the Donnan phase and the adjacent aqueous one. This sets up an ion concentration gradient, so a Donnan potential is created between the center of the Donnan phase and the bulk of the solution next to it. The sign of the electrical potential in the Donnan phase relative to the surrounding electrolyte solution is the same as the sign of the charge of the immobile ions. For example, because of the presence of dissociated carboxyl groups, the electrical potential in the cell wall is negative with respect to an external solution. Membranes also generally act as charged Donnan phases. In addition, Donnan phases occur in the cytoplasm, where the immobile charges are due to large proteins and other large polymers (RNA, DNA) that have many carboxyl and phosphate groups from which protons can dissociate, leaving the macromolecules with a net negative charge. Because the net electrical charge attracts counterions, locally higher osmotic pressures can occur. Actually, all small ions in the immediate vicinity of a Donnan phase are affected, including H^+, i.e., the local pH.

---

11. Such an attraction of positively charged species to the cell wall can increase the local concentration of solutes to about 600 mol m^{-3} (0.6 M), and consequently a greater osmotic pressure can exist in the cell wall than in surrounding aqueous solutions (see Richter and Dainty, 1989; Sentenac and Grigon, 1981).

Donnan phase
(immobile charges)

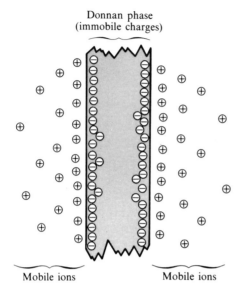

Figure 3.5

Spatial distribution of positively charged mobile ions (⊕) occurring on either side of a Donnan phase in which are embedded immobile negative charges (⊖).

Mobile ions   Mobile ions

Figure 3.5 illustrates a negatively charged, immobile Donnan phase with mobile, positively charged ions on either side. For example, layers containing cations are often formed in the aqueous solutions on each side of biological membranes, which generally act as Donnan phases with a net negative charge at physiological pH's.[12] At equilibrium no net movement of the ions occurs, so the chemical potentials of each of the mobile ions (e.g., K^+, Na^+, Cl^-, Ca^{2+}) have the same values up close to the Donnan phase as they do in the surrounding aqueous phase. The electrical potential difference (Donnan potential) can then be calculated by assuming constancy of the chemical potential. But this is exactly the same principle that we used in deriving the Nernst potential (Eq. 3.5), $E_{N_j} = E^{II} - E^{I} = (RT/z_j F) \ln (a_j^I/a_j^{II})$, between two aqueous compartments. In fact, because the argument again depends on the constancy of the chemical potential, the equilibrium distribution of any ion from the Donnan phase to the aqueous phase extending away from the barrier must satisfy the Nernst potential (E_{N_j}) for that particular ion.

The electrical potential difference known as the Donnan potential can also be regarded as a special case of a diffusion potential. We can assume that the mobile ions are initially in the same region as the immobile ones. In time, some of the mobile ions will tend to diffuse away. This tendency, based on thermal agitation,

---

12. The electrical potential drops or Donnan potentials on either side of a membrane (see Fig. 3.5) are in opposite directions and are *assumed* to cancel each other when a diffusion potential across the membrane is calculated.

causes a slight charge separation, thus setting up an electrical potential difference between the Donnan phase and the bulk of the adjacent solution. For the case of a single species of mobile cations and the anions fixed in the membrane (both assumed to be monovalent), the diffusion potential across that part of the aqueous phase next to the membrane can be described by Equation 3.10, $E^{II} - E^{I} = [(u_- - u_+)/(u_+ + u_-)](RT/F) \ln (c^{II}/c^{I})$, which we derived for diffusion toward regions of lower chemical potential in a solution. Fixed anions have zero mobility $(u_- = 0)$; hence $(u_- - u_+)/(u_+ + u_-)$ here is $-u_+/u_+$, or -1. Equation 3.10 then becomes $E^{II} - E^{I}$ equals $-(RT/F) \ln (c^{II}/c^{I})$, which is the Nernst potential (Eq. 3.5) for monovalent cations $[-\ln(c^{II}/c^{I}) = \ln (c^{I}/c^{II})]$. Thus, the Donnan potential can also be regarded as a type of diffusion potential occurring as the mobile ions tend to diffuse away from the immobile charges of opposite sign, which remain fixed in the Donnan phase.

CHARACTERISTICS OF CROSSING MEMBRANES

Entry of solutes into cells is crucial for many aspects of plant functioning, from nutrient uptake by roots to cell elongation during growth to the opening of stomatal pores that allows gas exchange with the environment. Solutes cross membranes either passively by diffusing toward regions of lower chemical potential or actively where energy is needed for the movement. Specifically, *active transport* implies that energy derived from metabolic processes is used to move a solute across a membrane toward a region of higher chemical potential. There are three different aspects to this description of active transport: a supply of energy, movement, and an increase in chemical potential. Although it is not really necessary, the expression "active transport" has conventionally been restricted to the case of movement in the energetically uphill direction, and we will follow such a restriction here.

By itself, a difference in chemical potential of a certain species across a membrane does not necessarily imply that active transport of that species is occurring. For example, if the solute does not penetrate the membrane, the solute is unable to attain equilibrium across it, and μ_j would not be expected to be the same on the two sides. For ions actually moving across some membrane, the ratio of the influx to the efflux of a particular ionic species provides information that helps indicate whether or not active transport is taking place. A simple but often effective approach for determining whether fluxes are active or passive is to remove possible energy sources. For photosynthesizing plant tissue, this can mean comparing the fluxes in the light with those in the dark. (In addition, we should check whether the permeability of the membrane changes or other alterations occur upon illumination.) Compounds or treatments that disrupt metabolism also can be useful for ascertaining whether metabolic energy is being used for the active transport of various solutes.

We will begin our discussion by showing how active transport can directly affect membrane potentials. We then compare the temperature dependencies of metabolic reactions with those for diffusion processes across a barrier to show that a marked enhancement of solute influx caused by increasing the temperature does not necessarily indicate that active transport is taking place. Next, we will consider a much more reliable criterion for deciding whether fluxes are passive or not—namely, the Ussing–Teorell, or flux ratio, equation. We then examine a specific case where active transport is involved, calculate the energy required, and finally speculate on why K^+ and Cl^- are actively transported into plant cells and Na^+ is transported out.

Electrogenicity

One of the possible consequences of actively transporting a certain ionic species into a cell or organelle is the development of an excess of electrical charge inside the membrane-surrounded entity. If for some ionic species the active transport process involves an accompanying ion of opposite charge or an equal release of a similarly charged ion, the total charge in the cell or organelle is unaffected. However, if the charge of the actively transported ion is not directly compensated for, the process is *electrogenic*; i.e., it tends to generate an electrical potential difference across the membrane. An electrogenic uptake of an ion that produces a net transport of charge into some cell or organelle will thus affect its membrane potential (see Blaustein and Lieberman, 1984; Higinbotham and Anderson, 1974; Hope, 1971; MacRobbie, 1977; Spanswick, 1981). We can appreciate this effect by referring to Equation 3.1 ($Q = C \Delta E$), which indicates that the difference in electrical potential across a membrane, ΔE, equals Q/C, where Q is the net charge enclosed within the cell or organelle and C is the membrane capacitance. The initial movement of net charge across a membrane by active transport leads to a fairly rapid increase or decrease in the potential difference across the membrane.

To be specific, we will consider a spherical cell of radius r into which an electrogenic influx of chloride is caused by active transport, $J^{in}_{a.t._{Cl}}$. The amount of charge transported in time t across the surface of the sphere (area $= 4\pi r^2$) is $J^{in}_{a.t._{Cl}} 4\pi r^2 t$. This active uptake of Cl^- increases the internal concentration of negative charge by the amount moved in divided by the cellular volume, or $J^{in}_{a.t._{Cl}} 4\pi r^2 t/(4\pi r^3/3)$, which is $3J^{in}_{a.t._{Cl}} t/r$. Let us suppose that $J^{in}_{a.t._{Cl}}$ is 10 nmol

m^{-2} s^{-1} and that the cell has a radius of 30 μm. In 1 s the concentration of Cl^- actively transported in would be

$$c_{a.t._{Cl}} = \frac{3 J^{in}_{a.t._{Cl}} t}{r} = \frac{(3)(10 \times 10^{-9} \text{ mol m}^{-2} \text{ s}^{-1})(1 \text{ s})}{(30 \times 10^{-6} \text{ m})}$$

$$= 1.0 \times 10^{-3} \text{ mol m}^{-3}$$

Assuming a membrane capacitance of 10 mF m^{-2}, we calculated on p. 113 that such a cell has 1.0×10^{-3} mol m^{-3} (1 μM) uncompensated negative charge when the interior is 100 mV negative with respect to the external solution. If no change were to take place in the other ionic fluxes, the electrogenic uptake of Cl^- into this cell would cause its interior to become more negative at the rate of 100 mV s^{-1}. The nonsteady-state charging of the membrane capacitance is indeed a rapid process.

The initiation of an electrogenic process causes an adjustment of the passive fluxes across the membrane. In particular, the net charge actively brought in is very soon electrically compensated by appropriate passive movements of it and other ions into or out of the cell. The actual electrical potential difference across the membrane then results from the diffusion potential caused by these new passive fluxes plus a steady-state contribution from the electrogenic process involving the active transport of some charged species. We can represent the electrical potential difference generated by the active transport of species j, $E_{a.t._j}$, as follows:

$$E_{a.t._j} = z_j F J_{a.t._j} R_j^{memb} \tag{3.20}$$

where z_j is the charge number of species j, F is Faraday's constant, $J_{a.t._j}$ is the flux density of species j caused by active transport, and R_j^{memb} is the membrane resistance for the specific pathway along which species j is actively transported. $F J_{a.t._j}$ in Equation 3.20—which is really a form of Ohm's law—is the charge flux density and can have units of (C mol^{-1}) (mol m^{-2} s^{-1}), or C m^{-2} s^{-1}, which is ampere m^{-2}, or current per unit area. Hence, R_j^{memb} can be expressed in ohm m^2, so that $z_j F J_{a.t._j} R_j^{memb}$ in Equation 3.20 can have units of ampere ohm, i.e., V, the proper unit for electrical potentials.

For many plant cells, R_H^{memb} is from 2 to 20 ohm m^2 for the active transport of H^+ ions (protons) out across the plasmalemma (Spanswick, 1981). How large is $E_{a.t._H}$ given by Equation 3.20 for such a proton "pump"? The electrical potential difference created by the active transport of 20 nmol m^{-2} s^{-1} of H^+ across a typical membrane resistance for H^+ of 10 ohm m^2 is

$$E_{a.t._{\text{H}}} = (1)(9.65 \times 10^4 \text{ C mol}^{-1})(-20 \times 10^{-9} \text{ mol m}^{-2} \text{ s}^{-1})(10 \text{ ohm m}^2)$$

$$= -1.9 \times 1.0^{-2} \text{ V} \quad (-19 \text{ mV})$$

We note that because the active transport of H^+ is out of the cell in the case considered, $J_{a.t._{\text{H}}}$ is negative, which leads to a decrease in membrane potential by the active transport process ($E_{a.t._{\text{H}}} < 0$). Measurements in the presence and the absence of metabolic inhibitors have indicated that an electrogenic efflux of H^+ can sometimes lead to an effect of more than 50 mV on E_M for plant cells (Higinbotham and Anderson, 1974; Spanswick, 1981). On the other hand, R_j^{memb} for ions such as K^+, Na^+, and Cl^- generally is only 0.1 to 1 ohm m² (Hope, 1971), so active transport of these ions does not generate much of an electrical potential drop. In any case, the actual electrical potential difference across a membrane can be obtained by adding the potential drop caused by active transport of uncompensated charge (Eq. 3.20) and that caused by passive fluxes (predicted by the Goldman equation, Eq. 3.19).

Boltzmann Energy Distribution and Q_{10}, a Temperature Coefficient

In general, most metabolic reactions are markedly influenced by temperature, whereas processes such as light absorption are insensitive to temperature. What temperature dependence should we expect for diffusion? Can we decide whether the movement of some solute into a cell is by active transport or by passive diffusion once we know how the fluxes depend on temperature? To answer such questions, we need an expression describing the distribution of energy among the molecules as a function of temperature to determine what fraction of the molecules has the requisite energy for a particular process. In aqueous solutions the relevant energy is generally the kinetic energy of motion of the species involved. Hence, we will begin by relating distribution of kinetic energy among molecules to temperature. The topics introduced here are important for a basic understanding of many aspects of biology—all the way from biochemical reactions to the consequences of light absorption.

    Very few molecules possess extremely large kinetic energies. In fact, the probability that a molecule has a given kinetic energy E decreases exponentially as E increases. The precise statement of this is known as the *Boltzmann energy distribution,* which describes the frequency with which specific kinetic energies are possessed by molecules at equilibrium at a given temperature (see Bull, 1964;

Edsall and Wyman, 1958; Van Holde, 1985). A convenient form for the Boltzmann energy distribution (Fig. 3.6) at temperature T is as follows:

$$n(E) = n_{total}e^{-E/kT} \qquad \text{molecule basis} \qquad (3.21a)$$

where $n(E)$ is the number of molecules possessing an energy of E or more out of the total number of molecules, n_{total}, and k is Boltzmann's constant. The quantity $e^{-E/kT}$, which by Equation 3.21a equals $n(E)/n_{total}$, is the *Boltzmann factor*. Equation 3.21a and Figure 3.6 indicate that the number of molecules with an energy of zero or greater, $n(0)$, equals $n_{total}e^{-0/kT}$ or $n_{total}e^{-0}$, which is simply n_{total}, the total number of molecules present.

Because of collisions based on thermal motion, energy is continually being gained or lost by individual molecules in a random fashion. Hence, a wide range of kinetic energies is possible, although very high energies are less probable (see

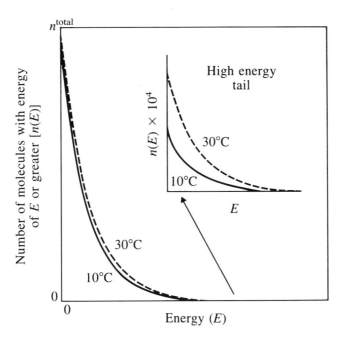

Figure 3.6

Boltzmann energy distributions at 10°C (———) and 30°C (— — —). The inset is a continuation of the right-hand portion of the graph with the scale of the abscissa (E) unchanged and that of the ordinate [$n(E)$] expanded by 10^4. The difference between the two curves is extremely small, except at high energies; although very few molecules are in this "high-energy tail," there are many more such molecules at the higher temperature.

Eq. 3.21a). As the temperature is raised, not only does the average energy per molecule become higher, but also the relative number of molecules in the "high-energy tail" of the exponential Boltzmann distribution increases substantially (see Fig. 3.6).

So far in this section we have considered energy per *molecule*, whereas in many cases it is more convenient to consider energy per *mole*. To change the Boltzmann distribution from a molecule to a mole basis, we multiply Boltzmann's constant k [energy/(molecule K)] by Avogadro's number N (molecules/mole), which gives us the gas constant R [energy/(mole K)]; i.e., R equals kN.[13] If $n(E)$ and n_{total} are numbers of moles and E is energy per mole, we simply replace k in the Boltzmann energy distribution (Eq. 3.21a) by R:

$$n(E) = n_{total}e^{-E/RT} \qquad \text{mole basis} \qquad (3.21b)$$

For diffusion across a membrane, the appropriate Boltzmann energy distribution indicates that the number of molecules with a kinetic energy of U or greater per mole resulting from velocities in some particular direction is proportional to $\sqrt{T}e^{-U/RT}$ (see Davson and Danielli, 1952). A minimum kinetic energy (U_{min}) is often necessary to diffuse past some barrier or to cause some specific reaction. In such circumstances, any molecule with a kinetic energy of U_{min} or greater has sufficient energy for the particular process. For the Boltzmann energy distribution appropriate to this case, the number of such molecules is proportional to $\sqrt{T}e^{-U_{min}/RT}$. (These expressions having the factor $\sqrt{T}$ actually only apply to diffusion in one dimension, e.g., for molecules diffusing across a membrane.) At a temperature 10°C higher, the number is proportional to $\sqrt{(T + 10)}e^{-U_{min}/[R(T+10)]}$. The ratio of these two quantities is called the Q_{10}, or *temperature coefficient* of the process[14]:

$$Q_{10} = \frac{\text{rate of process at } T + 10°C}{\text{rate of process at } T} = \sqrt{\frac{T + 10}{T}} \, e^{10U_{min}/[RT(T+10)]} \qquad (3.22)$$

---

13. In this text we will use two analogous sets of expressions: (1) molecule, mass of molecule, photon, electronic charge, k, kT; and (2) mole, molar mass, mole of photons, Faraday's constant, R, RT (see App. I for numerical values of k, R, kT, and RT). As indicated above, a quantity in the second set, which is more appropriate for most of our applications, is Avogadro's number N (6.022×10^{23}) times the corresponding quantity in the first set. We also note that 1 electron volt (eV), an energy unit often used on atomic and molecular levels, equals 1.602×10^{-19} J and that 1 eV molecule^{-1} = 96.55 kJ mol^{-1} = 23.06 kcal mol^{-1} (see App. II).

14. To obtain the form of the exponential given in Equation 3.22, we note that

$$\frac{-U_{min}}{R(T + 10)} + \frac{U_{min}}{RT} = \frac{-U_{min}T + U_{min}(T + 10)}{RT(T + 10)} = \frac{10U_{min}}{RT(T + 10)}$$

A Q_{10} near unity is characteristic of those passive processes with no energy barrier to surmount, i.e., where U_{min} equals 0. On the other hand, most enzymatic reactions take place only when the reactants have a considerable kinetic energy, so such processes tend to be quite sensitive to temperature. A value of 2 or greater for Q_{10} is often considered to indicate the involvement of metabolism, as occurs for active transport of a solute into a cell or organelle. However, Equation 3.22 indicates that *any* process having an appreciable energy barrier can have a large temperature coefficient.

The Q_{10} for a particular process can be indicative of the minimum kinetic energy required (U_{min}), and vice versa. A membrane often represents an appreciable energy barrier for the diffusion of charged solutes—U_{min} for ions crossing passively can be 50 kJ mol^{-1} (11.9 kcal mol^{-1}, or 0.52 eV molecule^{-1}; Stein, 1967). By Equation 3.22, this would lead to the following temperature coefficient at 20°C:

$$Q_{10} = \sqrt{\frac{(303 \text{ K})}{(293 \text{ K})}} \, e^{(10 \text{ K})(50 \times 10^3 \text{ J mol}^{-1})/[(8.3143 \text{ J mol}^{-1}\text{K}^{-1})(293 \text{ K})(303 \text{ K})]}$$

$$= 1.02e^{0.68} = 2.01$$

Therefore, the passive uptake of this ion doubles with only a 10°C increase in temperature. Hence, a passive process can have a rather high Q_{10} if there is an appreciable energy barrier, implying that a large Q_{10} for ion uptake does not necessarily indicate active transport.

A kinetic energy of 50 kJ mol^{-1} or greater is possessed by only a small fraction of the molecules in the Boltzmann energy distribution (Eq. 3.21 b). For instance, at 20°C the Boltzmann factor is

$$e^{-E/RT} = e^{-(50 \times 10^3 \text{ J mol}^{-1})/[(8.3143 \text{ J mol}^{-1} \text{ K}^{-1})(293 \text{ K})]} = 1.2 \times 10^{-9}$$

As T is raised, the fraction of molecules in the high energy part of the Boltzmann distribution increases greatly (see Fig. 3.6). Many more molecules then have the requisite kinetic energy, U_{min}, and consequently can take part in the process being considered. In particular, at 30°C the Boltzmann factor becomes 2.4×10^{-9}. Hence, the Boltzmann factor for a U_{min} of 50 kJ mol^{-1} essentially doubles for a 10°C rise in temperature, consistent with our Q_{10} calculation for this case.

Activation Energy and Arrhenius Plots

The existence of an energy barrier requiring a minimum energy U_{min} to cross is related to the concept of *activation energy*, which refers to the minimum amount of energy necessary for some reaction to take place. In the case of a membrane, we can evaluate U_{min} by determining how the number of molecules diffusing across the membrane varies with temperature, e.g., by invoking Equation 3.22. The required kinetic energy (U_{min}) corresponds to the activation energy for crossing the barrier. For a chemical reaction, we can also experimentally determine how the process is influenced by temperature. If we represent the activation energy per mole by A, the rate constant for such a reaction should vary with temperature as follows:

$$\text{rate constant} = Be^{-A/RT} \tag{3.23}$$

where B is essentially a constant (see Castellan, 1983; Noggle, 1985).[15]

Equation 3.23 is referred to as the *Arrhenius equation*. It was originally proposed on experimental grounds by Arrhenius at the end of the 19th century and subsequently interpreted theoretically. A plot of the logarithm of the rate constant (or rate of some reaction) versus $1/T$ is commonly known as an *Arrhenius plot* (see Fig. 3.7); by Equation 3.23, ln (rate constant) equals ln $B - A/RT$. Hence, the slope of an Arrhenius plot ($-A/R$) can be used to determine the activation energy. An enzyme greatly increases the rate of the reaction that it catalyzes by reducing the value of the activation energy needed (see Eq. 3.23); i.e., many more molecules then have enough energy to get over the energy barrier separating the reactants from the products. Even so, many reactions of importance in biochemistry have large values for A and are therefore extremely sensitive to temperature. Thus, the related concepts of activation energy and energy barrier have many applications in biology.

Let us next consider the activation energies for certain cases of diffusion. For diffusion in water at 20°C, A is 17 to 21 kJ mol⁻¹ for solutes with molecular weights from 20 to over 1 000 (Stein, 1986). After replacing U_{min} by A in Equation 3.22, we can calculate that the Q_{10} for such diffusion is about 1.3. This is a substantial Q_{10}, which is a consequence of the appreciable thermal energy required for a solute to move through the semicrystalline order found in aqueous solutions due to the presence of hydrogen bonds (see Ch. 2). The activation energy for the passive efflux of K^+ from many cells is about 63 kJ mol⁻¹ near 20°C, which corresponds to a Q_{10} of 2.5 (Stein, 1967). We again conclude that a purely passive process, such as diffusion across membranes, can have a substantial temperature dependence.

---

15. Actually, B can depend on temperature. For diffusion in one dimension as discussed above, the number of molecules with an energy of at least U_{min} per mole is proportional to $\sqrt{T}e^{-U_{min}/RT}$; B is then proportional to $\sqrt{T}$.

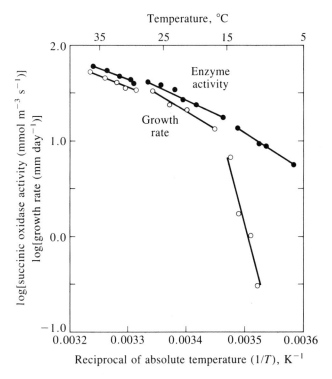

Figure 3.7
Arrhenius plots of mitochondrial succinate oxidase activity and growth rate of the hypocotyl plus radicle of *Vigna radiata*. [Data are replotted from J. K. Raison and E. A. Chapman, *Australian Journal of Plant Physiology* 3:291–299 (1976). Used by permission.]

Arrhenius plots have been used to identify those processes that change markedly over the range of temperatures that injure chilling-sensitive plants such as corn, cotton, cucumber, rice, soybean, and tomato (see P. L. Steponkus in Lange *et al.*, 1981; Lyons *et al.*, 1979; Simon, 1974). Such plants are often severely injured by exposure to temperatures that are low but far above freezing, e.g., near 10°C; prolonged exposure can even result in death. Visible symptoms include wilting, surface pitting of the leaves, and loss of chlorophyll. Damage at the cellular level caused by chilling can be manifested by loss of cytoplasmic streaming, metabolic dysfunction leading to the accumulation of toxic products, and enhanced membrane permeability. Figure 3.7 suggests that the temperature dependence of the rate of growth of the chilling-sensitive *Vigna radiata* (mung bean) may change near 15°C and near 28°C. The activation energy for a mitochondrial enzyme (succinate oxidase) also may change at these two temperatures

(Fig. 3.7), as apparently does the organization of membrane lipids in both mitochondria and chloroplasts (see Raison and Chapman, 1976). In fact, transitions in the physical properties of membranes (such as a change to a more fluid state), which can occur over a range of temperatures, may underlie the changes in slope seen in Figure 3.7. Such changes in the membranes can affect the catalytic properties of enzymes located in them, the permeability of solutes, and the general regulation of cellular metabolism, which then obviously would affect plant growth.[16]

Ussing–Teorell Equation

We now consider ways of distinguishing between active and passive fluxes that are more reliable than determining Q_{10}'s. One of the most useful physicochemical criteria for deciding whether a particular ionic movement across a membrane is active or passive is the application of the *Ussing–Teorell*, or *flux ratio*, equation. For ions moving passively, this expression shows how the ratio of the influx to the efflux depends on the internal and the external concentrations of that species and on the electrical potential difference across the membrane. If the Ussing–Teorell equation is satisfied, passive movements can account for the observed flux ratio, so active transport of the ions need not be invoked. We can readily derive this expression by considering how the influx and the efflux could each be determined experimentally, as the following arguments indicate.

For measuring the unidirectional inward component, or influx, of a certain ion (J_j^{in}), the plant cell or tissue can be placed in a solution containing a radioactive isotope of species j. Initially, none of the radioisotope is inside the cells, so the internal specific activity for this isotope equals zero at the beginning of the experiment. (As for any radioisotope study, only some of the molecules of species j are radioactive. This particular fraction is known as the *specific activity,* and it must be determined for both c_j^o and c_j^i in the present experiment.) Because originally none of the radioisotope is inside, its initial unidirectional outward

---

16. Apparent breaks in Arrhenius plots for enzyme activity (e.g., Fig. 3.7) most directly reflect changes in protein activation energies, not membrane phase transitions, which are best identified using X-ray diffraction or microcalorimetry. No change in the slope of an Arrhenius plot can occur at a phase transition (e.g., the diffusion of a small solute may be unaffected by the rearrangement of membrane lipids), and sharp breaks may occur in the absence of membrane phase transitions when temperature affects the conformation of a protein. Complicated processes such as growth involve many enzyme-catalyzed reactions and usually lead to continuous changes with temperature that cannot be analyzed with a single rate constant (see J. A. Berry and J. K. Raison, M. Aragno, and P. L. Steponkus in Lange *et al.*, 1981).

component, or efflux (J_j^{out}), is zero, and the initial net flux density (J_j) of the isotope indicates J_j^{in}. From Equation 3.15, this influx of the radioisotope can be represented by $(K_j u_j z_j FE_M/\Delta x)$ $[1/(e^{z_j FE_M/RT} - 1)]c_j^o$. After the isotope has entered the cell, some of it will start coming out. Therefore, only the initial flux will give an accurate measure of the influx of the radioisotope of species j.

Once the radioactivity has built up inside to a substantial level, we may remove the radioisotope from the outside solution. The flux of the isotope is then from inside the cells to the external solution. In this case, the specific activity for c_j^o equals zero, and c_j^i determines the net flux of the radioisotope. By Equation 3.15, this efflux of the isotope of species j differs in magnitude from the initial influx only by having the factor c_j^o replaced by $c_j^i e^{z_j FE_M/RT}$, the quantities in the first two parentheses remaining the same. The ratio of these two flux densities— each of which can be separately measured—takes the following relatively simple form:

$$\frac{J_j^{in}}{J_j^{out}} = \frac{c_j^o}{c_j^i e^{z_j FE_M/RT}} \qquad (3.24)$$

Equation 3.24 was independently derived by both Ussing and Teorell in 1949 (see Lüttge and Higinbotham, 1979; Teorell, 1949; Ussing, 1949) and is known as the *Ussing–Teorell equation* and as the *flux ratio equation*. It is strictly valid only for ions moving passively without interacting with other substances that may also be moving across the membrane. Our derivation uses Equation 3.15, which gives the passive flux of some charged species across a membrane in response to differences in its chemical potential. Equation 3.15 considers only one species at a time, so possible interactions between the fluxes of different species are not included in Equation 3.24. The Ussing–Teorell equation can thus be used to determine whether the observed influxes and effluxes are passive (i.e., responses to the chemical potentials of the ions on the two sides of a membrane), or whether additional factors such as interactions between species or active transport are involved. For example, when active transport of species j into a cell is taking place, the actual J_j^{in} would be the passive unidirectional flux density (i.e., the one predicted by Eq. 3.24) *plus* the influx due to active transport. After one further comment about Equation 3.24, we will consider a specific application of the Ussing–Teorell equation to evaluate membrane fluxes.

The ratio of the influx of species j to its efflux, as given by Equation 3.24, can be related to the difference in its chemical potential across a membrane, $\mu_j^o - \mu_j^i$. This difference causes the flux ratio to differ from unity, and we will return to it to estimate the minimum amount of energy needed for active transport of that ionic species across the membrane. After taking logarithms of both sides of the Ussing–Teorell equation (Eq. 3.24) and multiplying by RT, we obtain the following equalities:

$$RT \ln \frac{J_j^{in}}{J_j^{out}} = RT \ln \frac{c_j^o}{c_j^i} - z_j F E_M$$

$$= RT \ln a_j^o + z_j F E^o - RT \ln a_j^i - z_j F E^i$$

$$= \mu_j^o - \mu_j^i \tag{3.25}$$

where the membrane potential E_M has been replaced by $E^i - E^o$, in keeping with our previous convention. The derivation is restricted to the case of constant $\gamma_j (\gamma_j^o = \gamma_j^i)$, so $\ln (c_j^o/c_j^i)$ equals $\ln (\gamma_j^o c_j^o/\gamma_j^i c_j^i)$, or $\ln (a_j^o/a_j^i)$, which is $\ln a_j^o - \ln a_j^i$. Finally, the $\overline{V}_j P$ term in the chemical potential is ignored for these charged species—actually, we need only assume that $\overline{V}_j P^o$ equals $\overline{V}_j P^i$—so that μ_j then equals $\mu_j^* + RT \ln a_j + z_j F E$, where μ_j^* has the same value on the two sides of the membrane. Thus, $\mu_j^o - \mu_j^i$ is $RT \ln a_j^o + z_j F E^o - RT \ln a_j^i - z_j F E^i$, as is indicated in Equation 3.25.

A difference in chemical potential of species j across a membrane causes the flux ratio of the passive flux densities to differ from unity (see Fig. 3.8), a conclusion that follows directly from Equation 3.25. When μ_j^o equals μ_j^i, the influx balances the efflux, so no net passive flux density of species j occurs across the membrane ($J_j = J_j^{in} - J_j^{out}$ by Eq. 3.15). This condition ($\mu_j^o = \mu_j^i$) is also described by Equation 3.4, which was used to derive the Nernst equation. In fact, in such a case of equality of the chemical potentials, the electrical potential difference across the membrane is the Nernst potential, as given by Equation 3.5: $E_{N_j} = (RT/z_j F) \ln (a_j^o/a_j^i)$. Thus, when E_M equals E_{N_j} for some species, J_j^{in} equals J_j^{out}, and no net passive flux density of that ion is expected across the membrane, nor is any energy expended in moving the ion from one side of the membrane to the other. When Equations 3.24 and 3.25 are not satisfied for some species, such ions are not moving across the membrane passively, or perhaps not moving independently from other fluxes. One way this may occur is for the various fluxes to be interdependent, a condition describable by irreversible thermodynamics. Another way is through active transport of the ions, whereby energy derived from metabolism is used to move solutes to regions of higher chemical potential.

Example of Active Transport

The above criteria for deciding whether or not active transport of certain ions is taking place can be illustrated by using data obtained with the large internodal

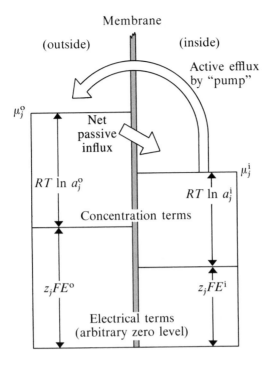

Membrane

(outside)   (inside)

Active efflux
by "pump"

μ_j^o

Net
passive
influx

$RT \ln a_j^o$

$RT \ln a_j^i$

μ_j^i

Concentration terms

$z_j FE^o$ $z_j FE^i$

Electrical terms
(arbitrary zero level)

Figure 3.8

Diagram illustrating the situation for an ion not in equilibrium across a membrane. Because μ_j^o is greater than μ_j^i, there is a net passive flux density into the cell. In the steady state, this net influx is balanced by an equal efflux caused by active transport of species j out of the cell.

cells of *Nitella translucens* (Table 3.1). All the parameters appearing in the Ussing–Teorell equation have been measured for Na$^+$, K$^+$, and Cl$^-$ using *Nitella*. For experimental purposes, this fresh water alga is often placed in a dilute aqueous solution containing 1 mM NaCl, 0.1 mM KCl, plus 0.1 mM CaCl$_2$ (1 mM = 1 mol m^{-3}), which establishes the values for all three c_j^o's (this solution is not unlike the pond water in which *Nitella* grows and is often referred to as "artificial pond water"). The concentrations of Na$^+$, K$^+$, and Cl$^-$ measured in the cytosol, the c_j^i's, are given in the third column of Table 3.1. Assuming that activities can be replaced by concentrations, we can calculate the Nernst potential across the plasmalemma from these concentrations by using Equation 3.5, $E_{N_j} = (58.2/z_j) \log (c_j^o/c_j^i)$ in mV (the numerical factor is 58.2 because the measurements were at 20°C). We thus get

$$E_{N_{Na}} = \frac{(58.2 \text{ mV})}{(1)} \log \frac{(1.0 \text{ mM})}{(14 \text{ mM})} = -67 \text{ mV}$$

Similarly, E_{N_K} is -179 mV and $E_{N_{Cl}}$ is 99 mV (Table 3.1). Direct measurement of the electrical potential difference across the plasmalemma (E_M) gives -138 mV, as indicated earlier in discussing the Goldman equation. Because E_{N_j} differs from E_M in all three cases, none of these ions is in equilibrium across the plasmalemma of *Nitella*.

A difference between E_M and E_{N_j} for a particular species indicates departure from equilibrium for that ionic species; it also tells us in which compartment μ_j is higher. Specifically, if the membrane potential is algebraically more negative than the calculated Nernst potential, the chemical potential in the inner aqueous phase (here the cytosol) is lower for a cation (Fig. 3.8) but higher for an anion, compared with the values in the external solution—consider the effect of z_j in the electrical term of the chemical potential, $z_j FE$. Because E_M (-138 mV) is more negative than $E_{N_{Na}}$ (-67 mV), Na^+ is at a lower chemical potential in the cytosol than outside in the external solution. Analogously, we find that K^+ (with $E_{N_K} = -179$ mV) has a higher chemical potential inside, and Cl^- (with $E_{N_{Cl}} = 99$ mV) has a much higher chemical potentia! inside. If these ions can move across the plasmalemma, this suggests an active transport of K^+ and Cl^- into the cell and an active extrusion of Na^+ from the cell, as is schematically indicated in Figure 3.9.

We can also consider the movement of the various ions into and out of *Nitella* in terms of the Ussing–Teorell equation to help determine whether active transport needs to be invoked to explain the fluxes. The Ussing–Teorell equation predicts that the quantity on the right-hand side of Equation 3.24 ($c_j^o/c_j^i e^{z_j FE_M/RT}$) equals the ratio of influx to efflux of the various ions, if the ions are moving passively in response to gradients in their chemical potential. Using the values given in Table 3.1 and noting that RT/F is 25.3 mV at 20°C (App. I), we find that this ratio for Na^+ is:

$$\frac{J_{Na}^{in}}{J_{Na}^{out}} = \frac{(1.0 \text{ mM})}{(14 \text{ mM})e^{(1)(-138 \text{ mV})/(25.3 \text{ mV})}} = 17$$

Similarly, the expected flux ratio is 0.20 for K^+ and 0.000085 for Cl^- (values given in Table 3.1, column 5). However, the observed influxes in the light equal the effluxes for each of these three ions (Table 3.1, columns 6 and 7). Equal influxes and effluxes are quite reasonable for mature cells of *Nitella*, which are essentially in a steady-state condition. On the other hand, if J_j^{in} equals J_j^{out}, the flux ratios given by Equation 3.24 are not satisfied for Na^+, K^+, or Cl^-. In fact, as suggested by Figure 3.9, active transport of K^+ and Cl^- in and Na^+ out accounts for the marked deviations from the Ussing–Teorell equation for *Nitella*.

As we mentioned earlier, another approach for studying active transport is to

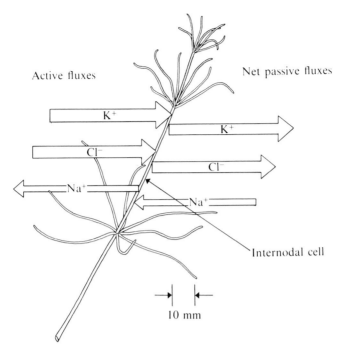

Active fluxes

Net passive fluxes

K⁺

K⁺

Cl⁻

Cl⁻

Na⁺

Na⁺

Internodal cell

10 mm

Figure 3.9

For the steady-state condition in the light the three active fluxes across the plasmalemma of the large internodal cell of *Nitella* are balanced by net passive K^+ and Cl^- effluxes and a net passive Na^+ influx.

remove the supply of energy. In the case of *Nitella,* cessation of illumination causes an appreciable decrease in the Na^+ efflux, the K^+ influx, and the Cl^- influx (last two columns of Table 3.1). But these are the three fluxes that are toward regions of higher chemical potential for the particular ions involved, so we may reasonably expect all three to be active. On the other hand, some fluxes remain essentially unchanged upon placing the cells in the dark (values in the last two columns of Table 3.1 refer to the fluxes soon after extinguishing the light, not the steady-state fluxes). For instance, the Na^+ influx and the K^+ efflux are initially unchanged when the *Nitella* cells are transferred from the light to the dark; i.e., these unidirectional fluxes toward lower chemical potentials do not depend on energy derived from photosynthesis. [For reasons that are unclear, the energetically downhill efflux of Cl^- apparently increases in the dark. This J^{out}_{Cl} is not included in Table 3.1 (see Hope and Walker, 1975).]

The passive diffusion of ions toward regions of lower chemical potential helps create the electrical potential difference across a membrane, and active transport is instrumental in maintaining the asymmetrical ionic distributions that sustain the passive fluxes. Thus, the passive and the active fluxes are interdependent in the ionic relations of cells, and both are crucial for the generation of the observed diffusion potentials. Moreover, active and passive fluxes can occur simultaneously in the same direction. For example, we calculated that J_{Na}^{in} should equal 17 times J_{Na}^{out}, if both flux densities were passive ones obeying the Ussing–Teorell equation. Because J_{Na}^{in} is passive and equal to 5.5 nmol m^{-2} s^{-1}, we expect a passive efflux of Na$^+$ equaling $(5.5)/(17)$, or 0.3 nmol m^{-2} s^{-1}. Thus, the active component of the Na$^+$ efflux in the light may be $5.5 - 0.3$, or 5.2 nmol m^{-2} s^{-1}. At cessation of illumination, J_{Na}^{out} decreases from 5.5 to 1.0 nmol m^{-2} s^{-1} (Table 3.1). Extinguishing the light removes photosynthesis as a possible energy source for active transport, but respiration could still supply energy in the dark. This can explain why J_{Na}^{out} in the dark does not decrease all the way to 0.3 nmol m^{-2} s^{-1}, the value predicted for the passive efflux.

Energy for Active Transport

Suppose that the chemical potential of some species is higher outside than inside a cell, as is illustrated in Figure 3.8. The minimum amount of energy needed to transport a mole of that species from the internal aqueous phase on one side of some membrane to the external solution on the other is the difference in chemical potential of that solute across the membrane, $\mu_j^o - \mu_j^i$ (here, $\mu_j^o > \mu_j^i$). As we noted in considering Equation 3.25, the quantity $\mu_j^o - \mu_j^i$ for ions is $RT \ln (a_j^o/a_j^i) - z_j F E_M$. Because the Nernst potential E_{N_j} is $(RT/z_j F) \ln (a_j^o/a_j^i)$ (Eq. 3.5), we can express the difference in chemical potential across the membrane as follows:

$$\mu_j^o - \mu_j^i = z_j F E_{N_j} - z_j F E_M \qquad (3.26a)$$

$$= z_j F (E_{N_j} - E_M)$$

or

$$\mu_j^i - \mu_j^o = z_j F (E_M - E_{N_j}) \qquad (3.26b)$$

Using the Nernst potentials of Na$^+$, K$^+$, and Cl$^-$ for *Nitella translucens* and the value of E_M given in Table 3.1, we can calculate $z_j (E_M - E_{N_j})$ for transporting these ions across the plasmalemma of this alga. Such a quantity is $(+1)[(-138$ mV$) - (-67$ mV$)]$ or -71 mV for Na$^+$, $+41$ mV for K$^+$, and $+237$ mV for

Cl⁻. By Equation 3.26, these values for $z_j(E_M - E_{N_j})$ mean that Na^+ is at a higher chemical potential in the external bathing solution, whereas K^+ and Cl^- are at higher chemical potentials inside the cell, as we concluded above (see also Figs. 3.8 and 3.9). By Equation 3.26a, the minimum energy required to transport actively or "pump" Na^+ out across the plasmalemma of the *Nitella* cell is

$$\mu_{Na}^o - \mu_{Na}^i = (1)(9.65 \times 10^{-2} \text{ kJ mol}^{-1} \text{ mV}^{-1})[-67 \text{ mV} - (-138 \text{ mV})]$$

$$= 6.9 \text{ kJ mol}^{-1}$$

Similarly, to pump K^+ inward requires 4.0 kJ mol⁻¹. The active extrusion of Na^+ from certain algal cells is linked to the active uptake of K^+, ATP being implicated as the energy source for this coupled exchange process (see Lüttge and Higinbotham, 1979). As we will discuss later (Ch. 6), the hydrolysis of ATP under biological conditions usually releases at least 40 kJ mol⁻¹ (10 kcal mol⁻¹). For the case of a *Nitella* cell this is more than sufficient energy per mole of ATP hydrolyzed to pump one mole of Na^+ out and one mole of K^+ in. The transport of Cl^- inward takes a minimum of 23 kJ mol⁻¹ according to Equation 3.26, which is a fairly large amount of energy. Although the mechanism for actively transporting Cl^- into *Nitella* or other plant cells is not fully understood, exchanges with OH^- or cotransport with H^+ may be involved (see Lüttge and Pitman, 1976; MacRobbie, 1977; Serrano, 1985). The involvement of proton chemical potential differences across membranes in chloroplast and mitochondrial bioenergetics will be discussed in Chapter 6.

Speculation on Active Transport

The active uptake of K^+ and Cl^- together with an active extrusion of Na^+, as for *Nitella,* actually occurs for many plant cells (Gutknecht and Dainty, 1968; Hedrich and Schroeder, 1989; Lüttge and Higinbotham, 1979; Serrano, 1985). We might ask, why does a cell actively transport K^+ and Cl^- in and Na^+ out? Although no definitive answer can be given to such a question, we shall *speculate* on possible reasons, based on the principles that we have been considering.

Let us imagine that a membrane-bounded cell containing negatively charged proteins is placed in an NaCl solution, possibly reflecting primeval conditions when life on earth originated. When Na^+ and Cl^- are both in equilibrium across a membrane, E_M equals $E_{N_{Cl}}$ and $E_{N_{Na}}$. Using concentrations (instead of activities) in Equation 3.5, log (c_{Na}^o/c_{Na}^i) then equals $-\log (c_{Cl}^o/c_{Cl}^i)$, or $c_{Na}^o/c_{Na}^i = c_{Cl}^i/c_{Cl}^o$, and hence $c_{Na}^o c_{Cl}^o = c_{Na}^i c_{Cl}^i$. For electroneutrality in an external solution containing only NaCl, c_{Na}^o equals c_{Cl}^o. Because $a^2 = bc$ implies that $2a \leqslant b + c$,[17] we conclude that $^o + {}^o \leqslant {}^i + {}^i$. But the proteins, which cannot diffuse across the membrane, also make a contribution to the internal osmotic pressure (Π^i), so Π^i is greater than Π^o. When placed in an NaCl solution, water therefore tends to enter such a membrane-bounded cell containing proteins, causing it to swell without limit. An outwardly directed active transport of Na^+ would lower c_{Na}^i and thus prevent excessive osmotic swelling of primitive cells.

An energy-dependent uptake of solutes into a plant cell tends to increase Π^i, which can lead to a rise in P^i. This higher internal hydrostatic pressure favors cell enlargement and consequently cell growth. For a plant cell surrounded by a cell wall, we might therefore expect an active transport of some species into the cell, e.g., Cl^-. (Animal cells do not have to push against a cell wall in order to enlarge and do not generally have an active uptake of Cl^-.) Enzymes have evolved that operate efficiently when exposed to relatively high concentrations of K^+ and Cl^-. In fact, many actually require K^+ for their activity, so an inwardly directed K^+ pump is probably necessary for metabolism as we know it. The presence of a substantial concentration of such ions ensures that electrostatic effects adjacent to a Donnan phase can largely be screened out—otherwise, a negatively charged substrate, e.g., an organic acid or a phosphorylated sugar, might be electrostatically repelled from the catalytic site on an enzyme (proteins are generally negatively charged at cytosolic pH's). Once active transport has set up certain concentration differences across a membrane, the membrane potential is an inevitable consequence of the tendency of such ions to diffuse passively toward regions of lower chemical potential. The roles that such diffusion potentials actually play in the physiology of plant cells are open to question. It is known, however, that they are essential for the transmission of electrical impulses in excitable cells of animals and certain plants.

MECHANISMS FOR CROSSING MEMBRANES

Although active transport is clearly of common occurrence, its mechanism—including the means whereby metabolic energy is used—remains uncertain. The possible involvement of a "carrier" molecule in the active transport of solutes across plant cell membranes was first suggested by Osterhout in the 1930's (see Osterhout, 1935). A carrier can selectively bind certain molecules and then carry

---

17. To show this, consider that $0 \leqslant (\sqrt{b} - \sqrt{c})^2 = b + c - 2\sqrt{bc}$, or $2\sqrt{bc} \leqslant b + c$; hence, if $a^2 = bc$, then $2a \leqslant b + c$. Here, we let $c_{Na}^o = c_{Cl}^o = a$, $c_{Na}^i = b$, and $c_{Cl}^i = c$.

them across the membrane. Carriers provide a cell with the specificity or selectivity needed to control the entry and exit of the various types of solutes encountered. Thus, certain metabolites can be specifically taken into a cell, and photosynthetic and waste products can be selectively moved out across the membranes. At the organelle level, such selectivity plays a key role in maintaining cellular compartmentation. At the organ level, the active transport of certain inorganic nutrients into epidermal cells in the root allows a plant to obtain and accumulate these solutes from the soil. Even though the mechanism for binding and moving solutes through membranes is still speculative, the carrier concept has found widespread application in the interpretation of experimental observations (see Epstein, 1972; Higinbotham, 1974; Lüttge and Higinbotham, 1979; Stein, 1986; Yagi and Pullman, 1987).

A great proliferation in terminology describing how solutes cross membranes has occurred, and new conceptual insights have greatly stimulated research efforts. Carriers are often referred to as transporters, or simply porters. A *symporter* is a porter that causes two different substances to move in the same direction across a membrane, and an *antiporter* causes them to move in opposite directions. Proton (H^+) fluxes appear to be involved with most symporters and antiporters in plant membranes (see Sze, 1985). Also, transmembrane channels with complex and specific properties are increasingly being implicated in the movement of ions across various plant membranes.

Carriers, Porters, Channels, and Pumps

Most researchers agree that transporters are generally proteins. Small proteins could bind some substance on one side of a membrane, thermally diffuse across the membrane, and then somehow release the substance on the other side. Such mobile carriers could bind a single substance, or they could bind two different substances, like the proton-solute symporter portrayed in Figure 3.10a. Candidates for transport by a proton symport in plants include inorganic ions such as Na^+ and metabolites such as sugars and amino acids (see Reinhold and Kaplan, 1984). Many substances apparently move in pores or *channels,* which can be membrane-spanning proteins. Such channels could have a series of binding sites, where the molecule or molecules transported must go from site to site through the membrane (Fig. 3.10b). As another alternative, the substance to be transported might first bind to a site accessible from one side of the membrane only. Following a conformational change of the protein involved, the substance might subsequently be exposed to the solution on the other side of the membrane (Fig. 3.10c). For mechanisms as speculative as these, there are a number of ways that

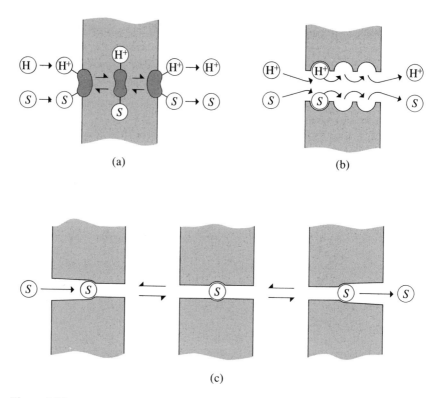

Figure 3.10

Hypothetical structures indicating possible mechanisms for transporters: (a) mobile carrier or porter acting as a symporter for protons (H^+) and some transported solute (S); (b) series of binding sites in a channel across a membrane, acting as a symporter for H^+ and S; and (c) three sequential conformations of a channel, leading to unidirectional movement of solute.

metabolic energy could be involved—e.g., using ATP to cause protein conformational changes or for active transport of H^+ to maintain an H^+ chemical potential difference across the membrane so that solute transport could be coupled to the passive, energetically downhill flow of protons.

Channels can allow many more solutes to cross membranes per unit time than can carriers. Individual carrier proteins have maximal processing rates of 10^4 to 10^5 transported solutes per second, whereas an open channel can allow 10^6 to 10^8 or more ions to cross a membrane per second (Hedrich and Schroeder, 1989; Tester, 1990). Even though many solute molecules can cross the membrane in a channel, flowing passively in the energetically downhill direction, the channel is still selective for a particular type of ion, possibly because of specific binding sites that depend on ion size and/or charge (Fig. 3.10b). Moreover, a so-called "patch-clamp" technique (Neher and Sakmann, 1976) has allowed the recording of electrophysiological responses of single ion channels. Specifically, a glass micropipette with a tip diameter of about 1 μm is pressed against some membrane (not through the membrane as in Fig. 3.4), a slight suction is applied to the micropipette such that a small piece of the membrane seals to its tip, and the voltage is then fixed or "clamped" across this patch of membrane (see Hedrich and Schroeder, 1989; Satter and Moran, 1988). Opening of a channel or channels for such a patch-clamp preparation leads to ion movements, which can be detected as a small current by a sensitive electronic amplifier. Indeed, the opening and closing of even a single channel mediated by the conformational changes of a single protein can be detected.

Potassium conducting channels are apparently involved in the pulvinus motor cells controlling leaf movement for *Samanea saman,* whose leaves tend to move diurnally, and in the opening and the closing of stomatal pores mediated by K^+ uptake and release across the plasmalemma of guard cells, which is a universal phenomenon among plants. The opening of channels is visualized to occur by the opening of "gates," which can be regulated by the membrane potential. In particular, as E_M rises above -40 mV, the gates open so that K^+ can readily cross the membranes of pulvinus motor cells or guard cells. In such cases μ_K is higher in the cells, so K^+ is released from the cells. When E_M for the plasmalemma of guard cells becomes more negative than -100 mV, K^+ channels open, but in this case μ_K is lower in the cells and so K^+ enters. The ensuing entry of an accompanying anion and then water (diffusing toward regions of higher Π, a process called *osmosis*) apparently leads to the swelling of guard cells and the opening of stomatal pores (discussed in Ch. 8). Potassium channels can have a frequency of about one or more channels per μm^2 of membrane surface area. Cellular control can be exerted on the opening of such K^+ channels, as high concentrations of cytosolic Ca^{2+} (above 3×10^{-4} mol m^{-3} or 0.3 μM) can inhibit channel opening. Other ion channels in plant membranes are specific for Ca^{2+} or Cl^-. Besides being sensitive to the electrical potential drop across a membrane, some channels seem to open upon stretching of a membrane (for further details on channels in plant membranes, see Hedrich and Schroeder, 1989; Satter and Moran, 1988; Schroeder and Hedrich, 1989; Tester, 1990).

Channels are apparently involved in plant processes in which large amounts of charged solutes rapidly cross membranes. Thus, channels have been implicated in the functioning of chloroplast thylakoids (discussed in Ch. 5) and guard cells (Ch. 8), in the motion of leaflets in plants sensitive to mechanical stimuli, and in the solar tracking of leaves (Ch. 7). Also, many plant cells are excitable and can transmit action potentials, a process in which ion channels are undoubtedly involved. In addition, ion channels are involved in the long-term maintenance of specific ion concentrations in plant cells.

Because of the principles of electroneutrality and electrogenicity discussed above, electrically uncompensated transport of charged solutes can have major impacts on the membrane potential, which in turn can control many of the properties of ion channels. Thus, an H^+-extruding ATPase (a porter protein that is linked to the hydrolysis of ATP), which is referred to as an H^+ or *proton pump*, can rapidly influence E_M for the plasmalemma and hence control the opening of K^+ and other channels. A blue-light-activated H^+ pump has been identified in the plasmalemma of guard cells that is apparently involved in stomatal opening (see Schroeder and Hedrich, 1989). The hyperpolarization (negative shift of E_M) caused by outwardly directed H^+ pumps in the plasmalemma can also subsequently affect symporters using cotransport of H^+ and various solutes (Fig. 3.10A). Also, an inwardly directed H+ pump in the tonoplast tends to lower the pH and to raise the electrical potential in the vacuole relative to values in the cytosol. Indeed, the most prevalent active transport processes in higher plant cells may be the electrogenic proton pumps taking H^+ out of the cytosol across the plasmalemma and across the tonoplast into the vacuole, leading to cytosolic pH's near 7 and vacuolar pH's below 6 (Sze, 1985). The tonoplast H^+-ATPase, which may transport 2 H^+'s per ATP hydrolyzed (Hedrich and Schroeder, 1989), can lead to passive ion uptake into the vacuole, an important process for plant growth. Thus, H^+ pumps depending on ATP to move protons to regions of higher chemical potential can generate changes in the electrical potential across membranes that in turn affect many cellular processes (see Tazawa *et al.*, 1987; Sze, 1985). Various actions of plant hormones also most likely involve H^+ pumps.

Michaelis–Menten Formalism

When an ion of some species is attached to a particular transporter, another similar ion (of the same or a different species) competing for the same binding site cannot also be bound. For example, the similar monovalent cations K^+ and Rb^+ appear to bind in a competitive fashion to the same site on a transporter. For some cells, the same carrier might transport Na^+ out of the cell and K^+ in—the so-called sodium-potassium pump alluded to above. Ca^{2+} and Sr^{2+} may compete with each other for binding sites on another common carrier. Two other divalent cations, Mg^{2+} and Mn^{2+}, are apparently transported by a single carrier that is different from the one for Ca^{2+} and Sr^{2+}. The halides (Cl^-, I^-, and Br^-) may also be transported by a single carrier.

One of the most important variables in the study of carrier-mediated uptake is the external concentration. As the external concentration of a transported solute increases, the rate of uptake eventually reaches an upper limit. We may then presume that all binding sites on the carriers for that particular solute have become filled or saturated. In particular, the rate of active uptake of species j, J_j^{in}, is often proportional to the external concentration of that solute, c_j^o, over the lower range of concentrations, but as c_j^o is raised a maximum rate, $J_{j\;max}^{in}$, is eventually reached. We can describe this kind of behavior by

$$J_j^{in} = \frac{J_{j\;max}^{in}\; c_j^o}{K_j + c_j^o} \qquad (3.27a)$$

where the constant K_j characterizes the affinity of the carrier used for species j when crossing a particular membrane and is expressed in the units of concentration. For the uptake of many ions into roots and other plant tissues, $J_{j\;max}^{in}$ is 30 to 300 nmol m^{-2} s^{-1}. Often, two different K_j's are observed for the uptake of the same ionic species into a root. The lower K_j is generally between 6 and 100 μM (6 and 100 mmol m^{-3}), which is in the concentration range of many ions in soil water, and the other K_j can be above 10 mM (Epstein, 1972).

Equation 3.27a is similar in appearance to the Michaelis–Menten equation describing enzyme kinetics in biochemistry, $v = v_{\;\text{\tiny max}}s/(K_M + s)$ (see Epstein, 1972; Smith et al., 1983; Stryer, 1988; Zubay, 1988). The substrate concentration s in the latter relation is analogous to c_j^o in Equation 3.27a, and the enzyme reaction velocity v is analogous to J_j^{in}. The term in the Michaelis–Menten equation equivalent to K_j in Equation 3.27a is the substrate concentration for half-maximal velocity of the reaction, K_M (the Michaelis constant). The lower is K_M, the greater is the reaction velocity at low substrate concentrations. Likewise, a low value for K_j indicates that the ion or other solute is more readily bound to some carrier and then transported across the membrane. In particular, Equation 3.27a describes a rectangular hyperbola (Fig. 3.11a), a class of equations that has been used to describe many adsorption and other binding phenomena. When c_j^o is small relative to K_j, J_j^{in} is approximately equal to $J_{j\;max}^{in}c_j^o/K_j$ by Equation 3.27a. The influx of species j is then not only proportional to the external concentration, as already indicated, but also inversely proportional to K_j. Hence, a low K_j, representing a high affinity for a carrier, indicates that species j is actually favored or selected for active transport into the cell, even when its external concentration is relatively low. We note that Equation 3.27a even applies to ions moving in channels where solute binding also can occur (Läuger, 1973). Hence, the Michaelis–Menten formalism has found widespread application in describing movement of solutes across membranes.

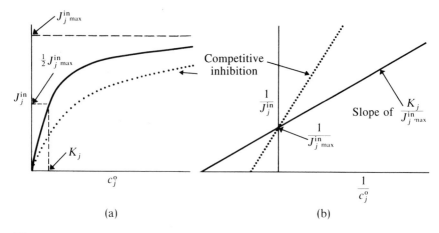

Figure 3.11

Relationship between the external solute concentration (c_j^o) and the rate of influx (J_j^{in}) for active uptake according to a Michaelis–Menten type of kinetics, as given by Equation 3.27: (a) linear plot, and (b) double-reciprocal plot.

The two most common ways of graphing data on solute uptake that fit Equation 3.27a are illustrated in Figure 3.11. Figure 3.11a shows that when the external concentration of species j, c_j^o, is equal to K_j, then J_j^{in} equals $\frac{1}{2}J_{j\,max}^{in}$, as we can see directly from Equation 3.27a; i.e., J_j^{in} is then $J_{j\,max}^{in}K_j/(K_j + K_j)$, or $\frac{1}{2}J_{j\,max}^{in}$. Thus, K_j is the external concentration at which the rate of active uptake is half-maximal—in fact, the observed values of K_j are convenient parameters for describing the uptake of various solutes. If two different solutes compete for the same site on some carrier, J_j^{in} for species j will be decreased by the presence of the second solute, which is known as *competitive inhibition* of species j. In the case of competitive inhibition, the asymptotic value for the active influx, $J_{j\,max}^{in}$, is not affected, because in principle we can raise c_j^o high enough to obtain the same maximum rate for the active uptake of species j. But the half-maximum rate occurs at a higher concentration, so the apparent K_j is raised if a competing solute is present (see Fig. 3.11a).

Often the experimental data are plotted such that a linear relationship is obtained when Equation 3.27a is satisfied for the active uptake of species j. Taking reciprocals of both sides of Equation 3.27a, we note that

$$\frac{1}{J_j^{in}} = \frac{K_j + c_j^o}{J_{j\,max}^{in}\,c_j^o} = \frac{K_j}{J_{j\,max}^{in}}\,\frac{1}{c_j^o} + \frac{1}{J_{j\,max}^{in}} \qquad (3.27b)$$

When $1/J_j^{in}$ is plotted against $1/c_j^o$ (Fig. 3.11b), Equation 3.27b yields a straight line with a slope of $K_j/J_j^{in}{}_{max}$ and an intercept on the ordinate of $1/J_j^{in}{}_{max}$. This latter method of treating the experimental results has found widespread application in studies of solute uptake by plant tissues, especially into roots. Because K_j increases and $J_j^{in}{}_{max}$ remains the same for competitive inhibition, the presence of a competing species causes the slope of the line (i.e., $K_j/J_j^{in}{}_{max}$) to be greater, whereas the intercept on the y-axis ($1/J_j^{in}{}_{max}$) is unchanged (see Fig. 3.11b).

Certain compounds structurally unrelated to species j can inhibit the functioning of the carriers. A *noncompetitive* inhibitor does not bind to the site used for transporting species j across the membrane, and K_j is unaffected. Because a noncompetitive inhibitor lowers the maximum rate of active influx ($J_j^{in}{}_{max}$), both the slope and the y-axis intercept in Figure 3.11b are changed. Finally, an *uncompetitive* inhibitor does not affect the binding of the solute, but rather inactivates the complex after the solute has been bound; the slope in a double-reciprocal plot is unchanged, but the intercept is increased (see Smith *et al.*, 1983; Zubay, 1988).

Facilitated Diffusion

The basis of the mathematical form of Equation 3.27 for describing uptake is the competitive binding of solutes to a limited number of specific sites. In other words, active processes involving metabolic energy do not have to be invoked; if a solute were to *diffuse* across the membrane only when bound to a carrier, the expression for the influx could also be Equation 3.27. This passive entry of a solute mediated by a carrier is termed *facilitated diffusion*.

Because facilitated diffusion is so important in biology and yet is often misunderstood, we will briefly elaborate upon it. Certain molecules passively enter cells more readily than would be expected from consideration of their molecular structure or from observations with analogous substances, so some mechanism is apparently *facilitating* their entry. The net flux density is still toward lower chemical potentials and hence is still in the same direction as ordinary diffusion.[18] To help explain facilitated diffusion, transporters are proposed to act as shuttles for a net passive movement of the specific molecules

---

18. The term "diffusion" usually refers to net thermal motion toward regions of lower concentration. It is used here with a broader meaning—namely, net motion toward regions of lower chemical potential.

across the membrane toward regions of lower energy. Instead of the usual diffusion across the barrier—based on random thermal motion of the solutes—transporters select out and bind certain molecules and then release them on the other side of the membrane without an input of metabolic energy. Such facilitation of entry may also be regarded as a special means of lowering the activation energy needed for the solute to cross the energy barrier represented by the membrane. Thus, transporters facilitate the influx of solutes in the same way that enzymes facilitate biochemical reactions.[19]

There are certain general characteristics of a facilitated diffusion system. As already mentioned, the net flux is toward lower chemical potentials. (According to the usual definition, active transport is in the energetically uphill direction; active transport may employ the same carriers as are used in facilitated diffusion.) Facilitated diffusion causes fluxes to be larger than those expected for ordinary diffusion. Furthermore, the transporters can exhibit a high degree of selectivity; i.e., they can be specific for certain types of molecules, while not binding closely related species. In addition, carriers in facilitated diffusion become saturated when the external concentration of the species transported is raised sufficiently, a behavior consistent with Equation 3.27. Finally, because carriers can exhibit competition phenomena, the flux density of a species entering a cell by facilitated diffusion can be reduced when structurally similar molecules are added to the external solution. (Such molecules would compete for sites on the carrier and thereby reduce the binding and the subsequent transfer of the original species into the cell.)

For convenience, we have been discussing facilitated diffusion *into* a cell, but exactly the same principles apply for exit, as well as for fluxes at the organelle level. Let us assume that a transporter for K^+ exists in the membrane of a certain cell and that it is being used as a shuttle for facilitated diffusion. Not only would the carrier lead to an enhanced net flux density toward the side with the lower chemical potential, but also both the unidirectional fluxes J_K^{in} and J_K^{out} could be increased over the values predicted for ordinary diffusion. This increase in the unidirectional fluxes by a carrier is often called *exchange diffusion*. In such a case, the molecules are interacting with a membrane component—namely, the carrier—and hence the Ussing–Teorell equation—Equation 3.24, $J_j^{in}/J_j^{out} = c_j^o/(c_j^i e^{z_j FE_M/RT})$—is not obeyed because it does not consider interactions with other substances. In fact, observation of departures from predictions of the Ussing–

---

19. As is true for all carrier mechanisms, the exact molecular details have not been worked out. We are thus using the term "carrier" or "transporter" in a phenomenological or operational sense.

Teorell equation is often how cases of exchange diffusion are actually discovered.[20]

Both active and passive fluxes across the cellular membranes can occur concomitantly, but these movements depend on concentrations in rather different ways. For passive diffusion, the unidirectional component J_j^{in} is proportional to c_j^o, as indicated by Equation 1.8 for neutral solutes $[J_j = P_j(c_j^o - c_j^i)]$ and by Equation 3.15 for ions. This proportionality strictly applies only over the range of external concentrations for which the permeability coefficient is essentially independent of concentration, and the membrane potential must not change in the case of charged solutes. Nevertheless, ordinary passive influxes do tend to be proportional to the external concentration, whereas an active influx or the special passive influx known as facilitated diffusion—either of which can be described by a Michaelis–Menten type of formalism—shows saturation effects at the higher concentrations. Moreover, facilitated diffusion and active transport exhibit selectivity and susceptibility to competition, whereas ordinary diffusion does not.

PRINCIPLES OF IRREVERSIBLE THERMODYNAMICS

So far we have been using classical thermodynamics—though, it may have been noticed, often somewhat illegitimately. For example, let us consider Equation 2.23 $(J_{V_w} = L_w \Delta\Psi)$. Because there is a difference in chemical potential represented here by a change in water potential, we expect a net (and irreversible) flow of water from one region to another—obviously not an equilibrium situation. Strictly speaking, however, classical thermodynamics is concerned solely with equilibria, not with movement. Indeed, classical thermodynamics might have been better named "thermostatics." Thus, we have frequently been involved in a kind of hybrid enterprise—appealing to classical thermodynamics for the driving forces, and using nonthermodynamic arguments and analogies to discuss fluxes. One of the objectives of irreversible thermodynamics is to help legitimize the arguments. But, as we shall see, legitimizing them brings in new ideas and considerations.

---

20. The term "exchange diffusion" has another usage in the literature—namely, to describe the carrier-mediated movement of some solute in one direction across a membrane in exchange for a *different* solute being transported in the opposite direction. Again, the Ussing–Teorell equation is not obeyed.

Irreversible thermodynamics uses the same parameters as classical thermo-dynamics—namely, temperature, pressure, free energy, activity, and so on. But these quantities are strictly defined for macroscopic amounts of matter only in equilibrium situations. How can we use them to discuss processes *not* in equi-librium, the domain of irreversible thermodynamics? This dilemma immediately circumscribes the range of validity of the theory of irreversible thermodynamics: it can deal only with "slow" processes or situations not very far from equilibrium, for only in such circumstances can equilibrium-related concepts such as tem-perature and free energy retain their validity—at least approximately. We have to assume from the outset that we can talk about and use classical thermodynamics parameters even in nonequilibrium situations.

Another refinement we will introduce is to recognize that the movement of one species may affect the movement of a second species. A particular flux of some solute may interact with another flux by way of collisions, each species flowing under the influence of its own force. For example, water, ions, and other solutes moving through a membrane toward regions of lower values for their respective chemical potentials can exert a frictional drag on each other. The magnitude of the flux of a solute may then depend on whether water is also flowing. In this way, the fluxes of various species across a membrane become interdependent. Stated more formally, the flux of a solute is not only dependent on the negative gradient of its own chemical potential—which is the sole driving force we have recognized up to now—but it may also be influenced by the gradient in the chemical potential of water. Again using Equation 2.23 ($J_{V_w} = L_w \, \Delta\Psi$) as an example, we have considered that the flow of water depends only on the difference in its own chemical potential between two locations and have thus far ignored any coupling to concomitant fluxes of solutes.

A quantitative description of interdependent fluxes and forces is given by irreversible thermodynamics, a subject that treats nonequilibrium situations such as those actually occurring under biological conditions.[21] In this brief introduction to irreversible thermodynamics we will emphasize certain underlying principles and then derive the *reflection coefficient*. To keep the analysis manageable, we will restrict our attention to isothermal conditions, which approximate many biological situations where fluxes of water and solutes are considered (for further details see Caplan and Essig, 1983; Dainty, 1963; Kedem and Katchalsky, 1958; Keizer, 1987; Prigogine, 1967).[22]

---

21. Nonequilibrium and irreversible are related, because a system in a *nonequilibrium* situation left isolated from external influences will spontaneously and *irreversibly* move toward equilibrium.

22. Onsager received the 1968 Nobel Prize and Prigogine received the 1977 Nobel Prize in chemistry for their contributions to irreversible thermodynamics.

Fluxes, Forces, and Onsager Coefficients

In our previous discussion of fluxes, the driving force leading to the flux density of species j, J_j, was the negative gradient in its chemical potential, $-\partial\mu_j/\partial x$. Irreversible thermodynamics takes a more general view—namely, the flux of species j depends not only on $-\partial\mu_j/\partial x$ but also potentially on *any* other force occurring in the system, such as the negative gradient in the chemical potential of some other species. A particular force, X_k, can likewise influence the flux of any species. Thus, the various fluxes become interdependent, or coupled, because they can respond to changes in any of the forces. Another premise of irreversible thermodynamics is that J_j is linearly dependent on the various forces, so we can treat only those cases that are not too far from equilibrium. Even with this simplification, the algebra often becomes cumbersome, owing to the coupling of the various forces and fluxes.

Using a linear combination of all the forces, we represent the flux density of species j by

$$J_j = \sum_k^n L_{jk}X_k = L_{j1}X_1 + L_{j2}X_2 + \cdots + L_{jj}X_j + \cdots + L_{jn}X_n \quad (3.28)$$

where the summation $\sum_k^n$ is over all forces (all n X_k's), and the L_{jk}'s are referred to as the *Onsager coefficients,* or the phenomenological coefficients, in this case for conductivity. The first subscript on these coefficients (j on L_{jk}) identifies the flux density J_j that we are considering; the second subscript (k on L_{jk}) designates the force, such as the gradient in chemical potential of species k. Each term, $L_{jk}X_k$, is thus the partial flux density of species j due to the particular force X_k. The individual Onsager coefficients in Equation 3.28 are therefore the proportionality factors indicating what contribution each force X_k makes to the flux density of species j. Equation 3.28 is sometimes referred to as the *phenomenological equation.* Phenomenological equations are used to describe observable phenomena without regard to explanations in terms of atoms or molecules. For instance, Ohm's law and Fick's laws are also phenomenological equations. They too assume linear relations between forces and fluxes.

A convenient relationship exists between the various phenomenological coefficients, namely, L_{jk} equals L_{kj}, which is known as the *reciprocity relation.* Such an equality of cross-coefficients was derived in 1931 by Onsager from statistical considerations utilizing the principle of "detailed balancing" (Onsager, 1931a,b). The argument involves microscopic reversibility; i.e., for local equilibrium, any molecular process and its reverse will be taking place at the same average rate in that region. The Onsager reciprocity relation means that the proportionality coefficient giving the flux density of species k caused by the force on species j equals the proportionality coefficient giving the flux density of j caused by the force on k. (Strictly speaking, conjugate forces and fluxes must be used, as they will in the case below.) The fact that L_{jk} equals L_{kj} can be further appreciated by considering Newton's third law—equality of action and reaction. For example, the frictional drag exerted by a moving solvent on a solute is equal to the drag exerted by the moving solute on the solvent. The pairwise equality of cross-coefficients given by the Onsager reciprocity relation reduces the number of coefficients needed to describe the interdependence of forces and fluxes in irreversible thermodynamics and consequently leads to a simplification in solving the sets of simultaneous equations.

Water and Solute Flow

As a specific application of the principles just introduced, we will consider the important coupling of water and solute flow. The driving forces for the fluxes are the negative gradients in chemical potential, which we will assume to be proportional to the differences in chemical potential across some barrier, here considered to be a membrane. In particular, we will represent $-\partial\mu_j/\partial x$ by $\Delta\mu_j/\Delta x$, which in the present case is $(\mu_j^o - \mu_j^i)/\Delta x$. (For convenience, the thickness of the barrier, Δx, will be incorporated into the coefficient multiplying $\Delta\mu_j$ in the flux equations.) To help keep the algebra relatively simple, the development will be carried out for a single nonelectrolyte. The fluxes are across a membrane permeable to both water (w) and the single solute (s), thereby removing the restriction in Chapter 2, where membranes permeable only to water were considered. Using Equation 3.28, we can represent the flux densities by the following linear combination of the differences in chemical potential:

$$J_w = L_{ww}\,\Delta\mu_w + L_{ws}\,\Delta\mu_s \qquad (3.29)$$

$$J_s = L_{sw}\,\Delta\mu_w + L_{ss}\,\Delta\mu_s \qquad (3.30)$$

Equations 3.29 and 3.30 allow for the possibility that each of the flux densities may depend on the differences in both chemical potentials, $\Delta\mu_w$ and $\Delta\mu_s$. Four phenomenological coefficients are used in these two equations. But by the Onsager reciprocity relation, L_{ws} equals L_{sw}. Thus, three different coefficients (L_{ww}, L_{ws}, and L_{ss}) are needed to describe the relationship of these two flux densities to the two driving forces. This is in contrast to Equation 3.6, $J_j = u_j c_j(-\partial\mu_j/$

∂x), where each flux density depends on but one force; accordingly, only two coefficients are then involved in describing J_w and J_s.[23]

To obtain more convenient formulations for the fluxes of water and the solute, we generally express $\Delta\mu_w$ and $\Delta\mu_s$ in terms of the differences in the osmotic pressure and the hydrostatic pressure, $\Delta\Pi$ and ΔP, because it is usually easier to measure ΔP and $\Delta\Pi$ than $\Delta\mu_w$ and $\Delta\mu_s$. The expression for $\Delta\mu_w$ is straightforward; the only possible ambiguity is in deciding on the algebraic sign. In keeping with the usual conventions for this specific case, $\Delta\mu_w$ is the chemical potential of water on the outside minus that on the inside, $\mu_w^o - \mu_w^i$. From Equation 2.12 ($\mu_w = \mu_w^* - \overline{V}_w\Pi + \overline{V}_w P + m_w gh$), $\Delta\mu_w$ is given by

$$\Delta\mu_w = -\overline{V}_w\Delta\Pi + \overline{V}_w\Delta P \qquad (3.31)$$

where $\Delta\Pi$ here equals $\Pi^o - \Pi^i$ and ΔP is $P^o - P^i$ ($\Delta h = 0$ across a membrane).

To express $\Delta\mu_s$ in terms of $\Delta\Pi$ and ΔP, we will first consider the activity term, $RT \ln a_s$. The differential $RT\, d(\ln a_s)$ equals $RT\, da_s/a_s$, or $RT\, d(\gamma_s c_s)/(\gamma_s c_s)$. When γ_s is constant, this latter quantity becomes $RT\, dc_s/c_s$. By Equation 2.10 ($\Pi_s = RT \sum_j c_j$), $RT\, dc_s$ equals $d\Pi$ for a dilute solution of a single solute. Hence, $RT\, d(\ln a_s)$ can be replaced by $d\Pi/c_s$ as a useful approximation. In expressing the difference in chemical potential across a membrane we are interested in macroscopic changes, not in the infinitesimal changes given by differentials. To go from differentials to differences, $RT\, d(\ln a_s)$ becomes RT ($\Delta \ln a_s$) and so $d\Pi/c_s$ can be replaced by $\Delta\Pi/\overline{c}_s$, where $\overline{c}_s$ is essentially the mean concentration of solute s, in this case across the membrane. Alternatively, we can simply define $\overline{c}_s$ as that concentration for which $RT(\Delta \ln a_s)$ exactly equals $\Delta\Pi/\overline{c}_s$. In any case, we can replace the term $RT(\Delta \ln a_s)$ in $\Delta\mu_s$ by the equivalent term, $\Delta\Pi/\overline{c}_s$. By Equation 2.4, μ_s equals $\mu_s^* + RT \ln a_s + \overline{V}_s P$ for a neutral species, and the difference in its chemical potential across a membrane, $\Delta\mu_s$, becomes

$$\Delta\mu_s = \frac{1}{\overline{c}_s}\,\Delta\Pi + \overline{V}_s\,\Delta P \qquad (3.32)$$

---

23. If the solute were a salt dissociable into two ions, we would have to consider three components and three forces, $\Delta\mu_w$, $\Delta\mu_+$, and $\Delta\mu_-$. This would lead to nine coefficients in the three flux equations (for J_w, J_+, and J_-). Invoking the Onsager reciprocity relation, we then have six different phenomenological coefficients to describe the movement of water, a cation, and its accompanying anion.

The two expressions representing the driving forces, $\Delta\mu_w$ (Eq. 3.31) and $\Delta\mu_s$ (Eq. 3.32), are thus expressed as functions of the same two pressure differences, $\Delta\Pi$ and ΔP, which are experimentally more convenient to measure.

Flux Densities, L_P, and σ

Now that we have appropriately expressed the chemical potential differences of water and the solute, we direct our attention to the fluxes. Expressed in our usual units, the flux densities J_w and J_s are the moles of water and solute, respectively, moving across one square metre of membrane surface in a second. A quantity of considerable interest is the volume flux density J_V, which is the rate of movement of the total *volume* of both water and solute across unit area of the membrane; J_V has the units of volume per unit area per unit time, e.g., $m^3\ m^{-2}\ s^{-1}$, or $m\ s^{-1}$.

The molar flux density of species j (J_j) in $mol\ m^{-2}\ s^{-1}$ times the volume occupied by each mole of it ($\bar{V}_j$) in $m^3\ mol^{-1}$ gives the volume flow for that component (J_{V_j}) in $m\ s^{-1}$. Hence, the total volume flux density is

$$J_V = \sum_j J_{V_j} = \sum_j \bar{V}_j J_j \qquad (3.33a)$$

For solute and water both moving across a membrane, J_V is the volume flow of water plus that of solute per unit area, which in the case of a single solute can be represented as

$$J_V = \bar{V}_w J_w + \bar{V}_s J_s \qquad (3.33b)$$

It is generally simpler and more convenient to measure the total volume flux density (such as that given in Eq. 3.33) than one of the component volume flux densities ($J_{V_j} = \bar{V}_j J_j$). For instance, we can often determine the volumes of cells or organelles under different conditions and relate any changes in volume to J_V. The volume flux density of water J_{V_w} (used in Ch. 2, Eq. 2.23) is simply $\bar{V}_w J_w$.

Although straightforward, the algebraic substitutions necessary to incorporate the various forces and fluxes (Eqs. 3.29–3.32) into the volume flow (Eq. 3.33b) lead to a rather cumbersome expression for J_V as a function of ΔP and $\Delta\Pi$ (see Nobel, 1983). Hence, the conventional approach, introduced by Kedem and Katchalsky in 1958 and adopted by essentially all subsequent treatments of irreversible thermodynamics, is to change to a more convenient set of conjugate forces and fluxes. A discussion of the criteria for deciding whether a particular force-flux pair is conjugate would take us into a consideration of the dissipation function and the rate of entropy production, topics outside the scope of this text. It turns out that the total volume flux density, J_V, is conjugate to ΔP, and the diffusional flux density, J_D, is conjugate to $\Delta\Pi$, where J_D can be represented as follows:

$$J_D = J_s/\bar{c}_s - J_w\bar{V}_w$$

$$\cong J_s/\bar{c}_s - J_w/c_w$$

$$= v_s - v_w \tag{3.34}$$

To obtain the second line, we note that $\bar{V}_w c_w$ is essentially unity for a dilute aqueous solution ($\bar{V}_w c_w + \bar{V}_s c_s = 1$ for a solution of a single solute, where $\bar{V}_s c_s \ll 1$ defines a dilute solution); hence $J_w\bar{V}_w$ is approximately equal to J_w/c_w. To obtain the bottom line in Equation 3.34, we note that the average speed of species j, v_j, is J_j/c_j (see Eq. 3.6 and Fig. 3.3).

The diffusional flux density (Eq. 3.34) is the difference between the mean velocities of solute and water. In mass flow (such as that described by Poiseuille's law, Eq. 9.10), v_s equals v_w, so J_D is then zero; such flow is independent of $\Delta\Pi$ and depends only on ΔP. On the other hand, let us consider $\Delta\Pi$ across a membrane that greatly restricts the passage of some solute relative to the movement of water, i.e., a barrier that acts as a differential filter; v_s would then be considerably less than v_w, so J_D would have a nonzero value in response to its conjugate "force," $\Delta\Pi$. Thus, J_D expresses the tendency of the solute relative to water to diffuse in response to a gradient in osmotic pressure.

Next, we will use J_V and J_D to express the linear interdependence of conjugate forces and fluxes in irreversible thermodynamics (Eq. 3.28, $J_j = \Sigma_j L_{jk}X_k$):

$$J_V = L_P \Delta P + L_{PD} \Delta\Pi \tag{3.35}$$

$$J_D = L_{DP} \Delta P + L_D \Delta\Pi \tag{3.36}$$

where the subscripts in both equations are those generally used in the literature. [Although these subscripts—P referring to pressure and D to diffusion—are not consistent with the L_{jk} convention, four coefficients are still needed to describe the dependence of two fluxes on their conjugate forces; sometimes L_P is replaced by L_{VP} (see Starzak, 1984).] By the Onsager reciprocity relation, which is applicable in the present case of conjugate forces and fluxes, L_{PD} equals L_{DP}, and again only three different coefficients are needed to describe the movement of water and a single solute across a membrane.

L_P is the *hydraulic conductivity coefficient*. It describes the mechanical filtration capacity of a membrane or other barrier; i.e., when $\Delta\Pi$ is zero, L_P relates the total volume flux density, J_V, to the hydrostatic pressure difference, ΔP. When ΔP is zero, Equation 3.36 indicates that a difference in osmotic pressure leads to a diffusional flow characterized by the coefficient L_D. Membranes also generally exhibit a property called *ultrafiltration*, whereby they offer different resistances to the passage of the solute and water. For instance, in the absence of an osmotic pressure difference ($\Delta\Pi = 0$), Equation 3.36 indicates that there will be a diffusional flux density equal to $L_{DP} \Delta P$. Based on Equation 3.34, v_s is then different from v_w, which would result if a membrane restricted the passage of solute more than that of water. Thus, the phenomenological coefficient L_{DP} helps describe the relative ease with which solute crosses a membrane compared with water. Such a property of relative selectivity by a barrier is embodied in the *reflection coefficient*, σ, defined by Staverman (1951) as

$$\sigma = -L_{DP}/L_P = -L_{PD}/L_P \qquad (3.37)$$

When a membrane is nonselective, both water and the solute move across it at the same velocity, i.e., v_s equals v_w and $\Delta\Pi$ is zero. (Admittedly, the idea of a solution containing a single solute is not realistic from a biological point of view, but it is convenient for illustrating the minimum value for σ.) If v_s equals v_w, then J_D must be zero by Equation 3.34. For this to be true for any ΔP and a zero $\Delta\Pi$, Equation 3.36 indicates that L_{DP} must be zero; hence the reflection coefficient is also zero in this case ($\sigma = -L_{DP}/L_P$, Eq. 3.37). Thus, σ is zero when the membrane does not select between solute and solvent. At the opposite extreme, the solute does not cross the membrane ($v_s = 0$); hence J_D is $-v_w$ ($J_D = v_s - v_w$, Eq. 3.34). When J_s is zero, J_V equals $\overline{V}_w J_w$ (Eq. 3.33b), which is simply v_w (see Eq. 3.34). Therefore, J_V equals $-J_D$ when the solute does not cross the membrane, and $L_P \Delta P + L_{PD} \Delta\Pi$ equals $-L_{DP} \Delta P - L_D \Delta\Pi$ by Equations 3.35 and 3.36. Because this is true for any ΔP, L_P must equal $-L_{DP}$, so $-L_{DP}/L_P$ then equals unity. The reflection coefficient is therefore unity ($\sigma = -L_{DP}/L_P$, Eq. 3.37) when the solute does not cross the membrane, indicating that all of the solute molecules in that case are *reflected* by the barrier.

Using the definition of σ (Eq. 3.37), we can rewrite Equation 3.35 to obtain the following form for the total volume flux density:

$$J_V = L_P \Delta P - L_P \sigma \Delta\Pi \qquad (3.38)$$

$$= L_P (\Delta P - \sigma \Delta\Pi)$$

We note that L_P is essentially the same as and hence usually replaces L_w, the water conductivity coefficient that we introduced in Chapter 2—e.g., in Equations 2.23 and 2.28. In the absence of a hydrostatic pressure difference across a membrane, the volume flux density J_V equals $-L_P \sigma \, \Delta\Pi$ by Equation 3.38. Thus, the magnitude of the dimensionless parameter σ determines the volume flux density expected in response to a difference in osmotic pressure across a membrane. It is this use that is most pertinent in biology.

Many different solutes can cross a membrane under usual conditions. Each such species j can be characterized by its own reflection coefficient, σ_j, for that particular membrane. The volume flux density given by Equation 3.38 can then be generalized to

$$J_V = L_P \left(\Delta P - \sum_j \sigma_j \, \Delta\Pi_j \right) \qquad (3.39)$$

where $\Delta\Pi_j$ is the osmotic pressure difference across the membrane for species j (e.g., $\Delta\Pi_j = RT \, \Delta c_j$ by Eq. 2.10). Although interactions with water are still taken into account, the generalization represented by Equation 3.39 introduces the assumption that the solutes do not interact with each other as they cross a membrane. Moreover, J_V in Equation 3.39 refers only to the movement of neutral species—otherwise we would also need a current equation to describe the flow of charge. Nevertheless, Equation 3.39 is a useful approximation of the actual situation describing the multicomponent solutions encountered by cells, and we will use it as the starting point for our general consideration of solute movement across membranes. Before discussing such movement, however, let us again consider the range of values of reflection coefficients.

Values of Reflection Coefficients

A reflection coefficient characterizes some particular solute interacting with a specific membrane. In addition, σ_j depends on the solvent on either side of the membrane—water is the only solvent we will consider here. Two extreme conditions can describe the passage of solutes: impermeability, which leads to the maximum value of unity for the reflection coefficient, and nonselectivity, where σ_j is zero. A reflection coefficient of zero may describe the movement of a solute across a very coarse barrier (one with large pores) that cannot distinguish or select between solute and solvent molecules, or it may refer to the passage through a membrane of a molecule very similar in size and structure to water itself. Impermeability describes the limiting case where water can cross some membrane, but the solute cannot.

Let us consider the rather realistic situation of ΔP equaling zero across a membrane bathed on either side by aqueous solutions. By Equation 3.38 [$J_V = L_P(\Delta P - \sigma \, \Delta\Pi)$], the volume flux density (J_V) is then simply $-L_P\sigma \, \Delta\Pi$. For a solute having a reflection coefficient equal to zero for that particular membrane, the volume flux density would be zero. By Equation 3.33b [$J_V = \overline{V}_w J_w + \overline{V}_s J_s$], a zero J_V implies that $\overline{V}_w J_w$ equals $-\overline{V}_s J_s$. In words, the volume flux density of water must be equal and opposite to the volume flux density of the solute to result in no net volume flux density. When ΔP is zero but $\Delta\Pi$ is nonzero, the absence of a net volume flux density across a membrane permeable to both water and the single solute indicates that the reflection coefficient for that solute is zero. This condition, σ equaling zero, occurs when the volume of water flowing toward the side with the higher Π is balanced by an equal volume of solute diffusing across the membrane in the opposite direction toward the side where the solute is less concentrated (lower Π). Such a situation of zero volume flux density anticipates the concept of a "stationary state" to be introduced in the next section.

In their interactions with biological membranes, solutes exhibit properties ranging all the way from freely penetrating ($\sigma_j = 0$), indicating nonselectivity by the membrane, to being unable to penetrate ($\sigma_j = 1$), indicating membrane impermeability. Substances retained in or excluded from plant cells have reflection coefficients quite close to unity for the cellular membranes. For instance, the σ_j's for sucrose and amino acids are usually near 1.0 for plant cells. Methanol and ethanol enter cells very readily and have reflection coefficients of approximately 0.3 for the plasmalemmas of some *Chara* and *Nitella* internodal cells (Dainty and Ginzburg, 1964; Lüttge and Higinbotham, 1979). On the other hand, σ_j can essentially equal zero for solutes crossing porous barriers, such as those presented by cell walls, or for molecules penetrating very readily across membranes, such as D_2O (^2H_2O). Just as for a permeability coefficient, the reflection coefficient of a species is the same for traversal in either direction across a membrane.

For many small neutral solutes not interacting with carriers in a membrane, the reflection coefficients are correlated with the partition coefficients (see Diamond and Wright, 1969; Wright and Diamond, 1969). For example, when K_j for nonelectrolytes is less than about 10^{-4}, σ_j is generally close to one. Thus, compounds that do not readily enter the lipid phase of a membrane (low K_j) also do not cross the membrane easily (σ_j near 1). When the partition coefficient is one or greater, the solutes can enter the membrane in appreciable amounts, and σ_j is generally close to zero. Considering the intermediate case, the reflection

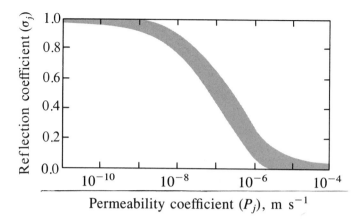

Figure 3.12

Correlation between the reflection coefficients for a series of nonelectrolytes (determined using rabbit gallbladder epithelium) and the permeability coefficients for the same compounds (measured by R. Collander using *Nitella mucronata*). Most of the measurements are in the area indicated. [The curve is adapted from J. M. Diamond and E. M. Wright, *Annual Review of Physiology 31*:581–646 (1969). Used by permission.]

coefficient of some species j can be near 0.5 for a small nonelectrolyte having a K_j of about 0.1 for the membrane lipids, although individual molecules differ depending on their molecular weight, branching, and atomic composition. Neglecting frictional effects with other solutes, we see that the intermolecular interactions affecting partition coefficients are similar to those governing the values of reflection coefficients, and therefore the permeability coefficient of a solute ($P_j = D_j K_j / \Delta x$) is correlated with its reflection coefficient for the same membrane. In fact, as Figure 3.12 illustrates, there is even a correlation between the reflection coefficients of a series of nonelectrolytes determined with animal membranes and the permeability coefficients of the same substances measured for plant membranes.[24] Thus, as the permeability coefficient goes from very small to very large values, the reflection coefficient decreases from unity (describing relative impermeability) to zero (for the opposite extreme of nonselectivity). In the next section we will specifically consider some of the consequences of reflection coefficients' differing from unity for solutes crossing cellular and organelle membranes.

Osmotic pressures play a key role in plant physiology, so σ_j's are important parameters for quantitatively describing the solute and water relations of plants.

---

24. Exceptions occur for certain solutes, e.g., those that hydrogen-bond to membrane components.

In particular, the incorporation of reflection coefficients allows the role of osmotic pressure to be precisely stated. For Poiseuille's law [$J_V = -(r^2/8\eta)\,\partial P/\partial x$, Eq. 9.11], which can adequately describe movement in the xylem and the phloem—as well as in veins, arteries, and household plumbing—the flow is driven by the hydrostatic pressure gradient. An alternative view of the same situation is that σ equals zero for the solutes. In that case, osmotic pressures would have no direct effect on the movement described by Poiseuille's law; i.e., we have a case of nonselectivity. At the opposite extreme of impermeability, σ is unity, so Equation 3.38 becomes $J_V = L_P(\Delta P - \Delta\Pi) = L_P\,\Delta\Psi$ (recall that the water potential, Ψ, can be $P - \Pi$, Eq. 2.13a). But this is similar to Equation 2.23, $J_{V_w} = L_w\,\Delta\Psi$, which we obtained when only water fluxes were considered. This correspondence is not unexpected, because when the solutes are nonpenetrating J_V is the same as J_{V_w} and L_w equals L_P. The real usefulness of reflection coefficients comes when σ_j is not at one of its two extremes of zero and unity. For such cases—intermediate between nonselectivity and impermeability—the volume flux density does not depend on the full osmotic pressure difference across the barrier, but it would be invalid to completely ignore the osmotic contribution of species j.

SOLUTE MOVEMENT ACROSS MEMBRANES

We can profitably reexamine certain aspects of the movement of solutes into and out of cells and organelles by using the more general equations developed from irreversible thermodynamics. One particularly important situation amenable to relatively uncomplicated analysis occurs when the total volume flux density J_V is zero, an example of a *stationary state*. This stationary state, in which the volume of the cell or organelle does not change over the time period of interest, can be brought about by having the net volume flux density of water in one direction across the membrane equal the net volume flux density of solutes in the opposite direction. A stationary state is therefore not the same as a steady state or an equilibrium condition for the cell or organelle; it represents a situation occurring only at a particular time or under some special experimental arrangement. In fact, μ_j can depend on both position and time for a stationary state, but only on position for a steady state, and on neither position nor time at equilibrium. When a net amount of water is moving into a cell, $\Delta\mu_w$ is nonzero (we are not at equilibrium) and, in general, μ_w^i will be increasing with time (we are not even in a steady state). But we still might have a stationary state, where the volume is not changing. Our restriction here to cases of zero net volume flux density considerably simplifies the algebra and emphasizes the role played by reflection coefficients. Moreover, a stationary state of no volume change often characterizes the experimental situations under which the Boyle–Van't Hoff relation (Eqs. 2.15 and 2.18) or the expression describing incipient plasmolysis (Eq. 2.19) is invoked. Thus, the derivation of both of these relationships will be reconsidered in terms of irreversible thermodynamics, and we will discuss the role of reflection coefficients.

What is the relationship between the internal and the external P's and Π's in the case of a stationary state? To answer this, we must first express the consequences of a stationary state in symbols. Our point of departure is Equation 3.39, $J_V = L_P(\Delta P - \Sigma_j \sigma_j \Delta\Pi_j)$, where the stationary state condition of zero net volume flux density ($J_V = 0$) leads to the following equalities: $0 = \Delta P - \Sigma_j \sigma_j \Delta\Pi_j = P^o - P^i - \Sigma_j \sigma_j(\Pi_j^o - \Pi_j^i)$. Here, the osmotic pressures indicate the effect of solutes on water activity, whereas in the general case both osmotic contributions from solutes (Π_s, here represented by Π_j for species j) and matric pressures resulting from the presence of interfaces (τ) might occur. Volume measurements for osmotic studies involving incipient plasmolysis or the Boyle–Van't Hoff relation are generally made when the external solution is at atmospheric pressure ($P^o = 0$) and when there are no external interfaces ($\tau^o = 0$). The stationary state condition of J_V equal to zero then leads to

$$\sum_j \sigma_j\Pi_j^o = \sigma^o\Pi^o = \sum_j \sigma_j\Pi_j^i + \tau^i - P^i \qquad (3.40)$$

where the possibilities of interfacial interactions and hydrostatic pressures within the cell or organelle are explicitly recognized by the inclusion of τ^i and P^i. Equation 3.40 applies when the solutes are capable of crossing the barrier, as when molecules interact with real—not idealized—biological membranes. Equation 3.40 also characterizes the external solutes by a mean reflection coefficient, σ^o, and the total external osmotic pressure, $\Pi^o = \Sigma_j \Pi_j^o$, the latter being relatively easy to measure (σ^o equals $\Sigma_j \sigma_j\Pi_j^o/\Sigma_j \Pi_j^o$, so it represents a weighted mean, with the weights based on the osmotic pressure of each solute species j, Π_j^o).

The Influence of Reflection Coefficients on Incipient Plasmolysis

In Chapter 2 we used classical thermodynamics to derive the condition for incipient plasmolysis ($\Pi_{plasmolysis}^o = \Pi^i$, Eq. 2.19), which occurs when the internal pressure P^i inside a plant cell just becomes zero. The derivation assumed equilibrium of water, i.e., equal water potential, across a membrane impermeable to solutes. But the assumptions of water equilibrium and impermeability are often not valid. We can now remedy this situation using an approach based on irreversible thermodynamics.

Measurements of incipient plasmolysis can be made when there is zero volume flux density ($J_V = 0$) and for a simple external solution ($\tau^o = 0$) at atmospheric pressure ($P^o = 0$). In this case, Equation 3.40 is the appropriate expression from irreversible thermodynamics, instead of the less realistic condition of water equilibrium used previously. For this stationary state condition, we obtain the following expression describing incipient plasmolysis ($P^i = 0$) when the solutes can cross the cell membrane:

$$\sigma^o \Pi^o_{\text{plasmolysis}} = \sum_j \sigma_j \Pi^i_j + \tau^i \qquad (3.41)$$

Because the value of σ^o depends on the particular external solutes present, Equation 3.41 (a corrected version of Eq. 2.19) indicates that the external osmotic pressure Π^o at incipient plasmolysis can vary with the particular solute placed in the solution surrounding the plant cells. Let us suppose that solute i cannot penetrate the membrane, so σ_i equals 1, a situation often true for sucrose. We will suppose that another solute, j, can enter the cells ($\sigma_j < 1$), as is the case for many small nonelectrolytes. If we are at the point of incipient plasmolysis for each of these two solutes as the sole species in the external solution, $\sigma_i \Pi^o_{i \text{ plasmolysis}}$ must equal $\sigma_j \Pi^o_{j \text{ plasmolysis}}$ by Equation 3.41. But solute i is unable to penetrate the membrane ($\sigma_i = 1$). Hence we obtain the following relationships:

$$\sigma_j = \frac{\sigma_i \Pi^o_{i \text{ plasmolysis}}}{\Pi^o_{j \text{ plasmolysis}}} = \frac{\Pi^o_{i \text{ plasmolysis}}}{\Pi^o_{j \text{ plasmolysis}}}$$

$$= \frac{\text{effective osmotic pressure of species } j}{\text{actual osmotic pressure of species } j} \qquad (3.42)$$

where the effective and actual osmotic pressures will be discussed presently.

Equation 3.42 suggests a straightforward way of describing σ_j. Because (by supposition) species j can cross the cell membrane, σ_j must be less than one. Therefore, by Equation 3.42, $\Pi^o_{j \text{ plasmolysis}}$ is greater than $\Pi^o_{i \text{ plasmolysis}}$, where the latter refers to the osmotic pressure of the nonpenetrating solute at the point of incipient plasmolysis. In other words, a higher external osmotic pressure is needed to cause plasmolysis if that solute is able to enter the plant cell. The "actual osmotic pressure" indicated in Equation 3.42 is defined either by Equation 2.7, $\Pi = -(RT/\overline{V}_w) \ln a_w$, where $\Pi = \Sigma_j \Pi_j$, or by Equation 2.10, $\Pi_j = RTc_j$. When the membrane is impermeable to the solute, σ_j equals one, and the apparent (effective) osmotic pressure of species j equals the actual Π_j^o. Effective osmotic pressure takes into consideration the fact that many solutes can cross biological membranes ($\sigma_j < 1$), and hence that the $\Delta\Pi_j$ effective in leading to a net volume flux density is reduced from its actual value—see Equation 3.39, $J_V = L_P(\Delta P - \Sigma_j \sigma_j \Delta\Pi_j)$. In summary, the reflection coefficient of species j indicates how effectively the osmotic pressure of that solute can be exerted across a particular membrane or other barrier.

We can use the condition of incipient plasmolysis to evaluate specific reflection coefficients. It is experimentally difficult to replace one external solution by another with no changes in the tissue taking place, or with none of the previous solution adhering to the cell (see Dainty, 1963). Also, although it is easy in principle and involves only the use of a light microscope, determination of when the plasmalemma just begins to pull away from the cell wall is a rather subjective judgment. Nevertheless, the use of Equation 3.42 provides a simple way of considering individual reflection coefficients for various solutes entering plant cells.

To indicate relationships between reflection coefficients, osmotic pressures, and plasmolysis, let us consider Figure 3.13. The cell in the upper left of the figure is at the point of incipient plasmolysis ($P^i = 0$) for a nonpenetrating solute ($\sigma^o = 1.0$) in the external solution ($\Pi^o = A$, where A is a constant); i.e., the plasmalemma is just beginning to pull away from the cell wall. Equation 3.41 indicates that $\sigma^i\Pi^i$ then also equals A. If we place the cell in a second solution containing a penetrating solute ($\sigma^o < 1$), Equation 3.42 indicates that the external solution must have a higher osmotic pressure for the cell to remain at the point of incipient plasmolysis. For instance, for a second solute with a σ_j of 0.5, the external osmotic pressure at the point of incipient plasmolysis is $2A$ (see Fig. 3.13b). Thus, when the external solute can enter, Π^o is less effective in balancing the internal osmotic pressure or in leading to a flow of water (see Eq. 3.42).

On the other hand, if Π^o were $2A$ for a nonpenetrating external solute, extensive plasmolysis of the same cell would occur, as is illustrated in Figure 3.13c. Because $\sigma^o\Pi^o$ is $2A$ in this case, Equation 3.40 ($\sigma^o\Pi^o = \sigma^i\Pi^i - P^i$, when $\tau^i = 0$) indicates that $\sigma^i\Pi^i$ must also be $2A$, so essentially half of the internal water has left the cell. Finally, if the reflection coefficient were 0.5 and the external osmotic pressure were A, $\sigma^o\Pi^o$ would be $\frac{1}{2}A$, and we would not be at the point of incipient plasmolysis. In fact, the cell would be under turgor with an internal hydrostatic pressure equal to $\frac{1}{2}A$, at least until the concentration of the penetrating solute begins to build up inside. We must therefore take into account the reflection coefficients of external (and internal) solutes to describe conditions at the point of incipient plasmolysis and, by extension, to predict the direction and the magnitude of volume fluxes across membranes.

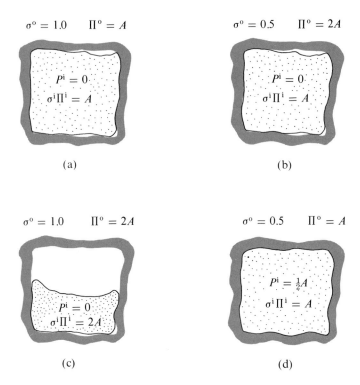

Figure 3.13

Diagrammatic cell showing a cell wall (shaded region) and a plasmalemma (thin line) for various external osmotic pressures: (a) point of incipient plasmolysis in the presence of a nonpenetrating solute (for clarity of showing the location of the plasmalemma, a slight amount of plasmolysis is indicated), (b) point of incipient plasmolysis with a penetrating solute, (c) extensive plasmolysis, and (d) cell under turgor.

Extension of the Boyle–Van't Hoff Relation

In Chapter 2 we derived the Boyle–Van't Hoff relation assuming that the water potential was the same on both sides of the cellular or organelle membrane under consideration. Not only were equilibrium conditions imposed on water, but we implicitly assumed that the membrane was impermeable to the solutes. However, zero net volume flux density ($J_V = 0$) is a better description of the experimental situations where the Boyle–Van't Hoff relation is applied. This condition of no volume change during the measurement is another example of a stationary state, so the Boyle–Van't Hoff relation will be reexamined from the point of view of irreversible thermodynamics. In this way we can remove two of the previous restrictions: equilibrium for water and impermeability of solutes (see Nobel, 1969).

When molecules can cross the membranes bounding cells or organelles, the reflection coefficients for both internal and external solutes should be included in the Boyle–Van't Hoff relation. Because σ^o is less than one when the external solutes can penetrate, the effect of the external osmotic pressure on J_V is reduced. Likewise, the reflection coefficients for solutes within the cell or organelle can lessen the contribution of the internal osmotic pressure of each solute. Replacing Π_j^i by $RTn_j^i/(\overline{V}_w n_w^i)$ (Eq. 2.10) in Equation 3.40 and dividing by σ^o leads to the following Boyle–Van't Hoff relation for the stationary state condition ($J_V = 0$ in Eq. 3.39):

$$\Pi^o = RT \frac{\sum_j \sigma_j n_j^i}{\sigma^o \overline{V}_w n_w^i} + \frac{\tau^i - P^i}{\sigma^o} \tag{3.43}$$

We note that the reflection coefficients of a membrane for both internal and external solutes enter into this extension of the expression relating volume and external osmotic pressure.

As indicated in Chapter 2, the quantity $V - b$ in the conventional Boyle–Van't Hoff relation, $\Pi^o(V - b) = RT\Sigma_j \varphi_j n_j$ (Eq. 2.15), can be identified with $\overline{V}_w n_w^i$. Comparing Equation 2.15 with Equation 3.43, the osmotic coefficient of species j, φ_j, can be equated to σ_j/σ^o as an explicit recognition of the permeation properties of solutes, both internal and external. Indeed, failure to recognize the effect of reflection coefficients on φ_j has led to misunderstandings of osmotic responses.

As we discussed in Chapter 2, the volume of pea chloroplasts (as well as other organelles and many cells) responds linearly to $1/\Pi^o$ (Fig. 2.9), indicating that $\tau^i - P^i$ in such organelles may be negligible compared with the external osmotic pressures used. To analyze experimental observations, $\sigma^o\Pi^o$ can be replaced by $\sigma_x\Pi_x^o + \alpha$, where Π_x^o is the contribution to the external osmotic pressure of solute x whose reflection coefficient (σ_x) is being considered, and α is the sum of $\sigma_j\Pi_j^o$ for all other external solutes. We can represent $RT\Sigma_j \sigma_j n_j^i$ by β, and replace $\overline{V}_w n_w^i$ by $V - b$. Making these substitutions into Equation 3.43, we obtain the following relatively simple form for testing osmotic responses in the case of penetrating solutes:

$$\sigma_x\Pi_x^o + \alpha = \frac{\beta}{V - b} \tag{3.44}$$

If we vary Π_x^o and measure V, we can then use Equation 3.44 to obtain the reflection coefficients for various nonelectrolytes in the external solution.

Reflection Coefficients of Chloroplasts

When the refinements introduced by reflection coefficients are taken into account, we can use osmotic responses of cells and organelles to describe quantitatively the permeability properties of their membranes. As a specific application of Equation 3.44, we note that the progressive addition of hydroxymethyl groups in a series of polyhydroxy alcohols causes the reflection coefficients to increase steadily from 0.00 to 1.00 for pea chloroplasts (Table 3.2). In this regard, the lipid: water solubility ratio decreases in going from methanol to ethylene glycol to glycerol to erythritol to adonitol; i.e., the partition coefficient K_x decreases. Because the permeability coefficient P_x equals $D_x K_x / \Delta x$ (Eq. 1.9), we expect a similar decrease in P_x as hydroxymethyl groups are progressively added. Figure 3.12 clearly shows that, as the permeability coefficient decreases, the reflection coefficient generally increases. Consequently, the increase in σ_x of alcohols as —CHOH groups are added can be interpreted as simply a lowering of K_x. (As

we go from methanol to adonitol, D_x also decreases, perhaps by a factor of 3, whereas K_x decreases about a thousandfold, so changes in K_x are the predominant influence on P_x and σ_x in this case.) The reflection coefficients of six-carbon polyhydroxy alcohols, such as sorbitol and mannitol, are unity for pea chloroplasts. This indication of relative impermeability suggests that these compounds would serve as suitable osmotica in which to suspend chloroplasts, as is indeed the case. From these examples, we see that the rather esoteric concepts of irreversible thermodynamics can be applied in a relatively simple manner to gain insights into the physiological attributes of membranes.

Table 3.2
Reflection coefficients of chloroplasts from *Pisum sativum* for alcohols. The reflection coefficients here apply to the pair of membranes surrounding the organelles, this being the barrier to solute entry or exit encountered in a plant cell (see Wang and Nobel, 1971).

Substance	σ_x	Substance	σ_x
Methanol	0.00	Adonitol	1.00
Ethylene glycol	0.40	Sorbitol	1.00
Glycerol	0.63	Mannitol	1.00
meso-Erythritol	0.90	Sucrose	1.00

Solute Flux Density

Our final objective in this chapter is to obtain an expression for the solute flux density, J_s, that takes into consideration the coupling of forces and fluxes of irreversible thermodynamics. Using Equations 3.33b and 3.34, we note that

$$J_V + J_D = J_w\overline{V}_w + J_s\overline{V}_s + J_s/\overline{c}_s - J_w\overline{V}_w$$

$$= J_s\left(\overline{V}_s + \frac{1}{\overline{c}_s}\right) \cong J_s/\overline{c}_s \qquad (3.45)$$

where the last step applies to a dilute solution ($\overline{V}_s\overline{c}_s \ll 1$, or $\overline{V}_s \ll 1/\overline{c}_s$). After multiplying both sides of Equation 3.45 by $\overline{c}_s$, and using Equation 3.36 for J_D, we obtain:

$$J_s \cong \overline{c}_s J_V + \overline{c}_s(L_{DP} \Delta P + L_D \Delta\Pi)$$

$$= \overline{c}_s J_V + \overline{c}_s L_{DP} \frac{(J_V - L_{PD} \Delta\Pi)}{L_P} + \overline{c}_s L_D \Delta\Pi$$

$$= \overline{c}_s J_V \left(1 + \frac{L_{DP}}{L_P}\right) + \overline{c}_s \Delta\Pi\left(L_D - \frac{L_{DP}L_{DP}}{L_P}\right) \qquad (3.46)$$

where Equation 3.35 is used to eliminate ΔP. Upon combining the factors multiplying $\Delta\Pi$ into a new coefficient for solute permeability, ω_s, and using the previous definition of σ (Eq. 3.37), we can rewrite the last line of Equation 3.46 as:

$$J_s = \overline{c}_s(1 - \sigma)J_V + \omega_s \Delta\Pi \qquad (3.47)$$

Equation 3.47 indicates that J_s not only depends on $\Delta\Pi$, as would be expected on the basis of classical thermodynamics, but also that the solute flux density can be affected by the overall volume flux density, J_V. In particular, the classical expression for J_s for a neutral solute is $P_j \Delta c_j$ (Eq. 1.8), which equals (P_j/RT) $\Delta\Pi_j$ using the Van't Hoff relation (Eq. 2.10, $\Pi_s = RT\Sigma_j c_j$). Thus, ω_s is analogous to P_j/RT of the classical thermodynamic description. The classical treatment indicates that J_s is zero if $\Delta\Pi$ is zero. On the other hand, when $\Delta\Pi$ is zero, Equation 3.47 indicates that J_s equals $\overline{c}_s(1 - \sigma)J_V$, so P_j may not always be an adequate parameter by which to describe the permeability of species j. Finally, in the case of a single solute, movement across a membrane is fully characterized by the three experimentally convenient parameters introduced by irreversible thermodynamics: L_P, σ, and ω_s.

Problems

3.1 At the beginning of this chapter we calculated that an average concentration of 1 mmol m^{-3} (1 μM) excess monovalent anions can lead to a -100 mV potential change across the surface of a spherical cell 30 μm in radius. (a) If the same total amount of charge were all concentrated in a layer 3 nm thick at the surface of the sphere, what would be its average concentration there? (b) If 10^7 sulfate ions are added inside the sphere, what is the new potential difference across the surface? (c) Approximately how much electrical work in joules is required to transport the 10^7 sulfate ions across the surface of the cell?

3.2 For purposes of calculation, let us assume that an external solution is 1 mol m^{-3} (1 mM) KCl, and the solution inside a cell is 160 mol m^{-3} KCl at 20°C. (a) If K$^+$ is in equilibrium across the membrane and activity coefficients are unity, what is the electrical potential difference across the membrane? (b) If K$^+$ is in equilibrium and the mean activity coefficients are calculated from Equation 3.3, what would the membrane potential be? (c) If 1 mol m^{-3} K$_3$ATP—which fully dissociates to 3K$^+$ and ATP^{-3}—is added inside the cell, and if K$^+$–ATP^{-3} can be considered to act as an ion pair, what are γ_{K-ATP} and a_{ATP}?

3.3 Consider a cell with a membrane potential E_M of -118 mV at 25°C. Suppose that the external solution contains 1 mM KCl, 0.1 mM NaCl, and 0.1 mM MgCl$_2$, and the internal concentration of K$^+$ is 100 mM, that of Ca^{2+} is 1 mM, and that of Mg^{2+} is 10 mM (1 mM $=$ 1 mol m^{-3}). Assume that activity coefficients are unity. (a) Are K$^+$ and Mg^{2+} in equilibrium across the membrane? (b) If Na$^+$ and Ca^{2+} are in equilibrium, what are their concentrations in the two phases? (c) If Cl$^-$ is 177 mV away from equilibrium, such that the passive driving force on it is outward, what is its Nernst potential, and what is the internal concentration of Cl$^-$? (d) What is J_{Cl}^{in}/J_{Cl}^{out} for passive fluxes? (e) What are $\Delta\mu_{Cl}$ and $\Delta\mu_{Mg}$ across the membrane?

3.4 A 10 mM KCl solution at 25°C is placed outside a cell formerly bathed in 1 mM KCl (1 mM $=$ 1 mol m^{-3}). (a) Assuming that some of the original solution adheres to the cell and that the ratio of mobilities (u_{Cl}/u_K) is 1.04, what diffusion potential would be present? (b) Assume that equilibrium is reached in the bathing solution after a sufficient lapse of time. The membrane may contain many carboxyl groups (—COOH) whose H$^+$'s will dissociate. The ensuing negative charge will attract K$^+$, and its concentration near the membrane may reach 200 mM. What type of and how large a potential would be associated with this situation? (c) Suppose that 10 mM NaCl is also in the external solution (with the 10 mM KCl), and that internally there is 100 mM K$^+$, 10 mM Na$^+$, and 100 mM Cl$^-$. Assume that P_{Na}/P_K is 0.20 and P_{Cl}/P_K is 0.01. What diffusion potential would be expected across the membrane? (d) What would E_M be if P_{Cl}/P_K were 0.00? If P_{Na}/P_K and P_{Cl}/P_K were both 0.00? Assume that other conditions are as in (c).

3.5 Consider an illuminated spherical spongy mesophyll cell 40 μm in diameter containing 50 spherical chloroplasts 4 μm in diameter. (a) Some monovalent anion produced by photosynthesis has a steady state net flux density out of the chloroplasts of 10 nmol m^{-2} s^{-1}. If this photosynthetic product is not changed or consumed in any of the cellular compartments, what is the net passive flux density out of the cell in the steady state? (b) If the passive flux density of the above substance into the cell at 25°C is 1 nmol m^{-2} s^{-1}, what is the difference in its chemical potential across the cellular membrane? (c) Suppose that, when the cell is placed in the dark, the influx and the efflux both become 0.1 nmol m^{-2} s^{-1}. If the plasmalemma potential is -118 mV (inside negative) and the same concentration occurs on the two sides of the membrane, what can be said about the energetics of the two fluxes? (d) If 1 ATP is required per ion transported, what is the rate of ATP consumption in (c)? Express your answer in μmol s^{-1} per m^3 of cellular contents.

3.6 The energy of activation for crossing biological membranes can represent the energy required to break hydrogen bonds between certain nonelectrolytes and the solvent water; e.g., to enter a cell the solute must first dissolve in the lipid phase of the membrane, and thus the hydrogen bonds with water must be broken. (a) What will be the Q_{10} for the influx of a solute that forms one H-bond per molecule with water if we increase the temperature from 10°C to 20°C? (b) How many hydrogen bonds would have to be ruptured per molecule to account for a Q_{10} of 3.2 under the conditions of (a)?

3.7 Suppose that transporters in the plasmalemma can shuttle K$^+$ and Na$^+$ into a cell. We will let the Michaelis constant K_j be 0.010 mM for the K$^+$ transporter and 1.0 mM for the Na$^+$ transporter (1 mM = 1 mol m^{-3}), and the maximum influx of either ion is 10 nmol m^{-2} s^{-1}. (a) What is the ratio of influxes, J_K^{in}/J_{Na}^{in}, when the external concentration of each ion is 0.010 mM? (b) What is J_K^{in}/J_{Na}^{in} when the external concentration of each ion is 100 mM? (c) If the entry of K$^+$ is by facilitated diffusion only, what is the rate of K$^+$ entry when c_K^o is 0.1 mM and ATP is being hydrolyzed at the rate of 10 mmol m^{-2} s^{-1}?

3.8 Consider a cell whose membrane has a hydraulic conductivity coefficient L_P of 10^{-12} m s^{-1} Pa^{-1}. Initially, no net volume flux density occurs when the cell is placed in a solution having the following composition: sucrose (Π_j^o = 0.2 MPa, σ_j = 1.00), ethanol (0.1 MPa, 0.30), and glycerol (0.1 MPa, 0.80). The external solution is at atmospheric pressure, and P^i is 0.5 MPa. Inside the cell the osmotic pressure caused by glycerol is 0.2 MPa, sucrose and ethanol are initially absent, and other substances having an osmotic pressure of 1.0 MPa are present. (a) What is the mean reflection coefficient for the external solution? (b) What is the mean reflection coefficient for the internal solutes other than glycerol? (c) Suppose that some treatment makes the membrane nonselective for all solutes present. Is there then a net volume flux density? (d) If another treatment makes the membrane impermeable to all solutes present, what would be the net volume flux density?

3.9 Consider a cell at the point of incipient plasmolysis in an external solution containing 0.3 molal sucrose, a nonpenetrating solute. The concentration of glycine that just causes plasmolysis is 0.4 molal. Assume that no water enters or leaves the cell during the plasmolytic experiments. (a) What is the reflection coefficient of glycine for the cellular membrane? (b) Suppose that chloroplasts isolated from such a cell have the same osmotic responses as in Problem 2.5. What is the volume of such chloroplasts *in vivo*? Assume that activity coefficients are unity and that the temperature is 20°C. (c) Suppose that chloroplasts are isolated in 0.3 molal sucrose, which has a reflection coefficient of 1.00 for the chloroplasts. If 0.1 mole of glycine is then added per kg of water in the isolation medium, and if the chloroplast volume is 23 μm^3, what is the reflection coefficient of glycine for the chloroplast membranes? (d) What is the external concentration of glycerol ($\sigma_j = 0.60$) in which the chloroplasts have the same initial volume as in 0.3 molal sucrose? What is the chloroplast volume after a long time in the glycerol solution?

References

Blaustein, M. P., and M. Lieberman, eds. 1984. *Electrogenic Transport: Fundamental Principles and Physiological Implications. Society of General Physiologists Series*, Vol. 38. Raven, New York.

Briggs, G. E., A. B. Hope, and R. N. Robertson. 1961. *Electrolytes and Plant Cells*. Davis, Philadelphia, Pennsylvania.

Bull, H. B. 1964. *An Introduction to Physical Biochemistry*. Davis, Philadelphia, Pennsylvania.

Caplan, S. R., and A. Essig. 1983. *Bioenergetics and Linear Nonequilibrium Thermodynamics: The Steady State*. Harvard University Press, Cambridge, Massachusetts.

Castellan, G. W. 1983. *Physical Chemistry*, 3rd ed. Addison-Wesley, Reading, Massachusetts.

Dainty, J. 1962. Ion transport and electrical potentials in plant cells. *Annual Review of Plant Physiology 13*:379–402.

Dainty, J. 1963. Water relations of plant cells. *Advances in Botanical Research 1*:279–326.

Dainty, J., and B. Z. Ginzburg. 1964. The reflection coefficient of plant cell membranes for certain solutes. *Biochimica et Biophysica Acta 79*:129–137.

Davson, H., and J. F. Danielli. 1952. *The Permeability of Natural Membranes*, 2nd ed. Cambridge University Press, Cambridge, England.

Diamond, J. M., and E. M. Wright. 1969. Biological membranes: The physical basis of ion and nonelectrolyte selectivity. *Annual Review of Physiology 31*:581–646.

Edsall, J. T., and J. Wyman. 1958. *Biophysical Chemistry*, Vol. I. Academic Press, New York.

Epstein, E. 1972. *Mineral Nutrition of Plants: Principles and Perspectives*. Wiley, New York.

Fuller, E. N., P. D. Schettler, and J. C. Giddings. 1966. A new method for prediction of binary gas-phase diffusion coefficients. *Industrial & Engineering Chemistry 58*:19–27.

Goldman, D. E. 1943. Potential, impedance, and rectification in membranes. *Journal of General Physiology 27*:37–60.

Gutknecht, J., and J. Dainty. 1968. Ionic relations of marine algae. *Oceanography and Marine Biology, An Annual Review 6*:163–200.

Hedrich, R., and J. I. Schroeder. 1989. The physiology of ion channels and electrogenic pumps in higher plants. *Annual Review of Plant Physiology and Plant Molecular Biology 40*:539–569.

Higinbotham, N. 1974. Conceptual developments in membrane transport, 1924–1974. *Plant Physiology 54*:454–462.

Higinbotham, N., and W. P. Anderson. 1974. Electrogenic pumps in higher plant cells. *Canadian Journal of Botany 52*:1011–1021.

Hodgkin, A. L., and B. Katz. 1949. The effect of sodium ions on the electrical activity of the giant axon of the squid. *Journal of Physiology 108*:37–77.

Hope, A. B. 1971. *Ion Transport and Membranes*. Butterworths, London; University Park Press, Baltimore, Maryland.

Hope, A. B., and N. A. Walker. 1975. *The Physiology of Giant Algal Cells*. Cambridge University Press, Cambridge, England.

Kedem, O., and A. Katchalsky. 1958. Thermodynamic analysis of the permeability of biological membranes to nonelectrolytes. *Biochimica et Biophysica Acta 27*:229–246.

Keizer, J. 1987. *Statistical Thermodynamics of Nonequilibrium Processes*. Springer-Verlag, New York.

Läuger, P. 1973. Ion transport through pores: A rate-theory analysis. *Biochimica et Biophysica Acta 311*:423–441.

Lange, O. L., P. S. Nobel, C. B. Osmond, and H. Ziegler, eds. 1981. *Physiological Plant Ecology. Encyclopedia of Plant Physiology, New Series*, Vol. 12A. Springer-Verlag, Berlin.

Lauffer, M. A. 1989. *Motion in Biological Systems*. Liss, New York.

Lüttge, U., and N. Higinbotham. 1979. *Transport in Plants*. Springer-Verlag, New York.

Lüttge, U., and M. G. Pitman. 1976. *Transport in Plants II, Part A, Cells; Part B, Tissues and Organs. Encyclopedia of Plant Physiology, New Series*, Vol. 2. Springer-Verlag, Berlin.

Lyons, J. M., D. Graham, and J. K. Raison, eds. 1979. *Low Temperature Stress in Crop Plants: The Role of the Membrane*. Academic Press, New York.

MacRobbie, E. A. C. 1962. Ionic relations of *Nitella translucens*. *Journal of General Physiology 45*:861–878.

MacRobbie, E. A. C. 1977. Functions of ion transport in plant cells and tissues. *International Review of Biochemistry, Plant Biochemistry II*, D. H. Northcote, ed. *13*, 211–247.

Neher, E., and B. Sakmann. 1976. Single-channel currents recorded from membrane of denervated frog muscle fibers. *Nature 260*:799–802.

Nobel, P. S. 1969. The Boyle–Van't Hoff relation. *Journal of Theoretical Biology 23*:375–379.

Nobel, P. S. 1974. *Introduction to Biophysical Plant Physiology*. Freeman, San Francisco, California.

Nobel, P. S. 1983. *Biophysical Plant Physiology and Ecology*. Freeman, New York.

Noggle, J. H. 1985. *Physical Chemistry*. Little, Brown, Boston, Massachusetts.

Onsager, L. 1931a. Reciprocal relations in irreversible processes. I. *Physical Review 37*:405–426.

Onsager, L. 1931b. Reciprocal relations in irreversible processes. II. *Physical Review 38*:2265–2279.

Osterhout, W. J. V. 1935. How do electrolytes enter the cell? *Proceedings of the National Academy of Sciences of the United States of America 21*:125–132.

Prigogine, I. 1967. *Thermodynamics of Irreversible Processes*, 3rd ed. Wiley (Interscience), New York.

Pytkowicz, R. M. 1979. *Activity Coefficients in Electrolyte Solutions.* CRC Press, Cleveland, Ohio.

Raison, J. K., and E. A. Chapman. 1976. Membrane phase changes in chilling-sensitive *Vigna radiata* and their significance to growth. *Australian Journal of Plant Physiology 3*:291–299.

Reinhold, L., and A. Kaplan. 1984. Membrane transport of sugars and amino acids. *Annual Review of Plant Physiology 35*:45–83.

Richter, C., and J. Dainty. 1989. Ion behavior in plant cell walls. I. Characterization of the *Sphagnum russowii* cell wall ion exchanger. *Canadian Journal of Botany 67*:451–459.

Satter, R. L., and N. Moran. 1988. Ionic channels in plant cell membranes. *Physiologia Plantarum 72*:816–820.

Schroeder, J. I., and R. Hedrich. 1989. Involvement of ion channels and active transport in osmoregulation and signaling of higher plant cells. *Trends in Biochemical Sciences 14*:187–192.

Sentenac, H., and C. Grigon. 1981. A model for predicting ionic equilibrium concentrations in cell walls. *Plant Physiology 68*:415–419.

Serrano, R. 1985. *Plasma Membrane ATPase of Plants and Fungi.* CRC Press, Boca Raton, Florida.

Simon, E. W. 1974. Phospholipids and plant membrane permeability. *New Phytologist 73*:377–420.

Smith, E. L., R. L. Hill, I. R. Lehman, R. J. Lefkowitz, P. Handler, and A. White. 1983. *Principles of Biochemistry*, 7th ed., *General Aspects*. McGraw-Hill, New York.

Smith, F. A., and J. A. Raven. 1979. Intracellular pH and its regulation. *Annual Review of Plant Physiology 30*:289–311.

Spanswick, R. M. 1981. Electrogenic ion pumps. *Annual Review of Plant Physiology 32*:267–289.

Spanswick, R. M., and E. J. Williams. 1964. Electrical potentials and Na, K, and Cl concentrations in the vacuole and cytoplasm of *Nitella translucens. Journal of Experimental Botany 15*:193–200.

Starzak, M. E. 1984. *The Physical Chemistry of Membranes.* Academic Press, Orlando, Florida.

Staverman, A. J. 1951. The theory of measurement of osmotic pressure. *Recueil des Travaux Chimiques des Pays-Bas 70*:344–352.

Stein, W. D. 1967. *The Movement of Molecules across Cell Membranes.* Academic Press, New York.

Stein, W. D. 1986. *Transport and Diffusion across Cell Membranes.* Academic Press, Orlando, Florida.

Stryer, L. 1988. *Biochemistry*, 3rd ed. Freeman, New York.

Sze, H. 1985. H^+-translocating ATPases: Advances using membrane vesicles. *Annual Review of Plant Physiology 36*:175–208.

Tazawa, M., T. Shimmen, and T. Mimura. 1987. Membrane control in the Characeae. *Annual Review of Plant Physiology 38*:95–117.

Teorell, T. 1949. Membrane electrophoresis in relation to bio-electrical polarization effects. *Archives des Sciences Physiologiques 3*:205–219.

Tester, M. 1990. Tansley Review No. 21. Plant ion channels: Whole cell and single channel studies. *New Phytologist 114*:305–340.

Ussing, H. H. 1949. The distinction by means of tracers between active transport and diffusion. *Acta Physiologica Scandinavica 19*:43–56.

Van Holde, K. E. 1985. *Physical Biochemistry*, 2nd ed. Prentice-Hall, Englewood Cliffs, New Jersey.

Vorobiev, V. N. 1967. Potassium ion activity in the cytoplasm and the vacuole of cells of *Chara* and *Griffithsia*. *Nature 216*:1325–1327.

Wang, C.t., and P. S. Nobel. 1971. Permeability of pea chloroplasts to alcohols and aldoses as measured by reflection coefficients. *Biochimica et Biophysica Acta 241*:200–212.

Wright, E. M., and J. M. Diamond. 1969. Patterns of non-electrolyte permeability. *Proceedings of the Royal Society of London, Series B 172*:227–271.

Yagi, K., and B. Pullman, eds. 1987. *Ion Transport through Membranes*. Academic Press, Orlando, Florida.

Zubay, G. 1988. *Biochemistry*, 2nd ed. Macmillan, New York.

CHAPTER **4**

Light

Through a series of nuclear reactions taking place within the sun, mass is converted into energy in accordance with Einstein's famous relation, $E = mc^2$. By such conversion of mass to energy, the sun maintains an extremely high surface temperature and thus radiates a great amount of energy into space. Some of this radiant energy is incident on the earth, only a small fraction of which is absorbed by plants. This absorption initiates a flow of energy through the biosphere (all living things and that portion of the earth that they inhabit).

The first step in the utilization of sunlight for this energy flow is the conversion of its radiant energy into various forms of chemical energy by the primary processes of photosynthesis. This chemical energy may then be stored in the plants, mainly in the form of carbohydrates. The stored energy may later be acquired by animals—directly by herbivores, indirectly by carnivores, or both directly and indirectly by omnivores like us. However the energy is acquired, its ultimate source is the solar radiation trapped by photosynthesis. Without this continuous energy input from the sun, living organisms would drift toward equilibrium and hence death.

Sunlight also regulates certain activities of plants and animals by acting as a trigger. The energy to carry out these activities is supplied by metabolic reactions, not directly by the light itself. Examples of light acting as a trigger include vision, phototaxis, phototropism, and the phytochrome regulation of various plant processes.

In this chapter we will be primarily concerned with the physical nature of light and the mechanism of light absorption by molecules. We will discuss how molecular states excited by light absorption can promote endergonic (energy-requiring) reactions or be dissipated by other de-excitation processes. In Chapter 5 we will consider the photochemistry of photosynthesis and in Chapter 6 the bioenergetics of energy conversion, especially that taking place in organelles. In Chapter 7 we will demonstrate how the net energy input by radiation is dissipated by a leaf.

As an introduction to the topic of light, let us next consider certain historical developments in the understanding of its nature (for details see Ditchburn, 1976; Resnick and Halliday, 1977 and 1986; Seliger and McElroy, 1965; Sobel, 1987; Waldman, 1983). In 1666 Newton showed that a prism could disperse white light into many different colors, suggesting that such radiation is a mixture of many components. Soon thereafter, Huygens proposed that the propagation of light through space could be by wave motion. In the early 1800's Young attributed interference properties to the wave character of light. But a wave theory of light was not generally accepted until about 1850, when Foucault demonstrated that light travels more slowly in a dense medium such as water than in a rarefied medium such as air, one of the predictions of the wave theory. In 1887 Hertz discovered that light striking the surface of a metal could cause the release of electrons from the solid—the "photoelectric" effect. But he also found that wavelengths above a certain value could not eject any electrons at all, no matter what the total energy in the light beam. This important result was contrary to the then-accepted wave theory of light. In an important departure from wave theory, Planck proposed in 1901 that radiation was particle-like; i.e., light was describable as consisting of discrete packets, or quanta, each of a specific energy. In 1905 Einstein explained the photoelectric effect of Hertz as a special example of the particle nature of light, indicating that the absorption of a photon of sufficiently short wavelength by an electron in the metal could supply enough energy to cause the ejection or release of that electron, whereas if the wavelengths were longer, then the individual photons were not energetic enough to eject any electrons. The intriguing wave-particle duality of light has subsequently become described in a consistent manner through the development of quantum mechanics. Both wave and particle attributes of light are necessary for a complete description of radiation, and we will consider both aspects in this chapter. (Although this text does not require a background in quantum mechanics, some knowledge of this field is essential for a comprehensive understanding of light.)

WAVELENGTH AND ENERGY

Light is often defined as that electromagnetic radiation perceivable by the human eye. Although such a definition may be strictly correct, the word *light* is frequently used to refer to a wider range of electromagnetic radiation. In this section we will discuss the range of electromagnetic radiation important in biology, including the subdivisions into various wavelength intervals. The wavelength of light will be related to its energy. After noting various conventions used to describe radiation, we will briefly consider some of the characteristics of solar radiation reaching the earth.

Light Waves

The regular and repetitive changes in the intensity of the minute electric and magnetic fields indicate the passage of a light wave (see Fig. 4.1). Light can travel in a solid (e.g., certain plastics), in a liquid (water), in a gas (air), and even in a vacuum (the space between the sun and the earth's atmosphere). One way to characterize light is by its wavelength—the distance between successive points of the same phase, such as between two successive peaks of a wave (Fig. 4.1). A wavelength, then, is the distance per *cycle* of the wave. The preferred

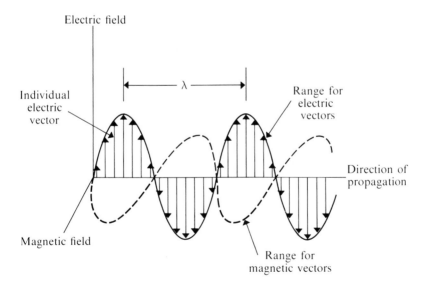

Figure 4.1

Light can be represented by an electromagnetic wave corresponding to oscillations of the local electric and magnetic fields. The oscillating electric vectors at a particular instant in time are indicated by arrows in the diagram. A moment later, the entire pattern of electric and magnetic fields will shift in the direction of propagation of the wave. A wavelength (λ) is the distance between two successive points of the same phase along the wave.

unit for wavelengths of light is the nanometer (nm)—10^{-9} m—and we will use it in this text (1 nm = 0.001 μm = 10 Å).

The wavelength regions of major interest in biology are the ultraviolet, the visible, and the infrared (see Table 4.1). Wavelengths below about 400 nm are referred to as ultraviolet (UV)—meaning on the other side of or beyond violet in the sense of having a shorter wavelength. The lower limit for UV is somewhat arbitrary, often near 10 nm, but very little solar radiation occurs at wavelengths less than 150 nm. The visible region extends from approximately 400 nm to 740 nm and is subdivided into various bands such as blue, green, and red (Table 4.1). These divisions are based on the subjective color experienced by humans. The infrared (IR) region has wavelengths longer than those of the red end of the visible spectrum, extending up to approximately 100 μm.

Besides wavelength, we can also characterize a light wave by its frequency of oscillation, ν, and by the magnitude of its velocity of propagation, v (i.e., v is the speed of light). These three quantities are related as follows:

$$\lambda \nu = v \tag{4.1}$$

where λ is the wavelength. In a vacuum the speed of light for all wavelengths is a constant, generally designated as c, which experimentally equals 299 800

Table 4.1

Definitions and characteristics of the various wavelength regions of light. The ranges of wavelengths leading to the sensation of a particular color are somewhat arbitrary and vary with individuals. Both frequencies and energies in the table refer to the particular wavelength indicated in column 3 for each wavelength interval. Wavelength magnitudes are those in a vacuum.

Color	Approximate wavelength range (nm)	Representative wavelength (nm)	Frequency (cycles s^{-1}, or hertz)	Energy (kJ mol^{-1})
Ultraviolet	below 400	254	11.80×10^{14}	471
Violet	400 to 425	410	7.31×10^{14}	292
Blue	425 to 490	460	6.52×10^{14}	260
Green	490 to 560	520	5.77×10^{14}	230
Yellow	560 to 585	570	5.26×10^{14}	210
Orange	585 to 640	620	4.84×10^{14}	193
Red	640 to 740	680	4.41×10^{14}	176
Infrared	above 740	1 400	2.14×10^{14}	85

km s^{-1}, or about 3.00×10^8 m s^{-1} (for measurements of the speed of light, see Resnick and Halliday, 1977 and 1986; Sobel, 1987; Waldman, 1983). Light passing through a medium other than a vacuum has a speed less than c. For example, light with a wavelength of 589 nm in a vacuum is decreased in speed by 0.03% by air, 25% by water, and 40% by dense flint glass. The wavelength undergoes a decrease in magnitude equal to the decrease in the speed of propagation in these various media, because the unchanging property of a wave propagating through different media is the frequency. We also note that ν is the frequency of the oscillations of the local electric and magnetic fields of light (see Fig. 4.1).

For light, $\lambda_{vacuum}\nu$ equals υ_{vacuum} by Equation 4.1, where υ_{vacuum} is the constant c. Therefore, if we know the wavelength in a vacuum, we can calculate the frequency. In fact, the wavelengths given for light generally refer to values in a vacuum, as is the case for columns 2 and 3 in Table 4.1 (λ's in air differ only slightly from the magnitudes listed). As a specific example, let us select a wavelength in the blue region of the spectrum, e.g. 460 nm. By Equation 4.1, the frequency of this blue light is

$$\nu = \frac{(3.00 \times 10^8 \text{ m s}^{-1})}{(460 \times 10^{-9} \text{ m cycle}^{-1})} = 6.52 \times 10^{14} \text{ cycles s}^{-1}$$

Because ν does not change from medium to medium, it is often desirable to describe light by its frequency, as has been done in column 4 of Table 4.1.[1]

Energy of Light

In addition to its wave-like properties, light also exhibits particle-like properties, as in the case of the photoelectric effect mentioned above. Thus, light can act as if it were divided (or quantized) into discrete units, which we call *photons*. The light energy (E_λ) carried by a photon is

$$E_\lambda = h\nu = hc/\lambda_{vacuum} \qquad \text{photon basis} \qquad (4.2a)$$

where h is a fundamental physical quantity called *Planck's constant*. By Equation 4.2a, a photon of light has an energy directly proportional to its frequency and inversely proportional to its wavelength in a vacuum. For most applications in

---

1. For purposes of calculation, the wavelength of light is generally expressed in nm, not nm cycle^{-1}, and ν is considered to have units of s^{-1} (not cycles s^{-1}).

this book, we will describe light by its wavelength λ. To emphasize that the light energy of a photon depends on its wavelength, we have used the symbol E_λ in Equation 4.2a. A *quantum* (plural: quanta) refers to the light *energy* carried by a photon; i.e., $h\nu$ represents a quantum of electromagnetic energy. The terms "photon" and "quantum" are sometimes used interchangeably, a usage that is generally clear but not strictly correct.

The introduction of the constant h by Planck in the early 1900's represented a great departure from the accepted wave theory of light. It substantially modified the classical equations describing radiation and provided a rational basis for determining the energy of a given number of photons. Because frequency has the units of time^{-1}, Equation 4.2a indicates that the dimensions of Planck's constant are energy $\times$ time, e.g., 6.626×10^{-34} J s (see App. I). We note that hc is 1 240 eV nm (App. I), so we can readily calculate the energy per photon of blue light at 460 nm using Equation 4.2a:

$$E_\lambda = \frac{(1\ 240\ \text{eV nm})}{(460\ \text{nm})} = 2.70\ \text{eV}$$

Instead of energy per photon, we are usually more interested in the energy per Avogadro's number N (6.022×10^{23}) of photons (N photons is sometimes called an einstein, but the correct SI unit is mole). On a mole basis, Equation 4.2a becomes

$$E_\lambda = Nh\nu = Nhc/\lambda_{\text{vacuum}} \qquad \text{mole basis} \qquad (4.2b)$$

Let us consider blue light of 460 nm, which has a frequency of 6.52×10^{14} cycles s^{-1} (Table 4.1). Using Equation 4.2b, we calculate that the energy per mole of 460-nm photons is

$$E_\lambda = (6.022 \times 10^{23}\ \text{mol}^{-1})(6.626 \times 10^{-34}\ \text{J s})(6.52 \times 10^{14}\ \text{s}^{-1})$$
$$= 260\ \text{kJ mol}^{-1}$$

Alternatively, we can calculate the energy by dividing Nhc (119 600 kJ mol^{-1} nm) by the wavelength in a vacuum (see Table 4.1).

Quanta of visible light represent relatively large amounts of energy. The hydrolysis of ATP, the main currency for chemical energy in biology, generally yields about 40 to 50 kJ mol^{-1} under physiological conditions (see Ch. 6), whereas, as we just calculated, blue light has about five to six times as much energy per mole of photons. Quanta of ultraviolet light represent even higher energies than do those of visible light; e.g., 254-nm photons have 471 kJ mol^{-1} of radiant energy (Table 4.1). This is greater than the carbon-carbon bond energy of 348 kJ mol^{-1} or the oxygen-hydrogen bond energy of 463 kJ mol^{-1}. The high quantum energy of UV radiation underlies its mutagenic and bactericidal action, because it is energetic enough to cause disruption of certain covalent bonds.

The photoelectric effect, for which light leads to the ejection of electrons from the surface of a metal, clearly illustrates the distinction between light energy and the energy of its photons. Millikan found that photons with at least 175 kJ mol^{-1}, representing wavelengths of 683 nm or less, can lead to a photoelectric effect for sodium (see Resnick and Halliday, 1977 and 1986).[2] For wavelengths beyond 683 nm, however, no matter how much light energy is absorbed, electrons are not ejected from the surface. Hence, a photon of a specific minimum energy can be necessary for a certain reaction. Measurement of the total light energy does not indicate how many photons are involved or what their individual energies are, unless the wavelength distribution is known.

Absorption of radiation by an atom or a molecule leads to a more energetic state of the absorbing species. Such energetic states can also be produced by collisions resulting from the random thermal motion of the molecules. The higher is the temperature, the greater is the average kinetic energy of the atoms and molecules, and the greater is the probability of achieving a relatively energetic state by collision. The number of molecules having a particular kinetic energy can often be approximated by the Boltzmann energy distribution. By Equation 3.21b [$n(E) = n_{\text{total}}e^{-E/RT}$], the fraction of atoms or molecules having a kinetic energy of molecular Brownian motion of E or greater at thermal equilibrium equals the Boltzmann factor $e^{-E/RT}$, where we have expressed energy on a mole basis (see Ch. 3, footnote 13). Based on the kinetic theory of gases, the *average* kinetic energy of translational motion for molecules in a gas phase is $(3/2)RT$, which equals 3.72 kJ mol^{-1} at 25°C. What fraction of the molecules exceeds this average energy? The Boltzmann factor becomes $e^{-(3RT/2)/(RT)}$, or $e^{-1.5}$, which is 0.22, so 22% of the molecules have kinetic energies that are higher than the average. At what temperature would 44% have such kinetic energies? The Boltzmann factor then equals 0.44, so T would be

$$T = \frac{1}{-\ln(0.44)} \frac{E}{R} = \frac{1}{(0.82)} \frac{(3\,720 \text{ J mol}^{-1})}{(8.3143 \text{ J mol}^{-1} \text{ K}^{-1})}$$

$$= 546 \text{ K} \qquad (273°C)$$

---

2. For his experimental work on the photoelectric effect, Millikan was awarded the Nobel Prize in physics in 1923. Others whose contributions to the understanding of radiation led to Nobel prizes in physics include Wien in 1911 for heat radiation laws, Planck in 1918 for the quantum concept, Stark in 1919 for spectral properties, Einstein in 1921 in part for discovering the photoelectric effect, Bohr in 1922 for atomic radiation, Franck and Hertz in 1925 for atom-electron interactions, and Pauli in 1945 for atomic properties.

Raising the temperature clearly increases the fraction of molecules with higher energies. However, temperature can be raised to only a limited extent under biological conditions without causing cellular damage.

Let us next consider a kinetic energy E of 260 kJ mol^{-1}, as is possessed by blue light of 460 nm. The fraction of molecules possessing at least this energy is $e^{-E/RT}$, which equals $e^{-(260 \text{ kJ mol}^{-1})/(2.48 \text{ kJ mol}^{-1})}$, or only 3×10^{-46} at 25°C! For comparison, the total number of atoms in the entire biomass of all living organisms is about 3×10^{41}. The chance that a particular molecule can gain the equivalent of 260 kJ mol^{-1} by means of thermal collisions is therefore vanishingly small. Hence, the absorption of blue light can lead to energetic states that otherwise would simply not occur at temperatures encountered by plants and animals. Absorption of the relatively high quantum energy of light thus promotes the attainment of very improbable energetic states, a key point in the understanding of photobiology.

Illumination, Photon Flux Density, and Irradiance

For many purposes in studying plants, we need to know the total amount of incident light. There are three common classes of instruments for measuring such fluxes: (1) photometers, which measure the available illuminating power, a quantity related to the wavelength sensitivity of the human eye; (2) quantum meters, which measure the number of photons; and (3) radiometers, which measure the total energy of the radiation (see Pearcy, 1989).

By definition, photometers do not appreciably respond to radiation in the infrared or the ultraviolet. They are "light" meters in the sense that they mimic human vision; i.e., they respond to photons in the visible region, in the same way as the light meter on a camera. A candle is a unit of luminous intensity, originally based on a standard candle or lamp. The present international unit is called a *candela* (sometimes still referred to as a "candle"), which was previously defined as the total light intensity of 1.67 mm^2 of a *blackbody* radiator (one that radiates maximally) at the melting temperature of pure platinum (2 042 K). In 1979 the candela was redefined as the luminous intensity of a monochromatic source with a frequency of 5.40×10^{14} cycles s^{-1} (λ of 555 nm) emitting 0.01840 J s^{-1} or 0.01840 W (1/683 W steradian^{-1}, where W is the abbreviation for watt and steradian is the unit for solid angle[3]). The total light emitted in all

---

3. A solid angle is the three-dimensional counterpart of the more familiar planar angle. It can be interpreted as a region of space defined by a series of lines radiating from a point, the lines forming a smooth surface (such as a conical surface viewed from its apex). The region delimits a certain surface area A on a sphere of radius r; the solid angle in steradians then equals A/r^2. For instance, a hemisphere viewed from its center subtends $2\pi \, r^2/r^2$ or 2π steradians.

directions by a source of 1 candela is 4π lumens; 1 lumen m^{-2} is the photometric illuminance unit, *lux*. Hence, 1 lux is the luminous flux density (illuminance) 1 m from a spherically uniform source of 1 candela, which is pretty dim (similar to the lighting in a movie theatre during the picture); indeed, the minimum illumination recommended for reading in libraries and most public conveyances is 300 lux (Kaufman and Haynes, 1981). Measurement in lux is adequate for certain purposes where human vision is involved, but it is not appropriate for studies with plants (see McCree, 1981).

For many purposes in plant studies, it is important to know the photon flux density. For instance, the rate of photosynthesis depends on the rate of absorption of photons (not on the rate of absorption of energy). Some instruments are sensitive only to photons that might be useful photosynthetically (e.g., wavelengths from 400 to 700 nm), the so-called *photosynthetically active radiation* (PAR) or the *photosynthetic photon flux density* (PPFD). Photon flux density is expressed in μmol m^{-2} s^{-1}.

Radiometers—e.g., thermocouples, thermopiles, or thermistors that have been treated so as to absorb all wavelengths—respond to radiant energy and thus are sensitive to irradiation in the ultraviolet and infrared as well as in the visible. Readings are expressed in energy per unit area and time, e.g., J m^{-2} s^{-1}, which is W m^{-2} (conversion factors for radiometric units are given in App. II). If the irradiance or radiant energy flux density at a specified wavelength is measured in radiometric units, the value can be converted to a photon flux density by using the energy carried by individual photons (see Eq. 4.2).[4] In general, radiometric units, photometric units, and photon flux densities cannot be unambiguously interconverted (unless the frequency distribution is known), although empirical relations exist among them for certain light sources (see Thimijan and Heins, 1983).

In choosing a lamp for a controlled-environment chamber used for growing plants, both the total energy emitted and the wavelength distribution of the

---

4. Various terms are used to describe electromagnetic radiation. *Emittance* is the flux density at the emitting surface, e.g., radiant emittance (expressed in J m^{-2} s^{-1} = W m^{-2}). *Illuminance* is the luminous flux density or "illumination" at some surface (lux). *Intensity* refers to a property of a source; e.g., light intensity designates the rate of light emission for a photometric source (lumens per unit solid angle, or candelas), and radiant intensity is the power emitted per unit solid angle (W steradian^{-1}). *Irradiance* (commonly termed *irradiation*) is the radiant energy flux density received on some surface (W m^{-2}). *Radiance* is the rate of radiant energy emmission (power emitted) per unit solid angle per unit area (W steradian^{-1} m^{-2}). *Radiant flux* is the radiant energy emitted or received per unit time (W). *Radiant flux density* is the radiant flux per unit area (W m^{-2}). The term *fluence* is sometimes used for amount per area, so radiant energy per unit area is the energy fluence (J m^{-2}). *Fluence rate* is therefore the fluence per unit time; e.g., the photon fluence rate (mol m^{-2} s^{-1}) is the same as photon flux density. (See Bell and Rose, 1981; Gates, 1980; McCree, 1981; Monteith, 1973.)

Figure 4.2

Wavelength distributions of the sun's photons incident on the earth's atmosphere and its surface. The curve for the solar irradiation on the atmosphere is an idealized one based on Planck's radiation distribution formula. The spectral distribution and the amount of solar irradiation reaching the earth's surface depend on clouds, other atmospheric conditions, altitude, and the sun's angle in the sky. The pattern indicated by the lower curve is appropriate at sea level on a clear day with the sun overhead (for further details see Bickford and Dunn, 1972; Gates, 1980; Monteith, 1973; and Seliger and McElroy, 1965).

photons should be considered. Tungsten lamps are comparatively poor sources of visible radiation, because about 90% of their radiant energy is emitted in the infrared—the actual amount depends on the operating temperature of the filaments—creating appreciable cooling problems when large numbers of such lamps are used. Typical fluorescent lamps emit only about 10% of their energy in the IR, so cooling problems are less severe than with tungsten lamps. On the other hand, the wavelength distribution of the sun's photons in the visible region (Fig. 4.2) is matched far better by tungsten lamps than by fluorescent ones. A close match can be important, because the wavelength distribution influences the relative amounts of the two forms of phytochrome discussed at the end of this chapter (see Bickford and Dunn, 1972; Haynes, 1981; Kaufman and Hart, 1988; Seliger and McElroy, 1965; Tibbitts and Kozlowski, 1979).

Sunlight

Essentially all energy for life originates in the form of electromagnetic radiation from the sun. In radiometric units the radiant flux density of solar irradiation (irradiance) perpendicularly incident on the earth's atmosphere—the "solar constant"—is about 1,368 W m^{-2} (see Willson and Hudson, 1988).[5] (In Ch. 6 we will consider the solar constant in terms of the annual photosynthetic yield, and in Ch. 7 in terms of the energy balance of a leaf.) Based on the solar constant and averaged for all latitudes, the mean total daily amount of radiant energy from the sun incident on a horizontal surface just outside the earth's atmosphere is 29.6 MJ m^{-2} day^{-1}, of which atmospheric conditions such as clouds permit an average of only 58% or 17.0 MJ m^{-2} day^{-1} to reach the earth's surface (at mid latitudes the amount at the earth's surface on a cloudless day in the summer can be about 30 MJ m^{-2} day^{-1}). The instantaneous irradiance in the visible region at noon on a cloudless day with the sun overhead can be 420 W m^{-2} (total irradiance of about 850 W m^{-2}; equivalent to about 100 000 lux). If we represent sunlight by yellow light of 570 nm, which by Table 4.1 carries 210 kJ mol^{-1}, then 420 W m^{-2} in the visible region from the sun is a photon flux density of about (420 J m^{-2} s^{-1})/(210 000 J mol^{-1}), or 2.0×10^{-3} mol m^{-2} s^{-1} (2 000 μmol m^{-2} s^{-1}) at the earth's surface.

Figure 4.2 shows the relative number of the sun's photons impinging on the earth's atmosphere and reaching its surface as a function of wavelength. About 5% of the photons incident on the earth's atmosphere are in the ultraviolet (below 400 nm), 28% in the visible, and 67% in the infrared (beyond 740 nm). Most of the ultraviolet fraction of sunlight incident on the atmosphere is prevented from reaching the surface of the earth by ozone (O_3) present in the stratosphere, 20 to 30 km above the earth's surface (see Gates, 1980).[6] Ozone absorbs some visible radiation (e.g., near 600 nm) and effectively screens out the shorter ultraviolet rays by absorbing strongly below 300 nm. Much of the infrared from the sun is absorbed by atmospheric water vapor and CO_2 (see Fig. 4.2). Water absorbs strongly near 900 nm and 1 100 nm, and above 1 200 nm, having a major infrared absorption band at 1 400 nm (1.4 μm). Although the amount of water vapor in the air varies with latitude, longitude, and season, the mean vapor concentration is equivalent to a path of liquid water approximately 20 mm thick (see Gates, 1980; Seliger and McElroy, 1965). The substantial absorption of UV and IR by atmospheric gases causes the solar irradiation at the earth's surface to have a larger fraction in the visible region than that incident on the outer

---

5. The solar constant varies from the average by up to $\pm 3.5\%$ because the earth's orbit is elliptical. The value given is for the mean distance between the earth and the sun (the earth is closest to the sun on 3 January, at 1.471×10^8 km, and furthest away on 4 July, at 1.521×10^8 km). There are additional variations in solar irradiation based on changes in solar activity, as occur for sun spots, which lead to the 11-year solar cycle.

6. The absorption of sunlight by this O_3 leads to a pronounced heating of the upper atmosphere.

atmosphere. In the example in Figure 4.2, about 2% of the photons at the earth's surface are in the ultraviolet, 45% in the visible, and 53% in the infrared.

The radiation environment in water is quite different from that on land. For instance, absorption by water causes most of the IR in sunlight to be removed after penetrating less than 1 m in lakes or oceans. This, coupled with greater scattering at shorter wavelengths, causes a greater fraction of the photons to be in the visible region at greater depths. But water attenuates the visible region too, so even at the wavelengths for greatest penetration (approximately 500 nm) in the clearest oceans, only about 1% of the solar photon flux density incident on the water surface penetrates to 200 m (see Jerlov, 1976). Light is considerably attenuated with depth in typical freshwater lakes and reservoirs, where substances in the water reduce the flux density of the wavelengths penetrating most readily (near 550 nm) to 1% of the surface values generally by a depth of 10 m (see Kirk, 1983).

The wavelength distribution of photons reaching the earth's surface profoundly influences life. For example, the substantial absorption of UV by ozone reduces the potential hazard of mutagenic effects caused by this short-wavelength irradiation. On the other hand, prior to the advent of much ozone in the upper atmosphere, UV irradiation from the sun would have been a potent factor affecting genetic processes. Even now, exposure to the UV in sunlight can inhibit photosynthesis and decrease leaf expansion (for a discussion of plant responses to UV radiation, see Caldwell, 1981). As another example, the peak near 680 nm for photons reaching the earth's surface (Fig. 4.2) coincides with the red absorption band of chlorophyll. Vision also utilizes the wavelength region where most of the sunlight reaches the earth. Thus, selective pressure must have favored the evolution of photochemical systems able to use the most abundant wavelengths while avoiding the too energetic UV and the too weak IR.

Planck's and Wien's Formulae

The shape of the curve depicting the wavelength distribution of photons incident upon the earth's atmosphere can be closely predicted using *Planck's radiation distribution formula*. This expression indicates that the relative *photon* flux density per unit wavelength interval is proportional to $\lambda^{-4}/(e^{hc/\lambda kT} - 1)$, where T is the temperature of the radiation source. Such a formula applies exactly to a perfectly efficient emitter, a so-called *blackbody*. A blackbody is a convenient idealization describing an object that absorbs all wavelengths—it is uniformly "black" at all wavelengths—and emits in accordance with Planck's radiation distribution formula. This formula is a good approximation for describing radiation from the sun—T is the surface temperature (about 5 800 K)—so it was used to obtain the upper curve in Figure 4.2. Also, the radiation from a tungsten lamp of a few hundred watts can be fairly well described by Planck's radiation distribution formula using a T of 2 900 K; i.e., the curve for the relative photon flux density from a tungsten lamp has the same shape as the solid line in Figure 4.2, although it is shifted toward longer wavelengths, because the temperature of a tungsten filament is less than that of the sun's surface (see Bickford and Dunn, 1972; Kaufman and Haynes, 1981; Seliger and McElroy, 1965). Planck's radiation distribution formula indicates that any object with a temperature greater than 0 K will emit electromagnetic radiation.

If we know the temperature of a blackbody, we can predict at what wavelength the radiation from it will be maximal. To derive such an expression, we differentiate Planck's radiation distribution formula with respect to wavelength and set the derivative equal to zero. The relation thus obtained is known as *Wien's displacement law*:

$$\lambda_{max}T = 3.6 \times 10^6 \text{ nm K} \qquad \text{photon basis} \qquad (4.3a)$$

where λ_{max} is the wavelength position for maximum *photon* flux density and T is the surface temperature of the source. Because the surface of the sun is about 5 800 K, Wien's displacement law (Eq. 4.3a) predicts that

$$\lambda_{max} = \frac{(3.6 \times 10^6 \text{ nm K})}{(5 800 \text{ K})} = 620 \text{ nm}$$

as the upper line in Figure 4.2 indicates. For a tungsten lamp operating at a temperature of 2 900 K (half the temperature of the sun's surface), the position for maximum photon flux density shifts to 1 240 nm (a doubling of λ_{max} compared with the sun), consistent with tungsten lamps emitting primarily infrared radiation. From Wien's displacement law, the λ_{max} for a maximum photon flux density from a body at 298 K occurs at 1.2×10^4 nm, which is 12 μm, i.e., far into the IR. Plants emit such infrared radiation, which is a crucial aspect of their overall energy balance, as we will consider in Chapter 7.

When considering the number of photons available for absorption by pigment molecules, as is relevant for discussing photosynthesis, we generally use the spectral distribution of *photons* per unit wavelength interval (see Fig. 4.2). On the other hand, for applications such as describing the energy gain by leaves exposed to sunlight, we are usually more interested in the spectral distribution of *energy* per unit wavelength interval. To recast Figure 4.2 on an energy basis, we need to divide each point on the curves by its wavelength—the ordinate then becomes "Relative energy flux density per unit wavelength interval." Planck's radiation distribution formula indicates that the relative *energy* flux density per unit wavelength interval is proportional to $\lambda^{-5}(e^{hc/\lambda kT} - 1)$. Wien's displacement law for maximum *energy* output is

$$\lambda_{max}T = 2.9 \times 10^6 \text{ nm K} \qquad \text{energy basis} \qquad (4.3b)$$

(For further details see Ditchburn, 1976; Gates, 1980; Resnick and Halliday, 1977 and 1986.)

ABSORPTION OF LIGHT BY MOLECULES

In our discussion of light utilization we will first consider the absorption process itself. Only light that is actually absorbed can produce a chemical change, a principle embodied in the Grotthuss–Draper law of photochemistry. This is true whether radiant energy is to be converted to some other form and then stored or is to be used as a trigger. Another important principle of photochemistry is the Stark–Einstein law, which specifies that each absorbed photon activates only one molecule. Einstein further postulated that all the energy of the light quantum is transferred to a single electron during the absorption event, resulting in the movement of this electron to a higher energy state. To help understand light absorption, we should first consider some of the properties of electrons. We will discuss the fate of the excited electron in the next section. [See Alberty (1983), Calvert and Pitts (1966), Cowan and Drisko (1976), Depuy and Chapman (1972), Turro (1978), and Wayne (1988) for a more thorough treatment of the topics in this section.]

Role of Electrons in Absorption Event

From a classical viewpoint, an electron is a charged particle that can move in some orbit around an atomic nucleus. Its energy depends both on the location of the orbit in space and on how fast the electron moves in its orbit. The increase in energy of an electron upon absorbing a photon could be used either to transfer that electron into an orbit that lies at a higher energy than the original orbit or to cause the electron to move more rapidly about the nucleus than it did before the light arrived. The locations of various possible electron orbits and the speeds for electrons moving in them are both limited to certain discrete, or "allowed," values, a phenomenon that has been interpreted by quantum mechanics. Hence,

Table 16.1. A comparison of photometric and radiant energy measurements (adapted from Thimijan and Heins, 1983)

Photometric		Radiant Energy (400-700 nm)			
Foot-candles	Lux	$\mu mol\ s^{-1}m^{-2}$		WM^{-2}	
		Daylight	HPS	Daylight	HPS
300	3,240	60	39.5	13.1	7.9
400	4,320	80	52.7	17.5	10.6
500	5,400	100	65.9	21.9	13.2
600	6,480	120	79.0	26.3	15.8
700	7,560	140	92.2	30.6	18.5
800	8,640	160	105.4	35.0	21.2
900	9,720	180	118.5	39.4	23.8
1000	10,800	200	132.7	43.8	26.6

Conversion Factors

1. Foot-candle to lux — multiple foot-candle readings by 10.8.

2. $\mu mol\ s^{-1}m^{-2}$ ($\mu E\ s^{-1}m^{-2}$) to Wm^{-2} PAR (400-700 nm) divide $\mu mol\ s^{-1}m^{-2}$ reading by the following factor for each lamp or light source.

sun and sky, daylight	4.57
high pressure sodium	4.98
metal halide	4.59
cool and white fluorescent	4.59

3. Lux to $\mu mol\ s^{-1}m^{-2}$ PAR (400-700 nm) divide lux reading by the following factor for each lamp or light source.

sun and sky, daylight	54
high pressure sodium	82
metal halide	71
cool white fluorescent	74

Supplementary Radiation With HPS Lamps

The lamp source for commercial rose lighting has focused on high pressure sodium vapor lamps because of their efficiency in converting electrical power to radiant energy. Thirty six percent of the power input from HPS lamps is converted to radiant energy within the range of 400-850 nanometers, whereas the conversion efficiency for mercury vapor and metal halide lamps is 13 and 30%, respectively (Cathey and Campbell, 1980). Furthermore, the maintenance cost is lower as the burning life of HPS lamps is 50% longer than mercury vapor or fluorescent lamps.

the energy of an electron in an atom or molecule can change only by certain specific amounts. Light of the appropriate wavelength will have the proper energy to cause the electron to move from one possible energetic state to another. For light absorption to occur, therefore, the energy of a photon as given by Equation 4.2 must equal the difference in energy between some allowed excited state of the atom or molecule and the initial state, the latter usually being the ground (lowest energy) state.

During light absorption an interaction occurs between the electromagnetic field of the light and some electron. Because electrons are charged particles, they experience a force when they are in an electric field. The oscillating electric field of light (Fig. 4.1) thus represents a periodic driving force acting on electrons. This electric field—a vector having a certain direction in space, such as along the vertical axis in Figure 4.1—causes or induces the electrons to move. If the frequency of the electromagnetic radiation causes a large sympathetic oscillation or beating of some electron, that electron is said to be in *resonance* with the light wave. Such a resonating electron corresponds to an *electric dipole* (local separation of positive and negative charge in the molecule), as the electron is forced to move first in a certain direction and then in the opposite one in response to the oscillating electric field of light. The displacement of the electron back and forth requires energy—in fact, it may take the entire energy of the photon, in which case the quantum is captured or absorbed. The direction and the magnitude of the induced electric dipole will depend on the resisting, or restoring, forces on the electron provided by the rest of the molecule. These restoring forces depend on the other electrons and the atomic nuclei in the molecule; consequently, they are not the same in different types of molecules. Therefore, the electric dipoles that can be induced in a particular molecule are characteristic of that molecule, which helps explain why each molecular species has its own unique absorption spectrum.

The probability that light will be absorbed depends on both its wavelength and the relative orientation of its electromagnetic field with respect to the possible induced oscillations of the electrons in the molecules. Absorption of a photon without ejection of an electron from the absorbing species can take place only if the following two conditions exist: (1) the photon has the proper energy to get to a discrete excited state of the molecule, i.e., has a specific wavelength (see Eq. 4.2); and (2) the electric field vector associated with the light (Fig. 4.1) has a component parallel to the direction of some potential electric dipole in the molecule so that an electron can be induced to oscillate. In other words, the electric field of light must be able to exert a force on some electron. It must exert this force in the direction of a potential electric dipole so that the electron can be induced to move and thus be able to accept the energy of the photon.

The probability for absorption is proportional to the square of the cosine of the angle between the electric field vector of light and the direction of the induced electric dipole in the molecule. Within these limits set by the wavelength and the orientation, light energy can be captured by the molecule, placing it in an excited state.

Electron Spin and State Multiplicity

Light absorption is affected by the arrangement of electrons in an atom or a molecule, which depends among other things on a property of the individual electrons known as their *spin*. We can view each electron as a charged particle spinning about an axis in much the same way as the earth spins about its axis. Such rotation has an angular momentum, or spin, associated with it. The magnitude of the spin of all electrons is the same; but because spin is a vector quantity, it can have different directions in space. For an electron, only two orientations are found—the spin of the electron is aligned either parallel or antiparallel to the local magnetic field, i.e., either in the same direction or in the opposite direction as the magnetic field. Even in the absence of an externally applied magnetic field, such as that of the earth or some electromagnet, a local internal magnetic field is provided by both the moving charges in the nucleus and the motion of the electrons. A magnetic field therefore always exists with which the electron spin can be aligned.

Angular momentum and hence spin have units of energy $\times$ time. It proves convenient to express the spin of electrons in units of $h/(2\pi)$, where h is Planck's constant, which itself has units of energy $\times$ time, e.g., J s. In units of $h/(2\pi)$, the projection along the magnetic field of the spin for a single electron is either $+\frac{1}{2}$ (e.g., when it is parallel to the local magnetic field), or $-\frac{1}{2}$ (when the spin is in the opposite direction, or antiparallel). The net spin of an atom or molecule is the vector sum of the spins of all the electrons, each individual electron having a spin of either $+\frac{1}{2}$ or $-\frac{1}{2}$. The magnitude of this net spin is given the symbol S, an extremely important quantity in spectroscopy.

To discuss the spectroscopic properties of various molecules, we introduce the *spin multiplicity*. The spin multiplicity of an electronic state is defined as $2S + 1$, where S is the magnitude of the net spin for the whole atom or molecule. For example, if S equals zero—so the spin projections of all the electrons taken along the magnetic field cancel each other—then $2S + 1$ is one, and the state is called a *singlet*. On the other hand, if S equals one, the state is a *triplet* ($2S + 1$ then equals 3). Singlets and triplets are the two most important spin multiplicities encountered in biological systems. When referring to an absorbing species, the spin multiplicity is generally indicated by S for singlet and T for triplet, as we will do here. When S equals $\frac{1}{2}$, as could occur if there were an odd number of electrons in a molecule, $2S + 1$ is 2; such *doublets* occur for free radicals (molecules with a single unpaired electron—such molecules consequently are generally quite reactive).

Electrons are found only in certain "allowed" regions of space; the particular locus in which some electron can move is referred to as its *orbital*. In the 1920's Pauli noted that, when an electron is in a given atomic orbital, a second electron having its spin in the same direction is excluded from that orbital. This led to the enunciation of the *Pauli exclusion principle* of quantum mechanics: when two electrons are in the same orbital, their spins must be in opposite directions. When a molecule has all of its electrons paired in orbitals with their spins in opposite directions, the total spin of the molecule is zero ($S = 0$), and the molecule is in a singlet state ($2S + 1 = 1$; see Fig. 4.3a). The ground, or unexcited, state of most molecules is such a singlet; i.e., all the electrons are in pairs in the lowest energy orbitals. When some electron is excited to an unoccupied orbital, two spin possibilities exist. The spins of the two electrons (which are now in different orbitals) may be in opposite directions, as they were when paired in the ground state (Fig. 4.3b); this electronic configuration is still

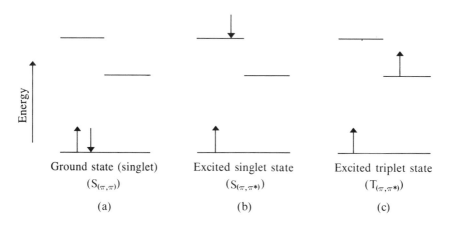

Ground state (singlet)   Excited singlet state   Excited triplet state
$(S_{(\pi,\pi)})$ $(S_{(\pi,\pi*)})$ $(T_{(\pi,\pi*)})$

(a)   (b)   (c)

Figure 4.3

Effect of light on a pair of electrons in a molecular orbital. The arrows indicate the directions of the electron spins with respect to the local magnetic field. (a) In the ground state, the two electrons in a filled orbital have their spins in opposite directions. (b) The absorption of a photon can cause the molecule to go to an excited *singlet* state where the spins of the electrons are still in opposite directions. (c) In an excited *triplet* state, the spins of the two electrons are in the same direction.

a singlet state. Or the two electrons may have their spins in the same direction—
a triplet state (Fig. 4.3c). (Because the electrons are in different orbitals, their
spins can be in the same direction without violating the Pauli principle.) An
important rule—first enunciated by Hund based on empirical observations and
later explained using quantum mechanics—is that the level with the greatest spin
multiplicity has the lowest energy. Thus, an excited triplet state is lower in
energy than is the corresponding excited singlet state, as is illustrated in Figure
4.3c.

Molecular Orbitals

For a discussion of the light absorption event involving the interaction of an
electromagnetic wave with some electron, it is easiest to visualize the electron
as a small point located at some specific position in the atom or molecule. For
a classical description of the energy of an electron, we can imagine the electron
moving in some fixed trajectory or orbit about the nucleus. However, to describe
the role of electrons in binding atoms together to form a molecule, it is most
convenient to imagine the electron as spatially distributed like a cloud of negative
charge surrounding the nuclei of adjacent atoms in the molecule. In this last
description, involving probability considerations introduced by quantum mechanics, we say that the electrons are located in *molecular orbitals*. Such considerations lead us into a different way of looking at things. Our common
experience yields relations such as Newton's laws of motion, which describe
events on a scale much larger than atomic or molecular. In contrast, the main
application of quantum mechanics involves molecular, atomic, and subatomic
dimensions. At this scale our intuition often fails us. Moreover, many things
that otherwise seem to be absolute, such as the position of an object, turn out
to be describable only on a relative basis or as a *probability*. We cannot say that
an electron is located at such and such a place. Instead, we must be satisfied
with knowing only the probability of finding an electron in some region about
a nucleus.

A logical way to begin our discussion of molecular orbitals is to consider the
probabilities of finding electrons in given regions about a single atomic nucleus.
The simplest atom is that of hydrogen, which has a single electron. If we were
to determine the probability of finding that electron in the various regions of
space about its nucleus, we would find that it spends most of its time fairly close
to the nucleus (within about 0.1 nm), and that the probability distribution is
spherically symmetrical in space. In other words, the chance of finding the
electron is the same in all directions about the nucleus. In atomic theory, this

spherically symmetrical distribution of the electron about an H nucleus is called an s orbital, and the electron is referred to as an s electron.

The next simplest atom is that of helium; He has two s electrons moving in the s orbital about its nucleus. By the Pauli exclusion principle these two electrons must have their spins in opposite directions. What happens for lithium, which contains three electrons? Two of its electrons are in the same type of orbital as that of He. This orbital is known as the K shell and represents the spherically symmetrical orbital closest to the nucleus. The third electron is excluded from the K shell but occurs in another orbital with spherical symmetry whose probability description indicates that on the average its electrons are farther away from the nucleus than is the case for the K shell. Electrons in this new orbital are still s electrons, but they are in the L shell. Beryllium, which has an atomic number of four, can have two s electrons in its K shell and two more in its L shell. What happens when we have five electrons moving about a nucleus, as for an uncharged boron atom? The fifth electron is in an orbital that is not spherically symmetrical about the nucleus. Instead, the probability distribution has the shape of a dumbbell, although it is still centered about the nucleus (similar in appearance to the distributions depicted in Fig. 4.4b). This new orbital is referred to as a p orbital and electrons in it are called p electrons. The probability of finding a p electron in various regions of space is greatest along some axis through the nucleus and vanishingly small at the nucleus itself.

How do all these various atomic orbitals relate to the spatial distribution of electrons in molecules? A molecule contains more than one atom (except for "molecules" like helium or neon), and certain of the electrons can move between the atoms—this interatomic motion is crucial for holding the molecule together. Fortunately, the spatial localization of electrons in molecules can be described using suitable linear combinations of the spatial distributions represented by various atomic orbitals centered about the nuclei involved. In fact, molecular orbital theory is concerned with giving the correct quantum-mechanical or wave-mechanical description of the probability of finding electrons in various regions of space in *molecules* by using the already established descriptions for the probability of finding electrons in *atomic* orbitals.

Some of the electrons in molecules are localized about a single nucleus, and others are *delocalized,* or shared between nuclei. For the delocalized electrons, the combination of the various atomic orbitals used to describe the spatial positions of the electrons in three dimensions is consistent with a sharing of the electrons between adjacent nuclei. In other words, the molecular orbitals of the delocalized electrons spatially overlap more than one nucleus. This sharing of electrons is responsible for the chemical bonds that prevent the molecule from separating into its constituent atoms; i.e., the negative electrons moving between the positive nuclei hold the molecule together by attracting the nuclei of different atoms. Moreover, these delocalized electrons are the ones usually involved in light absorption by molecules.

The lowest energy molecular orbital is a σ orbital, which is analogous in symmetry properties to an s atomic orbital.[7] Nonbonding, or lone-pair, electrons contributed by atoms like oxygen or nitrogen occur in n orbitals and retain their atomic character in the molecule. These n electrons are essentially physically separate from the other electrons in the molecule and do not take part in the bonding between nuclei. We shall devote most of our attention to electrons in π molecular orbitals, which are the molecular equivalent of p electrons in atoms. These π electrons are delocalized in a bond joining two or more atoms. They are of prime importance in light absorption. In fact, most photochemical reactions and spectroscopic properties of biological importance result from the absorption of photons by the π electrons.[8]

The excitation of a π electron by light absorption can lead to an excited state of the molecule in which the electron moves into a π^* orbital, the asterisk referring to an excited, or high-energy, molecular orbital. The probability distributions for electrons in both π and π^* orbitals are illustrated in Figure 4.4 (the circumscribed regions indicate where the electrons are most likely to be found). Figure 4.4a shows a π orbital delocalized between two nuclei; the same clouds of negative charge electrostatically attract both nuclei. Such sharing of electrons in the π orbital helps join the atoms together, so we refer to this type of molecular orbital as *bonding*. As shown in Figure 4.4b, electrons in a π^* orbital do not help join atoms together; rather, they tend to decrease bonding between atoms in the molecule, because the clouds of negative charge in adjacent atoms are located in a position to repel each other. A π^* orbital is therefore referred to as *antibonding*. The decrease in bonding in going from a π to a π^* orbital results in a less stable (higher energy) electronic state for the molecule, so a π^* orbital is at a higher energy than a π orbital.

In photochemistry, the energy required to move an electron from the attractive (bonding) π orbital to the antibonding π^* orbital is obtained by the absorption of a photon of the appropriate wavelength. For molecules such as chlorophylls and carotenoids, which we will consider later, the π^* orbitals are often only a few hundred kJ mol^{-1} higher in energy than are the corresponding π orbitals.

---

7. The spatial distribution of the electron cloud for a σ orbital is actually cylindrically symmetrical about the internuclear axis for the pair of atoms involved in the σ bond. It can be constructed by linear combinations of s atomic orbitals.

8. With respect to orbital nomenclature, π and σ are the Greek transliterations for p and s, respectively, and n derives from nonbonding.

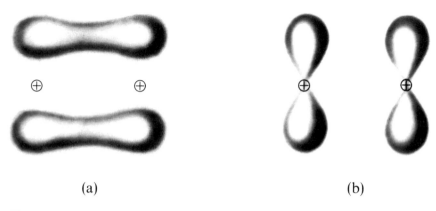

(a)   (b)

Figure 4.4

Typical π and π^* orbitals, indicating the spatial distribution about the nuclei ($\oplus$) where the greatest probability of finding the electrons occurs: (a) π orbital (bonding), and (b) π^* orbital (antibonding).

In such molecules, the absorption of visible light corresponding to photons with energies from 160 to 300 kJ mol^{-1} (Table 4.1) can lead to the excitation of π electrons into the π^* orbitals.

Photoisomerization

Light energy can cause molecular changes known as *photoisomerizations*. The three main types of photoisomerization are (1) *cis-trans* isomerization about a double bond; (2) a double bond shift; and (3) a molecular rearrangement involving changes in carbon-carbon bonds, e.g., ring cleavage or formation. *Cis-trans* photoisomerization—which illustrates the consequences of the different spatial distribution of electrons in π and π^* orbitals—makes possible the generally restricted rotation about a double bond.[9] A double bond has two π electrons; light absorption leads to the excitation of one of them to a π^* orbital. The attraction between the two carbon atoms caused by the remaining π electron is mostly canceled by the antibonding, or repulsive, contribution from this π^* electron (see Fig. 4.4). Hence, the molecular orbitals of the original two π

---

9. In the *trans* configuration, the two large groups are on opposite sides of the double bond between them $\left(\begin{smallmatrix} R \\ \diagdown \\ \diagup \end{smallmatrix} C{=}C \begin{smallmatrix} \diagup \\ \diagdown \\ R' \end{smallmatrix} \right)$, and in the *cis* form they are on the same side.

electrons that prevented rotation about the double bond have been replaced by an electronic configuration permitting relatively easy rotation about the carbon-carbon axis. Let us consider the absorption of light by a molecule in the *cis* form. When the excited π^* electron drops back to a π orbital, the two large groups can be on the same side (the original *cis* isomer) or on opposite sides (the *trans* isomer) of the double bond. Because the two isomers can have markedly different properties, such use of light to trigger their interconversion can have profound importance in biology. For example, light can cause the photo-isomerization of a *cis* isomer of retinal,[10] yielding a *trans* isomer. This is the basic photochemical event underlying vision (see Cowan and Drisko, 1976).

Light Absorption by Chlorophyll

We will use chlorophyll to help illustrate some of the terms introduced above describing the absorption of light by molecules. (Chlorophyll will be considered in more detail in Ch. 5.) The principal energy levels and electronic transitions of chlorophyll are presented in Figure 4.5. Chlorophyll is a singlet in the ground state, as are nearly all pigments of importance in biology. When a photon is absorbed, some π electron is excited to a π^* orbital. If this excited state is a singlet, it is represented by $S_{(\pi, \pi^*)}$ (Figs. 4.3 and 4.5). The first symbol in the subscript is the type of electron (here a π electron) that has been excited to the antibonding orbital indicated by the second symbol (here π^*). We will represent the ground state by $S_{(\pi, \pi)}$, indicating that no electrons are then in excited, or antibonding, orbitals. If the orientation of the spin of the excited π electron becomes reversed during excitation, it would be in the same direction as the spin of the electron that remained in the π orbital (see Fig. 4.3c). (Each filled orbital contains two electrons whose spins are in opposite directions.) In this case, the net spin of the molecule in the excited state would be 1, i.e., a triplet. The excited triplet state of chlorophyll is represented by $T_{(\pi, \pi^*)}$ in Figure 4.5.

Chlorophyll has two principal excited singlet states that differ considerably in energy. One of these states, designated $S^a_{(\pi, \pi^*)}$ in Figure 4.5, can be reached by absorption of red light, e.g., a wavelength of 680 nm. The other state, $S^b_{(\pi, \pi^*)}$, involves a π^* orbital that lies higher in energy and is reached by absorption of blue light, e.g., 430 nm. The electronic transitions caused by the absorption of photons are indicated by solid vertical arrows in Figure 4.5, and the vertical distances correspond approximately to the differences in energy involved. Excitation of a singlet ground state to an excited triplet state is usually only about

---

10. Retinal is a carotenoid attached to the lipoprotein opsin; the complex is referred to as rhodopsin.

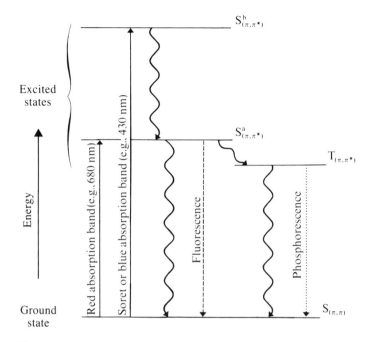

Figure 4.5

Energy level diagram indicating the principal electronic states and some of the transitions of chlorophyll. Straight vertical lines represent the absorption of light; wavy lines indicate radiationless transitions, for which the energy is eventually released as heat; broken lines indicate those de-excitations accompanied by radiation.

10^{-5} times as probable as going to an excited singlet state, so the transition from $S_{(\pi,\pi)}$ to $T_{(\pi,\pi^*)}$ has not been indicated for chlorophyll in Figure 4.5. To go from $S_{(\pi,\pi)}$ to $T_{(\pi,\pi^*)}$, the energy of an electron must be substantially increased *and* the orientation of its spin must be simultaneously reversed. The coincidence of these two events is very improbable, so very few chlorophyll molecules are directly excited to $T_{(\pi,\pi^*)}$ from the ground state by the absorption of light.

We will now briefly compare the two most important excitations in photo-biology—the transitions of n and π electrons to π^* orbitals. The n electrons have very little spatial overlap with other electrons, and they tend to be higher in energy than the π electrons, which are reduced in energy (stabilized) by being delocalized over a number of nuclei. Therefore, the excitation of an n electron to a π^* orbital generally takes less energy than the transition of a π electron to the same antibonding orbital (see Fig. 4.6). Hence, an $S_{(n,\pi^*)}$ state usually occurs at a lower energy—a longer wavelength is required for its excitation—than does the analogous $S_{(\pi,\pi^*)}$. Another difference is that π and π^* orbitals may overlap

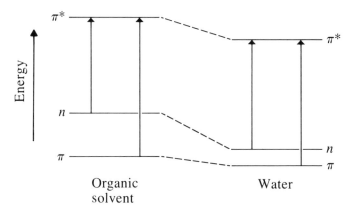

Figure 4.6

Influence of the solvent on the energies of n, π, and π^* orbitals. The indicated n-to-π^* transition takes more energy in water than in an organic solvent, and the π-to-π^* transition takes less.

spatially (see Fig. 4.4), but n and π^* orbitals generally do not. Consequently, the n to π^* transition is not as favored, or probable, as the π to π^* one. Thus, transitions to $S_{(\pi,\pi^*)}$ states tend to dominate the absorption properties of a molecular species (as evidenced, for example, by an absorption spectrum) compared with excitations yielding $S_{(n,\pi^*)}$ states.

As our final topic in this section, we will consider the effect of molecular environment on n, π, and π^* orbitals (Fig. 4.6). The n electrons generally interact strongly with water, e.g., by participating in hydrogen bonding, and therefore the energy level of n electrons is considerably lower in an aqueous environment than in an organic solvent. The energy of π^* electrons is also lowered by water, but to a lesser extent than for n electrons. The π orbitals are physically deepest within the molecule and hence are the least affected by the solvent. Thus, a transition from an n to a π^* orbital takes more energy in water than in an organic solvent (Fig. 4.6). On the other hand, the π to π^* transition takes less energy in an aqueous environment than in an organic one. As an example, let us consider the transition of chlorophyll to an excited singlet state in different solvents. The excitation to either excited singlet state takes less energy in an aqueous environment than it does in an organic solvent; i.e., when chlorophyll is in water, the required photons have a longer wavelength than when the chlorophyll is dissolved in an organic solvent like acetone. This lower energy requirement in water is consistent with our statement that the transition for chlorophyll corresponds to the promotion of a π electron to a π^* orbital. orbital.

DE-EXCITATION

The primary processes of photochemistry involve the light absorption event, which we have already discussed, together with the subsequent de-excitation reactions. We can portray such transitions on an energy level diagram, as in Figure 4.5 for chlorophyll. In this section we discuss the various de-excitation processes, including a consideration of their rate constants and lifetimes.

One characteristic of the various excitation and de-excitation processes is the time needed for the transitions. A useful estimate of the time it takes for the absorption of a photon is the time required for one cycle of the light wave to pass an electron. This time is the distance per cycle of the wave divided by the speed with which light travels, or λ/v, which equals $1/v$ by Equation 4.1. Hence, the time required for the absorption of a photon is approximately equal to the reciprocal of the frequency of the light, $1/v$, which is the time necessary for one complete oscillation of the electromagnetic field. During one oscillation the electric vector of light can induce an electron to move first in one direction and then in the opposite one, thereby setting up a beating or resonating of the electron. To be specific, let us consider blue light with a wavelength of 460 nm in a vacuum. From its frequency (given in Table 4.1), we can calculate that the time for one cycle is $1/(6.52 \times 10^{14} \text{ s}^{-1})$, or 1.5×10^{-15} s. Light absorption is indeed an extremely rapid event!

Times for de-excitation reactions are usually expressed in *lifetimes*. A lifetime is the time required for the number of molecules in a given state to decrease to $1/e$, or 37%, of the initial number. Lifetimes are extremely convenient for describing first-order processes because the initial species in such processes decay (disappear) exponentially with time (see Eq. 4.9 and also App. III for examples of first-order processes). A *half-time,* the time necessary for the number of species in a given state to decrease by 50%, is occasionally used to describe de-excitation processes; for an exponential decay, one half-time is ln 2, or 0.693, times the size of a lifetime.

Fluorescence, Phosphorescence, and Radiationless Transitions

As we would expect, the excess energy of the excited state can be dissipated by a number of competing reactions. One of the ways in which de-excitation can occur is by the emission of light known as *fluorescence* (indicated by a dashed straight line in Fig. 4.5 for the principal transitions of chlorophyll). Fluorescence describes the electromagnetic radiation emitted when a molecule goes from an excited singlet state to a singlet ground state. Fluorescence lifetimes for most organic molecules range from 10^{-9} to 10^{-6} s (see Turro, 1978).[11] The properties of chlorophyll fluorescence are extremely important for an understanding of the primary events of photosynthesis, as we shall see in Chapter 5.

De-excitation of an excited state often occurs without the emission of any radiation. Such transitions are termed nonradiative, or *radiationless* (indicated by wavy lines in Fig. 4.5). For a radiationless transition to a lower energy excited state or back to the ground state, the energy of the absorbed photon is eventually converted to heat, which is passed on by collisions with the surrounding molecules. Radiationless transitions can be extremely rapid from some excited singlet state to a lower energy excited singlet state in the same molecule. For example, the radiationless transition from $S^b_{(\pi,\pi^*)}$ to $S^a_{(\pi,\pi^*)}$ indicated for chlorophyll in Figure 4.5 takes about 10^{-12} s (see Clayton, 1965). This transition is so rapid that hardly any fluorescence can be emitted from $S^b_{(\pi,\pi^*)}$, so no fluorescence emission from the upper excited singlet state of chlorophyll is indicated in Figure 4.5. The excited singlet state lying at the lower energy, $S^a_{(\pi,\pi^*)}$, can decay to the ground state by a radiationless transition. As another alternative, $S^a_{(\pi,\pi^*)}$ can go to $T_{(\pi,\pi^*)}$, also by a radiationless transition (Fig. 4.5). In fact, excited triplet states in molecules are mainly formed by radiationless transitions from excited singlet states at higher energies. The de-excitation of $T_{(\pi,\pi^*)}$ to $S_{(\pi,\pi)}$ can be radiationless, or it can be by radiation known as *phosphorescence* (dotted straight line in Fig.4.5), which we will consider next.

Phosphorescence is the electromagnetic radiation that can accompany the transition of a molecule from an excited triplet state to a ground state singlet. Because we go from a triplet to a singlet state, the net spin of the molecule must change during the emission of this radiation. The lifetimes for phosphorescence are usually from 10^{-3} to 10 s (Calvert and Pitts, 1966; Wayne, 1988), which are long compared with those for fluorescence (10^{-9} to 10^{-6} s). These relatively long times for de-excitation occur because there is both the transition from one electronic state to another and the simultaneous reversal of the electron spin—the coincidence of these two events is rather improbable. In fact, the above-mentioned low probability of forming $T_{(\pi,\pi^*)}$ from $S_{(\pi,\pi)}$ by light absorption has the same physical basis as the long lifetime for de-excitation of the excited triplet state back to the ground state by emitting phosphorescence.

Another, although usually minor, way for an excited molecule to emit radiation is by "delayed fluorescence" (see Clayton, 1980). Specifically, a molecule in the relatively long-lived excited triplet state $T_{(\pi,\pi^*)}$ can sometimes be supplied

---

11. The lifetime of an excited state indicates the time course for de-excitation, not the time required for the de-excitation event per se; e.g., the time required for the emission of a photon is essentially the same as for its capture.

enough thermal energy by collisions to put it into the higher energy excited singlet state $S_{(\pi,\pi^*)}$. Subsequent radiation, as the molecule goes from this excited singlet state to the ground state $S_{(\pi,\pi)}$, has the characteristics of fluorescence. However, it is considerably *delayed* after light absorption compared with normal fluorescence because the excitation spent some time as $T_{(\pi,\pi^*)}$.

Competing Pathways for De-excitation

Each excited state has not only a definite energy but also a specific lifetime, the length of which depends on the particular processes competing for the de-excitation of that state. In addition to fluorescence, phosphorescence, and the radiationless transitions that we have just introduced, the excitation energy can also be transferred to another molecule, putting this second molecule into an excited state while the originally excited molecule returns to its ground state. In another type of de-excitation process, an excited (energetic) electron may leave the molecule that absorbed the photon; this occurs for certain excited chlorophyll molecules that we will discuss later. The excited state of pivotal importance in photosynthesis is the lower excited singlet state of chlorophyll, indicated by $S^a_{(\pi,\pi^*)}$ in Figure 4.5. We will use this state to illustrate some of the possible competing reactions leading to the de-excitation of an excited singlet state, $S_{(\pi,\pi^*)}$ (see Cowan and Drisko, 1976; Seliger and McElroy, 1965).

The absorbed quantum can be reradiated as electromagnetic energy, $h\nu$, causing the excited molecule $S_{(\pi,\pi^*)}$ to drop back to its ground state $S_{(\pi,\pi)}$:

$$S_{(\pi,\pi^*)} \xrightarrow{k_1} S_{(\pi,\pi)} + h\nu \qquad (4.4)$$

Such fluorescence exponentially decays with time after the exciting light is removed, indicating a first-order process with a rate constant k_1 in Equation 4.4 (first-order rate constants have units of s^{-1}).[12] The fluorescence lifetime of $S_{(\pi,\pi^*)}$ is typically about 10^{-8} s. This fluorescence lifetime would be the actual lifetime of $S_{(\pi,\pi^*)}$, if no other competing de-excitation processes occurred. When the energy of the absorbed quantum is dissipated as fluorescence, no photochemical work can be done. Therefore, the fluorescence lifetime is an upper time limit within which any biologically useful reactions can be driven by the

---

12. For a first-order process, the rate of disappearance of the excited state, $dS_{(\pi,\pi^*)}/dt$, is linearly proportional to the amount of $S_{(\pi,\pi^*)}$ present at any time. First-order processes are further discussed in Appendix III.

lowest excited singlet state of a molecule; i.e., if the reactions take longer than the fluorescence lifetime, most of the absorbed energy will already have been dissipated.

The next de-excitation processes that we will consider are the radiationless transitions by which $S_{(\pi,\pi*)}$ eventually dissipates its excess electronic energy as heat. Two different states can be reached by radiationless transitions from $S_{(\pi,\pi*)}$:

$$S_{(\pi,\pi*)} \xrightarrow{k_2} S_{(\pi,\pi)} + \text{heat} \tag{4.5}$$

$$S_{(\pi,\pi*)} \xrightarrow{k_3} T_{(\pi,\pi*)} + \text{heat} \tag{4.6}$$

Radiationless transitions such as those indicated in Equations 4.5 and 4.6 involve de-excitations in which the excess energy is often first passed on to other parts of the same molecule. This causes the excitation of certain vibrational modes for other pairs of atoms within the molecule—vibrations like those we will discuss in conjunction with the Franck–Condon principle (see Fig. 4.7). This energy, which has become distributed over the molecule, is subsequently dissipated by collisions with other molecules in the randomizing interchanges that are the basis of temperature. When an excited molecule directly returns to its ground state by a radiationless transition (Eq. 4.5), all of the radiant energy of the absorbed light is eventually converted into the thermal energy of motion of the surrounding molecules. In Equation 4.6 only some of the excess electronic energy appears as the quantity designated "heat," which in that case represents the difference in energy between $S_{(\pi,\pi*)}$ and $T_{(\pi,\pi*)}$.

As for fluorescence, radiationless transitions also obey first-order kinetics. Although the dissipation of excitation energy as heat in a transition to the ground state (Eq. 4.5) is photochemically wasteful in that no biological work is performed, the transition to $T_{(\pi,\pi*)}$ indicated by Equation 4.6 can be quite useful. The lowest excited triplet state usually lasts 10^4 to 10^8 times longer than does $S_{(\pi,\pi*)}$, which allows time for many more intermolecular collisions. Because each collision increases the opportunity for a given reaction to occur, $T_{(\pi,\pi*)}$ can be an important excited state in photobiology.

The absorption of light can lead to a photochemical reaction initiated by a molecule other than the one that actually absorbed the photon. This phenomenon suggests that electronic excitation can be transferred between molecules, resulting in the excitation of one and the de-excitation of the other. For instance, the excitation energy of $S_{(\pi,\pi*)}$ might be transferred to a second molecule, represented in the ground state by $S_{2(\pi,\pi)}$:

$$S_{(\pi,\pi*)} + S_{2(\pi,\pi)} \xrightarrow{k_4} S_{(\pi,\pi)} + S_{2(\pi,\pi*)} \tag{4.7}$$

This second molecule thereby becomes excited, indicated by $S_{2(\pi,\pi*)}$, and the molecule that absorbed the quantum becomes de-excited and is returned to its ground state. Such transfer of electronic excitation from molecule to molecule underlies the energy migration among the pigments involved in photosynthesis (see Ch. 5). We will assume that Equation 4.7 represents a first-order reaction, as it does for the excitation exchanges between chlorophyll molecules *in vivo* (in certain cases, Eq. 4.7 can represent a second-order reaction; i.e., $dS_{(\pi,\pi*)}/dt$ then equals $-k'\ S^2_{(\pi,\pi*)}$).

As another type of de-excitation process, $S_{(\pi,\pi*)}$ can take part in a *photochemical* reaction. For example, the excited π^* electron can be donated to a suitable acceptor:

$$S_{(\pi,\pi*)} \xrightarrow{k_5} D_{(\pi)} + e^* \tag{4.8}$$

where the ejected electron is represented by e^*. The electron removed from $S_{(\pi,\pi*)}$ is replaced by another one donated from some other compound; $D_{(\pi)}$ in Equation 4.8, which represents a doublet because one of the π orbitals contains an unpaired electron, then goes back to its original ground state $S_{(\pi,\pi)}$. (We will discuss electron energy and donation in Chs. 5 and 6.) Photochemical reactions of the form of Equation 4.8 serve as the crucial link in the conversion of radiant energy into chemical or electrical energy. Indeed, Equation 4.8 can be used to represent the photochemical reaction taking place at the special chlorophyll molecules P_{680} and P_{700} that we will discuss later.

Lifetimes

Equations 4.4 through 4.8 represent five competing pathways for the de-excitation of the excited singlet state, $S_{(\pi,\pi*)}$. They must all be considered when predicting the lifetime of $S_{(\pi,\pi*)}$. Assuming that each of these de-excitation processes is first-order and that no reaction leads to the formation of $S_{(\pi,\pi*)}$, then the disappearance of the excited singlet state satisfies the following relation:

$$-\frac{dS_{(\pi,\pi*)}}{dt} = (k_1 + k_2 + k_3 + k_4 + k_5)S_{(\pi,\pi*)} \tag{4.9}$$

where the various k_j's in Equation 4.9 are the rate constants for the five individual decay reactions (Eqs. 4.4 through 4.8). After multiplying both sides by $dt/S_{(\pi,\pi^*)}$ and then integrating Equation 4.9 (see App. III), taking exponentials of both sides followed by rearrangement leads to the following expression for the time dependence of the number of molecules in the excited singlet state:

$$S_{(\pi,\pi^*)_t} = S_{(\pi,\pi^*)_0} e^{-(k_1+k_2+k_3+k_4+k_5)t} \tag{4.10}$$

where $S_{(\pi,\pi^*)_0}$ represents the number of molecules in the excited singlet state when the illumination ceases ($t = 0$), and $S_{(\pi,\pi^*)_t}$ is the number of excited singlet states remaining at a subsequent time t. Relations such as Equation 4.10— showing the amount of some state remaining at various times after illumination or other treatment—are extremely important for describing processes with first-order rate constants.

Because the lifetime of an excited state is the time required for the number of excited molecules to decrease to $1/e$ of the initial value, $S_{(\pi,\pi^*)_t}$ in Equation 4.10 equals $(1/e)\, S_{(\pi,\pi^*)_0}$ when t equals the lifetime τ, i.e.,

$$S_{(\pi,\pi^*)_\tau} = e^{-1} S_{(\pi,\pi^*)_0} = S_{(\pi,\pi^*)_0} e^{-(k_1+k_2+k_3+k_4+k_5)t} \tag{4.11}$$

which leads to the following relationship:

$$(k_1 + k_2 + k_3 + k_4 + k_5)\tau = 1 \tag{4.12}$$

Equation 4.12 indicates that the greater is the rate constant for *any* particular de-excitation process, the shorter will be the lifetime of the excited state.

Equation 4.12 can be generalized to include all competing reactions, giving the following expression for the lifetime:

$$\frac{1}{\tau} = k = \sum_j k_j = \sum_j \frac{1}{\tau_j} \tag{4.13}$$

where k_j is the first-order rate constant for the jth de-excitation process and τ_j is its lifetime ($\tau_j = 1/k_j$). Also, τ is the lifetime of the excited state, and k in Equation 4.13 is the overall rate constant for its decay. Using Equation 4.13 we can reexpress Equation 4.10 as follows:

$$S_{(\pi,\pi^*)_t} = S_{(\pi,\pi^*)_0} e^{-kt} = S_{(\pi,\pi^*)_0} e^{-t/\tau} \tag{4.14}$$

Equations 4.12 and 4.13 indicate that, when more than one de-excitation process is possible, τ is less than the lifetime of any individual competing reaction acting alone. In other words, because each de-excitation reaction is independent of the others, the observed rate of decay of an excited state is faster than deactivation by any single competing reaction acting by itself.

If the rate constant for a particular reaction is much larger than for its competitors, the excited state becomes de-excited predominantly by that process. As an example, we will consider an excited triplet state of a molecule that shows delayed fluorescence. Suppose that the lifetime for phosphorescence, τ_P, is 10^{-2} s, in which case k_P is 100 s^{-1}. When sufficient thermal energy is supplied, $T_{(\pi,\pi*)}$ can be raised in energy to an excited singlet state, which could then emit "delayed" fluorescence if $S_{(\pi,\pi*)}$ decays to the ground state by emitting electromagnetic radiation. Suppose that the rate constant ($k_{T*\to S*}$) for the transition from $T_{(\pi,\pi*)}$ to $S_{(\pi,\pi*)}$ is 20 s^{-1}. By Equation 4.13, k for these two competing pathways is 100 s^{-1} + 20 s^{-1} or 120 s^{-1}, which corresponds to a lifetime of $1/(120$ s$^{-1})$, or 0.008 s. Suppose now that another molecule, which can readily take on the excitation of $T_{(\pi,\pi*)}$ of the original species, is introduced into the solution—$k_{transfer}$ might be 10^4 s^{-1}. Because of the relatively large rate constant, such a molecule "quenches" the phosphorescence and delayed fluorescence originating from $T_{(\pi,\pi*)}$ (i.e., its decay pathway predominates over the competing processes), so it is generally referred to as a *quencher*. For the three pathways indicated, the overall rate constant is 100 s^{-1} + 20 s^{-1} + 10^4 s^{-1}, or essentially 10^4 s^{-1}. The de-excitation here is dominated by the quencher, because $k_{transfer} \gg k_P + k_{T*\to S*}$.

Quantum Yields

We often use a *quantum yield* (or *quantum efficiency*), Φ_i, to describe the de-excitation processes following absorption of light. Here Φ_i represents the fraction of molecules in some excited state that will decay by the ith de-excitation reaction out of all the possible competing pathways:

$$\Phi_i = \frac{\text{number of molecules using }i\text{th de-excitation reaction}}{\text{number of excited molecules}} \tag{4.15}$$

$$= \frac{k_i}{\sum_j k_j} = \frac{\tau}{\tau_i}$$

Equation 4.15 indicates that the rate constant for a particular pathway determines what fraction of the molecules in a given excited state will use that de-excitation process. Hence, k_i determines the quantum yield for the ith de-excitation pathway. From Equation 4.13 and the definition of τ_j given above, we can also indicate such competition among pathways using lifetimes (see Eq. 4.15). The shorter the lifetime for a particular de-excitation pathway, the larger will be the fraction of the molecules using that pathway, and hence the higher will be its quantum yield. Finally, by Equations 4.13 and 4.15, the sum of the quantum yields for all the competing de-excitation pathways, $\Sigma_i\ \Phi_i$, equals unity.

To illustrate the use of Equation 4.15, let us consider the quantum yield for chlorophyll fluorescence, Φ_{Fl}. The fluorescence lifetime τ_{Fl} of the lower excited singlet state of chlorophyll in ether is 1.5×10^{-8} s, and the observed lifetime τ for de-excitation of this excited state in ether is 0.5×10^{-8} s (Clayton, 1965). By Equation 4.15, the expected quantum yield for fluorescence is $(0.5 \times 10^{-8}$ s$)/(1.5 \times 10^{-8}$ s$)$, or 0.3, which is consistent with the observed Φ_{Fl} of 0.33 for the fluorescence de-excitation of chlorophyll in ether.

ABSORPTION SPECTRA AND ACTION SPECTRA

The absorption of radiation causes a molecule to go from its ground state to some excited state in which one of the electrons enters an orbital of higher energy. We have so far considered that both the ground state and the excited states occur at *specific* energy levels, as is indicated by the horizontal lines in Figure 4.5 for the case of chlorophyll. This leads to a consideration of whether only a very limited number of wavelengths are absorbed. For instance, are 430 nm and 680 nm the only wavelengths absorbed by chlorophyll (Fig. 4.5)? We will find that each electronic energy level is divided or split into various discrete sublevels that differ in energy. The largest splitting is due to *vibrational sublevels*. Vibrational sublevels of specific electronic states affect the wavelengths of light involved in photosynthesis and other photochemical processes.

Vibrational sublevels are the result of the vibration of atoms in a molecule. This motion determines the sum of the kinetic and potential energies of the atoms, and consequently it affects the total energy of the molecule. We can describe this atomic oscillation by the accompanying change in the internuclear distance. Therefore, we will refer to an *energy level diagram* indicating the range of positions, or trajectories, taken on by the vibrating nuclei (Fig. 4.7). The trajectories of such nuclear vibrations are quantized; i.e., only certain specific vibrations can occur. As a consequence of this quantization of the trajectories of the vibrating nuclei, only discrete energies are possible for the vibrational sublevels of a given state.

At temperatures normally encountered by plants, essentially all molecules are in the ground or unexcited state. Moreover, these molecules are nearly all in the lowest vibrational sublevel of the ground state—another consequence of the

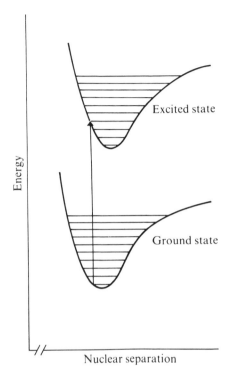

Figure 4.7

Energy curves for the ground state and an excited state showing the various vibrational sublevels. The vertical arrow represents a transition that is caused by the absorption of a photon and that is consistent with the Franck–Condon principle.

Boltzmann energy distribution (see Eq. 3.21). The absorption of a photon can cause a transition of the molecule from the lowest vibrational sublevel of the ground state to one of the vibrational sublevels of the excited state. The actual sublevel reached depends on the energy of the absorbed photon. For reasons that should become clear as we proceed in the discussion of vibrational sublevels, the *probability* that a photon will be absorbed also depends on its energy. Consideration of this absorption probability as a function of wavelength leads to an *absorption spectrum* for that particular molecule. The *effect* or consequence of light absorption on some process, when presented as a function of wavelength, leads to an *action spectrum*. At the end of this section we will discuss action spectra for a phytochrome-mediated reaction.

Vibrational Sublevels in an Energy Level Diagram

Various vibrational sublevels of both ground and excited states of a molecule are schematically indicated in Figure 4.7. In principle, such an energy level diagram can be prepared for any pigment, e.g., chlorophyll, carotenoid, or phytochrome. Energy level diagrams are extremely useful for predicting which electronic transitions are most likely to accompany the absorption of light. They can also be used to help explain why certain wavelengths predominate in the absorption process.

The abscissa in Figure 4.7 represents the distance between a pair of nuclei vibrating back and forth with respect to each other and can be obtained by imagining one nucleus to be situated at the origin of the coordinate system while the position of the other nucleus (in the same molecule) is plotted relative to this origin. The ordinate represents the total energy of the electrons plus the pair of nuclei. Figure 4.7 shows that the excited state is at a higher energy than the ground state and indicates that the two states are split into many vibrational sublevels differing in energy (see Seliger and McElroy, 1965; Turro, 1978; Wayne, 1988).

Because internuclear and interelectronic repulsive forces act against the electrostatic attraction between nuclei and shared electrons, only a certain range of internuclear separations can occur for a particular bond in a molecule. As internuclear separation decreases, the nuclei repel each other more and more, and the energy of the molecule increases. Moreover, the clouds of negative charge representing electrons localized on each nucleus have a greater overlap as internuclear separation decreases, resulting in interelectronic repulsion and likewise an increase in molecular energy. These effects account for the steep rise of the energy curves in Figure 4.7 as nuclear separation becomes less (left-hand side of the figure). At the other extreme, the delocalized (bonding) electrons shared by the two nuclei resist an unlimited increase in nuclear separation, which diminishes the attractive electrostatic interaction between nuclei and electrons; such an increase in internuclear distance corresponds to a stretching of the chemical bond. Thus, the energy curve in Figure 4.7 also rises as the internuclear distance becomes greater (right-hand side of the figure). Because of these two opposing tendencies, the range of possible nuclear separations is confined to an *energy trough*. It is within these potential energy curves—one for the ground state and another for the excited state—that the trajectories of the nuclear vibrations occur.

A horizontal line in Figure 4.7 represents the range of nuclear separations corresponding to a given vibrational sublevel; i.e., nuclei vibrate back and forth along the distance indicated by a horizontal line. As is evident in the figure, both ground and excited states have many vibrational sublevels differing in energy. For the upper vibrational sublevels of a given state, the nuclei vibrate over longer distances (i.e., more extensive range of nuclear separations in Fig. 4.7), which also corresponds to higher vibrational energies. Because the excited state has an electron in an antibonding orbital, it usually has a greater mean internuclear separation than does the ground state. This increase in bond length is shown in Figure 4.7 by a slight displacement to greater nuclear separations, i.e., to the right, for the upper curve.

The direction of nuclear motion is reversed at the extremities of the vibrational pathways (horizontal lines in Fig. 4.7), so the velocity of the nuclei must be zero at these turning points. As the turning point at either end of the oscillation range is approached, the nuclei begin to slow down and eventually stop before reversing their direction of motion. Consequently, the nuclei spend most of their time at or near the extreme ends of their trajectory. Therefore, a photon is most likely to arrive at the molecule when the nuclei are at or near the extremes of their vibrational range.[13] Because of our probability consideration, the electronic transition resulting from the absorption of a photon, represented by the vertical arrow in Figure 4.7, has been initiated from one of the ends of the nuclear oscillation range for the lowest vibrational sublevel of the ground state. This arrow begins from the lowest sublevel because nearly all ground state molecules are in the lowest vibrational sublevel at the temperatures encountered in plants, which, as we noted above, is a consequence of the Boltzmann energy distribution.

During the light absorption event, the energy of the photon is transferred to some electron in the molecule. Because the molecular orbital describing the trajectory of an electron in the excited state has a small, but finite, probability of spatially overlapping with the nuclei, an interaction is possible between the excited electron and the nuclei over which it is delocalized. Such interactions do occur and generally take place over a rather short time period (less than 10^{-13} s). An interaction between an energetic electron and the nuclei can cause the excitation of nuclei to higher energy vibrational states. In fact, the transition represented by the arrow in Figure 4.7 corresponds to both the excitation of an electron leading to an excited state of the molecule and the subsequent excitation of the nuclei to some excited vibrational sublevel. Thus, part of the energy of the photon is rapidly passed on to nuclear vibrations. The length of the arrow in Figure 4.7 is proportional to the light energy (or quantum) added to the molecule and therefore represents the energy distributed to the nuclei plus that remaining with the excited electron. Next, we will demonstrate which vibrational sublevel of the excited state has the highest probability of being reached by the excitation process.

---

13. Actual quantum-mechanical calculations beyond the scope of this text lead to somewhat different conclusions, especially for the lowest vibrational sublevel (see Calvert and Pitts, 1966; Turro, 1978).

The Franck–Condon Principle

Franck and Condon in 1926 enunciated a principle, based mainly on classical mechanics, to help rationalize the various bands in absorption spectra and fluorescence emission spectra (Condon, 1926; Franck, 1926). We will direct our attention to the nuclei to discuss the effect of the quantized modes of nuclear vibration (see the energy level diagram in Fig. 4.7). The Franck–Condon principle states that the nuclei change neither their *separation* nor their *velocity* during those transitions for which the absorption of a photon is most probable (see Seliger and Mc Elroy, 1965; Turro, 1978). We can use this principle to predict which vibrational sublevels of the excited state are most likely to be involved in the electronic transitions accompanying light absorption.

First we will consider the part of the Franck–Condon principle stating that nuclei do not change their position for the most probable electronic transitions caused by light absorption. Because a vertical line in Figure 4.7 corresponds to no change in nuclear separation, the absorption of a photon has been indicated by a vertical arrow. This condition of constant internuclear distance during absorption can be satisfied most often when the nuclei are moving slowly or are actually stopped at the extremes of their oscillation range. Thus, the origin of the arrow indicating an electronic transition in Figure 4.7 is at one of the turning points of the lowest vibrational sublevel of the ground state, and the tip is drawn to an extremity of one of the vibrational sublevels in the excited state (the fourth vibrational sublevel for the particular case illustrated).

The other condition embodied in the Franck–Condon principle is that a photon has the greatest chance of being absorbed when the velocity (a vector) of the vibrating nuclei does not change. In other words, absorption is maximal when the nuclei are moving in the same direction and at the same speed in both the ground and the excited state. Again, this condition has the greatest probability of being met when the nuclei are moving slowly or not at all, as occurs at the turning points for a nuclear oscillation, because the nuclei spend the most time at a particular velocity under such conditions. Therefore, the most probable electronic transition represented in a diagram like Figure 4.7 is a vertical line that originates from one of the ends of the horizontal line representing the range of nuclear separations for the lowest vibrational sublevel of the ground state and terminates at the end of the vibrational trajectory for some sublevel of the excited state.

Another way to view the Franck–Condon principle is to consider that the light absorption event is so rapid that the nuclei do not have a chance to move during it. The absorption of a photon requires about 10^{-15} s. In contrast, the period for one nuclear vibration back and forth along an oscillation range (like those represented by horizontal lines in Fig. 4.7) is generally somewhat more than 10^{-13} s. Thus, the nuclei cannot move any appreciable distance during the time necessary for the absorption of a photon, especially when the nuclei are moving relatively slowly near the ends of their vibrational trajectory. Also, nuclear velocity would not change appreciably in a time interval as short as 10^{-15} s (see Turro, 1978; Wayne, 1988).

The time scale for nuclear oscillations has far-reaching consequences for energy dissipation. In particular, as the nuclei oscillate back and forth along their trajectories, they can interact with other nuclei. These encounters make possible the transfer of energy from one nucleus to another (within the same molecule or to adjacent molecules). Thus, the time for one cycle of a nuclear vibration, approximately 10^{-13} s, is an estimate of the time in which excess vibrational energy can be dissipated as heat by interactions with other nuclei. As excess energy is exchanged by such processes, the part of the molecule indicated in Figure 4.7 soon reaches the lowest vibrational sublevel of the excited state; these transitions within the same electronic state, e.g., $S_{(\pi,\pi^*)}$, are usually complete in about 10^{-12} s. Fluorescence lifetimes generally are on the order of 10^{-8} s, so an excited singlet state gets to its lowest vibrational sublevel before appreciable de-excitation can occur by fluorescence. The rapid dissipation of excess vibrational energy causes some of the energy of the absorbed photon to be released as heat. Therefore, flourescence is generally of lower energy (longer wavelength) than the absorbed light, as we will show for chlorophyll in the next chapter.

The Franck–Condon principle predicts the most likely transition caused by the absorption of light. But others do occur. These other transitions become statistically less probable the more that nuclear position or velocity changes during the absorption of the photon. Because transitions from the ground state to both higher and lower vibrational sublevels in the excited state occur with a lower probability than transitions to the optimal sublevel, the absorption of light is not as great at the wavelengths that excite the molecule to such vibrational sublevels. Moreover, some transitions begin from an excited vibrational sublevel of the ground state. We can calculate the fraction of the ground state molecules in the various vibrational sublevels from the Boltzmann energy distribution, $n(E) = n_{\text{total}} e^{-E/RT}$ (Eq. 3.21b). For example, $e^{-E/RT}$ is 0.22 for an E of 3.72 kJ mol^{-1}, the mean translational energy at 25°C.

For many molecules the vibrational sublevels of both the ground state and the excited state are 10 to 20 kJ mol^{-1} apart. Consequently, as the wavelength of incident light is increased or decreased from that for the most intense absorption, transitions involving other vibrational sublevels become important, e.g, at approximately 15 kJ mol^{-1} intervals. An energy difference of 15 kJ mol^{-1} between two photons corresponds to a difference in wavelength of about 40 nm near the middle of the visible region (green or yellow, Table 4.1). Such wavelength spacings can be seen by the various peaks near the major absorption bands of chlorophyll (Fig. 5.3) or the three peaks in the absorption spectra of typical carotenoids (Fig. 5.5). In summary, the amount of light absorbed is maximal at a certain wavelength corresponding to the most probable transition predicted by the Franck–Condon principle. Transitions from other vibrational sublevels of the ground state and to other sublevels of the excited state occur less frequently and help create an absorption spectrum that is characteristic of a particular molecule.[14]

Absorption Bands and Absorption Coefficients

Our discussion of light absorption has so far been primarily concerned with transitions from the ground state energy level to those of excited states, which we just expanded to include the occurrence of vibrational sublevels of the states. In addition, vibrational sublevels are subdivided into *rotational* states. In particular, the motion of the atomic nuclei within a molecule can be described by quantized rotational states of specific energy, leading to the subdivision of a given vibrational sublevel of a state into a number of rotational *sub*sublevels. The energy increments between rotational states are generally about 1 kJ mol^{-1} (approximately 3 nm in wavelength in the visible region). The further broadening of absorption lines because of a continuum of *translational* energies of the whole molecule is generally much less, often about 0.1 kJ mol^{-1}. Moreover, interactions with the solvent or other neighboring molecules can affect the distribution of electrons in a particular molecule and, consequently, can shift the position of the various energy levels. The magnitude of the shifts caused by intermolecular interactions varies considerably and can be 5 kJ mol^{-1} or more (Depuy and Chapman, 1972; Seliger and McElroy, 1965).

The photons absorbed in an electronic transition involving specific vibrational sublevels—including the range of energies due to the various rotational subsublevels and other shifts—give rise to an *absorption band*. These wavelengths represent the transition from a vibrational sublevel of the ground state to some

---

14. We note that electromagnetic radiation in the far infrared region does not have enough energy to cause an electronic transition, but its absorption can excite a molecule to an excited vibrational sublevel of the ground state. For instance, by Equation 4.2b ($E_\lambda = Nhc/\lambda_{vacuum}$) IR at 4 μm has an energy of (119 600 kJ mol^{-1} nm)/(4 000 nm) or 30 kJ mol^{-1}. Absorption of such radiation could excite a molecule to the third vibrational sublevel of the ground state when the vibration sublevels have a typical spacing of 15 kJ mol^{-1}.

vibrational sublevel of an excited state. A plot of the relative efficiency for light absorption as a function of wavelength is an *absorption spectrum*, which may include more than one absorption band. Such bands can represent transitions to or from different vibrational sublevels and to different excited states. The smoothness of the absorption bands of most pigment molecules indicates that a great range of photon energies can correspond to the transition of an electron from the ground state to some excited state (see absorption spectra in Figs. 4.11, 5.3, 5.5, and 5.7).[15] Because of the rather large effects that intermolecular interactions can have on electronic energy levels, we should always specify the solvent when presenting an absorption band or spectrum.

Absorption bands and spectra indicate how light absorption varies with wavelength. The absorption at a particular wavelength by a certain species is quantitatively described using an *absorption coefficient*, ε_λ (ε_λ is also referred to as an *extinction* coefficient). Because of its descriptive usefulness, we will now derive an expression incorporating ε_λ.

Let us consider a monochromatic beam of parallel light of flux density J. Because "monochromatic" refers to light of a single wavelength, J can be expressed as either photon or energy flux densities. Some of the light may be absorbed in passing through a solution, so the emerging beam will generally have a lower flux density—we will assume that scattering and reflection are negligible (an assumption that must be tested when absorption properties are determined). In a small path length dx along the direction of the beam, J decreases by dJ due to absorption by a species having a concentration c. Lambert is often credited with recognizing in 1768 that $-dJ/J$ is proportional to dx—actually, Bouguer had expressed this in 1729—and in 1852 Beer noted that $-dJ/J$ is proportional to c. Upon putting these two observations together, we obtain the following expression:

$$-\frac{dJ}{J} = k_\lambda c\ dx \qquad (4.16)$$

where k_λ is a proportionality coefficient referring to a particular wavelength, and the minus sign indicates that the flux density is decreased by absorption. We

---

15. To "sharpen" an absorption spectrum, the translational broadening of absorption bands can be reduced by substantially decreasing the temperature, e.g., by using liquid nitrogen (boiling point $= -196°C$) to cool the sample. Also, the reduction in temperature decreases the number of the absorbing molecules in excited vibrational sublevels and higher energy rotational sublevels of the ground state—we can predict the relative populations of these states from the Boltzmann factor (see Eq. 3.21).

can integrate Equation 4.16 across a solution of a given concentration, which leads to

$$-\int_{J_0}^{J_b} dJ/J = -\ln\frac{J_b}{J_0} = \int_0^b k_\lambda c \; dx = k_\lambda cb \qquad (4.17)$$

where J_0 is the flux density of the incident beam, and J_b is its flux density after traversing a distance b through the solution (see Fig. 4.8).

Equation 4.17 is usually recast into a slightly more convenient form. We can replace the natural logarithm by the common logarithm (ln = 2.303 log, App. II or III) and $k_\lambda/2.303$ by the absorption coefficient at a specific wavelength, ε_λ. Equation 4.17 then becomes

$$A_\lambda = \log\frac{J_0}{J_b} = \varepsilon_\lambda cb \qquad (4.18)$$

where A_λ is the absorbance (colloquially, the "optical density") of the solution at a particular wavelength. Equation 4.18 is generally referred to as Beer's law, although it is also called the Beer–Lambert law, the Lambert–Beer law, and even the Bouguer–Lambert–Beer law. When more than one absorbing species is present in a solution, we can generalize Equation 4.18 to give

$$A_\lambda = \log\frac{J_0}{J_b} = \sum_j \varepsilon_{\lambda_j} c_j b \qquad (4.19)$$

where c_j is the concentration of species j, and ε_{λ_j} is its absorption coefficient at wavelength λ (for additional information see Calvert and Pitts, 1966; Clayton, 1980; Eisenberg and Crothers, 1979; Williams and Fleming, 1987).

According to Beer's law, the absorbance at some wavelength is proportional to the concentration of the absorbing species, to its absorption coefficient at that wavelength, and to the optical path length b (Fig. 4.8). Values of ε_λ for organic compounds range up to over 10^4 m^2 mol^{-1} in the visible region.[16] If ε_λ at some wavelength is known for a particular species, we can determine its concentration from the measured absorbance at that wavelength by using Beer's law (Eq. 4.18).

---

16. For laboratory absorption studies, the optical path length b is often 1 cm and c_j is expressed in mol litre^{-1} (i.e., molarity), in which case ε_{λ_j} has units of litre mol^{-1} cm^{-1} and is referred to as the *molar absorption* (or *extinction*) *coefficient* (1 litre mol^{-1} cm^{-1} = 1 M^{-1} cm^{-1} = 10^{-3} mM^{-1} cm^{-1} = 10^3 cm^2 mol^{-1} = 10^{-1} m^2 mol^{-1}). The absorbing species is usually dissolved in a solvent that does not absorb at the wavelengths under consideration.

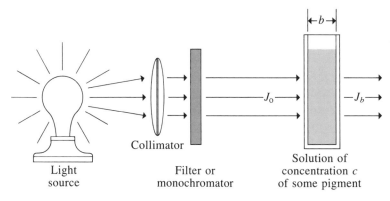

Figure 4.8

Quantities involved in light absorption by a solution as described by Beer's law, $\log(J_0/J_b) = \varepsilon_\lambda cb$ (Eq. 4.18).

As an application of Beer's law, we will estimate the average chlorophyll concentration in leaf cells. The palisade and spongy mesophyll cells in the leaf section portrayed in Figure 1.2 might correspond to an average thickness of chlorophyll-containing cells of about 200 μm. The maximum molar absorption coefficient ε_λ in the red or the blue bands of chlorophyll (see Figs. 4.5 and 5.3) is about 10^4 m^2 mol^{-1}. At the peaks of the absorption bands, about 99% of the incident red or blue light can be absorbed by chlorophyll in a leaf section such as the one illustrated in Figure 1.2. This corresponds to having an emergent flux density J_b equal to 1% of the incident flux density J_0, so the absorbance A_λ in Equation 4.18 equals $\log(100/1)$ or 2—for simplicity, we are ignoring absorption by pigments other than chlorophyll. Using Beer's law (Eq. 4.18), we find that the average chlorophyll concentration is

$$c = \frac{A_\lambda}{\varepsilon_\lambda b} = \frac{(2)}{(10^4 \ m^2 \ mol^{-1})(200 \times 10^{-6} \ m)}$$

$$= 1 \ mol \ m^{-3} \quad (1 \ \text{mM})$$

a value characteristic of the average chlorophyll concentration in the photosynthesizing cells of many leaves.

Chlorophyll is located only in the chloroplasts, which in turn occupy about 3% to 4% of the volume of mesophyll cell in the leaf of a higher plant. The average concentration of chlorophyll in chloroplasts is thus about 30 times higher than the above estimate of chlorophyll in a leaf, or approximately 30 mol m^{-3}. A typical light path across the thickness of a chloroplast (Fig. 1.9) is about 2 μm. Using Beer's law (Eq. 4.18), we find that the absorbance of a single chloroplast in the red or blue bands is about

$$A_\lambda = (10^4 \text{ m}^2 \text{ mol}^{-1})(30 \text{ mol m}^{-3})(2 \times 10^{-6} \text{ m})$$

$$= 0.6 = \log \frac{J_0}{J_b}$$

Hence, J_b equals J_0/(antilog 0.6), or $0.25J_0$. Therefore, approximately 75% of the incident red or blue light at the peak of absorption bands is absorbed by a single chloroplast, which helps explain why individual chloroplasts appear green under a light microscope.

Conjugation

Light absorption by molecules generally involves transitions of π electrons to excited states where the electrons are in π^* orbitals. These π electrons occur in double bonds; the more double bonds that there are in some molecule, the greater is the probability for light absorption by that species. Indeed, both the effectiveness in absorbing electromagnetic radiation and the wavelengths involved are affected by the number of double bonds in *conjugation,* where conjugation refers to the alternation of single and double bonds (e.g., C—C=C—C=C—C=C—C) along some part of the molecule. For the molecules that we will consider, this alternation of single and double bonds mostly involves C atoms, but it also includes N atoms and O atoms. The absorption coefficient ε_λ increases with the number of double bonds in conjugation, because more *delocalized* (shared) π electrons can then interact with light (each π bond contains two electrons, so the number of excitable electrons increases proportionally to the number of double bonds in conjugation). As the number of double bonds in the conjugated system increases, the absorption bands also shift to longer wavelengths.

To help understand why the number of double bonds in conjugation affects the wavelength position for an absorption band, let us consider the shifts in energy for the various orbitals as the number of π electrons in a conjugated system changes. The various π orbitals in a conjugated system occur at different energy levels, the average energy remaining about the same as for the π orbital in an isolated double bond not part of a conjugated system (Fig. 4.9). The more double bonds occurring in a conjugated system, the more π orbitals that there are in the conjugated system, and the greater is the energy range from the lowest to the highest energy π orbital. Because the *average* energy of the π orbitals in

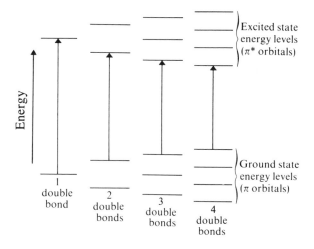

Figure 4.9

Effect on the energy levels of π and π^* orbitals as the number of double bonds in conjugation increases. The vertical arrows represent transitions caused by the absorption of photons—the required energy becomes less as the extent of conjugation increases.

a conjugated system does not markedly depend on the number of double bonds, the energy of the highest energy π orbital increases as the number of double bonds in conjugation increases. The π^* orbitals are similarly split into various energy levels (diagrammed in Fig. 4.9). Again, the range of energy levels about the mean for these π^* orbitals increases as the number of double bonds in the conjugated system increases. Consequently, the more π^* orbitals available in the conjugated system, the lower in energy will be the lowest of these (see Calvert and Pitts, 1966).

In Figure 4.9 we present transitions from the highest energy π orbital to the lowest energy π^* orbital for a series of molecules differing in the number of double bonds in conjugation. We notice that the more delocalized π electrons that there are in the conjugated system, the less is the energy required for a transition. This is illustrated in Figure 4.9 by a decrease in the length of the vertical arrow (which represents an electronic transition) as the number of double bonds in conjugation increases. Moreover, the most likely or probable electronic transition in this case is the one involving the least amount of energy; i.e., the excitation from the highest energy π orbital to the lowest energy π^* orbital is the transition that predominates.[17]

---

17. The transition probability depends on the spatial overlap between the quantum-mechanical wave functions describing the trajectories of the electrons in the two states, and such overlap is relatively large between the highest energy π orbital and the lowest energy π^* one (see Orchin and Jaffé, 1971).

The decrease in energy separation between the π orbitals and the π^* orbitals as the number of double bonds in conjugation increases (Fig. 4.9) accounts for the accompanying shift of the peaks of the absorption bands toward longer wavelengths. For example, an isolated double bond (C—C=C—C) generally absorbs maximally near 185 nm in the ultraviolet and has a maximum absorption coefficient of nearly 10^3 m^2 mol^{-1}. For two double bonds in conjugation (C—C=C—C=C—C), the maximum absorption coefficient doubles, and the wavelength position for maximum absorption shifts to about 225 nm, i.e., toward longer wavelengths. As the number of double bonds in conjugation in straight-chain hydrocarbons increases from three to five to seven to nine, the center of the absorption band for these hydrocarbons when they are dissolved in hexane shifts from approximately 265 to 325 to 375 to 415 nm, respectively (Calvert and Pitts, 1966; Levine, 1970). The maximum absorption coefficient is approximately proportional to the number of double bonds in the conjugated system, so it increases to almost 10^4 m^2 mol^{-1} for the hydrocarbon containing nine double bonds in conjugation. For molecules to absorb strongly in the visible region, a fairly extensive conjugated system of double bonds is generally necessary, as is indeed the case for pigments such as the chlorophylls and the carotenoids discussed in the next chapter.

Action Spectra

The relative *effectiveness* of various wavelengths in producing a specified response is of basic importance in photobiology and is presented in an *action spectrum*. An action spectrum is complementary to an absorption spectrum, the latter being the relative probability for the absorption of different wavelengths, e.g., ε_λ versus λ. When many different types of pigments are present, the action spectrum for a particular response can differ greatly from the absorption spectrum of the entire system. However, the Grotthuss–Draper law implies that an action spectrum should resemble the absorption spectrum of the substance that absorbs the light responsible for the specific effect or action being considered.

To obtain an action spectrum for some particular response, we could expose the system to the same photon flux density at each of a series of wavelength intervals and measure the resulting effect or action. The action could be the amount of O_2 evolved, the fraction of seeds germinating, or some other measured change. We could then plot the responses obtained as a function of their respective wavelength intervals to see which wavelengths are most effective in leading to that "action" (see Fig. 5.10). Another way to obtain an action spectrum is to plot the reciprocal of the number of photons required in the various wavelength intervals to give a particular response. If twice as many photons are needed at one wavelength compared with a second, the action spectrum has half the height at the first wavelength, and thus the effectiveness of various wavelengths can easily be presented (see Fig. 4.12). Using the latter approach, the photon flux density is varied until the response is the same for each wavelength inverval. This is an important point—if it is to be a true action spectrum, the action or effect measured must be linear with photon flux density for each of the wavelength intervals used; i.e., we must not be approaching light saturation. When we approach light saturation at certain wavelengths, the measured action per photon is less than it should be, compared with the values at other wavelengths. Consequently, an action spectrum, such as Figure 5.10, is flattened for those wavelengths where the response approaches light saturation. In the extreme case of light saturation at *all* wavelengths, the action spectrum is perfectly flat, because the response is then the same at each wavelength (see Clayton, 1980; Seliger and McElroy, 1965).

We can compare the action spectrum of some response with the absorption spectra of the various pigments suspected of being involved to see which pigment is responsible. If the measured action spectrum closely resembles the known absorption spectrum of some molecule, light absorbed by that molecule may be leading to the action considered. Examples where the use of action spectra have been important in understanding the photochemical aspects of plant physiology include the study of photosynthesis (Ch. 5) and investigations of the responses mediated by the pigment phytochrome, to which we now turn.

Absorption and Action Spectra of Phytochrome

Phytochrome is an important plant pigment that may be present in all eukaryotic photosynthetic organisms. It regulates photomorphogenic aspects of plant growth and development, such as seed germination, stem elongation, leaf expansion, formation of certain pigments, chloroplast development, and flowering. Many of these classical phytochrome effects can be caused by very low amounts of light; e.g., 500 μmol m^{-2} of red light (the photons in the visible for $\frac{1}{4}$ second of full sunlight) can saturate most of the processes, and 0.3 μmol m^{-2} can lead to half-saturation for a very sensitive one (Salisbury and Ross, 1991). High irradiance levels can also lead to effects mediated by phytochrome. We will direct our attention first to the structure of phytochrome and then to the absorption spectra for two of its forms. The absorption spectra will subsequently be compared with the action spectra obtained for the promotion as well as for the inhibition of seed germination of *Lactuca sativa* (lettuce).

(a)

Pr ⟶ (Red light, Sunlight) ⟵ (Far-red light, Canopy-filtered light) Pfr ⟶ Physiological response
e.g., promotion of seed germination
          inhibition of etiolation
            (excess stem elongation)
          promotion of leaf expansion
          inhibition of flowering

Reversion in dark

(b)

Figure 4.10

Phytochrome structure, interconversions, and associated physiological responses. (a) Possible structure for P$_r$, indicating the tetrapyrrole forming the chromophore and the protein attachment (see Furuya, 1987; Song, 1984). The convention for lettering the rings is also indicated. (b) Light and dark interconversions of phytochrome, indicating some of the reactions promoted by the physiologically active form, P$_{fr}$.

Phytochrome consists of a protein to which is covalently bound the *chromophore*, or light-absorbing part of the pigment (Fig. 4.10a). The chromophore is a *tetrapyrrole*, as occurs in the chlorophylls and the phycobilins that we will discuss in the next chapter. *Pyrrole* refers to a five-membered ring having four carbons, one nitrogen, and two double bonds, ⌷. The chromophore–protein bond (see Fig. 4.10a) may be an ester linkage to the propionic acid side chain on ring C and a thioether bond (to cysteine) to ring A. Light absorption may change a double bond near ring A, may affect the chromophore-protein binding, or may alter the three-dimensional structure of the protein and in any case evidently leads to an isomeric shift about the double bond between rings C and D; this photoisomerization may underlie the conversion of phytochrome to its physiologically active form (see Furuya, 1987; Lagarias, 1985; Song, 1984).

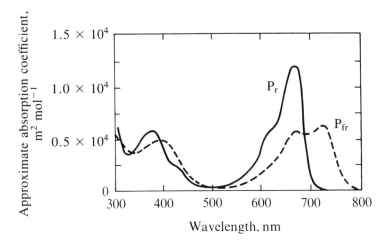

Figure 4.11

Absorption spectra for the red (P_r) and the far-red (P_{fr}) absorbing forms of phytochrome. (Data are replotted from W. L. Butler, S. B. Hendricks, and H. W. Siegelman, in *Chemistry and Biochemistry of Plant Pigments*, T. W. Goodwin, ed., Academic Press, London, 1965, pp. 197–210. Used by permission.) Note that the absorption spectra presented in this text were obtained at or near room temperature.

The chromophore of phytochrome is highly conjugated, as the structure for P_r in Figure 4.10a indicates (P stands for pigment and the subscript r indicates that it absorbs in the red region). In fact, P_r has eight double bonds in conjugation. Other double bonds are not in the conjugated system, because only those alternating with single bonds along the molecule are part of the main conjugation. For instance, the double bond at the top of pyrrole ring C is a branch, or cross-conjugation, to the main conjugation and only slightly affects the wavelength position for maximum absorption.[18] On the basis of the extensive conjugation, we would expect phytochrome to absorb in the visible region, as indeed it does.

The absorption spectra of two forms of phytochrome are presented in Figure 4.11. P_r has a major absorption band in the red, with a peak near 667 nm (the exact location varies with the plant species). Upon absorption of red light, P_r can be converted to a form having an absorption band in the far-red, P_{fr} (Figs. 4.10 and 4.11). Although maximum absorption for P_{fr} occurs near 725 nm,

---

18. Semiempirical rules exist for predicting the wavelength position of maximum absorption based on all double bonds that occur, including those in branches to the main conjugated system (see Calvert and Pitts, 1966; Williams and Fleming, 1987).

Figure 4.12

Action spectra for the promotion of lettuce seed germination and its reversal, or "inhibition." (Data are replotted from S. B. Hendricks and H. A. Borthwick, in *Chemistry and Biochemistry of Plant Pigments*, T. W. Goodwin, ed., Academic Press, London, 1965, pp. 405–436. Used by permission.)

appreciable absorption occurs in the near IR up to about 800 nm. In fact, the absorption bands of both pigments are rather broad, and many different wavelengths of light can be absorbed by each of them (Fig. 4.11). As is indicated in Figure 4.11, the maximum absorption coefficients are about 10^4 m^2 mol^{-1} for both P_r and P_{fr}. The plant pigments that we will consider in the next chapter have similar high values of ε_λ. We should also note that the absorption spectra for P_r and P_{fr} overlap considerably; e.g., 660 nm light can be readily absorbed by either pigment.

We will next examine the action spectra for seed germination responses with regard to the known absorption properties of phytochrome. Figure 4.12 indicates that a pigment absorbing in the red region promotes the germination of lettuce seeds. If we compare this action spectrum with the absorption spectra given in Figure 4.11, we see that the pigment absorbing the light that promotes seed germination most likely is the P_r form of phytochrome. This enhancement of seed germination by red light can be reversed by subsequent irradiation of the seeds with far-red light. In other words, the potential promotion of seed germination by red light can be stopped or at least reduced by irradiation with various longer wavelengths immediately following irradiation with the red light. The action spectrum for such "inhibition" of lettuce seed germination is also presented in Figure 4.12. This inhibition induced by far-red light requires fewest photons at wavelengths near 725 to 730 nm, similar to the position of the peak in the absorption spectrum of P_{fr} (Fig. 4.11). In summary, both the promotion of lettuce seed germination and the reversal of this promotion are apparently controlled by two forms of phytochrome, P_r and P_{fr}, which can be reversibly interconverted by light.[19]

To illustrate the ecological consequences of the phytochrome control of seed germination, let us consider a seed present on the surface of the soil under a dense canopy of leaves. Because of their chlorophyll, leaves in the canopy absorb red light preferentially to far-red light; thus very little red light compared with far-red light reaches the seed. The phytochrome in the seed hence occurs predominantly in the P_r form. However, if this seed were to become exposed to sunlight—as could happen because of a fire or the natural death of a shading tree—a larger fraction of the P_r would be converted to the active form, P_{fr} (see Fig. 4.10b). Seed germination would then proceed under light conditions favorable for photosynthesis and thus for growth.

Studies using action spectra have indicated that, on a photon basis, the maximum sensitivity is usually quite different for the opposing responses attributable to the two forms of phytochrome. (Because the expression of phytochrome action involves a multistep process including biochemical reactions, the two responses need not have the same sensitivity.) For lettuce seed germination, about 30 times more far-red photons (e.g., at 730 nm) are required to cause a 50% inhibition than the number of red photons (e.g., at 660 nm) needed to promote seed germination by 50%—compare the different scales used for the ordinates in Figure 4.12. Thus, ordinary sunlight is functionally equivalent to red light, because much of the phytochrome is converted to the active form, P_{fr}. Also, P_r has a much larger absorption coefficient in the red region than does P_{fr} (Fig. 4.11).

The P_{fr} form of phytochrome reverts spontaneously to P_r in the dark (Fig. 4.10b), except in most monocots. Reversion can take place in less than one hour and probably is not used by plants as a timing mechanism per se, although phytochrome may be involved in other aspects of photoperiodism (the influence of daylength or nightlength on plant processes). Also, the interconversions of phytochrome indicated in a simplified fashion in Figure 4.10b are actually multistep processes involving a number of intermediates. The mechanisms of action of phytochrome are not yet fully understood, although changes in membrane permeability, in the activity of membrane-bound enzymes, and in the transcription or translation of genetic information have all been implicated. Finally, as we might expect, individual plants vary considerably in how they amplify and use the information provided by the P_r–P_{fr} system (for further information on

---

19. Germination is not enhanced by red light for seeds of all species. Enhancement tends to be more common in small seeds rich in fat that come from wild plants (Salisbury and Ross, 1991).

phytochrome, see Furuya, 1987; Kendrick and Frankland, 1983; Salisbury and Ross, 1991; Song, 1984).

Our discussion of phytochrome leads us to the concept of a *photostationary state*. A photostationary state refers to the relative amounts of interconvertible forms of some pigment that occur in response to a particular steady illumination— such a state should more appropriately be called a "photosteady" state. When illumination is constant, the conversion of P_r to P_{fr} eventually achieves the same rate as the reverse reaction (see Fig. 4.10). The ratio of P_r to P_{fr} for this photostationary state depends on the absorption properties of each form of the pigment for the incident wavelengths (note that the absorption spectra overlap, Fig. 4.11), the number of photons in each wavelength interval, the kinetics of the competing de-excitation reactions, and the kinetics of pigment synthesis or degradation. If the light quality or quantity were to change to another constant condition, we would shift to a new photostationary state, where P_r/P_{fr} would in general be different. Because the ratio of P_r to P_{fr} determines the overall effect of the phytochrome system, a description in terms of photostationary states can be useful for discussing the influence of this pigment.

Problems

4.1 Consider electromagnetic radiation with the indicated wavelengths in a vacuum. (a) If λ is 400 nm, how much energy is carried by 10^{20} photons? (b) If a mol of 1 800 nm photons is absorbed by 10^{-3} m^3 (1 litre) of water at 0°C, what would be the final temperature? Assume that there are no other energy exchanges with the external environment; the volumetric heat capacity of water averages 4.19×10^6 J m^{-3} °C^{-1} over the temperature range involved. (c) A certain filter, which passes all wavelengths below 600 nm and absorbs all those above 600 nm, is placed over a radiometric device. If the meter indicates 1 W m^{-2}, what is the maximum photon flux density in μmol m^{-2} s^{-1}? (d) What is the illuminance in (c) expressed in lux (lumens m^{-2})?

4.2 Consider electromagnetic radiation having a frequency of 0.9×10^{15} cycles s^{-1}. (a) The speed of the radiation is 2.0×10^8 m s^{-1} in dense flint glass. What are the wavelengths in a vacuum, in air, and in such glass? (b) Can such radiation cause an $S_{(\pi,\pi)}$ ground state to go directly to $T_{(\pi,\pi*)}$? (c) Can such radiation cause the transition of a π electron to a $\pi*$ orbital in a molecule having six double bonds in conjugation? (d) Electromagnetic radiation is often expressed in "wave numbers," which is the frequency divided by the speed of light in a vacuum, i.e., ν/c, which is $1/\lambda_{vacuum}$. What is the wave number in m^{-1} in the present case?

4.3 Suppose that the quantum yield for ATP formation—molecules of ATP formed/ numbers of excited chlorphyll molecules—is 0.40 at 680 nm, and the rate of ATP formation is 0.20 mol m^{-2} hour^{-1}. (a) What is the minimum photon flux density in μmol m^{-2} s^{-1} at 680 nm? (b) What is the energy flux density under the conditions of (a)? (c) If light of 430 nm is used, the ground state of chlorophyll, $S_{(\pi,\pi)}$, is excited to $S^b_{(\pi,\pi*)}$. Suppose that 95% of $S^b_{(\pi,\pi*)}$ goes to $S^a_{(\pi,\pi*)}$ in 10^{-12} s and that the rest of the upper excited singlet state returns to the ground state. What is the energy conversion efficiency of 430 nm light as an energy source for ATP formation compared with 680 nm light? (d) The hydrolysis of ATP to ADP and phosphate under physiological conditions can yield about 40 kJ of free energy mol^{-1}. What wavelength of light has the same amount of energy mol^{-1}?

4.4 Assume that some excited singlet state can become de-excited by three competing processes: (1) fluorescence (lifetime $= 10^{-8}$ s), (2) a radiationless transition to an excited triplet state (5×10^{-9} s), and (3) a radiationless transition to the ground state (10^{-8} s). (a) What is the lifetime of the excited singlet state? (b) What is the maximum quantum yield for all de-excitations leading directly or indirectly to electromagnetic radiation? (c) Suppose that the above molecule is inserted into a membrane, which adds a de-excitation pathway involving intermolecular transfer of energy from the excited singlet state (rate constant $= 10^{12}$ s^{-1}). What is the new lifetime of the excited singlet state?

4.5 The *cis*-isomer of some species has an absorption coefficient of 2.0×10^3 m^2 mol^{-1} at 450 nm, where the spectrophotometer has a photon flux density of 10^{17} photons m^{-2} s^{-1}. (a) What concentration of the *cis*-isomer will absorb 65% of the incident 450 nm light for a cuvette (a transparent vessel used in a spectrophotometer) with an optical path length of 10 mm? (b) What is the absorbance of the solution in (a)? What would be the absorbance if the flux density at 450 nm were halved? (c) Suppose that the 450 nm light caused a photoisomerization of the *cis*-isomer to the *trans*-isomer with a quantum yield of 0.50, the other de-excitation pathway being the return to the *cis*-form. If the *trans*-isomer did not absorb at 450 nm, what would be the initial rate of decrease of the *cis*-isomer and rate of change of absorbance at 450 nm under the conditions of (a)? (d) If ε_{450} for the *trans*-isomer were 10^3 m^2 mol^{-1}, and 450 nm light led to a photoisomerization of the *trans*-isomer with a quantum yield of 0.50 for forming the *cis*-isomer, what would be the ratio of *cis* to *trans* after a long time?

4.6 (a) Suppose that the spacing in wave numbers (see Problem 4.2) between vibrational sublevels for the transition depicted in Figure 4.7 is 1.2×10^5 m^{-1} and that the most probable absorption predicted by the Franck–Condon principle occurs at 500 nm (the main band). What are the wavelength positions of the satellite bands that occur for transitions to the vibrational sublevels just above and just below the one for the most probable transition? (b) Suppose that the main band has a maximum absorption coefficient of 5×10^3 m^2 mol^{-1}, and each satellite band has an ε_λ one-fifth as large. If 20% of the incident light is absorbed at the wavelengths of either of the satellite bands, what percentage is absorbed at the main band wavelength? (c) When the pigment is placed in a cuvette with an optical path length of 5 mm, the maximum absorbance is 0.3. What is the concentration?

4.7 A straight chain hydrocarbon has 11 double bonds in conjugation. Suppose that it has three absorption bands in the visible region, one at 450 nm ($\varepsilon_{450} = 1.0 \times 10^4$

m^2 mol^{-1}), one at 431 nm ($\varepsilon_{431} = 2 \times 10^3$ m^2 mol^{-1}), and a minor band near 470 nm ($\varepsilon_{470} \cong 70$ m^2 mol^{-1}). Upon cooling from 20°C to liquid helium temperatures, the minor band essentially disappears. (a) What is the splitting between vibrational sublevels in the excited state? (b) What transition could account for the minor band? Support your answer by calculation. (c) If the λ_{max} for fluorescence is at 494 nm, what transition is responsible for the 450 nm absorption band? (d) If the double bond in the middle of the conjugated system is reduced (by adding 2 H's so that it becomes a single bond) and the rest of the molecule remains unchanged, calculate the new λ_{max} for the main absorption band and its absorption coefficient. Assume that for every double bond added to the conjugated system, λ_{max} shifts by 25 kJ mol^{-1}, and that $\varepsilon_{\lambda_{max}}$ is directly proportional to the number of double bonds in conjugation.

References

Alberty, R. A. 1983. *Physical Chemistry*, 6th ed. Wiley, New York.

Bell, C. J., and D. A. Rose. 1981. Light measurement and the terminology of flow. *Plant, Cell and Environment* 4:89–96.

Bickford, E. D., and S. Dunn. 1972. *Lighting for Plant Growth*. Kent State University Press, Kent, Ohio.

Caldwell, M. M. 1981. Plant response to solar ultraviolet radiation. In *Physiological Plant Ecology*, O. L. Lange, P. S. Nobel, C. B. Osmond, and H. Ziegler, eds. *Encyclopedia of Plant Physiology, New Series*, Vol 12A. Springer-Verlag, Berlin. Pp. 169–197.

Calvert, J. G., and J. N. Pitts, Jr. 1966. *Photochemistry*. Wiley, New York.

Clayton, R. K. 1965. *Molecular Physics in Photosynthesis*. Blaisdell, New York.

Clayton, R. K. 1980. *Photosynthesis: Physical Mechanisms and Chemical Patterns*. Cambridge University Press, Cambridge, England.

Condon, E. 1926. A theory of intensity distribution in band systems. *Physical Review* 28:1182–1201.

Cowan, D. O., and R. L. Drisko. 1976. *Elements of Organic Photochemistry*. Plenum, New York.

Depuy, C. H., and O. L. Chapman. 1972. *Molecular Reactions and Photochemistry*. Prentice-Hall, Englewood Cliffs, New Jersey.

Ditchburn, R. W. 1976. *Light*, 3rd ed. Academic Press, London.

Eisenberg, D., and D. Crothers. 1979. *Physical Chemistry with Applications to the Life Sciences*. Benjamin/Cummings, Menlo Park, California.

Franck, J. 1926. Elementary processes of photochemical reactions. *Transactions of the Faraday Society* 21:536–542.

Furuya, M., ed. 1987. *Phytochrome and Photoregulation in Plants*. Academic Press, Tokyo.

Gates, D. M. 1980. *Biophysical Ecology*. Springer-Verlag, New York.

Hart, J. W. 1988. *Light and Plant Growth*. Unwin Hyman, London.

Jerlov, N. G. 1976. *Marine Optics*, 2nd ed. Elsevier, Amsterdam.

Kaufman, J. E., and H. Haynes, eds. 1981. *IES Lighting Handbook*, 2 vols. Illuminating Engineering Society of North America, New York.

Kendrick, R. E., and B. Frankland. 1983. *Phytochrome and Plant Growth*, 2nd ed. *The Institute of Biology's Studies in Biology*, No. 68. Arnold, London.

Kirk, J. T. O. 1983. *Light and Photosynthesis in Aquatic Ecosystems*. Cambridge University Press, Cambridge, England.

Lagarias, J. C. 1985. Progress in the molecular analysis of phytochrome. *Photochemistry and Photobiology 42*:811–820.

Levine, J. N. 1970. *Quantum Chemistry*, Vol. I, *Quantum Mechanics and Molecular Electronic Structure*. Allyn and Bacon, Boston, Massachusetts.

McCree, K. J. 1981. Photosynthetically active radiation. In *Physiological Plant Ecology*, O. L. Lange, P. S. Nobel, C. B. Osmond, and H. Ziegler, eds. *Encyclopedia of Plant Physiology, New Series*, Vol. 12A. Springer-Verlag, Berlin. Pp. 41–55.

Monteith, J. L. 1973. *Principles of Environmental Physics*. American Elsevier, New York.

Orchin, M., and H. H. Jaffé. 1971. *Symmetry, Orbitals, and Spectra*. Wiley (Interscience), New York.

Pearcy, R. W. 1989. Radiation and light measurements. In *Plant Physiological Ecology: Field Methods and Instrumentation*, R. W. Pearcy, J. Ehleringer, H. A. Mooney, and P. W. Rundel, eds. Chapman and Hall, London. Pp. 97–116.

Resnick, R., and D. Halliday. 1977 (Part One) and 1986 (Part Two). *Physics*, 3rd ed. Wiley, New York. See also D. Halliday and R. Resnick. 1988. *Fundamentals of Physics*, 3rd ed. Wiley, New York.

Salisbury, F. B., and C. W. Ross. 1991. *Plant Physiology*, 4th ed. Wadsworth, Belmont, California. In press.

Seliger, H. H., and W. D. McElroy. 1965. *Light: Physical and Biological Action*. Academic Press, New York.

Sobel, M. I. 1987. *Light*. University of Chicago Press, Chicago, Illinois.

Song, P.-S. 1984. Phytochrome. In *Advanced Plant Physiology*, M. B. Wilkins, ed. Pitman, London. Pp. 354–379.

Thimijan, R. W., and R. D. Heins. 1983. Photometric, radiometric, and quantum light units of measurement: A review of procedures for interconversion. *HortScience 18*:818–822.

Tibbitts, T. W., and T. T. Kozlowski, eds. 1979. *Controlled Environment Guidelines for Plant Research*. Academic Press, New York.

Turro, N. J. 1978. *Modern Molecular Photochemistry*. Benjamin/Cummings, Menlo Park, California.

Waldman, G. 1983. *Introduction to Light*. Prentice-Hall, Englewood Cliffs, New Jersey.

Wayne, R. P. 1988. *Principles and Applications of Photochemistry*. Oxford University Press, Oxford, England.

Williams, D. H., and I. Fleming. 1987. *Spectroscopic Methods in Organic Chemistry*, 4th ed. McGraw-Hill, London.

Willson, R. C., and H. S. Hudson. 1988. Solar luminosity variations in solar cycle 21. *Nature 332*:810–812.

CHAPTER 5

Photochemistry of Photosynthesis

Photosynthesis is the largest-scale synthetic process on earth. In one year about 0.7×10^{14} kg of carbon are fixed into organic compounds by photosynthetic organisms (this is often called the *net primary productivity*; see Hall and Rao, 1987; Lieth and Whittaker, 1975). This equals about 1% of the world's known reserves of fossil fuels (coal, gas, and oil), or 10 times the world's present annual energy consumption. The carbon source used in photosynthesis is the 0.04% CO_2 contained in the air (about 8×10^{14} kg carbon) and the CO_2 or HCO_3^- dissolved in lakes and oceans (about 400×10^{14} kg carbon; O'Neill, 1985). In addition to the organic compounds, another product of photosynthesis essential for all respiring organisms is O_2. At the present rate, the entire atmospheric content of O_2 is replenished by photosynthesis every few thousand years.

Photosynthesis is not a single reaction; rather, it is composed of many individual steps that work together with remarkably high overall efficiency. We can divide the process into three stages: (1) the photochemical steps, our primary concern in this chapter; (2) electron transfer, to which is coupled the formation of ATP, which we will consider in both this and the next chapter; and (3) the biochemical reactions involving the incorporation of CO_2 into carbohydrates. Figure 5.1 summarizes the processes involved and introduces the relative amounts of the various reactants and products taking part in the three stages of photosynthesis. The photochemical reactions, which are often referred to as the *primary events* of photosynthesis, lead to electron transfer along a sequence of molecules, resulting in the formation of NADPH and ATP.

Let us now consider the net chemical reaction for photosynthesis. In Figure

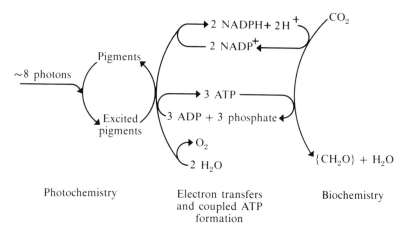

| Photochemistry | Electron transfers and coupled ATP formation | Biochemistry |

Figure 5.1

Schematic representation of the three stages of photosynthesis. The absorption of light can cause the excitation of photosynthetic pigments, which leads to the photochemical events where electrons are donated by special chlorophylls. The electrons are then transferred along a series of molecules, leading to the reduction of $NADP^+$ to NADPH; ATP formation is coupled to the electron transfer steps. The biochemistry of photosynthesis can proceed in the dark and requires 3 moles of ATP and 2 moles of NADPH per mole of CO_2 fixed into a carbohydrate, represented in the figure by $\{CH_2O\}$.

5.1, two H_2O's are indicated as reactants in the O_2 evolution step, and one H_2O is a product in the biochemical stage. Hence, the overall *net* chemical reaction describing photosynthesis is: CO_2 plus H_2O yields carbohydrate plus O_2. It is instructive to indicate the energy of each of the chemical bonds in these compounds, which leads to the following representation for the net photosynthetic reaction:

$$H \overset{463}{-\!\!\!-} O \overset{463}{-\!\!\!-} H + O \overset{800}{=\!\!\!=} C \overset{800}{=\!\!\!=} O \rightarrow$$

$$\overset{\tfrac{1}{2}(348)}{|}$$
$$H \overset{413}{-\!\!\!-} C \overset{350}{-\!\!\!-} O \overset{463}{-\!\!\!-} H + O \overset{498}{=\!\!\!=} O \qquad (5.1)$$
$$\underset{\tfrac{1}{2}(348)}{|}$$

where the numbers represent the various bond energies in kJ mol^{-1} (literature values for bond energies vary somewhat; see Atkins, 1986; Rabinowitch and Govindjee, 1969). A C—C bond, which occurs on two sides of the carbon in $\{CH_2O\}$, has a value of 348 kJ mol^{-1}, so $\tfrac{1}{2}(348)$ has been indicated in the appropriate places in Equation 5.1.

The formulation of photosynthesis in Equation 5.1 fails to do justice to the complexity of the reactions, but it is useful for estimating the amount of Gibbs free energy that is stored. We will begin by considering the energy of the chemical bonds. The total bond energy of the reactants in Equation 5.1 is 2 526 kJ mol^{-1} (463 + 463 + 800 + 800), and it is 2 072 kJ mol^{-1} for the products (413 + 348 + 350 + 463 + 498). Thus the reactants H_2O and CO_2 represent the lower energy (i.e., they are more "tightly" bonded), because 454 kJ mol^{-1} is necessary for the bond changes to convert them to the products $\{CH_2O\}$ plus O_2. This energy change actually represents the increase in enthalpy required, ΔH, although we are really more concerned here with the change in Gibbs free energy, ΔG (see Ch. 6 and App. IV). (For a reaction at constant temperature, ΔG equals $\Delta H - T \Delta S$, where S is the entropy; ΔG is about the same as ΔH for Eq. 5.1.) Although the actual ΔG per mole of C depends somewhat on the particular carbohydrate involved, 454 kJ is approximately the increase in Gibbs free energy per mole of CO_2 that reacts according to Equation 5.1. For instance, the Gibbs free energy released when glucose is oxidized to CO_2 and H_2O is 2 872 kJ mol^{-1} (686 kcal mol^{-1}) of glucose, or 479 kJ mol^{-1} of C. In discussing photosynthesis we will frequently use this ΔG, which actually refers to standard state conditions (25°C, pH 7, 1 molal concentrations, 1 atmosphere pressure).

About 8 photons are required in photosynthesis per CO_2 fixed and O_2 evolved (see Fig. 5.1). Red light at 680 nm corresponds to 176 kJ mol^{-1} (Table 4.1), so 8 moles of such photons have 1 408 kJ of radiant energy. Using this as the energy input and 479 kJ as the amount of energy stored per mole of CO_2 fixed, the efficiency of energy conversion by photosynthesis is (479 kJ/1 408 kJ)(100), or 34%. Actually, slightly more than 8 photons may be required per CO_2 fixed. Furthermore, the energy for wavelengths less than 680 nm, which are also used in photosynthesis, is higher than 176 kJ mol^{-1}. Both of these considerations lead to a calculated maximum efficiency for the utilization of absorbed energy of less than 34%. Nevertheless, photosynthesis is an extremely efficient energy conversion process, considering all the steps involved, each with its inherent energy losses.

Nearly all of the enzymes involved in the synthetic reactions of photosynthesis are also found in nonphotosynthetic tissue. Thus, the unique feature of photosynthesis is the conversion of radiant energy into chemical energy. The present chapter will emphasize the light absorption and the excitation transfer aspects of photosynthesis. We consider the structures and the absorption characteristics of the photosynthetic pigments and the means by which radiant energy is trapped, transferred, and eventually used. Thus, the emphasis is on the *photo* part of photosynthesis. We will examine the deductions concerning photosystems that derive from the early experiments on light absorption and will consider the structures of the various types of molecules involved in electron transport. In the next chapter, we will discuss energy conversion in an energetic context, paying particular attention to ATP and NADPH.

CHLOROPHYLL—CHEMISTRY AND SPECTRA

Chlorophylls represent the principal class of pigments responsible for light absorption in photosynthesis and are found in all photosynthetic organisms. There are a number of different types of chlorophyll, as Tswett demonstrated in 1906 using adsorption chromatography. For instance, approximately 1 g of the chlorophylls designated *a* and *b* is present per kg fresh weight of green leaves. The empirical formulas were first given by Willstätter, beginning in 1913; Fisher established the actual structures of various chlorophylls by 1940. These two investigators, as well as Woodward, who first synthesized chlorophyll *in vitro*, have all received the Nobel prize in chemistry for their studies on this important plant pigment (in 1915, 1930, and 1965, respectively). We will first consider the structure of chlorophyll *a* (abbreviated Chl *a*), then its absorption and fluorescence characteristics (for historical details see Goodwin, 1976; Vernon and Seeley, 1966).

Types and Structures

The various types of chlorophyll are identified by letters or by the taxonomic group of the plants in which they occur. The most important is Chl *a*. It has a molecular weight (relative molecular mass) of 893.5, and its structure is given in Figure 5.2. Chl *a* is found in all photosynthetic organisms except the green and the purple bacteria, i.e., in all species where O_2 evolution accompanies photosynthesis. It is a tetrapyrrole (see p. 236) having a relatively flat porphyrin "head" about 1.5 nm by 1.5 nm (15 Å by 15 Å) in the center of which a magnesium atom is coordinately bound. Attached to the head is a long-chain terpene alcohol, phytol, which acts like a "tail" about 2 nm in length containing 20 carbon atoms (Fig. 5.2). This tail provides a nonpolar region that helps bind the chlorophyll molecules to chlorophyll–protein complexes in the lamellar membranes, but it makes no appreciable contribution to the optical properties of chlorophyll in the visible region. The system of rings in the porphyrin head of Chl *a* is highly conjugated, having nine double bonds in conjugation (plus three other double bonds in branches to the main conjugated system). These alternating single and double bonds of the porphyrin ring provide many delocalized π electrons that can take part in the absorption of light.

Figure 5.2

Structure of Chl *a*, illustrating the highly conjugated porphyrin "head" to which is attached a phytol "tail." The convention for numbering the various rings is also indicated. The solid lines to Mg indicate a resonating form with shared electrons in the bonds, and the dashed lines indicate bonds with little electron sharing at that moment.

A number of other chlorophyll forms structurally similar to Chl *a* occur in nature. For instance, Chl *b* differs from Chl *a* by having a formyl group (—CHO) in place of a methyl group (—CH₃) on ring II. Chl *b* is found in virtually all land plants (including ferns and mosses), the green algae, and the Euglenophyta; the ratio of Chl *a* to Chl *b* in these organisms is usually about 3. Chl *b* is not essential for photosynthesis, as a barley mutant containing only Chl *a* carries out photosynthesis quite satisfactorily (Levine, 1969).

Another type is Chl *c*, which occurs in the dinoflagellates, diatoms, golden algae, and brown algae. The common chlorophyll of the green and the purple bacteria is bacteriochlorophyll *a*, or Bchl *a*. The purple photosynthetic bacteria contain Bchl *a* (or *b* in some species), and Bchl *a* plus Chlorobium chlorophyll occur in green photosynthetic bacteria. These bacterial pigments differ from green plant chlorophylls in that they contain two more hydrogens in the porphyrin ring. They also have different substituents around the periphery of the porphyrin ring. In addition, Chlorobium chlorophyll has the alcohol farnesol (15 C and three double bonds) in the place of phytol (20 C and one double bond). The pigment of principal interest in this text is Chl a (for further comments on the distribution of chlorophylls among various plants, see Goodwin, 1976; Jeffrey, 1981; Vernon and Seely, 1966).

Absorption and Fluorescence Emission Spectra

The absorption spectrum of Chl a has a blue band and a red band; hence the characteristic color of chlorophyll is green. The band in the blue part of the spectrum has a peak at 430 nm for Chl a in ether (Fig. 5.3). This band is known as the *Soret band,* which occurs in the UV, violet, or blue region for all tetra-pyrroles. We will designate the wavelength position for a maximum of the absorption coefficient in an absorption band by λ_{max}. Figure 5.3 indicates that the absorption coefficient at the λ_{max} for the Soret band of Chl a is just over 1.2 $\times$ 10^4 m^2 mol^{-1} (1.2 $\times$ 10^5 M^{-1} cm^{-1}; see footnote 16, Ch. 4). Such a high value is a consequence of the many double bonds in the conjugated system of

Figure 5.3

Absorption and fluorescence emission spectra of Chl a dissolved in ether. [Source: A. S. Holt and E. E. Jacobs, *American Journal of Botany 41*:710–717 (1954). Data reprinted by permission.]

the porphyrin ring of chlorophyll. Chl a has a major band in the red region with a λ_{max} at 662 nm when the pigment is dissolved in ether (Fig. 5.3).

Chl a also has a number of minor absorption bands. For instance, Chl a dissolved in ether has a small absorption band at 615 nm, which is 47 nm shorter than the λ_{max} of the main red band (Fig. 5.3). Absorption of light at 615 nm leads to an electronic transition requiring 14 kJ mol^{-1} more energy than the main band at 662 nm. This extra energy is similar to the energy spacing between vibrational sublevels. In fact, this small band on the shorter wavelength (higher energy) side of the red band corresponds to electrons going to the vibrational sublevel in the excited state immediately above the sublevel for the λ_{max} at 662 nm—an aspect to which we will return shortly.

Although chlorophyll absorbs strongly in both the red and the blue, the fluorescence emission is essentially all in the red region (Fig. 5.3). This is because the upper singlet state of chlorophyll excited by blue light ($S^b_{(\pi,\pi*)}$ in Fig. 4.5) is extremely unstable and goes to the lower excited singlet state $S^a_{(\pi,\pi*)}$ in about 10^{-12} s, i.e., before any appreciable blue fluorescence can take place.[1] Because of such rapid energy degradation by a radiationless transition, photons absorbed by the Soret band of chlorophyll are no more effective for photosynthesis than are the lower energy photons absorbed in the red region. We can observe the red fluorescence of chlorophyll accompanying light absorption by the Soret band if we illuminate a leaf with blue or shorter-wavelength light in a darkened room. With a light microscope we can see the red fluorescence emanating from individual chloroplasts in the leaf's cells when using such exciting light (the red fluorescence is often masked by scattering when using red exciting light, so shorter wavelengths are used in most fluorescence studies).

On the basis of the discussion in the previous chapter, we can reason that the transition having a λ_{max} at 662 nm in the absorption spectrum for Chl a dissolved in ether corresponds to the excitation of the molecule from the lowest vibrational sublevel of the ground state to some vibrational sublevel of the lower excited state. We can use the Boltzmann factor [$n(E)/n_{total} = e^{-E/RT}$, Eq. 3.21b] to estimate the fraction of chlorophyll molecules in the first excited vibrational sublevel of the ground state. Because RT is 2.48 kJ mol^{-1} at 25°C (App. II), and the distance between vibrational sublevels is about 14 kJ mol^{-1} for chlorophyll, the Boltzmann factor equals $e^{-(14 \text{ kJ mol}^{-1})/(2.48 \text{ kJ mol}^{-1})}$, or $e^{-5.65}$, which is 0.0035. Therefore, only about one in 300 chlorophyll molecules are normally in the first excited vibrational sublevel of the ground state when light arrives.

---

1. $S^b_{(\pi,\pi*)}$ and $S^a_{(\pi,\pi*)}$ are distinct excited electronic states; each has its own energy curve in a diagram like Figure 4.7.

Consequently, the absorption of a photon most likely occurs when chlorophyll is in the lowest vibrational sublevel of the ground state.

In Chapter 4 we argued that fluorescence generally occurs from the lowest vibrational sublevel of the excited singlet state. In other words, any excess vibrational energy is usually dissipated before the rest of the energy of the absorbed photon is reradiated as fluorescence. But Figure 5.3 shows that the wavelength region for most of the fluorescence is nearly coincident with the red band in the chlorophyll absorption spectrum.[2] In particular, the λ_{max} for fluorescence occurs at 666 nm, which is only 1 kJ mol^{-1} lower in energy than the λ_{max} of 662 nm for the red band in the absorption spectrum. The slight shift, which is much less than the distance between vibrational sublevels of 14 kJ mol^{-1} for chlorophyll, is most likely due to the loss of some rotational energy (rotational subsublevels of a vibrational sublevel are generally about 1 kJ mol^{-1} apart). Thus, the transition from the lowest vibrational sublevel of the ground state up to the lower excited state (the red absorption band) has essentially the same energy as a transition from the lowest vibrational sublevel of that excited state down to the ground state (the red fluorescence band). The only way for this to occur is to have the lowest vibrational sublevels of both the ground state and the excited state involved in each of the transitions. Hence, the red absorption band corresponds to a transition of the chlorophyll molecule from the lowest vibrational sublevel of the ground state to the lowest vibrational sublevel of the lower excited state, as is depicted in Figure 5.4.

The participation of the lowest vibrational sublevels of both the ground state and the lower excited state of Chl a in the major red band can also be appreciated by considering the minor band adjacent to the major red band in both the absorption spectrum and the fluorescence emission spectrum (see Figs. 5.3 and 5.4). The shorter wavelength absorption band at 615 nm in ether—14 kJ mol^{-1} higher in energy than the 662 nm band—corresponds to a transition to the first excited vibrational sublevel in the lower excited state. De-excitations from the lowest vibrational sublevel of the lower excited state to excited vibrational sublevels of the ground state would correspond to fluorescence at wavelengths greater than 700 nm. In fact, a small band near 728 nm in the fluorescence emission spectrum of Chl a (Fig. 5.3) occurs about 62 nm on the long wavelength side of the main fluorescence band, indicating an electronic transition with 15 kJ mol^{-1} less energy than the 666 nm band. This far-red band most likely corresponds to fluorescence emitted as the chlorophyll molecule goes from the lowest vibrational sublevel of the lower excited state to the first excited vibrational sublevel

---

2. The difference in energy or wavelength between absorption and fluorescence bands is often called the Stokes shift.

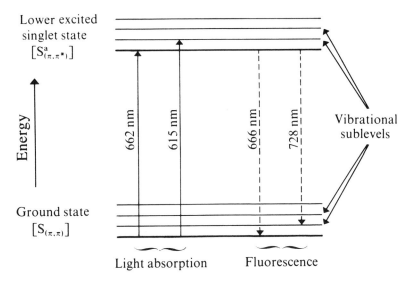

Figure 5.4

Energy level diagram, indicating the vibrational sublevels of the ground state $[S_{(\pi,\pi)}]$ and the lower excited singlet state $[S^a_{(\pi,\pi*)}]$ of Chl a. Solid vertical lines indicate the absorption of light by Chl a dissolved in ether; dashed lines represent fluorescence at the specified wavelengths. The lengths of the arrows are proportional to the amounts of energy involved in the various transitions.

of the ground state (Fig. 5.4). In summary, we note that: (1) excitations from excited vibrational sublevels of the ground state are uncommon, which is a reflection of the Boltzmann energy distribution; (2) fluorescence from excited vibrational sublevels of an excited state is also uncommon, because radiationless transitions to the lowest vibrational sublevel are so rapid; and (3) transitions to excited vibrational sublevels of the ground state can be quite significant (see Fig. 5.4 for Chl a).

Absorption *in Vivo*—Polarized Light

Chl a exhibits a number of spectroscopically different forms *in vivo*. The various values of λ_{max} for Chl a result from interactions between the chlorophylls and the surrounding molecules, such as the proteins and the lipids in the lamellae as well as adjacent water molecules. In fact, probably all Chl a is associated with proteins in chlorophyll–protein complexes. Hydrophobic interactions among phytol tails of adjacent chlorophylls and perhaps with hydrophobic regions in the protein help stabilize these chlorophyll–protein complexes. Because of the interactions of the porphyrin ring with the other molecules in the complex, and especially with the polar amino acids of the protein, the red bands for Chl *a in vivo* are shifted toward longer wavelengths than for Chl *a* dissolved in ether (λ_{max} at 662 nm), e.g., 670 to 680 nm. This is an example of the pronounced effect that the solvent or other neighboring molecules can have in determining the positions of the electronic energy levels of a pigment. The red absorption band of Chl *b in vivo* occurs as a "shoulder" on the short wavelength side of the Chl *a* red band, usually near 650 nm, and its Soret band occurs at slightly longer wavelengths than for Chl *a*.

A small amount of Chl *a* occurs in a special site that plays a particularly important role in photosynthesis. These Chl *a*'s have λ_{max}'s at approximately 680 nm and 700 nm and are referred to as P_{680} and P_{700}, respectively (P indicating pigment). P_{700} is apparently a dimer of Chl *a* molecules (i.e., 2 Chl *a*'s acting as a unit). P_{680} may also be a dimer, although some evidence indicates that it is a single Chl *a* (Amesz, 1987; Clayton, 1980; Gregory, 1989).

We can define the *bandwidth* of an absorption band as the difference in energy between photons on the two sides of the band at wavelengths where the absorption has dropped to half of that for λ_{max}. Such bandwidths of the red absorption bands of the various Chl *a*'s *in vivo* are fairly narrow—often about 10 nm at 20°C (Brown, 1972). At 680 nm a bandwidth of 10 nm is equivalent to 3 kJ mol^{-1}; i.e., a photon having a wavelength of 675 nm has an energy 3 kJ mol^{-1} greater than a photon with a wavelength of 685 nm. An energy of 3 kJ mol^{-1} is smaller than the spacing between vibrational sublevels of 14 kJ mol^{-1} for Chl *a*. Thus, a bandwidth of 3 kJ mol^{-1} results mainly from the rotational and the translational broadening of an electronic transition to a single vibrational sublevel of the excited state of Chl *a*.

The absorption of polarized light by chlorophyll *in vivo* can provide information on the orientations of individual chlorophyll molecules (see Clayton, 1980; Goedheer, 1972; Vernon and Seely, 1966). (*Polarized* means that the oscillating electric vector of light, Fig. 4.1, is in some specified direction.) The electronic transition of chlorophyll to the excited singlet state responsible for the red absorption band has its electric dipole in the plane of the porphyrin ring— actually, there are two dipoles in the plane in mutually perpendicular directions. Polarized light of the appropriate wavelength with its oscillating electric vector parallel to one of the dipoles is therefore preferentially absorbed by chlorophyll— recall that the probability for absorption is proportional to the square of the cosine of the angle between the induced dipole and the electric field vector of light (see Ch. 4). Absorption of polarized light indicates that the porphyrin rings of a few percent of the Chl *a* molecules, perhaps including P_{680}, are nearly parallel to the plane of the chloroplast lamellae. However, most of the chlorophyll molecules have their porphyrin heads randomly oriented in the internal membranes of chloroplasts.

Observations on the degree of polarization of fluorescence following the absorption of polarized light can tell us whether the excitation has been transferred from one molecule to another. If the same chlorophyll molecules that absorbed polarized light later emit photons when they go back to the ground state, the fluorescence would be polarized to within a few degrees of the direction of the electric vector of the incident light. However, the chlorophyll fluorescence following absorption of polarized light by chloroplasts is not appreciably polarized. This *fluorescence depolarization* indicates that the excitation energy has been transferred from one chlorophyll molecule to another so many times that the directional aspect becomes randomized, i.e., the chlorophyll molecule emitting fluorescence is randomly aligned relative to the chlorophyll that absorbed the polarized light.

When unpolarized light is incident on chloroplast lamellae that have been oriented in some particular direction, the observed fluorescence is polarized. The plane of polarization is similar to the plane of the membranes, indicating that the emitting chlorophyll molecules have their porphyrin rings in about the same orientation as the membrane. But the porphyrin rings of the absorbing chlorophyll molecules are randomly oriented. Again, we must conclude that the excitation has been transferred from the absorbing to the emitting molecule.

OTHER PHOTOSYNTHETIC PIGMENTS

Besides chlorophyll, other molecules in photosynthetic organisms also absorb light in the visible region. These molecules pass their electronic excitations on to Chl a (or to Bchl a,); they are often referred to as auxiliary or *accessory* pigments. In addition to Chl b and Chl c, two groups of accessory pigments important to photosynthesis are the *carotenoids* and the *phycobilins*. These two classes of accessory pigments can absorb yellow or green light, wavelengths for which absorption by chlorophyll is not appreciable.

Fluorescence studies have provided valuable information on the sequence of excitation transfer to and from the accessory pigments. For example, light absorbed by carotenoids, phycobilins, and Chl b leads to the fluorescence of Chl a. However, light absorbed by Chl a does not lead to the fluorescence of the accessory pigments, suggesting that excitation energy is not transferred from Chl a to the accessory pigments. Thus, accessory pigments can increase the photosynthetic use of white light by absorbing at wavelengths where Chl a

absorption is low; the excitations are then transferred to Chl a before the photochemical reactions take place.

Carotenoids

Carotenoids are found in essentially all green plants, algae, and photosynthetic bacteria (see Goodwin, 1980; Isler, 1971). In fact, the dominant pigments for plant leaves are the chlorophylls, which absorb strongly in the red and the blue regions, and the carotenoids, which absorb mostly in the blue and somewhat in the green region of the spectrum. The predominant colors reflected or transmitted by leaves are therefore green and yellow. In the autumn, chlorophyll in the leaves of deciduous plants can bleach and is usually not replaced, thereby greatly reducing absorption in the red and the blue regions. The remaining carotenoids absorb only in the blue and the green regions, leading to the well-known fall colors of such leaves, namely, yellow, orange, and red. Animals apparently do not synthesize carotenoids. Hence, brightly colored birds such as canaries and flamingoes, as well as many invertebrates, obtain their yellow or reddish colors from the carotenoids in the parts of the plants that they eat.

Carotenoids are bound to the chlorophyll–protein complexes, of which there are a number of types in the lamellar membranes of chloroplasts (Fig. 1.9). Carotenoids also occur in organelles known as *chromoplasts,* which are about the size of chloroplasts and are often derived from them. Lycopene (red) is in tomato fruit chromoplasts, α- and β-carotenes (orange) occur in carrot root chromoplasts, and the various chromoplasts of flowers contain a great diversity of carotenoids (for further details on chromoplasts, see Kirk and Tilney-Bassett, 1978).

Carotenoids are 40-carbon terpenoids or isoprenoids. They are composed of 8 *isoprene* units, where isoprene is a 5-carbon compound having two double bonds (CH_2＝C CH＝CH_2). In many carotenoids, the isoprene units on one or

$\quad\quad\quad\quad\quad\quad$|

$\quad\quad\quad\quad\quad$C H_3

both ends of the molecule are part of six-membered rings. Carotenoids are about 3 nm long, and those involved in photosynthesis generally have nine or more double bonds in conjugation.

The wavelength position of the λ_{max} depends on the solvent, on the substitutions on the hydrocarbon backbone, and on the number of double bonds in the conjugated system. We can illustrate this latter point for carotenoids in n-hexane, where the central maxima of the three observed peaks in the absorption spectra are at 286 nm for 3 double bonds in conjugation, at 347 nm for 5, at 400 nm for 7, at 440 nm for 9, at 472 nm for 11, and at 500 nm for 13 double bonds in conjugation—the only change from molecule to molecule in this series is in the number of double bonds in conjugation (see Goodwin, 1976). Thus, the greater the degree of conjugation, the longer is the wavelength representing λ_{max}, as we discussed in Chapter 4. For the 9 to 12 double bonds occurring in the conjugated systems of photosynthetically important carotenoids, the maximum absorption coefficient is greater than 10^4 m^2 mol^{-1}.

The carotenoids that serve as accessory pigments for photosynthesis absorb strongly in the blue (425 to 490 nm, Table 4.1) and somewhat in the green (490 to 560 nm), usually having triple-banded spectra in the region from 400 to 540 nm. For β-carotene in hexane, the three bands are centered at 425, 451, and 483 nm, and another major carotenoid in plants, lutein, has peaks at 420, 447, and 477 nm when dissolved in ethanol (absorption spectra in Fig. 5.5). The three absorption bands in each spectrum are about 17 kJ mol^{-1} apart, a reasonable value for the spacing between adjacent vibrational sublevels. Hence, the triple-banded spectra characteristic of carotenoids most likely represent transitions to

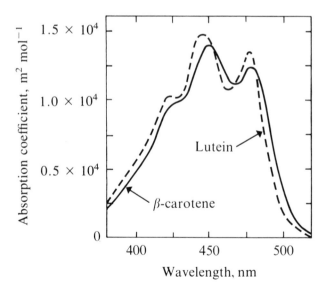

Figure 5.5

Absorption spectra for the two major carotenoids of green plants. [Data for β-carotene (in *n*-hexane) and lutein (in ethanol) are from F. P. Zscheile, J. W. White, Jr., B. W. Beadle, and J. R. Roach, *Plant Physiology* 17:331–346 (1942). Used by permission.]

three adjacent vibrational sublevels in the same excited state. It is not certain which vibrational sublevels are involved, in part because it has proved difficult to detect any carotenoid fluorescence. The spectra of the carotenoids *in vivo* are shifted about 20 to 30 nm toward longer wavelengths (lower energy) compared with absorption when the pigments are dissolved in hexane or ethanol.

Carotenoids are usually subdivided into the hydrocarbon ones, the *carotenes,* and the oxygenated ones, the *xanthophylls.* The major carotene in green plants is β-carotene (absorption spectrum in Fig. 5.5, structure in Fig. 5.6); α-carotene is also abundant. The xanthophylls exhibit a much greater structural diversity than do the carotenes, because the added oxygen can be in hydroxy, keto, epoxy, or methoxy groups. The three most abundant xanthophylls in green plants are lutein (absorption spectrum in Fig. 5.5, structure in Fig. 5.6), violaxanthin, and neoxanthin, in that order, with cryptoxanthin and zeaxanthin less frequently encountered. The major carotene of algae is again β-carotene, and lutein is the most common xanthophyll, although great variation in the type and the amount of xanthophylls is characteristic of algae. For instance, golden algae, diatoms, and brown algae contain considerable amounts of the special xanthophyll fucoxanthin (Fig. 5.6), which functions as the main accessory pigment in these organisms. The distribution and the types of carotenoids in plants have evolutionary implications and taxonomic usefulness.

In addition to functioning as accessory pigments for photosynthesis, carotenoids are also important for protecting photosynthetic organisms from the destructive photooxidations that can occur in the presence of light, O_2, and certain pigments. In particular, light absorbed by chlorophyll can lead to excited states of O_2. These highly reactive states can damage chlorophyll, but their interactions with carotenoids prevent harmful effects to the organism. Because photosynthesis in the green and the purple bacteria does not lead to O_2 evolution, it can proceed in the absence of carotenoids. For instance, a mutant of *Rhodopseudomonas spheroides* lacking carotenoids performs photosynthesis in a normal manner in the absence of O_2; when O_2 is introduced in the light, the bacteriochlorophyll becomes photo-oxidized and the bacteria are killed, a sensitivity not present in related strains containing carotenoids. On the other hand, cyanobacteria, algae, and higher plants produce O_2 as a photosynthetic product, so they must contain carotenoids to survive in the light. In addition, daily cycles in specific carotenoids can occur, as β-carotene and certain xanthophylls are oxidized during the daytime and then reduced back to the original form at night (Demmig-Adams *et al.,* 1989). Because such oxidations tend to increase with time, the fraction of carotenoids in the form of xanthophylls generally increases in leaves as the growing season progresses (Czeczuga, 1987; Kirk and Tilney-Bassett, 1978).

β-carotene

Lutein

Fucoxanthin

Phycoerythrobilin

Phycocyanobilin

Figure 5.6

Structure of five important accessory pigments. The phycobilins (lower two structures) occur covalently bound to proteins; i.e., they are the chromophores for phycobiliproteins.

Phycobilins

The other main accessory pigments important in photosynthesis are the *phycobilins*. Lemberg in the 1920s termed these molecules "phycobilins" because they occur in algae (red algae and blue-green algae, the latter now referred to as cyanobacteria; *phyco* is derived from the Greek for seaweed) but they structurally resemble bile pigments. Like the chlorophylls, the phycobilins are tetrapyrroles. However, the four pyrroles in the phycobilins occur in an open chain, as is the case for phytochrome, not in a closed porphyrin ring, as is the case for the chlorophylls. Phycobilins have a molecular weight (relative molecular mass) of 586. They occur covalently bound to proteins with molecular masses of 30 to 35 kDa. These assemblies containing 300 to 800 phycobilins are organized into *phycobilisomes*, which are about 40 nm in diameter and are associated with the outer (stromal) surfaces of lamellar membranes in cyanobacteria and red algae, where they function as the main accessory pigments (Gantt, 1981; Gregory, 1989).

Phycobilins generally have their major absorption bands from 500 to 650 nm,

Figure 5.7

Absorption spectra of phycoerythrin from a red alga and phycocyanin from a cyanobacterium. (Data are replotted from C. Ó hEocha, in *Chemistry and Biochemistry of Plant Pigments*, T. W. Goodwin, ed., Academic Press, London, 1965, pp. 175–196. Used by permission.)

with a relatively small Soret band in the UV (Fig. 5.7). These pigments are higher in concentration in many cyanobacteria and red algae than are the chlorophylls and are responsible for the color of certain species. The main phycobilins are phycocyanobilin and phycoerythrobilin (structures in Fig. 5.6). Phycoerythrobilin plus the protein to which it is covalently attached is called *phycoerythrin*.[3] Phycoerythrin is soluble in aqueous solutions, so we can obtain absorption spectra for it under conditions similar to those *in vivo*. Phycoerythrin is reddish, because it absorbs green and has at least one main band between 530 and 570 nm (see absorption spectrum in Fig. 5.7). It occurs throughout the red algae and in some cyanobacteria. *Phycocyanin* (phycocyanobilin plus protein) appears bluish, because it absorbs strongly from 610 to 660 nm (Fig. 5.7). It is the main phycobilin in the cyanobacteria and also is found in the red algae. As is the case for other pigments, the greater is the number of double bonds in conjugation in the phycobilins, the longer are the wavelengths for λ_{max}. For example, phycoerythrobilin has seven double bonds in the main conjugated system and absorbs maximally in the green region of the spectrum; phycocyanobilin has nine such double bonds and its λ_{max} occurs in the red (see the structures of these compounds in Fig. 5.6). The maximum absorption coefficients of both phycobilins exceed 10^4 m^2 mol^{-1} (see Fig. 5.7).

As we indicated in Chapter 4, both the quantity and the quality of radiation change with depth in water, with wavelengths near 500 nm penetrating the deepest. For instance, only about 10% of the blue and the red parts of the spectrum penetrate to 50 m in clear water, so chlorophyll would not be a very useful light-harvesting pigment at that depth (see Fig. 5.3 for a chlorophyll absorption spectrum). Although there are many exceptions, changes in the spectral quality (relative amounts of various wavelengths) with depth affect the distribution of photosynthetic organisms according to their pigment types. The predominant accessory pigment in green algae is Chl *b*, which absorbs mainly in the violet (400 to 425 nm; see Table 4.1) and the red (640 to 740 nm). Green algae as well as sea grasses and fresh water plants grow in shallow water, where the visible spectrum is little changed from that of the incident sunlight. Fucoxanthin (Fig. 5.6) is the major accessory pigment in brown algae, such as the kelps, and it absorbs strongly in the blue and the green regions (425 to 560 nm), helping to extend the range of such plants downward to over 20 m. Marine red algae can occur at even greater depths (e.g., 100 m), and their phycoerythrin absorbs the green light (490 to 560 nm) that penetrates to such distances. Changes

---

3. The phycobilins are covalently bound to their proteins (referred to as *apoproteins*) to form phycobiliproteins, whereas chlorophylls and carotenoids are joined to their apoproteins by weak bonds such as H-bonds and hydrophobic interactions.

in spectral quality can also induce changes in the synthesis of biliproteins within a given organism. For instance, green light induces the synthesis of the green-absorbing phycoerythrin, and red light leads to the synthesis of the red-absorbing phycocyanin (see Fig. 5.7) in certain cyanobacteria and red algae (for further details see Glazer, 1977; Jeffrey, 1981).

Interestingly enough, only two types of pigments appear to be involved in all the known photochemical reactions in plants. These are the carotenoids and the tetrapyrroles, the latter class including the chlorophylls, the phycobilins, and phytochrome. The maximum absorption coefficients for the most intense absorption bands are slightly over 10^4 m^2 mol^{-1} in each case. Cytochromes, which are involved in the electron transport reactions in chloroplasts and mitochondria, are also tetrapyrroles (considered later in this chapter). Table 5.1 summarizes the relative frequency of the main types of photosynthetic pigments that we have been discussing.

Table 5.1

Approximate relative amounts and locations of photosynthetic pigments. Data are expressed per 600 chlorophylls and are for representative leaves of green plants (except for the phycobilins) growing at moderate sunlight. Photosystems and the light-harvesting antennae are discussed later in this chapter. (For further details see Anderson, 1986; Goodwin, 1976; Haehnel, 1984.)

Pigment	Number	Location
Chl a	450	Approximately 30% in Photosystem I, 20% in Photosystem II, and 50% in the light-harvesting antennae
Chl b	150	In light-harvesting antennae
P_{680}	~1.6	Trap for Photosystem II
P_{700}	~1.0	Trap for Photosystem I
Carotenoids	120	Most in light-harvesting antennae of Photosystem II
Phycobilins	500	Covalently bound to proteins situated on the outer surface of photosynthetic membranes in cyanobacteria and red algae; serve in light-harvesting antennae of Photosystem II

EXCITATION TRANSFERS AMONG PHOTOSYNTHETIC PIGMENTS

Chlorophyll is at the very heart of the primary events of photosynthesis. It helps to convert the plentiful radiant energy from the sun into chemical free energy that can be stored in various ways. In this section we will represent light absorption, excitation transfer, and the photochemical step as chemical reactions; this will serve as a prelude to a further consideration of certain molecular details of photosynthesis.

Pigments and the Photochemical Reaction

The first step in photosynthesis is light absorption by one of the pigments. The absorption event (discussed in Ch. 4) for the various types of photosynthetic pigments described in this chapter can be represented as follows:

$$\left. \begin{array}{c} \text{accessory pigment} \\ \text{or} \\ \text{Chl } a \\ \text{or} \\ \text{trap chl} \end{array} \right\} + h\nu \rightarrow \left\{ \begin{array}{c} \text{accessory pigment*} \\ \text{or} \\ \text{Chl } a* \\ \text{or} \\ \text{trap chl*} \end{array} \right. \tag{5.2}$$

where the asterisk refers to an excited state of the pigment molecule caused by the absorption of a light quantum, $h\nu$. Trap chl indicates a special type of Chl a (e.g., P_{680} or P_{700}) that occurs much less frequently than the other chlorophylls (see Table 5.1); we will consider its important excitation-trapping properties at the end of this section.

Because the photochemical reactions take place only at the trap chl molecules, the excitations resulting from light absorption by either the accessory pigments or the other Chl a's must be transferred to the trap chl before they can be used for photosynthesis. The relative rarity of trap chl compared with the other photosynthetic pigments means that it absorbs only a small fraction of the incident light. In fact, under natural conditions in green plants over 99% of the photons are absorbed by either the accessory pigments or Chl a. The migration of excitations from the initially excited species to the trap chl—the mechanism for which we will discuss below—can be represented as follows:

$$\text{accessory pigment*} + \text{Chl } a \rightarrow \text{accessory pigment} + \text{Chl } a* \tag{5.3}$$

$$\text{Chl } a* + \text{trap chl} \rightarrow \text{Chl } a + \text{trap chl*} \tag{5.4}$$

In other words, the direction of excitation transfer or migration is from the accessory pigments to Chl a (Eq. 5.3), and from Chl a to the special "trap" chlorophylls (Eq. 5.4) where the actual photochemical reactions take place. Hence, the overall effect of the steps described by Equations 5.2 through 5.4 is to funnel the excitations that are caused by the absorption of light to the trap chl.

A prerequisite for the conversion of radiant energy into a form that can be stored chemically is the formation of reducing and oxidizing species. The *reducing* (electron-donating) and the *oxidizing* (electron-accepting) species that result from light absorption must be fairly stable and located in such a way that they do not interact. (We will discuss the energetics of oxidation and reduction in Ch. 6.) If we denote the molecule that accepts an electron from the excited trap chl by A, this electron transfer step can be represented by

$$\text{trap chl*} + A \rightarrow \text{trap chl}^+ + A^- \tag{5.5}$$

where A^- indicates the reduced state of the acceptor and trap chl$^+$ means that the special chlorophyll has lost an electron. Equation 5.5 represents a *photochemical* reaction because the absorption of a light quantum (Eq. 5.2) has led to the transfer of an electron away from a special type of chlorophyll, representing a chemical change in that molecule. The electron removed from trap chl* (Eq. 5.5) can be replaced by one coming from a donor, D, which leads to the oxidation of this latter species, D^+, and the return of the trap chl to its unexcited state:

$$\text{trap chl}^+ + D \rightarrow \text{trap chl} + D^+ \tag{5.6}$$

The generation of stable reduced (A^-) and oxidized (D^+) intermediates can be considered to complete the conversion of light energy into chemical potential energy. Combining Equations 5.2 through 5.6 gives us the following relation for the net reaction describing the primary events of photosynthesis:

$$A + D + hv \rightarrow A^- + D^+ \tag{5.7}$$

The light-driven change in chemical free energy represented by the conversion of $A + D$ to $A^- + D^+$ (Eq. 5.7) eventually causes chemical reactions leading to the evolution of O_2 from water, the production of a reduced compound (NADPH), and the formation of high-energy phosphates (ADP + phosphate $\rightarrow$ ATP). Such a conversion of light energy into chemical energy represented by Equation 5.7 is the cornerstone of photosynthesis.

Resonance Transfer of Excitation

We have already mentioned a number of different examples of excitation transfer among photosynthetic pigments. For instance, light absorbed by the accessory pigments can lead to the fluorescence of Chl a. Also, studies on the absorption of polarized light by chlorophyll in $vivo$, where the resulting fluorescence is not polarized, provide further evidence that excitations can migrate from molecule to molecule before the energy is emitted as radiation. One case to consider is the transfer of excitation between identical molecules; e.g., the excitation of the lower excited singlet state of chlorophyll can be passed to a second chlorophyll molecule. This causes the deactivation of the originally excited molecule and the attainment of the lower excited singlet state in the second chlorophyll, a process described by Equation 4.7, $S_{(\pi,\pi*)} + S_{2(\pi,\pi)} \rightarrow S_{(\pi,\pi)} + S_{2(\pi,\pi*)}$. The most widely accepted mechanism for such exchange of electronic excitation between chlorophyll molecules is $resonance$ $transfer$ (also called inductive resonance, the Förster mechanism, or weak coupling), which we next consider qualitatively.

On the basis of our discussion in the previous chapter, we might expect that an excited molecule could induce an excited state in a second molecule in close proximity. In particular, the oscillating electric dipole representing the energetic electron in the excited state of the first molecule leads to a varying electric field. This field can cause a similar oscillation or resonance of some electron in a second molecule. A transfer of electronic excitation energy takes place when an electron in the second molecule is induced to oscillate. When excitation transfer is completed, the previously excited electron in the first molecule has ceased oscillating and some electron in the second molecule is now oscillating, leading to an excited state of that molecule. Resonance transfer of excitation between molecules is thus analogous to the process by which light is originally absorbed, because an oscillation of some electron in the molecule is induced by a locally varying electric field. Resonance transfer of excitation is most probable when there is the proper orientation between the electric dipole in the excited molecule and the potential dipole in the second molecule and the energy of the original dipole is appropriate, an aspect that we will consider next.

For resonance transfer of electronic excitation to occur, the energy available in the excited molecule must match the energy that can be accepted by a second molecule. The wavelengths for fluorescence indicate the energy of the excited singlet state of a molecule (at least after the very rapid radiationless transitions to the lowest vibrational sublevel of that excited state have occurred). Thus, although fluorescence itself is not involved in this type of excitation transfer, the fluorescence emission spectrum usually gives the range of energies available for transfer to a second molecule. The range of wavelengths of light that can sympathetically induce an oscillation of some electron in a second molecule is given by the absorption spectrum of that molecule (Ch. 4), and therefore the absorption spectrum shows the energies that can be accepted by a molecule. As might be expected from these two considerations, the probability for resonance transfer is high when the overlap in wavelength between the fluorescence band for the excited oscillator (available energy) and the absorption band of an unexcited oscillator (acceptable energy) in a neighboring molecule is large. Because the overlap in the red region between the absorption spectrum and the fluorescence emission spectrum of Chl a is very great (Fig. 5.3), excitations can be efficiently exchanged between Chl a molecules by resonance transfer. Figure 5.8 illustrates the various energy considerations involved in the case of resonance transfer of excitation between two dissimilar molecules.

The probability for resonance transfer of electronic excitation decreases as the distance between the two molecules increases. If chlorophyll molecules were

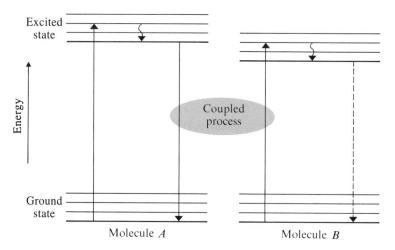

Figure 5.8

Resonance transfer of excitation from molecule A to molecule B. After light absorption by molecule A, a radiationless transition occurs to the lowest vibrational sublevel of its excited state. Next, resonance transfer of the excitation takes place from A to B, causing the second molecule to go to an excited state, while molecule A returns to its ground state. Following a radiationless transition to the lowest vibrational sublevel in the excited state, fluorescence can then be emitted by molecule B as it returns to its ground state. Based on the energy level diagrams (which include the vibrational sublevels for each of these two different pigments), we can conclude that generally the excitation rapidly decreases in energy after each intermolecular transfer between dissimilar molecules.

uniformly distributed in three dimensions in the lamellar membranes of chloroplasts, they would have a center-to-center spacing of approximately 2 nm, an
intermolecular distance over which resonance transfer of excitation can readily
occur (resonance transfer is effective up to about 10 nm for chlorophyll). Thus,
both the spectral properties of chlorophyll and its spacing in the lamellar membranes are conducive to an efficient migration of excitation from molecule to
molecule by resonance transfer (for a further discussion of resonance transfer,
see Barber, 1977; Clayton, 1980).

Transfers of Excitation between Photosynthetic Pigments

In addition to the transfer from one Chl a molecule to another, excitations can
also migrate by resonance transfer from the accessory pigments to Chl a. The
transfers of excitation among Chl a's can be nearly 100% efficient (i.e, $k_{transfer}$
$\gg$ rate constants for competing pathways; see p. 221), whereas the fraction of
excitation transfers between dissimilar molecules is quite variable. For instance,
the transfer of excitation from β-carotene to Chl a is very efficient in certain
algae, but the fraction of excitations transferred from most xanthophylls to Chl
a is generally low. The transfer of excitation from α-carotene and lutein to
chlorophyll usually is intermediate; e.g., about 40% of the excitations of these
carotenoids may be transferred to Chl a. Photons absorbed by the carotenoid
fucoxanthin—found in the golden algae, diatoms, and brown algae—can approach 100% efficiency of excitation transfer to Chl a. Also, most excitations
of phycobilins and Chl b can be transferred to Chl a. In red algae, 90% of the
electronic excitations produced by the absorption of photons by phycoerythrin
can be passed on to phycocyanin and then to Chl a; these transfers require about
4×10^{-10} s each (for efficiencies of excitation transfers between pigments, see
Clayton, 1980; Goedheer, 1972; Rabinowitch and Govindjee, 1969). We will
next consider the direction for excitation transfer between various photosynthetic
pigments and then the times involved for intermolecular excitation transfers of
chlorophyll.

Some energy is generally lost by each molecule to which the excitation is
transferred. Any excess vibrational or rotational energy is usually dissipated
rather rapidly as heat (see Fig. 5.8). Therefore, the λ_{max} for each type of pigment
involved in the sequential steps of excitation transfer tends to become longer in
the direction in which the excitation migrates. In particular, the fluorescence
emission spectrum of some molecule—which, for transfer to take place efficiently, must appreciably overlap the absorption spectrum of the molecule to
which the excitation is resonantly transferred—occurs at longer wavelengths than

the absorption spectrum of that molecule (see Fig. 5.3 for Chl a). Therefore, for a second molecule to become excited by resonance transfer, it should have an absorption band at longer wavelengths (lower energy) than the absorption band for the molecule from which it receives the excitation. Thus, the direction for excitation migration by resonance transfer among photosynthetic pigments is usually toward those pigments with longer λ_{max}. We can appreciate this important aspect by considering Figure 5.8. If excitation to the second excited vibrational sublevel of the excited state is the most probable transition predicted by the Franck–Condon principle for each molecule, then the excitation of molecule A requires more energy than that of B (the pigment to which the excitation is transferred). Hence, λ_{max_A} must be less than λ_{max_B} in the two absorption spectra, consistent with our statement that the excitation migrates toward the pigment with the longer λ_{max}.

As a specific example of the tendency for excitations to migrate toward pigments with longer λ_{max} in their absorption spectra, we will consider the transfer of excitations from the accessory pigments to Chl a. In red algae and in some cyanobacteria, phycoerythrin has a λ_{max} at about 560 nm and passes excitation energy on to phycocyanin, which has an absorption maximum near 620 nm. This excitation can then be transferred to a Chl a with a λ_{max} near 670 nm. The biliprotein allophycocyanin absorbs maximally at 650 nm and apparently intervenes in the transfer of excitation between phycocyanin and Chl a. Because some of the excitation energy is generally dissipated as heat by each molecule (see Fig. 5.8), the excitation represents less energy (longer λ) after each pigment in the sequence. Consequently, the overall direction for excitation migration is essentially irreversible.

As the nuclei vibrate back and forth after the absorption of a photon by some electron, their collisions with other nuclei every 10^{-13} s or so can lead in such short times to the dissipation of any excess energy in the excited vibrational sublevels. In addition, the radiationless transition from the upper excited singlet to the lower excited singlet of Chl a—$S_{(\pi,\pi*)}^b$ to $S_{(\pi,\pi*)}^a$ in Figure 4.5—is completed within 10^{-12} s. The time for the transfer of the excitation between two Chl a molecules *in vivo* is somewhat longer—about 1 to 2×10^{-12} s. Thus, the originally excited chlorophyll molecule usually attains the lowest vibrational sublevel of the lower excited singlet state before the excitation is transferred to another molecule. The amount of energy resonantly transferred from one Chl a to another therefore generally corresponds to the energy indicated by the fluorescence emission spectrum (Fig. 5.3). An excitation representing this amount of energy can, in principle, be transferred many times by resonance transfer with essentially no further degradation of the energy.

In Chapter 4 we noted that an upper time limit within which processes involving excited singlet states must occur is provided by the kinetics of fluorescence de-excitation. The lifetime for chlorophyll fluorescence from the lower excited singlet state is about 1.5×10^{-8} s. Thus, time is available for approximately 10 000 transfers of excitation among the Chl a molecules—each transfer requiring about 10^{-12} s—before the loss of the excitation by the emission of fluorescence. The number of excitation transfers between Chl a molecules actually taking place is much less than this, for reasons that will become clear shortly.

Excitation Trapping

We have already introduced the special Chl a's, P_{680} and P_{700}. These pigments tend to absorb at longer wavelengths than do the other types of Chl a; hence, the excited singlet states in P_{680} and P_{700} tend to be at lower energies. It is energetically feasible for the other Chl a's to excite such trap chl's by resonance transfer, but P_{680} and P_{700} usually do not pass the excitation back; i.e., they rapidly lose some of their excitation energy (within 10^{-12} s), so they do not retain enough energy to re-excite the other Chl a's by resonance transfer. Therefore, the excited singlet states of other Chl a molecules can have their excitations readily passed on to the trap chl's, but not vice versa—analogous to the irreversibility of the migration of excitations from the accessory pigments to Chl a. The excitations resulting from the absorption of radiation by the various photosynthetic pigments are thereby funneled into P_{680} or P_{700}. Such collecting of excitations by one species is the net effect of Equations 5.2 through 5.4 (with the term "trap chl" replaced by P_{680} or P_{700}).

Because one of the trap chl's is present per approximately 250 chlorophylls (Table 5.1), on the average only a few hundred transfers are necessary to get an excitation from Chl a to P_{680} or P_{700}. Thus, the 10 000 transfers of excitation from one Chl a to another possible within the fluorescence lifetime do not occur. Because each excitation transfer takes approximately 10^{-12} s, 100 transfers would require about 10^{-10} s. In agreement with this, both calculations from mathematical models and ingenious experimentation have shown that over 90% of the excitations of Chl a can migrate to P_{680} or P_{700} in less than 10^{-9} s (Clayton, 1980; Rabinowitch and Govindjee, 1969).

The characteristics of fluorescence provide information on the lifetime of the excited singlet state of chlorophyll *in vivo* and thus on the time available for migration of excitations. Specifically, about 1% to 3% of the light absorbed by Chl a *in vivo* is lost by fluorescence. The amount reradiated depends on the competing de-excitation reactions and hence is higher at higher incident light levels for which the photochemical reactions become saturated. Thus, near full sunlight the quantum yield for fluorescence *in vivo*, Φ_{Fl}, is approximately 0.03. Equation 4.12 ($\Phi_i = \tau/\tau_i$) indicates that this quantum yield is equal to τ/τ_{Fl}, where τ is the lifetime of the excited singlet state and τ_{Fl} is its fluorescence lifetime. It is reasonable to assume that τ_{Fl} *in vivo* is similar to the fluorescence lifetime of Chl *a in vitro*, 1.5×10^{-8} s. Therefore, Equation 4.12 predicts a lifetime for the excited state of Chl *a in vivo* of $(0.03)(1.5 \times 10^{-8}$ s), or 0.5×10^{-9} s. This is another estimate of the average time necessary for the excitation to migrate to the trap chl (not all of the Chl *a* readily fluoresces *in vivo*, and hence this is only an approximate guide to the lifetime; see Barber, 1978).

P_{680} and P_{700} act as traps for excitations in chloroplast lamellae, and a special type of bacteriochlorophyll with a λ_{max} between 870 and 890 nm (depending on the species) acts in an analogous manner for the green and the purple bacteria. One of the useful features of such excitation traps is to have an excited singlet state lower in energy than the excited singlet states in the other pigment molecules. This lower energy state (longer λ_{max} for absorption) is a consequence of the molecular environment in which the chlorophyll molecules acting as excitation traps are located. Moreover, the longer λ_{max} ensures the directionality for the migration of excitations. Another characteristic of the trap chl's is their relative rarity. Thus, most of the photosynthetic pigments act as "antennae," or

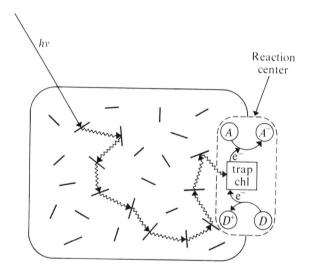

Figure 5.9

Schematic representation of a group of pigments that harvests a light quantum (*hv*) and passes the excitation on to a special trap chlorophyll. In the reaction center an electron (e^-) is transferred from the trap chl to some acceptor (A^- in the reduced form) and then replaced by another electron coming from a suitable donor (D^+ in the oxidized form).

light harvesters, which collect the radiation and channel the excitations toward the trap chl, as illustrated in Figure 5.9. Processing of the excitation originally caused by light takes place only at the trap chl's. This participation in the essentially irreversible, electron-transfer process is the crucial feature of an excitation trap.

When the excitation migrates to a trap such as P_{680} or P_{700}, this special Chl a goes to an excited singlet state, as would any other Chl a. Because the trap chl cannot readily excite other chlorophylls by resonance transfer, it might become de-excited by the emission of fluorescence. However, very little fluorescence from the trap chl's is observed *in vivo*. This is explained by the occurrence of a photochemical event (see Eq. 5.5, trap chl* $+ A \rightarrow$ trap chl$^+$ $+$ A^-); i.e., the rapid donation (within 10^{-10} s) of an electron to an acceptor could prevent the de-excitation of the trap chl's by fluorescence, which has a longer lifetime.

As we have just indicated, an excited trap chl can rapidly donate an electron to some acceptor molecule, which is part of the *photochemistry* of photosynthesis. The donation of the electron initiates the *chemical* reactions of photosynthesis and the subsequent storage of energy in stable chemical bonds. Moreover, once the trap chl has lost an electron, it can take on another one from some donor, indicated by D in Equation 5.6 (trap chl$^+$ $+ D \rightarrow$ trap chl $+ D^+$) and Figure 5.9. Thus, the photochemical reactions of photosynthesis lead to electron flow. An excitation trap such as P_{680} or P_{700} plays a key role in the conversion of radiant energy into forms of energy that are biologically useful. We generally refer to the trap chl plus A and D as a *reaction center* (illustrated in Fig. 5.9). The reaction center is the locus for the photochemistry of photosynthesis (Eq. 5.5).[4]

GROUPINGS OF PHOTOSYNTHETIC PIGMENTS

We have discussed the absorption of light by photosynthetic pigments and the ensuing transfers of excitation among these molecules, which leads us to a consideration of whether there are discrete units of such pigments acting together in some concerted fashion. Such an ensemble is presented in Figure 5.9, where the light-harvesting photosynthetic pigments greatly outnumber the special trap chl molecules, the latter occurring in a one-to-one relationship with suitable electron acceptors and donors. In this section we will consider whether photo-

---

4. The Nobel Prize in chemistry was awarded to Deisenhofer, Huber, and Michel in 1988 for unraveling the structure of the reaction center from the purple photosynthetic bacterium *Rhodopseudomonas viridis* using X-ray crystallography.

synthetic pigments are organized into functional groups. If they are, how many molecules make up one group? Are various groupings of pigments identical in terms of their function in photosynthesis?

Photosynthetic Units

At low light levels, one CO_2 can be fixed and one O_2 evolved for approximately every 8 photons absorbed by any of the photosynthetic pigments. Is one O_2 evolved for every 8 photons absorbed at *high* light levels? Data to answer this question were provided in 1932 by Emerson and Arnold, who exposed the green alga *Chlorella* to a series of repetitive intense flashes of light (see Clayton, 1980; Rabinowitch and Govindjee, 1969). These intense flashes excite practically all chlorophyll molecules and other photosynthetic pigments simultaneously. But the maximum yield in such experiments is only one evolved O_2 for approximately every 2 000 chlorophyll molecules. Assuming that each chlorophyll molecule absorbs one photon, 250 times more photons are needed than the 8 that are needed to produce one O_2 at low light levels. At low light levels sufficient time elapses between the arrival of individual photons for the excitations of the accessory pigments and Chl *a* to be efficiently collected in a trap chl and used for the chemical reactions of photosynthesis. At high light levels, however, many photosynthetic pigments become excited at the same time, and only one excited chlorophyll out of about 250 leads to any photochemical reaction. One possible interpretation is that 250 chlorophylls act together as a *photosynthetic unit*; when any one chlorophyll in this photosynthetic unit becomes excited by light and then has its excitation transferred to the reaction center, the concomitant excitation of other chlorophylls in that unit cannot be used for photosynthesis. However, the calculation only indicates the number of chlorophylls per reaction center, and should not be taken to imply structurally definable units.

The above conclusions can also be considered in terms of Figure 5.9, where the trap chl is shown interacting with the electron acceptor A and donor D. At high light levels, the rate-limiting step for photosynthesis is not light absorption, excitation transfer, or photochemistry (electron donation by trap chl*), but the subsequent steps leading to O_2 evolution and CO_2 fixation. A brief intense illumination thus leads to more excitations than can be processed by the electron transfer reactions and subsequent biochemical events. In the limit of a very intense flash exciting all photosynthetic pigments simultaneously, one excitation is processed by each reaction center, but all others are dissipated by various nonphotochemical de-excitation processes, such as those discussed in the previous chapter.

Excitation Processing

The electron excitation originally caused by the absorption of a photon can be processed by the chemical reactions leading to CO_2 fixation about once every 5 $\times$ 10^{-3} s (5 ms; see Gregory, 1989). This processing time has important consequences for both the efficiency of light use at different photon flux densities and the optimal number of chlorophylls per reaction center.

The highest photon flux density normally encountered by plants occurs when the sun is directly overhead on a cloudless day, in which case the photosynthetic photon flux density (PPFD) for wavelengths from 400 to 700 nm is about 2 000 μmol m^{-2} s^{-1} on a horizontal plane. The average chlorophyll concentration in chloroplasts is approximately 30 mol m^{-3} (p. 223), and in passing through a chloroplast 2 μm thick about 30% of the incident PPFD is absorbed. We can therefore estimate how often an individual chlorophyll molecule might absorb a photon. Specifically, (0.3) (2 000 $\times$ 10^{-6} mol photons m^{-2} s^{-1}) or 600 $\times$ 10^{-6} mol photons m^{-2} s^{-1} is absorbed by (30 mol chlorophyll m^{-3}) (2 $\times$ 10^{-6} m) or 60 $\times$ 10^{-6} mol chlorophyll m^{-2}, which is 10 mol photons (mol chlorophyll)$^{-1}$ s^{-1}. Thus, 10 photons per second are absorbed on the average by each chlorophyll molecule in a chloroplast exposed to full sunlight.

As we have just calculated, each chlorophyll in an unshaded chloroplast can absorb a photon on the average about once every 0.1 s. When there are 250 chlorophylls per reaction center, 12.5 of these molecules are excited every 5 ms (2 500 excitations s^{-1} $\times$ 0.005 s). However, because the average processing time per reaction center is about 5 ms, only one of these excitations can be used photochemically—the others are dissipated by nonphotochemical de-excitation reactions. Consequently, although the chemical reactions leading to CO_2 fixation operate at their maximum rate under such conditions of high PPFD, a large fraction of the electronic excitations caused by light absorption are not used for photosynthesis.

Full midday sunlight is seldom incident on a chloroplast under natural conditions, because chloroplasts are generally shaded by other chloroplasts in the same cell, by chloroplasts in other cells, and perhaps by overlying leaves. Furthermore, the amount of sunlight incident on a plant is much less at sunrise or sunset, during overcast periods, or during the winter than near noon on a clear day in the summer. For the sake of argument, let us consider that the PPFD incident on a chloroplast is one-tenth of that from the direct midday sun, namely 200 μmol m^{-2} s^{-1}. In this case, each chlorophyll in the chloroplast absorbs a photon once every second, which is a much longer time between the arrival of photons than the time required for processing an electronic excitation. However, when individual chlorophylls are excited every 1 s at this moderate illumination, one chlorophyll out of the 250 per reaction center would be excited on the average every 4 ms (1 s per 250 excitations means 1 000 ms/250 or 4 ms per single excitation). This excitation frequency is such that the photons can be efficiently used for photosynthesis. In other words, the photons are arriving at a rate such that the excitations produced by most of them can be used; moreover, the chemical reactions are working at their maximum capacity. Consequently, a reaction center, with its photochemistry and associated enzymatic reactions, will function very effectively at a moderate illumination.

What would happen if there were but one chlorophyll molecule per reaction center? If the chemical reactions required 5 ms as above, this single pigment molecule would be excited about once every second at a PPFD of 200 μmol m^{-2} s^{-1}, and the excitation could easily be processed by the chemical reactions. But the photochemical step plus the subsequent enzymatic reactions leading to CO_2 fixation would be working at only 0.5% of capacity—$(5 \times 10^{-3}$ s$)/(1$ s$)$, or 0.005, is the fraction of time they could be used. In other words, although all the absorbed photons would be used for photosynthesis, even the slowest of the chemical steps would be idle over 99% of the time.

Thus, given the 5 ms processing time for the chemical reactions, 250 chlorophylls per reaction center connected with the appropriate enzyme machinery provides a plant with a mechanism for efficiently handling the usual illuminations found in nature—both for harvesting the photons and for using the chemical reactions at a substantial fraction of their capacity. For instance, averaged over the earth's surface and the year, the mean PPFD reaching the ground during the daytime on clear days is about 800 μmol m^{-2} s^{-1}. The total area of all leaves divided by the total land area is about 4.3 (see Lieth and Whittaker, 1975; this ratio, called the leaf area index, will be discussed in Ch. 9), so the average PPFD on a leaf is somewhat less than 200 μmol m^{-2} s^{-1}. At this PPFD, which can also occur for chloroplasts in cells on the side away from the fully sunlit side of a leaf, the rate of photon absorption is well matched to the rate of excitation processing.

In addition to interspecific variations, the number of chlorophylls per reaction center can depend on the PPFD present during plant development. Some algae and leaves of land plants developing under low illumination can have over 700 chlorophylls per reaction center, whereas certain leaves developing under full sunlight can have as few as 100. The ratio is fairly low in bacteria, where there are 40 to 100 bacteriochlorophyll molecules per reaction center (Clayton, 1980).

Photosynthetic Action Spectra and Enhancement Effects

The electron excitations resulting from light absorption by any photosynthetic pigment can be transferred to a trap chl and thus lead to photochemical reactions.

The absorption spectrum of chloroplasts should therefore fairly well match the action spectrum for photosynthesis. However, the action spectrum for CO_2 fixation or O_2 evolution and the overall absorption spectrum for the photosynthetic pigments in the same organism do not always coincide. Among other things, there is a "red drop," whereby the photosynthetic action spectrum drops off much more rapidly in the red region beyond 690 nm than does the absorption spectrum for chlorophylls and other pigments (illustrated in Fig. 5.10 for the case of a green alga).

Emerson demonstrated in 1957 that the relatively low photosynthetic efficiency of *Chlorella* in far-red light—a red drop like that in Figure 5.10—can be increased by simultaneously using light of a shorter wavelength along with the far-red light (Emerson, 1957). The photosynthetic rate with the two beams can be 30% to 40% greater than the sum of the rates of the far-red light and the shorter wavelength light used separately. We can also describe the enhancement in terms of quantum yield (defined on p. 221). For example, the quantum yield for O_2 evolution by *Chlorella* can be determined using 700 nm radiation in both the presence and the absence of 650 nm light. In such an experiment the quantum yield using the 700 nm radiation is higher when the 650 nm light is also present (see Myers, 1971). Such synergism, or enhancement, suggests that photosynthesis involves the cooperation of two distinct photochemical reactions. Light

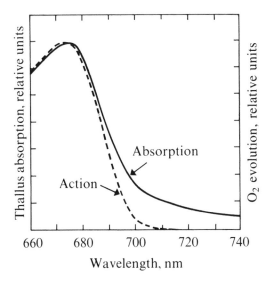

Figure 5.10

Absorption spectrum for an algal thallus and the action spectrum for its O_2 evolution, illustrating the "red drop" in photosynthesis. [Data for *Ulva taeniata* are from F. T. Haxo and L. R. Blinks, *Journal of General Physiology 33*:389–422 (1950). Used by permission.]

of wavelengths greater than 690 nm mainly powers only one of the two necessary reactions, and thus photosynthesis does not proceed at an appreciable rate. When shorter wavelengths are also used, however, the other necessary reaction takes place, resulting in a marked enhancement of the photosynthetic rate or quantum yield. Also, because photosynthesis is enhanced by adding a shorter wavelength to the far-red illumination, a smaller percentage of light is then reradiated as fluorescence.

Two Photosystems plus Light-Harvesting Antennae

We can use the photosynthetic enhancement effect to study the pigments associated with each of the two photochemical systems involved in photosynthesis. The system containing the pigments absorbing beyond 690 nm was discovered first and is referred to as Photosystem I, a terminology introduced by Duysens, Amesz, and Kamp in 1961 (see Duysens *et al.*, 1961). Much of the far-red absorption by Photosystem I is due to a type of Chl *a* with a λ_{max} near 680 nm. The special Chl *a* dimer, P_{700}, is found exclusively in Photosystem I. Therefore, light above 690 nm is absorbed mainly by this long wavelength form of Chl *a* and P_{700} in Photosystem I.

An action spectrum for the enhancement of photosynthesis in the presence of a constant irradiation with wavelengths longer than 690 nm, which are absorbed by Photosystem I, should indicate which pigments are in the other system, Photosystem II. For example, the action spectrum for such photosynthetic enhancement represented in Figure 5.11 for the red alga *Porphyridium cruentum* resembles the absorption spectrum of phycoerythrin (λ_{max} near 540 nm; see Fig. 5.7). A marked increase in photosynthesis in the cyanobacterium *Anacystis nidulans* occurs when far-red light is supplemented by light absorbed by phycocyanin (see Fig. 5.11 and the absorption spectrum for phycocyanin in Fig. 5.7). It is reasonable to conclude, therefore, that in such organisms the phycobilins funnel their excitations mainly into Photosystem II. Studies using photosynthetic enhancement also indicate that the excitations of Chl *b* preferentially go to Photosystem II, as do those of fucoxanthin in the brown algae. In a related experimental approach, action spectra for the evocation of Chl *a* fluorescence indicate that light absorbed by the phycobilins and Chl *b* leads to the fluorescence of Chl *a* in Photosystem II, perhaps including fluorescence by P_{680}. But the results have not always been clear-cut, because photons absorbed by Chl *b* and the phycobilins sometimes lead to photochemical reactions powered by P_{700}.

Steady progress has occurred in the isolation and chemical identification of the photosystems as well as in the recognition that certain photosynthetic pigments in other distinct complexes can transfer excitations to either Photosystem

Figure 5.11

Action spectra for the enhancement of O_2 evolution. Cells were exposed to a constant far-red illumination (wavelengths beyond 690 nm that would produce little O_2 evolution) plus a specific photon flux density at the various wavelengths indicated on the abscissa. The ordinate represents the rate of O_2 evolution for the two beams acting together minus that produced by the shorter wavelength acting alone. The pigment absorbing the light leading to enhancement of O_2 evolution in *Porphyridium cruentum* is phycoerythrin, and phycocyanin is responsible in *Anacystis nidulans*—see absorption spectra in Figure 5.7. [Source: R. Emerson and E. Rabinowitch, *Plant Physiology 35*:477–485 (1960). Data used by permission.]

I or II. These latter pigments are referred to as being in a *light-harvesting antenna*. For instance, probably all of the Chl *b* occurs in chlorophyll–protein complexes that are part of light-harvesting antennae (Tables 5.1 and 5.2).

Each photosystem is composed of various protein subunits with which are associated specific pigments. For instance, the core of Photosystem I consists of a tetramer of 70 kDa polypeptides to which can be bound about 130 molecules of Chl *a* (Table 5.2). This supramolecular organization contains a reaction center with P_{700} and also interacts with a chlorophyll–protein complex that acts as a light-harvesting antenna (Table 5.2). Similarly, Photosystem II is composed of various protein subunits, 50 Chl *a* molecules, and a reaction center containing P_{680}. This photosystem interacts with two light-harvesting complexes, a tightly bound one and a slightly larger peripheral one (Table 5.2). The peripheral light-

Table 5.2

Chlorophyll composition of the photosystems and light-harvesting antennae. Data are for chloroplasts from leaves of plants growing under moderate sunlight (see Glazer and Melis, 1987). Other data indicate up to four different light-harvesting antennae associated with Photosystem I, two different Photosystems II, and up to five different light-harvesting antennae for Photosystem II; also, the chlorophylls measured per polypeptide depends on the isolation technique (see Gregory, 1989; Thornber *et al.*, 1990; Zuber, 1987).

Component	Approximate number of chlorophyll molecules
Photosystem I	
Core	130 Chl *a*
Light-harvesting antenna	60 Chl *a*, 20 Chl *b*
Photosystem II	
Core	50 Chl *a*
Intrinsic light-harvesting antenna	45 Chl *a*, 35 Chl *b*
Peripheral light-harvesting antenna	65 Chl *a*, 55 Chl *b*

harvesting complex apparently can become disconnected from Photosystem II, at which time it may transfer excitations to Photosystem I. In particular, when absorption of photons is markedly in excess of what can be processed by Photosystem II, some of the light-harvesting complex associated with Photosystem II becomes detached, apparently after phosphate groups are added that can lead to electrostatic repulsion (see Fork and Satoh, 1986). Movement of such antenna chlorophylls away from Photosystem II and their ability to pass excitations on to Photosystem I help create a better balance in photon processing by the two photosystems. The phycobiliproteins (organized into phycobilisomes in cyanobacteria and red algae, which have no Chl *b*) are also part of a light-harvesting complex, in this case one interacting with Photosystem II. Only a single photosystem occurs in the green and the purple bacteria, which oxidize primarily organic acids or inorganic sulfur-containing compounds instead of water (hence, O_2 evolution does not accompany their photosynthesis); again, a special light-harvesting complex is involved in collecting excitations and directing them toward a trap Bchl in the core.

We note that the stoichiometry of the components varies, as Photosystem II is about 40% to 90% more abundant than is Photosystem I for plants growing in full sunlight, and the relative amounts of various pigments in the complexes can also be influenced by the illumination level during development (see Glazer and Melis, 1987). Photosystem I is apparently excluded from most, if not all, of the appressed membrane regions where lamellae are stacked into grana (Fig. 1.9). Thus the stromal lamellae are greatly enriched in Photosystem I, which also apparently occurs in the exposed end membranes and the margins of the grana. Chloroplasts in plants or parts of plants developing in the shade tend to have more photosynthetic pigments per unit volume, more thylakoids, and a greater stacking of thylakoids into grana than do chloroplasts developing under high illumination. Consequently, the Photosystem I–enriched lamellae exposed to the stroma increase from about 40% of the lamellar membranes for plants under low illuminations to 60% under high illuminations (see Haehnel, 1984). The increase in the ratio of Photosystem I to Photosystem II with increasing availability of light leads to an increase in the ratio of Chl *a* to Chl *b,* as Photosystem I and its associated light-harvesting antenna have a higher fraction of their chlorophyll as Chl *a* than does Photosystem II with its antennae (Table 5.2). Such relative increases of Photosystem I and Chl *a* with increasing illumination can be seen between plants, between leaves on a tree, and even between cells in a single leaf (see Anderson, 1986).

ELECTRON FLOW

The photochemical reaction of photosynthesis involves the removal of an electron from an excited state of the special chlorophyll that acts as an excitation trap. The movement of the electron from this trap chl to some acceptor begins a series of electron transfers that can ultimately lead to the reduction of $NADP^+$. The oxidized trap chl, which has lost an electron, can accept another from some donor, as in the steps leading to O_2 evolution. Coupled to the electron transfer reactions in chloroplasts is the formation of ATP, a process known as "photophosphorylation." In this section we will consider some of the components of chloroplasts involved in accepting and donating electrons; a discussion of the energetics of such processes will follow in the next chapter (for further details see Amesz, 1987; Gregory, 1989; Kamen, 1963).

The various steps of photosynthesis occur over a wide range of times. For instance, the absorption of light and the transfer of excitation from both accessory pigments and Chl *a* molecules to a trap chl take from 10^{-15} to 10^{-9} s after the arrival of a photon. The photochemical event at the reaction center leads to the separation of an electron from the trap chl (Eq. 5.5), which causes a bleaching— a decrease in the absorption coefficient for wavelengths in the visible region— of this pigment. By observing the kinetics of this bleaching in the far-red region, we can determine the occurrence of photochemistry, which signals that the excitation caused by absorption of radiant energy has led to an increase in chemical energy. The donation of an electron to the oxidized trap chl (Eq. 5.6) usually occurs 10^{-7} to 10^{-4} s after the arrival of a photon and restores the original spectral properties of the trap chl. The ensuing electron flow to some components can last into the millisecond range. Finally, the overall processing time for a reaction center plus the associated enzymes is about 5 ms per excitation.

Electron Flow Model

In 1937 Hill demonstrated that isolated chloroplasts, when placed in an aqueous solution in the presence of a suitable exogenous electron acceptor, could evolve oxygen in the light, a process that subsequently has become known as the Hill reaction. Oxygen evolution proceeded in the absence of CO_2, suggesting that CO_2 fixation and O_2 evolution were separate processes, contrary to the prevailing belief. Using ^{18}O-labeled H_2O and ^{18}O-labeled CO_2 in different experiments, Ruben and Kamen showed in 1941 that the evolved O_2 comes from water and not from CO_2. Subsequent studies have elucidated many of the steps intervening between O_2 evolution and CO_2 fixation in photosynthesis (see Barber, 1977; Gregory, 1989; Kamen, 1963).

Our point of departure for the present discussion of electron flow in photosynthesis will be to outline the overall process. Although the complete mechanism of this water oxidation is still unclear, it can be represented as follows:

$$2H_2O \rightarrow O_2 + 4H \qquad (5.8)$$

where H in Equation 5.8 represents $H^+ + e^-$. CO_2 fixation into a carbohydrate involves the reduction of carbon, with four hydrogen atoms being required per carbon atom:

$$CO_2 + 4H \rightarrow \{CH_2O\} + H_2O \qquad (5.9)$$

where $\{CH_2O\}$ represents a general carbohydrate, as indicated previously (Fig. 5.1). The movement of the reductant H in Equations 5.8 and 5.9 can conveniently be followed by tracing the flow of the electrons ($H = H^+ + e^-$), which explains why our topic here is electron flow and not hydrogen movement (reducing a compound is chemically equivalent to adding electrons, and oxidation is the removal of electrons).

Some pathway for photosynthetic electron flow is needed so that the H's indicated as products in Equation 5.8 can become the H's required as reactants in Equation 5.9. This electron transfer pathway involves both Photosystems I and II as well as a number of other components. The oxidant involved with O_2 evolution (Eq. 5.8) is provided by Photosystem II. The primary oxidant is trap chl$^+$, which leads to the oxidation of water (see Eq. 5.6, trap chl$^+$ + D → trap chl + D^+). The reductant required for carbon reduction (Eq. 5.9) is produced by the excited trap chl in Photosystem I (see Eq. 5.5, trap chl* + A → trap chl$^+$ + A^-). These two photosystems are linked by a chain of components along which a transfer of electrons occurs, as is illustrated in Figure 5.12.

For each photon absorbed by any of the accessory pigments or Chl a's whose excitations are funneled into a reaction center, one electron can be removed from its trap chl. Because 4 electrons are involved per O_2 derived from water, the evolution of this molecule of O_2 requires the absorption of 4 photons by Photosystem II or the light-harvesting antenna feeding into it (see Eq. 5.8 and Fig. 5.12). An additional 4 photons whose excitations arrive at the trap chl of Pho-

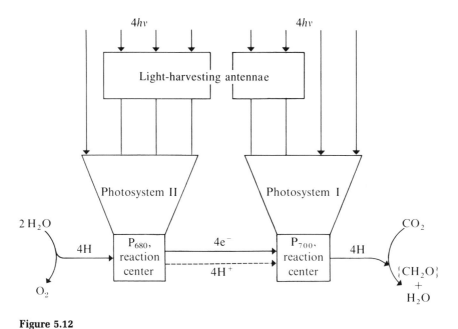

Figure 5.12

Schematic model for a series representation of the two photosystems of photosynthesis, indicating the stoichiometry of various factors involved in the reduction of a CO_2 molecule to a carbohydrate ($\{CH_2O\}$). Some of the photons (hv) are captured by the accessory pigments and Chl a in the light-harvesting antennae; these excitations are then fed into the two photosystems, but primarily to Photosystem II.

tosystem I are required for the reduction of the 2 molecules of $NADP^+$ necessary for the subsequent reduction of 1 CO_2 molecule (Eq. 5.9 and Fig. 5.12). Hence, 8 photons are needed for the evolution of 1 molecule of O_2 and the fixation of 1 molecule of CO_2. (In the next chapter we will consider how many photons are used to provide the ATP's required per CO_2 fixed.) Although there are alternatives to the series representation (Fig. 5.12) proposed by Hill and Bendall in 1960, the scheme takes into consideration the results of many investigators and has generally become accepted as an overall description of electron flow in chloroplast grana (see Hill and Bendall, 1960). After introducing the concept of redox potential in the next chapter, we will portray the energetics of the series representation (see Fig. 6.4, which includes many of the components that we will discuss next).

Components of the Electron Transfer Pathway

We shall now turn our attention to the specific molecules that act as electron acceptors or donors in chloroplasts. A summary of the characteristics of the better-known components of the pathway is presented in Table 5.3. We will begin our discussion by considering the photochemistry at the reaction center of Photosystem II, and then proceed with a consideration of the various substances in the sequence in which they are involved in electron transfer along the pathway from Photosystem II to Photosystem I. We will conclude by considering the fate of the excited electron in Photosystem I.

The electron removed from the excited trap chl in the reaction center of Photosystem II is replaced by one coming from water in the process that leads to O_2 evolution (Eq. 5.8). Using very brief flashes of light, Joliot showed in the 1960s that essentially no O_2 evolution accompanied the first or even the second flash (see Barber, 1977, 1978; Clayton, 1980; Haehnel, 1984). If the 4 electrons involved come from four different Photosystem II's, then a single intense flash should cause O_2 evolution, because four Photosystem II's would have been excited and the 4 electrons coming from each O_2 could be accepted. If two photosystems were involved, with each one accepting 2 electrons sequentially, then two flashes should lead to O_2 evolution. In fact, every fourth flash leads to substantial O_2 evolution, indicating that a single Photosystem II is responsible for the 4 electrons involved in the evolution of each O_2 molecule; i.e., four consecutive photochemical acts in a particular Photosystem II are required before a molecule of O_2 can be evolved. The times for the first three steps are about 0.2, 0.5, and 1 ms, respectively. The last step, which leads to the release of O_2 *inside* a thylakoid, takes about 1 ms. (In an illuminated leaf, there are many O_2-

Table 5.3

Representative properties of some of the components involved with electron transfer in chloroplasts. The frequency of components is per 600 chlorophylls for plants growing under moderate sunlight (see Table 5.1), and the order presented is in the sequence for electron flow (see Anderson, 1986; Gregory, 1989; Haehnel, 1984). We will discuss redox potentials in Chapter 6.

Name	Molecular mass (kDa)	Approximate number per 600 chlorophylls	Numbers of electrons accepted or donated per molecule	Approximate midpoint redox potential (V)	Comment
P_{680}	1 or 2 × 0.893	1.6	1	0.90	A Chl a or a Chl a dimer; acts as the trap of Photosystem II
Plastoquinone A	0.748	10	2	0.11	Located in membrane; acts as a mobile pool accepting electrons from Photosystem II and donating them to the cytochrome b_6–f complex
Cyt b_6	23	2	1	−0.05 and −0.17	The two hemes are bound to the same polypeptide; part of cytochrome b_6–f complex
Cyt f	34	1	1	0.35	Part of the cytochrome b_6–f complex; λ_{max} for α band at 554 nm
Plastocyanin	11	2	1	0.37	Blue protein (reduced form is colorless) that accepts electrons from the cytochrome b_6–f complex and donates them to Photosystem I; contains one Cu; soluble in aqueous solutions, but occurs in the thylakoid lumen
P_{700}	2 × 0.893	1.0	1	0.45	Probably a Chl a dimer; acts as the trap of Photosystem I
Ferredoxin	11	1 to 2	1	−0.42	Nonheme protein with two Fe and two S; accepts electrons from Photosystem I by way of intermediates; soluble in aqueous solutions
Ferredoxin–NADP reductase	37	1	—	—	An enzyme containing one flavin adenine dinucleotide per molecule; bound to outside of lamellae
$NADP^+$–NADPH	0.743 to 0.744	30	2	−0.32	Soluble in aqueous solutions

evolving loci, and the four steps are in different stages at any one time, thus making O_2 evolution appear to be continuous.) Manganese is apparently involved in the O_2-evolving process, presumably while bound to a protein. Four Mn's occur per reaction center in Photosystem II, and their progressive experimental removal leads to a stoichiometric reduction in the O_2-evolving ability. Chloride is also apparently necessary for O_2 evolution (see Anderson, 1986; Brudvig *et al.*, 1989).

The electron from trap chl* of Photosystem II is transferred by a series of molecules making up the photosynthetic *electron transfer chain,* a term describing the pathway from Photosystem II to Photosystem I (see Fig. 5.12). The electron removed from P_{680} is very rapidly transferred to *pheophytin* (in about 2×10^{-12} s); pheophytin is chlorophyll (Fig. 5.2) without the central magnesium, indicating that it is a highly conjugated porphyrin absorbing in the blue region (Soret band) and the red region (another chlorophyll molecule may intervene between P_{680} and pheophytin). From pheophytin, the electron rapidly (in 1 to 2 $\times 10^{-10}$ s) moves to a quinone (generally referred to as Q_A), the latter step helping to stabilize the charge separation and thus preventing the electron from going back to P_{680}^+. The electron is later transferred to a second quinone (Q_B); when P_{680} is excited again, another electron is passed on to this second quinone (see Andréasson and Vänngård, 1988; Glazer and Melis, 1987).

Chloroplast lamellae contain different types of quinone. A quinone becomes a semiquinone when one hydrogen atom ($H^+ + e^-$) is added (a semiquinone is a free radical because of the presence of an unpaired electron), and a hydroquinone when two are added (a hydroquinone is also called a hydroquinol, or simply a quinol):

quinone   semiquinone   hydroquinone

There is great variation among quinones because of substituents attached to the ring; e.g., plastoquinone A is

Like chlorophyll, plastoquinone A has a nonpolar terpenoid or isoprenoid tail, which may stabilize the molecule at the proper location in the lamellar membranes of chloroplasts. When donating or accepting electrons, plastoquinones have characteristic absorption changes in the UV near 250 to 260 nm, 290 nm, and 320 nm, that can be monitored to study their electron transfer reactions.[5] The plastoquinones involved in photosynthetic electron transport are divided into two categories: (1) two plastoquinones, apparently bound to proteins, that rapidly receive single electrons from P_{680}; and (2) a mobile group or pool of about 10 plastoquinones that subsequently receives two electrons (plus two H^+) from the second of these plastoquinones (all of these quinones are located in the lamellar membranes). From the plastoquinone pool, electrons move to the cytochrome b_6–f complex (see Haehnel, 1984).

Cytochromes are extremely important components of electron transfer pathways in chloroplasts and mitochondria. They have three absorption bands in the visible region: the α, β, and γ bands. (Absorption of light by cytochromes apparently plays no part in photosynthesis.) In 1925 Keilin described three different types of cytochrome based on the spectral position of their α (long) wavelength band. Cytochromes of the a type have a λ_{max} for the α band from 600 to 605 nm, b types near 560 nm, and c types near 550 nm. All cytochromes have β bands near 515 to 530 nm. The main short wavelength band, the γ or Soret band, usually has a λ_{max} between 415 and 430 nm (for further information on cytochromes, see Dolphin, 1979; B.T. Storey in Davies, 1980; Trebst and Avron, 1977).

Cytochromes consist of an iron-containing tetrapyrrole or porphyrin known as *heme* (Fig. 5.13), which is bound to a protein. The various cytochromes differ in both the substituents around the periphery of the porphyrin ring and the protein to which the chromophore is attached. Cyt f is a c-type cytochrome[6] occurring in chloroplasts and contains the chromophore indicated in Figure 5.13. Many different hemoproteins of the Cyt b type are found in plants, two occurring in chloroplasts (Cyt b_{559}, which occurs associated with Photosystem II, and another one with a λ_{max} at about 563 nm, which is part of the cytochrome b_6–f complex, Table 5.3). These various b cytochromes appear to have the same chromophore

---

5. Plastoquinone refers to a quinone found in a plastid such as a chloroplast; these quinones have various numbers of isoprenoid residues, such as nine for plastoquinone A, the most common plastoquinone in higher plants.

6. f from *frons*, the Latin for leaf. The first three cytochromes studied by Keilin were designated Cyt a, Cyt b, and Cyt c. Additional ones were indicated by subscripts. For instance, Cyt f from higher plants is also known as Cyt c_1, because it resembles the absorption properties of the first additional c-type cytochrome identified by Keilin. Cyt b_6, which is often referred to as Cyt b_{563}, has absorption properties similar to the sixth additional b-type cytochrome identified.

Figure 5.13

The iron-containing porphyrin known as *heme,* the chromophore for cytochrome *c.*

attached to different proteins, and thus their individual absorption properties are probably due to changes in the protein. Like the cytochromes, chlorophylls are tetrapyrroles (compare the structure of Chl *a* in Fig. 5.2), but they have a Mg atom in the center of the porphyrin ring, whereas cytochromes have an Fe. Furthermore, the acceptance or donation of an electron by a cytochrome involves a transition between the two states of its iron, Fe^{2+} and Fe^{3+}, whereas the electron removed from chlorophyll in the photochemical reactions of photosynthesis is one of the π electrons in the conjugated system of the porphyrin ring.

Instead of the usual valence bonds, the metal atoms in chlorophylls and cytochromes should be presented in terms of the six coordinate bonds described by ligand-field (molecular-oribtal) theory. Fe has one coordinate bond to each of the four N's in the porphyrin ring (the two solid and the two broken lines emanating from Fe in Fig. 5.13; also see Fig. 5.2), one to an N in the imidazole side chain of a histidine, and the sixth to another histidine for *b*-type cytochromes in chloroplasts and to a lysine or a methionine in Cyt *f* (see Cramer *et al.*, 1987)[7]; these amino acids occur in the protein to which the chromophore is bound (one bond is above and one is below the plane of the porphyrin in Fig. 5.13). The donation of an electron by the ferrous form of Cyt *c*, ferrocytchrome *c* (Fig.

---

7. For *a*-type cytochromes occurring in mitochondria, the sixth position is not bound to an amino acid but rather interacts with O_2, Cu, or other ligands.

5.13), causes the iron to go to the ferric state, Fe^{3+}. The extra positive charge on the iron in ferricytochrome c can either attract anions such as OH^- or be delocalized to adjacent parts of the molecule. The six coordinate bonds remain in ferricytochrome c, and the conjugation in the porphyrin ring is only moderately changed from that in ferrocytochrome c. Hence, the extensive bleaching of chlorophyll following the loss of an electron from its porphyrin ring does not occur with electron donation by the iron atom in cytochrome. Instead, the transition from Fe^{2+} to Fe^{3+} leads to less of a change in the absorption bands, because electrons in the ring conjugation are influenced only secondarily. Nevertheless, such changes in absorption upon the acceptance or the donation of electrons by cytochromes can be detected and are important for monitoring electron flow *in vivo*.

The removal of an electron from a cytochrome (oxidation) causes its three absorption bands to become less intense, broader, and to shift toward shorter wavelengths. The absorption coefficients at λ_{max} in the reduced form are about 3×10^3 m^2 mol^{-1} for the α band, somewhat less for the β band, and over 10^4 m^2 mol^{-1} for the γ (Soret) band. Upon oxidation, the ε_λ for the α band decreases just over 50%, and smaller fractional changes generally occur in the absorption coefficients of the β and the γ bands (λ_{max} is also at shorter wavelengths in the oxidized form). Such absorption changes permit a study of the kinetics of electron transfer while the cytochrome molecules remain embedded in the internal membranes of chloroplasts or mitochondria. By monitoring the spectral changes of the chloroplast cytochromes we can observe that the electron movement along the electron transfer chain between Photosystem II and Photosystem I (Figs. 5.12 and 6.4) occurs 2×10^{-3} s after light absorption by some pigment molecule in Photosystem II.

The cytochrome b_6–f complex is a supramolecular protein unit embedded in the chloroplast lamellar membranes. It is composed of various polypeptides, including a polypeptide that binds cytochrome f and another that binds two cytochrome b_6's (Table 5.3), as well as iron-plus-sulfur-containing protein (the Rieske-Fe S center) and a bound plastoquinone. The protein plastocyanin (Table 5.3) carries electrons from the cytochrome b_6–f complex to Photosystem I. Plastocyanin occurs associated with the inner side of the thylakoid membranes, i.e., the side toward the lumen and away from the stroma (Fig. 1.9). The electrons from plastocyanin, which are apparently donated and accepted by copper atoms, can be accepted by the trap chl of Photosystem I, P_{700}, if the latter is in the oxidized form (see Eq. 5.6, trap $chl^+ + D \rightarrow$ trap chl $+ D^+$). The photochemical change in P_{700} can be followed spectrophotometrically, because the loss of an electron causes a bleaching of both its Soret and red absorption bands; the subsequent acceptance of an electron restores the original spectral properties.

Specifically, P_{700} is bleached (oxidized) by light absorbed by Photosystem I and then restored (reduced) following the absorption of photons by Photosystem II. Thus, the electron removed from P_{700} is replaced by one coming from Photosystem II by means of the electron transfer chain (Figs. 5.12 and 6.4). Far-red light (above 690 nm) absorbed by Photosystem I leads to an oxidation of Cyt f, indicating that an electron from Cyt f can be donated to P_{700} in about 3×10^{-4} s (see Haehnel, 1984).

Because Photosystem II tends to occur in the grana and Photosystem I in the stromal lamellae, the intervening components of the electron transport chain may have to diffuse over considerable distances in the lamellar membranes to link the two photosystems. We can examine such diffusion using the time–distance relationship that we derived in Chapter 1 (Eq. 1.6, $x_e^2 = 4D_j t_e$). In particular, the diffusion coefficient for plastocyanin in a membrane can be about 3×10^{-12} $m^2\ s^{-1}$ and about the same in the lumen of the thylakoids, unless diffusion of plastocyanin is physically restricted in the lumen by the appression of the membranes (see Haehnel, 1984). For such a D_j, in 3×10^{-4} s (the time for electron transfer from the cytochrome b_6–f complex to P_{700}), plastocyanin could diffuse about $[(4)\ (3 \times 10^{-12}\ m^2\ s^{-1})\ (3 \times 10^{-4}\ s)]^{1/2}$ or 60 nm, indicating that this complex probably occurs in the lamellae in relatively close proximity to its electron acceptor, Photosystem I. Plastoquinone is smaller and hence would diffuse more readily than plastocyanin, and a longer time (2×10^{-3} s) is apparently necessary to move electrons from Photosystem II to the cytochrome b_6–f complex; hence, these two components can be separated by much greater distances than are the cytochrome b_6–f complex and Photosystem I.

The electron from P_{700}^* reduces the iron-containing protein, ferredoxin, via a series of at least five intermediates (see Andréasson and Vänngård, 1988). Electron transfer to the first component, which may be a form of chlorophyll, is rapid (about 3×10^{-12} s); the next component may be a quinone (reached in about 35×10^{-12} s), followed by three centers containing iron and sulfur. Finally, the electron reaches ferredoxin, which contains two irons in the ferric state that interact with two sulfur atoms (each Fe also interacts with S's in two cysteines); the acceptance of an electron by ferredoxin can be viewed as reducing one of the ferric atoms to the ferrous form. By means of the enzyme ferredoxin–NADP reductase, two molecules of ferredoxin reduce one molecule of $NADP^+$ to yield NADPH (we will discuss $NADP^+$ and NADPH in the next chapter).

Types of Electron Flow

Three different types of photosynthetic electron transfer, or flow, occur—noncyclic, pseudocyclic, and cyclic—each one depending on the compound to which electrons are transferred from ferredoxin (see Trebst and Avron, 1977). In *noncyclic* electron flow, electrons coming originally from water are used to reduce $NADP^+$. The overall noncyclic electron flow is as follows: an electron from water goes to the trap chl^+ of Photosystem II; there it replaces a donated electron, which moves along the electron transfer chain to the oxidized P_{700} in Photosystem I; the electron from P_{700} moves to ferredoxin and then to $NADP^+$. Such noncyclic electron flow follows essentially the same pathway as the reductant H moving from left to right in Figure 5.12 (see also Fig. 6.4).

Electrons from ferredoxin may also reduce O_2, which yields H_2O_2 and eventually H_2O ($O_2 + 2e^- + 2H^+ \rightleftharpoons H_2O_2 \rightleftharpoons H_2O + \frac{1}{2}O_2$).[8] Because equal amounts of O_2 are evolved at Photosystem II and then consumed using reduced ferredoxin in a separate reaction, such electron flow is termed *pseudocyclic* (illustrated in Fig. 6.4). There is no net O_2 change accompanying pseudocyclic electron flow, although it is not a cycle in the sense of having electrons cyclically traverse a certain pathway. It is not yet clear to what extent pseudocyclic electron flow occurs *in vivo*, although such electron flow can readily be demonstrated with isolated chloroplasts (see H. Gimmler in Trebst and Avron, 1977).

In *cyclic* electron flow, the electron from the reduced form of ferredoxin moves back to the electron transfer chain between Photosystems I and II via the cytochrome b_6–f complex (including the involvement of a plastoquinone) and eventually reduces an oxidized P_{700}. Cyclic electron flow does not involve Photosystem II at all, so it can be caused by far-red light absorbed only by Photosystem I—a fact that is often exploited in experimental studies. When far-red light absorbed by Photosystem I is used, cyclic electron flow can occur but noncyclic does not, so no NADPH is formed and no O_2 is evolved (cyclic electron flow can lead to the formation of ATP, as is indicated in the next chapter). When light absorbed by Photosystem II is added to cells exposed to far-red illumination, both CO_2 fixation and O_2 evolution can proceed, and photosynthetic enhancement is achieved. Treatment of chloroplasts or plant cells with the O_2-evolution inhibitor DCMU [3-(3,4-dichlorophenyl)-1,1-dimethyl urea] also leads to only cyclic electron flow; DCMU therefore has many applications in the laboratory and is also an effective herbicide, because it markedly inhibits photosynthesis. Cyclic electron flow may be more common in stromal lamellae, because they have predominantly Photosystem I activity.

As indicated in Table 5.3, P_{680}, P_{700}, the cytochromes, plastocyanin, and

---

8. The light-dependent consumption of O_2, as occurs when electrons from ferredoxin or from one of the iron-plus-sulfur-containing intermediates of Photosystem I move to O_2, is termed the Mehler reaction.

ferredoxin accept or donate only one electron per molecule. These electrons interact with $NADP^+$ and the plastoquinones, both of which transfer two electrons at a time. The two electrons that reduce plastoquinone apparently come sequentially from the same Photosystem II. It is less clear how the two electrons move from a plastoquinone to the one-electron carrier Cyt f. The enzyme ferredoxin–NADP reductase is apparently involved in matching the one-electron chemistry of ferredoxin to the two-electron chemistry of NADP. Both the pyridine nucleotides and the plastoquinones are considerably more numerous than are other molecules involved with photosynthetic electron flow (Table 5.3), which has important implications for the electron transfer reactions. Moreover, $NADP^+$ is soluble in aqueous solutions and so can diffuse to the ferredoxin–NADP reductase, where two electrons are transferred to it to yield NADPH (ferredoxin and plastocyanin are also soluble in aqueous solutions).

Photophosphorylation

Three ATP molecules are generally required for the reductive fixation of one CO_2 molecule into a carbohydrate (see Fig. 5.1). Such ATP is produced by *photophosphorylation*; i.e., light absorbed by the photosynthetic pigments in the lamellar membranes leads to a flow of electrons, to which is coupled the phosphorylation of ADP. We will consider the energetics of this dehydration of ADP plus phosphate to yield ATP in the next chapter.

Photophosphorylation was first demonstrated in cell-free systems in 1954. Frenkel, working with bacterial chromatophores, and Arnon, Allen, and Whatley, using broken spinach chloroplasts, observed ATP formation in the light (Arnon *et al.*, 1954; Frenkel, 1954). The enzymes are localized in or on the lamellar membranes, and the energy transfer steps are very sensitive to perturbation of the membranes. Moreover, none of the molecular species (ADP, ATP, and phosphate) can be readily determined quantitatively *in vivo*. Specifically, the interconversions of these compounds in the chloroplasts cannot be monitored by measuring changes in spectral properties, a technique that is quite successful for studying the acceptance or donation of electrons by cytochromes and trap chl's. Furthermore, all three molecules (ATP, ADP, and phosphate) take part in many different biochemical reactions. Nevertheless, considerable progress has been made in understanding the relationship between ATP formation and proton (H^+) transport across membranes. In the next chapter we will reconsider ATP formation coupled to electron flow in both chloroplasts and mitochondria, after some of the underlying energy concepts have been introduced.

Vectorial Aspects of Electron Flow

Because the electron flow components are associated with membranes, the possibility exists for chemical asymmetries to develop. In fact, the electrons and their associated protons are moved in specific directions in space by the processes we have been considering, causing the flows to have a vectorial nature (see Fig. 5.14).

The chlorophyll–protein complexes are oriented in the lamellar membranes in such a way that the electron transfer steps at the reaction centers lead to an outward movement of electrons. For instance, the electron donated by Photosystem II is moved from the lumen side to the stromal side of a thylakoid (see Figs. 1.9 and 5.14). The electron that is donated back to the trap chl (P_{680}) comes from H_2O, leading to the evolution of O_2 by Photosystem II (Eq. 5.8). The O_2 and the H^+ from this reaction are released inside the thylakoid (Fig. 5.14). Because O_2 is a small neutral molecule, it readily diffuses out across the

Figure 5.14

Schematic representation of reactions occurring at the photosystems and certain electron transfer components, emphasizing the vectorial or unidirectional flows developed in the thylakoids of a chloroplast. Outwardly directed electron movements occur in the two photosystems (PS I and PS II), where the electron donors are on the inner side and the electron acceptors are on the outer side of the membrane. Light-harvesting complexes (LHC) act as antennae for these photosystems. The plastoquinone pool (PQ) and the cytochrome b_6–f complex (Cyt b_6–f) occur in the membrane, whereas plastocyanin (PC) occurs on the lumen side and ferredoxin–NADP reductase (FNR), which catalyzes electron flow from ferredoxin (FD) to $NADP^+$, occurs on the stromal side of the thylakoids. Protons (H^+) are produced in the lumen by the oxidation of water and are transported into the lumen accompanying electron (e^-) movement along the electron transfer chain. (See Anderson, 1986; Gregory, 1989.)

lamellar membranes into the chloroplast stroma. But the proton (H^+) carries a charge and hence has a low partition coefficient for the membrane, so it does not readily move out of the thylakoid lumen.

The electron excited away from P_{680} in Photosystem II eventually reaches a quinone in that photosystem that accepts two electrons and also picks up two protons (H^+) from the stroma (Fig. 5.14). This quinone transfers its two electrons and two protons to a mobile plastoquinone in the plastoquinone pool occurring in the lamellar membranes, and the mobile plastoquinone in turn interacts with the cytochrome b_6–f complex. The cytochrome b_6–f complex is responsible for the final step in the vectorial transport of protons from the stroma to the lumens of the thylakoids; also it delivers electrons to plastocyanin, which occurs on the inner side of the lamellar membranes. Electrons from plastocyanin move to the reaction center of Photosystem I. Via a photochemical event, the trap chl of Photosystem I (P_{700}) donates an electron that eventually reaches ferredoxin, which occurs on the outer side of the thylakoid (Fig. 5.14). Ferredoxin, which is soluble in aqueous solutions, diffuses to ferredoxin–NADP reductase (a flavoprotein, Table 5.3), where two electrons are accepted by $NADP^+$, yielding NADPH. The flavoprotein is bound on the outer side of the lamellar membranes, so NADPH is formed in a region where subsequent biochemical reactions can utilize this crucially important molecule.

Let us now recapitulate the accomplishments of the various processes described above. O_2 is evolved inside the thylakoid and readily diffuses out. The protons from the O_2-evolving step plus those transported by the cytochrome b_6–f complex are released in the thylakoid lumen, where the membranes prevent their ready escape. In cyclic electron flow, electrons from P_{700} move to ferredoxin and thence to the cytochrome b_6–f complex, which also causes protons to be delivered from the stroma to the lumens of the thylakoids. The accumulation of protons inside the thylakoid, together with the transfer of electrons out, raises the electrical potential inside with respect to outside and also increases the internal concentration of protons, thus setting up a chemical potential gradient capable of doing work. In fact, this proton chemical potential gradient is energetically coupled to the formation of ATP (photophosphorylation), as we will see in the next chapter.

Problems

5.1 A spherical spongy mesophyll cell is 40 μm in diameter and contains 50 spherical chloroplasts that are 4 μm in diameter. Assume that such cells contain 1 g chlorophyll

kg^{-1}, that the cell is 90% water by weight, and that the cellular density is 1 000 kg m^{-3} ($= 1.00$ g cm^{-3}). (a) What volume fraction of the cell is occupied by chloroplasts? (b) If the CO_2 fixation rate is 100 mmol (g chlorophyll)$^{-1}$ hour^{-1}, how long does it take to double the dry weight of the cell? Assume that CO_2 and H_2O are the only substances entering the cell. (c) If the ratio Chl a/Chl b is 3, what is the mean molecular weight of chlorophyll? (d) Assuming that the chlorophyll is uniformly distributed throughout the cell, what is the maximum absorbance by one cell in the red and the blue regions? Use absorption coefficients given in Figure 5.3.

5.2 Suppose that some pigment has 8 double bonds in conjugation and has a single absorption band with a λ_{max} at 580 nm, which corresponds to a transition to the fourth vibrational sublevel of the excited state (see Fig. 4.7). A similar pigment has 10 double bonds in conjugation, which causes the lowest vibrational sublevel of the excited state to move down in energy by 20 kJ mol^{-1} and the lowest vibrational sublevel of the ground state to move up in energy by 20 kJ mol^{-1} compared with the corresponding levels in the other molecule. Assume that the splitting between vibrational sublevels remains at 10 kJ mol^{-1} and that the most likely transition predicted by the Franck–Condon principle for this second molecule is also to the fourth vibrational sublevel. (a) What is the shortest λ_{max} for fluorescence by each of the two molecules? (b) Can either or both molecules readily pass their excitation on to Chl a in vivo? (c) Can the absorption of blue light by Chl a lead to excitation of either of the pigments? Give your reasoning.

5.3 Let us approximate chloroplasts by short cylinders 4 μm in diameter and 2 μm thick (i.e., 2 μm along the cylinder axis), which contain 20 mol chlorophyll m^{-3}. The chloroplasts are exposed to 40 W m^{-2} of 675 nm light parallel to the axis of the cylinder. Assume that ε_{675} is 0.60×10^4 m^2 mol^{-1} for the chlorophylls. (a) What is the absorbance at 675 nm for the chlorophyll in a single chloroplast? What is the fraction of the incident light absorbed? (b) How many μmol photons m^{-2} s^{-1} of 675 nm light will be absorbed in passing through a single chloroplast? How many chlorophyll molecules participate in this absorption? (c) Assume that a photosynthetic unit has 250 chlorophyll molecules and that 0.01 s is needed to process each excitation. How often are chlorophyll molecules excited on the average and what fraction of the absorbed photons can be processed? (d) How many moles of O_2 m^{-2} s^{-1} are evolved for each chloroplast? Assume that 8 photons are needed to evolve 1 molecule of O_2. (e) What would be the answers to (c) for a chloroplast shaded by three overlying chloroplasts?

5.4 Chloroplasts corresponding to 10 mmol of chlorophyll m^{-3} of solution are suspended in a cuvette with a 10 mm light path. The rate of O_2 evolution is proportional to PPFD up to 10 μmol photons absorbed m^{-2} s^{-1}, which gives 10^{-4} mol m^{-3} (10^{-7} M) O_2 evolved s^{-1}. The maximum O_2 evolution rate under very high PPFD is 5×10^{-4} mol m^{-3} s^{-1}. For very brief and intense flashes of light, the O_2 evolution is 5×10^{-6} mol m^{-3} per flash. (a) Using the above data, how many photons are required per O_2 evolved? (b) How many chlorophyll molecules occur in a photosynthetic unit? (c) How much time is required for the processing of an excitation by a photosynthetic unit? (d) An "uncoupler" is a compound that decreases the ATP formation coupled to photosynthetic electron flow. When such a compound is added to chloroplasts incubated at a high photon flux density, the O_2 evolution rate eventually becomes less than a control without the uncoupler. Explain.

5.5 Suppose that the absorbance of pea chloroplasts in a cuvette with a 10 mm light path is 0.1 at 710 nm and 1.0 at 550 nm. Assume that chlorophyll is the only species absorbing at 710 nm and that no chlorophyll absorbs at 550 nm. Suppose that no CO_2 is fixed when either 550 nm or 710 nm light is used alone, but that both together lead to CO_2 fixation. (a) Is any ATP formation caused by the 550 nm or by the 710 nm light? (b) What type of pigments are absorbing at 550 nm? Are they isoprenoids or tetrapyrroles? (c) If equal but low incident photon flux densities are simultaneously used at both 550 nm and 710 nm, what is the maximum quantum yield for CO_2 fixation for each beam? (d) The initial bleaching of P_{700} at 700 nm leads to a decrease in absorbance of 10^{-5} in 10^{-6} s. What is the minimum number of moles of photons per unit area absorbed by Photosystem I that could account for this? Assume that ε_{700} is 0.8×10^4 m^2 mol^{-1} for the trap chl.

References

Amesz, J., ed. 1987. *Photosynthesis*. Elsevier, Amsterdam.

Anderson, J. M. 1986. Photoregulation of the composition, function, and structure of thylakoid membranes. *Annual Review of Plant Physiology 37*:93–136.

Andréasson, L.-E., and T. Vänngård. 1988. Electron transport in photosystems I and II. *Annual Review of Plant Physiology and Plant Molecular Biology 39*:379–411.

Arnon, D. I., M. B. Allen, and F. R. Whatley. 1954. Photosynthesis by isolated chloroplasts. *Nature 174*:394–396.

Atkins, P. W. 1986. *Physical Chemistry*, 3rd ed. Freeman, New York.

Barber, J., ed. 1977. *Primary Processes of Photosynthesis. Topics in Photosynthesis*, Vol. 2. Elsevier, Amsterdam.

Barber, J. 1978. Biophysics of photosynthesis. *Reports on Progress in Physics 41*:1157–1199.

Brown, J. S. 1972. Forms of chlorophyll *in vivo*. *Annual Review of Plant Physiology 23*:73–86.

Brudvig, G. W., W. F. Beck, and J. C. de Paula. 1989. Mechanism of water oxidation. *Annual Review of Biophysics and Biophysical Chemistry 18*:25–46.

Clayton, R. K. 1980. *Photosynthesis: Physical Mechanisms and Chemical Patterns*. Cambridge University Press, Cambridge, England.

Cramer, W. A., M. T. Black, W. R. Widger, and M. E. Girvin. 1987. Structure and function of photosynthetic cytochrome b–c_1 and b_6–f complexes. In *The Light Reactions*, J. Barber, ed. Elsevier, Amsterdam. Pp. 447–493.

Czeczuga, B. 1987. Carotenoid contents in leaves grown under various light intensities. *Biochemical Systematics and Ecology 15*:523–527.

Davies, D. D., ed. 1980. *Metabolism and Respiration*. In *The Biochemistry of Plants: A Comprehensive Treatise*, P. K. Stumpf and E. E. Conn, eds., Vol. 2. Academic Press, New York.

Demmig-Adams, B., W. W. Adams III, K. Winter, A. Meyer, U. Schreiber, J. S. Pereira, A. Krüger, F.-C. Czygan, and O. L. Lange. 1989. Photochemical efficiency of photosystem II, photon yield of O_2 evolution, photosynthetic capacity, and carotenoid composition during the midday depression of net CO_2 uptake in *Arbutus unedo* growing in Portugal. *Planta 177*:377–387.

Dolphin, D., ed. 1979. *The Porphyrins*, 7 vols. Academic Press, New York.

Duysens, L. M. N., J. Amesz, and B. M. Kamp. 1961. Two photochemical systems in photosynthesis. *Nature 190*:510–511.

Emerson, R. 1957. Dependence of yield of photosynthesis in long-wave red on wavelength and intensity of supplementary light. *Science 125*:746.

Fork, D. C., and K. Satoh. 1986. The control by state transitions of the distribution of excitation energy in photosynthesis. *Annual Review of Plant Physiology 37*:335–361.

Frenkel, A. 1954. Light-induced phosphorylation by cell-free preparations of photosynthetic bacteria. *Journal of the American Chemical Society 76*:5568–5569.

Gantt, E. 1981. Phycobilisomes. *Annual Review of Plant Physiology 32*:327–347.

Glazer, A. N. 1977. Structure and molecular organization of the photosynthetic accessory pigments of cyanobacteria and red algae. *Molecular and Cellular Biochemistry 18*:125–140.

Glazer, A. N., and A. Melis. 1987. Photochemical reaction centers: Structure, organization, and function. *Annual Review of Plant Physiology 38*:11–45.

Goedheer, J. C. 1972. Fluorescence in relation to photosynthesis. *Annual Review of Plant Physiology 23*:87–112.

Goodwin, T. W. ed. 1976. *Chemistry and Biochemistry of Plant Pigments*, 2nd ed. Academic Press, London.

Goodwin, T. W. 1980. *The Biochemistry of the Carotenoids*, Vol. 1, *Plants*, 2nd ed. Chapman and Hall, London.

Gregory, R. P. F. 1989. *Biochemistry of Photosynthesis*, 3rd ed. Wiley, Chichester, England.

Haehnel, W. 1984. Photosynthetic electron transport in higher plants. *Annual Review of Plant Physiology 35*:659–693.

Hall, D. O., and K. K. Rao. 1987. *Photosynthesis*, 4th ed. Arnold, London.

Hill, R., and F. Bendall. 1960. Function of the two cytochrome components in chloroplasts: A working hypothesis. *Nature 186*:136–137.

Isler, O., ed. 1971. *Carotenoids*. Birkhäuser Verlag, Basel, Switzerland.

Jeffrey, S. W. 1981. Responses to light in aquatic plants. In *Physiological Plant Ecology*, O. L. Lange, P. S. Nobel, C. B. Osmond, and H. Ziegler, eds. *Encyclopedia of Plant Physiology, New Series*, Vol. 12A. Springer-Verlag, Berlin. Pp. 249–276.

Kamen, M. D. 1963. *Primary Processes in Photosynthesis*. Academic Press, New York.

Kirk, J. T. O., and R. A. E. Tilney-Bassett. 1978. *The Plastids*, 2nd ed. Elsevier, Amsterdam.

Levine, R. P. 1969. The analysis of photosynthesis using mutant strains of algae and higher plants. *Annual Review of Plant Physiology 20*:523–540.

Lieth, H., and R. H. Whittaker, eds. 1975. *Primary Productivity of the Biosphere. Ecological Studies*, Vol. 14. Springer-Verlag, New York.

Myers, J. 1971. Enhancement studies in photosynthesis. *Annual Review of Plant Physiology 22*:289–312.

O'Neill, P. 1985. *Environmental Chemistry*. Allen & Unwin, London.

Rabinowitch, E., and Govindjee. 1969. *Photosynthesis*. Wiley, New York.

Thornber, J. P., D. T. Morishige, S. Anandan, and G. F. Peter. 1990. Chlorophyll–carotenoid–proteins of higher plant thylakoids. In *The Chlorophylls*, H. Scheer, ed. *CRC UniScience Series*. CRC Press, Boca Raton, Florida. In press.

Trebst, A., and M. Avron, eds. 1977. *Photosynthesis I: Photosynthetic Electron Transport and Photophosphorylation. Encyclopedia of Plant Physiology, New Series*, Vol. 5. Springer-Verlag, Berlin.

Vernon, L. P., and G. R. Seely, eds. 1966. *The Chlorophylls*. Academic Press, New York.

Zuber, H. 1987. The structure of light-harvesting pigment–protein complexes. In *The Light Reactions*, J. Barber, ed. Elsevier, Amsterdam. Pp. 197–259.

CHAPTER 6

Bioenergetics

Throughout this book we have considered various aspects of energy in biological systems. The concept of chemical potential was introduced in Chapter 2 and then applied to the specific case of water. In Chapter 3 we used this thermodynamic approach to discuss the movement of ions. We also considered the use of energy for the active transport of substances toward higher chemical potentials. Chapter 4 dealt with the absorption of light, an event that is followed by various de-excitation reactions for the excited states of the molecules. The photochemistry of photosynthesis (Ch. 5) involves the conversion of such electromagnetic energy into forms that are biologically useful. This last aspect—namely, the production and use of various energy currencies in biological systems—is the topic of the present chapter.

Two energy "currencies" produced in chloroplasts following the trapping of radiant energy are ATP and NADPH. These substances represent the two main classes of energy-storage compounds associated with the electron transfer pathways of photosynthesis and respiration. We can appreciate the importance of ATP by noting that about 100 mol of ATP ($\cong 50$ kg, as the molecular weight of ATP is 507.2) are hydrolyzed in the synthesis of 1 kg dry weight ($\cong 10$ kg wet weight) of many microorganisms. Because only about 10^{-2} mol of ATP typically occurs per kg dry weight of cells, a great turnover of ATP is necessary both to synthesize new tissue and to maintain mature cells in a state far from equilibrium (see Forrest and Walker, 1971; Payne, 1970; Penning de Vries, 1972).

In this chapter we will first examine energy storage in terms of the chemical potential changes accompanying the conversion of a set of reactants into their products. This consideration of the Gibbs free energy allows us to determine the amount of chemical energy that a particular reaction can store or release. We will then evaluate the energy-carrying capacity of ATP in terms of the energetics of its formation and hydrolysis. NADPH can be regarded as possessing electrical energy, the particular amount depending on the oxidation-reduction potential of the system with which it interacts. After considering ATP and NADPH as individual molecules, we will place them in their biological context—namely, as part of the bioenergetic scheme of chloroplasts and mitochondria.

GIBBS FREE ENERGY

Using the appropriate thermodynamic relations, we can calculate the energy changes that accompany biological reactions. Two conditions that are often met by physiological processes greatly simplify these calculations. First, most biological reactions take place at *constant temperature*; i.e., they are isothermal. Second, processes in cells or tissues generally take place at *constant pressure*. These two special conditions make the Gibbs free energy, G, a very convenient function for describing energetics in biology, because the decrease in G under these conditions equals the maximum amount of energy available for work. (See Ch. 2 for an introduction to G, and App. IV for a mathematical presentation of the Gibbs free energy.) Biological use of free energy, or "work," takes a number of different forms—from muscular movement to chemical synthesis and active transport. The difference in Gibbs free energy between two states can be used to predict the spontaneous direction for a reaction and to indicate how much energy the transition makes available for performing work. Biologists are generally more interested in such changes in free energy than in the absolute amount, which must be defined relative to some arbitrary level.

The spontaneous direction for a reaction at constant T and P is toward a minimum of the Gibbs free energy of the system; minimum Gibbs free energy is achieved at equilibrium (see Fig. 6.1). In principle, a spontaneous process can be harnessed to do work; the reversal of a spontaneous reaction requires an input of free energy (see Fig. 2.5). As we will elaborate below, light can be harnessed to produce a free energy source that causes the phosphorylation of ADP and the reduction of $NADP^+$. These two processes are prime examples of energy-requiring reactions that are at the very heart of chloroplast bioenergetics.

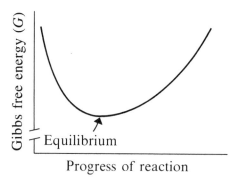

Figure 6.1

Relation between progress of a reaction and the Gibbs free energy of the system (G). When the product concentration is low (abscissa values near the origin), the reaction will spontaneously proceed in its forward direction toward lower values of G. The Gibbs free energy for a reaction attains a minimum at equilibrium; ΔG is then zero for the reaction proceeding a short distance in either direction. For high product concentrations (right-hand side of the abscissa), ΔG to drive the reaction further in the forward direction is positive, indicating that a free energy input is then needed. When ΔG in the forward direction is positive, ΔG for the reverse reaction is negative. The absolute value of G is arbitrary (e.g., $G = \Sigma_j\, n_j \mu_j$, where each μ_j contains an unknown constant, μ_j^*), so the ordinate is interrupted in the figure.

The concept of free energy was introduced in Chapter 2 in presenting the chemical potential μ_j. The chemical potential is actually the partial molal Gibbs free energy with respect to that species, i.e., $\mu_j = (\partial G/\partial n_j)_{T,P,E,h,n_i}$ (Eq. IV.9, App. IV). We must therefore consider the Gibbs free energy of an entire system consisting of many components to define the chemical potential of species j. In turn, G depends on each of the species present, an appropriate expression being

$$G = \sum_j n_j \mu_j \tag{6.1}$$

where n_j is the number of moles of species j in some system, μ_j is the chemical potential of species j (Eq. 2.4), and the summation is over all of the species present (see App. IV, where Eq. 6.1 is derived). The Gibbs free energy is expressed relative to some arbitrary zero level—the arbitrariness in the baseline for G in Equation 6.1. is a consequence of the μ_j^* included in each μ_j (see Fig. 6.1). We will find that the Gibbs free energy as represented by Equation 6.1 is in a very useful form for our application of free energy relations to bioenergetics.

Chemical Reactions and Equilibrium Constants

From the thermodynamic point of view, we are interested in the overall change in free energy for an individual reaction—or perhaps a sequence of reactions. Let us consider a general chemical reaction for which A and B are the reactants and C and D are the products:

$$aA + bB \rightleftharpoons cC + dD \qquad (6.2)$$

where a, b, c, and d are the numbers of moles of the various species taking part in the reaction.

How much energy is stored (or released) when the reaction proceeds a certain extent in either direction? More specifically, what is the change in Gibbs free energy for the reaction in Equation 6.2 proceeding in the forward direction, with a moles of A and b moles of B reacting to give c moles of C and d moles of D? For most applications, we can consider that the chemical potentials of the species involved are constant; in other words, we are concerned with a hypothetical change in the Gibbs free energy were the reaction to take place at certain concentrations and under other fixed conditions. Using Equation 6.1 ($G = \Sigma_j n_j \mu_j$), we can express the change in the Gibbs free energy for such a reaction as

$$\Delta G = -a\mu_A - b\mu_B + c\mu_C + d\mu_D \qquad (6.3)$$

where Δn_j is positive for a product and negative for a reactant ($\Delta G = \Sigma_j \Delta n_j \mu_j$ when all μ_j are constant). Equation 6.3 indicates that the free energy change for a chemical reaction is the Gibbs free energy of the products minus that of the reactants.

To transform Equation 6.3 into a more useful form, we need to incorporate expressions for the chemical potentials of the species involved. The chemical potential of species j was presented in Chapter 2, where μ_j is a linear combination of various terms: $\mu_j = \mu_j^* + RT \ln a_j + \bar{V}_j P + z_j FE + m_j gh$ (Eq. 2.4).[1]

---

1. μ_j^* is a constant, a_j is the activity of species j, $\bar{V}_j$ is its partial molal volume, P is the pressure in excess of atmospheric, z_j is its charge number, F is Faraday's constant, E is the electrical potential, m_j is its mass per mole, and h is the vertical position in the gravitational field.

Substituting such chemical potentials of A, B, C, and D into Equation 6.3, and collecting similar terms, we obtain

$$\Delta G = -a\mu_A^* - b\mu_B^* + c\mu_C^* + d\mu_D^*$$

$$+ RT(-a \ln a_A - b \ln a_B + c \ln a_C + d \ln a_D)$$

$$+ P(-a\overline{V}_A - b\overline{V}_B + c\overline{V}_C + d\overline{V}_D)$$

$$+ FE(-az_A - bz_B + cz_C + dz_D)$$

$$+ gh(-am_A - bm_B + cm_C + dm_D) \tag{6.4}$$

Let us next simplify this equation. The constant terms, $-a\mu_A^* - b\mu_B^* + c\mu_C^* + d\mu_D^*$, can be replaced by ΔG^*, a quantity that we will evaluate below. If the volume of the products, $c\overline{V}_C + d\overline{V}_D$, is the same as that of the reactants, $a\overline{V}_A + b\overline{V}_B$, the factor multiplying P in equation 6.4 is zero. In any case, the value of $P \Sigma_j n_j \overline{V}_j$ for biochemical reactions is usually relatively small for pressures encountered in plant cells, so we will not retain these terms here.[2] The algebraic sum of the terms making up the factor multiplying FE is zero because no charge is created or destroyed by the reaction given in Equation 6.2; i.e., the total charge of the products, $cz_C + dz_D$, equals that of the reactants, $az_A + bz_B$. Because no mass is created or destroyed by the reaction, the factor multiplying gh is also zero.

We can now convert Equation 6.4 to a relatively simple form. Using ΔG^*, the constancies of charge and mass, and the generally valid assumption that $P(-a\overline{V}_A - b\overline{V}_B + c\overline{V}_C + d\overline{V}_D)$ is negligible, the change in Gibbs free energy for the reaction in Equation 6.2 proceeding in the forward direction becomes

$$\Delta G = \Delta G^* + RT \ln \frac{(a_C)^c(a_D)^d}{(a_A)^a(a_B)^b} \tag{6.5}$$

where various properties of logarithms presented in Appendix III have been used to obtain the form indicated. Equation 6.5 is a general expression giving the Gibbs free energy stored or released by a chemical reaction.

---

2. Ignoring $P\Sigma_j n_j\overline{V}_j$ may not be valid for reactions occurring under very high pressures (e.g., deep in the ocean), especially if volume changes are suspected, such as can occur when two oppositely charged species react to form a neutral one (see Zimmerman, 1970).

The quantity ΔG^* in Equation 6.5 has a special value and meaning. At equilibrium, the argument of the logarithm in Equation 6.5, $[(a_C)^c(a_D)^d]/[(a_A)^a(a_B)^b]$, is the equilibrium constant K of the reaction given by Equation 6.2. Furthermore, at constant T and P the Gibbs free energy achieves a minimum at equilibrium (Fig. 6.1), and it does not change for a conversion of reactants to products under such conditions; i.e., ΔG equals zero for a chemical reaction proceeding a short distance in either direction at equilibrium under these conditions. Equation 6.5 thus indicates that $\Delta G^* + RT \ln K$ is zero at equilibrium, so

$$\Delta G^* = -RT \ln K \tag{6.6}$$

where K is the equilibrium constant of the reaction; i.e., K equals the equilibrium value for $[(a_C)^c(a_D)^d]/[(a_A)^a(a_B)^b]$. Hence ΔG^* depends rather simply on the equilibrium constant.

Let us next assume that the reactants, A and B, and the products, C and D, initially all have unit activity. Thus $[(a_C)^c(a_D)^d]/[(a_A)^a(a_B)^b]$ is one, so the logarithm of the activity term is zero. Equation 6.5 then indicates that, for a moles of A plus b moles of B reacting to form c moles of C plus d moles of D under these conditions, the Gibbs free energy change ΔG is ΔG^*. If the equilibrium constant K for the reaction is greater than 1, ΔG^* is negative ($\Delta G^* = -RT \ln K$, Eq. 6.6), so the reaction in the forward direction is spontaneous for this case of unit activity of reactants and products. For $K > 1$ we would expect the reaction to proceed in the forward direction, because at equilibrium the products are then favored over the reactants—in the sense that $[(a_C)^c(a_D)^d] > [(a_A)^a(a_B)^b]$ at equilibrium. On the other hand, if K is less than one, ΔG^* is positive, and such a reaction would not proceed spontaneously in the forward direction for the given initial condition of unit activity of all reactants and products.

Let us now consider the units of G and ΔG. G is an extensive variable; i.e., it depends on the extent or size of the system and is obtained by summing its values throughout the whole system (see App. IV). Specifically, Equation 6.1 ($G = \Sigma_j\, n_j \mu_j$) indicates that the Gibbs free energy is the sum, over all species present, of the number of moles of species j (n_j, an extensive variable) times the energy/mole of species j (μ_j, an intensive variable, i.e., a quantity that can be measured at some point in a system such as is the case for T, P, $\overline{V}_j$, and μ_j). Hence, G has the dimensions of energy. On the other hand, Equation 6.5 suggests that G has the same units as RT—namely, energy *per mole* (a logarithm is dimensionless). To help us out of this apparent dilemma, let us reconsider the conventions used in Equation 6.2, $aA + bB \rightleftharpoons cC + dD$. We usually write a chemical reaction using the smallest possible integers for a, b, c, and d, not the actual number of moles reacting. In fact, in the equations describing nearly all biochemical reactions, either or both a and c are unity, e.g., ADP + phosphate $\rightleftharpoons$ ATP + H_2O. Another convention is to express the Gibbs free energy change per mole of the species on which attention is being focused, e.g., per mole of a certain reactant ($\Delta G/a$) or per mole of a certain product ($\Delta G/c$). So, when a or c represents one mole, ΔG has the same magnitude, whether as energy or as energy/mole. When we use actual values for ΔG or ΔG^* to describe chemical reactions, we will always indicate on what basis we are using the Gibbs free energy, e.g., "ΔG per mole of ATP formed."

What are the numerical values of ΔG^* per mole of reactant A or product C for K's of 100 and 0.01? Because RT is 2.48 kJ mol^{-1} at 25°C and ln equals 2.303 log, an equilibrium constant of 100 corresponds to a $\Delta G^*/a$ or $\Delta G^*/c$ given by Equation 6.6 of $-(2.48$ kJ mol$^{-1})(2.303)$ log (100), or -11.4 kJ mol^{-1} for a moles of reactant A or c moles of product C, and a K of 0.01 leads to a $\Delta G^*/a$ or $\Delta G^*/c$ of $+11.4$ kJ mol^{-1}. In the former case Equation 6.5 indicates that 11.4 kJ of energy per mole of the reactant or product (assuming a or c is 1 mole) is released, and in the latter case the same amount of energy per mole is required, when the reaction proceeds in the forward direction starting with unit activity of all reactants and products.

Interconversion of Chemical and Electrical Energy

To help understand how chemical energy can be converted into electrical energy, and vice versa, we must re-examine the properties of both chemical reactions and the movement of charged species. Let us first consider a chemical reaction such as the dissociation of sodium chloride: NaCl $\rightleftharpoons$ Na$^+$ + Cl$^-$. Although two charged species are produced upon dissociation of NaCl, no overall change in the electrical components of the chemical potentials of Na$^+$ plus Cl$^-$ occurs. In other words, the electrical term $z_j FE$ for Na$^+ (z_{Na} = +1)$ is balanced by an opposite change in the electrical component of μ_{Cl} ($z_{Cl} = -1$). Next, let us consider the following type of reaction: Ag$_s$ $\rightleftharpoons$ Ag$^+$ + e$^-$, i.e., the dissociation of solid silver to an ion plus an electron. Again, no net change in the overall electrical contribution to the two chemical potentials occurs for the dissociation as written. However, the production of an electron opens up various other possibilities, because electrons can be conducted to regions where the electrical potential may be different. Such reactions, in which electrons are produced in one region and then conducted to regions of different electrical potential, allow for the interconversion of chemical and electrical energy. The electron-producing and electron-consuming reactions are referred to as *electrode,* or *half-cell,* re-actions and occur in batteries as well as in the electron transfer chains located in chloroplast and mitochondrial membranes.

To elaborate on energy conversion, let us consider mixing ferrous (Fe^{2+}) and cupric (Cu^{2+}) ions in an aqueous solution; we will assume that the common anion is Cl^-. A chemical reaction occurs in which the products are ferric (Fe^{3+}) and cuprous (Cu^+) ions. Because this is a spontaneous process, the Gibbs free energy decreases; also, the reaction is exothermic, as heat is evolved. Next, let us consider the electrode reaction, $Fe^{2+} \rightleftharpoons Fe^{3+} + e^-$. If the electrons produced in such a half-cell initially containing only Fe^{2+} can be *conducted* by a wire to the Cu^{2+} ions (Fig. 6.2), another electrode reaction can occur in a second beaker initially containing only Cu^{2+}—namely, $Cu^{2+} + e^- \rightleftharpoons Cu^+$. Except for heat evolution, the net result in the solutions is the same as occurs by mixing Fe^{2+} and Cu^{2+}. When the electrons move in the conductor, however, they can be used to do various types of electrical work, e.g., powering a direct-current

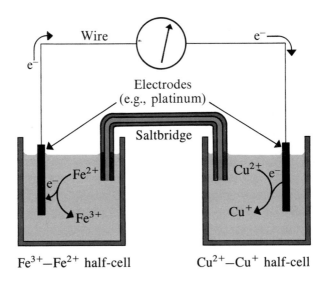

Fe^{3+}–Fe^{2+} half-cell   Cu^{2+}–Cu$^+$ half-cell

Figure 6.2

Two half-cells, or redox couples, connected by a wire and a saltbridge to complete the electrical circuit. Electrons donated by Fe^{2+} to one electrode are conducted by the wire to the other electrode where they reduce Cu^{2+}. Both electrodes (couples) are necessary before electrons can flow. A saltbridge, which provides a pathway along which ions can move and so helps maintain electroneutrality by avoiding the buildup of charge in either half-cell, often contains agar and KCl; the latter minimizes the diffusion potentials at the junctions between the saltbridge and the solutions in the beakers (see Ch. 3).

electrical motor or a light bulb. Such an arrangement provides a way of converting the change in Gibbs free energy of the two spontaneous half-cell reactions into electrical energy that can be used for performing work. Indeed, the important component for obtaining electrical work from the two half-cell reactions is the conducting pathway between them (for a further discussion of half-cells, see Castellan, 1983; MacInnes, 1961; Morris, 1968; Weyer, 1968).

When electrons are moved to a lower electrical potential ($\Delta E < 0$), e.g., by using the chemical energy in a battery, their electrical energy ($z_j FE$, where $z_j = -1$ for an electron) increases. We can use this increase in the electrical energy of the electrons to power a chemical reaction when the electrons subsequently move spontaneously to higher E. In photosynthesis, light energy is used to move electrons toward lower electrical potentials, thereby setting up a spontaneous flow of electrons in the other direction using the electron transfer components that we introduced in the previous chapter. This latter, energetically downhill, spontaneous electron movement is harnessed to drive the photophosphorylation reaction, ADP + phosphate $\rightleftharpoons$ ATP + H_2O, in the forward direction and thereby to store chemical energy.

Now let us consider the interconversion of chemical and electrical energy in more formal terms. Suppose that n moles of electrons ($z_j = -1$) are transferred from one region to another where the electrical potential differs by ΔE, e.g., from one half-cell to another. As we noted in Chapter 3, the charge carried by a mole of protons is Faraday's constant (F); hence the total charge moved in the present case is $-nF$. Electrical work is expressed as the charge transported times the electrical potential difference through which it moves (ΔE). The change in the electrical energy of n moles of electrons therefore is $-nF \Delta E$. This can be converted to an equal change in Gibbs free energy (ΔG):

$$\Delta G = -nF \Delta E \qquad (6.7)$$

We note that n is usually expressed with respect to the compound of interest, for instance, 2 moles of electrons are used to reduce one mole of $NADP^+$; because n is dimensionless, ΔG has the units of energy mol^{-1}, just as for ΔG of chemical reactions.

According to Equation 6.7, the amount of Gibbs free energy that can be stored or released is directly proportional to the difference in electrical potential across which the electrons move. Moreover, this equation indicates that the flow of electrons toward more positive electrical potentials ($\Delta E > 0$) corresponds to a decrease in free energy ($\Delta G < 0$); thus, this movement proceeds spontaneously. We should emphasize that *two* half-cells are necessary to get a ΔE and thus a ΔG for electron transfer (see Fig. 6.2). We will apply these free energy considerations to the energetics of electrons moving from molecule to molecule in the electron transfer chains of chloroplasts and mitochondria.

Redox Potentials

Many organic compounds involved in photosynthesis accept or donate electrons (see Table 5.3). The negatively charged electrons spontaneously flow toward more positive electrical potentials, which are measured by *redox potentials* in the case of the components involved with electron flow in chloroplast lamellae or the inner membranes of mitochondria. Redox potentials are a measure of the relative chemical potential of electrons accepted or donated by a particular type of molecule. The oxidized plus the reduced forms of each electron transfer component can be regarded as an electrode, or half-cell. Such a half-cell can interact with other electron-accepting and electron-donating molecules in the membrane, in which case the electrons spontaneously move toward the component with the higher redox potential.

We will begin by writing a chemical reaction describing a general electrode (half-cell) reaction. We can represent the acceptance or the donation of electrons by some species as follows:

$$\text{oxidized form} + qe^- \rightleftharpoons \text{reduced form} \qquad (6.8)$$

where q is a dimensionless parameter indicating the number of electrons transferred per molecule; and *oxidized* and *reduced* refer to different forms of the same species—e.g., $NADP^+$ is an oxidized form and NADPH represents the corresponding reduced form (q is 2 in this case). Like any other chemical reaction, an oxidation–reduction reaction such as Equation 6.8 has a change in Gibbs free energy associated with it when the reactants are converted to products. Thus oxidation–reduction, or "redox," reactions can be described by the relative tendency of the redox system, or *couple* (the oxidized plus the reduced forms of the compound), to proceed in the forward direction, which for Equation 6.8 means accepting electrons.

It is more useful to describe redox reactions in terms of relative electrical potentials instead of the equivalent changes in Gibbs free energy. The electrons in Equation 6.8 come from or go to some other redox couple, and whether or not the reaction in Equation 6.8 will proceed in the forward direction depends on the relative electrical potentials of these two couples. It is therefore convenient to assign a particular electrical potential to a system accepting or donating electrons, a value known as its *redox potential*. This oxidation–reduction potential can then be compared with that of another couple to predict the direction for spontaneous electron flow when the two couples are allowed to interact—electrons spontaneously move toward higher redox potentials. The redox potential of species j, E_j, is defined as follows:

$$E_j = E_j^* - \frac{RT}{qF} \ln \frac{(\text{reduced}_j)}{(\text{oxidized}_j)} \tag{6.9}$$

where E_j^* is an additive constant, q is the number of electrons transferred (the same q as in Eq. 6.8), and (reduced_j) and (oxidized_j) refer to the activities of the two different redox states of species j. Equation 6.9 indicates that the oxidation–reduction potential of a particular redox couple is determined by the ratio of the reduced to the oxidized form plus an additive constant, a quantity that we consider next.

An electrical circuit is formed when two electrodes (half-cells) are connected and pathways for electron flow are provided (see Fig. 6.2). Because the sum of the electrical potential drops (voltage changes) going completely around such a circuit is zero, we can determine the half-cell potential on an absolute basis for a particular electrode if the potential of some standard reference electrode is known. By international agreement, the E_j^* of a hydrogen half-cell ($\frac{1}{2}H_{2\text{gas}} \rightleftharpoons$ $H^+ + e^-$) is arbitrarily set equal to zero for an activity of hydrogen ions of 1 molal equilibrated with hydrogen gas at a pressure of one atmosphere, i.e., $E_H^* = 0$. Fixing the zero level of the electrical potential for the hydrogen half-cell removes the arbitrary nature of redox potentials for all half-cells, because the redox potential for any species can then be determined relative to that of the hydrogen electrode. We will replace E_j^* in Equation 6.9 by $E_j^{*,H}$ to emphasize the convention of referring electrode potentials to the standard hydrogen electrode.[3]

According to Equation 6.9, the larger is the ratio $(\text{reduced}_j)/(\text{oxidized}_j)$, the more negative the redox potential becomes. Because electrons are negatively charged, a lower E_j corresponds to higher energies for the electrons. Thus, the further Equation 6.8 is driven in the forward direction, the more energy will be required to reduce species j. Likewise, the larger is $(\text{reduced}_j)/(\text{oxidized}_j)$, the higher will be the electrical energy of the electrons that the reduced form of that couple can then donate. When (reduced_j) equals (oxidized_j), E_j equals $E_j^{*,H}$ by Equation 6.9. This $E_j^{*,H}$ is commonly referred to as the *midpoint redox potential* (see the values in Table 5.3 for the $E_j^{*,H}$ of some components involved with

---

3. The hydrogen half-cell is not very convenient for routine laboratory usage—indeed, 1 molal H^+ and 1 atmosphere H_2 can be rather dangerous. Hence, secondary standards are used, e.g., mercury/mercurous (calomel) or silver/silver chloride electrodes, which have midpoint redox potentials of 0.244 V and 0.222 V, respectively (for measurement details see Allen and Holmes, 1986).

electron transfer in chloroplasts). For certain purposes, knowledge of the mid-point redox potentials may be sufficient, as we will show later in this chapter (for more details on redox reactions, see Morris, 1968; Smith *et al.*, 1983; Stryer, 1988).

BIOLOGICAL ENERGY CURRENCIES

In photosynthesis photons are captured, initiating an electron flow leading both to the production of NADPH and to a coupled process whereby ATP is formed. Light energy is thereby converted into chemical energy in the for-mation of a phospho-anhydride (ATP) in an aqueous environment. More-over, it is converted into electrical energy by providing a reduced compound (NADPH) under oxidizing conditions. ATP and NADPH are the two energy storage compounds, or "currencies," that we will consider in this section. Both can occur as ions, both can readily diffuse around within a cell or organelle, and both can carry appreciable amounts of energy under biological condi-tions. In addition to its use in processes such as active transport and muscle contraction, the chemical energy stored in ATP is also used in certain bio-synthetic reactions involving the formation of anhydrous links, or bonds, in the aqueous milieu of a cell. The relatively high atmospheric levels of O_2 ensure that appreciable amounts of this strong oxidizing agent will be pre-sent in most biological systems; a reduced compound like NADPH is thus an important currency for energy storage. We will discuss these two compounds in turn, after briefly considering the difference between ATP and NADPH as energy currencies.

Redox couples are assigned a relative electrical energy, whereas chemical reactions have a specific chemical energy. In a chemical reaction certain reactants are transformed into products, and the accompanying change in Gibbs free energy can be calculated. This change in chemical energy need not depend on any other chemical species. For instance, if the concentrations—strictly speaking, the chemical activities—of ADP, phosphate, and ATP as well as certain other con-ditions (e.g., temperature, pH, Mg^{2+} concentration, and ionic strength) are the same in different parts of an organism, then the Gibbs free energy released upon the hydrolysis of a certain amount of ATP to ADP and phosphate will be the same in each of the different locations. But an oxidation–reduction couple must donate electrons to, or accept electrons from, another redox system, and the change in electrical energy depends on the difference in the redox potential between the two couples. Thus, the amount of electrical energy released when NADPH is oxidized to $NADP^+$ depends on the redox potential of the particular couple with which NADPH interacts.

ATP—Structure and Reactions

To help us understand the bioenergetics of chloroplasts and mitochondria, we need to know how much energy is stored in ATP, that is, the difference between its chemical potential and that of the reactants (ADP and phosphate) used in its formation. We must then look for reactions that have a large enough free energy decrease to drive the ATP synthesis reaction in the energetically uphill direction; this will lead us to a consideration of the energetics of electron flow in organelles—topics that we will discuss in the next two sections. Our immediate concern is with the following: (1) the chemical reaction describing ATP formation, (2) the associated change in Gibbs free energy for that reaction, and (3) the implications of the substantial amount of energy storage in ATP.

ADP, ATP, and phosphate can all occur in a number of different charge states in aqueous solutions. Moreover, all three compounds can interact with other species, notably Mg^{2+} and Ca^{2+}. Thus, many different chemical reactions can be used for describing ATP formation. A predominant reaction occurring near neutral pH in the absence of divalent cations is as follows:

(6.10)

where adenosine is adenine esterified to the $1'$ position of the sugar ribose, i.e.,

Figure 6.3

Structures of three molecules important in bioenergetics. The dissociations and bindings of protons indicated in the figure are appropriate near pH 7. Note the similarity between the molecules.

The attachment of adenosine to the phosphates in ADP and ATP—and in NADP$^+$ as well as in FAD—is by means of an ester linkage (Ch. 1) with the hydroxymethyl group on the 5′ position of the ribose moiety (see Fig. 6.3).

Equation 6.10 indicates a number of features of ATP production. For instance, the formation of ATP from ADP plus phosphate is a dehydration; the reversal of Equation 6.10, in which the phospho-anhydride is split with the incorporation of water, is known as ATP *hydrolysis*. Because Equation 6.10 contains H$^+$, the equilibrium constant depends on pH, $-\log(a_{H^+})$. Moreover, the fractions of ADP, phosphate, and ATP in various states of ionization depend on the pH. Near pH 7 about half of the ADP molecules are doubly charged and half are triply charged, the latter form being indicated in Equation 6.10.[4] Likewise, ATP

---

4. For simplicity, we are ignoring the charge due to the extra proton bound to an adenine nitrogen, which gives that part of the ADP and ATP molecules a single positive charge at pH 7.

at pH 7 is about equally distributed between the forms with charges of -3 and -4. Because of their negative charges in aqueous solutions, both ADP and ATP can readily bind positive ions, especially divalent cations such as Mg^{2+} or Ca^{2+}. A chelate is formed such that Mg^{2+} or Ca^{2+} is held electrostatically between two negatively charged oxygen atoms on the same molecule (consider the many $-O^-$'s occurring on the chemical structures indicted in Eq. 6.10). Also, inorganic phosphate can interact electrostatically with Mg^{2+} and other divalent cations, further increasing the number of complexed forms of ADP, ATP, and phosphate that are possible.

The activities (or concentrations) of a species in all of its ionization states and complexed forms are generally summed to obtain the total activity (or concentration) of that species. The number of relations and equilibrium constants needed to describe a reaction like ATP formation is then reduced to one; i.e., a separate equilibrium constant is not needed for every possible combination of ionization states and complexed forms of all the reactants and products. Using this convention, we can replace Equation 6.10 and many others like it, which also describe ATP formation, by the following general reaction for the phosphorylation of ADP:

$$ADP + phosphate \rightleftharpoons ATP + H_2O \qquad (6.11)$$

We will return to ATP formation as represented by Equation 6.11 after briefly commenting on two important conventions used in biochemistry. First, most equilibrium constants for biochemical reactions are defined at a specific pH, usually pH 7 ($a_{H^+} = 10^{-7}$ M). At constant pH the activity of H^+ does not change. Thus, H^+ need not be included as a reactant or a product in the expression for the change in Gibbs free energy (Eq. 6.5). In other words, the effect of H^+ in relations such as Equation 6.10 is incorporated into the equilibrium constant, which itself generally depends on pH. Second, biological reactions such as ATP formation usually take place in aqueous solutions where the concentration of water does not change appreciably. (The concentrations of other possible reactants and products are much, much less than that of water.) Hence, the a_{H_2O} term coming from relations like Equation 6.11 is also usually incorporated into the equilibrium constant. We can illustrate these points concerning a_{H^+} and a_{H_2O} by specifically considering ATP formation as described by Equation 6.10. For the reaction as written, the equilibrium constant K is equal to $[(a_{ATP})(a_{H_2O})]/[(a_{ADP})(a_{phosphate})(a_{H^+})]$; hence, $(a_{ATP})/[(a_{ADP})(a_{phosphate})]$ equals $(a_{H^+})K/(a_{H_2O})$. For a dilute aqueous solution, a_{H_2O} is essentially constant during the reaction; therefore, $(a_{H^+})K/(a_{H_2O})$—which is conventionally called the equilibrium constant in biochemistry—has a fixed value at a given pH, e.g., at pH 7, $(a_{H^+})K/(a_{H_2O}) = (10^{-7} \text{ M})K/(a_{H_2O}) = K_{pH\ 7}$.

Next, we will specifically consider the equilibrium constant for ATP formation under biological conditions. Using the above conventions for H^+ and H_2O, an equilibrium constant for Equation 6.11 at pH 7($K_{pH\ 7}$) is

$$K_{pH\ 7} = \frac{[\text{ATP}]}{[\text{ADP}][\text{phosphate}]} \cong 5 \times 10^{-6} \text{ M}^{-1} \text{ at } 25°C \qquad (6.12)$$

where the total concentration of each species involved is indicated in brackets; i.e., it is experimentally more convenient to measure $K_{pH\ 7}$ using concentrations [indicated by brackets] instead of activities (indicated by parentheses).[5]

When activities of ions ($a_j = \gamma_j c_j$, Eq. 2.5) are replaced by concentrations [c_j], the resulting equations for equilibrium constants or free energy changes apply only to a particular ionic strength ($\frac{1}{2}\Sigma_j\ c_j z_j^2$; see p. 115), because the activity coefficients of ions (γ_j) can markedly depend on ionic strength (see Eq. 3.3). The value for $K_{pH\ 7}$ given in Equation 6.12 is suitable for ionic strengths close to 0.2 M (200 mol m^{-3}), which is an ionic strength that can occur *in vivo* (a 0.05 M increase or decrease in ionic strength changes K in the opposite direction by about 10%). The magnitude of the equilibrium constant for ATP formation also depends on the concentration of Mg^{2+}. The value of $K_{pH\ 7}$ in Equation 6.12 is appropriate for 10 mM Mg^{2+}, a concentration similar to that in many plant cells (a 5 mM increase or decrease in Mg^{2+} changes K in the opposite direction by about 10% to 20%). An equilibrium constant also depends on temperature, $K_{pH\ 7}$ for ATP formation increasing 1% to 5% per °C. A large effect on K is produced by pH, which may not be near 7 and often is unknown in a cell. The equilibrium constant for ATP formation increases about 3-fold as the pH is lowered 1 unit from pH 7 and decreases 7-fold as it is raised one unit (see Alberty, 1968; D. O. Hall in Barber, 1976; Rosing and Slater, 1972).

Gibbs Free Energy Change for ATP Formation

The energetics of a reaction such as ATP formation is summarized by its Gibbs free energy change, ΔG. For a general chemical reaction, Equation 6.5 indicates

---

5. Because the equilibrium constant for ATP formation is quite small (see Eq. 6.12) and is sensitive to temperature, pH, Mg^{2+}, and ionic strength, measured values of K vary considerably. Actually, instead of K, the standard Gibbs free energy, ΔG^*, is usually determined ($\Delta G^* = -RT$ ln K, Eq. 6.6). Values of ΔG^* for ATP formation range from 28 to 45 kJ mol^{-1}, but a value near 30 kJ mol^{-1} seems most likely at pH 7, 25°C, 10 mM Mg^{2+}, and an ionic strength of 0.2 M (see D. O. Hall in Barber, 1976; Rosing and Slater, 1972).

that ΔG equals $\Delta G^* + RT \ln \{[(a_C)^c(a_D)^d]/[(a_A)^a(a_B)^b]\}$, where ΔG^* is $-RT$ $\ln K$ (Eq. 6.6). For the present case, the reactant A is ADP, B is phosphate, the product C is ATP, and the equilibrium constant is given by Equation 6.12. Therefore, our sought-after free energy relationship describing ATP formation is

$$\Delta G = -RT \ln (K_{\text{pH } 7}) + RT \ln \frac{[\text{ATP}]}{[\text{ADP}][\text{phosphate}]} \qquad (6.13a)$$

which at pH 7 and 25°C becomes

$$\Delta G \cong 30 + 5.71 \log \frac{[\text{ATP}]}{[\text{ADP}][\text{phosphate}]} \qquad \text{kJ (mol ATP)}^{-1} \quad (6.13b)$$

where $\ln$ equals 2.303 $\log$, 2.303 RT is 5.71 kJ mol^{-1} at 25°C (App. I), and $-(5.71$ kJ mol$^{-1}) \log (5 \times 10^{-6})$ is 30 kJ mol^{-1} (7.2 kcal mol^{-1}).

It is apparent from Equation 6.13 that ATP usually does not tend to form spontaneously, because the Gibbs free energy change for the reaction is generally quite positive. In fact, the energy required for the phosphorylation of ADP is rather large compared with the free energy changes for most biochemical reactions. Stated another way, much energy can be stored by converting ADP plus phosphate to ATP. Although ATP in an aqueous solution is *thermodynamically unstable*, in that its hydrolysis can release a considerable amount of Gibbs free energy, it still can last for a long enough time in cells to be an important energy currency; i.e., it is *kinetically* stable. In particular, ATP is generally not hydrolyzed very rapidly unless the appropriate enzymes necessary for its use in certain biosynthetic reactions or other energy-requiring processes are present (for long-term energy storage, plants use carbohydrates like the polysaccharide starch).

We will next estimate the changes in Gibbs free energy that might be expected for photophosphorylation under physiological conditions. For purposes of calculation, we will assume that in unilluminated chloroplasts the concentration of ADP is 2.2 mM, phosphate is 10 mM, and ATP is 0.2 mM (1 mM = 1 mol m^{-3}). From Equation 6.13b, the free energy change required to form ATP then is

$$\Delta G = 30 + 5.71 \log \frac{(0.2 \times 10^{-3} \text{ M})}{(2.2 \times 10^{-3} \text{ M})(10 \times 10^{-3} \text{ M})}$$

$$= 35 \text{ kJ (mol ATP)}^{-1}$$

The change in Gibbs free energy required is positive, indicating that energy must be supplied to power photophosphorylation. Moreover, the energy necessary depends in a predictable way on the concentrations of the reactants and the product. After a certain period of time in the light, ATP may increase to 2.2 mM with a concomitant decrease in ADP to 0.2 mM and in phosphate to 8 mM. The free energy required for photophosphorylation under these conditions is 48 kJ (mol ATP)$^{-1}$. Consequently, the further photophosphorylation goes to completion, the greater is the energy required to form more ATP.

The high energy of ATP relative to ADP plus phosphate for usual physiological concentrations is not the property of a single bond but of the local configuration in the ATP molecule, a point that we can appreciate by considering the phosphorus atoms in ADP, ATP, and phosphate. Phosphorus is in group V of the third period of the periodic table and has five electrons in its outermost shell. It can enter into a total of five bonds with four oxygen atoms, the bonding to one O being a double bond (consider structures in Eq. 6.10). In inorganic phosphate all four bonds are equivalent, so four different structures for phosphate exist in resonance with each other. The terminal P of ADP has only three resonating forms, because one of the O's is connected to a second phosphorus atom and does not assume a double bond configuration (see Eq. 6.10). When inorganic phosphate is attached to this terminal P of ADP to form ATP, a resonating form is lost both from the ADP and from the phosphate. Configurations having more resonating structures are in general more probable or stable (lower in energy), so energy must be supplied to form ATP from ADP plus phosphate with an accompanying loss of two resonating forms.

Let us next quantitatively examine some of the ways in which ATP can be used as a free energy currency. Each of our four examples will relate to a different variable term in the chemical potential ($\mu_j = \mu_j^* + RT \ln a_j + \overline{V}_j P + z_j FE + m_j gh$, Eq. 2.4). To transfer a mole of a neutral compound against a 10-fold increase in activity requires 2.303 RT, or 5.7 kJ of Gibbs free energy at 25°C, and it takes 11.4 kJ for a 100-fold increase and 17.1 kJ to move in the direction of a 1 000-fold increase in activity. To move a monovalent cation from one side of a membrane to the other, where it has the same concentration but the electrical potential is 0.1 V higher, requires 9.6 kJ mol^{-1} ($F = 96.49$ kJ mol^{-1} V^{-1}, App. I). ATP usually supplies at least 40 kJ mol^{-1} when hydrolyzed and can act as the Gibbs free energy source for the active transport of solutes across membranes toward regions of higher chemical potential. ATP is also the free energy currency for the contraction of muscles. The ATP-driven contraction of the muscles surrounding the left ventricle of the human heart can increase the blood pressure within it by 20 kPa (0.2 bar or 150 mm Hg). This increases the chemical potential of the water in the blood (i.e., the $\overline{V}_w P$ term), which causes the blood to flow out to the aorta and then to the rest of the circulatory system toward lower hydrostatic pressures. Pressure-driven flow is an efficient way to move fluids; e.g., it takes only 0.02 kJ of Gibbs free energy to increase the pressure of 10^{-3}

m^3 (1 litre) of water by 20 kPa. As an example of gravitational work that can be mediated by ATP, the increase in Gibbs free energy as a 50 kg person climbs up 100 m is 49 kJ. Because of its large free energy release upon hydrolysis— about 40 to 50 kJ mol^{-1} (10 to 12 kcal mol^{-1}), and up to 60 kJ mol^{-1} in exceptional circumstances—and because of its convenient form as a relatively abundant ion, ATP is an extremely useful cellular energy currency.

NADP$^+$–NADPH Redox Couple

Another class of energy storage compounds consists of redox couples such as NADP$^+$–NADPH. The reduced form, NADPH, is produced by noncyclic electron flow in chloroplasts. Photosynthesis in bacteria makes use of a different redox couple, NAD$^+$–NADH. The reduced member of this latter couple causes an electron flow in mitochondria and an associated formation of ATP. NAD is nicotinamide adenine dinucleotide and differs from NADP by not having a phosphate esterified to the 2′ hydroxy group of the ribose in the adenosine part of the molecule (see Fig. 6.3). Our present discussion will focus on the NADP$^+$– NADPH couple, but the same arguments and also the same midpoint redox potential apply to the NAD$^+$–NADH couple (see Morris, 1968; Smith *et al.*, 1983; Stryer, 1988).

The reduction of a molecule of NADP$^+$ involves its acceptance of two electrons. Only the nicotinamide portion (illustrated in Eq. 6.14) of NADP$^+$ is involved in accepting the electrons. The actual half-cell reaction describing this reduction is

(6.14)

where R represents a ribose attached at its 1C position to nicotinamide and at its 5C position by a pyrophosphate bridge to an adenosine having the 2′ hydroxy group of its ribose moiety esterified to an additional phosphate (see Fig. 6.3). Adenosine less the ribose is called adenine, hence the name of nicotinamide adenine dinucleotide phosphate, or NADP. The reduction of NADP$^+$ involves the transfer of two electrons to the nicotinamide ring, plus the attachment of one H$^+$ to the *para* position (top of the ring for the NADPH indicated in Eq. 6.14). That is, *two* electrons are accepted by the NADP$^+$ molecule during its reduction, although one of the *two* accompanying protons is not attached to the reduced form, as Equation 6.14 indicates.

A particular half-cell reaction such as Equation 6.14 can accept or donate electrons. We quantitatively describe this by the oxidation–reduction (redox) potential for that reaction, as expressed by Equation 6.9 $\{E_j = E_j^{*,H} - (RT/qF)$ ln [(reduced$_j$)/(oxidized$_j$)]$\}$. We will use (NADPH) to represent the activity of all of the various ionization states and complexed forms of the reduced nicotinamide adenine dinucleotide phosphate, and (NADP$^+$) has an analogous meaning for the oxidized species of the NADP$^+$–NADPH couple. For redox reactions of biological interest the midpoint (standard) redox potential is generally determined at pH 7. By using Equation 6.9, in which the number q of electrons transferred per molecule reduced is 2, we can express the oxidation–reduction potential of the NADP$^+$–NADPH couple (Eq. 6.14) as follows:

$$E_{\text{NADP}^+\text{-NADPH}} = E_{\text{pH 7}}^{*,H} - \frac{RT}{2F} \ln \frac{(\text{NADPH})}{(\text{NADP}^+)} \tag{6.15a}$$

which at 25°C and pH 7 becomes

$$E_{\text{NADP}^+\text{-NADPH}} = -0.32 - 0.030 \log \frac{(\text{NADPH})}{(\text{NADP}^+)} \quad \text{V} \tag{6.15b}$$

where in the second equation ln has been replaced by 2.303 log, and the numerical value for 2.303 RT/F of 0.0592 V at 25°C (App. I) has been used. The midpoint redox potential, $E_{\text{pH 7}}^{*,H}$, for the NADP$^+$–NADPH couple is -0.32 V (Table 5.3), a value achieved when (NADPH) equals (NADP$^+$).

To determine whether electrons spontaneously flow toward or away from the NADP$^+$–NADPH couple, we must compare its redox potential with that of some other redox couple. As indicated above, electrons spontaneously flow toward more positive redox potentials, whereas energy must be supplied to move electrons in the energetically uphill direction of algebraically decreasing redox potentials. We will next examine the redox potentials of the various redox couples involved in electron flow in chloroplasts, then in mitochondria.

CHLOROPLAST BIOENERGETICS

In Chapter 5 we introduced various molecules that are involved with electron transfer in chloroplasts, together with a consideration of the sequence of electron flow between components. Now that the concept of redox potential has been presented, we will resume our discussion of electron transfer in chloroplasts. We will compare the midpoint redox potentials of the various redox couples, not only to help understand the direction of spontaneous electron flow, but also to see the important role of light absorption in changing the redox properties of the trap chl. Also, we will consider how ATP formation is coupled to electron flow.

Redox Couples

Although the ratio of reduced to oxidized forms of species j affects its redox potential $\{E_j = E_j^{*,H} - (RT/qF) \ln [(\text{reduced}_j)/(\text{oxidized}_j)]$, Eq. 6.9 $\}$, the actual activities of the two forms are generally unknown for biological studies *in vivo*. Moreover, the value of the local pH (which can affect $E_j^{*,H}$) is also generally unknown. Consequently, we usually compare midpoint (standard) redox potentials determined at pH 7 to predict the direction for spontaneous electron flow in lamellar membranes of chloroplasts. We will follow such an approach here and will assume that free energy is required to transfer electrons to a compound with a more negative *midpoint* redox potential, whereas the reverse process can go on spontaneously.

The absorption of a photon can markedly affect the redox properties of a pigment molecule. An excited molecule such as trap chl* has an electron in an antibonding orbital (see Ch. 4). It takes less energy to remove such an electron from the excited molecule than when that molecule is in its ground state. Thus, the electronically excited molecule is a better electron donor (reducing agent) than is one in the ground state, inasmuch as it has a considerably more negative redox potential. Once the electron is removed, this oxidized molecule (trap chl$^+$) becomes a very good electron acceptor (oxidizing agent). Such electron acceptance, trap chl$^+$ + D $\rightarrow$ trap chl + D$^+$ (Eq. 5.6), involves the ground state of the chlorophyll molecule, which has a rather positive redox potential (see Table 5.3 and Fig. 6.4). The electronic state of trap chl that donates an electron is therefore an excited state with a negative redox potential, and the ground state with its rather positive redox potential can readily accept an electron. In short, the absorption of light energy transforms trap chl's from couples with positive redox potentials to couples with negative redox potentials. In this section we

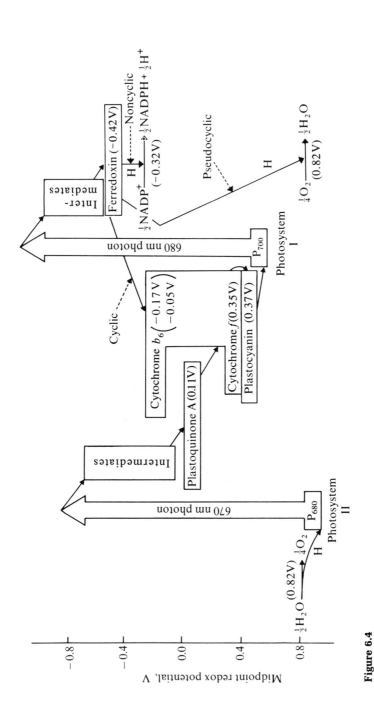

Figure 6.4

Energy aspects of photosynthetic electron flow. The lengths of the arrows emanating from the trap chl's of Photosystems I and II represent the increases in chemical potential of the electrons that occur upon absorption of red light near the λ_{max}'s of the Chl a's. The diagram also shows the various midpoint redox potentials of the couples involved (data from Table 5.3) and the three types of electron flow mediated by ferredoxin.

will estimate the redox potential spans at each of the two photosystems in chloroplasts and then diagram the overall pattern of photosynthetic electron flow.

Following light absorption by Chl a or an accessory pigment feeding into Photosystem II, the excitation migrates to P_{680}, where an electron transfer reaction takes place (trap chl* $+ A \rightarrow$ trap chl$^+ + A^-$, Eq. 5.5). The electron removed from P_{680}^* is replaced by one coming from water, which results in O_2 evolution as described by Equation 5.8. (For energetic considerations, we write Eq. 5.8 as $H_2O \rightleftharpoons \frac{1}{2}O_2 + 2H$.) The water–oxygen system represents a half-cell reaction with a midpoint redox potential of 0.82 V at 25°C, pH 7, and an O_2 pressure of one atmosphere (see Fig. 6.4). Water oxidation and the accompanying O_2 evolution follow spontaneously after the photochemistry at the reaction center of Photosystem II has led to P_{680}^+. Thus, the required oxidant for water—i.e., P_{680}^+ or some intermediate oxidized by it—must have a redox potential more positive than 0.82 V for the electron to move energetically downhill from water to the trap chl$^+$ in the reaction center of Photosystem II. As we indicated in Table 5.3, the redox potential of P_{680} in the ground state is about 0.90 V.

The electron removed from P_{680}^* goes to pheophytin, from which it moves to two intermediate quinones and then to the plastoquinone pool. The midpoint redox potentials are -0.61 V for pheophytin, -0.05 to -0.25 V for the first plastoquinone (Q_A), about 0.10 V for the second one (Q_B), and 0.11 V for plastoquinone A (see Glazer and Melis, 1987; Gregory, 1989; Rich and Moss, 1987; Velthuys, 1987). As expected, the negatively charged electron spontaneously moves toward higher redox potentials, in this case from pheophytin to intermediate quinones to plastoquinone A. The electrical potential span in Photosystem II is thus from the redox potential of 0.90 V for the ground state of P_{680} to -0.61 V for the pheophytin couple, or -1.51 V overall.

The energy required to move an electron in the energetically uphill direction toward lower redox potentials in Photosystem II is supplied by a light quantum (Fig. 6.4). In particular, Photosystem II can be excited by 670-nm light (as well as by other wavelengths, this value being near the λ_{max} for the red band of its Chl a). From Equation 4.2a ($E = hc/\lambda_{vacuum}$) and the numerical value of hc (1 240 eV nm, App. I), we can calculate that the energy of 670-nm light is (1 240 eV nm)/(670 nm), or 1.85 eV per photon. Such a photon could move an electron across 1.85 V, which is more than sufficient energy to cross the redox potential span of 1.51 V estimated for Photosystem II.[6]

---

6. Actually, the amount of absorbed light energy and the accompanying changes in free energy available for decreasing the redox potential are not the same; i.e., the increase in internal energy U upon light absorption is generally not equal to the change in G (see Eq. IV.4a, App. IV, $G = U + PV - TS$). The magnitude of the increase in Gibbs free energy caused by the absorption of a photon by chlorophyll depends on the level of illumination as well as on the various pathways competing for the de-excitation of trap chl* (see Knox, 1977; Ross and Calvin, 1967).

We can make a similar analysis of the energetics for Photosystem I, where the trap chl is P_{700}. The redox potential span across which electrons are moved is from the midpoint redox potential of 0.45 V for the P_{700}–P_{700}^+ couple (Table 5.3) to about -0.90 V for the monomeric form of chlorophyll that acts as the first acceptor (Gregory, 1989; Malkin, 1987). Thus, the electrical potential span in Photosystem I is about 1.35 V. A photon at 680 nm, which is the λ_{max} for the red band of Chl a in Photosystem I, has an energy of 1.82 eV, which is ample energy to move an electron from P_{700} to the first acceptor in Photosystem I. From the first acceptor, the electron spontaneously moves to a quinone (midpoint redox potential possibly near -0.80 V), then to three Fe-S centers (midpoint redox potentials of -0.73, -0.59, and -0.53 V, respectively), and then to ferredoxin (Table 5.3). From ferredoxin to the next component in the noncyclic electron flow sequence, $NADP^+$, electrons go from -0.42 V to -0.32 V (midpoint redox potentials of the couples; see Fig. 6.4). Again, moving toward higher redox potentials is energetically downhill for electrons, so this step leading to the reduction of the pyridine nucleotide follows spontaneously from the reduced ferredoxin—the enzyme ferredoxin–NADP reductase (see Table 5.3) catalyzes this process.

In noncyclic electron flow, two electrons originating in the water–oxygen couple with a midpoint redox potential of 0.82 V are moved to the redox level of -0.32 V for the reduction of one molecule of the $NADP^+$–NADPH couple. Because a midpoint redox potential of -0.32 V is more negative than most encountered in biology, NADPH can spontaneously reduce most other redox systems; reduced pyridine nucleotides are therefore an important energy currency. Moving electrons from 0.82 V to -0.32 V requires a considerable expenditure of free energy, which helps to explain why light, with its relatively large amount of energy (see Table 4.1), is needed. We can calculate the actual free energy change for the overall process using Equation 6.7 ($\Delta G = -nF \, \Delta E$):

$$\Delta G = -(2)(96.5 \text{ kJ mol}^{-1} \text{ V}^{-1})(-0.32 \text{ V} - 0.82 \text{ V})$$
$$= 220 \text{ kJ mol}^{-1}$$

for the overall movement of two moles of electrons along the pathway for noncyclic electron flow—a process that leads to the reduction of one mole of $NADP^+$.

The incorporation of CO_2 into a carbohydrate during photosynthesis requires 3 ATP's and 2 NADPH's (see Fig. 5.1). Using these energy currencies, CO_2 fixation could energetically proceed in the absence of light, so the steps of the reductive pentose cycle are often referred to as the *dark reactions* of photosynthesis (actually, because several of the enzymes are light-activated, not much CO_2 fixation would occur in the dark). The hydrolysis of ATP in chloroplasts can release about 50 kJ mol^{-1}; we have just stated that 220 kJ are required to reduce a mole of NADP$^+$ using electrons coming from water (this is actually the Gibbs free energy change between standard states, because midpoint redox potentials were used in our calculations); and the increase in Gibbs free energy per mole of CO_2 incorporated into a carbohydrate during photosynthesis is 479 kJ (p. 247). Using these numbers, we can estimate the efficiency for free energy storage by the dark reactions. Dividing the energy stored per mole of CO_2 fixed, 479 kJ, by the energy input, (3 mol ATP)(50 kJ mol^{-1} ATP) + (2 mol NADPH)(220 kJ mol^{-1} NADPH), we find that the efficiency is [(479 kJ)/(590 kJ)](100), or 81%! The dark reactions of photosynthesis are indeed extremely efficient.

Figure 6.4 incorporates the midpoint redox potentials of various components involved with photosynthetic electron transfer discussed both here and in Chapter 5 (see Table 5.3). The direction for spontaneous electron flow to higher midpoint redox potentials is downward (toward the bottom of Fig. 6.4); the absorption of light quanta with their relatively large energies corresponds to moving electrons vertically upward to higher energy. The role played by ferredoxin at the crossroads of cyclic, noncyclic, and pseudocyclic electron flow is also illustrated in Figure 6.4.

H$^+$ Chemical Potential Differences Caused by Electron Flow

We indicated in Chapter 5 that the components involved with electron flow are so situated in the lamellar membranes of chloroplasts that they lead to a vectorial or unidirectional movement of electrons and protons (see Fig. 5.14). We will now return to this theme and focus on the gradients in H$^+$ (protons) thus created. The difference in the chemical potential of H$^+$ from the inside to the outside of a thylakoid in the light acts as the energy source to drive photophosphorylation. This was first clearly recognized in the 1960's by Mitchell, who received the 1978 Nobel Prize in chemistry for his enunciation of what has become known as the *chemiosmotic* hypothesis for interpreting the relationship between electron flow, proton movements, and ATP formation (see Clayton, 1980; Mitchell, 1979; Nicholls, 1982; Witt, 1979).

Figure 6.5 indicates that the O_2-evolution step and the electron flow mediated by the plastoquinones and the cytochrome b_6–f complex lead to an accumulation of H$^+$ inside a thylakoid in the light. This causes the internal H$^+$ concentration c_H^i or activity a_H^i to increase. These steps are dependent on the light-driven electron flow, which leads to electron movement out across the thylakoid in each of the photosystems (see Fig. 5.14). Such movements of electrons out and protons

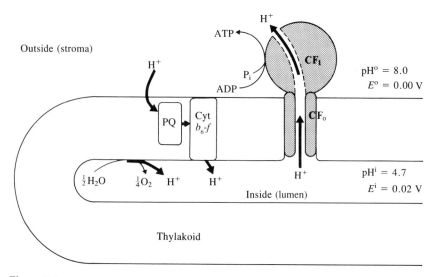

Figure 6.5

Energetics and directionality of the coupling between electron flow and ATP formation in chloroplasts, emphasizing the role played by H^+ (see also Fig. 5.14). The O_2 evolution from H_2O and electron flow via plastoquinones (PQ) and the cytochrome b_6–f complex (Cyt b_6–f) lead to H^+ accumulation inside the thylakoid. This H^+ can move back out through a hydrophobic channel (CF$_o$) and another protein factor (CF$_1$), leading to ATP formation.

in can increase the electrical potential inside the thylakoid (E^i) relative to that outside (E^o), allowing an electrical potential difference to develop across a thylakoid membrane. By the definition of chemical potential ($\mu_j = \mu_j^* + RT \ln a_j + z_j FE$, Eq. 2.4 with the pressure and gravitational terms omitted; see p. 110), the difference in chemical potential of H^+ across a membrane is

$$\mu_H^i - \mu_H^o = RT \ln a_H^i + FE^i - RT \ln a_H^o - FE^o \qquad (6.16a)$$

Incorporating the definitions of pH (pH $= -\log a_H = -\ln a_H / 2.303$) and E_M ($E_M = E^i - E^o$) into Equation 6.16a, we obtain

$$\mu_H^i - \mu_H^o = -2.303 \, RT \, \text{pH}^i + 2.303 \, RT \, \text{pH}^o + FE_M \qquad (6.16b)$$

which, using numerical values from Appendix I, at 25°C becomes

$$\mu_H^i - \mu_H^o = 5.71 \, (\text{pH}^o - \text{pH}^i) + 96.5 \, E_M \qquad \text{kJ mol}^{-1} \qquad (6.16c)$$

where E_M is in volts.

According to the chemiosmotic hypothesis (which might, more appropriately, be termed a "transmembrane hypothesis"), the ATP reaction is driven in the energetically uphill forward direction by protons moving out of the thylakoids in their energetically downhill direction. Based on the amount of ATP formed in the dark and the accompanying efflux of protons, the stoichiometry is apparently 3 H^+'s per ATP (Strotmann and Lohse, 1988). Because the formation of ATP in chloroplasts usually requires 40 to 50 kJ (mol ATP)$^{-1}$, as indicated above, the difference in chemical potential of H^+ must be at least (40 to 50)/3 or 13 to 17 kJ (mol H^+)$^{-1}$ to drive the reaction energetically using 3 H^+'s per ATP. By Equation 6.16c, such an energy difference corresponds to a pH difference of (13 to 17)/(5.71) or 2.3 to 3.0 pH units, or a difference in electrical potential of (13 to 17)/(96.5) or 0.13 to 0.18 V. In turn, if the proton chemical potential gradient is established by coupling to electron flow, then an energetically uphill movement of protons of at least 13 to 17 kJ mol^{-1} requires an energetically even larger downhill flow of electrons. We will next examine the evidence and the thermodynamic requirements for the various steps from electron flow to ATP formation.

Evidence for Chemiosmotic Hypothesis

One of the most striking pieces of evidence in support of the chemiosmotic hypothesis was obtained in the 1960s by Jagendorf and Uribe (Jagendorf and Uribe, 1966). When chloroplast lamellae were incubated in a solution at pH 4— in which case pH^i presumably attained a value near 4—and then rapidly transferred to a solution with a pH^o of 8 containing ADP and phosphate, the lamellae were capable of leading to ATP formation *in the dark*. Approximately 100 ATP's could be formed per Cyt f. When the difference in pH across the membrane was 2.5 or less, essentially no ATP was formed by the chloroplast lamellae. This is in close agreement with the energetic argument presented above, where a minimum ΔpH of 2.3 to 3.0 pH units was required to lead to ATP formation. Also, if the pH of the external solution is gradually increased from 4 to 8 in the dark (over a period of tens of seconds), the protons "leak" out across the lamellar membranes, ΔpH is relatively small, and no ATP is formed.

The electrical term in the chemical potential of H^+ also can power ATP formation (see Gräber *et al.*, 1977; Witt, 1979). For instance, when an E_M of 0.20 V was artifically created across lamellar membranes, ATP formation could be induced in the dark. This is consistent with our prediction that an electrical potential difference of at least 0.13 to 0.18 V is necessary. In chloroplast thylakoids, E_M in the light appears to be fairly low, e.g., near 0.02 V in the steady state (see Fig. 6.5). However, the electrical term seems to be the main contributor to $\Delta\mu_H$ for the first 1 to 2 seconds after chloroplasts are exposed to a high photosynthetic photon flux density (PPFD). Also, the electrical component of the H^+ chemical potential difference can be much larger for the chromatophores (see Ch. 1) of certain photosynthetic bacteria; in chromatophores from the purple photosynthetic bacterium *Rhodopseudomonas spheroides*, E_M can be 0.20 V in the light in the steady state.

When chloroplasts are illuminated, electron flow commences, which causes μ_H^i within the thylakoids to increase relative to μ_H^o. We would expect a delay before $\Delta\mu_H$ given by Equation 6.16 is large enough to lead to ATP formation. Indeed, a lag of a fraction of a second to seconds can be observed before photophosphorylation commences at low PPFD's, and the lag can be decreased by increasing the PPFD (see Avron, 1978). We can also reason that a gradient in the chemical potential of H^+ will affect the movement of other ions. For instance, the light-induced uptake of H^+ into the thylakoids is accompanied by a release of Mg^{2+}, which can cause the stromal concentration of this ion to increase by 10 mm (see Barber, 1976). This released Mg^{2+} can activate various enzymes involved with CO_2 fixation in the stroma, indicating that the ionic readjustments following light-dependent proton movements can act as a cellular control for biochemical reactions.

We can imagine various ways of dissipating the chemical potential gradient of H^+ across the lamellar membranes and thus uncoupling electron flow from ATP formation. Compounds that accomplish this are called *uncouplers*. For instance, neutral weak bases (e.g., methyl amine, CH_3NH_2; or ammonia, NH_3) can readily diffuse into the thylakoids and there combine with H^+. This lowers a_H^i and raises pH^i. Moreover, the protonated base ($CH_3NH_3^+$ or NH_4^+) cannot readily diffuse back out, because it is now charged. The uncoupler nigericin competitively binds H^+ and K^+, and can exchange K^+ outside for H^+ inside, which also tends to lower a_H^i (because nigericin can stoichiometrically lead to H^+ movement one way and K^+ the other, it acts as an H^+/K^+ antiporter; see Ch. 3). Detergents can remove certain membrane components, thus making the thylakoids leaky to H^+ and other ions, which also dissipates the $\Delta\mu_H$. Such studies further show the importance of the H^+ chemical potential difference in leading to ATP formation.

Coupling of Flows

We will next reconsider the vectorial aspects of the coupling of proton and electron flow (Figs. 5.14 and 6.5) and examine the associated energetics. Also, we will discuss the structures involved in the coupling of ATP formation to proton flow. Finally, we will consider the stoichiometry of the various flows with respect to the ATP and NADPH requirements of CO_2 fixation.

Let us start with the O_2-evolution step, $\frac{1}{2}H_2O \rightleftharpoons \frac{1}{4}O_2 + H^+ + e^-$ (essentially Eq. 5.8). To obtain H^+ inside thylakoids from the O_2 evolving step, we need: (1) H_2O inside the thylakoids, which can readily diffuse in from the stroma; (2) an oxidant with a redox potential more positive than the 0.82 V of the $H_2O–O_2$ couple, which is supplied by the $P_{680}^+–P_{680}$ couple in Photosystem II with a midpoint redox potential of 0.90 V (Table 5.3); (3) a pathway for removing electrons, which is provided by Photosystem II, because it moves electrons from the inner to the outer side of the thylakoid membrane (see Fig. 5.14); and (4) removal of O_2 from inside the thylakoid, which readily occurs by outward diffusion of this small neutral molecule. Thus, the asymmetrical nature of the reaction center of Photosystem II, together with the known properties of membranes, allowing small neutral molecules to cross easily while retarding the penetration of charged species, leads to an accumulation of H^+ inside thylakoids owing to the light-dependent, O_2-evolution step.

We argued above that a decrease in the chemical potential of H^+ of 13 to 17 kJ (mol H^+)$^{-1}$ is needed to cause ADP plus phosphate to react to form ATP if 3 H^+'s are used per ATP. One H^+ is transferred from the stroma to inside a thylakoid for each electron traversing the electron transport pathway from a plastoquinone in Photosystem II to Cyt f. The decrease in energy of this electron, as it spontaneously flows along the electron transport pathway toward higher redox potentials, supplies the energy to move an H^+ toward its higher chemical potential inside the thylakoid. The minimum change in redox potential to provide the Gibbs free energy to drive such H^+ movement can be calculated from Equation 6.7, $\Delta G = -nF \, \Delta E$ ($F = 96.49$ kJ mol^{-1} V^{-1}, App. I) as follows:

$$\Delta E = \frac{(-13 \text{ to } -17 \text{ kJ mol}^{-1})}{-(1)(96.49 \text{ kJ mol}^{-1} \text{ V}^{-1})}$$

$$= 0.13 \text{ to } 0.18 \text{ V}$$

The midpoint redox potentials are about 0.10 V for the second plastoquinone in Photosystem II (Q_B), where H^+ is probably first bound (Fig. 5.14), 0.11 V for plastoquinone A, and 0.35 V for Cyt f (Table 5.3 and Fig. 6.4). Hence, sufficient energy appears to be available from the energetically downhill movement of e^- to couple to and hence drive the energetically uphill movement of H^+ into the thylakoid lumen.

Coupling between the H^+ movements across the thylakoid membranes associated with electron flow and ATP formation occurs via an ATP synthetase, also referred to as an ATPase (because it can catalyze the reverse reaction leading to ATP hydrolysis) as well as a coupling factor. As illustrated in Figure 6.5, the ATP synthetase has two components: (1) a protein factor that occurs on the stromal side of a thylakoid, which can bind ADP, P_i, and ATP (labeled CF_1 in Fig. 6.5); and (2) another protein factor that is hydrophobic and hence occurs in the thylakoid membrane, through which H^+ can pass (labeled CF_o). The chemical nature of the complexes has been extensively studied and is found to be similar for ATP synthetases in chloroplasts, mitochondria, and bacterial membranes (see Futai et al., 1989).[7] The complex CF_1 is readily dislodged from the thylakoids and is soluble in water, where it can catalyze ATP hydrolysis. It consists of five subunits, two of which may occur in triplicate, with an overall molecular mass of about 400 kDa. Approximately one CF_1, which comprises about 10% of the thylakoid protein, occurs per Photosystem I (just as for Photosystem I, CF_1 is absent where the thylakoids stack together to form grana). Upon removal of CF_1, CF_o remains in the thylakoid membrane as a channel through which passive proton movement can readily occur. CF_o most likely has four subunits, some of which apparently span the thylakoid membrane, and an overall molecular mass of about 100 kDa. Although the H^+ moving through CF_o can come from the lumen of a thylakoid (Fig. 6.5), protons (H^+) can also diffuse along domains on the inner side of the thylakoid membrane, such as via H^+-binding sites on proteins or phospholipids in the membrane (see Dilley et al., 1987). Indeed, the pathway for H^+ movement, the manner in which 3 H^+'s are involved per ATP formed, and other details concerning the catalysis mechanism for the chloroplast ATP synthetase currently are active areas of research.

A matter related to the coupling of various flows across the thylakoids is the relative amounts of ATP produced and $NADP^+$ reduced in chloroplasts. Three ATP's and two NADPH's are needed per CO_2 photosynthetically fixed in the majority of plants, which are referred to as C_3 plants because CO_2 is incorporated into ribulose-1,5-diphosphate to yield two molecules of 3-phosphoglyceric acid, a *three carbon* compound (Fig. 8.8). Four to 6 ATP's (depending on which of three different decarboxylating enzymes is involved) and 2 NADPH's are required per CO_2 fixed in C_4 plants, where the first photosynthetic products are *four carbon* organic acids (e.g., oxaloacetic acid). (We will briefly discuss C_3 and C_4 plants in Ch. 8.) If the absorption of 8 photons leads to the processing

---

7. F_o and F_1 were originally studied in mitochondria and when analogous complexes were found in chloroplasts, they were designated CF_o and CF_1. F_1 was the first of a series of proteinaceous factors involved with oxidative phosphorylation that were isolated from mitochondria by Racker and others in the 1960s. Later the hydrophobic factor F_o, which makes the ATPase activity of F_1 sensitive to oligomycin and hence has "o" as a subscript, was isolated from mitochondrial membranes (much literature inadvertently uses zero as the subscript on F_o).

of 4 excitations in each of the two photosystems, then one O_2 can be evolved and 4 H^+'s produced inside a thylakoid by Photosystem II, 4 H^+'s would be delivered from the stroma to the thylakoid lumen by the plastoquinones plus the cytochrome b_6–f complex and 2 NADPH's would be produced by the overall noncyclic electron flow (see Figs. 5.12 and 6.4). Although this is sufficient NADPH to fix one CO_2 in C_3 plants, the 3 ATP's require 9 H^+'s moving out across the thylakoid instead of the 8 generated. Thus cyclic, or perhaps pseudocyclic, electron flow could lead to the other H^+ (as we indicated in Ch. 5, no $NADP^+$ reduction accompanies either of these types of electron flow). For instance, cyclic electron flow takes an electron from ferredoxin via a plastoquinone to the cytochrome b_6–f complex, from which the electron moves via plastocyanin back to Photosystem I; accompanying the electron movement, an H^+ is transferred from the stroma to the thylakoid lumen. Hence, 9 photons may be required to produce the 3 ATP's and 2 NADPH's needed per CO_2. On the other hand, some evidence exists for an additional H^+ being transported from the stroma to the thylakoid lumen via the cytochrome b_6–f complex accompanying noncyclic electron flow (see Gregory, 1989). So far, the precise quantum requirement and actual electron flow involvement for CO_2 fixation are experimentally unresolved.

MITOCHONDRIAL BIOENERGETICS

The activities of chloroplasts and mitochondria are related in a number of ways, as is clear from Figure 6.6. For instance, the O_2 evolved by photosynthesis can be consumed during respiration, and the fate of CO_2 for the two processes is just the opposite. Moreover, ATP formation is coupled to electron flow in both organelles; in mitochondria the electron flow is from a reduced pyridine nucleotide to the oxygen–water half-cell, and in chloroplasts it is in the opposite direction (see Fig. 6.6). A few to many thousands of mitochondria occur in a given plant cell, their frequency tending to be somewhat less in cells where chloroplasts are abundant. In photosynthetic tissue at night, and at all times in the nongreen tissues of a plant, oxidative phosphorylation in mitochondria is the predominant supplier of ATP for the cells.

Electron Flow Components—Redox Potentials

As with chloroplast membranes, various compounds in mitochondrial membranes accept and donate electrons. These electrons originate from biochemical cycles

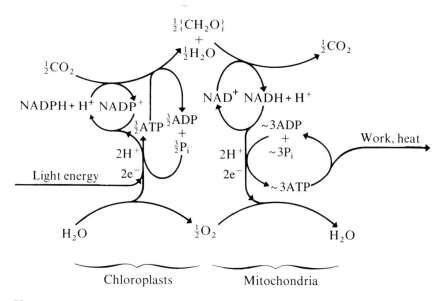

Figure 6.6

Schematic representation of the interrelationship between components involved in chloroplast and mitochondrial bioenergetics.

occurring in the cytosol as well as in the mitochondrial matrix (Fig. 1.8)—most come from the tricarboxylic acid (Krebs) cycle, which leads to the oxidation of pyruvate and the reduction of NAD^+ within mitochondria. Certain principal components for mitochondrial electron transfer and their midpoint redox potentials are indicated in Figure 6.7, where the spontaneous electron flow to higher redox potentials is toward the bottom of the figure. As for photosynthetic electron flow, only a relatively few types of compounds are involved in electron transfer in mitochondria—namely, pyridine nucleotides, flavoproteins, quinones, cytochromes, and the water–oxygen couple (some iron-plus-sulfur containing centers or clusters also occur).

The reduced compounds that introduce electrons directly into the mitochondrial electron transfer chain are NADH and succinate ($COOHCH_2CH_2COOH$), the latter passing two hydrogens ($H = H^+ + e^-$) on to FAD (see Fig. 6.7). FAD is flavin adenine dinucleotide, and consists of riboflavin (vitamin B_2) bound by a pyrophosphate bridge to adenosine (Fig. 6.3). Upon accepting two electrons and two protons—one H going to each N with a double bond in the riboflavin part of the molecule—FAD is reduced to $FADH_2$. The FAD–$FADH_2$ couple is usually bound to a protein, in which case the combination is referred to as a *flavoprotein* (ferredoxin–NADP reductase, Table 5.3, is a flavoprotein involved

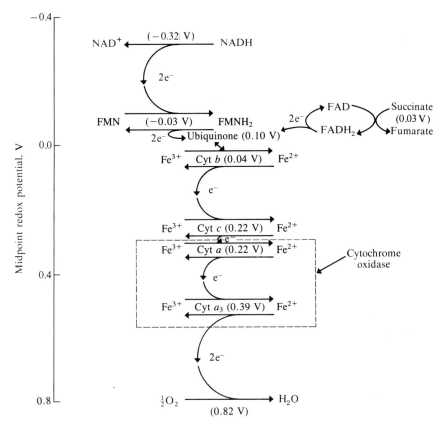

Figure 6.7
Components of the mitochondrial electron transport chain with midpoint redox potentials in parentheses (see Ikuma, 1972; Lehninger, 1982; Smith *et al.*, 1983; and Stryer, 1988).

with photosynthetic electron transfer). We note that a flavoprotein containing flavin mononucleotide (FMN) as the prosthetic group also occurs in mitochondria, where FMN is riboflavin phosphate:

Mitochondria contain ubiquinone, which differs from plastoquinone A (p. 284) by having two methoxy groups in place of the methyl groups on the ring, and 10 instead of 9 isoprene units in the side chain. A c-type cytochrome, referred to as Cyt c_1 in animal mitochondria, intervenes just before Cyt c, and there is a b-type cytochrome in plant mitochondria that is involved with an electron transfer that bypasses cytochrome oxidase on the way to O_2. The cytochrome oxidase complex contains two Cyt a plus two Cyt a_3 molecules and copper on an equimolar basis with the heme (see Fig. 5.13). Both the Cu and the Fe of the heme of Cyt a_3 are involved with the reduction of O_2 to H_2O (see Wikström *et al.*, 1981). Cytochromes a, b, and c are in roughly equal amounts in mitochondria (the ratios vary somewhat with plant species; see Ikuma, 1972), and flavoproteins are about 4 times, ubiquinones 7 to 10 times, and pyridine nucleotides 10 to 30 times more abundant than individual cytochromes. Likewise, in chloroplasts the quinones and pyridine nucleotides are much more abundant than are the cytochromes (see Table 5.3).

Most of the components involved in electron transport in mitochondria are contained in four supramolecular protein complexes that traverse the inner mitochondrial membrane (see Hatefi, 1985). The first complex, which has a molecular mass of approximately 800 kDa (contributed by about 25 different polypeptides) and contains FMN and various iron-sulfur clusters as active sites, transfers electrons from NADH to ubiquinone (Fig. 6.7). The second complex, which has a molecular mass of 140 kDa and contains FAD, various iron-sulfur clusters, and Cyt b_{560}, transfers electrons from succinate also to ubiquinone. Ubiquinone functions as a pool of two electron carriers, analogous to the function of plastoquinone A in the lamellar membranes of chloroplasts, which accepts electrons from complexes I and II and delivers them to the third protein complex.[8] This complex III, which has a molecular mass of 250 kDa and contains Cyt b_{562}, Cyt b_{566}, Cyt c_1, an iron-sulfur cluster, and a quinone, transfers electrons to Cyt c (Fig. 6.7). In turn, the pool of Cyt c molecules, which are soluble in aqueous solutions, passes electrons to the fourth complex (also called cytochrome oxidase), which has a molecular mass of about 170 kDa and contains Cyt a, Cyt a_3, and copper atoms as active sites. Complex IV delivers electrons to oxygen, which acts as the terminal electron acceptor in mitochondria.

---

8. Complex III is analogous to the cytochrome b_6–f complex of chloroplasts, both with respect to contents (two Cyt b's, one Cyt c, an Fe-S protein, and a quinone) and function within the membranes (e.g., interaction with a quinol; the isolated cytochrome b_6–f complex can also pass electrons to Cyt c as well as to its natural electron acceptor, plastocyanin). It is also structurally and functionally analogous to a supramolecular protein complex in bacteria.

Oxidative Phosphorylation

ATP formation coupled to electron flow in mitochondria is generally called *oxidative phosphorylation*. Because electron flow involves reduction as well as oxidation, more appropriate names are "respiratory phosphorylation" and "respiratory-chain phosphorylation," terms that are also more consistent with photophosphorylation for the ATP formation occurring in photosynthesis. As with photophosphorylation, the mechanism of oxidative phosphorylation is not yet fully understood in molecular terms. Processes like phosphorylation accompanying electron flow are intimately connected with membrane structure, and thus they are much more difficult to study than are the biochemical reactions taking place in solutions. Coupling between electron flow and ATP formation in mitochondria was previously thought to be via specific chemical components, the electron flow possibly inducing conformational changes facilitating the synthesis of ATP. A chemiosmotic mechanism is now generally accepted, and we will discuss some of its characteristics at the end of this section after examining the energetics of electron flow (for further details, see Douce and Neuburger, 1989; Kim *et al.*, 1987; Smith *et al.*, 1983; Stryer, 1988).

Experiments with isolated mitochondria have shown that the number of ATP's produced per pair of electrons used to reduce O_2 depends on the particular compound introducing the electrons into the electron transfer pathway. For instance, when a pair of electrons moves from NADH to O_2, 3 ATP molecules may be produced. Substrates, such as succinate, lead to a reduction of a flavoprotein whose oxidation leads to the reduction of ubiquinone and the phosphorylation of perhaps only 2 ADP's per pair of electrons moving along the mitochondrial electron transfer chain.[9] This suggests that a "coupling site" leading to ATP formation exists between NADH and ubiquinone (see Fig. 6.7), where for animal mitochondria a redox potential span of 0.31 V can occur (the actual ΔE, not the difference in midpoint redox potentials; see Smith *et al.*, 1983). Using Equation 6.7 ($\Delta G = -nF \Delta E$), we can calculate that a redox potential increase of 0.31 V for a pair of electrons can lead to a free energy change of $-(2 \text{ mol}) (96.49 \text{ kJ mol}^{-1} \text{ V}^{-1}) (0.31 \text{ V})$ or -60 kJ per 2 moles of electrons, which is sufficient energy to synthesize a mole of ATP. Such an energetic argument, however, does not indicate the coupling mechanism.

Let us next examine the coupling between electron flow and ATP formation

---

9. Although most researchers agree that NADH oxidation can lead to one more ATP produced than can succinate oxidation, certain evidence indicates that only about 2.5 ATP's are produced per pair of electrons moving from NADH to O_2 (see Ferguson, 1986; Stoner, 1987).

in mitochondria according to the chemiosmotic hypothesis, in which the intermediate is the electrochemical gradient of protons (H^+). Accompanying electron flow in mitochondria, H^+ is transported from the matrix on the inner side of the inner membrane to the space between the limiting membranes but outside the inner membrane (Fig. 6.8). Certain electron flow components are apparently so situated in the membranes that they can carry out this vectorial movement. As just indicated, movement of H^+ is associated with the protein complex I that oxidizes NADH; the number of H^+'s moved has been variously indicated to be 2 to 4 per NADH, with the latter number more likely (see P. Mitchell in Kim *et al.*, 1987; Nicholls, 1982; S. Papa in Kim *et al.*, 1987; Skulachev, 1988). Transport of H^+ outward from the matrix side also occurs via protein complex III, where 2 to 4 H^+'s can be transported per pair of electrons traversing the electron transport chain. Complex IV (cytochrome oxidase) may transport 4 H^+'s out, although again such values are not universally agreed upon (Fig. 6.8

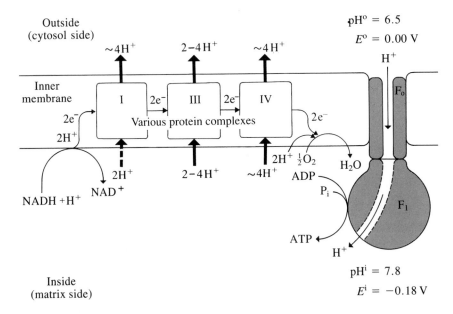

Figure 6.8

Schematic representation of certain electron flow and ATP synthesis components in the inner mitochondrial membrane, emphasizing the directional flows of H^+. The stoichiometry of H^+ per pair of electrons for the protein complexes discussed in the text is tentative. The H^+ moved out toward higher μ_H accompanying electron flow along the respiratory chain can move back through a hydrophobic channel (F_o) and another protein factor attached to the inside of the inner membrane (F_1), leading to ATP synthesis.

summarizes these possibilities). We also note that 2 H^+'s are necessary for the reduction of $\frac{1}{2}O_2$ to H_2O, and these protons could also be taken up from the internal solution (Fig. 6.8). In any case, the pH is higher and the electrical potential is lower in the inner region (Fig. 6.8) than they are outside the inner membrane, opposite to the polarity found for chloroplasts (Fig. 6.5).

The transport of protons out of the matrix leads to a difference in the H^+ chemical potential across the inner mitochondrial membrane. Using Equation 6.16c and the values in Figure 6.8, we can calculate that $\mu_H^i - \mu_H^o = (5.71)$ $(6.5 - 7.8) + (96.5)(-0.18 - 0.00) = -7.4 - 17.4$ or -25 kJ (mol H^+)$^{-1}$, indicating that the H^+ chemical potential is lower on the inside. In some cases where ATP formation occurs, E_M is about -0.15 V and $pH^o - pH^i$ is about -0.5, in which case 17 kJ (mol H^+)$^{-1}$ is available for use in ATP formation from the chemical potential difference of H^+ across the inner mitochondrial membrane. We also note that for chloroplasts most of the $\Delta\mu_H$ is due to the pH term, whereas for mitochondria the electrical term is generally more important.

As with chloroplasts, we can uncouple ATP formation from electron flow by adding compounds that dissipate the H^+ chemical potential difference. In addition to the types of compounds mentioned in connection with chloroplast bioenergetics, studies with the ionophore valinomycin have proved useful. Valinomycin acts like an organic ring with a hydrophilic center through which K^+ and NH_4^+ can readily pass but Na^+ and H^+ cannot, and it can thus provide a selective channel when it is embedded in a membrane. If the antiporter nigericin, which facilitates $K^+ - H^+$ exchange, is added together with the ionophore valinomycin, protons tend to move back into the matrix through the antiporter, thereby diminishing the ΔpH without affecting E_M, while the ionophore causes K^+ entry, thereby collapsing the electrical potential difference. As $\Delta\mu_H$ is thus dissipated, ATP formation ceases.

ATP formation is apparently coupled to the energetically downhill H^+ movement back into the mitochondrial matrix through a hydrophobic protein factor in the inner membrane (F_o; see footnote 7, this chapter) and a protein factor (F_1) that protrudes into the matrix from the inner side of the inner membrane (Fig. 6.8). These two factors make up the mitochondrial ATP synthetase, a supramolecular protein complex with an overall molecular mass of about 500 kDa (see Hatefi, 1985). In particular, F_1, which looks like a knob about 9 nm in diameter, has a molecular mass of approximately 380 kDa and consists of five polypeptides (two probably occurring in triplicate; see Boyer, 1989), just like CF_1; when removed from F_o, to which it is bound by a short stalk, F_1 can lead to hydrolysis of ATP in an aqueous solution. F_o, which has a molecular mass of just over 100 kDa, consists of at least four different polypeptide subunits. When F_1 is removed, the inner membrane becomes leaky to H^+, suggesting that the hydrophobic F_o is a channel or transporter for protons, just as is CF_o for chloroplasts. The formation of one ATP apparently requires the movement of 3 H^+'s through the mitochondrial ATP synthetase ($F_o + F_1$, Fig. 6.8), just as for the chloroplastic ATP synthetase ($CF_o + CF_1$, Fig. 6.5), although the H^+ motion is inward for the inner mitochondrial membrane and outward for the chloroplast thylakoid.

ATP is produced in the mitochondrial matrix but is usually needed in the cytosol. As we mentioned in Chapter 1, the outer mitochondrial membrane is quite permeable to small solutes including succinate, ADP, and ATP; a channel-forming protein is responsible for this high permeability (Colombini, 1987). On the other hand, specific porters are necessary for moving heavily charged solutes such as ATP across the inner mitochondrial membrane. In fact, an ADP/ATP antiporter exists in the inner membrane, which serves to replenish the internal ADP pool as well. Phosphate is also needed for ATP synthesis, and it enters by a P_i/OH^- antiporter (the concentration of OH^- is relatively high in the matrix) or, perhaps, an H^+/P_i symporter. In any case, moving OH^- out or H^+ in tends to dissipate the chemical potential difference of protons established across the inner mitochondrial membrane. In contrast, the ATP synthesized during photophosphorylation is produced where it is mainly utilized, namely, in the chloroplast stroma where CO_2 fixation occurs.

Let us next reconsider the ratio of protons to ATP. For the usual state of ionization, an ADP/ATP antiporter taking ATP into the mitochondrial matrix would be electrogenic (one more negative charge brought in than taken out). Also, a P_i/OH^- antiporter (or an H^+/P_i symporter) transporting HPO_4^{2-} would be electrogenic but in the opposite direction. Because these two porters must operate at the same rate to sustain ATP formation in the mitochondrial matrix, no overall effect on the membrane potential is expected. However, the extra proton required for these porters per ATP synthesized raises the proton requirement to 4 H^+'s per ATP formed in the mitochondrial matrix. The 12 H^+'s possibly moved out across the inner mitochondrial membrane when a pair of electrons move from NADH to O_2 (see Fig. 6.8) is consistent with 4 H^+'s per ATP together with the long-standing observation of 3 ATP's formed per NADH oxidized. However, only 2.5 ATP's may be formed per NADH, which is consistent with the possibility of 10 H^+'s per pair of electrons (Fig. 6.8) and 4 H^+'s per ATP. Such matters are still unresolved (for further details, see Ferguson, 1986; Stoner, 1987; Wohlrab, 1986).

As with chloroplasts, many questions concerning electron flow and the coupled ATP formation in mitochondria remain unanswered. The first part of the mitochondrial electron transfer chain has a number of two-electron carriers (NAD^+, FMN, and ubiquinone) that must interact with the cytochromes (one-electron carriers). Do the two electrons from NADH move in sequence to the same Cyt b or to two different Cyt b's? How many Cyt a_3's are involved per NADH? Actually, the reduction of O_2 apparently involves 4 electrons coming sequentially from the same Cyt a_3. Of perhaps more interest is deciding on the actual mechanism of ATP formation (see Boyer, 1989). Among other things, this involves reexamining the ratio of electrons to protons and the ratio of protons to ATP, where there are many similarities between chloroplasts and mitochondria but also some perplexing differences.

ENERGY FLOW IN THE BIOSPHERE

The foregoing discussion of the way organisms interconvert energy on an organelle level sets the stage for a consideration of bioenergetics in a broader context. We will begin with certain biochemical aspects and then discuss the overall flow of energy from the sun through the biosphere. We will consider photosynthetic efficiency as well as transfer of energy from plants to animals. This material will serve as a transition between the molecular and cellular levels, considered up to now, and the organ and organism levels of the succeeding three chapters.

In Chapter 4 we indicated that the radiation input of the sun to the earth's atmosphere is 1 368 W m^{-2} (the "solar constant"). Some of the radiant energy is incorporated into the free energy increases needed to form ATP and NADPH in chloroplasts. In turn, these energy currencies lead to the reductive fixation of CO_2 into a carbohydrate in photosynthesis (Fig. 5.1). In the same photosynthetic cells, in other plant cells, and in animal cells, the carbohydrates formed during photosynthesis can serve as the free energy source for respiration, which leads to the generation of ATP by oxidative phosphorylation.

When used as a fuel in respiration, the carbohydrate glucose is first broken down into two molecules of pyruvate in the cytosol. Pyruvate enters the mitochondria and is eventually oxidized to CO_2 and H_2O by the tricarboxylic acid cycle. One mole of glucose is conventionally believed to lead to the formation of 36 moles of ATP.[10] As we noted above, the Gibbs free energy released by

---

10. Two NADH's are produced in the cytosol during the oxidation of glucose to pyruvate. Depending on the carrier used to transport these NADH's into the mitochondria, a net of 2 or 3 ATP's can be formed per NADH. The figure of 36 ATP's per glucose assumes 2 ATP's for each such NADH (see Lehninger, 1982; Smith *et al.*, 1983). A more serious problem is the number of ATP's formed per NADH oxidized by the mitochondrial electron transfer pathway; if but 2.5 ATP's are formed (Ferguson, 1986; Stoner, 1987; Wohlrab, 1986), then only about 30 ATP's can be produced per glucose.

the complete oxidation of a mole of glucose is 2 872 kJ, and about 50 kJ may be required for the phosphorylation of one mole of ADP in mitochondria. Hence, the efficiency of the many-faceted conversion of Gibbs free energy from glucose to ATP may be [(36 mol ATP/mol glucose)(50 kJ/mol ATP)/(2 872 kJ/mol glucose)](100), or 63%, indicating that the Gibbs free energy in glucose can be efficiently mobilized to produce ATP. Such ATP is used by the cells to transport ions actively, to synthesize proteins, and to provide for growth and maintenance in other ways. We can readily appreciate that, if free energy were not constantly supplied to their cells, plants and animals would drift toward equilibrium and death.

One of the consequences of the flux of energy through the biosphere is the formation of rather complex and energetically improbable molecules such as proteins and nucleic acids. Such compounds represent a considerably greater amount of Gibbs free energy than does an equilibrium mixture containing the same relative amounts of the various atoms. (Equilibrium corresponds to a minimum in Gibbs free energy; see Fig. 6.1.) For instance, the atoms in the nonaqueous components of cells have an average of about 26 kJ mol^{-1} more Gibbs free energy than do the same atoms at equilibrium (see Morowitz, 1979). The Boltzmann energy distribution (Eq. 3.21b) predicts that at equilibrium the fraction of atoms with kinetic energy in excess of E is equal to $e^{-E/RT}$, which for 26 kJ mol^{-1} is only 0.000028 at 25°C. Thus, only a very small fraction of atoms would have a kinetic energy equal to the average enrichment in Gibbs free energy per atom of the nonaqueous components in cells. It is the flux of energy from the sun through plants and animals (see Fig. 6.6) that leads to such an energy enrichment in the molecules and that ensures that biological systems will be maintained in a state far from equilibrium, as is essential for life.

Incident Light—Stefan–Boltzmann Law

To help understand the energy available to the biosphere, we need to reconsider some properties of radiating bodies. In Chapter 4 we indicated that the distribution of radiant energy per unit wavelength interval is proportional to $\lambda^{-5}/(e^{hc/\lambda kT} - 1)$, where T is the temperature of the radiation source. This form of Planck's radiation distribution formula applies to an object that radiates maximally, a so-called blackbody. When blackbody radiation is integrated over all wavelengths, we can determine the maximum amount of energy radiated by an object. Appropriate integration of Planck's radiation distribution formula leads to the following expression:

$$\text{maximum radiant energy flux density} = \sigma T^4 \qquad (6.17)$$

where σ is a constant and T is in kelvins (temperature in °C + 273.15). Although Equation 6.17 can be derived from quantum-physical considerations developed by Planck in 1900, it was first proposed in the latter part of the 19th century. In 1879 Stefan empirically determined that the maximum radiation was proportional to the fourth power of the absolute temperature; in 1884 Boltzmann interpreted this in terms of classical physics. The coefficient of proportionality σ was deduced from measurements then available and has become known as the Stefan–Boltzmann constant, which equals 5.670×10^{-8} W m^{-2} K^{-4} (App. I). In the case where the object does not radiate as a blackbody, the radiant energy flux density at the surface of the radiator equals $e\sigma T^4$, where e is the *emissivity*. Emissivity depends on the surface material of the radiating body and achieves its maximum value of unity for a blackbody (for further details see Gates, 1980).

We will now estimate the amount of energy radiated from the sun's surface and how much of this is annually incident on the earth's atmosphere. Using Equation 6.17, which is known as the Stefan–Boltzmann law, and the effective surface temperature of the sun, about 5 800 K (p. 203), the rate of energy radiation per unit area of the sun's surface is

$$J_{\text{energy}} = (5.670 \times 10^{-8} \text{ W m}^{-2} \text{ K}^{-4})(5\ 800 \text{ K})^4 = 6.4 \times 10^7 \text{ W m}^{-2}$$

The entire output of the sun is about 3.84×10^{26} W, which leads to 1.21×10^{34} J year^{-1}. Because the amount incident on the earth's atmosphere is 1 368 W m^{-2} (p. 201) and the projected area of the earth is 1.276×10^{14} m^2, the annual energy input into the earth's atmosphere from the sun is (1 368 J m^{-2} s^{-1})(1.276 $\times$ 10^{14} m^2)(3.156 $\times$ 10^7 year^{-1}), or 5.51×10^{24} J year^{-1}.

Absorbed Light and Photosynthetic Efficiency

Only a small fraction of the sun's energy incident on the earth's atmosphere each year is actually absorbed by photosynthetic pigments, and only a small fraction of the absorbed energy is stored as chemical energy of the photosynthetic products (see Gates, 1980; Lieth and Whittaker, 1975; Nobel, 1974). Specifically, approximately 5% of the 5.51×10^{24} J annually incident on the earth's atmosphere is absorbed by chlorophyll or other photosynthetic pigments, leading to a radiant energy input into this part of the biosphere of about 2.8×10^{23} J year^{-1}. How much of this energy is annually stored in photosynthetic products? As we indicated in Chapter 5, a net of approximately 0.7×10^{14} kg of carbon are annually fixed by photosynthesis. For each mole of CO_2 (12 g carbon) incorporated into carbohydrate, approximately 479 kJ of Gibbs free energy are stored. The total amount of energy stored each year by photosynthesis is thus

$(0.7 \times 10^{17}$ g year$^{-1})(1$ mol/12 g$)(4.79 \times 10^5$ J mol$^{-1})$, or 2.8×10^{21} J year^{-1}. Hence, only about 1% of the radiant energy absorbed by photosynthetic pigments ends up stored by plant cells.

There are many different ways to represent the efficiency of photosynthesis. If we express it on the basis of the total solar irradiation incident on the earth's atmosphere $(5.51 \times 10^{24}$ J year$^{-1})$, it is only 0.051%. This low figure takes into consideration many places of low productivity, e.g., open oceans, polar icecaps, winter landscapes, and arid regions. Furthermore, not all solar radiation reaches the earth's surface, and much that does is in the infrared (see Fig. 4.2); the efficiency would be 0.089% if we considered only the solar irradiation reaching the ground. Nevertheless, even with the apparently very low overall energy conversion, the trapping of solar energy by photosynthesis is the only source of free energy used to sustain life.

What is the highest possible efficiency for photosynthesis? For low levels of *red* light, the conversion of radiant energy into the Gibbs free energy of photosynthetic products can be up to 34% in the laboratory (p. 247). Solar irradiation incident on plants spans many wavelengths, slightly less than half of this radiant energy being in the region that can be absorbed by photosynthetic pigments, 400 to 700 nm (see Figs. 4.2 and 7.2). If all the incident wavelengths from 400 to 700 nm (about half of the radiant energy) were absorbed by photosynthetic pigments, and 8 photons were required per CO_2 fixed, the maximum photosynthetic efficiency for the use of low levels of incident solar irradiation would be just under half of 34%, say 15%. But some sunlight is reflected or transmitted by plants (see Fig. 7.3), and some is absorbed by nonphotosynthetic pigments in the plant cells. Thus, the maximum photosynthetic efficiency for using incident solar energy under ideal conditions of temperature, water supply, and physiological status of plants in the field is closer to 10%.

Actual measurements of photosynthetic efficiency in the field have indicated that up to about 7% of the incident solar energy can be stored in photosynthetic products for a rapidly growing crop under ideal conditions. Usually, the photosynthetic photon flux density (PPFD) on the upper leaves of vegetation is too high for all excitations of the photosynthetic pigments to be used for the photochemistry of photosynthesis. The energy of many absorbed photons is therefore simply wasted as heat, especially when leaves of C_3 plants are exposed to a PPFD of more than 600 μmol m^{-2} s^{-1}. Hence, the maximum sustained efficiency for the conversion of solar energy into Gibbs free energy stored in photosynthetic products is often near 3% for a crop when averaged over a day in the growing season (see Gates, 1980; Loomis *et al.*, 1971). For all vegetation averaged over a year, the efficiency is about 0.5% of the incident solar irradiation. This is consistent with our previous statement that about 1% of the radiant energy absorbed by photosynthetic pigments is stored in the products of photosynthesis, because only about half of the solar irradiation incident on plants is absorbed by chlorophylls, carotenoids, and phycobilins.

Food Chains and Material Cycles

We will now consider the fate of the Gibbs free energy stored in photosynthetic products when animals enter the picture. We begin by noting that across each step in a food chain the free energy decreases, as required by the second law of thermodynamics. For instance, growing herbivores generally retain only 10% to 20% of the free energy of the ingested plant material, and a mature animal uses essentially all of its Gibbs free energy consumption just to maintain a state far from equilibrium. Growing carnivores will store about 10% to 20% of the free energy content of herbivores or other animals that they eat. Thus, a sizable loss in Gibbs free energy occurs for each link in a food chain, and as a consequence there are seldom more than four links, or steps, in a chain.

Although modern agriculture tends to reduce the length of our food chain for meat, humans still make a rather large demand on the Gibbs free energy available in the biosphere. The global average intake of free energy is about 10 MJ person^{-1} day^{-1} (2 400 kcal person^{-1} day^{-1}; see Gates, 1980). For a world population of 6 billion (projected for the year 1999; Hoffman, 1989), the annual consumption of Gibbs free energy for food would be 2.2×10^{19} J year^{-1}. (Humans also consume plants and animals for clothing, shelter, firewood, papermaking, and in many other ways.) Thus, our food consumption alone amounts to almost 1% of the 2.8×10^{21} J year^{-1} stored in photosynthetic products. If we were to eat only carnivores that ate herbivores with a 10% retention in Gibbs free energy across each step in the food chain, we would indirectly be responsible for the consumption of the entire storage of energy by present-day photosynthesis. Fortunately, we obtain most of our free energy requirements directly from plants. The average daily consumption in the United States is 13 MJ person^{-1}, of which just over 9 MJ comes from plants and just under 4 MJ from animals. In turn, the animals consumed in this case require an average Gibbs free energy input of 26 MJ of plant material, which corresponds to 15% efficiency in free energy retention across this link in the food chain.

The harnessing of solar radiation by photosynthesis starts the flow of Gibbs free energy through the biosphere. In addition to maintaining individual chemical reactions as well as entire plants and animals in a state far from equilibrium, the annual degradation of chemical energy to heat sets up various cycles (see Morowitz, 1979). We have already indicated some of these, e.g., O_2 is evolved in photosynthesis and then consumed by respiration, and CO_2 cycles in the reverse direction between these two processes (Fig. 6.6). There is also a cycling between ATP and ADP + phosphate at the cellular level, as well as the inevitable birth–death cycle of organisms. In addition, we can recognize a cycling of nitrogen, phosphorus, and sulfur in the biosphere. All of these material cycles can be regarded as consequences of the unidirectional flow of Gibbs free energy, which becomes less after each step along the way.

Problems

6.1 A reaction $A + B \rightleftharpoons C$ has a ΔG^* of -17.1 kJ mol^{-1} of reactant or product at 25°C (K in molality). Assume that activity coefficients are unity. In which direction will the reaction proceed under the following conditions? (a) The concentrations of A, B, and C are all 1 molal. (b) The concentrations of A, B, and C are all 1 millimolal. (c) The concentrations of A, B, and C are all 1 micromolal. (d) What is the equilibrium constant for the reaction?

6.2 Consider the following two half-cell reactions at 25°C:

$$A \rightleftharpoons A^+ + e^- \qquad \Delta G^* = 8.37 \text{ kJ mol}^{-1}$$
$$B \rightleftharpoons B^+ + e^- \qquad \Delta G^* = 2.93 \text{ kJ mol}^{-1}$$

Assume that the midpoint redox potential of the second reaction is $+0.118$ V and that all activity coefficients are unity. (a) If the redox potential of the B–B^+ couple is 0.000 V, what is the ratio of B^+ to B? (b) What is the midpoint redox potential of the A–A^+ couple? (c) Suppose that all reactants and products are initially 1 molal but that the couples are in separate solutions of equal volume. If the half-cells are electrically connected with a metal wire, what is the initial electrical potential difference between them, and in which direction do electrons flow? (d) If all reactants and products are initially 1 molal, what is the concentration of each at equilibrium in a single solution? (e) Qualitatively, how would the answer to (d) change if the initial conditions were identical to (c), but as the electrons flow through the wire they do electrical work?

6.3 Suppose that isolated chloroplasts are suspended in an aqueous medium initially containing 2 mM ADP, 5 mM phosphate, and no ATP (ignore any of these solutes originally in the chloroplasts). Assume that the temperature is 25°C and that the pH is 7. (a) What is the ATP concentration at equilibrium? (b) When the chloroplasts are illuminated, the ADP concentration decreases to 1 mM. What is the new concentration of ATP, and what is the change in Gibbs free energy for continued photophosphorylation? (c) If ferredoxin has a redox potential of -0.580 V and the activity of NADPH is 3% of that of NADP$^+$, what is the difference in redox potential between the two couples? (d) How much Gibbs free energy is available between the two couples in (c) when a pair of electrons moves between them? Is this enough for the continued formation of ATP under the conditions of (b)? (e) What difference in pH across a thylakoid membrane whose E_M is zero represents enough energy for ATP synthesis under the conditions of (b) when two protons are used per ATP? What if three protons are required per ATP?

6.4 Consider the following two mitochondrial cytochromes:

$$\text{Cyt } b \text{ (Fe}^{2+}) \rightleftharpoons \text{Cyt } b \text{ (Fe}^{3+}) + e^- \qquad E_b^{*,H} = 0.040 \text{ V}$$

$$\text{Cyt } c \text{ (Fe}^{3+}) + e^- \rightleftharpoons \text{Cyt } c \text{ (Fe}^{2+}) \qquad E_c^{*,H} = 0.220 \text{ V}$$

Assume that the temperature is 25°C, the chemical activity of Cyt b (Fe^{2+}) is 20% of that of the oxidized form, activity coefficients are equal to unity, and 40 kJ is required to form one mole of ATP in mitochondria. (a) What is the redox potential of Cyt b? (b) If the concentration of Cyt c (Fe^{2+}) is 1 mM, what is the concentration of ferricytochrome c such that the Cyt c couple can just transfer electrons back to Cyt b? (c) What is the redox potential of Cyt c such that one electron going from Cyt b to Cyt c represents the same energy as is required to form one ATP? (d) Assume that for each pair of electrons that ubiquinone delivers to Cyt b in a supramolecular protein complex, four protons are moved from the matrix side out across the inner mitochondrial membrane. If the proton concentration is the same on the two sides of the membrane, what difference in redox potential energetically corresponds to having the electrical potential 0.15 V higher on the inner side? (e) If the pH is the same on the two sides of the membrane and two protons move inward through the F_o-F_1 ATP synthetase per ATP, what is the minimum electrical potential difference across the inner mitochondrial membrane required to synthesize ATP by proton movement? (f) What is E_M in (e) if three protons are required per ATP? What is E_M if three H$^+$ are required and the energy losses (inefficiencies) are 30%?

References

Alberty, R. A. 1968. Effect of pH and metal ion concentration on the equilibrium hydrolysis of adenosine triphosphate to adenosine diphosphate. *Journal of Biological Chemistry* 243:1337–1343.

Allen, J. F., and N. G. Holmes. 1986. Electron transport and redox titration. In *Photosynthesis Energy Transduction: A Practical Approach*, M. F. Hipkins and N. R. Baker, eds. IRL Press, Oxford, England.

Avron, M. 1978. Energy transduction in photophosphorylation. *FEBS Letters* 96:225–232.

Barber, J., ed. 1976. *The Intact Chloroplast. Topics in Photosynthesis*, Vol. 1. Elsevier, Amsterdam.

Boyer, P. D. 1989. A perspective of the binding change mechanism for ATP synthesis. *FASEB Journal* 3:2164–2178.

Castellan, G. W. 1983. *Physical Chemistry*, 3rd ed. Addison-Wesley, Reading, Massachusetts.

Clayton, R. K. 1980. *Photosynthesis: Physical Mechanisms and Chemical Patterns.* Cambridge University Press, Cambridge, England.

Colombini, M. 1987. Regulation of the mitochondrial outer membrane channel, VDAC. *Journal of Bioenergetics and Biomembranes 19*:309–320.

Dilley, R. A., S. M. Theg, and W. A. Beard. 1987. Membrane–proton interactions in chloroplast bioenergetics: Localized proton domains. *Annual Review of Plant Physiology 38*:347–389.

Douce, R., and M. Neuburger. 1989. The uniqueness of plant mitochondria. *Annual Review of Plant Physiology and Plant Molecular Biology 40*:371–414.

Ferguson, S. J. 1986. The ups and downs of P/O ratios (and the question of non-integral coupling stoichiometries for oxidative phosphorylation and related processes). *Trends in Biochemical Sciences 11*:351–353.

Forrest, W. W., and D. J. Walker. 1971. The generation and utilization of energy during growth. *Advances in Microbial Physiology 5*:213–274.

Futai, M., T. Noumi, and M. Maeda. 1989. ATP synthetase (H^+-ATPase): Results by combined biochemical and molecular biological approaches. *Annual Review of Biochemistry 58*:111–136.

Gates, D. M. 1980. *Biophysical Ecology*. Springer-Verlag, New York.

Glazer, A.N., and A. Melis. 1987. Photochemical reaction centers: Structure, organization, and function. *Annual Review of Plant Physiology 38*:11–45.

Gräber, P., E. Schlodder, and H. T. Witt. 1977. Conformational change of the chloroplast ATPase induced by a transmembrane electric field and its correlation to phosphorylation. *Biochimica et Biophysica Acta 461*:426–440.

Gregory, R. P. F. 1989. *Biochemistry of Photosynthesis*, 3rd ed. Wiley, Chichester, England.

Hatefi, Y. 1985. The mitochondrial electron transport and oxidative phosphorylation system. *Annual Review of Biochemistry 54*:1015–1069.

Hoffman, M. S., ed. 1989. *The World Almanac and Book of Facts 1990*. World Almanac, New York.

Ikuma, H. 1972. Electron transport in plant respiration. *Annual Review of Plant Physiology 23*:419–436.

Jagendorf, A. T., and E. Uribe. 1966. ATP formation caused by acid–base transition of spinach chloroplasts. *Proceedings of the National Academy of Sciences of the United States of America 55*:170–177.

Kim, C. H., J. Tedeschi, J. J. Diwan, and J. C. Salerno, eds. 1987. *Advances in Membrane Biochemistry and Bioenergetics*. Plenum, New York.

Knox, R. S. 1977. Photosynthetic efficiency and exciton transfer and trapping. In *Primary Processes of Photosynthesis*, J. Barber, ed. Elsevier, Amsterdam. Pp. 55–97.

Lehninger, A. L. 1982. *Principles of Biochemistry*. Worth, New York.

Lieth, H., and R. H. Whittaker, eds. 1975. *Primary Productivity of the Biosphere. Ecological Studies*, Vol. 14. Springer-Verlag, New York.

Loomis, R. S., W. A. Williams, and A. E. Hall. 1971. Agricultural productivity. *Annual Review of Plant Physiology 22*:431–438.

MacInnes, D. A. 1961. *The Principles of Electrochemistry*. Dover, New York.

Malkin, R. 1987. Photosystem I. In *The Light Reactions*, J. Barber, ed. Elsevier, Amsterdam. Pp. 495–525.

Mitchell, P. 1979. Compartmentation and communication in living systems. Ligand conduction: A general catalytic principle in chemical, osmotic and chemiosmotic reaction systems. *European Journal of Biochemistry* 95:1–20.

Morowitz, H. J. 1979. *Energy Flow in Biology*. Ox Bow Press, Woodbridge, Connecticut.

Morris, J. G. 1968. *A Biologist's Physical Chemistry*. Addison-Wesley, Reading, Massachusetts.

Nicholls, D. G. 1982. *Bioenergetics: An Introduction to the Chemiosmotic Theory*. Academic Press, London.

Nobel, P. S. 1974. Free energy: The currency of life. In *Nature in the Round, A Guide to Environmental Science*, N. Calder, ed. Viking, New York. Pp. 157–167.

Payne, W. J. 1970. Energy yields and growth of heterotrophs. *Annual Review of Microbiology* 24:17–52.

Penning de Vries, F. W. T. 1972. Respiration and growth. In *Crop Processes in Controlled Environments*, A. R. Rees, K. E. Cockshull, D. W. Hand, and R. G. Hurd, eds. Academic Press, London. Pp. 327–347.

Rich, P. R., and D. A. Moss. 1987. The reactions of quinones in higher plant photosynthesis. In *The Light Reactions*, J. Barber, ed. Elsevier, Amsterdam. Pp. 421–445.

Rosing, J., and E. C. Slater. 1972. The value of $\Delta G°$ for the hydrolysis of ATP. *Biochimica et Biophysica Acta* 267:275–290.

Ross, R. T., and M. Calvin. 1967. Thermodynamics of light emission and free-energy storage in photosynthesis. *Biophysical Journal* 7:595–614.

Skulachev, V. P. 1988. *Membrane Bioenergetics*. Springer-Verlag, Berlin.

Smith, E. L., R. L. Hill, I. R. Lehman, R. J. Lefkowitz, P. Handler, and A. White. 1983. *Principles of Biochemistry*, 7th ed., *General Aspects*. McGraw-Hill, New York.

Stoner, C. D. 1987. Determination of the P/2e⁻ stoichiometries at the individual coupling sites in mitochondrial oxidative phosphorylation. Evidence for maximum values of 1.0, 0.5, and 1.0 at sites 1, 2, and 3. *Journal of Biological Chemistry* 262:10445–10453.

Strotmann, H., and D. Lohse. 1988. Determination of the H^+/ATP ratio of the H^+ transport-coupled reversible chloroplast ATPase reaction by equilibrium studies. *FEBS Letters* 229:308–312.

Stryer, L. 1988. *Biochemistry*, 3rd ed. Freeman, New York.

Velthuys, B. R. 1987. The Photosystem two reaction center. In *The Light Reactions*, J. Barber, ed. Elsevier, Amsterdam. Pp. 341–377.

Weyer, E. M., ed. 1968. Bioelectrodes. *Annals of the New York Academy of Sciences* 148:1–287.

Wikström, M., K. Krab, and M. Saraste. 1981. Proton-translocating cytochrome complexes. *Annual Review of Biochemistry* 50:623–655.

Witt, H. T. 1979. Energy conversion in the functional membrane of photosynthesis. Analysis by light pulse and electric pulse methods. The central role of the electric field. *Biochimica et Biophysica Acta* 505:355–427.

Wohlrab, H. 1986. Molecular aspects of inorganic phosphate transport in mitochondria. *Biochimica et Biophysica Acta* 853:115–134.

Zimmerman, A. M., ed. 1970. *High Pressure Effects on Cellular Processes*. Academic Press, New York.

CHAPTER 7

Temperature—Energy Budgets

We have already encountered many aspects of temperature. For instance, when introducing the special properties of water in Chapter 2, we noted that physiological processes generally take place within a fairly narrow temperature range. Such a range often dictates the boundaries of plant distributions, which can be quite apparent with respect to freezing limits. The water vapor content of air at saturation, as is nearly the case inside a leaf, is very temperature dependent (App I). Biochemical reactions usually exhibit a temperature optimum, although specific reactions in plants can acclimate to different temperature regimes. In Chapter 3 we discussed the Boltzmann energy distribution, Arrhenius plots, and Q_{10}, all of which involve the thermal energy of molecular motion. Light absorption (Ch. 4) causes molecules to attain states that are simply too improbable to be reached by collisions based on thermal energy. Transitions from some excited state to another one at lower energy or to the ground state can occur by radiationless transitions, where the energy is released as heat that is eventually shared by the surrounding molecules. The surface temperature of an object indicates both the wavelengths where radiation from it will be maximal (Wien's displacement law, Ch. 4) and also the total energy radiated (Stefan–Boltzmann law, Ch. 6).

The temperature of an object is the net result of *all* the ways that energy can enter or leave it. In this chapter we will examine these various ways, especially for leaves. We will then be able to predict the temperatures of leaves, as well as more massive plant parts, based on the ambient environmental conditions. We can also begin to understand the consequences of certain adaptations of plants to their environment and to recognize the experimental data needed for future refinements of our calculations.

We should emphasize at the outset that individual plants as well as environmental conditions vary tremendously. Thus, in this and the next two chapters we will be indicating an approach to the study of plant physiology and physiological ecology, rather than a compendium of facts suitable for all situations. However, certain basic features should become obvious. For instance, CO_2 uptake during photosynthesis is accompanied by a water efflux through the stomata. This water loss during transpiration serves to cool the leaf. Also, energy influxes are balanced against effluxes by changes in leaf temperature, which affect the amount of radiation emitted by the leaf as well as the heat conducted to the surrounding air. Another generality is that the temperatures of small leaves tend to be closer to those of the air than do the temperatures of large leaves. Because it is difficult to appreciate the relative contributions of the various factors without making calculations, we will use representative values for leaf and environmental parameters to describe the gas fluxes and the energy balance of leaves.

ENERGY BUDGET—RADIATION

The principle of the conservation of energy (the first law of thermodynamics) states that energy cannot be created or destroyed but only changed from one form to another. We will apply this principle to the energy balance of a leaf, which exists in an environment with many energy fluxes. We can summarize the various contributors to the energy balance of a leaf as follows:

absorbed solar irradiation, absorbed infrared irradiation from surroundings	emitted infrared radiation, heat convection, heat conduction, heat loss accompanying water evaporation	photosynthesis, other metabolism, leaf temperature changes	(7.1)
energy into leaf	− energy out of leaf	= energy storage by leaf	

Because heat moves *into* a leaf when its temperature is less than that of the surrounding turbulent air, Equation 7.1 describes the case in which the leaf temperature is greater than the temperature of the surroundings. Also, when water condenses onto a leaf, the leaf will gain heat. In such cases, the appropriate energy terms in Equation 7.1 change sign.

The various terms in Equation 7.1 differ greatly in magnitude. For instance, the energy storage terms generally are relatively small. As a basis for comparison, we will consider the average amount of solar irradiation incident on the earth's atmosphere (the term "irradiation" refers to incident radiation; see Ch. 4, footnote 4, for radiation terminology). This radiant flux density, the solar constant, is about 1 368 W m^{-2} (p. 201). The solar irradiation absorbed by an exposed leaf is often about half this during the daytime. Hence, any process under 7 W m^{-2} would correspond to less than 1% of the absorbed solar irradiation for full sunlight. What is the amount of energy stored by photosynthesis? A typical net rate of CO_2 fixation by a photosynthetically active leaf is 10 μmol m^{-2} s^{-1} (see Ch. 8). As indicated in Chapter 5, about 479 kJ of energy are stored per mole of CO_2 fixed into photosynthetic products. Hence, photosynthesis by leaves might store (10 $\times$ 10^{-6} mol m^{-2} s^{-1})(479 $\times$ 10^3 J mol^{-1}), or 5 W m^{-2}, which is only about 1% of the rate of absorption of solar irradiation under the same conditions. In some cases the rate of photosynthesis can be higher (see p. 451). Still, we can generally ignore the contribution of photosynthesis to the energy balance of a leaf. Other metabolic processes in a leaf, such as respiration and photorespiration, are usually even less important on an energy basis than is photosynthesis, so they too can generally be ignored.

We will now consider the amount of energy that can be stored because of changes in leaf temperature. For purposes of calculation, we will assume that a leaf has the high specific heat of water (4.19 kJ kg^{-1} °C^{-1} at 20°C), is 300 μm thick, and has an overall density of 700 kg m^{-3} (0.7 g cm^{-3})—a leaf is often 30% air by volume. Hence, the mass per unit leaf area in this case is (300 $\times$ 10^{-6} m)(700 kg m^{-3}), or 0.21 kg m^{-2}. If 7 W m^{-2} were stored by temperature changes in such a leaf, its temperature would rise at the rate of (7 J m^{-2} s^{-1})/ [(4 190 J kg^{-1} °C^{-1})(0.21 kg m^{-2})], or 0.008°C s^{-1} (0.5°C min^{-1}). Because this is a faster temperature change than that sustained for very long periods by leaves, we can assume that very little energy is stored (or released) in the form of leaf temperature changes. Hence, all three energy storage terms indicated in Equation 7.1 usually are relatively small for a leaf.

When the energy storage terms in Equation 7.1 are ignored, the remaining contributors to the energy balance of a leaf are either radiation or heat terms. We can then simplify our energy balance relation for a leaf as follows:

$$\underbrace{\begin{array}{c}\text{absorbed solar irradiation,}\\ \text{absorbed infrared irradiation}\\ \text{from surroundings}\end{array}}_{\text{energy into leaf}} \cong \underbrace{\begin{array}{c}\text{emitted infrared radiation,}\\ \text{heat convection,}\\ \text{heat conduction,}\\ \text{heat loss accompanying}\\ \text{water evaporation}\end{array}}_{\text{energy out of leaf}} \quad (7.2)$$

The heat conducted and convected from leaves is sometimes referred to as *sensible* heat, and that associated with the evaporation or the condensation of water is known as *latent* heat. In this section we will consider each of the terms in Equation 7.2, which have units of energy per unit area and time (e.g., J m^{-2} s^{-1}, which is W m^{-2}).

Solar Irradiation

Solar irridiation can reach a leaf in many different ways, the most obvious being *direct* sunlight. Alternatively, sunlight can be *scattered* by molecules and particles in the atmosphere before striking a leaf. Finally, both the direct and the scattered solar irradiation can be *reflected* by the surroundings toward a leaf.[1] In Figure 7.1 we summarize these various possibilities, which lead to six different ways by which solar irradiation can impinge on a leaf. The individual energy fluxes may involve the upper surface of a leaf, its lower surface, or perhaps both surfaces—in Figure 7.1 the direct solar irradiation (S^{direct}) is incident only on the upper surface of the leaf. To proceed with the analysis in a reasonable fashion, we will obviously need to make many simplifying assumptions and approximations (see Gates, 1980; Robinson, 1966; and van Wijk, 1966).

Some of the solar irradiation can be scattered or reflected from clouds before being incident on a leaf. On a cloudy day, the diffuse sunlight emanating from the clouds—S^{cloud}, or *cloudlight*—is substantial, whereas S^{direct} may be greatly reduced. For instance, a sky overcast by a fairly thin cloud layer 100 m thick might absorb or reflect away from the earth about 50% of the incident solar irradiation and diffusely scatter most of the rest toward the earth. A cloud 1 km thick will usually scatter somewhat less than 10% of the S^{direct} incident on it toward the earth (the actual numbers depend on the type and density of the clouds). Because the relative amounts of absorption, scattering, reflection, and transmission by clouds all depend on wavelength, cloudlight has a somewhat different wavelength distribution than does direct solar irradiation.

Solar irradiation scattered by air molecules and air-borne particles leads to S^{sky}, or *skylight*. Such scattering is generally divided into two categories: Rayleigh scattering due to molecules, which is approximately proportional to $1/\lambda^4$ (λ is the wavelength; see Ch. 4), and Mie scattering due to particles such as dust, which is approximately proportional to $1/\lambda$. Because of the greater scattering of the shorter wavelengths, skylight differs considerably from the wavelength dis-

---

1. Generally, the term *scattering* denotes the irregular changes in the direction of light caused by small particles or molecules; *reflection* refers to the change in direction of irradiation at a surface.

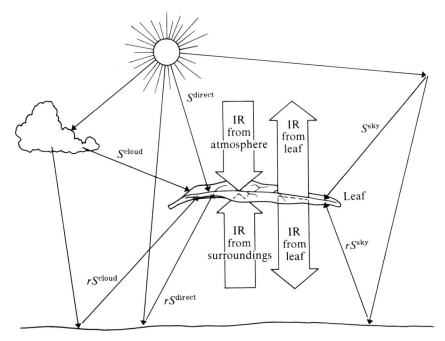

Figure 7.1

Schematic illustration of eight forms of radiant energy incident on an exposed leaf, including six that involve shortwave radiation from the sun and contain the letter S, and of the infrared radiation emitted from the leaf surfaces.

tribution for S^{direct}; in particular, S^{sky} is enriched in the shorter wavelengths (see Fig. 7.2). Specifically, the Rayleigh scattering causes a preponderance of the wavelengths below 500 nm to be scattered out of the direct solar beam (see Fig. 7.2) and to become skylight. This explains the blue color of the sky (and its skylight) during the daytime—and the red color of the sun at sunset (S^{direct} must then travel a greater distance through the earth's atmosphere, leading to more scattering of the shorter wavelengths). In terms of energy, S^{sky} can be up to 10% of S^{direct} for a horizontal leaf (Fig. 7.1) on a cloudless day when the sun is overhead (Fig. 7.2) and can exceed S^{direct} when the sun is near the horizon.

For convenience, we will refer to the direct sunlight plus the cloudlight and skylight as the *global irradiation, S.* Generally, $S^{\text{cloud}} + S^{\text{sky}}$ is referred to as diffuse shortwave irradiation, a readily measured quantity, whereas S^{cloud} and S^{sky} are often hard to measure separately. Thus, the global irradiation equals the direct plus the diffuse shortwave irradiation. The value of the global irradiation varies widely with the time of day, the time of year, latitude, altitude, and

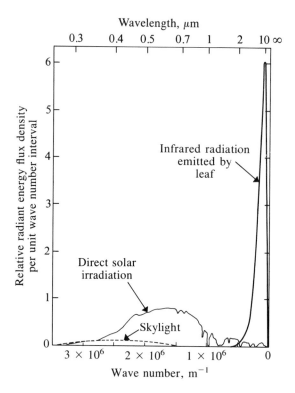

Figure 7.2

Wave number and wavelength distributions for direct solar irradiation, skylight, and the radiation emitted by a leaf at 25 °C. The areas under the curves indicate the total energy radiated; e.g., S^{direct} here is 840 W m^{-2}, and the IR emitted is 900 W m^{-2}.

atmospheric conditions, which must be taken into consideration when determining the energy balance of a leaf. As indicated above, the maximum solar irradiation incident on the earth's atmosphere is about 1 368 W m^{-2}. But, because of scattering and absorption of solar irradiation by atmospheric gases (Fig. 4.2), S on a cloudless day with the sun directly overhead in a dust-free sky is about 1 000 W m^{-2} at sea level.[2]

---

2. In the absence of clouds, S can be related to the solar constant (1 368 W m^{-2}; S_c) and the atmospheric transmittance τ (the fraction of sunlight transmitted when the sun is directly overhead), which ranges from 0.5 under hazy conditions at sea level to 0.8 for clear skies at higher elevations (see Ehleringer, 1989; Gates, 1980; Nobel, 1988). In particular, S equals $S_c \tau^{1/\sin \gamma} \sin \gamma$, where γ is the sun's altitude, or angle above the horizon; γ depends on the time of day, the time of year, and the latitude.

Sunlight may impinge on a leaf as direct solar irradiation, cloudlight, or skylight. These three components of global irradiation may first be reflected from the surroundings before striking a leaf (see Fig. 7.1). Although the reflected global irradiation can be incident on the leaf from all angles, for a horizontal exposed leaf it occurs primarily on the lower surface (Fig. 7.1). The reflected sunlight, cloudlight, and skylight often amount to 10% to 30% of the global irradiation. A related quantity is the fraction of the incident shortwave irradiation reflected from the earth's surface, termed the *albedo*, which averages about 0.60 for snow, 0.35 for dry sandy soil, 0.25 for dry clay, 0.10 for peat soil, 0.25 for most crops, and 0.15 for forests (see Rosenberg *et al.*, 1983).[3]

Each of these six forms of solar irradiation (reflected as well as direct forms of sunlight, cloudlight, and skylight) can have a different wavelength distribution. Because absorption depends on wavelength, the fraction of each one absorbed can in principle be different. Moreover, the fraction reflected also depends on wavelength. For simplicity, we will assume that the same absorptance (defined shortly) applies to S^{direct}, S^{cloud}, and S^{sky}, as well as to the reflected forms of these irradiations. We will also assume that the same reflectance applies to each component of the global irradiation. We can then represent the absorption of all forms of solar irradiation by a leaf as follows:

absorbed direct, scattered, and reflected solar irradiation

$$\cong a(S^{direct} + S^{cloud} + S^{sky}) + ar(S^{direct} + S^{cloud} + S^{sky})$$

$$= a(1 + r)S \tag{7.3}$$

where the *absorptance* a is the fraction of the global radiant energy flux density S absorbed by the leaf, and the *reflectance* r is the fraction of S reflected from the surroundings onto the leaf.[4]

Absorbed Infrared Irradiation

Besides the absorption of the various components of solar irradiation, additional *infrared*, or *thermal*, radiation is also absorbed by a leaf (see Eq. 7.2). Any object with a temperature above 0 K ("absolute zero") emits such thermal

---

3. The albedo can vary with the angle of incidence of the direct solar beam, being greater at smaller angles of incidence.

4. Absorptance is often called *absorptivity*, and reflectance is called *reflectivity*, especially when dealing with smooth surfaces of uniform composition.

radiation, including a leaf's surroundings as well as the sky. The peak in the spectral distribution of thermal radiation can be described by Wien's displacement law, which states that the wavelength for maximum emission of energy, λ_{max}, times the temperature of the emitting body, T, equals 2.9×10^6 nm K (Eq. 4.3b). Because the temperature of the surroundings is generally near 290 K, λ_{max} for radiation from them is close to $(2.9 \times 10^6$ nm K$)/(290$ K$)$, or 10 000 nm, which is 10 μm. Thus, the emission of thermal radiation from the surroundings occurs predominantly at wavelengths far into the infrared. Because of its wavelength distribution, we will also refer to thermal radiation as both infrared radiation and *longwave* radiation (over 99% of the radiant energy from the surroundings occurs at wavelengths longer than 4 μm, and over 98% of the solar or *shortwave* irradiation occurs at wavelengths shorter than this).

Most of the thermal radiation from the sky comes from H_2O, CO_2, and other molecules in the atmosphere that emit considerable radiation from 5 to 8 μm and above 13 μm. Moreover, the concentration of these gases varies, so the effective temperature of the sky, T^{sky}, as judged from its radiation, also varies. For instance, clouds contain much water in the form of vapor, droplets, or crystals, and this leads to a substantial emission of infrared radiation, so T^{sky} can be as high as 280 K on a cloudy day or night. On the other hand, a dry cloudless dust-free atmosphere might have a T^{sky} as low as about 220 K (see Munn, 1966; Robinson, 1966).

We will now consider the amount of thermal irradiation absorbed by a leaf. We will suppose that the infrared irradiation from the surroundings, acting as a planar source at an effective temperature of T^{surr}, is incident on the lower surface of the leaf, and that the upper surface of the leaf is exposed to the sky, which acts as a planar source with an effective temperature of T^{sky} (Fig. 7.1). In Chapter 6 we introduced the Stefan–Boltzmann law, which indicates that the amount of radiation emitted by a body depends markedly on its temperature (Eq. 6.17, maximum radiant energy flux density = σT^4). The Stefan–Boltzmann law predicts the maximum rate of energy radiation, i.e., the radiation emitted by a perfect radiator, the so-called blackbody (pp. 198, 336; see also Fig. 7.3). Here we will use *effective temperature* in the sense that the *actual* radiant energy flux density equals $\sigma(T_{effective})^4$. For instance, T^{sky} is not the temperature we would measure at some particular location in the sky, although $\sigma(T^{sky})^4$ equals the actual amount of radiant energy from the sky. By the Stefan–Boltzmann law, with effective temperatures to give the radiation emitted by the surroundings and the sky, the IR absorbed by a leaf is:

$$\text{infrared irradiation absorbed} = a_{IR}\sigma[(T^{surr})^4 + (T^{sky})^4] \qquad (7.4)$$

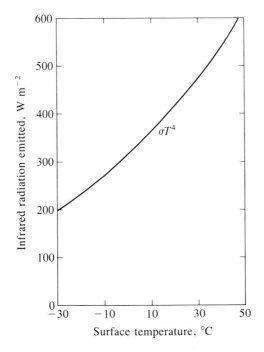

Figure 7.3

Rate of emission of infrared (longwave) radiation per unit area by a blackbody ($e_{IR} = 1.00$) as a function of its surface temperature.

where the absorptance a_{IR} is the fraction of the energy of the incident infrared irradiation absorbed by the leaf.

Emitted Infrared Radiation

Thermal radiation is also *emitted* by a leaf. Such radiation occurs at wavelengths far into the IR, because leaf temperatures, like those of its surroundings, are near 300 K. This is illustrated in Figure 7.2, where the emission of radiant energy from a leaf at 25°C is plotted both in terms of wavelength and wave number.[5] Using the wave-number scale makes it easier to illustrate the spectral distribution

---

5. Wave number, which we introduced in Problems 4.2 and 4.6, is the frequency of the radiation divided by the speed of light, i.e., v/c, which is $1/\lambda_{vacuum}$ ($\lambda_{vacuum}v = c$ by Eq. 4.1). Hence, wave number is proportional to energy (see Eq. 4.2a, $E_\lambda = hv = hc/\lambda_{vacuum}$).

of radiation from the sun and a leaf in the same figure; moreover, the area under the curves is proportional to the total radiant energy. The λ_{max} for sunlight is in the visible region; for the leaf it is in the infrared near 10 μm. Figure 7.2 also indicates that essentially all of the thermal radiation emitted by a leaf has wave numbers less than 5×10^5 m^{-1}, corresponding to IR wavelengths greater than 2 μm.

We will express the IR emitted by a leaf at a temperature T^{leaf} using the Stefan–Boltzmann law (Eq. 6.17), which describes the maximum rate of radiation per unit area. For the general emission case we incorporate a coefficient known as the *emissivity*, or *emittance*, *e*, which takes on its maximum value of unity for a perfect, or blackbody, radiator.[6] Because infrared radiation is emitted by both sides of a leaf (see Fig. 7.1), the factor 2 is necessary to describe its energy loss by thermal radiation:

$$\text{infrared radiation emitted} = 2\,e_{IR}\sigma(T^{leaf})^4 \tag{7.5}$$

As for our other flux density relations, Equation 7.5 is expressed on the basis of unit area for one side of a leaf. The substantial temperature dependency of emitted infrared radiation is indicated in Figure 7.3.

Values for a, a_{IR}, and e_{IR}

The parameters a, a_{IR}, and e_{IR} help determine the energy balance of a leaf. We will first consider how the absorptance of a leaf depends on wavelength and then discuss the leaf emittance for infrared radiation.

Figure 7.4 indicates that leaf absorptance at a particular wavelength, a_λ, varies considerably with the spectral region. For example, a_λ averages about 0.8 in the visible region (0.4 to 0.7 μm). Such relatively high fractional absorption by a leaf is due mainly to the photosynthetic pigments; the local minimum in a_λ near 0.55 μm is where chlorophyll absorption is relatively low (Fig. 5.3) and thus a larger fraction of the incident light is reflected or transmitted, leading to the green color (see Table 4.1) of leaves. Figure 7.4 indicates that a_λ is rather small

---

6. The actual radiant energy flux density equals $e\sigma(T_{actual})^4$, which is the same as $\sigma(T_{effective})^4$. We will use emissivites and actual temperatures to describe the energy radiated by leaves; effective temperatures are usually employed for thermal radiation from the surroundings and the sky, because their temperatures are difficult to measure or, indeed, hypothetical. Empirical equations incorporating the influence of air temperature, water vapor content, and clouds can also be used to predict the longwave radiation from the sky (see Campbell, 1977; Monteith, 1973).

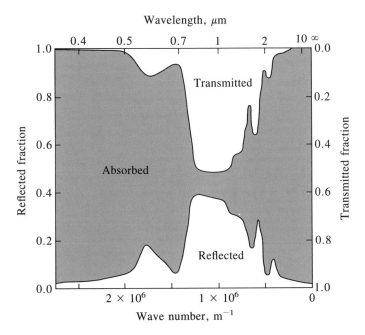

Figure 7.4

Representative fractions of irradiation absorbed, transmitted, and reflected by a leaf, as a function of wave number and wavelength. The sum $a_\lambda + r_\lambda + t_\lambda$ is unity. (See Gates, 1970; Woolley, 1971.)

from just beyond 0.7 μm up to nearly 1.2 μm. This is quite important for minimizing the energy input into a leaf, because much global irradiation occurs in this interval of the infrared. The fraction of irradiation absorbed becomes essentially unity for infrared irradiation beyond 2 μm. This does not lead to excessive heating of the leaves from absorption of global irradiation, because very little radiant emission from the sun occurs beyond 2 μm (see Fig. 7.2).

In our equations we have used two absorptances, a (in Eq. 7.3) and a_{IR} (in Eq. 7.4). In contrast to a_λ, these coefficients represent absorptances for a particular wavelength *range*. For example, a refers to the fraction of the incident solar energy absorbed (the wavelength distributions for direct sunlight and skylight are presented in Fig. 7.2). For most leaves, a is between 0.4 and 0.6, with 0.5 being typical. The shortwave absorptance can differ between the upper and the lower surfaces of a leaf, and also a tends to be lower for lower sun angles in the sky, because the shorter wavelengths are then scattered more (Fig. 7.2) and a_λ for the relatively enriched longer wavelengths is lower (Fig. 7.4; see

Gates, 1980). For the region 0.4 to 0.7 μm (designated "photosynthetic photon flux density," or PPFD in Ch. 4), the overall leaf absorptance is generally 0.75 to 0.90, as indicated above. Figure 7.4 shows that nearly all the IR irradiation beyond 2 μm is absorbed by a leaf. In fact, a_{IR} for leaves is usually from 0.95 to 0.98, and we will use a value of 0.96 for purposes of calculation.

Next, let us compare a_{IR} with e_{IR}. Because the emission of radiation is the reverse of its absorption, we should not be surprised that the same sort of electronic considerations that apply to the absorption of electromagnetic radiation (see Ch. 4) also apply to its emission. A good absorber of radiation is also a good emitter. In more precise language, the absorptance a_λ equals the emittance e_λ when they refer to the same wavelength (referred to as Kirchhoff's radiation law)—for a blackbody, $a_\lambda = e_\lambda = 1.00$ at all wavelengths (see Kreith and Bohm, 1986; Monteith, 1973). Because radiation from the surroundings and the sky occurs in essentially the same region of the infrared as that emitted by a leaf, e_{IR} is about the same as a_{IR}, e.g., 0.96.[7]

Net Radiation

We have now considered each of the terms that involve radiation in the energy balance of a leaf (Eq. 7.2). It is often convenient to single out these quantities and to refer to them as the *net radiation* balance for the leaf:

$$\text{net radiation} = \begin{array}{c} \text{absorbed solar irradiation} \\ + \\ \text{absorbed infrared irradiation} \\ \text{from surroundings} \end{array} - \begin{array}{c} \text{emitted infrared} \\ \text{radiation} \end{array} \quad (7.6a)$$

---

7. We can relate e_λ and a_λ to radiation quantities that we introduced in Chapter 4. The amount of radiant energy emitted by a blackbody per unit wavelength interval is proportional to $\lambda^{-5}/(e^{hc/\lambda kT} - 1)$, as predicted by Planck's radiation formula (p. 202). When we multiply this maximum radiation by e_λ at each of the wavelengths involved, we obtain the actual spectral distribution of the emitted thermal radiation. The absorptance a_λ is related to the absorption coefficient ε_λ. Equation 4.18 indicates that log J_0/J_b equals $\varepsilon_\lambda cb$, where c is the concentration of the absorbing species, b is the optical path length, J_0 is the incident flux density, and J_b is the flux density of the emergent beam when only absorption takes place (i.e., in the absence of reflection and scattering). The fraction of irradiation absorbed at a particular wavelength, $(J_0 - J_b)/J_0$, is the absorptance, a_λ. Thus, a_λ equals $1 - J_b/J_0$, which is $1 - 10^{-\varepsilon_\lambda cb}$, so the absorptance tends to be greater for wavelengths where the pigments absorb more (higher ε_λ), for thicker leaves, and for leaves with higher pigment concentrations.

We can also use Equations 7.3 through 7.5 to express the net radiation balance:

$$\text{net radiation} = \frac{a(1 + r)S}{a_{IR}\sigma[(T^{surr})^4 + (T^{sky})^4]} - 2e_{IR}\sigma\,(T^{leaf})^4 \qquad (7.6b)$$

Before continuing with our analysis of the energy balance of a leaf, we will examine representative values for each of the terms in the net radiation.

Examples for Radiation Terms

Let us consider a horizontal leaf exposed to full sunlight at sea level where the global irradiation S is 840 W m^{-2}. We will assume that the absorptance of the leaf for global irradiation a equals 0.60, and that the reflectance of the surroundings r is 0.20. By Equation 7.3, the direct plus the reflected sunlight, cloudlight, and skylight absorbed by the leaf is

$$a(1 + r)S = (0.60)(1.00 + 0.20)(840 \text{ W m}^{-2})$$

$$= 605 \text{ W m}^{-2}$$

To calculate the infrared irradiation absorbed by the leaf, we will let a_{IR} be 0.96, the temperature of the surroundings be 20°C, and the sky temperature be -20°C. Using Equation 7.4, with a Stefan–Boltzmann constant of 5.67×10^{-8} W m^{-2} K^{-4} (App. I), the absorbed IR is

$$a_{IR}\sigma[(T^{surr})^4 + (T^{sky})^4]$$

$$= (0.96)(5.67 \times 10^{-8} \text{ W m}^{-2} \text{ K}^{-4})[(293 \text{ K})^4 + (253 \text{ K})^4]$$

$$= 624 \text{ W m}^{-2}$$

Hence, the total irradiation load on the leaf is 605 W m^{-2} + 624 W m^{-2} or 1 229 W m^{-2}.

In the present case the rate of energy input per unit leaf area (1 229 W m^{-2}) is nearly the size of the solar constant (1 368 W m^{-2}). About half the total heat load on the leaf is contributed by the various forms of irradiation from the sun (605 W m^{-2}) and half by IR irradiation from the surroundings plus the sky (624 W m^{-2}). Because the sky generally has a much lower effective temperature for radiation than does the surroundings, the upper surface of an exposed leaf usually receives less IR than does the lower one: $a_{IR}\sigma(T^{sky})^4$ here is 223 W m^{-2}, and the IR absorbed by the lower surface of the leaf, $a_{IR}\sigma(T^{surr})^4$, is 401 W m^{-2} (see Fig. 7.3). Changes in the angle of an exposed leaf generally have only a moderate influence on the total irradiation load (changes of generally less than 300 W m^{-2}), because the IR, the scattered, and the reflected irradiation received by a leaf come from all angles. But leaf angle can still have important implications for the interaction of certain plants with their environment, as we will see near the end of this chapter.

To estimate the infrared radiation emitted by a leaf, we will let e_{IR} be 0.96 and the leaf temperature be 25°C. By Equation 7.5, the energy loss by thermal radiation then is

$$2e_{IR}\sigma(T^{leaf})^4 = (2)(0.96)(5.67 \times 10^{-8} \text{ W m}^{-2} \text{ K}^{-4})(298 \text{ K})^4$$

$$= 859 \text{ W m}^{-2}$$

In this case, the IR emitted by both sides of the exposed leaf is about 40% greater than the leaf's absorption of either solar or infrared irradiation. (Leaves shaded by other leaves are not fully exposed to T^{sky}, and for them IR emitted can be approximately equal to IR absorbed, a matter to which we will return later.)

The net radiation balance (see Eq. 7.6) for our exposed leaf is 1 229 W m^{-2} − 859 W m^{-2}, or 370 W m^{-2}. As we will discuss below, such excess energy is dissipated by conduction, convection, and the evaporation of water accompanying tranpiration. However, most of the energy input into a leaf is balanced by the emission of infrared radiation. The IR emitted in the present case, for example, amounts to (100)(859 W m^{-2})/(1 229 W m^{-2}), or 70% of the energy input from all sources of incident irradiation. The radiation terms for this exposed leaf are summarized in the top line of Table 7.1.

Because many conditions affect the net radiation balance for leaves, let us now consider some other examples. At an elevation of 2 000 m, S at noon on a cloudless day might be 1 050 W m^{-2}. Because of the higher incident global irradiation at 2 000 m, the leaf there will absorb 151 W m^{-2} more direct, scattered, and reflected solar irradiation than the leaf at sea level. The effective temperature of the sky is generally slightly lower at the higher elevation (-25°C vs. -20°C in Table 7.1), and we will assume that the surroundings are 10°C lower in temperature than for our example at sea level. The total irradiation input is then 1 311 W m^{-2}, which is 82 W m^{-2} greater than at sea level (see Table 7.1). For the net radiation to be the same in the two cases (370 W m^{-2}), the leaf at the

Table 7.1

Representatives values for the various terms in the net radiation balance of an exposed leaf. Equation 7.6 is used, taking a as 0.60 (except where indicated), r as 0.20, and both a_{IR} and e_{IR} as 0.96. (See text for interpretations.)

Condition	Global irradiation, S (W m^{-2})	Absorbed solar irradiation, $a(1+r)S$ (W m^{-2})	Temperature of surroundings (°C)	Sky temperature (°C)	Absorbed infrared, $a_{IR}\sigma[(T^{surr})^4 + (T^{sky})^4]$ (W m^{-2})	Leaf temperature (°C)	Emitted infrared, $2e_{IR}\sigma(T^{leaf})^4$ (W m^{-2})	Net radiation (W m^{-2})
Sea level on cloudless day	840	605	20	−20	624	25	859	370
2 000 m on cloudless day	1 050	756	10	−25	555	32	941	370
Silvery leaf (a = 0.50) at 2 000 m on cloudless day	1 050	630	10	−25	555	21	815	370
Sea level on cloudy night	0	0	1	1	614	1	614	0
Sea level on cloudless night	0	0	1	−20	530	−9	530	0
	0	0	1	−20	530	−1	596	−66

higher altitude must emit 82 W m^{-2} more thermal radiation than the leaf at sea level. As Table 7.1 indicates, this could be accomplished if T^{leaf} for the leaf at 2 000 m were 7°C higher than the one at sea level.

Certain plants have silvery or shiny leaves, which increases the amount of solar irradiation reflected. The fraction of S reflected by the leaf may increase from typical values of 0.1 or 0.2 (see Fig. 7.4) to 0.3 for silvery leaves, with an accompanying reduction in the absorptance from 0.6 to 0.5 or lower (see Gates, 1970). This reduction in absorptance can have a major influence on T^{leaf}. Other conditions remaining the same, a reduction of the absorptance a by only 0.1 can cause the leaf temperature to go from 32°C to 21°C (Table 7.1). Such a reduction in leaf temperature can have substantial effects on transpiration and photosynthesis. This can be particularly important for desert plants in hot environments. For instance, seasonal variations in pubescense (epidermal hairs) can increase the shortwave reflectance of leaves of the desert shrub *Encelia farinosa* produced in dry, hot periods by about 0.2 over the reflectance for leaves from cool, wet periods. The higher reflectance decreases T^{leaf} by about 5°C for this species; this leads to photosynthetic rates that are 10% to 60% higher during the dry, hot periods than they would otherwise be and at the same time conserves water, because transpiration rates average 23% less (see Ehleringer and Mooney, 1978).

Let us next consider the effect of clouds on leaf temperature at night. On a clear night at sea level the effective sky temperature might be −20°C, whereas for a heavy cloud cover it could be 1°C because of the IR emitted by the clouds (Table 7.1). If the temperature of the surroundings were 1°C in both cases, the infrared absorbed by the leaf, $a_{IR}\sigma[(T^{surr})^4 + (T^{sky})^4]$, would be 84 W m^{-2} lower on the clear night because of less IR from the cloudless sky (Table 7.1). For there to be no net gain or loss of energy by a leaf from all forms of radiation, i.e., $a_{IR}\sigma[(T^{surr})^4 + (T^{sky})^4] = 2e_{IR}\sigma(T^{leaf})^4$, the leaf temperature would have to be 1°C on the cloudy night and −9°C on the clear one (Table 7.1). Thus, on the clear night with the surroundings at 1°C, T^{leaf} could be considerably below freezing, so heat would be conducted from the surroundings to the leaf, raising its temperature to perhaps −1°C. Using this leaf temperature, we can calculate that there would be a net energy *loss* by radiation of 66 W m^{-2} on the clear night (see Table 7.1). A plant on a cloudless night may therefore freeze even though the temperatures of both air and surroundings are above 0°C, because the excess of IR emitted over that absorbed can lower the leaf temperature below the freezing point. Such freezing of leaves on clear nights is a severe problem for certain agricultural crops. A method for avoiding freezing damage employed for certain citrus and other orchards is to spray water on the trees. The release of the heat of fusion as this water freezes maintains the plant tissues at 0°C, which is above the freezing point of their cell sap (see Jones, 1983; Rosenberg *et al.*, 1983).

WIND–HEAT CONDUCTION AND CONVECTION

Now that we have considered the net radiation balance of a leaf, we will examine other ways that energy may be exchanged with the environment. For instance, heat can be conducted from one body to a cooler one in contact with it by molecular and/or electronic collisions. We will assume that in general a temperature difference exists across the air boundary layer adjacent to the leaf surface.[8] Heat can be *conducted* across this air layer by the random thermal collisions of the gas molecules. Heat convection, on the other hand, involves turbulent movement of a fluid, brought about for example by differences in pressure. There are two types of convection, *free* and *forced*. Free (natural) convection occurs when the heat transferred from a leaf causes the air outside the unstirred layer to warm, expand, and thus to decrease in density; this more buoyant warmer air then moves upward and thereby moves heat away from the leaf. Forced convection, caused by wind, can also remove the heated air outside the boundary layer. As the wind speed increases, more and more heat is dissipated by forced relative to free convection. However, even at a very low wind speed of only 0.1 m s^{-1}, forced convection dominates free convection as a means of heat loss from most leaves (0.10 m s^{-1} = 0.36 km h^{-1} = 0.22 mile h^{-1}). We can therefore generally assume that heat is *conducted* across the boundary layer adjacent to the leaf and then removed by forced *convection* in the surrounding turbulent air. We will shortly introduce certain dimensionless numbers that can help indicate whether forced or free convection should prevail. However, before discussing conduction and convection, we will examine some general characteristics of wind, paying particular attention to the air boundary layers adjacent to plant parts.

Wind—General Comments

Wind can influence plant growth, reproduction, and distribution, and can even cause death (see Nobel, 1981). It can mechanically deform plants and also can disperse pollen, seeds, disease organisms, and gaseous substances such as CO_2 and pollutants. Many effects of wind depend on the air boundary layers next to the aerial surfaces of a plant across which mass and heat exchanges occur with the environment.

---

8. Frictional interactions between a fluid and a solid phase moving with respect to each other lead to boundary layers of fluid adhering to the solid phase and across which heat and mass are exchanged. In Chapter 1 we considered unstirred layers of water adjacent to membranes (see Fig. 1.10), and in this section we will consider boundary layers of air adjacent to leaves.

Wind is caused by differences in air pressure that result from differential absorption of shortwave irradiation by the earth's surface and by clouds. On a macro scale, wind speed is affected by land surface features such as mountains and canyons, whereas about 10 m above a plant canopy wind can be mostly influenced by the plants, becoming completely arrested at their surfaces. In coastal regions wind speed can annually average as high as 7 m s^{-1} at 10 m above the ground, and in topographically flat inland areas 1 m s^{-1} is a typical mean wind speed (see Grace, 1977). Wind speeds are generally lower at night.

Whenever air is decelerated by an object such as a leaf or a branch, *form drag* occurs. When the air flow is entirely stopped, the force transmitted per unit area of surface perpendicular to the wind direction is $\frac{1}{2} \rho v^2$, where ρ is the air density (1.20 kg m^{-3} at 20°C, App. I) and v is the wind speed (the magnitude of the wind velocity, which is a vector). However, much of the air flow is directed around most objects, so the actual form drag is reduced from its maximum value. For increasing wind speeds, leaves with flexible petioles (leaf stalks) generally change their aspect such that a smaller area is presented perpendicular to the wind direction. If we designate the original area projected in the wind direction by A_{orig}, we can define a dimensionless drag coefficient c_d as follows:

$$c_d = \frac{\text{actual drag force}}{\frac{1}{2} \rho v^2 A_{orig}} \tag{7.7}$$

For relatively flat leaves that can align with the wind direction, c_d is generally 0.02 to 0.2, but it can be 0.5 to 1.0 for *bluff bodies* such as branches and tree trunks that block and hence substantially influence the air flow (see Jones, 1983; Grace, 1977; Monteith, 1973; Vogel, 1981). For flexible objects the drag coefficient as defined by Equation 7.7 can decrease as the wind speed increases (Vogel, 1989). In particular, the areas of leaves and small branches perpendicular to the wind direction can decrease up to threefold as the wind speed increases, leading to an aerodynamically more streamlined shape, as occurs for certain trees.

Near smooth surfaces air can flow in sheets, or laminae, nearly parallel to the surface. But turbulent flow, where air movement is not parallel and orderly, characterizes the wind regime near vegetation. In fact, the standard deviation of wind speed divided by the mean wind speed, the *turbulence intensity*, is often about 0.4 near vegetation, indicating that wind has considerable temporal variation (see Cionco, 1972). In such turbulent regimes air can be described as moving in packets or "eddies" (to be considered in Ch. 9).

Sites experiencing greater wind speeds often tend to have shorter vegetation, such as cushion plants in alpine tundra or the procumbent forms on coastal dunes (see Jaffe, 1980). Also, stem elongation can be reduced two- to threefold by high winds. On the other hand, environmental growth chambers tend to have low wind speeds (generally below 0.5 m s^{-1}), often leading to rather spindly plants that are unlike their field-grown counterparts. The retardation of stem elongation and the increase in girth caused by wind are mainly due to the development of shorter cells with thicker cell walls. Agronomic implications of wind-induced sway have long been recognized; e.g., increasing the spacing between nursery stock leads to sturdier trees of larger trunk diameter. Buttresses at the base of tree trunks and roots are more common on the windward side— such a location is more effective in resisting wind-caused upsetting moments than if the buttresses or enhanced root growth were on the leeward side, because the tensile strength of wood is greater than its compressional strength. A consequence of a prevailing wind direction are "flag trees," where branches occur mainly toward the leeward side. Many of these effects of wind on plant morphology have been shown to be hormonal responses (see Jaffe, 1985). At the extreme of sporadic high winds such as occur in gales (wind speeds of 17 to 21 m s^{-1}, which corresponds to 61 to 76 km h^{-1} or 38 to 47 miles h^{-1}), form drag causes stems to be permanently displaced from their upright position, a process termed "lodging" for various cereal crops. The possibility of lodging has fostered breeding efforts to develop wheat, barley, and oat genotypes with shorter, sturdier stems. Wind is also one of the major factors in the ecology of forests, forming gaps in the canopy by uprooting trees, creating special microhabitats by distributing leaf litter, and influencing succession by distributing pollen, spores, and seeds (see Nobel, 1981).

Air Boundary Layers

Wind speed affects the thickness of the air boundary layer next to a leaf or other plant part. Because boundary layers affect the heat exchange and hence temperature of the plant part, any plant process depending on temperature can be affected by wind speed. Also, any molecule entering or leaving a leaf in the gas phase must cross the boundary layer.

The boundary layer is a region dominated by the shearing stresses originating at some surface; such layers arise for any solid immersed in a fluid, e.g., a leaf in air (see Grace, 1977; Schlichting, 1979). Adjacent to the leaf is a laminar sublayer of air (Fig. 7.5), where movement is parallel to the leaf surface; air movement is arrested at the surface and has increasing speed at increasing distances. Diffusion perpendicularly away from the leaf surface is by molecular motion in the laminar sublayer. Farther from the surface, especially on the

Figure 7.5

Schematic illustration of originally nonturbulent air flowing over the top of a flat leaf, indicating the laminar sublayer (shorter straight arrows), the turbulent region, and the effective boundary layer thickness δ^{bl}. The arrows indicate the relative speed and the direction of the air movement.

downwind part of a leaf, the boundary layer becomes turbulent (Fig. 7.5). Here, heat and gas movements are eddy-assisted; i.e., the air swirls around in vortices and behaves as if it were moving in little units or packets. Instead of trying to describe the transfer processes in the laminar and the turbulent portions, both of which change in thickness across a leaf's surface, we use an effective or equivalent boundary layer thickness, δ^{bl}, averaged over the whole leaf surface (this is often referred to as the displacement air boundary layer). Actually, the boundary layer is somewhat thinner at the upwind or leading edge of a leaf than at its center (Fig. 7.5); as a consequence, the temperature there can differ from that in the center of the leaf. Because there is no sharp discontinuity of wind speed between the air adjacent to the leaf and the free airstream, the definition of boundary layer thickness is somewhat arbitrary, so it is generally defined operationally, e.g., in terms of the flux of heat across it, which we will consider shortly.

As we have indicated, δ^{bl} represents an *average* thickness of the unstirred air layer adjacent to a leaf (or to leaflets for a compound leaf). The main factors affecting δ^{bl} are the ambient wind speed and leaf size, with leaf shape exerting a secondary influence. Partly for convenience, but mainly because it has proved experimentally justifiable, we will handle the effect of leaf size on boundary layer thickness by the characteristic dimension l, which is the mean length of a leaf in the direction of the wind. Based on hydrodynamic theory for laminar flow adjacent to a flat surface as modified by actual observations under field conditions (see Pearman *et al.*, 1972), there is an approximate, but useful, expression for the average thickness of the boundary layer next to a flat leaf:

$$\delta^{bl}_{(mm)} = 4.0\sqrt{\frac{l_{(m)}}{v_{(m\ s^{-1})}}} \tag{7.8}$$

where $l_{(m)}$ is the mean length of the leaf in the downwind direction in m, $v_{(m\ s^{-1})}$ is the ambient wind speed in m s^{-1}, and $\delta^{bl}_{(mm)}$ is the average thickness of the boundary layer in mm (the factor 4.0 has units of mm s$^{-1/2}$, as we can deduce by using the indicated units for the three variables).

Instead of the numerical factor 4.0 in Equation 7.8, hydrodynamic theory predicts a factor closer to 6.0 for the "displacement," or effective, boundary layer thickness adjacent to a flat plate (see Cowan, 1968; Monteith, 1973; and especially Schlichting, 1979). However, wind tunnel measurements under high turbulence intensity, as well as field measurements, have indicted that 4.0 is actually a more suitable factor for leaves (or 3.7; see Gates, 1980). This divergence from theory relates to the relatively small size of leaves, their irregular shape, leaf curl, leaf flutter, and, most importantly, to the high turbulence intensity occurring under field conditions. Moreover, the dependency of δ^{bl} on $l^{0.5}$, which applies to large flat surfaces, does not always give the best fit to the data; wind tunnel measurements for various leaf shapes and sizes give values for the exponent of 0.3 to 0.5 (see Gates and Papian, 1971). Thus, various shape considerations, including the leaf dimension perpendicular to the wind direction, can affect δ^{bl}. In addition, instead of the dependence on $v^{-0.5}$ indicated in Equation 7.8, actual measurements with leaves are best described by exponents of -0.5 to -0.7. Consequently, Equation 7.8 must be regarded as only a useful approximation for indicating how the average boundary layer thickness varies with leaf size and wind speed.

Table 7.2
Thickness in mm of the boundary layer adjacent to a leaf, $\delta^{bl}_{(mm)}$, for a variety of leaf sizes and wind speeds. Calculations were made assuming that $\delta^{bl}_{(mm)} = 4.0\sqrt{l_{(m)}/v_{(m\ s^{-1})}}$ (Eq. 7.8), where $l_{(m)}$ is the mean leaf length in the wind direction in m and $v_{(m\ s^{-1})}$ is the ambient wind speed in m s^{-1}.

$l_{(m)}$	$v_{(m\ s^{-1})}$						
	0.10	0.28	0.45	1.00	2.78	4.47	10.00
0.002	0.57	0.34	0.27	0.179	0.107	0.085	0.057
0.01	1.26	0.76	0.60	0.40	0.24	0.189	0.126
0.05	2.8	1.69	1.33	0.89	0.54	0.42	0.28
0.2	5.7	3.4	2.7	1.8	1.07	0.85	0.57
0.5	8.9	5.3	4.2	2.8	1.70	1.34	0.89

Note that 1 km h^{-1} equals 0.278 m s^{-1}, and 1 mile h^{-1} equals 0.447 m s^{-1}.

Average wind speeds generally range from approximately 0.1 m s^{-1} up to about 10 m s^{-1}—exposed leaves commonly experience wind speeds near 1 m s^{-1}. Because the thickness of the air boundary layer enters into many calculations of heat and gas fluxes for leaves, the magnitudes of δ^{bl} for a wide variety of wind speeds and leaf dimensions are presented in Table 7.2.

Boundary Layers for Bluff Bodies

Although relatively flat leaves can be described by the boundary layer considerations presented above (Fig. 7.5 and Eq. 7.8), many plant parts such as stems, branches, inflorescences, fruits, and even certain leaves (e.g., the tubular leaves of onion, *Allium cepa*) represent three-dimensional objects. Air flow is intercepted by such bluff bodies and forced to move around them. Here, we will consider two particular shapes, cylinders and spheres. In the next subsection we will present heat flux equations for cases of cylindrical and spherical symmetry as well as for flat leaves.

On approximately the upwind half of a cylinder, a laminar boundary layer develops (Fig. 7.6). It is analogous to the laminar sublayer for flat plates (Fig. 7.5), and air movements in it can be theoretically analyzed (see Kreith and Bohm, 1986; Schlichting, 1979). On the downwind portion of the cylinder, the air flow becomes turbulent, can be opposite to the wind in direction, and in general is quite complicated. Nevertheless, an effective boundary layer thickness

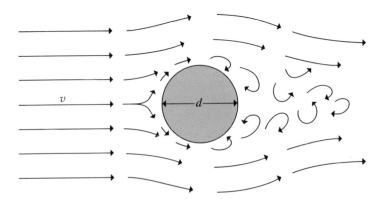

Figure 7.6

Schematic illustration of air flow around a cylinder. Flow can be laminar on the upwind half, but turbulence develops on the downwind side.

can be estimated (see Nobel, 1974). For turbulence intensities appropriate to field conditions, $\delta^{bl}_{(mm)}$ in mm can be expressed as follows:

$$\delta^{bl}_{(mm)} = 5.8 \sqrt{\frac{d_{(m)}}{v_{(m\ s^{-1})}}} \qquad \text{cylinder} \qquad (7.9)$$

where $d_{(m)}$ is the cylinder diameter in m. As is the case for flat leaves, the boundary layer is thinner for smaller objects or at higher wind speeds.

A similar analysis for the effective average boundary layer around a sphere under turbulent intensities appropriate to field conditions leads to

$$\delta^{bl}_{(mm)} = 2.8 \sqrt{\frac{d_{(m)}}{v_{(m\ s^{-1})}}} + \frac{0.25}{v_{(m\ s^{-1})}} \qquad \text{sphere} \qquad (7.10)$$

where $d_{(m)}$ is here the diameter of the sphere in m (See Nobel, 1975). Equation 7.10 has been successfully used to predict boundary layer thicknesses across which heat transfer occurs for the approximately spherical fruit of grape, *Vitis vinifera* (see Smart and Sinclair, 1976).

Heat Conduction/Convection Equations

Now that we have considered the average boundary layer thickness for objects of various shapes, let us return to a consideration of convective heat exchange, where the heat is first conducted across the boundary layer and then convected away in the moving airstream. For the one-dimensional case, heat flow by conduction equals $-K\ \partial T/\partial x$, where K is the thermal conductivity coefficient (e.g., $W\ m^{-1}\ °C^{-1}$) and $\partial T/\partial x$ is the temperature gradient (this relation is sometimes referred to as Fourier's heat-transfer law). Because heat can be conducted across the boundary layers on either side of a leaf, we need to incorporate the factor 2 to describe the total rate of heat flux by conduction per unit area of one side of a leaf. (For convenience, we will assume that the boundary layers on the two sides are of equal thickness, δ^{bl}.) The amount of heat conducted across the boundary layers and convected away from a leaf per unit time and area therefore is

$$J^C_H = -2K^{air}\frac{\partial T}{\partial x}$$

$$= 2K^{air}\frac{(T^{leaf} - T^{ta})}{\delta^{bl}} \qquad (7.11)$$

where J_H^C is the rate of heat conduction per unit area (e.g., W m^{-2}), K^{air} is the thermal conductivity coefficient of air, T^{leaf} is the leaf temperature, and T^{ta} is the temperature of the turbulent air outside a boundary layer of thickness δ^{bl}. Because heat is conducted from the solid surface of the leaf across the adjacent unstirred air, J_H^C does not depend on whether the stomata are open or closed— we have a planar source at a specific temperature, T^{leaf}, from which heat is conducted across the boundary layer to the outside turbulent air at T^{ta}. The heat flux density in Equation 7.11 is considered positive when heat goes from the leaf to the surrounding air. Of course, heat is conducted into the leaf when T^{leaf} is less than T^{ta}, in which case J_H^C is negative.

All of our flux equations used so far in this book have been for one-dimensional cases. Because we have introduced the average thickness of the boundary layer for cylinders (Eq. 7.9) and spheres (Eq. 7.10), let us also consider the appropriate fluxes for such cases, which can have many biological applications. For the case of cylindrical symmetry, thermal properties can change in the radial direction perpendicularly away from the cylinder axis but not with angle around the cylinder or length along its axis. The heat flux density at the cylinder's surface for such cylindrical symmetry is

$$J_H^C = \frac{K^{air}(T^{surf} - T^{ta})}{r \ln\left(\dfrac{r + \delta^{bl}}{r}\right)} \qquad \text{cylinder} \qquad (7.12)$$

where r is the cylinder radius, T^{surf} is its surface temperature, δ^{bl} is calculated by Equation 7.9, and the other quantities have the same meaning as for the one-dimensional case, Equation 7.11. For the case of spherical symmetry, where properties vary only in the radial direction and not with any angle, the heat flux density at the sphere's surface for conduction across the boundary layer and then convection in the surrounding turbulent air is

$$J_H^C = \frac{(r + \delta^{bl})K^{air}(T^{surf} - T^{ta})}{r\delta^{bl}} \qquad \text{sphere} \qquad (7.13)$$

where r is the radius of the sphere.

As we can see, the flux density for heat conduction across the air boundary layer is proportional to $K^{air}(T^{surf} - T^{ta})$ for all the shapes considered. Values of the thermal conductivity coefficient for dry air at various temperatures are given in W m^{-1} °C^{-1} in Appendix I. Because the conduction of heat in a gas phase is based on the random thermal motion of the molecules, the composition of air, such as the content of water vapor, can influence the value of K^{air}. For

K^{air} at 20°C is 1% *less* for 100% relative humidity than it is for dry air. Air can hold considerably more water vapor at 40°C, where K^{air} is 2% lower for water-saturated air than that of 0% relative humidity.

Dimensionless Numbers

In many studies on heat and gas fluxes, relationships between parameters are expressed in terms of dimensionless numbers. This facilitates comparisons between objects of the same shape but different sizes and for different wind speeds; i.e., dimensionless numbers allow application of data to different, but geometrically similar, situations (see Campbell, 1977; Gates, 1980; Incropera and DeWitt, 1985; Monteith, 1973; Schlichting, 1979; Vogel, 1981). For instance, dimensionless numbers were used to determine the effects of wind speed on δ^{bl} for the various shapes presented above (Nobel, 1974, 1975). Dimensionless numbers can also be used to study boundary layer phenomena and flow characteristics for water (Wheeler and Neushul, 1981)—water speeds can exceed 10 m s^{-1} in intertidal regions owing to wave action, although they are usually from 0.01 to 0.2 m s^{-1} in coastal regions of lakes and oceans. We will here consider three dimensionless numbers that are particularly important for discussing heat fluxes.

Before presenting these dimensionless numbers, we will indicate one common convention for describing heat conduction across boundary layers, a convention that is invariably used for objects of irregular shape and is often used for geometrically regular objects. Instead of expressions involving δ^{bl} (e.g., Eq. 7.11 for flat plates, Eq. 7.12 for cylinders, and Eq. 7.13 for spheres), the following relation is used to describe the heat flux density across the air boundary layer:

$$J_H^C = h_c(T^{surf} - T^{ta}) \tag{7.14}$$

where h_c is called the heat convection coefficient (or the convective heat-transfer coefficient; Eq. 7.14 is known as Newton's law of cooling). Upon comparing Equations 7.14 and 7.11 and noting that h_c generally refers to unit surface area of one side of a leaf [hence, the total heat flux density is $2h_c(T^{surf} - T^{ta})$ for a flat leaf], we find that h_c equals K^{air}/δ^{bl} for flat leaves (slightly more complex relations hold for cylinders and spheres). Even when the boundary layer thickness cannot be determined analytically, e.g., for the irregular shapes of cacti with their surface ribbing and projecting spines, Equation 7.14 can still be used to relate the convective heat exchange to the temperature difference between the air and the plant part. Appropriate units of h_c are W m^{-2} °C^{-1}.

Now that we have introduced the heat convection coefficient, we will proceed to define our first dimensionless number, the *Nusselt number*. We will represent the size of a particular plant part by a characteristic dimension d, which for a flat plate is the quantity l in Equation 7.8 and for a cylinder or sphere is the diameter. This leads us to

$$\text{Nusselt number} = \text{Nu} = \frac{h_c d}{K^{\text{air}}} \tag{7.15}$$

$$= \frac{d}{\delta^{\text{bl}}}$$

where in the second line we have used our above comment on h_c for flat leaves ($h_c = K^{\text{air}}/\delta^{\text{bl}}$); i.e., Nu relates the characteristic dimension to the boundary layer thickness, and thus Nusselt numbers are useful for describing heat transfer occurring across boundary layers.

Next, we will introduce a dimensionless number that describes flow characteristics—e.g., it can help indicate whether the flow will be laminar or turbulent. This quantity indicates the ratio of inertial forces (due to momentum, which tends to keep things moving) to viscous forces (due to frictional interactions, which tend to slow things down), and it is known as the *Reynolds number*:

$$\text{Reynolds number} = \text{Re} = \frac{vd}{\nu} \tag{7.16}$$

where v is the magnitude of the fluid velocity moving past an object of characteristic dimension d and ν is the kinematic viscosity (ordinary viscosity divided by density) for the fluid (1.51×10^{-5} m^2 s^{-1} for dry air at 20°C, App. I). At low Reynolds numbers viscous forces dominate inertial forces and the flow tends to be laminar, whereas at high Reynolds numbers (above about 10^4 for plant parts; see Gates, 1980) the flow becomes turbulent owing to the dominance of inertial forces. For a cylindrical plant part such as a branch 0.1 m in diameter, turbulence sets in for wind speeds above about 1.5 m s^{-1}. Although air movement is different below compared with above such critical Reynolds numbers, Equations 7.8 through 7.13 are satisfactory for most applications to plants in either flow regime.[9]

---

9. For flat leaves under field conditions, analysis by dimensionless numbers indicates that Nu = 0.97 Re$^{0.5}$ (see Pearman *et al.*, 1972; Gates, 1980). Hence, d/δ^{bl} equals $0.97(vd/\nu)^{1/2}$ by Equations 7.15 and 7.16, so δ^{bl} is $(1/0.97)(\nu/vd)^{1/2}d = (1/0.97)(1.53 \times 10^{-5}$ m^2 s$^{-1})^{1/2}(d/v)^{1/2} = 4.0 \times 10^{-3}$ $\sqrt{d/v}$ m s$^{-1/2}$, which is essentially Equation 7.8.

At very low wind speeds and large values of $T^{surf} - T^{ta}$, free convection can dominate forced convection for large objects. In such cases, the Reynolds number should be replaced in heat flow studies by the *Grashof number*, which takes into account buoyant forces. Specifically, the Grashof number Gr indicates the tendency of a parcel of air to rise or fall, and thus it describes the tendency for free convection. In fact, Gr represents the ratio of buoyant times inertial forces to the square of the viscous forces:

$$\text{Grashof number} = \text{Gr} = \frac{g\beta\,\Delta T\,d^3}{v^2} \qquad (7.17)$$

where g is the gravitational acceleration, β is the coefficient of volumetric thermal expansion (i.e., the fractional change in volume with temperature at constant pressure, which equals $1/T$ for an essentially perfect gas such as air, e.g., $\beta = 3.4 \times 10^{-3}\ °C^{-1}$ at 20°C), and ΔT is the temperature difference from the object's surface to the ambient air ($T^{surf} - T^{ta}$).

Using the Grashof and the Reynolds numbers, we can indicate whether forced or free convection dominates in a particular case. Because Re equals inertial forces/viscous forces and Gr equals buoyant × inertial forces/(viscous forces)^2, Re^2/Gr equals inertial forces/buoyant forces. Thus, Re^2/Gr reflects forced convection/free convection. Experiments show that forced convection accounts for nearly all heat transfer when Re^2/Gr > 10, free convection accounts for nearly all heat transfer when Re^2/Gr < 0.1, and the intervening region has mixed convection; i.e., both forced and free convection should then be considered, especially for Re^2/Gr near 1. Using Equations 7.16 and 7.17, we obtain:

$$\frac{Re^2}{Gr} = \frac{\left(\dfrac{vd}{v}\right)^2}{\left(\dfrac{g\beta\,\Delta T\,d^3}{v^2}\right)} = \frac{v^2}{g\,\beta\,\Delta T\,d}$$

$$= \frac{v^2}{(9.8\ \text{m s}^{-2})(3.4 \times 10^{-3}\ °C^{-1})\,\Delta T\,d}$$

$$= (30\ \text{s}^2\,°C\,\text{m}^{-1})\,\frac{v^2}{\Delta T\,d} \qquad (7.18)$$

Considering $v^2/(\Delta T\,d)$ in Equation 7.18, we note that inertial forces and hence forced convection become more important for higher wind speeds, smaller temperature differences, and smaller objects. For a ΔT of 5°C and a d of 0.1 m, Equation 7.18 indicates that Re^2/Gr equals 1 when the wind speed is

$$v = \left[\frac{(5°C)(0.1 \text{ m})}{(30 \text{ s}^2 \text{ °C m}^{-1})} \right]^{1/2} = 0.1 \text{ m s}^{-1}$$

Thus, for wind speeds greater than 0.1 m s^{-1}, forced convection will dominate free convection when ΔT is 5°C and d is 0.1 m. This domination of forced convection over free convection will occur for most of our applications (see Leuning, 1987).

Examples of Heat Conduction/Convection

Let us now calculate the heat conduction across the air boundary layer for a leaf at 25°C when the surrounding turbulent air is at 20°C. We will consider a leaf with a mean length in the wind direction of 0.10 m (10 cm) and a wind speed of 0.8 m s^{-1}. From Equation 7.8, the boundary layer thickness is

$$\delta^{bl}_{(mm)} = 4.0 \sqrt{\frac{(0.10 \text{ m})}{(0.8 \text{ m s}^{-1})}} = 1.4 \text{ mm}$$

which is a rather typical value for a leaf. Using Equation 7.13 and K^{air} of 0.0259 W m^{-1} °C^{-1} (appropriate for 20°C to 25°C; App. I), we can calculate the heat flux density for conduction across the boundary layer:

$$J^C_H = \frac{(2)(0.0259 \text{ W m}^{-1} \text{ °C}^{-1})(25°C - 20°C)}{(1.4 \times 10^{-3} \text{ m})}$$

$$= 190 \text{ W m}^{-2}$$

A leaf at 25°C at sea level on a sunny day could have a net radiation balance of 370 W m^{-2} (see top line of Table 7.1). In the present case, just over half of this energy input by radiation is dissipated by conduction of heat across the boundary layers on each side of the leaf (190 W m^{-2} total), followed by forced convection in the surrounding turbulent air outside the boundary layers.

A leaf with a thick boundary layer can have a temperature quite different from that of the surrounding air, because air is a relatively poor conductor of heat. Specifically, K^{air} is relatively low compared with the thermal conductivity coefficients for liquids and most solids. A large leaf in a low wind might have a boundary layer 4 mm thick (see Table 7.2 or Eq. 7.8 for the wind speeds and leaf sizes implied by this). If J^C_H and K^{air} are the same as in the previous paragraph, where the difference between leaf and turbulent air temperatures is 5°C when the boundary layer is 1.4 mm thick, then $T^{leaf} - T^{ta}$ will be (4 mm/1.4 mm) (5°C), or 14°C, for a δ^{bl} of 4 mm. Thus, the combination of large leaves and low wind speeds favors a large drop in temperature across the boundary layers. On the other hand, a small leaf in a moderate wind can have a δ^{bl} of 0.2 mm. For the same J_H^C and K^{air} as above, $T^{leaf} - T^{ta}$ is (0.2 mm/1.4 mm)(5°C), or 0.7°C, for this thin boundary layer. Hence, small leaves tend to have temperatures quite close to that of the air, especially at moderate to high wind speeds. This close coupling between leaf and air temperatures for small leaves can keep the leaf temperature low enough for optimal photosynthesis (often 30°C to 35°C) in hot, sunny climates. Also, the lower is the leaf temperature, the lower is the concentration of water vapor in the pores of the cell walls of mesophyll cells, and consequently less water then tends to be lost in transpiration (Ch. 8), an important consideration in arid and semi-arid regions.

We can apply the same heat conduction analysis to massive structures such as the stems of cacti. Let us consider a cylindrical stem 0.25 m in diameter exposed to the sun on a cloudless day at a high elevation of 2 000 m and calculate the air temperature expected if we use the radiation terms presented in Table 7.1, line 2 (net radiation balance of 370 W m^{-2} for a surface temperature of 32°C). Cacti are Crassulacean acid metabolism (CAM) plants (to be discussed in Ch. 8), so the stomata tend to be closed during the daytime, making transpirational cooling then minimal. Consequently, if we ignore heat conduction and heat storage in the stem, the net radiation will be balanced by heat conduction across the boundary layer. For an ambient wind speed of 1.0 m s^{-1}, the boundary layer thickness (Eq. 7.9) is 2.9 mm. Assuming cylindrical symmetry (Eq. 7.12), we can then calculate the temperature of the surrounding turbulent air such that a net radiation input of 370 W m^{-2} is just balanced by heat conduction across the boundary layer:

$$T^{ta} = T^{surf} - \frac{J_H^C \, r \, \ln\left(\dfrac{r + \delta^{bl}}{r}\right)}{K^{air}}$$

$$= 32°C - \frac{(370 \text{ W m}^{-2})(0.125 \text{ m}) \ln\left(\dfrac{0.125 \text{ m} + 0.0029 \text{ m}}{0.125 \text{ m}}\right)}{(0.0253 \text{ W m}^{-1} \text{ °C}^{-1})}$$

$$= -10°C$$

Actually, we would not expect the stem surface temperature T^{surf} to be 42°C above T^{ta}—in fact, our assumption that no heat is stored or conducted in the stem is incorrect for a massive cactus, an aspect that we will reconsider toward the end of this chapter.

LATENT HEAT—TRANSPIRATION

Evaporation of water is a cooling process. Water evaporates at the air-liquid interfaces along the pores on the cell walls of mesophyll, epidermal, and guard cells (see Fig. 1.2) and then diffuses out of a leaf. Thus, transpiration represents a means of heat loss by a leaf (Eq. 7.2). A leaf can have a latent heat gain if dew or frost condenses onto it, as we will discuss shortly.

Heat Flux Density Accompanying Transpiration

We will represent the flux density of water vapor diffusing out of a leaf during transpiration by J_{wv}. If we multiply the amount of water leaving per unit time and per unit leaf area, J_{wv}, by the energy necessary to evaporate a unit amount of water at the temperature of the leaf, H_{vap}, we obtain the heat flux density accompanying transpiration, J_H^T:

$$J_H^T = J_{wv} H_{vap} = \frac{H_{vap} D_{wv} \, \Delta c_{wv}^{total}}{\Delta x^{total}} = \frac{H_{vap} D_{wv} (c_{wv}^e - c_{wv}^{ta})}{\Delta x^{total}} \qquad (7.19)$$

where Fick's first law (Eqs. 1.1 and 1.8) has been used to express J_{wv} in terms of the diffusion coefficient for water vapor, D_{wv}, and the total drop in water vapor concentration, Δc_{wv}^{total}, over some effective total distance, Δx^{total}. In turn, Δc_{wv}^{total} equals the water vapor concentration at the sites of evaporation within a leaf, c_{wv}^e, minus the value in the turbulent air just outside the boundary layer, c_{wv}^{ta}. J_H^T and J_{wv} in Equation 7.19 are conventionally expressed based on the area of one side of a leaf, and Δx^{total} is usually incorporated into a resistance or a conductance. In the next chapter we will discuss in detail the effective distance, resistance, and conductance for the water vapor flux.

How much of the heat load on a leaf is dissipated by the evaporation of water during transpiration? For an exposed leaf of a typical mesophyte during the daytime, J_{wv} is about 4 mmol m^{-2} s^{-1} (p. 423). In Chapter 2 we noted that water has a high heat of vaporization, e.g., 44.0 kJ mol^{-1} at 25°C (values at other temperatures are given in App. I). By Equation 7.19, the heat flux density out of the leaf by transpiration then is

$$J_H^T = (4 \times 10^{-3} \text{ mol m}^{-2} \text{ s}^{-1})(44.0 \times 10^3 \text{ J mol}^{-1}) = 180 \text{ W m}^{-2}$$

For the leaf described in the top line of Table 7.1, a heat loss of 180 W m^{-2} by evaporation dissipates slightly under half of the net radiation balance (370 W m^{-2}), the rest of the energy input being removed by heat conduction across the boundary layer followed by forced convection (see Fig. 7.7a). A heat loss accompanying transpiration reduces leaf temperatures during the daytime. Although such latent heat losses can benefit a plant, evaporation and its associated cooling are an inevitable consequence of gas exchange by leaves, where opening of the stomata is necessary for substantial rates of CO_2 uptake.

Heat Flux Density for Dew or Frost Formation

So far we have regarded c_{wv}^e as greater than c_{wv}^{ta}, in which case there is a net loss of water from the leaf and a consequent dissipation of heat. When the

Leaf temperature	25°C	−1°C
Air temperature	20°C	1°C
Relative humidity	50%	92%
Wind speed	0.8 m s^{-1}	0.4 m s^{-1}
Water flux density	4 mmol m^{-2} s^{-1} transpired	0.2 mm frost in 8 h
	(a)	(b)

Figure 7.7

Energy budget for an exposed leaf: (a) at midday, and (b) at night with frost. The flux densities in W m^{-2} are indicated for net radiation (R); conduction/convection (C), also referred to as sensible heat; and latent heat (L).

turbulent air is warmer than the leaf, however, the water vapor concentration in the turbulent air may be greater than that in the leaf (in Ch. 2 we noted that the water vapor concentration and partial pressure at saturation increase rapidly with temperature; also, see values for P_{wv}^* and c_{wv}^* in App. I). If c_{wv}^{ta} is greater than the water vapor concentration in the leaf, then a net diffusion of water vapor occurs toward the leaf. This can cause c_{wv} at the leaf surface to increase, and it may reach c_{wv}^*, the saturation value. If c_{wv}^{ta} is greater than this c_{wv}^*, dew—or frost, if the leaf temperature is below freezing—can form as water vapor diffuses toward the leaf and then condenses onto its surface, which is cooler than the turbulent air.[10]

Condensation of water vapor leads to a heat *gain* by a leaf. Because water condensation is just the reverse of the energy-dissipating process evaporation, the heat gain per unit amount of water condensed is the heat of vaporization of water at the temperature of the leaf, H_{vap}. If the condensation is on the leaf surface, the diffusion step is across the air boundary layers of thickness δ^{bl} that are present on either side of a leaf. We can then modify Equation 7.19 to the following form to describe the rate of heat gain per unit area accompanying the water vapor condensation that leads to dew formation:

$$J_H^{dew} = \frac{2H_{vap}D_{wv}(c_{wv}^{ta} - c_{wv}^{leaf*})}{\delta^{bl}} \qquad (7.20)$$

where c_{wv}^{leaf*} is the saturation concentration of water vapor at the temperature of the leaf. The factor 2 is necessary because water vapor diffuses toward the leaf across the boundary layer on each side (as usual, we will assume that these layers are of equal average thickness δ^{bl}).

The temperature to which the turbulent air must be reduced at constant pressure for it to become saturated with water vapor is known as the *dew point*, or dew temperature, $T_{dew\ point}^{ta}$. When T^{leaf} is below $T_{dew\ point}^{ta}$, the turbulent air contains more water vapor (c_{wv}^{ta}) than the air at the leaf surface can hold (c_{wv}^{leaf*}). Water vapor then diffuses toward the leaf, which can lead to dew formation (Eq. 7.20). If T^{leaf} is below 0°C and less than $T_{dew\ point}^{ta}$, the water that condenses onto the leaf surface will freeze. Under such conditions we should replace H_{vap} in Equation 7.20 by the heat of sublimation, H_{sub}, at that particular leaf temperature to describe the heat gain by frost formation.

---

10. Condensation resulting from water emanating from the soil is sometimes called "distillation," the term "dew" then being reserved for water coming from the air above.

Examples of Frost and Dew Formation

As an example of nighttime frost formation, we will consider a leaf with boundary layers 2.0 mm thick and a temperature of $-1°C$ when the surrounding air is at $1°C$ and 92% relative humidity. The leaf can be the same leaf considered above (p. 372), where we estimated a boundary layer thickness of 1.4 mm during the daytime; at night, the wind speed is generally lower—if v were halved, then the boundary layer would be 41% thicker, or 2.0 mm, by Equation 7.8. D_{wv} is 2.13×10^{-5} m^2 s^{-1} at $0°C$, H_{sub} is 51.0 kJ mol^{-1} at $-1°C$, and c_{wv}^* is 0.249 mol m^{-3} at $-1°C$ and 0.288 mol m^{-3} at $1°C$ (App. I). By Equation 7.20, the rate of heat gain per unit area by frost formation is:

$$
\begin{aligned}
J_H^{frost} &= (2)(51.0 \times 10^3 \text{ J mol}^{-1})(2.13 \times 10^{-5} \text{ m}^2 \text{ s}^{-1}) \\
&\quad \times \frac{[(0.92)(0.288 \text{ mol m}^{-3}) - (0.249 \text{ mol m}^{-3})]}{(2.0 \times 10^{-3} \text{ m})} \\
&= 17 \text{ W m}^{-2}
\end{aligned}
$$

Because T^{leaf} is here less than T^{ta}, heat is conducted *into* the leaf from the air. By Equation 7.11 [$J_H^C = 2K^{air}(T^{leaf} - T^{ta})/\delta^{bl}$, where K^{air} is 0.0243 W m^{-1} $°C^{-1}$ at $0°C$; Appl. I], the heat conduction across the boundary layer is (2) (0.0243 W m^{-1} $°C^{-1}$)($-1°C - 1°C$)/(2.0×10^{-3} m) or -49 W m^{-2}, where the minus sign indicates that heat is conducted into the leaf. The bottom line of Table 7.1 indicates that this leaf can lose 66 W m^{-2} by net radiation. Thus, the heat inputs from frost formation (17 W m^{-2}) and heat conduction (49 W m^{-2}) balance the energy loss by net radiation (see Fig. 7.7b).

How long would it take to form a layer of frost 0.1 mm thick on each side of a leaf for the above conditions? Because J_H^{frost} equals $J_{wv} H_{sub}$ (see Eq. 7.19), the rate of water deposition per unit area (J_{wv}) in kg m^{-2} s^{-1} is J_H^{frost} divided by H_{sub} in J kg^{-1} (2.83 MJ kg^{-1} at $-1°C$, App. I). This J_{wv} divided by the density of ice (ρ_{ice}; 917 kg m^{-3}) gives the rate of frost accumulation in m s^{-1}. To accumulate 0.1 mm of ice on each side of a leaf therefore takes

$$
\begin{aligned}
t &= \frac{\text{thickness}}{J_{wv}/\rho_{ice}} = \frac{(0.2 \text{ mm}) \rho_{ice}}{J_H^{frost}/H_{sub}} \\
&= \frac{(0.2 \times 10^{-3} \text{ m})(917 \text{ kg m}^{-3})}{(17 \text{ J s}^{-1} \text{ m}^{-2})/(2.83 \times 10^6 \text{ J kg}^{-1})} \\
&= 3.1 \times 10^4 \text{ s} \quad (8.6 \text{ h})
\end{aligned}
$$

Dew formation can be as great as 0.5 mm per night, which can be an important source of water in certain regions for part of the year (see Jones, 1983; Levitt, 1980; Oke, 1987; Rosenberg *et al.*, 1983).[11] For the lichen *Ramalina maciformis* in the Negev Desert, dew sufficient to lead to photosynthesis the next day occurs on over half of the nights, and the annual dewfall can be about 30 mm (see Kappen *et al.*, 1979). Dew or frost formation is favored on cloudless nights and when the relative humidity of the surrounding turbulent air is close to 100%. As we mentioned above, T^{leaf} tends to be further from T^{ta} for large leaves than for small ones. On a cloudless night when T^{leaf} is less than T^{surr} and T^{ta} (see Table 7.1), the larger exposed leaves generally dip below $T^{\text{ta}}_{\text{dew point}}$ sooner than do the small ones, and hence dew or frost tends to form on the larger leaves first. For convenience, we have assumed that the leaf has a uniform temperature. In fact, however, the boundary layer tends to be thinner at the edges of a leaf (Fig. 7.5), so T^{leaf} is somewhat closer to T^{ta} at the edges than at the center. Thus, dew or frost generally forms first at the center of a leaf, where δ^{bl} is larger and the temperature slightly lower than at the leaf edges.

SOIL

Soil acts as an extremely important component in the energy balance of plants. For instance, shortwave irradiation can be reflected from its surface, it is the source of longwave radiation that can correspond to a temperature considerably different from that of the surrounding air, and heat can be conducted to or from stems in their region of contact with the soil. Also, considerable energy can be stored by the soil, contrary to the case for most leaves (see Eqs. 7.1 and 7.2). Although the soil surface can have large daily oscillations in temperature, the soil temperature at moderate depths of 1 m or so can be fairly steady, even from season to season.

Thermal Properties

Soil has a relatively high heat capacity. To raise the temperature of 1 kg of dry sand by 1°C takes about 0.8 kJ, with similar values for dry clay or loam. Because the density of soil solids is about 2 600 kg m^{-3} and soil is about half pores by

---

11. At 10°C, a layer of liquid water 0.5 mm thick corresponds to the entire content of water vapor in a column of saturated air (see App. I) that is 53 m tall. Hence, considerable downward movement of water vapor is necessary to have dew formation over large areas, or much of the water vapor must emanate from the ground (often termed distillation; see Oke, 1987).

volume, the overall density of dry soil is about 1 300 kg m^{-3}—values generally range from 1 200 kg m^{-3} for dry loam, slightly higher for clay, and up to 1 500 kg m^{-3} for dry sand. Thus, the volumetric heat capacity at constant pressure, C_P, of dry soil is about 1.1 MJ m^{-3} °C^{-1}. For a moist loam containing 20% water by volume (water has a C_P of 4.18 MJ m^{-3} °C^{-1} at 20°C), C_P^{soil} is about 1.9 MJ m^{-3} °C^{-1} (see van Wijk, 1966).

The relatively high heat capacity of soil means that a considerable amount of energy can be involved with its temperature changes. For instance, if the soil temperature of the upper 0.4 m of a moist soil with a C_P^{soil} of 2.0 MJ m^{-3} °C^{-1} increases by an average of 2°C during the daytime, then (0.4 m)(2° C)(2.0 MJ m^{-3} °C^{-1}), or 1.6 MJ, can be stored per m^2 of ground. Nearly all the heat stored in the soil during the daytime is released that night. Generally, the soil temperature at depths below 0.4 m changes less than 0.5°C during a day or night, although annually it will vary considerably more, as we will indicate below.

Although it has a substantial volumetric heat capacity, soil does not have a very high thermal conductivity coefficient, K^{soil}. Heat is therefore not conducted very readily in a soil, where the heat flux density by conduction is

$$J_H^C = -K^{soil} \frac{\partial T}{\partial z} \tag{7.21}$$

Heat can of course be conducted in all directions in the soil, instead of only vertically, as we will consider here (see Eq. 7.21). J_H^C can be expressed in W m^{-2}, $\partial T/\partial z$ in °C m^{-1}, and therefore K^{soil} in W m^{-1} °C^{-1}, just as for K^{air}. K^{soil} depends on the soil water content; replacement of soil air (a relatively poor heat conductor) by water will increase K^{soil}. For instance, K^{soil} can vary from 0.2 W m^{-1} °C^{-1} for a dry soil to 2 W m^{-1} °C^{-1} for a wet one, as the addition of water forms bridges between the soil particles (see Fig. 9.8) and thereby increases the thermal conductivity. For comparison, K^{water} is 0.60 W m^{-1} °C^{-1} and K^{air} is 0.026 W m^{-1} °C^{-1} near 20°C (see App. I; Rose, 1966; van Wijk, 1966).

During the daytime the surface of the ground can be considerably warmer than the underlying layers of the soil, which leads to heat conduction into the soil. Because the soil exposed to the turbulent air tends to be drier than the underlying layers, the thermal conductivity coefficient can be lower near the soil surface. For the upper part of a fairly moist sandy loam, K^{soil} may therefore be 0.6 W m^{-1} °C^{-1} and $\partial T/\partial z$ may be -100°C m^{-1} (at least for the upper 0.05 m or so; z is here considered positive into the soil). Using Equation 7.21, the heat flux density by conduction into the soil is

$$J_H^C = -(0.6 \text{ W m}^{-1} \text{ °C}^{-1})(-100\text{°C m}^{-1}) = 60 \text{ W m}^{-2}$$

This heat conducted into the soil could lead to the daytime heat storage calculated above, 1.6 MJ m^{-2}, in $(1.6 \times 10^6$ J m$^{-2})/(60$ J m^{-2} s$^{-1})$, or 2.67×10^4 s, which is about 7 h.

Soil Energy Balance

The components of the energy balance for the soil surface are similar to those for leaves (see Eqs. 7.1 and 7.2). However, we must also take into consideration the heat conducted into the soil itself (Eq. 7.21), which leads to a gradual temperature change of its upper layers.

The absorption and emission of radiation usually takes place in the upper few millimetres of the soil. Using Equations 7.3 though 7.6, we can generally represent the net radiation balance for the soil surface by $aS + a_{IR}\sigma(T^{surr})^4 - e_{IR}\sigma (T^{soil})^4$, where the values of all parameters are those at the soil surface. (If the soil is exposed directly to the sky, T^{surr} should be replaced by T^{sky} for the incident IR.) For a soil exposed to direct sunlight, the net energy input by radiation can be quite large—the temperature of the uppermost part of the soil can exceed 70°C in a desert.

Let us now consider the flux for heat conduction between the overlying air and the soil, as well as the heat flux accompanying water evaporation in the soil. The heat conducted across the relatively still air next to the soil surface is equal to $-K^{air} \partial T/\partial z$, where K^{air} is the thermal conductivity coefficient of air at the local temperature (see Eq. 7.11, which indicates that the heat conducted from both sides of a leaf is $-2K^{air} \partial T/\partial x$). The heat conducted within the soil can be calculated using Equation 7.21 ($J_H^C = -K^{soil} \partial T/\partial z$). For a given layer of soil where the water vapor flux density changes by ΔJ_{wv}, the heat loss accompanying water evaporation (or heat gain accompanying condensation) equals $\Delta J_{wv} H_{vap}$, where H_{vap} is the heat necessary to evaporate a unit amount of water at the local soil temperature (see Eq. 7.19, $J_H^T = J_{wv} H_{vap}$). Except in the upper few millimetres, the main energy flux in the soil is for heat conduction, not for radiation or for phase changes of water. Because of the large heat capacity of water, its movement in the soil can also represent an important means of heat movement.

Variations in Soil Temperature

Because the energy flux in the soil is often mainly by heat conduction, we can fairly readily estimate the soil temperature at various depths, although complications arise because of the heterogeneous nature of soil as well as the many types of plant cover. To obtain some idea of daily and annual temperature variations, we will assume that the volumetric heat capacity (C_P^{soil}) and the thermal conductivity coefficient (K^{soil}) are both constant with depth, and we will ignore water movement in the soil. Moreover, we will assume that the soil surface temperature varies sinusoidally around an average value $\overline{T}^{\text{surf}}$, with a daily or annual amplitude of ΔT^{surf}, a useful approximation that can be tested. We then obtain the following relation for the temperature T at time t and depth z (see Arya, 1988; Campbell, 1977; de Vries, 1975; Rosenberg et al., 1983; van Wijk, 1966):

$$T = \overline{T}^{\text{surf}} + \Delta T^{\text{surf}} e^{-z/d} \cos\left(\frac{2\pi t}{p} - \frac{2\pi t_{\max}}{p} - \frac{z}{d}\right) \qquad (7.22)$$

where the damping depth d, which is the depth in the soil where the variation in temperature has been damped down to $1/e$ of the value at the soil surface, equals $(pK^{\text{soil}}/\pi C_P^{\text{soil}})^{1/2}$; and p is the period (24 h, or 8.64×10^4 s, for a daily variation and 365 times longer for an annual variation). We note that z equals zero at the surface and is considered positive downward, and that the soil surface has its maximum temperature, $\overline{T}^{\text{surf}} + \Delta T^{\text{surf}}$, when t equals $t_{\max}$.[12]

Let us next estimate the depths where the variation in soil temperature would be only $\pm 1°C$, in one case daily and in another annually. As above, we will use a C_P^{soil} of 2 MJ m^{-3} °C^{-1} and a K^{soil} of 0.6 W m^{-1} °C^{-1}. The damping depth for the daily case then is

$$d = \left[\frac{(8.64 \times 10^4 \text{ s})(0.6 \text{ J s}^{-1} \text{ m}^{-1} \text{ °C}^{-1})}{(\pi)(2 \times 10^6 \text{ J m}^{-3} \text{ °C}^{-1})}\right]^{1/2}$$
$$= 0.091 \text{ m}$$

and it is 1.7 m for the annual case. From Equation 7.21 the daily variation in temperature at depth z is $\pm \Delta T^{\text{surf}} e^{-z/d}$. The amplitude of the daily variation in soil surface temperature about its mean for bare soil is often 15°C (i.e., $\overline{T}_{\max}^{\text{surf}} - \overline{T}_{\min}^{\text{surf}} = 30°C$), and the annual amplitude for variations in average daily surface temperatures is generally somewhat less (e.g., 10°C). The depth where the daily variation in temperature is $\pm 1°C$ is then

$$z = -d \ln\left(\frac{1°C}{\Delta T^{\text{surf}}}\right) = -(0.091 \text{ m}) \ln\left(\frac{1°C}{15°C}\right)$$
$$= 0.25 \text{ m}$$

---

12. Equation 7.22 is of the form $y = A + B \cos \alpha$, where A is the average value of y and B is the amplitude of the variation about the mean.

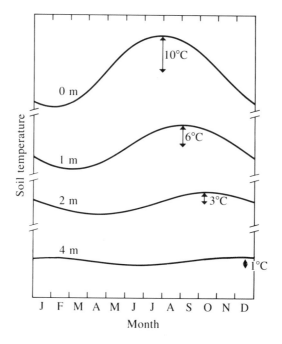

Figure 7.8

Simulated annual variation in soil temperatures at the indicated depths. The average daily temperature at the soil surface was assumed to vary sinusoidally, with a maximum on August 1 and an annual amplitude of 10°C. See text for other conditions.

which is consistent with our previous statement that soil temperatures generally change less than 0.5°C daily at about 0.4 m. On an annual basis, the ±1°C variation would occur at a depth of 3.9 m (Fig. 7.8).

The factor z/d occurring in the cosine in Equation 7.22 indicates that the peak of the temperature "wave" arrives later at greater depths in the soil. This peak occurs when the cosine equals 1, which corresponds to $2\pi t/p - 2\pi t_{max}/p = z/d$ (cos 0 = 1). The speed of movement of the wave, $\Delta z/\Delta t$, is thus $2\pi d/p$. Figure 7.8 shows this effect at various depths on an annual basis. For instance, at 4 m the peak temperature occurs $4\frac{1}{2}$ months later than at the surface.

FURTHER EXAMPLES OF ENERGY BUDGETS

So far in this chapter our discussion of the energy fluxes of leaves has been for those in exposed location. Although their temperatures depend on the net effect of all the energy fluxes that we have discussed, many observations and calculations indicate that exposed leaves when sunlit tend to be above air temperature for T^{ta} up to about 30°C and below T^{ta} for air temperatures above about 35°C (see Gates, 1980). This primarily reflects the increasing importance of infrared radiation emission as leaf temperature rises [see Eq. 7.5, $J_{IR} = 2e_{IR}\sigma(T^{leaf})^4$] and the increase with temperature of the water vapor concentration in the leaves, which affects transpiration (to be discussed in the next chapter). Such influences of energy budget terms often lead to temperatures for exposed leaves that are more favorable for photosynthesis than is the ambient air temperature. We can readily extend our analysis to include leaves shaded by overlying ones. Also, our brief description of the thermal properties of soil serves as an introduction to the energy balance of certain plant parts where heat storage and heat conduction within the tissue are important. We will conclude this section with some comments on the time constants for changes in leaf temperature.

Leaf Shape and Orientation

Leaf sizes and shapes vary tremendously, which can have important consequences for leaf temperature (see Gates, 1980; Taylor, 1975). Leaves developing in full sunlight tend to have smaller areas when mature than do leaves on the same plant that develop in the shade—"sun" leaves generally range from 20% to over 80% less in surface area than do "shade" leaves. When shade leaves are placed in exposed locations, their larger size leads to thicker boundary layers (Eq. 7.8), less convective heat loss (Eq. 7.11), and consequently greater differences from air temperature than for sun leaves. For leaves above air temperature, this can lead to high transpiration rates, because the water vapor concentration at saturation depends more or less exponentially on temperature, as indicated in Chapter 2 (also see Fig 8.5 and App. I). Moreover, the amount of CO_2 photosynthetically fixed per unit of water transpired can be higher for sun leaves than for shade leaves in exposed sunlit locations and for shade leaves than for sun leaves in shaded locations based on model calculations (Parkhurst and Loucks, 1972) and observations on a desert shrub, *Hyptis emoryi* (Smith and Nobel, 1977).

Lobing and dissection tend to decrease the effective length across a leaf in the direction of the wind and hence to reduce δ^{bl}, with a consequent enhancement of convective exchange. For instance, the heat convection coefficient h_c (Eq. 7.14) increases with the depth of serrations on model leaves (Gottschlich and Smith, 1982). The greater lobing often observed for sun leaves compared with shade leaves reduces the heating of sun leaves above air temperature (Vogel,

1970). Also, heat convection can be approximately twice as great for a pinnate leaf as for a simple leaf of the same area (Balding and Cunningham, 1976).

Certain plants, especially those exposed to intense shortwave irradiation, have vertically oriented leaves, e.g., willow, many species of *Eucalyptus,* and certain chaparral and desert shrubs. Over the course of a day, vertical leaves often intercept nearly as much shortwave irradiation as do horizontal leaves, but they intercept less at midday, when air temperatures tend to be high. Higher leaf temperatures at midday lead to greater transpiration for a given stomatal opening and possibly to temperatures above those photosynthetically optimal. Also, leaves generally become more vertical upon wilting, thereby reducing their interception of shortwave irradiation for higher sun angles. As we indicated above (p. 358), shortwave plus longwave irradiation absorbed by a leaf at midday could decrease by up to 300 W m^{-2} if a leaf were rotated from horizontal to vertical. For the exposed horizontal leaf that we have considered (radiation terms in the top line of Table 7.1, convection and transpiration in Fig. 7.7a), such rotation would decrease its temperature from 25°C to 19°C, or by 6°C. Consistent with this are observations that rotating the leaves of *Cercis canadensis* (redbud) and *Erythrina berteroana* from vertical to horizontal at midday increases their temperatures by 2°C to 6°C (see Gates, 1980).

Leaf temperature can be influenced by seasonal differences in leaf orientation and by diurnal solar "tracking" movements, as occurs for cotton, other species in the Malvaceae, many clovers, and certain desert annuals (see Ehleringer and Forseth, 1989). Leaves of *Malvastrum rotundifolium,* a winter annual inhabiting warm deserts but growing during the cool part of the year, track the sun so well that the leaf surface is within 20° of being perpendicular to the direct solar beam throughout the day. As well as leading to better light interception, this tracking raises the leaf temperature to values more optimal for photosynthesis (Mooney and Ehleringer, 1978). The mechanism for solar tracking involves changes in hydrostatic pressure induced by the absorption of blue light by photoreceptors in the leaf veins or the pulvinus, perhaps via steps similar to those for stomatal opening discussed in the next chapter, leading to changes in leaf orientation (see Koller, 1986).

Shaded Leaves within Plant Communities

Let us next consider a shaded leaf at the same temperature as its surroundings. As we indicated above, the infrared radiation absorbed by such a leaf is the same as the IR emitted by it (when $a_{IR} = e_{IR}$ and $T^{leaf} = T^{surr}$). The net radiation is then due solely to the various forms of solar radiation that reach the leaf.

Because the transmission by leaves is fairly high from 0.7 to 2 μm (Fig. 7.4), much of the solar radiation reaching a shaded leaf is in a region not useful for photosynthesis. The leaf absorptance in this range is lower than for the solar irradiation incident on an exposed leaf (Fig. 7.4). For purposes of calculation, we will assume that S on the shaded leaf is 70 W m^{-2}, that its absorptance a is 0.30, and that the reflectance r of the surroundings has the rather high value of 0.40 because of considerable reflection of radiation by the other leaves within the plant community. By Equations 7.3 and 7.6, the net radiation balance is then $(0.30)(1.00 + 0.40)(70$ W m$^{-2})$, or 30 W m^{-2}.

Our shaded leaf has a rather low rate of photosynthesis because the amount of radiation in the visible region reaching it is fairly small. Moreover, the stomata generally are partially closed at lower illumination levels, which increases the stomatal resistance and further decreases photosynthesis (Ch. 8). Three factors tend to reduce the flux of water vapor out of a shaded leaf: the increase in stomatal resistance at the lower light levels, which is the main factor reducing transpiration; a lower wind speed for a protected than for an exposed leaf, which leads to a thicker air boundary layer; and a concentration of water vapor in the turbulent air that is generally higher than at the top of the plant canopy. Instead of the water vapor flux density of 4 mmol m^{-2} s^{-1} that can occur for an exposed leaf of a typical mesophyte (p. 423), J_{wv} for the shaded leaf might be only 0.7 mmol m^{-2} s^{-1}. By Equation 7.19 ($J_H^T = J_{wv}H_{vap}$), the heat dissipation by the latent heat term is then 30 W m^{-2}. No heat is conducted across the boundary layers if the leaf is at the same temperature as the surrounding air. Thus, the heat loss by water evaporation (30 W m^{-2}) here balances the energy gain from the absorption of shortwave irradiation for this shaded leaf.

Heat Storage

We indicated at the beginning of this chapter that very little energy is stored in the temperature changes of leaves. However, massive plant parts, such as tree trunks, can store considerable energy, as can the soil. We will represent the heat storage rate (e.g., in J s^{-1}, or W) as follows:

$$\text{heat storage rate} = C_P V \frac{\Delta T}{\Delta t} \tag{7.23}$$

where C_P is the volumetric heat capacity (e.g., in J m^{-3} $°C^{-1}$; see values introduced above for soil), and V is the volume that undergoes a change in temperature ΔT in the time interval Δt.

As indicated above, heat storage can be important for any massive plant part. To model its energy balance, such a part can be divided into a number of isothermal subvolumes, which are generally referred to as *nodes* in heat transfer studies (see Kreith and Bohm, 1986). This approach has been applied to the energy balance of massive stems of cacti, which have been divided into about 100 nodes for the barrel-shaped *Ferocactus acanthodes* (Fig. 7.9) and over 200 nodes for the tall, columnar *Carnegiea gigantea,* or saguaro (Lewis and Nobel, 1977; Nobel, 1980). The stem is divided into surface nodes, which have no volume and hence no heat storage, plus interior nodes, which have volume but no radiation, boundary-layer, or latent-heat terms (Fig. 7.9). These interior nodes are thus involved with changes in temperature leading to heat storage (Eq. 7.23)

(a)

(b)

(c)

Figure 7.9

System of nodes or subvolumes used in energy balance studies on the barrel cactus, *Ferocactus acanthodes:* (a) vertical section indicating the division of the stem into various levels, (b) horizontal section indicating the surface nodes (1 through 8) and interior nodes (9 through 17), and (c) three-dimensional representation of certain nodes. (Adapted from Lewis and Nobel, 1977; used by permission.)

and heat conduction to or from surface nodes as well as to or from other interior nodes (describable by equations of the form of Eqs. 7.11 and 7.21).

Once the stem has been divided into nodes, an energy balance for each node can be calculated using the various environmental parameters that we described above—greater precision requires a greater number of nodes. The analysis is complicated, in part because the various contributors depend on temperature in different ways. Specifically, shortwave and longwave absorption is independent of temperature, longwave emission depends on T^4, conduction depends on a temperature gradient or difference (e.g., $T^{\text{surf}} - T^{\text{ta}}$), saturation water vapor content—which can affect the latent heat term—varies approximately exponentially with temperature, and heat storage depends on $\Delta T/\Delta t$. When these various energy terms for all the nodes are incorporated into a simulation model, or when cactus temperatures are directly measured in the field, parts of the cactus stem facing the sun are found to range up to 15°C above air temperature for stems approximately 0.25 m in diameter, not the 42°C calculated when heat storage and heat conduction in the tissue are ignored (p. 373). Moreover, parts facing away from the sun can be below air temperature, and time lags of a few hours are observed in the heating of the center of the stem.

We indicated near the beginning of this chapter that consumption or production of energy by metabolic processes can generally be ignored in the energy budget of a leaf. An interesting situation in which metabolic heat production must be reckoned with occurs for the inflorescences of many members of the Araceae (*Arum* family), for which high respiratory rates can substantially raise the inflorescence temperature and lead to considerable heat storage. An extreme example is presented by the 2- to 9-g inflorescence or spadix of *Symplocarpus foetidus* (eastern skunk cabbage). By consuming O_2 at the same rate as an active mammal of the same size (heat production of about 0.10 W g^{-1}), the tissue temperature can be 15°C to 35°C above ambient air temperatures of -15°C to 15°C for at least 2 weeks (Knutson, 1974).

Time Constants

A matter related to heat storage is the time constant for temperature changes. Analogous to our use of lifetimes in Chapter 4, we will define a time constant τ as the time required for the change of surface temperature from some initial value T_0^{surf} to within $1/e$ (37%) of the overall change to a final value approached asymptotically (T_∞^{surf}):

$$T^{\text{surf}} - T_\infty^{\text{surf}} = (T_0^{\text{surf}} - T_\infty^{\text{surf}})e^{-t/\tau} \qquad (7.24)$$

If we ignore transpiration and assume uniform tissue temperatures, then the time constant is

$$\tau = \frac{(V/A)\, C_P}{4 e_{\mathrm{IR}} \sigma (T^{\mathrm{surf}} + 273)^3 + K^{\mathrm{air}}/\delta^{\mathrm{bl}}} \tag{7.25}$$

where V is the volume of a plant part having total surface area A (V/A indicates the mean depth for heat storage, which is half its thickness for a leaf), C_P is the heat capacity per unit volume, and T^{surf} is expressed in °C. For a rapidly transpiring leaf, the time constant is about 50% less than indicated by Equation 7.25 (see Gates, 1980; and especially Monteith, 1981).

Let us next estimate the time constants for a leaf and a cactus stem. If we consider a 300-μm-thick leaf at 25°C with a volumetric heat capacity that is 70% as large as that of water and a boundary layer thickness of 1.4 mm, by Equation 7.25 the time constant is

$$\tau =$$

$$\frac{(150 \times 10^{-6}\ \mathrm{m})(0.7)(4.2 \times 10^6\ \mathrm{J\ m^{-3}\ °C^{-1}})}{(4)(0.96)(5.67 \times 10^{-8}\ \mathrm{J\ s^{-1}\ m^{-2}\ K^{-4}})(298\ \mathrm{K})^3 + \dfrac{(0.0259\ \mathrm{J\ s^{-1}\ m^{-1}\ °C^{-1}})}{(1.4 \times 10^{-3}\ \mathrm{m})}}$$

$$= 18\ \mathrm{s}$$

Thus, 63% of the overall change in leaf temperature occurs in only 18 s, indicating that such leaves respond rapidly to variations in environmental conditions, consistent with our statement (p. 347) that very little heat can be stored by means of temperature changes of such leaves. On the other hand, stems of cacti can store appreciable amounts of heat. For the stem portrayed in Figure 7.9, V/A is about 0.05 m, and C_P is about 90% of C_P^{water} (cactus stems generally have a much smaller volume fraction of intercellular air spaces than the 30% typical of a leaf). Assuming the other factors are the same as in the above calculation for a leaf, τ is then 7.8×10^3 s (2.2 h) for the cactus stem. Indeed, massive stems do have very long time constants for thermal changes.

Among other effects, long thermal time lags help avoid overheating of tree trunks as well as of seeds and roots in the soil during rapidly moving fires. For instance, temperatures near the soil surface average about 300°C during fires in different ecosystems, but at 0.1 m below the soil surface they rarely exceed 50°C (see Rundel, 1981). Energy budget analyses can provide information on the multitude of physiological and ecological processes involving temperature— from frost to fire to photosynthesis.

Problems

7.1 An exposed leaf at 10°C has an a_{IR} and e_{IR} of 0.96, while a is 0.60 and r is 0.10. Suppose that the effective temperature for radiation is 2°C for a cloudy sky and -40°C for a clear sky. (a) If the absorbed IR equals the emitted IR for the leaf, what are the temperatures of the surroundings for a clear sky and for a cloudy one? (b) What are the λ_{max}'s for emission of radiant energy by the leaf and the surroundings for (a)? (c) If the global irradiation is 700 W m^{-2} and the temperature of the surroundings is 9°C on a clear day, how much radiation is absorbed by the leaf? (d) Under the conditions of (c), what percentage of the energy input by absorbed irradiation is dissipated by the emission of thermal radiation? (e) Assume that the clouds block out the sunlight and the skylight. Let S^{cloud} be 250 W m^{-2} on the upper surface of the leaf and 15% as much be reflected onto the lower surface. If the temperature of the surroundings is 9°C, what is the net radiation for the leaf?

7.2 Consider a circular leaf at 25°C that is 0.12 m in diameter. The ambient wind speed is 0.80 m s^{-1}, and the ambient air temperature is 20°C. (a) What is the mean distance across the leaf in the direction of the wind? (b) What is the boundary layer thickness? What would δ^{bl} be, assuming that the mean distance is the diameter? (c) What is the heat flux density conducted across the boundary layer? (d) If the net radiation balance for the leaf is 300 W m^{-2}, what is the transpiration rate such that the leaf temperature remains constant?

7.3 Suppose that the global radiation absorbed by the ground below some vegetation averages 100 W m^{-2}. We will assume that the bulk of the vegetation is at 22°C, the top of the soil is at 20°C, and that both emit like ideal blackbodies. (a) What is the net radiation balance for the soil? (b) Suppose that there are 4 plants/m^2 of ground and that their average stem diameter is 3 cm. If the thermal conductivity coefficient of the stem is the same as that of water, and the temperature changes from that of the bulk of the vegetation to that of the ground in 0.8 m, what is the rate of heat conduction in W down each stem? What is the average value of such J_H^C per m^2 of the ground? (c) Suppose that the 4 mm of air immediately above the ground acts like a boundary layer and that the air temperature at 4 mm is 21°C. What is the rate of heat conduction from the soil into the air? (d) What is J_H^C into the soil in the steady state if 0.3 mmol m^{-2} s^{-1} of water evaporates from the upper part of the soil where the radiation is absorbed? (e) If K^{soil} is the same as K^{water}, what is $\partial T/\partial z$ in the upper part of the soil?

7.4 Let us consider a spherical cactus 0.2 m in diameter with essentially no stem mass below ground. The surface of the cactus averages 25.0°C and is 50% shaded by spines. Assume that the ambient wind speed is 1.0 m s^{-1}, the ambient air temperature is 20.0°C, the global irradiation with the sun overhead is 1 000 W m^{-2}, the effective temperature of the surroundings (including the sky) is -20°C, $a_{IR} = e_{IR} = 0.97$, $r_{surr} = 0.00$, $a_{spine} = 0.70$, and the spines have no transpiration or heat storage. (a) What is the mean boundary layer thickness for the stem and the heat conduction across it? (b) What is the stem heat convection coefficient? (c) Assuming that the spines can be represented by cylinders 1.2 mm in diameter, what is their boundary layer thickness? (d) If the net radiation balance averaged over the spine surface is due entirely to shortwave irradiation (i.e., $IR_{absorbed} = IR_{emitted}$), and if the maximum shortwave irradiation measured perpendicular to the cylinder surface is 100 W m^{-2}, what is the mean T^{spine} to within 0.1°C? Ignore spine heat conduction to the stem and assume that spines do not transpire. (e) What is the absorbed-minus-emitted longwave radiation at the stem surface in the presence and the absence of spines? (f) Assume that 30% of the incident shortwave is absorbed by the stem surface for the plant with spines. What is the net energy balance averaged over the stem surface? What is the hourly change in mean tissue temperature? Assume that the volumetric heat capacity is 80% of that of water, and ignore transpiration.

References

Arya, S. P. 1988. *Introduction to Micrometeorology.* Academic Press, San Diego, California.

Balding, F. R., and G. L. Cunningham. 1976. A comparison of heat transfer characteristics of simple and pinnate leaf models. *Botanical Gazette 137*:65–74.

Campbell, G. S. 1977. *An Introduction to Environmental Biophysics.* Springer-Verlag, Heidelberg.

Cionco, R. M. 1972. Intensity of turbulence within canopies with simple and complex roughness elements. *Boundary-Layer Meteorology 2*:453–465.

Cowan, I. R. 1968. Mass, heat and momentum exchange between stands of plants and their atmospheric environment. *Quarterly Journal of the Royal Meteorological Society 94*:523–544.

deVries, D. A. 1975. Heat transfer in soils. In *Heat and Mass Transfer in the Biosphere,* Part 1, *Heat and Mass Transfer in Processes in Plant Environment,* D. A. de Vries and N. H. Afgan, eds. Halsted Press, Wiley, New York. Pp. 5–28.

Ehleringer, J. R. 1989. Temperature and energy budgets. In *Plant Physiological Ecology: Field Methods and Instrumentation,* R. W. Pearcy, J. Ehleringer, H. A. Mooney, and P. W. Rundel, eds. Chapman and Hall, London. Pp. 117–135.

Ehleringer, J. R., and I. N. Forseth. 1989. Diurnal leaf movements and productivity in canopies. In *Plant Canopies: Their Growth, Form and Function,* G. Russell, B. Marshall, and P. G. Jarvis, eds. Cambridge University Press, Cambridge, England. Pp. 129–142.

Ehleringer, J. R., and H. A. Mooney. 1978. Leaf hairs: Effects on physiological activity and adaptive value to a desert shrub. *Oecologia 37*:183–200.

Gates, D. M. 1970. Physical and physiological properties of plants. In *Remote Sensing.* National Academy of Sciences, Washington, D.C. Pp. 224–252.

Gates, D. M. 1980. *Biophysical Ecology.* Springer-Verlag, New York.

Gates, D. M., and L. E. Papian. 1971. *Atlas of Energy Budgets of Plant Leaves.* Academic Press, New York.

Gottschlich, D. E., and A. P. Smith. 1982. Convective heat transfer characteristics of toothed leaves. *Oecologia 53*:418–420.

Grace, J. 1977. *Plant Responses to Wind*. Academic Press, London.

Incropera, F. P., and D. P. DeWitt. 1985. *Fundamentals of Heat and Mass Transfer*, 2nd ed. Wiley, New York.

Jaffe, M. J. 1980. Morphogenetic responses of plants to mechanical stimuli or stress. *BioScience 30*:239–243.

Jaffe, M. J. 1985. Wind and other mechanical effects in the development and behavior of plants, with special emphasis on the role of hormones. In *Hormonal Regulation of Development III*, R. P. Pharis and D. M. Reid, eds. *Encyclopedia of Plant Physiology, New Series*, Vol. 11. Springer-Verlag, Berlin. Pp. 444–484.

Jones, H. G. 1983. *Plants and Microclimate*. Cambridge University Press, Cambridge, England.

Kappen, L., O. L. Lange, E.-D. Schulze, M. Evenari, and U. Buschbom. 1979. Ecophysiological investigations on lichens of the Negev Desert. VI. Annual course of the photosynthetic production of *Ramalina maciformis* (Del.) Bory. *Flora 168*:85–108.

Knutson, R. M. 1974. Heat production and temperature regulation in eastern skunk cabbage. *Science 186*:746–747.

Koller, D. 1986. Yearly review: The control of leaf orientation by light. *Photochemistry and Photobiology 44*:819–826.

Kreith, F., and M. S. Bohm. 1986. *Principles of Heat Transfer*, 4th ed. Harper & Row, New York.

Leuning, R. 1987. Leaf temperatures during radiation frost part II. A steady state theory. *Agricultural and Forest Meteorology 42*:135–155.

Levitt, J. 1980. *Responses of Plants to Environmental Stresses*, 2nd ed., Vol II, *Water, Radiation, Salt, and Other Stresses*. Academic Press, New York.

Lewis, D. A., and P. S. Nobel. 1977. Thermal energy exchange model and water loss of a barrel cactus, *Ferocactus acanthodes*. *Plant Physiology 60*:609–616.

Monteith, J. L. 1973. *Principles of Environmental Physics*. American Elsevier, New York.

Monteith, J. L. 1981. Coupling of plants to the atmosphere. In *Plants and Their Atmospheric Environment*, J. Grace, E. D. Ford, and P. G. Jarvis, eds. Blackwell, Oxford, England. Pp. 1–29.

Mooney, H. A., and J. R. Ehleringer. 1978. The carbon gain benefits of solar tracking in a desert annual. *Plant, Cell and Environment 1*:307–311.

Munn, R. E. 1966. *Descriptive Micrometeorology*. Academic Press, New York.

Nobel, P. S. 1974. Boundary layers of air adjacent to cylinders. Estimation of effective thickness and measurements on plant material. *Plant Physiology 54*:177–181.

Nobel, P. S. 1975. Effective thickness and resistance of the air boundary layer adjacent to spherical plant parts. *Journal of Experimental Botany 26*:120–130.

Nobel, P. S. 1980. Morphology, surface temperatures, and northern limits of columnar cacti in the Sonoran desert. *Ecology 61*:1–7.

Nobel, P. S. 1981. Wind as an ecological factor. In *Physiological Plant Ecology*, O. L. Lange, P. S. Nobel, C. B. Osmond, and H. Ziegler, eds. *Encyclopedia of Plant Physiology, New Series*, Vol. 12A. Springer-Verlag, Berlin. Pp. 475–500.

Nobel, P. S. 1988. *Environmental Biology of Agaves and Cacti.* Cambridge University Press, New York.

Oke, T. R. 1987. *Boundary Layer Climates*, 2nd ed. Methuen, London.

Parkhurst, D. F., and O. L. Loucks. 1972. Optimal leaf size in relation to environment. *Journal of Ecology 60*:505–537.

Pearman, G. I., H. L. Weaver, and C. B. Tanner. 1972. Boundary layer heat transfer coefficients under field conditions. *Agricultural Meteorology 10*:83–92.

Robinson, N. ed. 1966. *Solar Radiation.* Elsevier, Amsterdam.

Rose, C. W. 1966. *Agricultural Physics.* Pergamon, Oxford, England.

Rosenberg, N. J., B. L. Blad, and S. B. Verma. 1983. *Microclimate: The Biological Environment*, 2nd ed. Wiley, New York.

Rundel, P. W. 1981. Fire as an ecological factor. In *Physiological Plant Ecology*, O. L. Lange, P. S. Nobel, C. B. Osmond, and H. Ziegler, eds. *Encyclopedia of Plant Physiology, New Series*, Vol. 12A. Springer-Verlag, Berlin. Pp. 501–538.

Schlichting, H. 1979. *Boundary–Layer Theory*, 7th ed. McGraw-Hill, New York.

Smart, R. E., and T. R. Sinclair. 1976. Solar heating of grape berries and other spherical fruits. *Agricultural Meteorology 17*:241–259.

Smith, W. K., and P. S. Nobel. 1977. Temperature and water relations for sun and shade leaves of a desert broadleaf, *Hyptis emoryi. Journal of Experimental Botany 28*:169–183.

Taylor, S. E. 1975. Optimal leaf form. In *Perspectives of Biophysical Ecology*, D. M. Gates and R. B Schmerl, eds. Springer-Verlag, Heidelberg. Pp. 73–86.

van Wijk, W. R., ed. 1966. *Physics of Plant Environment*, 2nd ed. North-Holland, Amsterdam.

Vogel, S. 1970. Convective cooling at low airspeeds and the shapes of broad leaves. *Journal of Experimental Botany 21*:91–101.

Vogel, S. 1981. *Life in Moving Fluids; the Physical Biology of Flow.* Willard Grant Press, Boston, Massachusetts.

Vogel, S. 1989. Drag and reconfiguration of broad leaves in high winds. *Journal of Experimental Botany 40*:941–948.

Wheeler, W. W., and M. Neushul. 1981. The aquatic environment. In *Physiological Plant Ecology*, O. L. Lange, P. S. Nobel, C. B. Osmond, and H. Ziegler, eds. *Encyclopedia of Plant Physiology, New Series*, Vol. 12A. Springer-Verlag, Berlin. Pp. 229–247.

Woolley, J. T. 1971. Reflectance and transmittance of light by leaves. *Plant Physiology 47*:656–662.

CHAPTER 8

Leaves and Fluxes

In this chapter we will draw upon previously introduced topics to discuss gas fluxes for leaves. Specifically, we will reconsider the transpiration of water and the photosynthetic fixation of carbon dioxide. The driving forces for such fluxes are differences in CO_2 and H_2O concentrations or mole fractions. We will quantitatively discuss the movement of gases into and out of leaves in terms of the various resistances or conductances involved. The use of resistances to describe gas fluxes quantitatively was first applied to leaves by Brown and Escombe in 1900 (Brown and Escombe, 1900). Over the years, consideration of resistance networks has become very convenient for specifying which parts of the pathway are most limiting for photosynthesis or transpiration. Recently, use of two distinctly different forms of conductance has become more popular, especially for describing transpiration, and we will consider both forms. We will see that greater stomatal opening may be an advantage for photosynthesis, but can result in excessive transpiration. Consequently, a benefit/cost index, such as the amount of CO_2 fixed per unit of water lost, can be important for evaluating ecological aspects of gas exchange. Figure 8.1 indicates how the flux densities of water vapor and CO_2 can be measured for a leaf. Although we will usually be referring to leaves, the discussion of gas fluxes is also applicable to stems, flower petals, and other plant parts.

Figure 8.1

Experimental arrangement for the measurement of leaf transpiration and photosynthesis. The water vapor and CO_2 content of the gas entering a transparent chamber enclosing a leaf is compared with that leaving. If the water vapor concentration goes from 0.6 mol m^{-3} entering to 1.0 mol m^{-3} leaving for a gas flow rate of 10^{-5} m^3 s^{-1} (10 cm^3 s^{-1}), then J_{wv} for a leaf of area 10^{-3} m^2 (10 cm^2) would be (1.0 mol m^{-3} − 0.6 mol m^{-3})(10^{-5} m^3 s^{-1})/(10^{-3} m^2), or 0.004 mol m^{-2} $^{-1}$. [See Field *et al.* (1989) and Leuning (1983) for refinements of measurement technique and analysis.]

RESISTANCES AND CONDUCTANCES—TRANSPIRATION

The resistances and conductances we will discuss in this section are those encountered by water vapor as it diffuses from the pores in the cell walls of mesophyll cells or other sites of water evaporation out into the turbulent air surrounding a leaf. We will define these quantities for the intercellular air spaces, the stomata, the cuticle, and the boundary layer next to a leaf. As explained later in this chapter, CO_2 diffuses across the same gaseous-phase resistances or conductances as does water vapor, and also across a number of other components in the liquid phases of mesophyll cells.

Throughout this book we have used equations of the following general form: flux density = proportionality coefficient × force (e.g., pp. 11, 121). The proportionality coefficients in such flux density expressions are measures of *conductivity*. To represent force, instead of using the gradient in chemical potential—which, as we noted in Chapter 3, is a very general force—we often use quantities that are more convenient experimentally, such as differences in concentration (consider Fick's first law, Eq. 1.1; see pp. 11 and 124), partial pressure, or mole fraction. Such "forces," however, do not have the proper units for force: so, to be correct, the proportionality coefficient is not the conductivity, but rather the *conductance*:

$$\text{flux density} = \text{conductivity} \times \text{force} \qquad (8.1a)$$

$$\text{flux density} = \text{conductance} \times \text{"force"} \qquad (8.1b)$$

We might equally well choose an alternative form for the relation between forces and fluxes: flux density = force/resistivity, where *resistivity* is the reciprocal of conductivity. In turn, we can define a *resistance,* which is the reciprocal of conductance:

$$\text{flux density} = \frac{\text{force}}{\text{resistivity}} \qquad (8.1c)$$

$$\text{flux density} = \frac{\text{"force"}}{\text{resistance}} \qquad (8.1d)$$

To help understand the difference between resistivity and resistance, we will consider the electrical usage of these terms. Electrical resistivity is a fundamental physical property of a material; e.g., such resistivities are tabulated in suitable handbooks. Electrical resistance, on the other hand, describes a particular component in an electrical circuit, i.e., a particular piece of material. We can readily measure the magnitude of resistance in the laboratory or purchase a resistor of known resistance in an electronics store. Electrical resistivity, ρ, generally has the units of ohm m, whereas electrical resistance, $R = \rho \, \Delta x / A$ (see p. 121), is expressed in ohms. Thus, besides differing conceptually (compare Eq. 8.1c and d), resistance and resistivity differ in their units. In fact, much of our attention in this chapter will be devoted to the units for variables such as resistance, because many different systems are in use in plant physiology, ecology, meteorology, soil science, and other related disciplines.

We will initially consider both the resistance and the conductance of a particular component, and we will present the expressions for both resistance and conductance for series versus parallel arrangements of components. Because a flux density is directly proportional to conductance (Eq. 8.1b) but inversely proportional to resistance (Eq. 8.1d), conductance terminology is often more convenient for discussing plant responses to environmental factors. Also, resistance has no upper limit—i.e., it varies from some minimal value to infinity—whereas conductance varies between zero and some maximum value. However, resistance terminology can be easier to use when a substance must cross a series of components in sequence, such as CO_2 diffusing across the cell wall, plasmalemma, cytosol, and chloroplast membranes. To help familiarize the reader with the various conventions found in the literature, we will use both systems, emphasizing conductances for transpiration and resistances for photosynthesis.

Boundary Layer Adjacent to Leaf

As a starting point for our discussion of gas fluxes across air boundary layers, let us consider the one-dimensional form of Fick's first law of diffusion, $J_j = -D_j \, \partial c_j / \partial x$ (Eq. 1.1). As in Chapter 1, we will replace the concentration gradient by the difference in concentration across some distance. In effect, we are considering cases not too far from equilibrium, so the flux density depends linearly on the force, and the force can be represented by the difference in concentration. The distance that interests us here is the one occurring across the air boundary layer adjacent to the surface of a leaf, δ^{bl} (see Ch. 7 for a discussion of boundary layers and equations for their thickness). Consequently, Fick's first law assumes the following form for the diffusion of species j:

$$J_j = -D_j \frac{\partial c_j}{\partial x} = D_j \frac{\Delta c_j}{\Delta x} = D_j \frac{\Delta c_j^{bl}}{\delta^{bl}}$$

$$= g_j^{bl} \, \Delta c_j^{bl} = \frac{\Delta c_j^{bl}}{r_j^{bl}} \tag{8.2}$$

Equation 8.2 shows how the net flux density of species j depends on its diffusion coefficient D_j and on the difference in its concentration Δc_j^{bl} across a distance δ^{bl} of the air. The net flux density J_j is toward regions of lower c_j, which is apparent from the negative sign associated with the concentration gradient and otherwise is incorporated into the definition of Δc_j in Equation 8.2; we will specifically consider the diffusion of water vapor and CO_2 toward lower concentrations in this chapter. Also, we will assume that the same boundary layer thickness δ^{bl} derived for heat transfer (Eqs. 7.8 through 7.13) applies for mass transfer, an example of the similarity principle. Outside δ^{bl} is a region of air turbulence, where we will assume that the concentrations of gases are the same as in the bulk atmosphere (an assumption that we will remove in the next chapter). Equation 8.2 indicates that J_j equals Δc_j^{bl} multiplied by a conductance, g_j^{bl}, or divided by a resistance, r_j^{bl}, as we will consider next.

The air boundary layers on both sides of a leaf represent important barriers in the pathways for the entry of CO_2 and the exit of H_2O, as was clearly shown by Raschke in the 1950s (see Raschke, 1956). Movement of gas molecules across these layers is by diffusion in response to differences in concentration. Using Equation 8.2, we can represent the conductance and the resistance of a boundary layer of air as follows:

$$g_j^{bl} = \frac{J_j}{\Delta c_j^{bl}} = \frac{D_j}{\delta^{bl}} = \frac{1}{r_j^{bl}} \tag{8.3}$$

Because the SI unit for the diffusion coefficient D_j is $m^2\ s^{-1}$ (see Ch. 1) and the thickness of the boundary layer δ^{bl} is in m, g_j^{bl} is in $(m^2\ s^{-1})/(m)$, or $m\ s^{-1}$ (values are often expressed in $mm\ s^{-1}$), and r_j^{bl} is in $s\ m^{-1}$. J_j is expressed per unit leaf area, so g_j^{bl} and r_j^{bl} also relate to unit area of a leaf.

We note that D_j is a fundamental measure of conductivity (values available in suitable handbooks) describing the diffusion of species j in a given medium. On the other hand, δ^{bl} characterizes a particular situation, because the thickness of the boundary layer depends on wind speed and leaf size (see Ch. 7, e.g., Eq. 7.8). Thus, r_j^{bl} as defined by Equation 8.3 describes a particular component of the path, analogous to the resistance (R) used in Ohm's law. Recalling Equation 1.9, $P_j = D_j K_j/\Delta x$, we recognize that D_j/δ^{bl} is the permeability coefficient for species j, P_j, as it diffuses across an air boundary layer of thickness δ^{bl}. Thus, g_j^{bl} as defined by Equation 8.3 is equal to P_j. When something readily diffuses across a boundary layer, P_j and g_j^{bl} are large and r_j^{bl} is small. Using resistances and conductances, we can describe gas fluxes into and out of leaves employing a number of relations originally developed for the analysis of electrical circuits (see Gaastra, 1959; J. L. Monteith in Evans, 1963; Lake, 1967).

We will next estimate values that might be expected for g_{wv}^{bl} and r_{wv}^{bl}—the conductance and the resistance, respectively—for water vapor diffusing across the boundary layer of air next to a leaf. In Chapter 7 we indicated that the boundary layer thickness in mm ($\delta_{(mm)}^{bl}$) for a flat leaf under field conditions is $4.0\sqrt{l_{(m)}/v_{(m\ s^{-1})}}$ (Eq. 7.10), where $l_{(m)}$ is the mean length of the leaf in the direction of the wind in m and $v_{(m\ s^{-1})}$ is the ambient wind speed in m s^{-1}. Let us consider a relatively thin boundary layer of 0.3 mm and a thick one of 3 mm (see Table 7.2 for the values of $l_{(m)}$ and $v_{(m\ s^{-1})}$ that this implies). For water vapor diffusing in air at 20°C, D_{wv} is 2.4 × 10^{-5} m^2 s^{-1} (App. I). Using Equation 8.3, we obtain for the thin boundary layer:

$$g_{wv}^{bl} = \frac{(2.4 \times 10^{-5}\ m^2\ s^{-1})}{(0.3 \times 10^{-3}\ m)} = 8 \times 10^{-2}\ m\ s^{-1} = 80\ mm\ s^{-1}$$

$$r_{wv}^{bl} = \frac{(0.3 \times 10^{-3}\ m)}{(2.4 \times 10^{-5}\ m^2\ s^{-1})} = 13\ s\ m^{-1}$$

For the thick boundary layer, g_{wv}^{bl} equals 8 mm s^{-1} and r_{wv}^{bl} equals 130 s m^{-1} (Table 8.1).[1] Usually, boundary layer conductances are larger and resistances are smaller than their respective values for diffusion along the stomatal pores, which we will examine next.

Stomata

As we indicated in Chapter 1, stomata help control the exit of water vapor from leaves and the entry of CO$_2$ into them. Although the epidermal cells occupy a much greater fraction of the leaf surface area than do the stomatal pores, the waxy cuticle covering the outer surface of the epidermal cells greatly reduces the water loss from their cell walls to the turbulent air surrounding a leaf, the *cuticular* transpiration. The usual pathway for water vapor leaving a leaf during transpiration is therefore through the stomata.

The stomatal aperture is controlled by the confirmation of the two guard cells surrounding a pore (see Figs. 1.2 and 8.2). These cells are generally kidney-

---

1. Instead of estimating the boundary layer conductance or resistance based on δ^{bl} calculated by Equation 7.8, which cannot account for all the intricacies of different leaf shapes, it is often more expedient to construct a filter-paper replica of the leaf; if this "leaf" is then moistened, the observed J_{wv} from it for a certain Δc_{wv}^{bl} will indicate g_{wv}^{bl} or r_{wv}^{bl} (see Eq. 8.2), because water vapor will cross only a boundary layer in this case.

Table 8.1

Summary of representative values of conductances and resistances for water vapor diffusing out of leaves. See text for specific calculations (conductances in mmol m^{-2} s^{-1} are from Eq. 8.8), as well Farquhar and von Caemmerer (1982), Körner *et al.* (1979), and Zelitch (1971).

Component condition	Conductance		Resistance	
	mm s^{-1}	mmol m^{-2} s^{-1}	s m^{-1}	m^2 s mol^{-1}
Boundary layer				
thin	80	3 200	13	0.3
thick	8	320	130	3
Stomata				
large area—open	19	760	53	1.3
small area—open	1.7	70	600	14
closed	0	0	∞	∞
mesophytes—open	4 to 20	160 to 800	50 to 250	1.3 to 6
xerophytes and trees—open	1 to 4	40 to 160	250 to 1 000	6 to 25
Cuticle				
crops	0.1 to 0.4	4 to 16	2 500 to 10 000	60 to 250
many trees	0.05 to 0.2	2 to 8	5 000 to 20 000	125 to 500
many xerophytes	0.01 to 0.1	0.4 to 4	10 000 to 100 000	250 to 2 500
Intercellular air spaces				
calculation	24 to 240	1 000 to 10 000	4.2 to 42	0.1 to 1
waxy layer				
typical	50 to 200	2 000 to 8 000	5 to 20	0.1 to 0.5
certain xerophytes	10	400	100	2.5
typical	40 to 100	1 600 to 4 000	10 to 25	0.2 to 0.6
Leaf (lower surface)				
crops—open stomata	2 to 10	80 to 400	100 to 500	2.5 to 13
trees—open stomata	0.5 to 3	20 to 120	300 to 2 000	8 to 50

shaped (dumbbell-shaped in grasses), may be 40 μm long, and, unlike ordinary epidermal cells, usually contain chloroplasts. When the guard cells are relatively flaccid, the stomatal pore is nearly closed, as is the case for most plants at night. Upon illumination, guard cells take up K$^+$, which may increase in concentration by 0.3 to 0.6 M in a cell (see Fischer, 1972; Jarvis and Mansfield, 1981; MacRobbie,

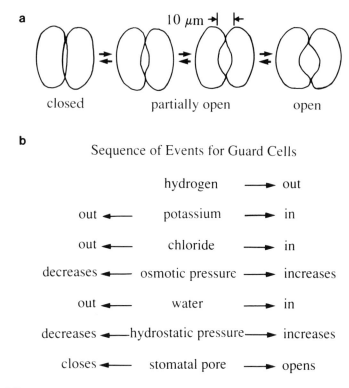

Figure 8.2

Schematic representation of the opening and closing movements for stomatal pores: (a) Pair of guard cells as viewed toward leaf surface; and (b) cellular events involved. Arrows to the right are for stomatal opening and those to the left for closing. (Modified from A. C. Gibson and P. S. Nobel. 1986. *The Cactus Primer*. Harvard University Press, Cambridge, Massachusetts. Used by permission.)

1988; Raschke, 1975). The K^+ uptake raises the internal osmotic pressure and thus lowers the internal water potential; water then spontaneously flows from the epidermal cells into the guard cells (Fig. 8.2). This water entry leads to an increase in the internal hydrostatic pressure of a pair of guard cells, which in turn causes them to expand and their opposed cell walls (those on either side of the pore) to become concave. As the kidney-shaped guard cells thus bow outward, an elliptical pore develops between the two cells. The formation of this pore is a consequence of the anisotropic properties of the cell wall surrounding each guard cell (see Aylor *et al.*, 1973; Sharpe *et al.*, 1987). The distance between the guard cells across the open pore (the pore "width") is generally 5 to 15 μm, and the major axis of the elliptical pore may be about 20 μm.

The water relations of the special epidermal cells immediately surrounding the guard cells, which are referred to as subsidiary cells, are also crucial for stomatal opening (see Cowan, 1972, 1977; Tyree and Yianoulis, 1980). For instance, solutes such as K^+ can move from the subsidiary cells to the guard cells, causing water to leave the subsidiary cells and their internal hydrostatic pressure to decrease while P^i of the guard cells increases. Once the stomatal pore has opened, water evaporates from the inner side of the guard cells and subsidiary cells and then diffuses out of the leaf. This lowers the water potential in the cell walls, which in turn can cause water to leave the protoplasts of the guard cells and subsidiary cells, thereby lowering their P^i and causing some stomatal closure (local water flow depends on the local Ψ and leads to different kinetics for the changes of P^i in guard cells versus subsidiary cells). The ensuing partial stomatal closure reduces the water loss from the leaf and hence reduces evaporation, thereby allowing P^i to build back up in the guard cells and subsidiary cells, leading to re-opening of the stomata. The resulting oscillation of stomatal pore aperture can have a period of about 30 to 60 min, although it is often damped out by changes in environmental factors such as wind and irradiation.

What controls the opening of stomatal pores? This question has proved difficult to answer, in part because a number of factors are involved (see Zeiger, 1983). An initial event apparently is the active H^+ extrusion from the guard cells, which lowers the electrical potential inside relative to outside (i.e., hyperpolarizes the membrane potential, E_M; see Ch. 3) as well as lowers the internal concentration of H^+ (i.e., raises the internal pH). The lowered E_M favors passive K^+ uptake, the latter most likely by K^+ channels that are opened by the hyperpolarization of the plasmalemma (Ch. 3; see also MacRobbie, 1988). After stomatal opening is initiated, the added K^+ within the guard cells is electrically balanced—partly by a Cl^- influx (perhaps via an OH^- antiporter or more likely by an H^+ symporter; see p. 157) and partly by the production of organic anions like malate in the guard cells. For most plants, malate is the main counterion for K^+, but Cl^- appears to be the main anion in those few species whose guard cells lack chloroplasts. For some plants stomatal opening increases with the light level only up to a photosynthetic photon flux density (PPFD) of about 200 μmol m^{-2} s^{-1}, but for others it may increase all the way up to full sunlight (2 000 μmol m^{-2} s^{-1}).

The degree of stomatal opening often depends on the CO_2 concentration in the guard cells, which reflects their own carbohydrate metabolism as well as the CO_2 level in the air within the leaf (see Cowan, 1977; Hall, 1982; Raschke, 1975). For instance, upon illumination, the CO_2 concentration in the leaf intercellular air spaces is decreased by photosynthesis, resulting in decreased CO_2 levels in the guard cells, which somehow triggers stomatal opening. CO_2 can then enter the leaf and photosynthesis can continue (to be discussed again near the end of this chapter). In the dark, respiration generally leads to relatively high CO_2 levels in the leaves, which triggers stomata to close. Lowering the CO_2 concentration in the ambient air can induce stomatal opening in the *dark* (again as a response to the low CO_2 level in the guard cells), which indicates that the energy for opening can be supplied by respiration. Stomata can also respond directly to light, which stimulates proton extrusion from guard cells, independent of the response to CO_2; the light responses involve both the absorption of PPFD by chloroplasts in the guard cells as well as another system that absorbs in the blue region and that is sensitive to low photon flux densities (see Zeiger, 1983). In addition, stomata often tend to close partially as the relative humidity of the ambient air decreases, another means of regulating water loss. If a constant water vapor concentration difference from inside the leaf to the surrounding air is maintained, stomata in the light generally tend to open with increasing temperature up to temperatures optimal for photophosphorylation, about 35°C; in the dark, the temperature-induced stomatal opening continues up to higher temperatures, possibly reflecting the higher optimal temperature for oxidative phosphorylation than for photophosphorylation. Stomatal movements can also be controlled by hormones, including those produced in the roots as well as in the leaves (see Schulze, 1986). For instance, abscisic acid increases in illuminated leaves during periods of water stress. This leads to stomatal closure, which conserves water, although at the expense of a decrease in photosynthesis. The halftimes for stomatal movements are generally 5 to 20 min, with closing usually more rapid than opening.

Currently, less is known about what triggers stomatal closure compared with stomatal opening (see MacRobbie, 1988; Schroeder and Hedrich, 1989). An initial event may be the opening of Ca^{2+} channels in the plasmalemma of the guard cells, causing Ca^{2+} entry into the cytosol and a depolarization of the plasmalemma. This depolarization and the increased cytosolic Ca^{2+} concentration apparently cause anion channels to open, allowing Cl^- and possibly malate to move out across the plasmalemma, leading to further depolarization. Such depolarization opens K^+ channels, so K^+ also passively moves out of the guard cells. The decreased osmotic pressure causes water to move out, which in turn causes the hydrostatic pressure to decrease, leading to stomatal closure (Fig. 8.2).

For leaves of dicots, stomata are usually more numerous on the lower surface than on the upper one. In many dicots, stomata may even be nearly absent from the upper epidermis. On the other hand, most grasses and certain other plants with vertically oriented leaves have approximately the same number of stomata per unit area on each side. A frequency of 40 to 300 stomata per mm^2 is representative for the lower surface of most leaves. The pores of the open stomata generally occupy 0.2% to 2% of the leaf surface area. Thus, the area for diffusion of gases through the stomatal pores in the upper or lower epidermis of a leaf, A^{st}, is much less than the leaf surface area, A (see Meidner and Mansfield, 1968).

Stomatal Conductance and Resistance

We can apply Fick's first law in the form $J_j = D_j \, \Delta c_j / \Delta x$ (Eq. 8.2) to describe the diffusion of gases through stomatal pores. Let the depth of the stomatal pore be δ^{st} (see Fig. 8.3); the concentration of species j changes by Δc_j^{st} along the distance δ^{st}. For the steady state and ignoring the cuticular pathway, the amount of species j moving per unit time toward or away from the leaf (J_j times the leaf area A) must equal the amount of species j moving per unit time through the stomata (the flux density within the stomata, $D_j \, \Delta c_j^{st} / \delta^{st}$, times the stomatal area A^{st} that occurs for leaf area A). Recognition of this extremely important constricting effect that the stomata have on the area available for the diffusion of gas molecules gives us the following relations:

$$J_j A = D_j \frac{\Delta c_j^{st}}{\delta^{st}} A^{st} \qquad (8.4a)$$

or

$$J_j = D_j \frac{\Delta c_j^{st}}{\delta^{st}} \frac{A^{st}}{A} = \frac{D_j na^{st}}{\delta^{st}} \Delta c_j^{st} \qquad (8.4b)$$

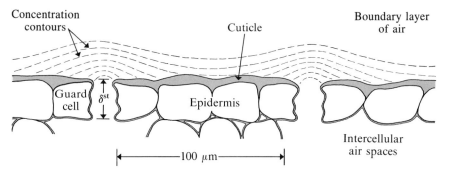

Figure 8.3

Concentration contours of water vapor outside open stomata.

where n is the number of stomata per unit area of the leaf and a^{st} is the average area per stomatal pore; i.e., na^{st} equals the fraction of the leaf surface area occupied by stomatal pores, A^{st}/A. The flux density J_j in Equation 8.4 refers to the rate of movement of species j *per unit area of the leaf*, a quantity that is considerably easier to measure than is the flux density within a stomatal pore.

The area available for water vapor diffusion abruptly changes from A^{st} to A at the leaf surface. On the other hand, the three-dimensional surfaces of equal concentration must fan out from each stomatal pore (Fig. 8.3). This geometrical aspect could introduce considerable complications, but fortunately we can still use a one-dimensional form for Fick's first law to describe gases moving across the boundary layer next to a leaf, although we need to make an "end correction" to allow for the diffusion pattern at the end of the stomatal pores. The distance between stomata is often about 100 μm, which is considerably less than the thicknesses of nearly all boundary layers (see Table 7.2). Thus, the three-dimensional concentration contours from adjacent stomata tend to overlap in the boundary layer (Fig. 8.3). Because of this, the concentration of water vapor varies only slightly in planes parallel to the leaf surface, but changes substantially in the direction perpendicular to the leaf surface, and therefore we can generally use a one-dimensional form of Fick's first law (see Cooke and Rand, 1980; Meidner and Mansfield, 1968; Penman and Schofield, 1951). However, the concentration patterns on both ends of a stomatal pore cause the pore to have an effective depth greater than δ^{st} by about the mean, or effective, "radius" of the pore, r^{st}.[2] Because stomatal pores are approximately elliptical, not circular, we will define r^{st} as follows: $\pi(r^{st})^2 = a^{st}$, where a^{st} is the effective area of the pore.

For our subsequent applications to transpiration and photosynthesis, we will define a stomatal conductance, g_j^{st}, and resistance, r_j^{st}, for the diffusion of species j. We have previously described analogous quantities for the boundary layer (Eq. 8.3) based on a form of Fick's first law for the diffusion of species j. Here we will use Equation 8.4 to define the conductance and the resistance for the diffusion of gases along the length of the stomata:

$$g_j^{st} = \frac{J_j}{\Delta c_j^{st}} = \frac{D_j na^{st}}{\delta^{st} + r^{st}} = \frac{1}{r_j^{st}} \tag{8.5}$$

---

2. Formulas for the end correction due to the three-dimensional nature of the concentration gradients at each end of a stomatal pore vary and are more complicated than that used here, r^{st} (see Cooke and Rand, 1980; Meidner and Mansfield, 1968; Monteith, 1973; Parlange and Waggoner, 1970). However, the differences among the various correction formulas are generally relatively small compared with δ^{st}.

Equation 8.5 incorporates the effective depth of the stomatal pore, $\delta^{st} + r^{st}$, where r^{st} is the mean pore radius just introduced. When the width of a stomatal pore is about 0.1 to 0.3 μm, as occurs when the pores are nearly closed, the mean free path for molecules diffusing in air is about the same as the dimensions of the opening.[3] In this case, the molecular interactions with the sides of the stomatal pore are important, and this affects the value of D_j. Even though we will ignore this interaction and make certain geometrical approximations for δ^{st}, r^{st}, and a^{st}, Equation 8.5 is still quite useful for estimating stomatal conductances and resistances (for a general treatment, including the effects of water vapor movements on the stomatal CO_2 conductance, see Field *et al.*, 1989; and especially Leuning, 1983).

Let us now calculate some values of the stomatal conductance for the diffusion of water vapor. We will consider air at 20°C for which D_{wv} is 2.4×10^{-5} m^2 s^{-1}. The stomatal conductance tends to be high when a large portion of the leaf surface area (e.g., $na^{st} = 0.02$) is occupied by open stomata of relatively short pore depth (e.g., $\delta^{st} = 20$ μm). If we assume that the mean radius r^{st} is 5 μm, Equation 8.5 predicts that the conductance is then

$$g^{st}_{wv} = \frac{(2.4 \times 10^{-5} \text{ m}^2 \text{ s}^{-1})(0.02)}{(20 \times 10^{-6} \text{ m} + 5 \times 10^{-6} \text{ m})}$$

$$= 1.9 \times 10^{-2} \text{ m s}^{-1} = 19 \text{ mm s}^{-1}$$

At the other extreme, the open stomata may occupy only 0.4% of the lower surface of a leaf ($na^{st} = 0.004$), and the pore depth may be rather large, e.g., 50 μm. Again assuming that r^{st} is 5 μm, the stomatal conductance calculated using Equation 8.5 is 1.7 mm s^{-1}, a rather small value for open stomata (Table 8.1).

The stomatal conductance is usually 2 to 20 mm s^{-1} for water vapor diffusing out through the open stomata of most mesophytes (r^{st}_{wv} of 50 to 500 s m^{-1}). In general, agricultural crops tend to have high values of g^{st}_{wv}, whereas for certain xerophytes and many trees with open stomata it may be only 1 mm s^{-1} (Table 8.1). Some xerophytes have sunken stomata leading to another conductance in series with g^{st}_{wv}, which slightly decreases the overall conductance for water loss by transpiration. Of course, as the stomatal pores close, the conductance decreases accordingly, because g^{st}_{wv} is proportional to the stomatal area by Equation

---

3. The mean free path of a gas molecule, which is the average distance such a molecule travels before colliding with another gas molecule, is about 0.07 μm at pressures and temperatures normally experienced by plants.

8.5. Because g_{wv}^{st} is essentially the only conductance in the whole diffusion pathway that is variable over a wide range, changes of the openings of the stomatal pores provide a plant with an important control mechanism for regulating the movement of gases into or out of the leaves.

Cuticle

Some water molecules diffuse across the waxy cuticle of the epidermal cells, a process called *cuticular transpiration*. This movement of H_2O across the cuticle probably involves liquid water as well as water vapor. We will identify a cuticular conductance, g_j^c, and resistance, r_j^c, for the diffusion of species j across the cuticle. These quantities are in parallel with the analogous quantities for the stomata—i.e., species j can leave the leaf either by crossing the cuticle or by going through the stomatal pores (see Fig. 1.2). The conductance for cuticular transpiration, g_{wv}^c, generally ranges from 0.05 to 0.3 mm s^{-1} for different species, although it can be considerably lower for xerophytes with thick cuticles that greatly restrict water loss (Table 8.1). Thus, g_{wv}^c is usually much smaller than is g_{wv}^{st} for open stomata. When the stomata are nearly closed (low g_{wv}^{st}), the cuticular transpiration can exceed the loss of water through the stomata. If the cuticle is mechanically damaged or develops cracks, as can occur for older leaves, g_{wv}^c can be raised, and more water will then move out of the leaf by this pathway.

Intercellular Air Spaces

Another conductance encountered by the diffusion of species j in plant leaves is that of the intercellular air spaces, g_j^{ias}. Because of the irregular shape of these air spaces, which generally account for about 30% of the leaf volume, g_j^{ias} is difficult to estimate accurately from geometrical considerations alone. However, the intercellular air spaces do act as an unstirred air layer across which substances must diffuse; hence we will describe g_j^{ias} by a relation similar to Equation 8.3 ($g_j^{bl} = D_j/\delta^{bl}$). As a first approximation, we can regard the thickness of the unstirred air layer in the present case as an average distance from the surfaces of the mesophyll cells within a leaf to the inner side of a stomatal pore, δ^{ias}. By analogy with Equations 8.3 and 8.5, we can express g_j^{ias} as

$$g_j^{ias} = \frac{D_j}{\delta^{ias}} = \frac{1}{r_j^{ias}} \qquad (8.6)$$

For convenience, we will combine a number of factors to get an effective distance, δ^{ias}. For instance, we will let δ^{ias} include the effective length of the air-filled part of the cell wall pores from the cell wall surface to the sites where the evaporation of water or the dissolving of CO_2 takes place. This length is greater than the actual distance along the pores, because we must correct for the decrease in the cross-sectional area available for diffusion, a decrease caused by the nongaseous parts of the cell wall. We will also include in δ^{ias} the effective length of the thin waxy layer that generally occurs on the surfaces of mesophyll cells. This thin waxy layer has a conductance of 50 to 200 mm s^{-1} for mesophytes, but may be much lower for certain xerophytes where cutinization of the mesophyll cells is appreciable (Table 8.1). In addition, δ^{ias} incorporates the fact that the entire cross section of the mesophyll region is not available for diffusion of gases, as the flow is constricted to the intercellular air spaces, which have a smaller cross-sectional area than the corresponding leaf area, i.e., $A^{ias}/A < 1$. Thus, that part of δ^{ias} referring to the intercellular air spaces per se equals the actual distance involved times A/A^{ias}. If the mesophyll region were one-third air by volume, A/A^{ias} would be three.

The effective length δ^{ias}, including all of the factors just enumerated, ranges from 100 μm to 1 mm for most leaves. Equation 8.6 indicates that the water vapor conductance across the intercellular air spaces then ranges from an upper limit of

$$ g_{wv}^{ias} = \frac{(2.4 \times 10^{-5} \text{ m}^2 \text{ s}^{-1})}{(100 \times 10^{-6} \text{ m})} = 0.24 \text{ m s}^{-1} = 240 \text{ mm s}^{-1} $$

for a δ^{ias} of 100 μm down to about 24 mm s^{-1} for a δ^{ias} of 1 mm ($D_{wv} = 2.4 \times 10^{-5}$ m^2 s^{-1} at 20°C; see App. I). Thus, the conductance of the intercellular air spaces is relatively large (the resistance is relatively small) compared with the other conductances encountered by gases diffusing into or out of leaves (Table 8.1).[4]

Fick's First Law and Conductances

The form of Fick's first law that we are using to describe the flux density of water vapor out of or CO_2 into a leaf is appropriate for a one-dimensional situation

---

4. CO_2 and water vapor can enter or leave cells along the length of the intercellular air spaces, which complicates the analysis somewhat (see Cooke and Rand, 1980; Parkhurst, 1977). In any case, g_j^{ias} generally is relatively large compared with the other conductances.

(e.g., $J_j = D_j \, \Delta c_j / \Delta x = g_j \, \Delta c_j = \Delta c_j / r_j$). The H_2O lost from a leaf during transpiration evaporates from the cell walls of mesophyll cells, the inner sides of guard cells, and the adjacent subsidiary cells. If the cell walls were uniform and wet, then most of the water would be expected to evaporate from the immediate vicinity of the stomatal pores (see Cooke and Rand, 1980; Cowan, 1977; Tanton and Crowdy, 1972; Tyree and Yianoulis, 1980). However, the waxy material that occurs on the cell walls within a leaf, especially on guard cells and other nearby cells, causes much of the water to evaporate from the mesophyll cells in the leaf interior (see Boyer, 1985; Nonami and Schulze, 1989). We can imagine that the water vapor moves in the intercellular air spaces toward the leaf surface by diffusing down planar fronts of successively lower concentration (see Fig. 8.3). Our imaginary planar fronts are parallel to the leaf surface, so the direction for the fluxes is perpendicular to the leaf surface. When we reach the inner side of a stomatal pore, the area for diffusion is reduced from A^{ias} to A^{st}. In other words, we are still discussing the advance of planar fronts in one dimension, but the flux is now constricted to the stomatal pores. (A small amount of water constituting the cuticular transpiration diffuses across the cuticle in parallel with the stomatal fluxes.) Finally, the movement of water vapor across the boundary layer at the leaf surface is again a one-dimensional diffusion process—in this case across a distance δ^{bl}. Thus, all fluxes of gases that we will consider here are moving perpendicular to the leaf surface, and thus a one-dimensional form of Fick's first law is usually appropriate. We generally express gas fluxes on the basis of unit leaf area. All conductances and resistances are also given per unit leaf area, as we mentioned above.

   We should next ask whether Δc_j is an accurate representation of the force for diffusion. Also, do the coefficients g_j and r_j relating flux densities to Δc_j change greatly as other quantities vary over physiological ranges? For instance, the coefficients could depend on concentration of species j or on temperature. Finally, are other quantities—such as changes in chemical potential $\Delta\mu_j$, changes in water potential $\Delta\Psi_{wv}$, or changes in partial pressure ΔP_j—more appropriate than Δc_j for flux calculations?

   We will begin by writing the equation for the chemical potential of species j in a gas phase, μ_j^{vapor}. By Equations 2.20 and IV.10 (in App IV), μ_j^{vapor} equals $\mu_j^* + RT \ln (P_j/P_j^*)$, where P_j is the partial pressure of species j in the gas phase and P_j^* is its saturation partial pressure at that temperature and one atmosphere of total pressure (we are ignoring the gravitational term of μ_j^{vapor}, because it does not influence the fluxes of gases into or out of leaves). At constant T the most general way to represent the force promoting the movement of species j in the x-direction is $-\partial\mu_j^{vapor}/\partial x$ (see p. 121), which becomes $-RT \, \partial[\ln (P_j/P_j^*)]/\partial x$ for our above representation of the chemical potential, and in turn equals $-(RT/$

P_j) $\partial P_j/\partial x$. For an ideal or perfect gas, $P_j V$ equals $n_j RT$. The total number of moles of species j divided by the volume (n_j/V) is the concentration of species j, c_j; hence P_j ($= n_j RT/V$) can be replaced by $c_j RT$. Making this substitution for P_j, and using the very general flux relation given by Equation 3.6 [$J_j = u_j c_j$ ($-\partial\mu_j/\partial x$)], we can express the flux density of gaseous species j as follows:

$$J_j = u_j c_j \left(\frac{-\partial\mu_j^{\text{vapor}}}{\partial x}\right) = -u_j c_j \frac{RT}{P_j}\frac{\partial P_j}{\partial x}$$

$$= -u_j \frac{\partial P_j}{\partial x} = -u_j RT \frac{\partial c_j}{\partial x} \tag{8.7}$$

$$= \frac{u_j RT}{\Delta x}\Delta c_j = \frac{D_j}{\Delta x}\Delta c_j = g_j \,\Delta c_j$$

where in the last line of the equation we have replaced the negative concentration gradient, $-\partial c_j/\partial x$, by an average concentration gradient, $\Delta c_j/\Delta x$. Also, we have replaced D_j by $u_j RT$ (see p. 124) and then used Equation 8.2, which indicates that g_j equals $D_j/\Delta x$. In essence, Equation 8.7 represents a thermodynamic derivation of Fick's first law for a gas phase (see p. 124).

Let us next examine the coefficients multiplying the various driving forces in Equation 8.7. First, we note that g_j equals $u_j RT/\Delta x$, so g_j is essentially independent of concentration—the very slight dependence of u_j or D_j on concentration can be ignored for gases. On the other hand, if we replace $\partial\mu_j^{\text{vapor}}/\partial x$ by $\Delta\mu_j^{\text{vapor}}/\Delta x$, Equation 8.7 becomes $J_j = u_j c_j \,\Delta\mu_j^{\text{vapor}}/\Delta x$. The factor multiplying $\Delta\mu_j^{\text{vapor}}$ depends directly on concentration, so in addition to knowing $\Delta\mu_j^{\text{vapor}}$, we would also need to specify c_j before calculating J_j. Moreover, c_j can vary near a leaf. Thus, $\Delta\mu_j^{\text{vapor}}$ is not used to represent the force in the relations for transpiration or photosynthesis. Because $\Delta\Psi_{wv}$ equals $\Delta\mu_{wv}/\overline{V}_w$ (Eq. 2.21), Equation 8.7 becomes $J_{wv} = (u_{wv} c_{wv} \overline{V}_w/\Delta x)\,\Delta\Psi_{wv}$ if differences in water potential are used to represent the force. Again, the coefficient depends on concentration, so using $\Delta\Psi_{wv}$ is not appropriate for describing transpiration. But Equation 8.7 also indicates that J_j could equal $(u_j/\Delta x)\,\Delta P_j$. This can lead us to an alternative formulation for flux relations in which conductance has less dependence on temperature and ambient pressure.

Instead of partial pressures, this alternative formulation uses mole fractions (see p. 71) to represent the driving force. We have already encountered a difference in mole fraction representing a driving force when we discussed water flow ($J_{V_w} = P_w \,\Delta N_w$, p. 99). For the case of ideal gases, which approximates situations of biological interest, Dalton's law of partial pressures indicates that the mole fraction of species j, N_j, equals P_j/P, where P is the total pressure. We can then modify Equation 8.7 as follows:

$$J_j = -u_j \frac{\partial P_j}{\partial x} = u_j \frac{\Delta P_j}{\Delta x} = \frac{u_j RT}{RT} \frac{P}{\Delta x} \frac{\Delta P_j}{P}$$

$$= \frac{D_j P}{RT \Delta x} \Delta N_j = g'_j \Delta N_j$$

(8.8)

where we have again incorporated the relationship between diffusion coefficients and mobility ($D_j = u_j RT$) to lead to the conductance g'_j.

Let us next see how the conductance in Equations 8.2 and 8.7 ($g_j = D_j/\Delta x$) and that introduced in Equation 8.8 [$g'_j = D_j P/(RT \Delta x) = g_j P/(RT)$] depend on environmental parameters like pressure and temperature. We have noted that, for gases, D_j depends inversely on pressure (p. 19) and approximately on temperature raised to the power 1.8 (p. 124); i.e.,

$$D_j \cong D_j^0 \left(\frac{T}{273}\right)^{1.8} \frac{P^0}{P}$$

(8.9)

where D_j^0 is the diffusion coefficient of species j at 273 K (0°C) and an ambient pressure of P^0 (often taken as 1 atmosphere).[5] As the pressure decreases, D_j increases proportionally (Eq. 8.9), as does g_j ($g_j = D_j/\Delta x$, Eq. 8.2). On the other hand, as pressure changes, g'_j is unchanged [$g'_j = D_j P/(RT \Delta x)$, Eq. 8.8]. As temperature increases, D_j and hence g_j increase approximately as $T^{1.8}$ (Eq. 8.9), whereas g'_j increases approximately as $T^{0.8}$. Thus, g'_j has no dependence on pressure and much less dependence on temperature than does g_j, and thus the former conductance is more appropriate for describing gas fluxes (see Cowan, 1977; Farquhar et al., 1978; Hall, 1982).

We can illustrate the difference between the two types of conductance by comparing a leaf with *constant* anatomical properties under various environmental conditions. If a leaf at sea level and 10°C were heated to 40°C, g_j would increase 20% even though the leaf's anatomical properties are unchanged. If the leaf were then transferred to 2 000 m, where the ambient air pressure is 22% lower than at sea level, g_j would increase 28% more. On the other hand, heating

---

5. Experimental errors occur in the measurement of D_j and its temperature dependence (see Fuller et al., 1966; Marrero and Mason, 1972; Reid et al., 1977). We will use $D_{wv} = 2.13 \times 10^{-5}$ m^2 s^{-1} and $D_{CO_2} = 1.33 \times 10^{-5}$ m^2 s^{-1} at 0°C and 1 atmosphere (0.1013 MPa); values at other temperatures are given in Appendix I.

from 10°C to 40°C would increase g_j' only 8%, and transferring to 2 000 m would not affect g_j'.

Because ΔN_j is dimensionless, g_j' has the same units as J_j, e.g., mol m^{-2} s^{-1}. At 1 atmosphere and 20°C, P/RT equals 41.6 mol m^{-3}. Thus, a conductance g_j of 1 mm s^{-1} then corresponds to a conductance g_j' of (1 mm s^{-1})(41.6 mol m^{-3}), or 41.6 mmol m^{-2} s^{-1}. In our discussion of transpiration and photosynthesis, we will give examples of the use of both forms of conductance and of resistance (see Table 8.1).

WATER VAPOR FLUXES ACCOMPANYING TRANSPIRATION

Here we will describe the flux of water vapor out of a leaf during transpiration in terms of the conductances and the resistances involved. We will represent the conductances and resistances using symbols (namely, $-\wedge\wedge\wedge-$) borrowed from electrical circuit diagrams. Typical values for the components will be used along with the resulting differences in water vapor concentration and mole fraction across them. Our analysis of water vapor fluxes will indicate the important control of transpiration exercised by stomata.

Conductance and Resistance Network

Water vapor that evaporates from cell walls of mesophyll cells or the inner side of leaf epidermal cells diffuses through the intercellular air spaces to the stomata and then into the outside air. We have already introduced the four components involved—two are strictly anatomical (intercellular air spaces and cuticle), one depends on anatomy and yet responds to metabolic as well as environmental factors (stomata), and one depends on leaf morphology and wind speed (boundary layer). Figure 8.4 summarizes the symbols and uses an electrical circuit analogy to illustrate how the four components can be arranged. We will analyze resistances and conductances for these components, some of which occur in series and some in parallel.

The conductance g_{wv}^{ias} and the resistance r_{wv}^{ias} include all parts of the pathway from the site of water evaporation to the leaf epidermis. The water can evaporate at the air-water interfaces of mesophyll cells, the inner side of epidermal cells (including guard cells), and even cells of the vascular tissue in a leaf before diffusing in the generally tortuous pathways of the intercellular air spaces. The water generally has to cross a thin waxy layer on the cell walls of the cells within a leaf, this layer often being as thick as 0.1 μm (see Esau, 1965). After crossing

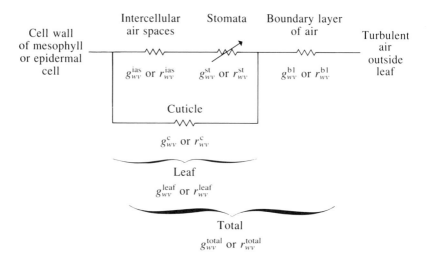

Figure 8.4

Conductances and resistances involved in water vapor flow accompanying transpiration, as arranged into an electrical circuit. The symbol for the stomatal component indicates that it is variable.

the waxy layer, the water vapor diffuses through the intercellular air spaces and then through the stomata (conductance $= g_{wv}^{st}$, resistance $= r_{wv}^{st}$) to reach the boundary layer adjacent to the leaf surface. Alternatively, water in the cell walls of leaf cells might move as a liquid to the cell walls on the cuticle side of epidermal cells, where it could evaporate and the vapor then diffuse across the cuticle (or water may move across the cuticle as a liquid) before reaching the boundary layer adjacent to the leaf. The pathway for such cuticular transpiration (conductance $= g_{wv}^{c}$, resistance $= r_{wv}^{c}$) is in parallel with the pathway for transpiration that occurs through the stomatal pores. For simplicity, we have not included in Figure 8.4 the intermediate case of water evaporating from the cell walls of mesophyll cells and then moving through the intercellular air spaces before crossing the waxy cuticle (the resistance of this pathway is indistinguishable from r_{wv}^{c}, because $r_{wv}^{c} \gg r_{wv}^{ias}$). The final component encountered by the diffusing water vapor is that of the boundary layer just outside the leaf (conductance $= g_{wv}^{bl}$, resistance $= r_{wv}^{bl}$).

We will first analyze the electrical circuit presented in Figure 8.4 in terms of resistances and then in terms of conductances. We begin by noting that the resistances r_{wv}^{ias} and r_{wv}^{st} occur in series. The total resistance of a group of resistors in series, r_{series}, is simply the sum of the individual resistances ($r_{series} = \Sigma_i r_i$, where r_i is the resistance of series resistor i). Thus, the resistance of the pathway from the site of water evaporation across the intercellular air spaces and through the stomata is $r_{wv}^{ias} + r_{wv}^{st}$. This resistance is in parallel with r_{wv}^{c}, the cuticular resistance. The reciprocal of the total resistance of a group of resistors in parallel is the sum of the reciprocals of the individual resistances ($1/r_{parallel} = \Sigma_i\ 1/r_i$, where r_i is the resistance of parallel resistor i). For two resistors in parallel, $r_{parallel}$ is equal to $r_1 r_2/(r_1 + r_2)$. Because ($r_{wv}^{ias} + r_{wv}^{st}$) and r_{wv}^{c} are in parallel, it is often convenient to express their combined resistance as the leaf resistance for water vapor, r_{wv}^{leaf}:

$$r_{wv}^{leaf} = \frac{(r_{wv}^{ias} + r_{wv}^{st})(r_{wv}^{c})}{r_{wv}^{ias} + r_{wv}^{st} + r_{wv}^{c}} \qquad (8.10)$$

When the stomata are open, the cuticular resistance (r_{wv}^{c}) is generally much larger than is the resistance in parallel with it, $r_{wv}^{ias} + r_{wv}^{st}$. In that case, the leaf resistance as given by Equation 8.10 is approximately equal to $r_{wv}^{ias} + r_{wv}^{st}$.

The leaf resistance is in series with that of the boundary layer, r_{wv}^{bl}. Thus the total resistance for the flow of water vapor from the site of evaporation to the turbulent air surrounding a leaf (r_{wv}^{total}) is $r_{wv}^{leaf} + r_{wv}^{bl}$ (see Fig. 8.4). We now must face the complication created by the presence of two leaf surfaces representing parallel pathways for the diffusion of water vapor from the interior of a leaf to the surrounding turbulent air. We will represent the leaf resistance for the pathway through the upper (adaxial) surface by $r_{wv}^{leaf_u}$ and that for the lower (abaxial) surface by $r_{wv}^{leaf_l}$. Each of these resistances is in series with that of an air boundary layer—$r_{wv}^{bl_u}$ and $r_{wv}^{bl_l}$ for the upper and the lower surfaces, respectively. The parallel arrangement of the pathways through the upper and lower surfaces of a leaf leads to the following comprehensive expression for the total resistance to the diffusion of water vapor from a leaf:

$$r_{wv}^{total} = \frac{(r_{wv}^{leaf_u} + r_{wv}^{bl_u})(r_{wv}^{leaf_l} + r_{wv}^{bl_l})}{r_{wv}^{leaf_u} + r_{wv}^{bl_u} + r_{wv}^{leaf_l} + r_{wv}^{bl_l}} \qquad (8.11)$$

Now let us reconsider the analysis of the electrical circuit using conductances. For a group of conductances in series, the reciprocal of the total conductance ($1/g_{series}$) is the sum of the reciprocals of the individual conductances ($1/g_{series} = \Sigma_i\ 1/g_i$), whereas the conductance of a group of conductances in parallel is the sum of the individual conductances ($g_{parallel} = \Sigma_i\ g_i$). Thus, the water vapor conductance of the intercellular air spaces and the stomata in series is $g_{wv}^{ias} g_{wv}^{st}/(g_{wv}^{ias} + g_{wv}^{st})$, and the water vapor conductance of the leaf (see Eq. 8.10 for resistance) is

$$g_{wv}^{leaf} = \frac{g_{wv}^{ias} g_{wv}^{st}}{g_{wv}^{ias} + g_{wv}^{st}} + g_{wv}^{c} \qquad (8.12)$$

Next, we note that g_{wv}^{leaf} is in series with a boundary layer conductance g_{wv}^{bl} and that the two sides of a leaf act as parallel conductances for moving water vapor from the interior of a leaf. We therefore obtain the following expression for the total conductance of a leaf with air boundary layers on each side:

$$g_{wv}^{total} = \frac{g_{wv}^{leaf_u} g_{wv}^{bl_u}}{g_{wv}^{leaf_u} + g_{wv}^{bl_u}} + \frac{g_{wv}^{leaf_l} g_{wv}^{bl_l}}{g_{wv}^{leaf_l} + g_{wv}^{bl_l}} \qquad (8.13)$$

Values of Conductances

We generally assume that the average thickness of the boundary layer is the same on the two sides of a leaf, in which case $g_{wv}^{bl_u}$ equals $g_{wv}^{bl_l}$. However, this does not simplify our analysis of gas fluxes very much, except in special cases. Usually the boundary layer conductance is much greater than that of the leaf. If, in addition, $g_{wv}^{leaf_l}$ is much greater than $g_{wv}^{leaf_u}$, as can occur for leaves with stomata primarily on the lower surface, then the diffusion of water vapor is mainly out through the lower surface of the leaf. This is often the case for deciduous trees. The total conductance for water vapor diffusion from the sites of evaporation inside the leaf to the turbulent air surrounding the leaf (g_{wv}^{total}) is then approximately $g_{wv}^{leaf_l} g_{wv}^{bl_l} / (g_{wv}^{leaf_l} + g_{wv}^{bl_l})$ (see Eq. 8.13). For the symmetrical case where $g_{wv}^{leaf_l}$ equals $g_{wv}^{leaf_u}$ and the boundary layers are of equal thickness (hence, $g_{wv}^{bl_l} = g_{wv}^{bl_u}$), the conductance is the same through either surface; g_{wv}^{total} is then double that through either surface, e.g., $2 g_{wv}^{leaf_l} g_{wv}^{bl_l} / (g_{wv}^{leaf_l} + g_{wv}^{bl_l})$ by Equation 8.13. This can occur for certain monocots—e.g., certain grasses, oats, barley, wheat, corn—which can have fairly equal stomatal frequencies on the two leaf surfaces. For many plants, two to four times more stomata occur on the lower surface of the leaves than on the upper one, so A^{st}/A is considerably larger for the lower surface. In such cases it may be necessary to use the rather cumbersome expression for g_{wv}^{total} given by Equation 8.13.

Let us next examine some representative values for the conductances encountered by water vapor as it diffuses out of leaves. For crop plants such as beet, spinach, tomato, and pea, g_{wv}^{ias} is generally 40 to 100 mm s^{-1} or 1 600 to 4 000 mmol m^{-2} s^{-1}, and for open stomata g_{wv}^{st} usually ranges from 5 to 20 mm s^{-1} or 200 to 800 mmol m^{-2} s^{-1}. Both of these conductances tend to be somewhat lower for many other agricultural crops, but the maximum $g_{wv}^{leaf_l}$ is still 2 to 10 mm s⁻¹ or 80 to 400 mmol m⁻² s⁻¹ for most cultivated plants. Compared with crop plants, the maximum g_{wv}^{leaf} is usually somewhat lower for leaves of deciduous trees and conifer needles (see Table 8.1). Of course, as the stomata close, g_{wv}^{st} and hence g_{wv}^{leaf} decrease accordingly. The minimum value for leaf water vapor conductance occurs for fully closed stomata and essentially equals the cuticular conductance (see Eq. 8.12). Although measured values of g_{wv}^{c} vary considerably, in part because of the difficulty in telling when stomata are fully closed, it may be as low as 0.1 mm s⁻¹ or 4 mmol m⁻² s⁻¹ for cultivated plants, twofold lower for trees, and 10-fold lower for certain xerophytes (Table 8.1).

Effective Lengths and Resistance

Let us next examine a simplified expression that is often adequate for describing the total resistance to diffusion of water vapor from the sites of evaporation in cell walls to the turbulent air surrounding a leaf. We will consider the case in which nearly all the water vapor diffusing from the leaf moves out across the lower epidermis and when cuticular transpiration is negligible. By Equations 8.10 and 8.11, the total resistance then is

$$
\begin{aligned}
r_{wv}^{total} &\cong r_{wv}^{leaf_l} + r_{wv}^{bl_l} \\
&= r_{wv}^{ias} + r_{wv}^{st_l} + r_{wv}^{bl_l} \\
&= \frac{1}{D_{wv}} \left(\delta^{ias} + \frac{\delta^{st} + r^{st}}{na^{st}} + \delta^{bl} \right)
\end{aligned}
\tag{8.14}
$$

where Equations 8.3, 8.5, and 8.6 have been used to obtain the bottom line (see Penman and Schofield, 1951).

When diffusion is described by a one-dimensional form of Fick's first law, Equation 8.14 indicates that each part of the pathway contributes its own effective length influencing the movement of water vapor out of the leaf. For instance, the effective length of the boundary layer is its thickness, δ^{bl}. The distance δ^{ias} includes the effective depth of the cell wall pores, the effective thickness of the waxy layer on the mesophyll cells, and the constriction on the region available for diffusion of gases within a leaf caused by the presence of mesophyll cells. A similar constricting effect greatly increases the effective length of the stomatal pores over the value of $\delta^{st} + r^{st}$. Specifically, Equation 8.14 indicates that the effective length of the stomatal pores is $(\delta^{st} + r^{st})/(na^{st})$, which is considerably greater than $\delta^{st} + r^{st}$ because na^{st} is much less than one (recall that $na^{st} = A^{st}/A$, so na^{st} is a dimensionless number indicating the *fraction* of the surface area occupied by stomatal pores). The large effective length of the stomatal pores caused by the factor $1/na^{st}$ is an alternative way of viewing the constricting effect of the stomata.

If we need to incorporate yet another series resistance for gaseous diffusion, we can include the effective length of the additional part of the pathway within the parentheses in Equation 8.14. For instance, the stomata in some xerophytes and conifers occur sunken beneath the leaf surface in cavities or crypts. We can estimate the effective length of this additional part of the pathway by using appropriate geometrical approximations. The effective length divided by the diffusion coefficient gives the resistance. Many leaves are covered by epidermal "hairs," which can be unicellular projections from the epidermal cells or multicellular appendages with the cells often occurring in a row. Such *trichomes* cause an additional air layer to be held next to the leaf surface. The thickness of this layer may equal the average distance that the hairs project perpendicularly away from the leaf surface (more than 1 mm for some leaves). This length can be added within the parenthesis in Equation 8.14 to calculate the overall resistance. Also, the extra resistance for water vapor diffusion caused by the hairs can be separately calculated as the thickness of the additional air layer divided by D_{wv}. Some water may evaporate from the surfaces of the epidermal hairs; this evaporation can lead to a substantial complication in the analysis of water vapor fluxes for leaves with numerous trichomes.

Leaf Water Vapor Concentrations and Mole Fractions

As we can see from relations like Equation 8.2 ($J_j = g_j \, \Delta c_j = \Delta c_j / r_j$), the conductances or resistances of the various parts of the pathway determine the drop in concentration across each component when the flux density is constant. Here we will apply this condition to a specific consideration of water vapor concentration and mole fraction in a leaf. We will also discuss the important effect of temperature on the water vapor content of air.

Let us represent the difference in water vapor concentration across the intercellular air spaces by Δc_{wv}^{ias}, that across the stomatal pores by Δc_{wv}^{st}, and that across the boundary layer adjacent to the leaf surface by Δc_{wv}^{bl}. Then the overall drop in water vapor concentration from the cell walls where the water evaporates to the turbulent air surrounding a leaf, Δc_{wv}^{total}, is

$$\Delta c_{wv}^{total} = c_{wv}^{e} - c_{wv}^{ta}$$

$$= \Delta c_{wv}^{ias} + \Delta c_{wv}^{st} + \Delta c_{wv}^{bl} \qquad (8.15)$$

where c_{wv}^e is the concentration of water vapor at the evaporation sites, presumably in the cell wall pores, and c_{wv}^{ta} is its value in the turbulent air surrounding the leaf. We also note that $\Delta c_{wv}^{ias} + \Delta c_{wv}^{st}$ is equal to Δc_{wv}^c, the difference in water vapor concentration from the cell walls of epidermal cells across the cuticle to the leaf surface, because the endpoints of the pathway are the same in each case (Fig. 8.4).

The rate of water vapor diffusion per unit leaf area, J_{wv}, is equal to the difference in water vapor concentration multiplied by the conductance across which Δc_{wv} occurs ($J_j = g_j \, \Delta c_j$, Eq. 8.2). In the steady state, when the flux density of water vapor and the conductance of each component are constant, this relation holds both for the overall pathway and any individual segment of it. Because some water evaporates from the cell walls of mesophyll cells along the pathway within the leaf, J_{wv} is actually not spatially constant in the intercellular air spaces. For simplicity, however, we generally assume that J_{wv} is unchanging from the mesophyll cell walls out to the turbulent air outside the leaf. When water vapor moves out mainly across the lower epidermis of the leaf, and when cuticular transpiration is negligible, we obtain the following relations in the steady state:

$$J_{wv} = g_{wv}^{total} \, \Delta c_{wv}^{total} = g_{wv}^{ias} \, \Delta c_{wv}^{ias}$$

$$= g_{wv}^{st} \, \Delta c_{wv}^{st} = g_{wv}^{bl} \, \Delta c_{wv}^{bl} \qquad (8.16)$$

Equation 8.16 indicates that the drop in water vapor concentration across a particular component is inversely proportional to its conductance when J_{wv} is constant (Δc_{wv} is then directly proportional to the resistance, as $J_{wv} = \Delta c_{wv}/r_{wv}$). Because g_{wv}^{ias} and g_{wv}^{bl} are generally larger than is g_{wv}^{st} when the stomata are open (Table 8.1), by Equation 8.16 the largest Δc_{wv} then occurs across the stomata.

The quantity c_{wv} is sometimes referred to as the *absolute* humidity. Its value in the turbulent air generally does not change very much during the day unless there is precipitation or other marked changes in the weather, whereas the *relative* humidity can vary greatly as the air temperature changes. For instance, air that is saturated with water vapor at 20°C (100% relative humidity) drops to 57% relative humidity when heated at constant pressure to 30°C. Most data for c_{wv} are expressed as mass of water/unit volume of air, and therefore we will use g m^{-3} as one of our units for water vapor concentration. A unit for c_{wv} that is often more appropriate is mol m^{-3}, and we will use it as well (see Fig. 8.5).

Equations 8.15 and 8.16 can be adapted to water vapor expressed as a mole fraction, N_{wv}, simply by replacing c with N throughout. For instance, N_{wv}^{total} equals $N_{wv}^e - N_{wv}^{ta}$, and J_{wv} equals $g_{wv}^{total} \, \Delta N_{wv}^{total}$ (we will here use the same symbol for

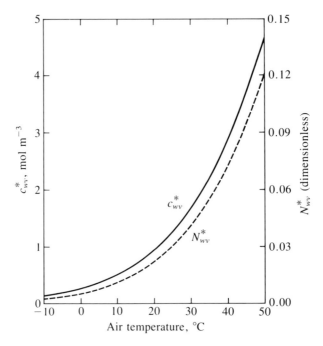

Figure 8.5

Variation in saturation values of water vapor concentration (c_{wv}^*) and mole fraction (N_{wv}^*) with temperature. Ambient pressure (needed to determine N_{wv}^*) was assumed to be 1 atmosphere (0.1013 MPa). Digital values are presented in Appendix I.

conductance in either system, the units being dictated by whether changes in concentration or mole fraction represent the driving force). Water vapor is generally the third most prevalent gas in air, although its mole fraction is relatively low compared with mole fractions for O_2 or N_2. Its saturation value is quite temperature dependent; e.g., N_{wv}^* equals 0.0231 at 20°C and 0.0419 at 30°C (Fig. 8.5). In fact, the water vapor content of air at saturation (represented by c_{wv}^*, N_{wv}^*, or P_{wv}^*) increases nearly exponentially with temperature.[6] Appendix I lists c_{wv}^*, N_{wv}^*, and P_{wv}^* at various temperatures from −30°C to 60°C.

---

6. It may be helpful for understanding this temperature dependence to note that the fraction of molecules in the upper part of the Boltzmann distribution (p. 143) that have a high enough energy to escape from the liquid increases exponentially with temperature.

Examples of Water Vapor Levels in a Leaf

We will now consider the actual relative humidities, concentrations, and mole fractions of water vapor in different parts of the transpiration pathway associated with a leaf. We will suppose that the leaf temperature is 25°C and that the turbulent air surrounding the leaf is at 20°C with a relative humidity of 50%, which correspond to a c_{wv}^{ta} of 0.48 mol m^{-3} and an N_{wv}^{ta} of 0.0115 (see Fig. 8.6 and App. I).

Let us begin by assuming that the water vapor in the pores of the cell walls of mesophyll cells is in equilibrium with the water in the cell wall and that $\Psi^{cell\ wall}$ is -1.0 MPa. Using Equation 2.21 and omitting the gravity term, we can calculate the corresponding % relative humidity (RH) at 25°C as follows:

$$\Psi^{cell\ wall} = \Psi_{wv}^{e} = \frac{RT}{\overline{V}_w} \ln \left(\frac{RH}{100} \right)$$

or

$$RH = 100 \; e^{\overline{V}_w \Psi^{cell\ wall}/RT}$$

$$= 100 \; e^{(-1.0 \text{ MPa}/137.3 \text{ MPa})} = 99.3\%$$

Thus, the air in the pores of the mesophyll cell walls is nearly saturated with water vapor. From Appendix I, c_{wv}^{*} equals 1.28 mol m^{-3} and N_{wv}^{*} equals 0.0313 at 25°C; therefore 99.3% relative humidity corresponds to a c_{wv}^{e} of (0.993)(1.28 mol m^{-3}), or 1.27 mol m^{-3}, and an N_{wv}^{e} of 0.0311. Hence, the difference in water vapor concentration from the mesophyll cell walls to the turbulent air is 1.27 mol m^{-3} $-$ 0.48 mol m^{-3}, or 0.79 mol m^{-3} ($\Delta c_{wv}^{total} = c_{wv}^{e} - c_{wv}^{ta}$, Eq. 8.15), and $N_{wv}^{e} - N_{wv}^{ta}$ is 0.0311 $-$ 0.0115, or 0.0196 (see Fig. 8.6).[7]

---

7. The condition $N_{wv}^{e} > N_{wv}^{ta}$ necessary for transpirational efflux of water vapor from a leaf leads to $N_{air}^{ta} > N_{air}^{e}$, where N_{air} represents the mole fraction of everything but water vapor—i.e., $N_{wv} + N_{air} = 1$ (there is a relatively small ΔN_{CO_2} resulting from CO_2 uptake during photosynthesis, which is mostly compensated for by a small ΔN_{O_2} acting in the opposite direction). Thus we expect a diffusion of air (mainly N_2) into a leaf, which can lead to slightly higher pressures in leaves (generally 0.1 to 1 kPa higher) compared to the ambient pressure outside (the continual supply of water vapor by evaporation inside the leaf and its removal in the outside turbulent air is necessary for the maintenance of this pressure difference). The internally elevated pressure can lead to mass flow within a plant, such as along the stems of the yellow waterlily, *Nuphar luteum*, where a ΔN_{wv} of

Equation 8.15 indicates that Δc_{wv}^{total} is the sum of three components, Δc_{wv}^{ias}, Δc_{wv}^{st}, and Δc_{wv}^{bl}. Likewise, ΔN_{wv}^{total} equals $\Delta N_{wv}^{ias} + \Delta N_{wv}^{st} + \Delta N_{wv}^{bl}$. The magnitude of each of these differences in water vapor level is inversely proportional to the conductance across which the drop occurs—e.g., $g_{wv}^{x} \Delta N_{wv}^{x} = g_{wv}^{total} \Delta N_{wv}^{total}$— and so ΔN_{wv}^{x} is larger when g_{wv}^{x} is smaller, where x refers to any series component in the pathway (see Eq. 8.16). Values of c_{wv}, N_{wv}, and relative humidity are given in Figure 8.6, which also indicates specific values of the conductances as well as the overall series conductance for the diffusion of water vapor from the sites of evaporation to the turbulent air ($1/g_{series} = \Sigma_i \, 1/g_i$). The largest drop in water vapor level occurs across the stomatal pores, because they have the smallest conductance in the present case. For instance, ΔN_{wv}^{st} equals $g_{wv}^{total} \Delta N_{wv}^{total}/g_{wv}^{st}$, which is (154 mmol m^{-2} s^{-1})(0.0196)/(200 mmol m^{-2} s^{-1}), or 0.0151 (see Fig. 8.6).

We note that a small drop in relative humidity, here from 99% in the cell walls of mesophyll cells to 95% at the inner side of the stomata, is necessary for the diffusion of water vapor across the intercellular air spaces. The greater is the fraction of water evaporating from near the guard cells, the smaller is the drop across the intercellular air spaces, although values near 95% agree with measurements (see Egorov and Karpushkin, 1988).[8] There is a fairly large humidity drop across the stomatal pores, such that the relative humidity is 47% at the leaf surface (Fig. 8.6). After crossing the stomatal pores, water vapor moves energetically *downhill* as it diffuses across the boundary layer from 47% relative humidity at 25°C to 50% relative humidity at 20°C in the turbulent air surrounding the leaf (Fig. 8.6). Because a temperature change is involved in this part of the pathway, we must express the driving force for the diffusion of water vapor in terms of the difference in concentration or mole fraction, not the change in relative humidity.

In this text we will generally assume that the leaf has a uniform temperature.[9] For instance, the leaf temperature is 25°C in the example presented, including the air in the intercellular spaces and that in the stomatal pores. The turbulent

---

0.01 is accompanied by an air pressure 0.2 kPa higher inside a young leaf (see Dacey, 1981). Mass flow of air can also occur in submerged parts of rice, *Oryza sativa*, as O_2 is consumed by respiration and the resulting respiratory CO_2 is dissolved in the surrounding water, leading to a local reduction in air pressure (see Raskin and Kende, 1985).

8. Similarly, a small drop in CO_2 concentration occurs across the intercellular air spaces during photosynthesis, which generally represents only 1% to 2% of the atmospheric CO_2 concentration (see Cooke and Rand, 1980; Parkhurst *et al.*, 1988; Sharkey, 1985).

9. Temperature differences of a few degrees C can develop across the width of a moderate-sized leaf at a wind speed of 1 m s^{-1}, reflecting differences in boundary layer thickness (p. 364) and spatial stomatal variation (see Clark and Wigley, 1975).

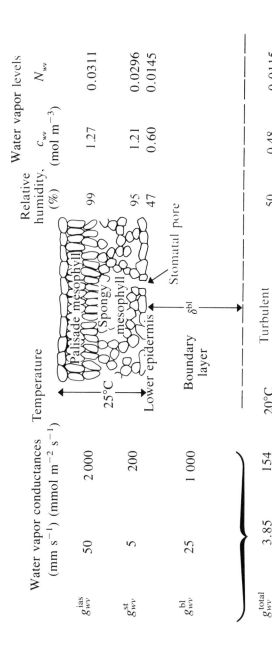

Figure 8.6

Representative values of quantities influencing the diffusion of water vapor out of a leaf. Conductances are given for the indicated parts of the pathway assuming water moves out only through the lower surface and ignoring cuticular transpiration.

air outside the leaf generally has a different temperature, here 20°C. Fick's first law (e.g., $J_j = D_j \, \Delta c_j / \Delta x$, Eq. 8.2) strictly applies only to isothermal situations. For instance, D_j depends on the absolute temperature (D_j is proportional to $T^{1.8}$; Eq. 8.9), so $\Delta(D_j c_j)$ may be nonzero and lead to a flux even when Δc_j is zero. Fortunately, even when there are temperature differences between leaves and the turbulent air, Fick's first law in the form of Equation 8.2 generally proves adequate for describing the fluxes of H_2O and CO_2, our primary concern in this chapter. However, flux relations based on differences in mole fraction (e.g., Eq. 8.8), which have a much lower dependence on temperature for their conductance, are preferred when there is a temperature difference from leaf to air, as is usually the case for plants.

Water Vapor Fluxes

Based on the quantities indicated in Figure 8.6, we can readily calculate the flux density of water vapor moving out of the lower side of a leaf. Specifically, J_{wv}^l equals $g_{wv}^{total} \, \Delta c_{wv}^{total}$ or $g_{wv}^{total} \, \Delta N_{wv}^{total}$, e.g.,

$$J_{wv}^l = (154 \text{ mmol m}^{-2} \text{ s}^{-1})(0.0311 - 0.0115)$$

$$= 3.0 \text{ mmol m}^{-2} \text{ s}^{-1}$$

Based on the decrease in water vapor concentration, J_{wv}^l is $(3.85 \text{ mm s}^{-1})(1.27 \text{ mol m}^{-3} - 0.48 \text{ mol m}^{-3})$, or $3.0 \text{ mmol m}^{-2} \text{ s}^{-1}$. If we express Δc_{wv}^{total} in mass per unit volume, then J_{wv}^l is the rate of movement of mass per unit leaf area; e.g., Δc_{wv}^{total} is here 14 g m^{-3}, so J_{wv}^l is $0.054 \text{ g m}^{-2} \text{ s}^{-1}$.

For simplicity, we have been considering the movement of water vapor across only the lower surface of the leaf, and we have ignored cuticular transpiration. Cuticular transpiration is generally small compared with the transpiration through open stomata that is in parallel with it; i.e., the drop in water vapor level across the cuticle is essentially the same as the drop across the stomata, whereas the stomatal conductance for open stomata is much greater than the cuticular conductance (see Table 8.1). We can add the cuticular transpiration to that through the stomata to get the total transpiration through one side of a leaf. To obtain the overall rate of water vapor diffusing out of both sides of a leaf, we could scale up J_{wv}^l calculated for the lower surface by an appropriate factor—the reciprocal of the fraction of transpiration through the lower surface. For instance, about 70% of the water loss in transpiration might be through the lower surface for a representative mesophyte. The J_{wv}^l of 3.0 mmol m^{-2} s^{-1} that we calculated in the previous paragraph would then become $(1/0.70)(3.0$ mmol m^{-2} s^{-1}), or 4.3 mmol m^{-2} s^{-1}, when both leaf surfaces are considered; i.e., $J_{wv}^l = 0.70 J_{wv}$ implies that $J_{wv} = J_{wv}^l/0.70$. Alternatively, we could use the actual g_{wv}^{total} as given by Equation 8.13, which considers the two leaf surfaces in parallel, or we could measure g_{wv}^{total} experimentally. This conductance times Δc_{wv} or ΔN_{wv}, as appropriate, would give J_{wv} through both surfaces, but expressed per unit area of one side of the leaf; i.e., the flux density would still be expressed on the same basis as are the other fluxes in this chapter.

Let us next relate the rate of water loss by transpiration to the water content of a leaf. For a 300-μm-thick leaf that contains 30% intercellular air spaces by volume, the nongaseous material corresponds to a thickness of $(0.70)(300$ μm), or 210 μm. Water typically comprises about 90% of a leaf's mass, so the water thickness would be about $(0.90)(210$ μm) or 190 μm. The density of water is 1 000 kg m^{-3}, which corresponds to 56 kmol m^{-3}, so the leaf has $(190 \times 10^{-6}$ m) $(56 \times 10^3$ mol m^{-3}) or 11 mol water m^{-2}. For the calculated J_{wv} of 4.3 mmol m^{-2} s^{-1}, this amount of water could be transpired in $(11$ mol m$^{-2})/(4.3 \times 10^{-3}$ mol m^{-2} s^{-1}) or 2 600 s, which is 43 min. Hence, such a transpiring leaf must be continually supplied with water, as we will discuss in the next chapter. For many cultivated plants and other mesophytes under these conditions ($T^{leaf} = 25°C$, $T^{ta} = 20°C$, relative humidity$^{ta} = 50\%$), J_{wv} for open stomata is from 1 to 5 mmol m^{-2} s^{-1}, so our example represents a slightly above-average transpiration rate. Because many systems of units are used for transpiration rates, conversion factors for the more common ones are summarized in Table 8.2.

Control of Transpiration

Let us now reconsider the values of the various conductances affecting the diffusion of water vapor through the intercellular air spaces, out the stomata, and across the boundary layer at the leaf surface. Usually g_{wv}^{ias} is relatively large, g_{wv}^{bl} is rarely less than 500 mmol m^{-2} s^{-1}, but g_{wv}^{st} is generally less than this value and decreases as the stomata close. Consequently, control for limiting transpiration usually rests with the stomata, not with the boundary layer or the intercellular air spaces. When g_{wv}^{st} is at least a few times smaller than is g_{wv}^{bl}, as generally occurs under field conditions, moderate changes in the ambient wind speed have relatively little effect on g_{wv}^{total}. However, the boundary layer conductance can be the main determinant of J_{wv} for the fruiting bodies of Basidiomycetes (fungi), which have no stomata. For instance, near the ground, where

Table 8.2

Conversion factors for some of the more common units used in expressing transpiration, CO_2 levels, and photosynthesis. Values were determined using quantities in Appendices I and II. To convert from one set to another, a quantity expressed in the units in the left column should be multiplied by the factor in the column of the desired units.

Transpiration units	$\text{mmol m}^{-2}\,\text{s}^{-1}$	$\text{mg m}^{-2}\,\text{s}^{-1}$
$\mu\text{mol H}_2\text{O cm}^{-2}\,\text{s}^{-1}$	10	180.2
$\text{mol H}_2\text{O m}^{-2}\,\text{h}^{-1}$	0.278	5.01
$\mu\text{g H}_2\text{O cm}^{-2}\,\text{s}^{-1}$	0.555	10
$\mu\text{g H}_2\text{O cm}^{-2}\,\text{min}^{-1}$	9.25×10^{-3}	0.1667
$\text{mg H}_2\text{O dm}^{-2}\,\text{min}^{-1}$	9.25×10^{-2}	1.667
$\text{g H}_2\text{O dm}^{-2}\,\text{h}^{-1}$	1.542	27.8
$\text{kg H}_2\text{O m}^{-2}\,\text{h}^{-1}$	15.42	278

CO_2 levels	mmol m^{-3}	mg m^{-3}	Pa	mole fraction $\times\,10^6$ (ppm by volume) ($\mu\text{litre litre}^{-1}$) ($\mu\text{bar bar}^{-1}$) ($\mu\text{mol mol}^{-1}$)
$\text{nmol CO}_2\text{ cm}^{-3}$, $\text{mmol CO}_2\text{ m}^{-3}$, μM	1	44.0	2.44	24.4
$\text{ng CO}_2\text{ cm}^{-3}$, $\text{mg CO}_2\text{ m}^{-3}$	0.0227	1	0.0554	0.554
$\mu\text{litre CO}_2\text{ litre}^{-1}$,* ppm CO_2 by volume	0.0410	1.806	0.1	1
μbar*	0.0410	1.806	0.1	1
Pa*	0.410	18.06	1	10

Photosynthesis units	$\mu\text{mol m}^{-2}\,\text{s}^{-1}$
$\text{ng CO}_2\text{ cm}^{-2}\,\text{s}^{-1}$	0.227
$\text{nmol CO}_2\text{ cm}^{-2}\,\text{s}^{-1}$	10
$\text{mg CO}_2\text{ m}^{-2}\,\text{s}^{-1}$	22.7
$\text{mg CO}_2\text{ dm}^{-2}\,\text{h}^{-1}$, $\text{kg CO}_2\text{ hectare}^{-1}\,\text{h}^{-1}$	0.631
$\text{kg carbohydrate hectare}^{-1}\,\text{h}^{-1}$	0.92
$\text{mm}^3\text{ CO}_2\text{ cm}^{-2}\,\text{h}^{-1}$	0.114

*To convert a volume/volume number like ppm or mole fraction to a mole/volume or a pressure unit, or vice versa, we need to know the temperature and the pressure; an air temperature of 20°C and a pressure of 0.1 MPa (1 bar) were used here. To adjust for other temperatures and pressures, the ideal gas law ($PV = nRT$) must be employed; e.g., 1 $\mu\text{mol mol}^{-1}$ of CO_2 at temperature T_x in K and pressure P_x in MPa is equal to $0.410\ (293.15/T_x)(P_x)\ \text{mmol m}^{-3}$.

the fruiting bodies occur and the wind speed is relatively low (generally below 0.2 m s^{-1}), g_{wv}^{bl} exerts the main control on transpiration for *Lycoperdon perlatum* and *Scleroderma australe* (see Nobel, 1975).

Stomata tend to close as a leaf wilts, a common response of plants to water stress. Assuming that g_{wv}^{st} decreases 20-fold, and using values presented in Figure 8.6, we note that g_{wv}^{total} would decrease about 15-fold, e.g., from 154 mmol m^{-2} s^{-1} for a g_{wv}^{st} of 200 mmol m^{-2} s^{-1} to 9.9 mmol m^{-2} s^{-1} for a g_{wv}^{st} of 10 mmol m^{-2} s^{-1} (this latter conductance is 100 times lower than g_{wv}^{bl} and 200 times lower than g_{wv}^{ias} in the present case). The decrease in g_{wv}^{total} causes transpiration to decrease to only about 6% of its former value, if we ignore the parallel pathway across the cuticle. When the stomata close tightly, only cuticular transpiration remains, which can have a conductance of 1 to 10 mmol m^{-2} s^{-1} (Table 8.1). In that case, cuticular transpiration accounts for all of the loss of water vapor from the leaves.

CO₂ CONDUCTANCES AND RESISTANCES

We will next consider the main function of a leaf, photosynthesis, in terms of the conductances and resistances encountered by CO_2 as it diffuses from the turbulent air, across the boundary layers next to a leaf, through the stomata, across the intercellular air spaces, into the mesophyll cells, and eventually into the chloroplasts. The situation is obviously more complicated than is the analogous one of the movement of water vapor during transpiration, because CO_2 not only must diffuse across the same components encountered by water vapor moving in the opposite direction,[10] but also must cross the cell wall of a mesophyll cell, the plasmalemma, part of the cytosol, the membranes surrounding a chloroplast, and some of the chloroplast stroma. We will specifically indicate the resistance of each component, resistances being somewhat easier to deal with than are conductances for the series of components involved in the pathway for CO_2 movement. We will summarize values for these resistances, as well as both forms of conductances, in a table.

---

10. Additional complications relate to the mass flows that occur, particularly through the stomata where the diffusing species can also interact with the walls of the pores. The flux of water vapor thus affects the CO_2 concentration gradient because of molecular collisions between H_2O and CO_2 moving in opposite directions as well as because of pressure driven (bulk) flow (see footnote, pp. 419–420; for an authoritative treatment, see Field *et al.*, 1989; Jarman, 1974; and especially Leuning, 1983).

Resistance and Conductance Network

Figure 8.7 illustrates the various conductances and resistances affecting CO_2 as it diffuses from the turbulent air surrounding a leaf up to the sites in the chloroplasts where it is incorporated into photosynthethic products. For simplicity, we

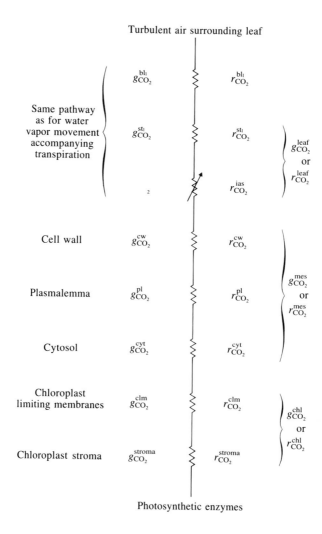

Figure 8.7

Principal conductances and resistances involved in the movement of CO_2 from the turbulent air surrounding a leaf, across the lower epidermis, and up to the enzymes involved in the fixation of CO_2 into photosynthetic products in the chloroplasts of mesophyll cells.

will initially restrict our attention to the diffusion of CO$_2$ into the leaf across the lower epidermis only. When the parallel pathways through the upper and the lower surfaces of a leaf are both important, we can readily modify our equations to handle the reduction in resistance, or increase in conductance, encountered between the turbulent air surrounding the leaf and the cell walls of its mesophyll cells (see Eqs. 8.11 and 8.13). We will also ignore the cuticular path for CO$_2$ entry into the leaf for the same sort of reason that we neglected this part of the pathway when discussing transpiration—namely $r_{CO_2}^c$ is generally considerably greater than is $r_{CO_2}^{ias} + r_{CO_2}^{st}$, the resistance in parallel with it.

The first three resistances encountered by CO$_2$ entering a leaf through the lower epidermis ($r_{CO_2}^{bl_l}$, $r_{CO_2}^{st_l}$, and $r_{CO_2}^{ias}$) have analogs in the case of transpiration. We can thus transfer Equation 8.14 to our present discussion, changing only the subscripts:

$$r_{CO_2}^{leaf_l} + r_{CO_2}^{bl_l} = \frac{1}{D_{CO_2}} \left(\delta^{ias} + \frac{\delta^{st} + r^{st}}{na^{st}} + \delta^{bl} \right) \qquad (8.17)$$

where $r_{CO_2}^{leaf_l}$ is the resistance of the intercellular air spaces plus the stomatal pores of the lower leaf surface to the diffusion of CO$_2$. We also note that $r_{CO_2}^{leaf_l} + r_{CO_2}^{bl_l}$ is analogous to r_{wv}^{total}, the *total* resistance encountered by water vapor (see Fig. 8.7).

The diffusion coefficient for CO$_2$, D_{CO_2}, is 1.51×10^{-5} m^2 s^{-1} in air at 20°C (App. I). This is smaller than D_{wv} (2.42×10^{-5} m^2 s^{-1}), primarily because CO$_2$ molecules are heavier and thus diffuse more slowly than do H$_2$O molecules (see p. 19). By Equations 8.14 and 8.17, $(r_{CO_2}^{leaf_l} + r_{CO_2}^{bl_l})/r_{wv}^{total}$ equals D_{wv}/D_{CO_2}, which is $(2.42 \times 10^{-5}$ m^2 s$^{-1})/(1.51 \times 10^{-5}$ m^2 s$^{-1})$, or 1.60. Consequently, CO$_2$ diffusing from the turbulent air up to the cell walls of mesophyll cells encounters 60% more resistance than does water vapor diffusing in the opposite direction over the same pathway. Likewise, the gas phase conductance is (100%)/(1.60) or only 62% as great for CO$_2$ as for water vapor.[11] Representative values for

---

11. Actually, movement across the boundary layer is partly by diffusion, where the ratio D_{wv}/D_{CO_2} applies, and partly by turbulent mixing (see Fig. 7.7), where molecular differences are obliterated. Thus, $r_{CO_2}^{bl}/r_{wv}^{bl}$ is intermediate between the extremes of 1.60 and 1.00, and indeed it is found to be $(D_{wv}/D_{CO_2})^{2/3}$, which is 1.37 (see Monteith, 1973; Schlichting, 1979). Although Equation 8.17 is generally satisfactory, certain situations may warrant replacing δ^{bl}/D_{CO_2} by $1.37 \, \delta^{bl}/D_{wv}$, which equals $0.86 \, \delta^{bl}/D_{CO_2}$. We also note that the effective boundary layer thicknesses for water vapor and heat transfer are quite similar, being within 7% for flat plates (see Monteith, 1973) and within 5% for cylinders and spheres (see Nobel, 1974, 1975).

CO_2 conductances and resistances are indicated in Table 8.4 for leaves of crops and trees with open stomata (see Table 8.1 for analogous values for water vapor).

As indicated in Figure 8.7, five additional resistances are involved in CO_2 flow compared with water vapor movement. The new components of the pathway are the nongaseous parts of the cell wall of a mesophyll cell (resistance $= r_{CO_2}^{cw}$), a plasmalemma ($r_{CO_2}^{pl}$), the cytosol ($r_{CO_2}^{cyt}$), then the chloroplast limiting membranes ($r_{CO_2}^{clm}$), and finally the interior of the chloroplasts ($r_{CO_2}^{stroma}$). For convenience we will divide these five resistances into two parts, the mesophyll resistance to CO_2, $r_{CO_2}^{mes}$, and the chloroplast resistance, $r_{CO_2}^{chl}$:

$$r_{CO_2}^{mes} = r_{CO_2}^{cw} + r_{CO_2}^{pl} + r_{CO_2}^{cyt} \qquad (8.18a)$$

$$r_{CO_2}^{chl} = r_{CO_2}^{clm} + r_{CO_2}^{stroma} \qquad (8.18b)$$

Mesophyll Area

The area of the cell walls of mesophyll cells across which CO_2 can diffuse is considerably larger than the surface area of the leaf. For the constricting effect caused by the stromata, we used A^{st}/A, the fraction of the leaf surface area occupied by stromatal pores. Here we will use the ratio A^{mes}/A to indicate the *increase* in area available for CO_2 diffusion, where A^{mes} is the total area of the cell walls of mesophyll cells that is exposed to the intercellular air spaces, and A is the area of one side of the same leaf.

Although A^{mes}/A varies with plant species as well as with leaf development, it is generally between 10 and 40 for mesophytes (see Björkman, 1981; El-Sharkawy and Hesketh, 1965; Esau, 1965; Nobel, 1980; Nobel and Walker, 1985). We can appreciate the large value of A^{mes}/A by examining Figures 1.2 and 8.6, which indicate that a considerable amount of air space and hence exposed cell walls occurs within a leaf; e.g., the palisade mesophyll is generally 10% to 40% air by volume, and the spongy mesophyll is 50% to 80% air. Although the spongy mesophyll region generally has a greater volume fraction of air, the palisade region usually has a greater total cell wall area exposed to the intercellular air spaces. Xerophytes tend to have a somewhat more highly developed palisade region than do mesophytes (in some cases, the spongy mesophyll cells are even absent in xerophytes), which leads to values of 20 to 50 for A^{mes}/A of many xerophytes.

To help appreciate the magnitude of A^{mes}/A, let us consider some geometrical idealizations. For a single layer of uniform spheres in an orthogonal array, A^{mes}/A

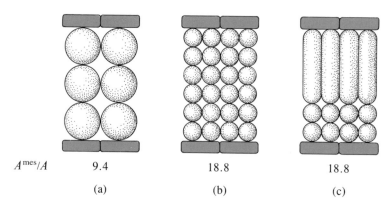

A^{mes}/A	9.4	18.8	18.8
	(a)	(b)	(c)

Figure 8.8

Representations of mesophyll cells showing how geometry affects A^{mes}/A. Spheres or cylinders with hemispherical ends in an orthogonal array lead to the indicated A^{mes}/A. The length of the lateral walls of the "palisade" cells in (c) is 6 times the radius.

is $4\pi r^2/(2r \times 2r)$, or π. Hence, three layers of spherical cells have an A^{mes}/A of 3π, or 9.4 (Fig. 8.8a). If the radius of the spheres were halved but the thickness of the array remained unchanged, then A^{mes}/A doubles (Fig. 8.8a versus b). In a more realistic representation of a leaf with a single palisade layer having cells four times as long as wide and two layers of spherical spongy cells, A^{mes}/A is 18.8, two-thirds of which is contributed by the palisade cells (Fig. 8.8c). Moreover, three-quarters of the exposed cell wall area of the palisade cells occurs on their lateral walls. To show that the relatively large area contribution of the lateral walls is quite typical, we will represent the palisade cells as cylinders of radius r and length l with hemispheres on each end. The total area of the lateral surface of the cylinder ($2\pi rl$) is generally greater than that of the two hemispherical ends ($4\pi r^2$), because l is usually considerably greater than r for such palisade cells—e.g., r may be 10 to 20 μm, whereas l is 30 to 100 μm for representative palisade cells. When such cylinders are packed together to form a layer of palisade cells, nearly the entire surface area of the lateral walls is exposed to the intercellular air spaces. This surface area and most of that of both ends of a palisade cell are available for the inward diffusion of CO_2.

The illumination condition under which a leaf develops can greatly influence the anatomy of its mesophyll region. Development in a dark or shaded environment can lead to a *shade* leaf, and differentiation under moderate to high illumination can lead to a *sun* leaf (see Esau, 1965). Besides being smaller in area, sun leaves usually are thicker and have a higher proportion of palisade cells than do shade leaves on the same plant. Moreover, their palisade cells are generally longer (larger l for the cylinders). Consequently, A^{mes}/A can be two to four times larger for sun leaves than for shade leaves on the same plant. For example, growing *Plectranthus parviflorus* at a PPFD of 17 μmol m^{-2} s^{-1} for 12-h days leads to thin leaves with an A^{mes}/A of 11, whereas a PPFD of 810 μmol m^{-2} s^{-1} leads to thick leaves with an A^{mes}/A of 50 (see Nobel *et al.*, 1975).

Although PPFD generally has the greatest influence, A^{mes}/A can also be influenced by changes in other environmental factors during leaf development (see Nobel, 1980; Nobel and Walker, 1985; Patterson, 1980). For instance, higher temperatures generally lead to smaller cells and up to a 40% higher A^{mes}/A. Reduced cell size generally accompanies water stress, but the influences on A^{mes}/A vary with species, ranging from no change to a 50% increase in A^{mes}/A. Higher salinities during leaf development generally lead to thicker leaves, which can be accompanied by a corresponding increase in cell dimensions with no change in A^{mes}/A, or sometimes by an increase in A^{mes}/A.

Resistance Formulation for Cell Components

The resistance to diffusion of a molecular species across a barrier equals the reciprocal of its permeability coefficient (see p. 397). We will let $P^j_{CO_2}$ be the permeability coefficient for CO_2 diffusion across barrier j. To express the resistance of a particular mesophyll or chloroplast component on a leaf area basis, we must also incorporate A^{mes}/A to allow for the actual area available for diffusion—the greater internal leaf area acts like more pathways in parallel and thus reduces the effective resistance. Because the area of the plasmalemma is about the same as that of the cell wall, and the chloroplasts generally occupy a single layer around the periphery of the cytosol (Fig. 8.9), the factor A^{mes}/A will be applied to all the diffusion steps of CO_2 with which we are concerned in mesophyll cells (all five individual resistances in Eq. 8.18). In other words, we are imagining for simplicity that the cell wall, the plasmalemma, the cytosol, and the chloroplasts are all in layers having essentially equal areas (Fig. 8.9). Thus, the resistance of any of the mesophyll or chloroplast components for CO_2 diffusion, $r^j_{CO_2}$, is reduced from $1/P^j_{CO_2}$ by the reciprocal of the same factor, A^{mes}/A:

$$r^j_{CO_2} = \frac{1}{A^{mes}/A} \frac{1}{P^j_{CO_2}} = \frac{A}{A^{mes} P^j_{CO_2}}$$

$$= \frac{A \, \Delta x^j}{A^{mes} D^j_{CO_2} K^j_{CO_2}} = \frac{1}{g^j_{CO_2}} \tag{8.19}$$

Intercellular air spaces

Figure 8.9

Schematic cross section near the periphery of a mesophyll cell, indicating the sequential occurrence of the anatomical components across which CO_2 diffuses from the intercellular air spaces to the carboxylation enzymes in the chloroplast stroma.

where the second line of this equation follows from the definition of a permeability coefficient, $P_j = D_j K_j / \Delta x$ (Eq. 1.9). In particular, Δx^j is the thickness of the jth barrier, $D^j_{CO_2}$ is the diffusion coefficient of CO_2 in it, and $K^j_{CO_2}$ is a suitably defined partition coefficient.

Partition Coefficient for CO₂

In Chapter 1 we introduced a partition coefficient to describe the ratio of the concentrations of some species in two adjacent phases. For instance, $K_{CO_2} c_{CO_2}$ could be the actual concentration of CO_2 in some region where concentrations are difficult to measure, c_{CO_2} is the equilibrium concentration of CO_2 in an adjacent region where it is readily measured, and K_{CO_2} is the partition coefficient. Similarly, we will express every $K^j_{CO_2}$ in the mesophyll cells as the actual concentration of all forms of CO_2 in component j divided by the concentration of CO_2 that would occur in an adjacent air phase at equilibrium, $c^{air}_{CO_2}$:

$$K^j_{CO_2} = \frac{\text{concentration of all forms of ``CO}_2\text{'' in barrier } j}{\text{equilibrium CO}_2 \text{ concentration in adjacent air phase}}$$

$$= \frac{c^j_{CO_2} + c^j_{H_2CO_3} + c^j_{HCO_3^-} + c^j_{CO_3^{2-}}}{c^{air}_{CO_2}} \tag{8.20}$$

Table 8.3
Influence of temperature and pH on partitioning of "CO$_2$" between an aqueous solution and an adjacent air phase. The partition coefficient for all three forms of "CO$_2$" at various pH's is for 20°C. Data are for an ionic strength of about 200 mmol m^{-3} and an air pressure of 0.1 MPa (see Stumm and Morgan, 1981).

Temperature (°C)	$\dfrac{c_{CO_2}^{water}}{c_{CO_2}^{air}}$	pH	$\dfrac{c_{CO_2}^{water} + c_{H_2CO_3}^{water} + c_{HCO_3^-}^{water}}{c_{CO_2}^{air}}$
0	1.65	4	0.91
10	1.19	5	0.96
20	0.91	6	1.48
30	0.71	7	6.6
40	0.58	8	58
50	0.50		

A concentration referred to as $c_{CO_2}^{air}$ will thus equal the actual concentration of all forms of CO$_2$ in component j divided by $K_{CO_2}^j$. This convention allows us to discuss fluxes in a straightforward manner, because CO$_2$ then diffuses toward regions of lower $c_{CO_2}^{air}$ regardless of the actual concentrations and partition coefficients involved. For instance, to discuss the diffusion of "CO$_2$" across a cell wall, we need to consider the partitioning of CO$_2$ between the air in the cell wall pores and the various species of CO$_2$ in the adjacent water within the cell wall interstices. $K_{CO_2}^{cw}$ is thus the actual concentration of CO$_2$ plus H$_2$CO$_3$, HCO$_3^-$, and CO$_3^{2-}$ in the cell wall water divided by the concentration of CO$_2$ in air in equilibrium with the cell wall water.

The concentrations of the various forms of "CO$_2$" present in an aqueous phase are temperature dependent and extremely sensitive to pH.[12] For instance, the equilibrium concentration of CO$_2$ dissolved in water divided by that of CO$_2$ in an adjacent gas phase, $c_{CO_2}^{water}/c_{CO_2}^{air}$, decreases over twofold from 10°C to 40°C (Table 8.3; the decreased solubility of CO$_2$ at higher temperatures is a characteristic of dissolved gases, which fit into the interstices of water, such space becoming less available as molecular thermal agitation increases with temperature). This partition coefficient is itself not very pH dependent, but the equi-

---

12. The concentrations also depend on the presence of other solutes, but this effect is presumably small for the cell wall water.

librium concentration of HCO_3^- in water relative to that of dissolved CO_2 is markedly affected by pH. In particular, CO_2 dissolved in an aqueous solution may interact with OH^- to form bicarbonate, which then associates with H^+ to form H_2CO_3: $CO_2 + OH^- + H^+ \rightleftharpoons HCO_3^- + H^+ \rightleftharpoons H_2CO_3$. Alternatively, CO_2 in water might form H_2CO_3, which then dissociates to HCO_3^- and H^+. The interconversions of CO_2 and H_2CO_3 are actually relatively slow unless a suitable catalyst, such as the enzyme carbonic anhydrase, is present (see Forster et al., 1969; Tashian and Hewett-Emmett, 1984). Because H^+ is involved in these reactions, its concentration, i.e., the pH, will affect the amount of HCO_3^- in solution, which in turn depends on the CO_2 concentration. The equilibrium concentration of H_2CO_3 is only about 1/400 of that of the dissolved CO_2, so our main concern will be with CO_2 and HCO_3^- (CO_3^{2-} is also not a major species until the pH exceeds 8, and at pH 8 it is only 1% of HCO_3^- at an ionic strength of 200 mmol m^{-3}, so CO_3^{2-} will be ignored here; see Stumm and Morgan, 1981).

Cell Wall Resistance

Let us begin our discussion of the newly introduced resistances by evaluating the components of $r_{CO_2}^{cw}$, the resistance encountered by CO_2 as it diffuses through the water-filled interstices of the cell wall from the interface with the intercellular air spaces on one side to the plasmalemma on the other side (Fig. 8.9). We will use Equation 8.19 to describe this resistance: $r_{CO_2}^{cw} = A \, \Delta x^{cw}/(A^{mes}D_{CO_2}^{cw} K_{CO_2}^{cw})$.

The distance across the barrier Δx^{cw} is the average thickness of the cell walls of the mesophyll cells. The diffusion coefficient for the gas CO_2 dissolved in water is 1.7×10^{-9} m^2 s^{-1} at 20°C (Table 1.1). However, the effective $D_{CO_2}^{cw}$ is probably lower by a factor of three or four, because the water-filled interstices represent slightly less than half the cell wall and their course through the cell wall is rather tortuous (see p. 36). Thus, $D_{CO_2}^{cw}$ may be about 5×10^{-10} m^2 s^{-1}. Besides moving as the dissolved gas, "CO_2" may also diffuse across the cell wall as H_2CO_3 or HCO_3^-. The diffusion coefficients of these two species in the cell wall are also most likely about 5×10^{-10} m^2 s^{-1}. However, the possible presence of H_2CO_3 and HCO_3^- makes the effective concentration of "CO_2" in the cell wall uncertain, which is why it is convenient to introduce $K_{CO_2}^{cw}$, the partition coefficient for CO_2 in the cell wall.

At 20°C, $(c_{CO_2}^{water} + c_{H_2CO_3}^{water} + c_{HCO_3^-}^{water})/c_{CO_2}^{air}$ (our definition for $K_{CO_2}^{cw}$, see Eq. 8.20) is about 1 from pH 4 to pH 6, but increases markedly above pH 7 (Table 8.3). The equilibrium value is affected by temperature in approximately the same way as the partition coefficient $c_{CO_2}^{water}/c_{CO_2}^{air}$ cited above; e.g., going from 20°C to 30°C it decreases by about 22%. We note that this appreciable temperature dependence of K_{CO_2} results in a similar temperature dependence of r'_{CO_2}—see Equation 8.19. Although the pH in the cell walls of mesophyll cells within a leaf is not known with certainty, it is probably under 6. Thus, at usual leaf temperatures, $K^{cw}_{CO_2}$ will be close to unity, the value we will use for calculation.

Let us now estimate a value for $r^{cw}_{CO_2}$. We will assume that the mesophyll cells have a typical cell wall thickness, Δx^{cw}, of 0.3 μm, that the diffusion coefficient in the cell walls for solutes like CO_2 or HCO_3^-, $D^{cw}_{CO_2}$, is 5×10^{-10} m² s⁻¹, and that $K^{cw}_{CO_2}$ is 1.0. The magnitude of $r^{cw}_{CO_2}$ also depends on the relative surface area of the mesophyll cells compared with the leaf area, a quantity that varies considerably. We will let A^{mes}/A be 20, a representative value for mesophytes. Using Equation 8.19, we can then calculate the resistance of the cell walls to the diffusion of CO_2:

$$r^{cw}_{CO_2} = \frac{(0.3 \times 10^{-6} \text{ m})}{(20)(5 \times 10^{-10} \text{ m}^2 \text{ s}^{-1})(1)} = 30 \text{ s m}^{-1}$$

This is a rather small value for a CO_2 resistance (see Table 8.4) and indicates that the cell walls of the mesophyll cells generally do not represent a major barrier to the diffusion of the various species of CO_2.

Table 8.4
Summary of representative values of conductances and resistances for CO_2 diffusing into leaves. Certain values are calculated in the text (see Evans et al., 1986; Longstreth et al., 1980; Šetlik, 1970; and references for Table 8.1).

Component	Conductance		Resistance	
	mm s⁻¹	mmol m⁻² s⁻¹	s m⁻¹	m² s mol⁻¹
Leaf (lower surface)—gas phase				
crops—open stomata	1.2 to 6	50 to 250	160 to 800	4 to 20
trees—open stomata	0.3 to 2	12 to 75	500 to 2 500	13 to 80
Cell wall	30	1 200	30	0.8
Plasmalemma	10	400	100	2.5
Cytosol	100	4 000	10	0.25
Mesophyll				
estimation	7	300	140	3.5
measurements—mesophytes	2.5 to 25	100 to 1 000	40 to 400	1 to 10
Chloroplast				
estimation	10	400	100	2.5
measurements	>5	>200	<200	<5

Plasmalemma Resistance

Let us now examine $r_{CO_2}^{pl}$, the resistance of the plasmalemma of mesophyll cells to diffusion of the various species of CO_2. Although we do not know the actual permeability coefficients of the plasmalemma of mesophyll cells for CO_2 and HCO_3^-, we expect them to be much lower for a charged species like HCO_3^- (see p. 33). For instance, $P_{HCO_3^-}^{pl}$ might be about 10^{-8} m s^{-1}. On the other hand, CO_2 is a small neutral linear molecule that enters cells extremely easily (see Forster *et al.*, 1969; Colman and Espie, 1985). The permeability coefficient of CO_2 entering plant cells is probably at least 2×10^{-4} m s^{-1}, which is somewhat higher than $P_{H_2O}^{pl}$ (p. 30), and may be 10^{-3} m s^{-1}—we will use 5×10^{-4} m s^{-1} for purposes of calculation. Using Equation 8.19 and a value of 20 for A^{mes}/A, the plasmalemma CO_2 resistance is

$$r_{CO_2}^{pl} = \frac{1}{(20)(5 \times 10^{-4} \text{ m s}^{-1})} = 100 \text{ s m}^{-1}$$

Based on the relative values for the two permeability coefficients, the resistance to the diffusion of HCO_3^- across the plasmalemma is over 10^4 times higher, namely 5×10^6 s m^{-1}. Because of the extremely high resistance for HCO_3^-, we conclude that bicarbonate does not *diffuse* across the plasmalemma at the rates necessary to sustain photosynthesis. But the resistance calculated for CO_2 (100 s m^{-1}) is relatively small, which suggests that diffusion of CO_2 can be adequate for moving this substrate of photosynthesis across the plasmalemma.

HCO_3^- or even CO_2 could be actively transported across the plasmalemma, or perhaps could cross by facilitated diffusion (p. 163). Facilitated diffusion would act as a low resistance pathway in parallel with the ordinary diffusion pathway and thus would reduce the effective resistance of the plasmalemma to either of these species. Unfortunately, the actual mechanism for CO_2 or HCO_3^- movement across the plasmalemmas of mesophyll cells is not known with certainty, although $r_{CO_2}^{pl}$ for diffusion of CO_2 is apparently low enough to account for the CO_2 fluxes actually observed.

Cytosol Resistance

The resistance of the cytosol of the mesophyll cells to the diffusion of CO_2 is small, because the distance is short, as the chloroplasts are located around the periphery of most mesophyll cells (Figs. 1.1 and 8.9). In particular, the average distance from the plasmalemma to the chloroplasts, Δx^{cyt}, is only 0.1 to 0.3 μm.

We will use a value of 0.2 μm for Δx^{cyt} when estimating $r_{CO_2}^{\text{cyt}}$ using Equation 8.19.

Apparently because of the presence of fibrous proteins in the cytosol, chloroplasts generally tend to remain stationary and quite close to the plasmalemma of a mesophyll cell. The diffusion coefficient of CO_2 in this region of the cytosol is somewhat less than its value in water at 20°C, 1.7×10^{-9} m² s⁻¹. For purposes of calculation, we will let $D_{CO_2}^{\text{cyt}}$ be 1×10^{-9} m² s⁻¹ for the various species of CO_2 diffusing from the plasmalemma to the chloroplasts. The value of $K_{CO_2}^{\text{cyt}}$ of mesophyll cells is not known, primarily because of the lack of knowledge about cytosolic pH. In any case, $K_{CO_2}^{\text{cyt}}$ cannot be much less than unity, and values greater than 1.0—the magnitude we will assume here—will not change our conclusions about the relative value of $r_{CO_2}^{\text{cyt}}$. As before, we will let A^{mes}/A be 20. Using Equation 8.19, we calculate that the cytosolic resistance for CO_2 is $(0.2 \times 10^{-6}$ m$)/[(20)(1 \times 10^{-9}$ m² s⁻¹$)(1)]$ or only 10 s m⁻¹, a very small value for a resistance to CO_2 diffusion (Table 8.4). Thus, the location of the chloroplasts around the periphery of a mesophyll cell is a "good" design in that it causes the resistance of this part of the pathway for CO_2 diffusion to be quite low.

Mesophyll Resistance

Let us now compare the measured values of the mesophyll resistance with the magnitudes just calculated for its three components. Most of the measurements of $r_{CO_2}^{\text{mes}}$ are rather indirect, but they do indicate that this resistance is generally 40 to 400 s m⁻¹ for mesophytes (Table 8.4). Based on our estimates of 30 s m⁻¹ for $r_{CO_2}^{\text{cw}}$, 100 s m⁻¹ for $r_{CO_2}^{\text{pl}}$, and 10 s m⁻¹ for $r_{CO_2}^{\text{cyt}}$, we predict a value of about 140 s m⁻¹ for $r_{CO_2}^{\text{mes}}$ (by Eq. 8.18a, $r_{CO_2}^{\text{mes}} = r_{CO_2}^{\text{cw}} + r_{CO_2}^{\text{pl}} + r_{CO_2}^{\text{cyt}}$). Thus, our estimate based on the *diffusion* of CO_2 across each barrier is consistent with the measured values of mesophyll resistance.

There are many assumptions and specific choices for parameter values involved in the calculation of $r_{CO_2}^{\text{mes}}$ here. For instance, we let A^{mes}/A be 20, whereas many leaves have values from 30 to 40; the latter ratios would reduce our $r_{CO_2}^{\text{mes}}$ to 70 to 90 s m⁻¹. The cell walls of some mesophyll cells are only 0.07 μm thick (see Longstreth *et al.*, 1980), which would decrease $r_{CO_2}^{\text{cw}}$ to less than 10 s m⁻¹. Permeability coefficients of the plasmalemmas of mesophyll cells for CO_2 so far have not been adequately measured. In this regard, P_j equals $D_j K_j/\Delta x$ (Eq. 1.9), where the diffusion coefficients of H_2O and CO_2 in the plasmalemma are probably about the same (within a factor of two of each other), the partition coefficient for CO_2 is most likely at least 10 times higher than $K_{H_2O}^{\text{pl}}$, and Δx^{pl} is the same for H_2O and CO_2. Because $P_{H_2O}^{\text{pl}}$ can be 10^{-4} m s⁻¹ (Ch. 1), our assumed value of 5×10^{-4} m s^{-1} for $P_{CO_2}^{pl}$ may be too low. By comparison, we noted in Chapter 1 (p. 30) that P_j for another small molecule, O_2, crossing erythrocyte membranes has an extremely high value of 0.3 m s^{-1}. A higher value for $P_{CO_2}^{pl}$ would decrease our estimate for $r_{CO_2}^{pl}$, and thus for $r_{CO_2}^{mes}$.

Chloroplast Resistance

The last two structural resistances encountered by the CO_2 used in photosynthesis are due to the chloroplasts ($r_{CO_2}^{chl} = r_{CO_2}^{clm} + r_{CO_2}^{stroma}$, Eq. 8.18b; see Fig. 8.9). As is the case for the plasmalemma, the resistance of the chloroplast limiting membranes, $r_{CO_2}^{clm}$, is extremely large for the diffusion of HCO_3^- and relatively low for the diffusion of CO_2. Numerous experiments indicate that the chloroplast limiting membranes are actually rather permeable to small solutes, so $r_{CO_2}^{clm}$ may be lower than $r_{CO_2}^{pl}$, which we estimated to be 100 s m^{-1}. Because Δx^{stroma} averages nearly 1 μm, whereas Δx^{cyt} is about 0.2 μm and the other pertinent parameters (A^{mes}/A, D_{CO_2}, K_{CO_2}) are approximately the same in the two cases, $r_{CO_2}^{stroma}$ is a few times larger than $r_{CO_2}^{cyt}$, which we calculated to be only 10 s m^{-1} (if the pH in the chloroplast stroma were near or above 7, this would increase $K_{CO_2}^{chl}$ above unity and reduce $r_{CO_2}^{stroma}$ accordingly). In any case, $r_{CO_2}^{stroma}$ is a relatively small resistance.

At present, all we can do is estimate a value of about 100 s m^{-1} for the resistance to the diffusion of CO_2 into chloroplasts and across their stroma. Measurement of $r_{CO_2}^{chl}$ *in vivo* is also difficult—analysis of available data indicates that most likely it is less than 200 s m^{-1}. Although active transport or facilitated diffusion of CO_2 or HCO_3^- into chloroplasts may occur and thus lower the effective resistance, the experimental values for $r_{CO_2}^{chl}$ are compatible with diffusion of CO_2 across the chloroplast limiting membranes (see Colman and Espie, 1985). All the resistances we have just discussed and their corresponding conductances, expressed in both types of units that we have considered, are summarized in Table 8.4.

CO₂ FLUXES ACCOMPANYING PHOTOSYNTHESIS

Now that we have discussed the CO_2 resistances and conductances, we are ready to examine CO_2 fluxes. We will do this for photosynthesis, a process that consumes CO_2, as well as for respiration and photorespiration, two processes that evolve CO_2. Our analysis will use an electrical circuit analogy so that we can represent the interrelationships between the various factors influencing net CO_2 uptake by a leaf.

Photosynthesis

The flux density of CO_2 into the chloroplasts represents the gross rate of photosynthesis per unit leaf area, $J^{ps}_{CO_2}$. In the steady state, $J^{ps}_{CO_2}$ is the same as the flux density entering the leaf, J_{CO_2}, corrected for any other reactions evolving or consuming CO_2, as we will consider later (most evidence indicates that CO_2, not HCO_3^-, is the substrate for photosynthesis in chloroplasts, and we will therefore focus our attention on CO_2). We will represent the average rate of photosynthesis per unit *volume* of the chloroplasts by v_{CO_2}, which can have units of mol CO_2 fixed m^{-3} s^{-1}. If chloroplasts have an average or effective thickness Δx^{chl} in a direction perpendicular to the plasmalemma (see Fig. 8.9), the rate of photosynthesis per unit leaf area $J^{ps}_{CO_2}$ can be represented as

$$J^{ps}_{CO_2} = A^{mes}\, v_{CO_2}\, \Delta x^{chl}/A \tag{8.21}$$

where the factor A^{mes}/A relates the chloroplast area to the leaf area. We can also derive Equation 8.21 by regarding the chloroplasts as a flat layer of average thickness Δx^{chl} occupying some area, e.g., A^{mes}. The rate of photosynthesis in this volume ($v_{CO_2}\, A^{mes}\, \Delta x^{chl}$) is equal to the gross CO_2 flux per unit leaf area times the leaf area corresponding to A^{mes}, i.e., $J^{ps}_{CO_2}\, A$.

Observations indicate that the gross rates of CO_2 fixation by leaves and by isolated chloroplasts are proportional to the CO_2 level over the lower range of its concentration and eventually reach an upper limit when the CO_2 concentration is raised sufficiently high. One way to describe such behavior is with a Michaelis–Menten type of expression:

$$v_{CO_2} = \frac{V_{max}\, c^{chl}_{CO_2}}{K_{CO_2} + c^{chl}_{CO_2}} \tag{8.22}$$

where V_{max} is the maximum rate of CO_2 fixation per unit volume and K_{CO_2} is essentially a Michaelis constant for CO_2 fixation, i.e., the value of $c^{chl}_{CO_2}$ at which v_{CO_2} equals $\frac{1}{2}V_{max}$ [see Eq. 3.27a, $J^{in}_j = J^{in}_{j\,max}\, c^o_j/(K_j + c^o_j)$, and the discussion that follows it]. Although convenient, using a Michaelis–Menten type of expression for the photosynthetic rate per unit volume may not always be justified for such a complicated series of reactions as photosynthesis (see P. Gaastra, in Evans, 1963; Tenhunen et al., 1980; Zelitch, 1971).

In Equation 8.22, V_{max} and, to some extent, K_{CO_2} depend on PPFD, temperature, and nutrient status. For instance, V_{max} is zero in the dark, because photosynthesis then ceases, and is directly proportional to PPFD up to about 50

μmol m^{-2} s^{-1}. If we continually increase the PPFD, V_{max} can reach an upper limit, its value for light saturation. This generally occurs by about 600 μmol m^{-2} s^{-1} for most C$_3$ plants, whereas photosynthesis in C$_4$ plants is usually not light-saturated even at full sunlight, 2 000 μmol m^{-2} s^{-1} (see pp. 326 and 462 for comments on C$_3$ and C$_4$ plants; see also Fig. 8.14 for responses of leaves of C$_3$ plants and a C$_4$ plant to PPFD). Photosynthesis is maximal at certain temperatures, often from 30°C to 40°C. V_{max} increases as the leaf temperature is raised to the optimum and then decreases with a further increase in temperature.

The values of V_{max} at light saturation and at the optimal temperatures for photosynthesis vary with the particular plant species, but they are generally from 2 to 10 mol m^{-3} s^{-1}. We can also estimate V_{max} from measurements of the maximum rates of CO$_2$ fixation by isolated chloroplasts. These maximum rates—which are sustained for rather short periods and are for optimal conditions—can be 100 mmol of CO$_2$ fixed (kg chlorophyll)$^{-1}$ s^{-1} [360 μmol (mg chlorophyll)$^{-1}$ h^{-1} in another common unit], which is equal to approximately 3 mol m^{-3} s^{-1} (1 kg chlorophyll is contained in about 0.035 m^3 of chloroplasts *in vivo*). The activity of the key enzyme for CO$_2$ fixation, ribulose-1,5-bisphosphate carboxylase/oxygenase, *in vitro* can have rates equivalent to 200 mmol (kg chlorophyll)$^{-1}$ s^{-1}. Usually, the estimates of V_{max} using isolated chloroplasts or enzymes are somewhat lower than its values determined for a leaf. Measurements using leaves generally indicate that K_{CO_2} is 5 to 15 mmol m^{-3}. For instance, K_{CO_2} can be 9 mmol m^{-3} at 25°C with a Q_{10} (see Ch. 3) of 1.8 (see Woodrow and Berry, 1988).

In Chapter 5 we noted that the processing time for CO$_2$ fixation was about 5 ms (p. 273). Eight photons are required, which are absorbed by approximately 2 000 chlorophyll molecules in chloroplasts where the chlorophyll concentration is about 30 mol m^{-3} (p. 232). Hence, the photosynthetic rate per unit volume of chloroplasts is

$$v_{CO_2} = \frac{\text{(concentration of CO}_2 \text{ fixation sites)}}{\text{(time for CO}_2 \text{ fixation at a site)}}$$

$$= \frac{(30 \text{ mol chlorophyll m}^{-3})/(2.0 \times 10^3 \text{ chlorophylls per site})}{(5 \times 10^{-3} \text{ s/CO}_2 \text{ at a site})}$$

$$= 3 \text{ mol CO}_2 \text{ m}^{-3} \text{ s}^{-1}$$

Under high PPFD, $c_{CO_2}^{chl}$ is generally limiting for photosynthesis, so this v_{CO_2} may be only 40% to 70% of the V_{max} at a particular temperature. Conversely, the 5 ms processing time may in large measure reflect the suboptimal CO$_2$ levels in the chloroplasts (these comments apply to C_3 plants; for C_4 plants $c_{CO_2}^{chl}$ is generally near the saturation value for photosynthesis, a topic to which we will return at the end of this chapter—see also Fig. 8.13). A related factor is the increase in atmospheric CO_2, primarily due to the burning of fossil fuels, from about 300 μmol mol^{-1} in 1900 to 353 μmol mol^{-1} in 1990, when the annual rate of CO_2 increase was nearly 2 μmol mol^{-1} (see Ramanathan, 1988; Waterman et al., 1989).[13] Other things being equal, this increase in $c_{CO_2}^{ta}$ will lead to a rise in $c_{CO_2}^{chl}$ and hence in v_{CO_2} (see Eq. 8.22).

Respiration and Photorespiration

So far, the only process involving CO_2 that we have considered in this chapter is photosynthesis. But we cannot neglect the CO_2 produced within mesophyll (and other) cells by respiration and photorespiration. If mitochondrial respiration in leaf cells were the same in the light as in the dark, when it can be readily measured, respiration would produce about 5% as much CO_2 as is consumed by photosynthesis at a moderate PPFD. In those plants that photorespire, the rate of light-stimulated production of CO_2 by photorespiration at moderate temperatures is often about 30% (range, 20% to 50%) of the rate of CO_2 fixation into photosynthetic products.

We have already considered respiration in Chapter 6, so we will next briefly comment on photorespiration (for further details, see Lorimer, 1981; Ogren, 1984; Osmond et al., 1980; Tolbert, 1981; Woodrow and Berry, 1988). Photorespiration is the uptake of O_2 and the evolution of CO_2 in the light resulting from glycolate synthesis in chloroplasts and subsequent glycolate and glycine metabolism in peroxisomes and mitochondria (Fig. 8.10). A crucial role is played by the enzyme ribulose-1,5-bisphosphate carboxylase/oxygenase (RuBisCO for

---

13. CO_2 levels are expressed in many units, parts per million by volume (ppm or ppmv) being commonly used, concise, clearly understood, but no longer generally accepted in the biological literature. Some data are still published on a volume basis using units of μlitre litre^{-1} or cm^3 m^{-3} (numerically equivalent to ppm). However, most data are currently expressed on a mole fraction basis, either as a dimensionless number or in μmol mol^{-1} (e.g., 350 $\times$ 10^{-6} and 350 μmol mol^{-1}, respectively) or on the basis of partial pressure. The latter is often in μbar bar^{-1} or μbar, although the more appropriate SI units are Pa MPa^{-1} and Pa, respectively. If pressure units are used for CO_2 (μbar or Pa), then the ambient atmospheric pressure should also be stated. By the ideal or perfect gas law ($PV = nRT$, where $P = \Sigma_i P_i$ and $V = \Sigma_i V_i$, the summations being over all gaseous species present), $V_{CO_2}/V_{total} = P_{CO_2}/P_{total} = n_{CO_2}/n_{total} = N_{CO_2}$, where N_{CO_2} is the mole fraction of CO_2, the unit that we will adopt. Another unit is mmol m^{-3}, the concentration of CO_2 (which we will also use), although both temperature and pressure should then be specified.

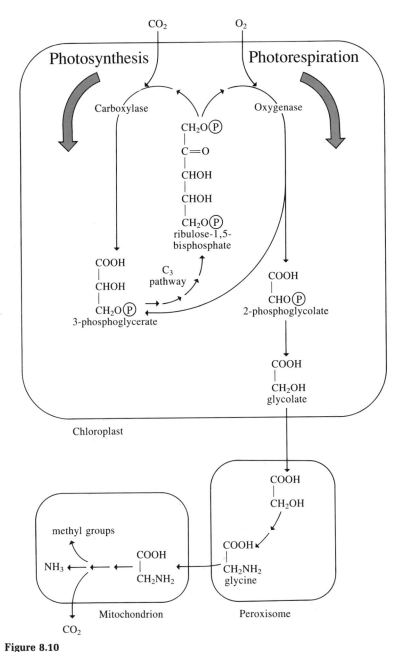

Figure 8.10

Schematic illustration of RuBisCO (ribulose-1,5-bisphosphate carboxylase/oxygenase) acting as the branch point for photosynthesis and photorespiration. All three of the organelles involved, but only a few of the biochemical steps, are indicated ($\textcircled{P}$ represents phosphate).

short), which has a molecular mass of about 540 kDa and consists of eight large subunits (53 kDa each) and eight small subunits (15 kDa; see Chapman *et al.*, 1987). RuBisCO constitutes about half of the soluble protein in the leaves of C_3 plants and perhaps one-sixth in the soluble protein in C_4 plants, making it the most abundant protein in the world. This carboxylase/oxygenase can interact with CO_2, leading to photosynthesis, or with O_2, leading to photorespiration (Fig. 8.10).[14]

In photorespiration, ribulose-1,5-biphosphate is split into 3-phosphogylcerate and 2-phosphoglycolate, the latter undergoing dephosphorylation in the chloroplasts and then entering the peroxisomes. Although CO_2 can be released by a decarboxylation of glycolate in the peroxisomes, the main product of glycolate metabolism is glycine, which then moves to the mitochondria where the CO_2 is released (Fig. 8.10). Because the generation of ribulose-1,5-bisphosphate depends on the C_3 photosynthetic pathway, photorespiration is influenced by PPFD and by temperature, although not quite in the same manner as is photosynthesis. For instance, photosynthesis usually doubles in going from 20°C to 30°C, whereas phororespiration often triples over this interval of leaf temperature; the oxygenase activity is apparently favored over the carboxylase activity with increasing temperature, and thus photorespiration increases at the expense of photosynthesis at higher temperatures. As far as CO_2 fixation is concerned, photorespiration apparently undoes what photosynthesis has done. We might then ask whether photorespiration benefits a plant—a question that so far has no entirely convincing answer.

Not all plants photorespire significantly. Many of these "nonphotorespiring" plants are tropical monocots, and all tend to have a leaf anatomy different from that of plants with high rates of photorespiration (Fig. 1.2 versus Fig. 8.11). Nonphotorespirers have a conspicuous group of chloroplast-containing cells surrounding the vascular bundles known as the *bundle sheaths* (the bundle sheaths in photorespirers tend to have smaller cells with few or no chloroplasts). Outside a bundle sheath in nonphotorespirers are mesophyll cells where CO_2 is fixed into 4-carbon dicarboxylic acids via a C_4 pathway elaborated by Hatch and Slack in the 1960s (see Hatch and Slack, 1970; O'Leary, 1982). In this C_4 pathway, CO_2 in the form of HCO_3^- reacts with phosphoenolypyruvate via the enzyme phos-

---

14. The competition for the same active site on RuBisCO by the two gases O_2 and CO_2, which thus act as alternative substrates, can be modeled by a modification of Equation 8.22 in which K_{CO_2} is replaced by $K_{CO_2}(1 + c_{O_2}^{chl}/K_{O_2})$ (see Farquhar *et al.*, 1980b; Farquhar and von Caemmerer, 1982; Tenhunen *et al.*, 1980; Woodrow and Berry, 1988).

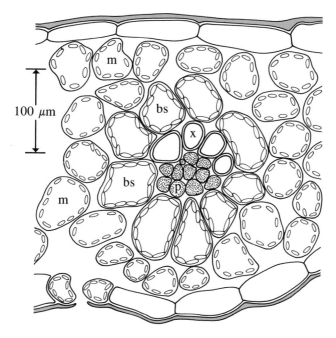

Figure 8.11

Schematic transverse section through a leaf of a C₄ plant, indicating a vascular bundle containing xylem (x) and phloem (p) cells, a concentric layer of bundle sheath cells (bs), and the surrounding mesophyll cells (m). Bundle sheath cells of C₄ plants appear more conspicuously green than do mesophyll cells, because the former generally contain more and/or larger chloroplasts (which are granaless in some types of C₄ plants).

phoenolpyruvate carboxylase located in the cytosol of the mesophyll cells.[15] The initial product is oxaloacetate, which is rapidly converted to malate and aspartate. For all chloroplasts in photorespiring (C₃) plants, and for the chloroplasts of the bundle sheath cells of C₄ plants, photosynthesis uses the ordinary C₃ pathway of the Calvin cycle—known since the 1940s—where CO₂ is incorporated into ribulose-1,5-bisphosphate, yielding two molecules of the 3-carbon compound, 3-phosphoglycerate (Fig. 8.10). Biochemical shuttles in C₄ plants move the 4-

---

15. CO_2 can diffuse across the plasmalemma and become hydrated to HCO_3^- by carbonic anhydrase, which occurs in the cytosol of mesophyll cells of C₄ plants (see Edwards and Walker, 1983). The Michaelis constant of phosphoenolpyruvate carboxylase for HCO_3^- is about 200 mmol m^{-3} (0.2 mM), which suggests that the cytosolic pH must be above 7 to get sufficient HCO_3^- formation (see Table 8.3) to match the CO₂ uptake rates of C₄ plants.

carbon compounds initially produced in the light, such as malate, from the mesophyll cells into the bundle sheath cells (Fig. 8.11). Decarboxylation of these compounds raises the CO_2 level in the bundle sheath cells to much higher levels than would be expected based on diffusion of CO_2 in from the atmosphere (to over 1 500 μmol CO_2 mol^{-1} in one estimate; Ehleringer and Björkman, 1977). Because of the high CO_2 level, the carboxylase activity of RuBisCO is dominant over (i.e., outcompetes) the oxygenase activity, so photosynthesis takes place in the bundle sheath cells with very little photorespiration.[16]

Comprehensive CO_2 Resistance Network

We can now develop an analytical framework within which to represent CO_2 fixation in photosynthesis and its evolution in respiration and photorespiration. The net flux of CO_2 into a leaf, J_{CO_2}, is a measure of the apparent (net) CO_2 assimilation rate by photosynthesis. The gross or "true" rate of photosynthesis, $J_{CO_2}^{ps}$, minus the rate of CO_2 evolution by respiration and photorespiration per unit leaf area, $J_{CO_2}^{r+pr}$, is related to J_{CO_2} as follows:

$$J_{CO_2} = J_{CO_2}^{ps} - J_{CO_2}^{r+pr} \tag{8.23}$$

Equation 8.23 summarizes the overall steady-state balance of CO_2 fluxes for leaves.

We will consider CO_2 fluxes and resistances for photosynthesis, respiration, and photorespiration using an electrical circuit analogy (Fig. 8.12). The sources of CO_2 for photosynthesis are the turbulent air surrounding a leaf (represented by the E battery in Fig. 8.12) and respiration plus photorespiration (the e battery). The E battery corresponds to the drop in CO_2 concentration (or mole fraction) from the turbulent air surrounding a leaf to the enzymes of photosynthesis inside chloroplasts, $c_{CO_2}^{ta} - c_{CO_2}^{chl}$, which represents the driving force for CO_2 diffusion. The batteries lead to currents that correspond to fluxes of CO_2; e.g., I, the current from the E battery, corresponds to J_{CO_2}, and i represents the flux density of CO_2 emanating from respiration and photorespiration, $J_{CO_2}^{r+pr}$. The current I crosses the resistances $r_{CO_2}^{bl}$, $r_{CO_2}^{leaf}$, and $r_{CO_2}^{mes}$ before being joined by i (see Fig. 8.12). The current i encounters $r_{CO_2}^{i}$, which is the resistance to the movement of CO_2 out of mitochondria and then across a short distance in the cytosol. Both I and i

---

16. Raising the external CO_2 level to 1 500 μmol mol^{-1} or so similarly would virtually eliminate photorespiration in C_3 plants (see Fig. 8.13); such CO_2 enrichment is common in greenhouses containing commercially valuable C_3 crops.

Figure 8.12

Electrical circuit indicating the resistances affecting photosynthesis, respiration, and photorespiration. The sources of CO_2 are the turbulent air surrounding the leaf (represented by the battery of electromotive force E) and respiration plus photorespiration (the e battery). The current I corresponds to the net CO_2 influx into the leaf (J_{CO_2}), i represents CO_2 evolution by respiration and photorespiration ($J_{CO_2}^{r+pr}$), and $I + i$ corresponds to gross photosynthesis ($J_{CO_2}^{ps}$). The voltages at various locations correspond to specific CO_2 concentrations; e.g., the voltage in the upper right corner (between $r_{CO_2}^{leaf_i}$ and $r_{CO_2}^{mes}$) corresponds to $c_{CO_2}^{ias}$, that in the middle right corner to $c_{CO_2}^{mito}$, and that in the lower left corner to $c_{CO_2}^{chl}$.

cross the resistance $r_{CO_2}^{chl}$, because CO_2 coming from the surrounding air, as well as that evolved in mitochondria by respiration and photorespiration, can be used for photosynthesis in the chloroplasts (this is a way of paraphrasing Eq. 8.23).

To analyze the electrical circuit in Figure 8.12, we will make use of Ohm's law ($\Delta E = IR$) and Kirchhoff's second law.[17] The latter rule—which is also known as the *loop theorem* and is a consequence of the conservation of energy—states that the overall change in electrical potential in going completely around a closed loop is zero. By considering a complete pathway around the left-hand part of the electrical circuit in Figure 8.12, we obtain the following relationship:

$$E - I(r_{CO_2}^{bl_i} + r_{CO_2}^{leaf_i} + r_{CO_2}^{mes}) - (I + i)(r_{CO_2}^{chl}) = 0 \qquad (8.24)$$

---

17. Kirchhoff's first law for electrical circuits states that the algebraic sum of the currents at any junction equals zero. For instance, at the junction in Figure 8.12 where current I meets current i, the current leaving that point equals $I + i$.

Let us now insert quantities into Equation 8.24 that relate to CO_2 exchange. E can be replaced by $c_{CO_2}^{ta} - c_{CO_2}^{chl}$, I by J_{CO_2}, and i by $J_{CO_2}^{r+pr}$. Upon moving the resistance terms to the opposite side of the equation, we obtain

$$c_{CO_2}^{ta} - c_{CO_2}^{chl} = J_{CO_2} (r_{CO_2}^{bl_1} + r_{CO_2}^{leaf_1} + r_{CO_2}^{mes} + r_{CO_2}^{chl}) + J_{CO_2}^{r+pr} r_{CO_2}^{chl} \quad (8.25)$$

Equation 8.25 indicates that, if respiration or photorespiration increase, then the net photosynthetic rate will decrease when other factors are unchanged.

Compensation Points

The atmospheric CO_2 concentration at which the CO_2 evolved by respiration and photorespiration is exactly compensated for by a CO_2 consumption in photosynthesis is known as the *CO_2 compensation point*. We can use Figure 8.12 and Equation 8.25 to demonstrate the CO_2 compensation point for photosynthesis in terms of forces and fluxes. If we continuously decrease $c_{CO_2}^{ta}$ for a leaf initially having a net uptake of CO_2, J_{CO_2} will decrease and eventually become zero when $c_{CO_2}^{ta} - c_{CO_2}^{chl}$ equals $J_{CO_2}^{r+pr} r_{CO_2}^{chl}$ (see Eq. 8.25). Thus, reducing the concentration of CO_2 in the turbulent air surrounding an illuminated leaf will cause the cessation of net CO_2 fixation when we reach the CO_2 compensation point.

The CO_2 compensation point is considerably higher for a C_3 plant than for a C_4 plant. At the compensation point, $c_{CO_2}^{ta} - c_{CO_2}^{chl}$ equals $J_{CO_2}^{r+pr} r_{CO_2}^{chl}$, and $J_{CO_2}^{r+pr}$ is larger when photorespiration is appreciable. Most C_4 plants—e.g., sugar cane, sorghum, maize, bermuda grass, Sudan grass, and *Amaranthus*—have CO_2 compensation points of 3 to 10 μmol CO_2 mol^{-1} in the turbulent air (Fig. 8.13). Most dicotyledons and temperate monocots are C_3 plants—e.g., cotton, tobacco, tomato, lettuce, oaks, maples, roses, wheat, and orchard grass—and have CO_2 compensation points of 40 to 100 μmol CO_2 mol^{-1} (Fig. 8.13). A few species, e.g., some species of *Mollugo, Moricandia,* and *Panicum,* appear to have intermediate CO_2 compensation points near 25 μmol CO_2 mol^{-1}, and shifts between C_3 and C_4 patterns can even occur during leaf development. The CO_2 compensation points generally increase with increasing temperature and decreasing PPFD, the values given being appropriate at 25°C when light is not limiting for photosynthesis (see Zelitch, 1971).

If we continuously reduce the amount of light incident on a leaf from the value for direct sunlight, we eventually reach a PPFD for which there is no net photosynthesis. This PPFD for which J_{CO_2} is zero is known as the *light compensation point* for photosynthesis. Because photorespiration depends on photosynthetic products, both photorespiration and gross photosynthesis decrease as

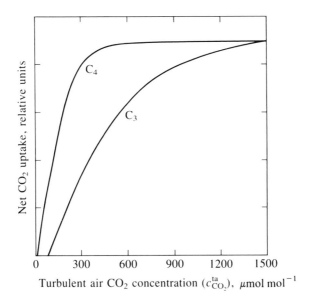

Turbulent air CO_2 concentration ($c_{CO_2}^{ta}$), μmol mol^{-1}

Figure 8.13

Dependence of net CO_2 uptake on external CO_2 level for leaves of representative C_3 and C_4 plants. C_3 plants require a higher $c_{CO_2}^{ta}$ at the CO_2 compensation point ($J_{CO_2} = 0$) and for CO_2 saturation than do C_4 plants. We note that because photosynthesis for C_4 plants is already saturated at present atmospheric CO_2 levels, elevated CO_2 generally will not enhance their photosynthetic rates, whereas the increasing atmospheric CO_2 levels will progressively increase photosynthesis for C_3 plants.

the PPFD is lowered. Hence, the light compensation point for leaves is approximately the same for C_3 and C_4 plants—at 20°C and 350 μmol CO_2 mol^{-1}, light compensation usually occurs at a PPFD of about 8 to 16 μmol m^{-2} s^{-1} for C_3 plants and 6 to 14 μmol m^{-2} s^{-1} for C_4 plants (lower values are for shade leaves or shade plants). The light compensation point, which occurs at less than 1% of a full sunlight PPFD of about 2 000 μmol m^{-2} s^{-1}, will be quite important in our consideration of plant canopies in the next chapter. For example, leaves shaded by many overlying leaves can actually be at (or below) the light compensation point when the exposed leaves have appreciable net rates of photosynthesis. Also, the uppermost leaves can reach the light compensation point on cloudy days or at sunset.

At either compensation point, J_{CO_2} is zero when $c_{CO_2}^{ta} - c_{CO_2}^{chl}$ equals $J_{CO_2}^{r+pr} r_{CO_2}^{chl}$ (see Eq. 8.25). For the light compensation point, $c_{CO_2}^{ta}$ is unchanged but $c_{CO_2}^{chl}$ increases owing to a decrease in CO_2 fixation by photosynthesis at the low PPFD. If we lower the PPFD below the light compensation point, J_{CO_2}

reverses its direction, which means a net flux of CO_2 out of the leaf. For instance, when $c_{CO_2}^{ta}$ equals $c_{CO_2}^{chl}$, J_{CO_2} equals $-J_{CO_2}^{r+pr} \, r_{CO_2}^{chl}/(r_{CO_2}^{bl_1} + r_{CO_2}^{leaf_1} + r_{CO_2}^{mes} + r_{CO_2}^{chl})$ by Equation 8.25, a conclusion that can also be reached by appl ing the loop theorem to the left-hand part of Figure 8.12 (note that E is zero when $c_{CO_2}^{ta} = c_{CO_2}^{chl}$). Thus, part of the respiratory plus photorespiratory flux density is then refixed in the chloroplasts and part comes out of the leaf. At night, J_{CO_2} becomes equal to $-J_{CO_2}^{r+pr}$, because $J_{CO_2}^{ps}$ is zero as gross photosynthesis stops upon cessation of illumination (see Eq. 8.23). The respiratory flow of CO_2 out of the leaf is then driven by the higher CO_2 concentration in the mitochondria than in the turbulent air, encountering the resistances $r_{CO_2}^{i}$, $r_{CO_2}^{mes}$, $r_{CO_2}^{leaf_1}$, and $r_{CO_2}^{bl_1}$, in that order (see Fig. 8.12). In particular, the E battery still corresponds to $c_{CO_2}^{ta} - c_{CO_2}^{chl}$, but $c_{CO_2}^{chl}$ is greater than $c_{CO_2}^{ta}$ at night, so the battery reverses its polarity. The condition $J_{CO_2}^{ps}$ equaling zero means no current through $r_{CO_2}^{chl}$ ($I = -i$), so applying the loop theorem to the left-hand part of Figure 8.12 yields $c_{CO_2}^{chl} - c_{CO_2}^{ta} = J_{CO_2}^{r+pr} (r_{CO_2}^{bl_1} + r_{CO_2}^{leaf_1} + r_{CO_2}^{mes})$ and to the right-hand part yields $c_{CO_2}^{mito} - c_{CO_2}^{chl} = J_{CO_2}^{r+pr} \, r_{CO_2}^{i}$. Adding these two relations yields $c_{CO_2}^{mito} - c_{CO_2}^{ta} = J_{CO_2}^{r+pr} (r_{CO_2}^{bl_1} + r_{CO_2}^{leaf_1} + r_{CO_2}^{mes} + r_{CO_2}^{i})$, which can be used to describe the efflux of respiratory CO_2 from leaves at night. Thus, our electrical circuit in Figure 8.12, and Equation 8.25 derived from it, are able to portray the CO_2 compensation point, the light compensation point, as well as the general interrelations of the fluxes of CO_2 for photosynthesis, photorespiration, and respiration in the light and the dark. Our discussion and Figure 8.12 have really been for C_3 plants—to apply an electrical circuit analog to C_4 and CAM plants, we need to consider the cytosolic location of the initial CO_2 fixing enzymes as well as the fate of the CO_2 released upon decarboxylation of the 4-carbon acids.

Fluxes of CO_2

Using Equation 8.25, we can relate the apparent or net rate of photosynthesis, J_{CO_2}, to the various resistances, respiration plus photorespiration, and the overall drop in CO_2 concentration, $\Delta c_{CO_2}^{total} = c_{CO_2}^{ta} - c_{CO_2}^{chl}$. Let us first rearrange Equation 8.25 into the following form:

$$J_{CO_2} = \frac{c_{CO_2}^{ta} - c_{CO_2}^{chl}}{r_{CO_2}^{bl_1} + r_{CO_2}^{leaf_1} + r_{CO_2}^{mes} + \left(1 + \dfrac{J_{CO_2}^{r+pr}}{J_{CO_2}}\right) r_{CO_2}^{chl}}$$

$$= \frac{\Delta c_{CO_2}^{total}}{r_{CO_2}^{total}}$$

(8.26)

where we have used our customary definition of resistance to obtain $r_{CO_2}^{total}$, the total effective resistance for CO_2 fixation—namely, resistance equals concentration drop divided by flux density (see Eq. 8.1d). We note that $r_{CO_2}^{total}$ depends on $J_{CO_2}^{r+pr}$, which is a consequence of the rather complicated electrical circuit (Fig. 8.12) needed to represent the various CO_2 components. Sometimes it may be convenient to rearrange Equation 8.25 in other ways, e.g., J_{CO_2} equals $(\Delta c_{CO_2}^{total} - J_{CO_2}^{r+pr} r_{CO_2}^{chl})/(r_{CO_2}^{bl_1} + r_{CO_2}^{leaf_1} + r_{CO_2}^{mes} + r_{CO_2}^{chl})$. This form clearly shows that J_{CO_2} is zero at the compensation points $(\Delta c_{CO_2}^{total} = J_{CO_2}^{r+pr} r_{CO_2}^{chl})$. Also, using Equation 8.23, we can manipulate the factor in Equation 8.26 containing $J_{CO_2}^{r+pr}$ as follows:

$$1 + \frac{J_{CO_2}^{r+pr}}{J_{CO_2}} = \frac{J_{CO_2} + J_{CO_2}^{r+pr}}{J_{CO_2}} = \frac{J_{CO_2}^{ps}}{J_{CO_2}} = \frac{J_{CO_2}^{ps}}{J_{CO_2}^{ps} - J_{CO_2}^{r+pr}} = \frac{1}{1 - \dfrac{J_{CO_2}^{r+pr}}{J_{CO_2}^{ps}}}$$

The appropriate form of this factor to use in Equation 8.26 depends on which ratio of fluxes is known.

Let us next consider specific values for the various parameters affecting net CO_2 uptake. We will use a CO_2 concentration of 360 µmol mol⁻¹ in the turbulent air, which corresponds to $(360)(0.0410)$, or 14.8 mmol CO_2 m⁻³ at 20°C and a pressure of 0.1 MPa (conversion factor in Table 8.2). Although we do not have reliable measurements of $c_{CO_2}^{chl}$, it may be about 8 mmol m⁻³ for a photorespiring plant at saturating PPFD. At 20°C, respiration plus photorespiration might be 30% as large as net photosynthesis. For purposes of calculation, we will let the gas phase resistance $r_{CO_2}^{bl_1} + r_{CO_2}^{leaf_1}$ be 400. s m⁻¹, let $r_{CO_2}^{mes}$ be 140 s m⁻¹, and let $r_{CO_2}^{chl}$ be 100 s m⁻¹ (see Table 8.4). Using Equation 8.26, we then calculate the net photosynthesis to be

$$J_{CO_2} = \frac{(14.8 \times 10^{-3} \text{ mol m}^{-3} - 8.0 \times 10^{-3} \text{ mol m}^{-3})}{400 \text{ s m}^{-1} + 140 \text{ s m}^{-1} + (1.0 + 0.3)(100 \text{ s m}^{-1})}$$

$$= \frac{(6.8 \times 10^{-3} \text{ mol m}^{-3})}{(670 \text{ s m}^{-1})} = 10.1 \text{ µmol m}^{-2} \text{ s}^{-1}$$

So far, we have considered CO_2 diffusing into a leaf only across its lower surface. In the general case, CO_2 can move in across its upper surface as well, which we can incorporate into our considerations by appropriately reducing the resistance $r_{CO_2}^{bl_1} + r_{CO_2}^{leaf_1}$ ($r_{CO_2}^{mes}$ and $r_{CO_2}^{chl}$ are unaffected when CO_2 diffuses in through both sides of a leaf). If 30% of the CO_2 diffused in through the upper side of a leaf, the effective resistance between the turbulent air and the surfaces of the mesophyll cells would be only 70% as great as $r_{CO_2}^{bl_1} + r_{CO_2}^{leaf_1}$.[18] For instance, $r_{CO_2}^{bl_1} + r_{CO_2}^{leaf_1}$ is here 400 s m^{-1}, so the resistance of this part of the pathway is reduced to $(0.70)(400$ s m$^{-1})$, or 280 s m^{-1}, if 30% of the CO_2 enters through the upper surface. This reduces $r_{CO_2}^{total}$ from 670 s m^{-1} to 550 s m^{-1}, which would raise J_{CO_2} to 12.4 μmol m^{-2} s^{-1} if $c_{CO_2}^{chl}$ were unchanged. Actually, $c_{CO_2}^{chl}$ must increase to lead to the higher photosynthetic rate (Eq. 8.22), offsetting most of the possible increase in J_{CO_2} caused by the lower gas-phase resistance (see Eq. 8.26). For example, if K_{CO_2} is 10 mmol m^{-3}, then simultaneously solving Equations 8.22 and 8.26 for $c_{CO_2}^{chl}$ (see App. III for the solution of a quadratic equation) indicates that $c_{CO_2}^{chl}$ will increase to 8.9 mmol m^{-3} (equivalent to about 220 μmol CO_2 mol^{-1}) and J_{CO_2} will be 10.7 μmol m^{-2} s^{-1} for the lower resistance condition.

Let us next consider what would happen to CO_2 uptake if the stomata provided no resistance to CO_2 entry whatsoever. Instead of a gas-phase resistance for CO_2 of 280 s m^{-1}, it might then be only 80 s m^{-1} for the two leaf surfaces acting in parallel. The $r_{CO_2}^{total}$ would thus be lowered to 350 s m^{-1}. Again simultaneously solving Equations 8.22 and 8.26 for $c_{CO_2}^{chl}$ with K_{CO_2} equal to 10 mmol m^{-3}, we find that J_{CO_2} would be 11.8 μmol m^{-2} s^{-1}. Thus, removing both epidermises completely would enhance CO_2 uptake by only 10%, indicating that the stomata do not greatly restrict the photosynthetic rate in the present case, although their pores occupy only a very small fraction of the leaf surface area. In summary, stomata exert major control on transpiratory water loss while reducing net CO_2 uptake only marginally (see also Farquhar and Sharkey, 1982; Long, 1985).

CO_2 Conductances

Our analysis for CO_2 fluxes could be carried out using conductances. Also, we could divide the CO_2 pathway into a gas-phase component from the turbulent air up to the mesophyll cells and a liquid-phase component representing the mesophyll cells. The drop in CO_2 concentration across the gas phase $\Delta c_{CO_2}^{gas}$ can be related to the CO_2 conductance for the gas phase $g_{CO_2}^{gas}$ and CO_2 concentrations as follows:

$$J_{CO_2} = g_{CO_2}^{gas} \, \Delta c_{CO_2}^{gas} = g_{CO_2}^{gas} \, (c_{CO_2}^{ta} - c_{CO_2}^{ias}) \tag{8.27}$$

18. See p. 422 for our discussion of the analogous situation in transpiration. Instead of knowing the relative flux densities through the two sides, we might know the actual resistances. We could then use Equation 8.11 (with CO_2 replacing wv as subscripts) to determine the resistance of the two leaf surfaces in parallel. Quite often the gas phase resistance for water vapor is measured for the two leaf surfaces together, in which case the gas phase CO_2 resistance is obtained by multiplying r_{wv}^{total} by D_{wv}/D_{CO_2}.

where $c_{CO_2}^{ias}$ is the CO_2 concentration in the intercellular air spaces. We note that $g_{CO_2}^{gas}$ equals $g_{wv}^{total}/1.60$, where g_{wv}^{total} can be defined by Equation 8.13 and 1.60 is the value of D_{wv}/D_{CO_2}. Under optimal photosynthetic conditions, $c_{CO_2}^{ias}$ is nearly twice as high for leaves of C_3 compared with C_4 plants (discussed below). Indeed, $c_{CO_2}^{ias}$ (or $N_{CO_2}^{ias}$) has proved to be an important parameter for evaluating photosynthesis at the mesophyll or liquid-phase level as well as for studying the regulation of stomatal opening.

We can identify a liquid-phase CO_2 conductance for the part of the pathway from the mesophyll cell walls up to the CO_2-fixation enzymes:

$$J_{CO_2} = g_{CO_2}^{liquid}(c_{CO_2}^{ias} - c_{CO_2}^{chl}) \qquad (8.28)$$

If respiration and photorespiration can be ignored, then $1/g_{CO_2}^{liquid}$ equals $1/g_{CO_2}^{mes} + 1/g_{CO_2}^{chl}$ (equivalently, $r_{CO_2}^{liquid} = r_{CO_2}^{mes} + r_{CO_2}^{chl}$). Otherwise, we could return to Figure 8.12 and note that $c_{CO_2}^{ias} - c_{CO_2}^{chl} = J_{CO_2}/g_{CO_2}^{mes} + J_{CO_2}^{ps}/g_{CO_2}^{chl} = J_{CO_2}[1/g_{CO_2}^{mes} + (J_{CO_2}^{ps}/J_{CO_2})/g_{CO_2}^{chl}]$, which in turn equals $J_{CO_2}/g_{CO_2}^{liquid}$ by Equation 8.28. This intermingling of conductances and fluxes again reflects the complication of having more than one source of CO_2 that can be fixed photosynthetically (see Eq. 8.25). If we are interested in the photosynthetic properties of the mesophyll cells themselves, we might wish to express the CO_2 conductance on the basis of mesophyll cell surface area:

$$g_{CO_2}^{cell} = \frac{g_{CO_2}^{liquid}}{A^{mes}/A} \qquad (8.29)$$

where $g_{CO_2}^{cell}$ is the cellular conductance for CO_2.

Range in Photosynthetic Rates

As we might expect, the net rates of photosynthesis vary considerably with plant species, temperature, PPFD and other conditions (Fig. 8.14; see also Šetlik, 1970; Zelitch, 1971). For instance, the maximum J_{CO_2} is often 5 to 10 μmol m⁻² s⁻¹ for the leaves of trees. Certain C_3 crop plants like sugar beet, soybean, and tobacco can have a J_{CO_2} of 20 to 25 μmol m⁻² s⁻¹ at saturating PPFD and temperatures near 30°C. For C_4 plants, J_{CO_2} tends to be larger, because $J_{CO_2}^{r+pr}$ is small and also the liquid-phase resistances ($r_{CO_2}^{mes}$ and $r_{CO_2}^{chl}$) are often relatively small (see Körner et al., 1979; Longstreth et al., 1980). Under optimal conditions of high PPFD and a leaf temperature of 35°C, J_{CO_2} can exceed 40 μmol m⁻² s⁻¹ for bermuda grass, maize, sorghum, sugar cane, and certain other C_4 plants (as well as a few C_3 species). An extremely high value of 67 μmol m⁻² s⁻¹ can

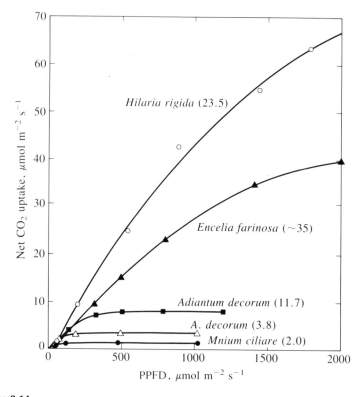

Figure 8.14

Photosynthetic responses to photosynthetic photon flux density for species differing in mesophyll surface area per unit leaf area. Curves were obtained at ambient CO_2 and O_2 concentrations, optimal temperatures, and the A^{mes}/A indicated in parentheses. [Sources: for the C_4 desert grass *Hilaria rigida*, P. S. Nobel, *Ecology 61*:252–258 (1980); for the C_3 desert composite *Encelia farinosa*, J. Ehleringer, O. Björkman, and H. A. Mooney, *Science 192*:376–377 (1976); and for the C_3 maidenhair fern *Adiantum decorum* and the C_3 moss *Mnium ciliare*, P. S. Nobel, *Physiologia Plantarum 40*:137–144 (1977).]

occur for the C_4 plant *Hilaria rigida* at full sunlight (Fig. 8.14) and for a few other species (see Longstreth *et al.*, 1980).

The influence of A^{mes}/A on J_{CO_2} deserves special comment. A^{mes}/A can be two or more times larger for sun leaves than for shade leaves on the same plant; this reduces $r_{CO_2}^{total}$ and thus enhances the maximal photosynthetic rates of sun leaves compared with shade leaves. Maximal rates of CO_2 uptake per unit area for C_3 plants can thus vary with A^{mes}/A for leaves on a single plant, on different plants of the same species, and sometimes even between species (Fig. 8.14). The large area of the cell walls of mesophyll cells that is available for CO_2 diffusion keeps $r_{CO_2}^{cw}$ and the other CO_2 resistances in the mesophyll cells at reasonable values.

For instance, if the mesophyll cells were tightly packed into a layer with no intervening air spaces, A^{mes}/A could equal 2.0. This occurs for the moss *Mnium ciliare,* whose leaves are one cell thick with the lateral walls completely touching; thus, the only area available for CO_2 to diffuse from the gas phase into the cells is their end walls, which have a total area twice that of one side of the leaf (see Nobel, 1980). Instead of a mesophyll resistance of 140 s m⁻¹ that we calculated for an A^{mes}/A of 20 (Table 8.4), $r^{mes}_{CO_2}$ would be 1 400 s m⁻¹ for an A^{mes}/A of 2. Thus, the evolution of a leaf anatomy with an abundance of mesophyll cell surface area leading to a large value for A^{mes}/A allows $r^{mes}_{CO_2}$ and $r^{chl}_{CO_2}$ to be fairly low, with a correspondingly high value for J_{CO_2} (see Fig. 8.14).

Many units are used to express photosynthesis and CO_2 fluxes for leaves. Conversion factors for some of the more common units are summarized in Table 8.2. For example, 16 mg CO_2 dm⁻² h⁻¹ corresponds to (16)(0.631), or 10 μmol m⁻² s⁻¹. The amount of chlorophyll per unit leaf area generally ranges from 0.2 to 0.8 g m⁻², with 0.4 to 0.5 g m⁻² being typical (see Björkman, 1981). Thus, 10 μmol m⁻² s⁻¹ might correspond to (10 μmol m⁻² s⁻¹)/(0.4 g chlorophyll m⁻²), or 25 μmol (g chlorophyll)⁻¹ s⁻¹, which equals 90 μmol CO_2 fixed (mg chlorophyll)⁻¹ h⁻¹.

Environmental Productivity Indices

From our above discussion we can surmise that water status, temperature, and PPFD all affect stomatal opening and the photosynthetic rates of leaves. Sometimes such environmental effects are incorporated into photosynthetic models by their influences on $g^{st}_{CO_2}$ (Eq. 8.5) or V_{max} (Eq. 8.22). At other times, how effects occur is merely implied by indicating their resulting influence on net CO_2 uptake. Indeed, indices have been proposed relating J_{CO_2} and the associated plant productivity to rainfall and the water status of the plants (see Le Houérou, 1984); to the ambient temperature, including the accumulated time that the temperature is above some minimum value (see Long and Woodward, 1988); and to the intercepted radiation (see Monteith, 1977; Jones, 1983). In the latter case, net CO_2 uptake depends on the PPFD absorbed times a conversion efficiency that in turn depends on water status and temperature.

Recognizing that various environmental factors can simultaneously limit net CO_2 uptake, an environmental productivity index (EPI) has been proposed to help predict daily net CO_2 uptake in the field based on J_{CO_2} for individual leaves (see Nobel, 1984, 1988):

$$\text{EPI} = \text{Water Index} \times \text{Temperature Index} \qquad (8.30)$$
$$\times \text{ PPFD Index}$$

where each of the component indices ranges from zero, when limitations by that factor abolish net CO_2 uptake, to unity, when that factor is optimal for net CO_2 uptake. To determine values for each component index, effects of individual factors on J_{CO_2} are determined over 24-h periods in the laboratory when the other environmental conditions are held constant. Then the values of each component index are calculated for the environmental conditions prevailing in the field to predict EPI, which hence represents the fraction of maximum net CO_2 uptake expected over the course of a day for a leaf under specific environmental conditions. This EPI times the maximum net CO_2 uptake measured in the laboratory for a 24-h period when the soil is wet (Water Index = 1.00), the air temperature is optimal for net CO_2 uptake (Temperature Index = 1.00), and the PPFD is saturating for photosynthesis (PPFD Index = 1.00) gives the daily CO_2 uptake under field conditions. Such measurements of maximum net CO_2 uptake can be made at any ambient CO_2 level and under the nutrient status pertinent to the field situation. Alternatively, a Nutrient Index can be incorporated into EPI to take into consideration the effect of soil elements on leaf net CO_2 uptake (see Nobel, 1988, 1989). Indeed, J_{CO_2} increases nearly linearly with leaf nitrogen content for a wide variety of plant species (see Field and Mooney, 1986).

Using EPI, net CO_2 uptake can be predicted at various times of the year, at various locations, and under various environmental conditions, such as under elevated atmospheric CO_2 levels and its associated suite of environmental changes (see Ramanathan, 1988) or as a function of elevation. For instance, net CO_2 uptake over 24-h periods is known for leaves of a common succulent of the Sonoran Desert, *Agave deserti*, for wet conditions as well as various durations of drought, for day/night temperatures encompassing the entire temperature range occurring in its native habitat, and from darkness up to full sunlight, so all three component indices of EPI (Eq. 8.30) can be calculated (see Nobel, 1984). In the summer, EPI for *A. deserti* increases fourfold from an elevation of 300 m to one of 1 300 m, primarily reflecting a threefold increase in rainfall, which raises the Water Index, and a nearly 10°C decrease in temperature, which raises the Temperature Index, as the warm temperatures at low elevations at this time of the year are considerably above those optimal for net CO_2 uptake (see Nobel and Hartsock, 1986). In the winter, EPI increases to mid elevations, reflecting an increase in the Water Index, and then decreases at higher elevations, reflecting the overriding importance of a decreasing Temperature Index at this cold time of year. The seasonal changes in EPI with elevation agree well with changes in productivity measured independently, indicating that a productivity index based on net CO_2 uptake determined in the laboratory, where one factor is varied at a time, can help interpret field CO_2 uptake and productivity when various environmental factors are changing simultaneously. Although secondary interactions do occur, such as a lower PPFD required for saturation of J_{CO_2} at suboptimal temperatures, EPI can be used to predict the major influences of climate on net CO_2 uptake in the field (see Nobel, 1988).

WATER-USE EFFICIENCY

As we have indicated, stomatal opening leading to the CO_2 uptake necessary for photosynthesis results in an inevitable loss of water. A useful parameter relating the two fluxes involved and showing the total CO_2 fixed (benefit) per unit water lost (cost) is the *water-use efficiency*, WUE:

$$\text{WUE} = \frac{\text{mass } CO_2 \text{ fixed}}{\text{mass } H_2O \text{ transpired}} \qquad \text{mass basis} \qquad (8.31a)$$

$$\text{WUE} = \frac{\text{mol } CO_2 \text{ fixed}}{\text{mol } H_2O \text{ transpired}} \qquad \text{mole basis} \qquad (8.31b)$$

A convenient unit for WUE on a mass basis, which is the more commonly used index, is g CO_2 (kg H_2O)$^{-1}$. A related quantity is the *transpiration ratio*, which is the reciprocal of the water-use efficiency and hence represents the water lost per CO_2 fixed.

Values of WUE

From the values of J_{CO_2} and J_{wv} calculated in this chapter, we can determine a WUE for the leaf of a representative C_3 mesophyte. Specifically, we obtained a J_{CO_2} of 10.7 μmol CO_2 fixed m^{-2} s^{-1} (p. 450) and a J_{wv} of 4.3 mmol H_2O transpired m^{-2} s^{-1} (p. 423). By Equation 8.31b, the water-use efficiency is

$$\text{WUE} = \frac{(10.7 \times 10^{-6} \text{ mol } CO_2 \text{ m}^{-2} \text{ s}^{-1})}{(4.3 \times 10^{-3} \text{ mol } H_2O \text{ m}^{-2} \text{ s}^{-1})} = 0.0025 \text{ } CO_2/H_2O$$

On a mass basis, this corresponds to a WUE of 6.1 g CO_2 (kg H_2O)$^{-1}$ (the molar masses of CO_2 and H_2O are 44.0 g mol^{-1} and 18.0 g mol^{-1}, respectively). We also note that the transpiration ratio in the above case is 400 H_2O/CO_2. This substantial water loss per CO_2 fixed is generally not a problem when plenty of water is available for transpiration. Plants in such environments often have a high g_{wv}^{total}, which leads to a somewhat higher $g_{CO_2}^{total}$ and somewhat higher rates of photosynthesis than for plants with a moderate g_{wv}^{total}.

Any loss of water can be potentially harmful for plants growing in arid regions, many of which have evolved a novel way of fixing CO_2 in a manner leading to a high water-use efficiency (see Osmond *et al.*, 1989). For example, many species in the family Crassulaceae, as well as other desert succulents, have their stomata closed during the daytime. This greatly reduces transpiration, but also essentially eliminates the net influx of CO_2 at that time. The stomata open at night, CO_2 then diffuses in, and it is fixed into malate (e.g., by carboxylation of phosphoenolpyruvate) and other organic acids. During the next day these organic acids are decarboxylated, and the released CO_2 is retained within the plant because of the closed stomata. This CO_2 is then fixed into photosynthetic products by means of the C_3 pathway. Plants with this CO_2 fixation mechanism are referred to as CAM (Crassulacean acid metabolism) plants, because such reactions were initially studied extensively in the Crassulaceae, although apparently first observed in the Cactaceae in 1804.

As just indicated, stomata for CAM plants tend to open at night, when leaf and air temperatures are lower than daytime values. The concentration of water vapor in the pores of the cell walls of chlorenchyma cells (c_{wv}^e) is then much lower, markedly reducing the rate of transpiration. For example, leaf temperatures of *Agave deserti* can be 25°C in the afternoon and 5°C at night (Fig. 8.15), leading to saturation water vapor concentrations of 23.1 g m^{-3} and 6.8 g m^{-3}, respectively (App. I). For air with a water vapor content of 4.0 g m^{-3}, which is fairly typical during the wintertime in the native habitat of *Agave deserti,* Δc_{wv}^{total} is $(23.1 - 4.0)/(6.8 - 4.0)$ or 7 times higher at 25°C than at 5°C, and therefore so is J_{wv} for the same degree of stomatal opening. Clearly, nocturnal stomatal opening can result in water conservation and hence a higher water-use efficiency. For the CAM plant *Agave deserti* on the day depicted in Figure 8.15, the WUE was 56 g CO_2 (kg H_2O)$^{-1}$, and it was 40 g CO_2 (kg H_2O)$^{-1}$ when averaged over a whole year (see Nobel, 1976)—both very high values.

Changes in the thickness of the air boundary layers adjacent to a leaf have a greater influence on the flux of water vapor than on the flux of CO_2. For instance, the total resistance for water vapor diffusion can equal $r_{wv}^{bl_1} + r_{wv}^{st_1} + r_{wv}^{ias}$ (Eq. 8.14), whereas $r_{CO_2}^{bl_1} + r_{CO_2}^{st_1} + r_{CO_2}^{ias}$ (Eq. 8.17) is generally just over half of the total resistance for CO_2 diffusion. Thus, changes in wind speed have a smaller fractional effect on $r_{CO_2}^{total}$ than on r_{wv}^{total}. Similarly, partial stomatal closure can appreciably reduce g_{wv}^{total}, but lead to a smaller fractional reduction in $g_{CO_2}^{total}$. Certain xerophytes have fairly low maximal values for g_{wv}^{st}—the maximal stomatal conductance is generally less than 2 mm s^{-1} for *Agave deserti* (see Fig. 8.15) compared with over 10 mm s^{-1} for many mesophytes (Table 8.1). Such reduced

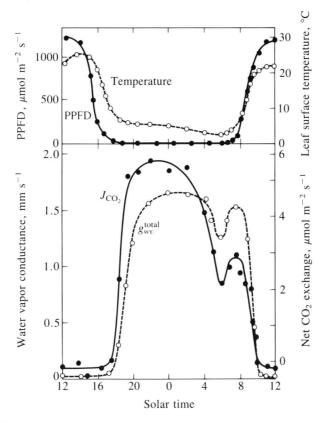

Figure 8.15

Photosynthetic photon flux density (on a horizontal surface), leaf surface temperature, water vapor conductance, and net CO_2 exchange for *Agave deserti* on clear winter days in the northwestern Sonoran Desert. (Modified from Nobel, 1976; used by permission.)

maximal stomatal conductance reduces transpiration to a greater degree than CO_2 uptake, with a consequent enhancement in water-use efficiency (see Eq. 8.27).

Even though photosynthesis and transpiration depend on environmental conditions, we can still make some generalizations about WUE for different types of plants. Specifically, WUE's averaged over a day for mature leaves are usually 1 to 3 g CO_2 (kg H_2O)$^{-1}$ for C_3 plants, 2 to 5 g CO_2 (kg H_2O)$^{-1}$ for C_4 plants, and 10 to 40 g CO_2 (kg H_2O)$^{-1}$ for CAM plants (see Osmond *et al.*, 1980; Szarek and Ting, 1975). C_4 plants have approximately double the WUE of C_3 plants, because C_4 plants tend to have lower gas-phase conductances (which conserves water with a relatively small negative effect on photosynthesis) and higher liquid-phase conductances (which affects photosynthesis positively) than do C_3 plants (see also Körner et al., 1979; Longstreth et al., 1980). However, maximizing WUE may not always be adaptive—e.g., water may not be limiting for an aquatic plant.

Elevational Effects on WUE

We would expect both transpiration and photosynthesis to be affected by elevation, because diffusion coefficients depend inversely on ambient (barometric) pressure $[D_j = D_j^0 (T/273)^{1.8} (P^0/P)$, Eq. 8.9] and also the partial pressures of water vapor and CO_2 generally decrease with elevation. At sea level, barometric pressure averages 0.101 MPa, and it averages 0.079 MPa at 2 000 m and about 0.054 MPa at 5 000 m. Thus, diffusion coefficients are nearly twice as large at 5 000 m as at sea level owing to this pressure change, which correspondingly increases the gas-phase conductances based on Δc (e.g., Eq. 8.2), although those based on ΔN (Eq. 8.8) are unchanged (see p. 410). For a typical lapse rate of $-5°C$ per km of elevation,[19] temperatures would decrease from 30°C at sea level to 5°C at 5 000 m, which by itself would decrease diffusion coefficients by 14% according to Equation 8.9. The partial pressure of CO_2 is reduced more or less in concert with the reduced barometric pressure; i.e., the mole fraction of CO_2 is approximately constant with elevation. The partial pressure of water vapor in the air also tends to decrease with elevation, and under isothermal conditions the driving force for water loss (both Δc_{wv} and ΔN_{wv}) increases, as does transpiration.[20]

Because of the interaction of many factors, especially the numerous temperature effects on both transpiration and photosynthesis, the effect of elevation on water-use efficiency is difficult to predict. When the turbulent mixing aspect in the leaf boundary layer is ignored (see footnote 11, this chapter), the higher D_{CO_2} in the gas phase at higher elevations tends to offset the lower P_{CO_2} as far as the CO_2 level in the intercellular air spaces is concerned. If we ignore changes

---

19. At the dry adiabatic lapse rate (9.8°C decrease in temperature per km increase in altitude), a rising parcel of dry air will cool by expansion due to the decrease in air pressure and will achieve the same temperature as the surrounding air—a case of neutral stability. That is, air movement is then neither favored nor retarded by buoyancy (see Monteith, 1973; Oke, 1987). Observed lapse rates are often $-5°C$ to $-6°C$ km^{-1}.

20. Because P_{wv}^* is essentially independent of P and N_{wv}^* equals P_{wv}^* /P, the mole fraction of water vapor in the leaf tends to increase as ambient pressure decreases, i.e., at higher elevations (actually, temperature decreases with elevation generally more than offset the effects of P changes on N_{wv}^*).

in stomatal aperture and temperature, the reduction in P_{CO_2} with elevation thus translates into smaller fractional decreases in $P_{CO_2}^{ias}$ and $c_{CO_2}^{ias}$, and $N_{CO_2}^{ias}$ is unchanged if J_{CO_2} is unchanged. Diffusion coefficients in the liquid phase are unaffected by barometric pressure. Hence, liquid-phase conductances expressed in mm s^{-1} are unaffected by the pressure changes with elevation, but those expressed in mmol m^{-2} s^{-1} are proportional to P (see p. 410) and hence decrease with elevation. Based on these changes in gas-phase and liquid-phase conductances, J_{CO_2} tends to decrease with elevation—the greater is the liquid-phase conductance relative to that of the gas phase, the less is the elevational effect (see Gale, 1972a). The lowering of temperature with elevation could have an even larger effect on photosynthesis, although the optimal temperature for photosynthesis can acclimate (usually by 2°C to 15°C) to match the average ambient temperature of the environment (see Berry and Björkman, 1980; Osmond *et al.*, 1980; Patterson, 1980).

As we discussed in Chapter 7, many factors affect leaf temperature, and an energy budget analysis can be used to calculate T^{leaf} and thus to indicate the effects of elevation on transpiration. Leaves at higher elevations generally experience a higher net radiation balance (more incident shortwave irradiation although somewhat less incident longwave irradiation; see Eq. 7.6) and lower air temperatures. For small leaves, which tend to be close to air temperature, transpiration generally decreases with elevation. For a 5 cm × 5 cm leaf in full sunlight, transpiration is unaffected at a lapse rate of -1°C km^{-1}, but decreases 25% at 1 000 m compared to sea level at a lapse rate of -5°C km^{-1} (Gale, 1972b). For large leaves in full sunlight, transpiration at typical lapse rates may decrease only slightly with elevation, because large leaves are further above air temperature at higher elevations (Smith and Geller, 1979). For temperature inversion conditions (increasing temperature with elevation), transpiration can actually increase with elevation (see Gale, 1972b). Because photosynthesis and transpiration can vary in so many ways with elevation, effects of elevation on WUE must be judged case by case.

Stomatal Control of WUE

To maximize water-use efficiency, stomatal opening must be synchronized with the capability for CO_2 fixation (see Cowan, 1977, 1981; Cowan and Farquhar, 1977; Farquhar *et al.*, 1978; Raschke, 1975). Stomatal opening can be regulated by the CO_2 level in the intercellular air spaces, a depletion of $N_{CO_2}^{ias}$ by photosynthesis leading to an increase in $g_{CO_2}^{st}$, which then lets more CO_2 into the leaf under a PPFD and other conditions favorable for photosynthesis. This is an example of a *feedback* system, as $N_{CO_2}^{ias}$ feeds a signal back to the stomata, which in turn leads to a change in $N_{CO_2}^{ias}$. Also, the PPFD may directly affect the metabolism of guard cells, which indeed contain chloroplasts and hence can utilize such radiation. This is an example of a *feedforward* system, as changes in stomatal aperture due to photosynthetic responses of chloroplasts feed forward (or anticipate) and adjust CO_2 entry into the leaf, thereby matching photosynthesis by the mesophyll region to environmental conditions.

Stomatal opening is also affected by the leaf water status. For instance, stomata tend to close as a leaf begins to wilt, especially after the leaf water potential drops below some threshold level and abscisic acid (ABA) is produced (as we mentioned on p. 402). In fact, ABA can induce stomatal closure even when $N_{CO_2}^{ias}$ favors opening. The water status thus affects stomatal opening and hence transpiration, which in turn feeds back onto the leaf water status. Stomatal opening is usually increased by higher N_{wv}^{ta}. This is another example of a feed-forward system, as it anticipates the effect of the ambient water vapor concentration on transpiration—e.g., higher N_{wv}^{ta} means less "force" leading to water loss from a leaf, so the stomata can open wider without leading to excessive transpiration.[21] These various processes regulating stomatal movements interact with each other—we will examine the consequences of this for gas exchange by leaves.

As a refinement on our consideration of water-use efficiency, we should consider what the optimal behavior of stomata might be over the course of a day. For instance, WUE is maximized by minimal stomatal opening, because transpiration is decreased more than photosynthesis by partial stomatal closure (see Eq. 8.31); i.e., J_{wv} changes proportionally more than does J_{CO_2} as g_{wv}^{st} changes (Fig. 8.16). However, this can lead to very little CO_2 uptake. Thus, a more pertinent consideration might be what is the maximum amount of CO_2 that can be taken up for a certain amount of water transpired. The amount of water lost depends on plant condition and environmental factors and should be considered over the course of a whole day. To help analyze the relationship between gas fluxes, curves showing J_{CO_2} versus J_{wv} can be drawn for any PPFD, temperature, wind speed, or relative humidity occurring for a given leaf during the day, the location of the curves varying but still conforming to the general shape illustrated in Figure 8.16. In addition to environmental factors, the location of the curves is influenced by leaf properties such as size, age, A^{mes}/A, and shortwave absorptance. Nevertheless, it has been hypothesized (see Cowan and Farquhar,

---

21. When N_{wv}^{ta} is low, the feedforward system can cause stomatal closure to such an extent that transpiration can actually be *less* under such conditions than at high humidity (Schulze *et al.*, 1972).

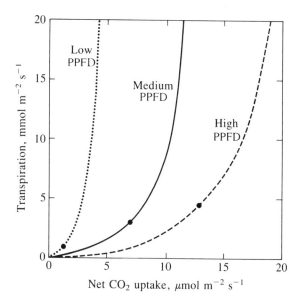

Figure 8.16

Relation between net photosynthesis (J_{CO_2}) and transpiration (J_{wv}) as stomatal conductance is varied. The three curves depict various PPFD levels, indicated as "low," "medium," and "high." The circles indicate where the slope $\partial J_{wv}/\partial J_{CO_2}$ is 1 000 H_2O/CO_2. Cuticular transpiration is ignored.

1977) that the stomata will open or close depending on the various feedback and feedforward processes in such a way that the following relation is obeyed:

$$\frac{\partial J_{wv}/\partial g^{st}_{wv}}{\partial J_{CO_2}/\partial g^{st}_{wv}} = \frac{\partial J_{wv}}{\partial J_{CO_2}} = \lambda = \text{constant} \qquad (8.32)$$

where $\partial J_{wv}/\partial g^{st}_{wv}$ and $\partial J_{CO_2}/\partial g^{st}_{wv}$ represent the sensitivity of transpiration and photosynthesis, respectively, to changes in stomatal conductance.

The solid circles in Figure 8.16 indicate the CO_2 and H_2O fluxes that can occur at different times of the day for a leaf with a λ of 1 000 H_2O/CO_2. At low PPFD, little stomatal opening occurs, so little water is used under conditions where the rate of photosynthesis inherently cannot be very high. Stomatal opening is much greater at high PPFD and thus both transpiration and photosynthesis then are greater, but the local slope of the J_{CO_2} versus J_{wv} curve is still the same for points of the same λ. Let us next consider what happens to the fluxes if we move along a curve away from one of the solid circles, e.g., the one for medium

PPFD (Fig. 8.16). If the stomatal opening changes so that J_{wv} increases by 1.0 mmol m^{-2} s^{-1}, then J_{CO_2} increases by 0.8 μmol m^{-2} s^{-1}. To lead to the same total transpiration for the day, we must decrease J_{wv} by the same amount at another time, which for simplicity we can also consider for the medium PPFD curve (note that all the solid circles occur for the same slope, 1 000 H$_2$O/CO$_2$). A decrease in J_{wv} of 1.0 mmol m^{-2} s^{-1} is accompanied by a decrease in J_{CO_2} of 1.4 μmol m^{-2} s^{-1}. When the effect of both changes is considered, we get the same total transpiration, but a lower net CO$_2$ uptake. In fact, the criterion expressed in Equation 8.32 leads to the maximum amount of CO$_2$ fixed for a particular amount of water transpired as well as to the minimum amount of water transpired for a particular amount of CO$_2$ fixed in a day (see Cowan and Farquhar, 1977). Thus, if stomata respond to keep $\partial J_{wv}/\partial J_{CO_2}$ constant, then the water-use efficiency of the leaf for the entire day is maximized.

The value of λ can change during the growth of a plant. For instance, λ can be small (e.g., 300 H$_2$O/CO$_2$) when water is in short supply. In such cases, constancy of λ requires that stomata close partially near midday, when temperatures are the highest and transpiration is potentially the greatest. Also, water stress conditions in general lead to higher ABA levels in the leaves and a lower λ. On the other hand, a large λ (e.g., 1 300 H$_2$O/CO$_2$) occurs when the plant is not under water stress, and no midday stomatal closure then takes place.

The proposed constancy of $\partial J_{wv}/\partial J_{CO_2}$ helps interpret the partial stomatal closure that occurs at midday when water is limiting, as well as the nocturnal stomatal closure, when PPFD is limiting. We can also use the constancy of $\partial J_{wv}/\partial J_{CO_2}$ to help interpret experiments in which a single environmental factor is varied, such as the driving force for water vapor loss, ΔN_{wv}. If the stomata maintained a constant J_{wv}, then changes in g_{wv}^{st} would be the inverse of changes in ΔN_{wv} (when cuticular transpiration is ignored). On the other hand, maintenance of constant J_{CO_2} as ΔN_{wv} is varied requires constancy of $g_{CO_2}^{st}$, which equals $g_{wv}^{st}/1.60$. In fact, however, varying ΔN_{wv} over a fourfold range for *Nicotiana glauca* (tobacco), *Corylus avellana* (hazel; Farquhar *et al.*, 1980a), and *Vigna unguiculata* (cowpea; Hall and Schulze, 1980) leads to stomatal behavior resulting in variation of both J_{wv} and J_{CO_2}, but $\partial J_{wv}/\partial J_{CO_2}$ is approximately constant. Consideration of other implications and determining the range of validity of the hypothesis represented by Equation 8.32 await future research.

C$_3$ versus C$_4$ Plants

We will next recapitulate some of the characteristics of C$_3$ and C$_4$ plants that we have introduced. After examining the influence of stomata on maximum photosynthetic rates under optimal conditions, we will predict effects on WUE for elevated levels of atmospheric CO_2.

The ecological advantages of the C_4 pathway are most apparent for plants in environments having high PPFD, high temperature, and limited water supply. C_4 plants are effective at high PPFD, as photosynthesis for leaves of many C_3 plants saturates below 600 μmol m^{-2} s^{-1}, but most C_4 plants have an increasing J_{CO_2} as the PPFD is raised up to 2 000 μmol m^{-2} s^{-1} (see Fig. 8.14). Optimal temperatures for net CO_2 uptake are usually 20°C to 35°C for C_3 plants but 30°C to 45°C for C_4 plants, which can be interpreted by considering ribulose-1,5-bisphosphate carboxylase/oxygenase; the CO_2-evolving photorespiration, which has very low rates in C_4 plants, becomes proportionally more important at higher temperatures and thus reduces the net CO_2 uptake at the higher temperatures for C_3 plants. The optimal temperature for photosynthesis is actually fairly variable, because it can change by 10°C in a matter of days (even for mature leaves), allowing for seasonal acclimation of photosynthetic performance. C_4 plants can more readily cope with limited water supply, because the gas-phase conductance for CO_2 is relatively lower and the liquid-phase CO_2 conductance is relatively higher for C_4 than for C_3 plants. Thus, C_4 plants tend to become dominant in deserts, grasslands, and certain subtropical regions, i.e., areas of high PPFD, high temperature, and limited water supply, as indicated above (CAM plants have an even higher WUE than do C_4 plants and achieve their greatest relative importance in regions of high PPFD and very limited water supply). Low rates of photorespiration and the associated high WUE allow C_4 plants to become very successful weeds—in fact, 8 out of the 10 agriculturally most noxious weeds use the C_4 pathway (see Holm *et al.*, 1977).

At 30°C and for an absorbed PPFD up to about 100 μmol m^{-2} s^{-1}, leaves of C_3 and C_4 plants can have a very similar quantum yield (approximately 0.053 mol CO_2/mol photons at an $N_{CO_2}^{ta}$ of 325 μmol mol^{-1}; Ehleringer and Björkman, 1977). As the temperature is raised, however, photorespiration increases relative to photosynthesis, so the quantum yield declines for C_3 plants but is essentially unchanged for C_4 plants. On the other hand, lowering the ambient O_2 level raises the quantum yield for C_3 (photorespiring) plants, because the oxygenase activity of RuBisCO (see Fig. 8.10) is then suppressed; such changes have little effect on C_4 plants until the O_2 level falls below about 2%, where mitochondrial respiration is affected.

CO_2 uptake by C_4 plants is CO_2-saturated at a relatively low CO_2 level in the intercellular air spaces. For instance, an $N_{CO_2}^{ias}$ equivalent to 150 μmol mol^{-1} usually leads to over 90% of the maximum J_{CO_2} for C_4 plants (Fig. 8.13), so increasing the ambient CO_2 level usually has little effect on the quantum yield. But the quantum yield for CO_2 fixation by C_3 plants progressively increases as the ambient CO_2 level is raised, and at an $N_{CO_2}^{ias}$ corresponding to 700 μmol mol^{-1} it approaches to within 10% of the value occurring when the ambient O_2 level is reduced 10-fold (about 0.081 mol CO_2/mol photons). Such raising of the CO_2 level is another way of favoring the carboxylase activity of RuBisCO (see footnote 16, this chapter). Also, the requirement for a high $N_{CO_2}^{ias}$ for C_3 plants is consistent with the K_{CO_2} of 5 to 15 mmol m^{-3} for CO_2 fixation by RuBisCO (noted on p. 439).

For a series of both C_3 and C_4 plants, the stomata open to a degree that gives an approximately constant CO_2 level in the intercellular air spaces, the level differing between plants representing the two photosynthetic pathways (Fig. 8.17; see Wong et al., 1979). A similar adjustment in stomatal conductance occurs as the PPFD increases for a particular plant (see Fig. 8.16). We must therefore conclude that stomata regulate the entry of CO_2 to match the photosynthetic capability of the mesophyll region. The slope of $g_{CO_2}^{st}$ versus J_{CO_2} gives the drop in CO_2 across the stomata; i.e., $J_{CO_2} = g_{CO_2}^{st} \Delta N_{CO_2}^{st}$, so the slope is $J_{CO_2}/$

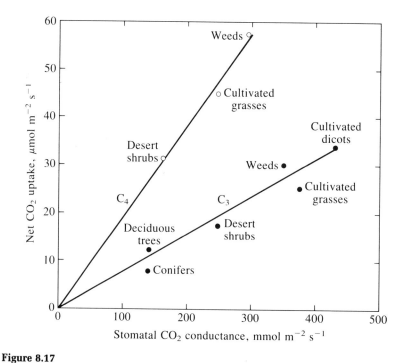

Figure 8.17

Relation between stomatal CO_2 conductance ($g_{CO_2}^{st}$) and net CO_2 uptake (J_{CO_2}) for various categories of C_3 and C_4 plants under optimal conditions and 340 μmol CO_2 mol^{-1}. (Data are from references cited in Tables 8.1 and 8.4.)

$g_{CO_2}^{st} = \Delta N_{CO_2}^{st}$. For the C_3 plants in Figure 8.17, the slope is 74×10^{-6} (i.e., 74 μmol mol^{-1}), and for the C_4 plants it is 189×10^{-6}. For an ambient CO_2 level of 340 μmol mol^{-1}, and ignoring the CO_2 drop across the boundary layer (about 10 to 30 μmol mol^{-1}), $N_{CO_2}^{ias}$ corresponds to 270 μmol mol^{-1} for C_3 plants and 150 μmol mol^{-1} for C_4 plants. [Changes in $N_{CO_2}^{ias}$ can occur, such as higher values at lower PPFD in the lower parts of a canopy (see Farquhar *et al.*, 1989).] Even though $N_{CO_2}^{ias}$ is lower for C_4 plants, it is still high enough to saturate their CO_2 fixation pathway. A higher $N_{CO_2}^{ias}$ brought about by a higher $g_{CO_2}^{st}$ does not benefit photosynthesis for C_4 plants, but the accompanying greater stomatal conductance would lead to more water loss. For a C_3 plant, photosynthesis does not approach saturation until $N_{CO_2}^{ias}$ exceeds 700 μmol mol^{-1} (see Fig. 8.13). However, opening stomata further than that required to maintain an $N_{CO_2}^{ias}$ of about 270 μmol mol^{-1} would not enhance photosynthesis very much, but it would considerably increase transpiration (see Fig. 8.16). Thus, the adjustment of stomatal opening to meet the conflicting demands of photosynthesis and transpiration, using feedback and feedforward control by microclimatic and leaf parameters, leads to a remarkable regulation that minimizes water loss while maximizing CO_2 uptake.

As a final topic in this chapter, let us consider what will happen to the water-use efficiency of C_3 and C_4 plants as the atmospheric level of CO_2 increases (see Morison, 1987). We will assume that other conditions such as temperature, water status, and nutrients are not severely limiting net CO_2 uptake. As the CO_2 level in the turbulent air increases, $N_{CO_2}^{ias}$ will increase, which will decrease stomatal opening and hence transpiration. For a doubling of atmospheric CO_2, transpiration should decrease 30% to 40% for leaves of both C_3 and C_4 plants. On the other hand, elevated CO_2 will increase the rate of photosynthesis for C_3 plants, approximately 30% for a doubling in atmospheric CO_2 levels, but should have no major effect on the photosynthetic rate of C_4 plants (Fig. 8.13). Thus, as atmospheric CO_2 levels become twice as high as the 1990 level in the latter half of the twenty-first century, the WUE should increase about 35% for C_4 plants and 75% for C_3 plants. Only the future will tell whether such predictions are borne out and what their ecological and agronomic ramifications will be.

Problems

8.1 Consider a leaf that is 0.5 mm thick with 64 stomata per mm^2. Approximate the stomatal opening by a rectangle that is 6 μm by 20 μm with a depth of 25 μm.

Assume that the leaf and air temperatures are both 20°C and the ambient air pressure is 1 atmosphere. (a) What are g_{wv}^{bl} (in mm s^{-1}) and r_{wv}^{bl} if the boundary layer is 0.8 mm thick? (b) What are na^{st} and the effective r^{st}? (c) What is the average flux density of water vapor within the stomatal pores compared with that across the boundary layer? (d) What is g_{wv}^{st} in mm s^{-1} and mmol m^{-2} s^{-1}? What are the values if the ambient air pressure is reduced to 0.9 atmosphere? (e) What is g_{wv}^{ias} in the two units if the effective path length in the intercellular air spaces equals the leaf thickness? (f) Suppose that each stoma is sunken in a cylindrical cavity 50 μm across and 100 μm deep. What additional resistance to water vapor diffusion does this provide?

8.2 Suppose that $g_{wv}^{bl_j}$ is 20 mm s^{-1}, $g_{wv}^{st_j}$ is 6 mm s^{-1}, g_{wv}^{c} is 0.1 mm s^{-1}, and g_{wv}^{ias} is 40 mm s^{-1}. (a) What is g_{wv}^{total} if water vapor diffuses out only across the lower epidermis of the leaf? (b) What are the three g_{wv}^{total}'s in (a) if the cuticular pathway is ignored, if the intercellular air spaces are ignored, and if both g_{wv}^{c} and g_{wv}^{ias} are ignored? (c) What is g_{wv}^{total} if the stomata in the upper epidermis have the same conductance as in the lower one? (d) What is g_{wv}^{total} if 28% of J_{wv} is through the upper epidermis? (e) Suppose that the leaf temperature is 30°C, the air in the cell wall pores where the water evaporates is at 99% relative humidity, and c_{wv}^{ta} is 7.5 g m^{-3}. What is J_{wv} through the lower epidermis? (f) What are g_{wv}^{total} in mmol m^{-2} s^{-1} and J_{wv} (in mmol m^{-2} s^{-1}) under the conditions of (e)? Assume that the air pressure is 1 atmosphere and that N_{wv}^{ta} is 0.0103. (g) Under the conditions of (e) and (f), what is the drop in water vapor concentration and mole fraction along the stomatal pores (ignore cuticular transpiration)?

8.3 Suppose that a shade leaf has a layer of tightly packed palisade mesophyll cells with rectangular sides that are externally 30 μm by 100 μm and with square ends 30 μm by 30 μm (the long dimension is perpendicular to the leaf surface). Suppose that there are two spherical spongy mesophyll cells (30 μm in diameter) under each palisade cell. Let the cell wall thickness of mesophyll cells be 0.2 μm, the mean distance from the plasmalemma to the chloroplasts be 0.1 μm, and the average distance that CO_2 diffuses in the chloroplasts before reaching the photosynthetic enzymes be 0.5 μm. (a) What is A^{mes}/A if essentially the entire surface area of the mesophyll cells is exposed to the intercellular air spaces? (b) Assume that a sun leaf on the same plant has two layers of palisade cells and half as many spongy mesophyll cells. If the dimensions of the cells are the same as for the shade leaf, what is A^{mes}/A for the sun leaf? (c) If $D_{CO_2}^{cw}$ is 5.0×10^{-10} m^2 s^{-1}, what is the maximum value for $r_{CO_2}^{cw}$ at 20°C for the shaded leaf? (d) If P_{CO_2} is 1.0×10^{-3} m s^{-1} for the plasmalemma and the chloroplast limiting membranes, what are $r_{CO_2}^{pl}$ and $r_{CO_2}^{clm}$ (shade leaf)? (e) If $D_{CO_2}^{cyt}$ and $D_{CO_2}^{stroma}$ are 1.0×10^{-9} m^2 s^{-1}, what are $r_{CO_2}^{cyt}$ and $r_{CO_2}^{stroma}$ (shade leaf)? Assume that the relevant partition coefficients for the various forms of CO_2 are unity. (f) What is the resistance to CO_2 diffusion from the intercellular air spaces to the photosynthetic enzymes for the sun leaf? Assume that $J_{CO_2}^{r+pr}$ is negligible and that $r_{CO_2}^{cw}$ has its maximal 20°C value.

8.4 Let us suppose that $c_{CO_2}^{ta}$ is 13 mmol m^{-3}, K_{CO_2} is 5 μM, $r_{CO_2}^{bl}$ is 60 s m^{-1}, $r_{CO_2}^{leaf_j}$ is 250 s m^{-1}, $r_{CO_2}^{mes}$ is 150 s m^{-1}, and $r_{CO_2}^{chl}$ is 100 s m^{-1} for a leaf that CO_2 enters only across the lower epidermis. (a) If the rate of gross photosynthesis is 4 mol CO_2 fixed m^{-3} s^{-1} when $c_{CO_2}^{chl}$ is 9 μM, what is V_{max}? (b) What is $c_{CO_2}^{chl}$ when v_{CO_2} is 90% of V_{max}? (c) If the rate of respiration plus photorespiration is 45% of that of gross photosynthesis, what are $r_{CO_2}^{total}$ and J_{CO_2}? Assume that $c_{CO_2}^{chl}$ is 9 μM. (d) Repeat (c) for a nonphotorespiring plant where the rate of respiration is 5% of $J_{CO_2}^{ps}$. Assume that $c_{CO_2}^{chl}$ is 7 μM. (e) Let us place a small transparent bag completely around a leaf of the nonphotorespiring plant. What would $c_{CO_2}^{chl}$ be if the CO_2 concentration in the bag in the steady state were 10 μmol mol^{-1}? Assume that all resistances and the rate of respiration are unchanged. (f) What is the concentration of CO_2 in the mitochondria at night for the nonphotorespiring plant? Let $r_{CO_2}^i$ be 500 s m^{-1}, and assume that the rate of respiration as well as the resistances remain the same as the daytime values. What is the mitochondrial c_{CO_2} at night if stomatal closure causes $r_{CO_2}^{leaf}$ to become 5 000 s m^{-1}?

8.5 Consider a sunlit leaf at 35°C with a g_{wv}^{bl} of 15 mm s^{-1}, stomata only in the lower epidermis, a g_{wv}^{ias} of 30 mm s^{-1}, and a $g_{CO_2}^{total}$ of 0.70 mm s^{-1}. Assume that the ambient air is at 30°C and 32% relative humidity, that cuticular transpiration is negligible and total transpiration is 5.0 mmol m^{-2} s^{-1}, and that the air in the intercellular air spaces reaches 100% relative humidity. (a) What are g_{wv}^{total} and g_{wv}^{st}? (b) What is the essentially immediate effect on g_{wv}^{total} and J_{wv} of decreasing the stomatal opening fourfold, as can occur during wilting? (c) What is the qualitative effect of the action in (b) on T^{leaf}? (d) Neglecting effects caused by leaf temperature, what are the percentage changes of photosynthesis and water-use efficiency caused by the action in (b)? Assume that $c_{CO_2}^{chl}$ is unchanged. (e) What is the essentially immediate effect on g_{wv}^{total} and J_{wv} of increasing the wind speed by fourfold? (f) What is the qualitative effect of the action in (e) on heat conduction across the boundary layer [$J_H^C = 2K^{air}(T^{leaf} - T^{air})/\delta^{bl}$, Eq. 7.11] and on T^{leaf}?

References

Aylor, D.E., J.-Y. Parlange, and A.D. Krikorian. 1973. Stomatal mechanics. *American Journal of Botany* 60:163–171.

Berry, J., and O. Björkman. 1980. Photosynthetic response and adaptation to temperature in higher plants. *Annual Review of Plant Physiology* 31:491–543.

Björkman, O. 1981. Responses to different quantum flux densities. In *Physiological Plant Ecology*, O. L. Lange, P. S. Nobel, C. B. Osmond, and H. Ziegler, eds. *Encyclopedia of Plant Physiology, New Series*, Vol. 12A. Springer-Verlag, Berlin. Pp. 57–107.

Boyer, J. S. 1985. Water transport. *Annual Review of Plant Physiology* 36:473–516.

Brown, H. T., and F. Escombe. 1900. Static diffusion of gases and liquids in relation to the assimilation of carbon and translocation in plants. *Philosophical Transactions of the Royal Society of London, Series B, 193*:223–291.

Chapman, M. S., S. W. Suh, D. Cascio, W. W. Smith, and D. Eisenberg. 1987. Sliding-layer conformational change limited by the quaternary structure of plant RuBisCO. *Nature 329*:354–356.

Clark, J. A., and G. Wigley. 1975. Heat and mass transfer from real and model leaves. In *Heat and Mass Transfer in the Biosphere, I, Transfer Processes in Plant Environment*, D. A. deVries and N. H. Afgan, eds. Halsted Press, Wiley, New York. Pp. 413–422.

Colman, B., and G. S. Espie. 1985. CO_2 uptake and transport in leaf mesophyll cells. *Plant, Cell and Environment 8*:449–457.

Cooke, J. R., and R. H. Rand. 1980. Diffusion resistance models. In *Predicting Photosynthesis for Ecosystem Models*, Vol. 1, J. D. Hesketh and J. W. Jones, eds. CRC Press, Boca Raton, Florida. Pp. 93–121.

Cowan, I. R. 1972. Oscillations in stomatal conductance and plant functioning associated with stomatal conductance: Observations and a model. *Planta 106*:185–219.

Cowan, I. R. 1977. Stomatal behavior and environment. *Advances in Botanical Research 4*:117–227.

Cowan, I. R. 1981. Coping with water stress. In *Biology of Australian Native Plants*, J. S. Pate and A. J. McCoomb, eds. University of Western Australia Press, Perth. Pp. 1–32.

Cowan, I. R., and G. D. Farquhar. 1977. Stomatal function in relation to leaf metabolism and environment. In *Integration of Activity in the Higher Plant*, D. H. Jennings, ed. Society for Experimental Biology Symposium No. 31. Cambridge University Press, Cambridge, England.

Dacey, J. W. H. 1981. Pressurized ventilation in the yellow waterlily. *Ecology 62*:1137–1147.

Edwards, G., and D. Walker. 1983. *C_3, C_4: Mechanisms, and Cellular and Environmental Regulation, of Photosynthesis*. University of California Press, Los Angeles.

Egorov, V. P., and L. T. Karpushkin. 1988. Determination of air humidity over evaporating surface inside a leaf by a compensation method. *Photosynthetica 22*:394–404.

Ehleringer, J., and O. Björkman. 1977. Quantum yields for CO_2 uptake in C_3 and C_4 plants: Dependence on temperature, CO_2, and O_2 concentration. *Plant Physiology 59*:86–90.

El-Sharkawy, M., and J. Hesketh. 1965. Photosynthesis among species in relation to characteristics of leaf anatomy and CO_2 diffusion resistances. *Crop Science 5*:517–521.

Esau, K. 1965. *Plant Anatomy*, 2nd. ed. Wiley, New York.

Evans, J. R., T. D. Sharkey, J. A. Berry, and G. D. Farquhar. 1986. Carbon isotope discrimination measured concurrently with gas exchange to investigate CO_2 diffusion in leaves of higher plants. *Australian Journal of Plant Physiology 13*:281–292.

Evans, L. T. 1963. *Environmental Control of Plant Growth*. Academic Press, New York.

Farquhar, G. D., and T. D. Sharkey. 1982. Stomatal conductance and photosynthesis. *Annual Review of Plant Physiology 33*:317–345.

Farquhar, G. D., and S. von Caemmerer. 1982. Modelling of photosynthetic response to environmental conditions. In *Physiological Plant Ecology*, O. L. Lange, P. S. Nobel, C. B. Osmond, and H. Ziegler, eds. *Encyclopedia of Plant Physiology, New Series*, Vol. 12B. Springer-Verlag, Berlin. Pp. 549–587.

Farquhar, G. D., D. R. Dubbe, and K. Raschke. 1978. Gain of the feedback loop involving carbon dioxide and stomata. Theory and measurement. *Plant Physiology 62*:406–412.

Farquhar, G. D., E.-D. Schulze, and M. Küppers. 1980a. Responses to humidity by stomata of *Nicotiana glauca* L. and *Corylus avellana* L. are consistent with the optimization of carbon dioxide uptake with respect to water loss. *Australian Journal of Plant Physiology* 7:315–327.

Farquhar, G. D., S. von Caemmerer, and J. A. Berry. 1980b. A biochemical model of photosynthetic CO_2 assimilation in leaves of C_3 species. *Planta* 149:78–90.

Farquhar, G. D., J. R. Ehleringer, and K. T. Hubick. 1989. Carbon isotope discrimination and photosynthesis. *Annual Review of Plant Physiology and Plant Molecular Biology* 40:503–507.

Field, C. B., J. T. Ball, and J. A. Berry. 1989. Photosynthesis: principles and field techniques. In *Plant Physiological Ecology: Field Methods and Instrumentation*, R. W. Pearcy, J. Ehleringer, H. A. Mooney, and P. W. Rundel, eds. Chapman and Hall, London. Pp. 209–253.

Field, C., and H. A. Mooney. 1986. The photosynthesis–nitrogen relationship in wild plants. In *On the Economy of Plant Form and Function*, T. J. Givnish, ed. Cambridge University Press, Cambridge, England. Pp. 25–55.

Fischer, R. A. 1972. Aspects of potassium accumulation by stomata of *Vicia faba*. *Australian Journal of Biological Sciences* 25:1107–1123.

Forster, R. E., J. T. Edsall, A. B. Otis, and F. J. W. Roughton, eds. 1969. CO_2: *Chemical, Biochemical, and Physiological Aspects*. National Aeronautics and Space Administration, Washington, D.C.

Fuller, E. N., P. D. Schettler, and J. C. Giddings. 1966. A new method for prediction of binary gas-phase diffusion coefficients. *Industrial & Engineering Chemistry* 58:19–27.

Gaastra, P. 1959. Photosynthesis of crop plants as influenced by light, carbon dioxide, temperature and stomatal diffusion resistance. *Mededelingen van de Landbouwhogeschool te Wageningen, Nederland* 59:1–68.

Gale, J. 1972a. Availability of carbon dioxide for photosynthesis at high altitudes: Theoretical considerations. *Ecology* 53:494–497.

Gale, J. 1972b. Elevation and transpiration: Some theoretical considerations with special reference to Mediterranean-type climate. *Journal of Applied Ecology* 9:691–702.

Hall, A. E. 1982. Mathematical models of plant water loss and plant water relations. In *Physiological Plant Ecology*, O. L. Lange, P. S. Nobel, C. B. Osmond, and H. Ziegler, eds. *Encyclopedia of Plant Physiology, New Series*, Vol. 12B. Springer-Verlag, Berlin. Pp. 231–261.

Hall, A. E., and E.-D. Schulze. 1980. Stomatal response to environment and a possible interrelation between stomatal effects on transpiration and CO_2 assimilation. *Plant, Cell and Environment* 3:467–474.

Hatch, M. D., and C. R. Slack. 1970. Photosynthetic CO_2-fixation pathways. *Annual Review of Plant Physiology* 21:141–162.

Holm, L. G., D. L. Plucknett, J. V. Pancho, and J. P. Herberger. 1977. *The World's Worst Weeds: Distribution and Biology*. University of Hawaii Press, Honolulu.

Jarman, P. D. 1974. The diffusion of carbon dioxide and water vapour through stomata. *Journal of Experimental Botany* 25:927–936.

Jarvis, P. G., and T. A. Mansfield, eds. 1981. *Stomatal Physiology*. Cambridge University Press, Cambridge, England.

Jones, H. G. 1983. *Plants and Microclimate*. Cambridge University Press, Cambridge, England.

Körner, C. H., J. A. Scheel, and H. Bauer. 1979. Maximum leaf diffusive conductance in vascular plants. *Photosynthetica 13*:45–82.

Lake, J. V. 1967. Respiration of leaves during photosynthesis. *Australian Journal of Biological Sciences 20*:487–499.

Le Houérou, H. N. 1984. Rain use efficiency: a unifying concept in arid-land ecology. *Journal of Arid Environments 7*:213–247.

Leuning, R. 1983. Transport of gases into leaves. *Plant, Cell and Environment 6*:181–194.

Long, S. P. 1985. Leaf gas exchange. In *Photosynthetic Mechanisms and the Environment*, J. Barber and N. R. Baker, eds. Elsevier, Amsterdam. Pp. 453–499.

Long, S. P., and F. I. Woodward, eds. 1988. *Plants and Temperature*. Symposium XXXXII of the Society for Experimental Biology. Company of Biologists, Cambridge, England.

Longstreth, D. J., T. L. Hartsock, and P. S. Nobel. 1980. Mesophyll cell properties for some C_3 and C_4 species with high photosynthetic rates. *Physiologia Plantarum 48*:494–498.

Lorimer, G. H. 1981. The carboxylation and oxygenation of ribulose 1,5-bisphosphate: The primary events in photosynthesis and photorespiration. *Annual Review of Plant Physiology 32*:349–383.

MacRobbie, E. A. C. 1988. Control of ion fluxes in stomatal guard cells. *Botanica Acta 101*:140–148.

Marrero, T. R., and E. A. Mason. 1972. Gaseous diffusion coefficients. *Journal of Physical and Chemical Reference Data 1*:3–118.

Meidner, M., and T. A. Mansfield. 1968. *Physiology of Stomata*. McGraw-Hill, New York.

Monteith, J. L. 1973. *Principles of Environmental Physics*. American Elsevier, New York.

Monteith, J. L. 1977. Climate and the efficiency of crop production in Britain. *Philosophical Transactions of the Royal Society of London, Series B, 281*:277–294.

Morison, J. I. L. 1987. Intercellular CO_2 concentration and stomatal response to CO_2. In *Stomatal Function*, E. Zeiger, G. D. Farquhar, and I. R. Cowan, eds. Stanford University Press, Stanford, California. Pp. 229–251.

Nobel, P. S. 1974. Boundary layers of air adjacent to cylinders. Estimation of effective thickness and measurements on plant material. *Plant Physiology 54*:177–181.

Nobel, P. S. 1975. Effective thickness and resistance of the air boundary layer adjacent to spherical plant parts. *Journal of Experimental Botany 26*:120–130.

Nobel, P. S. 1976. Water relations and photosynthesis of a desert CAM plant, *Agave deserti*. *Plant Physiology 58*:576–582.

Nobel, P. S. 1980. Leaf anatomy and water use efficiency. In *Adaptation of Plants to Water and High Temperature Stress*, N. C. Turner and P. J. Kramer, eds. Wiley, New York. Pp. 43–55.

Nobel, P. S. 1984. Productivity of *Agave deserti*: Measurement by dry weight and monthly prediction using physiological responses to environmental parameters. *Oecologia 64*:1–7.

Nobel, P. S. 1988. *Environmental Biology of Agaves and Cacti*. Cambridge University Press, New York.

Nobel, P. S. 1989. A nutrient index quantifying productivity of agaves and cacti. *Journal of Applied Ecology 26*:635–645.

Nobel, P. S., and T. L. Hartsock. 1986. Temperature, water, and PAR influences on predicted and measured productivity of *Agave deserti* at various elevations. *Oecologia 68*:181–185.

Nobel, P. S., and D. B. Walker. 1985. Structure of leaf photosynthetic tissue. In *Photosynthetic Mechanisms and the Environment*, J. Barber and N. R. Baker, eds. Elsevier, Amsterdam. Pp. 501–536.

Nobel, P. S., L. J. Zaragoza, and W. K. Smith. 1975. Relation between mesophyll surface area, photosynthetic rate, and illumination level during development for leaves of *Plectranthus parviflorus* Henckel. *Plant Physiology 55*:1067–1070.

Nonami, H., and E.-D. Schulze. 1989. Cell water potential, osmotic potential, and turgor in the epidermis and mesophyll of transpiring leaves. *Planta 177*:35–46.

Ogren, W. L. 1984. Photorespiration: Pathways, regulation, and modification. *Annual Review of Plant Physiology 35*:415–442.

Oke, T. R. 1987. *Boundary Layer Climates*, 2nd ed. Methuen, London.

O'Leary, M. H. 1982. Phosphoenolpyruvate carboxylase: An enzymologist's view. *Annual Review of Plant Physiology 33*:297–315.

Osmond, C. B., O. Björkman, and D. J. Anderson. 1980. *Physiological Processes in Plant Ecology. Toward a Synthesis with Atriplex*. Springer-Verlag, Berlin.

Osmond, C. B., W. W. Adams III, and S. D. Smith. 1989. Crassulacean acid metabolism. In *Plant Physiological Ecology: Field Methods and Instrumentation*, R. W. Pearcy, J. Ehleringer, H. A. Mooney, and P. W. Rundel, eds. Chapman and Hall, London. Pp. 255–280.

Parkhurst, D. 1977. A three-dimensional model for CO_2 uptake by continuously distributed mesophyll in leaves. *Journal of Theoretical Biology 67*:471–488.

Parkhurst, D. F., S.-C. Wong, G. D. Farquhar, and I. R. Cowan. 1988. Gradients of intercellular CO_2 level across the leaf mesophyll. *Plant Physiology 86*:1032–1037.

Parlange, J.-Y., and P. E. Waggoner. 1970. Stomatal dimensions and resistance to diffusion. *Plant Physiology 46*:337–342.

Patterson, D. T. 1980. Light and temperature adaptation. In *Predicting Photosynthesis for Ecosystem Models*, Vol. 1, J. D. Hesketh and J. W. Jones, eds. CRC Press, Boca Raton, Florida. Pp. 205–235.

Penman, H. L., and R. K. Schofield. 1951. Some physical aspects of assimilation and transpiration. *Symposia of the Society for Experimental Biology 5*:115–129.

Ramanathan, V. 1988. The greenhouse theory of climate change: A test by an inadvertent global experiment. *Science 240*:293–299.

Raschke, K. 1956. Über die physikalischen Beziehungen zwischen Wärmeübergangszahl, Strahlungsaustausch, Temperatur und Transpiration eines Blattes. *Planta 48*:200–238.

Raschke, K. 1975. Stomatal action. *Annual Review of Plant Physiology 26*:309–340.

Raskin, I., and H. Kende. 1985. Mechanism of aeration in rice. *Science 228*:327–329.

Reid, R. C., J. M. Prausnitz, and T. D. Sherwood. 1977. *The Properties of Gases and Liquids*, 3rd ed. McGraw-Hill, New York.

Schlichting, H. 1979. *Boundary-Layer Theory*, 7th ed. McGraw-Hill, New York.

Schroeder, J. I., and R. Hedrich. 1989. Involvement of ion channels and active transport in osmoregulation and signaling of higher plant cells. *Trends in Biochemical Sciences 14*:187–192.

Schulze, E.-D. 1986. Carbon dioxide and water vapor exchange in response to drought in the atmosphere and the soil. *Annual Review of Plant Physiology 37*:247–274.

Schulze, E.-D., O. L. Lange, U. Buschbom, L. Kappen, and M. Evenari. 1972. Stomatal responses to changes in humidity in plants growing in the desert. *Planta 108*:259–270.

Šetlik, I., ed. 1970. *Prediction and Measurement of Photosynthetic Productivity*. Pudoc, Centre for Agricultural Publishing and Documentation, Wageningen, The Netherlands.

Sharkey, T. D. 1985. Photosynthesis in intact leaves of C_3 plants: Physics, physiology and rate limitations. *The Botanical Review 51*:53–105.

Sharpe, P. J. H., H. Wu, and R. D. Spence. 1987. Stomatal mechanics. In *Stomatal Function*, E. Zeiger, G. D. Farquhar, and I. R. Cowan, eds. Stanford University Press, Stanford, California. Pp. 91–114.

Smith, W. K., and G. N. Geller. 1979. Plant transpiration at high elevations: Theory, field measurements, and comparisons with desert plants. *Oecologia 41*:109–122.

Stumm, W., and J. J. Morgan. 1981. *Aquatic Chemistry: An Introduction Emphasizing Chemical Equilibria in Natural Waters*, 2nd ed. Wiley, New York.

Szarek, S. R., and I. P. Ting. 1975. Photosynthetic efficiency of CAM plants in relation to C_3 and C_4 plants. In *Environmental and Biological Control of Photosynthesis*, R. Marcelle, ed. W. Junk, The Hague. Pp. 289–297.

Tanton, T. W., and S. H. Crowdy. 1972. Water pathways in higher plants. III. The transpiration stream within leaves. *Journal of Experimental Botany 23*:619–625.

Tashian, R. E., and D. Hewett-Emmett, eds. 1984. *Biology and Chemistry of the Carbonic Anhydrases*. New York Academy of Sciences, New York.

Tenhunen, J. D., J. D. Hesketh, and D. M. Gates. 1980. Leaf photosynthesis models. In *Predicting Photosynthesis for Ecosystem Models*, Vol. 1, J. D. Hesketh and J. W. Jones, eds. CRC Press, Boca Raton, Florida. Pp. 123–181.

Tolbert, N. E. 1981. Metabolic pathways in peroxisomes and glyoxysomes. *Annual Review of Biochemistry 50*:133–157.

Tyree, M. T., and P. Yianoulis. 1980. The site of water evaporation from sub-stomatal cavities, liquid path resistances and hydroactive stomatal closure. *Annals of Botany 46*:175–193.

Waterman, L. S., D. W. Nelson, W. D. Komhyr, T. B. Harris, K. W. Thoning, and P. P. Tans. 1989. Atmospheric carbon dioxide measurements at Cape Matatula, American Samoa, 1976–1987. *Journal of Geophysical Research 94*:14817–14829.

Wong, S. C., I. R. Cowan, and G. D. Farquhar. 1979. Stomatal conductance correlates with photosynthetic capacity. *Nature 282*:424–426.

Woodrow, I. E., and J. A. Berry. 1988. Enzymatic regulation of photosynthetic CO_2 fixation in C_3 plants. *Annual Review of Plant Physiology and Plant Molecular Biology 39*:533–594.

Zeiger, E. 1983. The biology of stomatal guard cells. *Annual Review of Plant Physiology 34*:441–475.

Zelitch, I. 1971. *Photosynthesis, Photorespiration, and Plant Productivity*. Academic Press, New York.

CHAPTER 9

Plants and Fluxes

In the previous chapter we analyzed gas fluxes for single leaves. We repeatedly used Fick's first law in the following form: flux density equals concentration (or mole fraction) difference divided by resistance, or, equivalently, conductance times concentration difference. This approach can be extended to cover some of the physiological aspects of an entire plant community. We will first describe certain fluxes in the air above the plants. Although the fluxes of water vapor and CO_2 in the air above the vegetation resemble diffusion, in that the net migration of these gases is toward regions of lower concentration, we are not dealing with the random thermal motion of molecules but rather with the random motion of relatively large packets of air in the turbulent region above the plants.

Our next task will be to discuss concentrations and fluxes *within* a plant community. When we begin to analyze the water vapor and CO_2 fluxes from the soil up to the tops of plants, we are immediately confronted by the great structural diversity among different types of vegetation. Each plant community has its own unique spatial patterns for water vapor and CO_2 concentration. The possible presence of many layers of leaves and the changing illumination also greatly complicate the analysis. In fact, even approximate descriptions of the gas fluxes within carefully selected plant communities involve complex calculations based on models incorporating numerous simplifying assumptions. To illustrate how CO_2 and water vapor concentrations and fluxes within a plant community can be handled in a practical situation, we will consider a cornfield as a specific example.

Three quarters of the water vaporized on land is transpired by plants. This water comes from the soil, and the soil also affects the CO_2 fluxes for vegetation. So, after we consider gas fluxes within a plant community, we will examine some of the hydraulic properties of soil. For instance, water in the soil is removed from larger pores before smaller ones. This removal decreases the soil conductivity for subsequent water movement, and a greater drop in water potential from the bulk soil up to a root is therefore necessary for a given water flux density.

Our final topic will be the flow of water in a continuous stream from the soil, to the root, into the root xylem, up to the leaves, and eventually out through the stomata into the atmosphere. As a useful first approximation, we can use the negative gradient of the water potential to represent the driving force for the flux across any segment where water moves as a liquid. We usually replace $-\partial\Psi/\partial x$ by $\Delta\Psi/\Delta x$. The greater the resistance—or, alternatively, the lower the conductance—the larger is the $\Delta\Psi$ required to maintain a given water flux across a particular component. However, $\Delta\Psi$ does not always represent the driving force on water. Furthermore, we must be prepared to recognize interactions between water movement in the xylem and in the other major transport system in plants, the phloem.

GAS FLUXES ABOVE THE LEAF CANOPY

When we considered the fluxes of H_2O and CO_2 for individual leaves in the last chapter, we assumed that outside the boundary layers on each side of a leaf a turbulent region occurs where both water vapor and CO_2 have uniform concentrations. Actually, as we will find, gradients in both CO_2 and H_2O exist within this turbulent region around plants. We will also find that the ambient wind speed is not constant but instead varies with distance above the vegetation.

Wind Speed Profiles

Because of frictional interactions between moving air and a leaf, the air immediately adjacent to a leaf surface is stationary. As we move short distances away from the leaf surface, there is a transition from laminar flow parallel to the leaf in the lower part of the boundary layer to turbulent flow with eddying motion (discussed in Ch. 7; see Fig. 7.5). The wind speed increases as we move still further away from the leaf (Fig. 9.1), often increasing logarithmically for

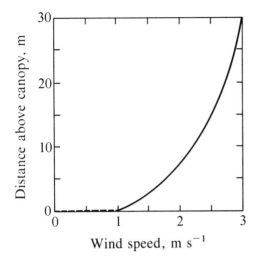

Figure 9.1

Change in wind speed with distance above a leaf at the top of a canopy. At the leaf surface v is 0, and at a distance δ^{bl} (on the order of mm) it is 1.0 m s^{-1}. At 0.5 m into the turbulent air, v increases to 1.1 m s^{-1}, and it can triple at 30 m above the canopy.

a few hundred meters above the leaf canopy.[1] Because the wind speed varies with distance in the turbulent air above vegetation, where to measure the ambient wind speed is somewhat uncertain. However, wind speed generally does not increase very much until we are at least 1 m above the foliage (see Fig. 9.1). Thus, wind speed measured about 0.2 m above the vegetation may be used as the ambient value, which is needed to calculate the boundary layer thickness for an exposed leaf at the top of the canopy (e.g., using Eq. 7.8).

Plants exert a frictional drag on moving air masses and thereby modify the local wind patterns (Fig. 9.1). The frictional interaction between trees and wind is quite different from that of a flexible crop such as wheat, which leads to different form drag (Eq. 7.7) and different wind patterns in the overlying turbulent air (see Grace, 1977; Oliver, 1971). Topographical features such as canyons or cliffs also affect the local wind speed profile.

The wind speed often increases approximately logarithmically with distance above a canopy (Fig. 9.1) and yet is influenced by properties of the plants. Hence, the following equation is often used to indicate the variation in wind speed v with distance above a large, horizontal, uniform canopy under stable atmospheric conditions:

$$v = \frac{v^*}{k} \ln \frac{z - d}{z_0} \tag{9.1}$$

---

1. The turbulent region generally extends 0.5 to 1.0 km above the earth's surface, above which more or less laminar flow occurs in the direction of the prevailing wind.

where v^* is termed the shearing or friction velocity, k is the von Karman constant (about 0.41), z is the height above the ground, d is the zero plane displacement, and z_0 is the roughness length (see Grace, 1977; Jones, 1983; Monteith, 1973; Rosenberg et al., 1983). Generally, d is about 70% of the canopy height, because most of the plant parts that produce form drag (Eq. 7.7) occur near that level. Although z_0 depends on the length of protuberances above the general canopy surface, it is often about 10% of the canopy height for dense vegetation and less for sparse vegetation. Equation 9.1 indicates that the extrapolated wind speed is zero at a height of $z_0 + d$ [the argument of the logarithm is then $(z_0 + d - d)/z_0$ or 1], although the actual wind speed at this height in the vegetation is nonzero, as we will indicate below (Eq. 9.1 applies only above the vegetation).

Flux Densities

A transpiring and photosynthesizing plant community as a whole can have a net vertical flux density of CO_2 (J_{CO_2}) toward it and a net vertical flux density of water vapor (J_{wv}) away from it into the turbulent air above the canopy. These flux densities are expressed per unit area of the ground or, equivalently, per unit area of the (horizontal) plant canopy. Each of the flux densities is related to the appropriate gradient. The vertical flux density of water vapor, for example, depends on the rate of change of the water vapor concentration in the turbulent air, c_{wv}^{ta}, with respect to distance, z, above the vegetation:

$$J_{wv} = -K_{wv} \frac{\partial c_{wv}^{ta}}{\partial z} \tag{9.2}$$

In Equation 9.2 we are again employing the relation, flux density equals a proportionality coefficient times a force, where the force here is the negative gradient of water vapor concentration. Because J_{wv} from the plant community can be expressed in mol m^{-2} s^{-1} and $\partial c_{wv}^{ta}/\partial z$ in mol m^{-4}, the coefficient K_{wv} in Equation 9.2 can have units of m^2 s^{-1}, the same as for diffusion coefficients. In fact, K_{wv} in Equation 9.2 is analogous to D_j in Fick's first law ($J_j = -D_j \partial c_j/\partial x$, Eqs. 1.1 and 8.2), except for one very important distinction: it does not reflect the random thermal motion of water vapor molecules, but rather the irregular swirling motion of packets, or eddies, of air in the turbulent region (see Fig. 9.2). This makes the coefficient much larger than for molecular motion.

Before further considering the random eddying motion of air packets, let us write down a general expression for the CO_2 flux density in the turbulent air

Figure 9.2

Schematic illustration of small packets or eddies of air swirling about in the turbulent region above vegetation. The eddies, which tend to increase in size with height, carry all molecules that they contain more or less as a unit. They are continually changing in size—breaking up, or coalescing with other eddies—making their actual size somewhat hypothetical.

above a canopy. By analogy with Equation 9.2, we can represent the flux density of CO_2 as

$$J_{CO_2} = -K_{CO_2} \frac{\partial c_{CO_2}^{ta}}{\partial z} \tag{9.3}$$

where K_{CO_2} is the "air packet" or *eddy* diffusion coefficient for CO_2 (K_j is also referred to as a transfer coefficient, an exchange coefficient, or a diffusivity coefficient). Similarly, the vertical flux density of O_2 in the turbulent air, J_{O_2}, can be equated to $-K_{O_2}\ \partial c_{O_2}^{ta}/\partial z$. We will let the positive direction for z be increasing altitude; consequently, the positive direction for a net flux density is from the plant canopy upward into the turbulent air.

Eddy Diffusion Coefficients

The eddy diffusion coefficients, K_{wv} and K_{CO_2}, unlike the ordinary diffusion coefficients, D_{wv} and D_{CO_2}, have the *same* value in a given situation. A small packet of air moves more or less as a unit, and thus carries with it all the H_2O, CO_2, and other molecules that it contains (see Fig. 9.2). Although we cannot really assign actual volumes to these eddies—which are constantly changing in size and shape because of shearing effects or coalescence with neighboring packets—they are indeed very large compared with intermolecular distances and contain enormous numbers of molecules.[2] The random motion of an air packet is caused by random fluctuations in pressure in local regions of the turbulent air. The eddying motions of the air packets promote a mixing, formally like the mixing due to diffusion, and thus lead to relations such as Equations 9.2 and 9.3. Besides their eddying motion, air packets have an average drift velocity represented by the local wind velocity. Pressure gradients over large distances cause the winds and the resulting horizontal drift of the air packets.

The values for K_{wv} and K_{CO_2} describing the "diffusion" of air packets vary with the wind speed above the canopy. Also, eddy diffusion coefficients are affected by the rates of change of both wind speed and air temperature with altitude. For instance, hot air tends to rise and become replaced by cooler air. Such buoyancy effects, which are encouraged when $\partial T/\partial z$ is steeply negative above the canopy, lead to more rapid mixing and higher values for K_j. The eddy diffusion coefficients usually are approximately proportional to the local wind speed. As the wind speed increases, turbulent mixing of the air is more likely, and thus K_j becomes larger. Because wind speed varies with height (see Fig. 9.1), and because K_j also depends on the gradient in wind speed, we often employ an eddy diffusion coefficient averaged over an appropriate distance to describe the vertical fluxes in some region of the turbulent air above the canopy. Moreover, the wind speed, its gradient, and the vertical temperature gradient all

---

2. The mean eddy size above vegetation has been approximated by $k(z - d)$, where k is the von Karman constant (about 0.41), z is the height above the ground, and d is the zero plane displacement, all parameters that appear in Equation 9.1 (see Thom, 1975).

vary during the day. Consequently, our value for K_j should also be averaged over a suitable time interval, e.g., an hour.

For a moderate wind speed of 2 m s^{-1}, the eddy diffusion coefficient is generally 0.05 to 0.2 m^2 s^{-1} just above the plant canopy. Under these conditions, K_j might be about 2 m^2 s^{-1} at 30 m above the canopy, and in excess of 5 m^2 s^{-1} at or above 300 m, where turbulent mixing is even greater (see van Wijk, 1966). By comparison, D_{wv} is 2.4 $\times$ 10^{-5} m^2 s^{-1} and D_{CO_2} is 1.5 $\times$ 10^{-5} m^2 s^{-1} in air at 20°C. Thus, K_j is 10^4 to 10^5 times larger in the turbulent air above the canopy than are these D_j's. The random motion of air packets is indeed much more effective than is the random thermal motion of molecules in moving H$_2$O and CO$_2$.

Because K_j increases rapidly with altitude as we move into turbulent regions with higher wind speeds, the steady-state concentration gradients become less steep with increasing height above the vegetation. Specifically, J_{wv} equals $-K_{wv}$ $\partial c_{wv}^{ta}/\partial z$ by Equation 9.2, and because K_{wv} increases with altitude, the absolute value of $\partial c_{wv}^{ta}/\partial z$ must become smaller at greater heights above the canopy. K_j may increase, for example, by a factor of 10 in the first 20 m above the vegetation, in which case the gradient in water vapor concentration would decrease 10-fold in this interval.

The equality of the eddy diffusion coefficients for different gaseous species results from the air packet and the molecules within it moving as a unit. In fact, K_j is often assumed to be the same for the transfer of gases, heat, and momentum (expressed in the same units), a relation that is often referred to as the *similarity principle*. K_j is therefore generally measured for the most convenient quantity in some situation and is then assumed to be the same (or at least similar) for all others (see Oke, 1987; Rosenberg *et al.*, 1983; van Wijk, 1966).

Resistance of Air above the Canopy

As in our use of resistances for gaseous diffusion in the previous chapter, we will identify a resistance to the flow of water vapor in the turbulent air by r_{wv}^{ta} and that for CO$_2$ by $r_{CO_2}^{ta}$. To derive such quantities, we will replace the negative gradient by the difference in concentration of species j, Δc_j^{ta}, across a given distance, Δz, in the turbulent air; i.e., Equation 9.2 becomes $J_{wv} = -K_{wv}$ $\partial c_{wv}^{ta}/\partial z = K_{wv} \Delta c_{wv}^{ta}/\Delta z = \Delta c_{wv}^{ta}/r_{wv}^{ta}$. Thus, by analogy with our previous definition of conductance and resistance ($g_j^{bl} = J_j/\Delta c_j^{bl} = D_j/\delta^{bl} = 1/r_j^{bl}$, Eq. 8.3), we can identify resistances from the flux density expressions in Equations 9.2

and 9.3. Because K_{wv} has the same value as K_{CO_2}, we obtain the following equalities:

$$
r_{wv}^{ta} = \frac{\Delta c_{wv}^{ta}}{J_{wv}} = \frac{\Delta z}{K_{wv}}
$$

$$
= \frac{\Delta z}{K_{CO_2}} = \frac{\Delta c_{CO_2}^{ta}}{J_{CO_2}} = r_{CO_2}^{ta} \tag{9.4}
$$

As with the analogous relations in Chapter 8 (e.g., Eqs. 8.3 and 8.5), Equation 9.4 describes the steady-state condition. Equation 9.4 indicates that r_{wv}^{ta} has the same value as $r_{CO_2}^{ta}$, as we would indeed expect based on the random motions of whole packets of air (also, $g_{wv}^{ta} = 1/r_{wv}^{ta} = 1/r_{CO_2}^{ta} = g_{CO_2}^{ta}$).

Let us now estimate the resistance of the turbulent air immediately above a plant canopy. For simplicity, we will let K_j average 1.0 m^2 s^{-1} for the first 30 m above the plants, a typical value in a moderate wind during the daytime. By Equation 9.4, the resistance over this 30-m interval then is

$$
r_{wv}^{ta} = r_{CO_2}^{ta} = \frac{(30 \text{ m})}{(1 \text{ m}^2 \text{ s}^{-1})}
$$

$$
= 30 \text{ s m}^{-1}
$$

Measured values for this resistance generally range from 20 to 40 s m^{-1} for moderate wind speeds, as do predicted values from computer analyses of r_j^{ta} using models incorporating the variation of K_j with altitude. Wind speeds, and therefore K_j, tend to be lower at night, so r_j^{ta} tends to be somewhat higher then than during the daytime (see J. L. Monteith in Evans, 1963).

Transpiration and Photosynthesis

As mentioned, we will express flux densities above plants per unit area of the ground or, equivalently, per unit area of the canopy. For many agricultural as well as ecological considerations, such a measure of the average transpiration or average photosynthesis of the whole plant community is far more useful than is the water vapor or CO_2 flux densities of an individual leaf. Environmental measurements in the turbulent air above vegetation can thus be a very important approach for measuring the overall rates of transpiration and photosynthesis, especially if the extent of similar plants is fairly large, as might occur for a cornfield or a grassland. Moreover, such measurements can generally be made without disturbing the plants or their leaves. On this large scale, however, we unavoidably lose sight of certain factors, such as the effect of stomatal opening or leaf size on the gas fluxes. Also, the turbulent air above the canopy is greatly influenced by the terrain as well as by the vegetation, so we must reckon with other factors not involved in our study of leaves. For example, both K_j and the gradients in water vapor or CO_2 depend on whether we are at the edge or the center of a field, whether and what types of trees are present, and whether the region is flat or hilly.

For simplicity, we are considering a one-dimensional situation where the net fluxes of water vapor and CO_2 occur only in the vertical direction above the canopy, as would be the case near the center of a large uniform plant community. Just as our assumption of a boundary layer of uniform thickness breaks down at the leading and the trailing edges of a leaf, we must also consider air packets transferring H_2O and CO_2 horizontally in and out at the sides of certain areas of vegetation. Such net horizontal transfer of various gases is referred to as *advection*. Instead of using Equations 9.2 through 9.4 to analyze net gas flux densities, we may have to use much more cumbersome three-dimensional equations to handle advection for small fields or individual plants.

As we considered in Chapter 8 (p. 455), J_{wv}/J_{CO_2} might be about 400 H_2O/CO_2 for a representative sunlit mesophytic leaf, and 200 for a photosynthetically efficient C_4 species such as corn. For an entire plant community, however, the water lost per CO_2 fixed is often considerably higher than for a single, well-illuminated leaf. In particular, J_{wv} measured above the canopy also includes water vapor coming from the soil and from leaves that are not well illuminated and that therefore contribute little to net photosynthesis. Some of the CO_2 taken up by the plant community is evolved by soil microorganisms, root cells, and leaves that do not receive much sunlight and so are below light compensation—all of which decrease the amount of CO_2 that needs to be supplied from above the plant canopy. These effects tend to raise J_{wv}/J_{CO_2} above the values for an exposed leaf. Although the absolute value of J_{wv}/J_{CO_2} above a leaf canopy depends on the ambient relative humidity and the physiological status of the plants, it is generally between 400 and 2 000 H_2O/CO_2 when averaged over a day in the growing season. Moreover, mainly because J_{CO_2} for C_4 plants is often about twice as large as for C_3 plants, the absolute value of J_{wv}/J_{CO_2} is lower and daily growth tends to be greater for C_4 plants (see Kozlowski, 1968, Vol. II; Kramer, 1983; Zelitch, 1971).

Values for Fluxes and Concentrations

We will use representative values of J_{CO_2} to calculate the decreases in concentration of CO_2 that might be expected over a certain vertical distance in the turbulent atmosphere. When net photosynthesis is occurring, the flux density of CO_2 is directed from the turbulent air down into the canopy. J_{CO_2} above the vegetation is then negative by our sign convention, which means that $c_{CO_2}^{ta}$ increases as we go vertically upward ($J_{CO_2} = -K_{CO_2} \, \partial c_{CO_2}^{ta}/\partial z$, Eq. 9.3). But this is just as we would expect if CO_2 is to be transferred downward toward the plants by the random motion of the eddies in the turbulent air.

A commonly observed value for J_{CO_2} above a plant canopy is -20 μmol m^{-2} s^{-1} at midday (see Loomis et al., 1971; Šetlik, 1970; Zelitch, 1971). For comparison, the flux density of CO_2 into an exposed *leaf* of a mesophyte at a moderate light level can be 11 μmol m^{-2} s^{-1} (p. 450). Using Equation 9.4 and a resistance of 30 s m^{-1} for the lower 30 m of the turbulent air ($r_{CO_2}^{ta}$), we calculate that the decrease in CO_2 concentration across this region would be[3]

$$\Delta c_{CO_2}^{ta} = J_{CO_2} r_{CO_2}^{ta} = (-20 \times 10^{-6} \text{ mol m}^{-2} \text{ s}^{-1})(30 \text{ s m}^{-1})$$

$$= -0.60 \text{ mmol m}^{-3}$$

Employing a conversion factor from Table 8.2, this $\Delta c_{CO_2}^{ta}$ corresponds to (0.60) (24.4), or 15 μmol mol^{-1} (a CO_2 mole fraction difference of 15×10^{-6}) at 20°C and an air pressure of 0.1 MPa (1 bar). Thus, CO_2, which might have a concentration near 360 μmol mol^{-1} well into the turbulent air, e.g., 30 m above vegetation, could be at $360 - 15$, or 345 μmol mol^{-1} just above the canopy.

For a rapidly photosynthesizing corn crop at noon, J_{CO_2} can be -60 μmol m^{-2} s^{-1}. For the above $r_{CO_2}^{ta}$, $\Delta c_{CO_2}^{ta}$ is -44 μmol mol^{-1} by Equation 9.4. Thus, the CO_2 concentration at the top of the canopy can be 316 μmol mol^{-1}, i.e., 44 μmol mol^{-1} lower than the 360 μmol mol^{-1} in the turbulent air tens of metres above the corn plants. In fact, measurement of $c_{CO_2}^{ta}$ at the canopy level indicates the net rate of photosynthesis by the plants. At night respiration occurs, but not photosynthesis, so vegetation then acts as a *source* of CO_2. Thus, the concentration of CO_2 just above the canopy at night is usually a few μmol mol^{-1} greater than it is higher up in the turbulent air, as we will see below.

The flux density of water vapor just above the canopy, which represents

---

3. By analogy with Ohm's law ($\Delta E = IR$), in this chapter we will use the form $\Delta c_j^{ta} = J_j r_j^{ta}$ for our calculations of concentration drops.

transpiration from the leaves plus evaporation from the soil, is often referred to as *evapotranspiration*. For fairly dense vegetation and a moist soil, this evapotranspiration is appreciable, generally amounting to 60% to 90% of the flux density of water vapor from an exposed water surface at the ambient air temperature. The daily evapotranspiration from a forest is often equivalent to a layer of water 3 to 5 mm thick, which averages 2 to 3 mmol m^{-2} s^{-1} (see Kozlowski, 1968, Vol. II; Kramer, 1983; Slatyer, 1967). At noon on a sunny day with a moderate wind, J_{wv} above a leaf canopy can be 7 mmol m^{-2} s^{-1}. Using Equation 9.4 ($\Delta c_{wv}^{ta} = J_{wv} r_{wv}^{ta}$) and employing our above value for r_{wv}^{ta} of 30 s m^{-1}, we note that the water vapor concentration decreases by (7 × 10^{-3} mol m^{-2} s^{-1})(30 s m^{-1}) or 0.21 mol m^{-3} over the first 30 m of the turbulent air. We indicated in Chapter 8 that the turbulent air immediately outside the boundary layer adjacent to a leaf contains 0.48 mol water m^{-3} when it is at 20°C and 50% relative humidity (Fig. 8.6). Our calculation indicates that c_{wv}^{ta} could drop by 0.21 mol m^{-3}, which means from 0.48 to 0.27 mol m^{-3} (28% relative humidity at 20°C), as we move 30 m upward into the turbulent air above the canopy. Such a marked decrease in absolute and relative humidity is indeed generally observed in the turbulent air above vegetation.

Because evapotranspiration is important both ecologically and agronomically, various methods have been developed for estimating it, including approaches based primarily on measurements made above the plant canopy. Just as for a leaf (Ch. 7), the main energy input into the plant canopy generally is from net radiation (see Eq. 7.6), and the main energy losses are from sensible heat (heat conduction across a boundary layer followed by convection to the turbulent air, see Eqs. 7.11–7.14) and latent heat loss accompanying water evaporation (Eq. 7.19). The ratio of the flux density of sensible heat to the air to the flux density of latent heat loss from the soil and the plant canopy is referred to as the *Bowen ratio,* based on work by Bowen in the 1920s (see Rosenberg *et al.,* 1983). When the evaporation rate is low, such as when water availability is limited, the Bowen ratio, β, tends to be high. For instance, β is about 0.1 for tropical oceans, 0.2 for tropical rain forests, 0.4 to 0.8 for temperate forests and grasslands, 2 to 6 for semi-arid regions, and 10 for deserts (see Oke, 1987). Incorporating β into the energy budget relation for an entire plant canopy allows the determination of the rate of evapotranspiration from measurements of the net radiation, the heat flux density into the soil, and the gradients in temperature and water vapor concentration above the canopy (see Jones, 1983). For example, evapotranspiration can be estimated to within a few percent of measured values using an energy budget analysis incorporating the Bowen ratio for grasses and other short crops as well as for forests (see Hutchinson and Hicks, 1985).

Figure 9.3

"Steam" rising from a leaf and a wooden fencepost that are rapidly heated by the sun after a rainstorm. Moisture-laden air next to the objects is swept in an eddying motion into a cooler region, where the water vapor condenses.

Condensation

What appears to be steam is often seen rising from leaves or other surfaces (Fig. 9.3) when the sun breaks through the clouds following a rainstorm or at sunrise following a night with a heavy dew. To analyze this phenomenon, we will assume that the sun warms up the leaves at the top of the canopy to 25°C and that the concentration of water vapor in the cell wall pores of their mesophyll cells is then 1.27 mol m^{-3} (see p. 419). Suppose that the air just outside the boundary layer adjacent to a leaf is at 22°C and has a high relative humidity of 95% just after the rainstorm; c_{wv}^{ta} is then 1.03 mol m^{-3} (c_{wv}^{*} = 1.08 mol m^{-3} at 22°C). Hence, water vapor will diffuse from the leaf, across the boundary layer, and into the turbulent air (Fig. 9.3). Now let us suppose that the air at a greater distance from the leaf is somewhat cooler, e.g., 20°C at 10 mm from the leaf. At 20°C c_{wv}^{*} is 0.96 mol m^{-3}. Thus, as the air with 1.03 mol water m^{-3} moves away from the unstirred layer adjacent to the leaf in an eddy, or air parcel, it will be cooled and some of its water vapor will condense, because c_{wv}^{ta} cannot exceed c_{wv}^{*} for the local air temperature. This condensation leads to the fog, or "steam," seen moving away from the plants into the surrounding cooler turbulent air (Fig. 9.3). As we move even further away to regions of lower water vapor concentration, the condensed water evaporates and so the steam disappears.

GAS FLUXES WITHIN PLANT COMMUNITIES

The pattern of concentrations and fluxes within the vegetation obviously depends on the particular plant community being considered (see Russell *et al.*, 1989).

Air flow within plant communities also is complicated by the three-dimensional architecture of the plants. For instance, wind speed does not necessarily decrease toward the ground—air in an open forest can "tunnel" under the branches and hence the wind speed can be greater there than further up in the canopy. We will not attempt to examine all types of vegetation, but instead we will focus on a cornfield representing a monospecific stand of high productivity (see Lemon *et al.*, 1971). The same general principles apply to other fairly uniform plant communities, but isolated plants provide special difficulty for analysis, because the gas concentrations and fluxes then vary in three dimensions.

Eddy Diffusion Coefficient and Resistance

Let us begin by considering how the eddy diffusion coefficient might vary within a plant community. Near the ground there is a boundary layer that can be rather thick, because the air there is generally quite still. In fact, K_j often averages 5 $\times$ 10^{-5} m^2 s^{-1} in the first 10 mm above the ground, a value only 2 to 3 times larger than are the diffusion coefficients of water vapor and CO_2 in air.[4] As we move up to the top of the canopy, the eddy diffusion coefficient increases, often more or less logarithmically with altitude in the upper part of many plant communities. It may reach a value of 0.2 m^2 s^{-1} at the top of a canopy in a moderate wind (Fig. 9.4). Because K_j is approximately proportional to wind speed, v^{wind} within the plant community varies in a manner similar to the variation described for the eddy diffusion coefficient (Fig. 9.4). For instance, the wind speed about 0.2 m above the ground might be 0.1 m s^{-1}, increasing to 2 m s^{-1} at the top of the canopy (see Millington and Peters, 1969).

Two aspects concerning K_j within the plant community deserve special emphasis. First, transfer of gaseous substances within the vegetation takes place by the random motion of relatively large parcels or eddies of air, just as in the turbulent region above the canopy. Second, because of frictional drag with the many leaves, branches, and other plant parts, the eddy diffusion coefficient within the vegetation is considerably less than in the air above the canopy.

We will now estimate the resistance of the turbulent air from the ground to the top of a corn crop that is 2 m in height (Fig. 9.4). To illustrate the relative contributions of various air layers, we will let K_j have specific average values for various height intervals above the ground (Table 9.1), consistent with the plot of K_j versus height in Figure 9.4. We can use Equation 9.4, $r_{wv}^{\text{ta}} = \Delta z / K_{wv}$

---

4. When K_j is of the same order of magnitude as diffusion coefficients, differences in movement between molecular species may become apparent. Effects of individual D_j's on K_j are noticeable only in the first 10 mm or so above the ground and they are ignored in the present example.

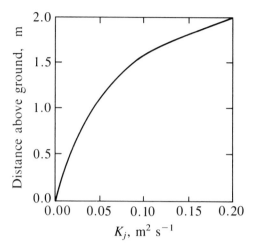

Figure 9.4

Idealized representation of the variation in the eddy diffusion coefficient within a uniform corn crop 2 m in height. The wind speed is 2 m s^{-1} at the top of the canopy.

$= r^{ta}_{CO_2}$, to estimate the resistance of each of the four air layers in series; e.g., r^{ta}_j for the lowermost layer is (0.01 m)/(5 × 10^{-5} m^2 s^{-1}), or 200 s m^{-1}. The total resistance from the ground up to the top of the canopy equals 200 + 90 + 50 + 10, or 350 s m^{-1} (summarized in Table 9.1). (In a sense, we are performing a numerical integration to determine the resistance.) Computer analyses using models describing the turbulent air within such a crop also indicate that the resistance is generally 300 to 400 s m^{-1}. Most of the resistance within a plant community is generally due to the relatively still air next to the ground. For instance, just over half (200 s m^{-1} out of 350 s m^{-1}) of the resistance for the 2 m pathway is provided by the lowest 0.01 m, whereas the entire upper half of the corn crop accounts for a resistance of only 10 s m^{-1} (see Table 9.1).

Water Vapor

A considerable amount of water can evaporate from the soil and move by air packets up through the vegetation. For instance, J_{wv} from a moist, intermittently illuminated soil, such as commonly occurs in a temperate forest, can be 0.2 to 1 mmol m^{-2} s^{-1}. (For comparison, 0.5 mmol m^{-2} s^{-1} corresponds to a depth of water of 0.8 mm/day or 280 mm/year.) If a flow of 0.5 mmol m^{-2} s^{-1} occurs across a resistance of 290 s m^{-1} to reach a distance 0.1 m above the ground, we can calculate using Equation 9.4 (by which $\Delta c^{ta}_{wv} = J_{wv} r^{ta}_{wv}$) that the drop in water vapor from the ground to this level is (0.5 × 10^{-3} mol m^{-2} s^{-1})(290 s m^{-1}) or 0.15 mol m^{-3}. At 20°C the saturation water vapor concentration is 0.96

Table 9.1

Summary of gas exchange parameters within a 2-m-tall corn crop at noon on a sunny day. Also indicated are the flux densities at ground level and just above the canopy. Bars over K_j, J_{wv}, and J_{CO_2} indicate values averaged over the height increment involved (see Fig. 9.1; Lemon et al., 1971).

Height above ground (m)	$\bar{K}_j$ (m² s⁻¹)	r_j^{ta} (s m⁻¹)	$\bar{J}_{wv}$ (mmol m⁻² s⁻¹)	Δc_{wv} (mol m⁻³)	$\bar{J}_{CO_2}$ (μmol m⁻² s⁻¹)	Δc_{CO_2} (mmol m⁻³)	Δc_{CO_2} (μmol mol⁻¹)
above canopy			7		−60		
1 to 2	1×10^{-1}	10	4	0.04	−30	−0.30	−7
0.1 to 1	2×10^{-2}	50	1	0.05	3	0.15	4
0.01 to 0.1	1×10^{-3}	90	0.5	0.05	2	0.18	4
0.00 to 0.01	5×10^{-5}	200	0.5	0.10	2	0.40	10
ground level			0.5		2		

mol m^{-3} (App. I), so a water vapor drop of 0.15 mol m^{-3} then corresponds to a 16% decrease in relative humidity. Thus, an appreciable drop in water vapor concentration can occur across the relatively still air near a moist soil under a canopy (Fig. 9.5).

Because of the addition of water vapor from transpiration by the leaves, J_{wv} increases as we move from the ground up through a corn crop. On a sunny day the water vapor flux density might be 1 mmol m^{-2} s^{-1} at 0.5 m, 2 at 1.0 m, 4 at 1.5 m, and 7 mmol m^{-2} s^{-1} at 2.0 m, the top of the canopy. (On a cloudy humid day, J_{wv} for a corn crop might be only 1 mmol m^{-2} s^{-1} at the top of the canopy.) If J_{wv} averages 1 mmol m^{-2} s^{-1} from 0.1 to 1.0 m above the ground, where the resistance is 50 s m^{-1} (Table 9.1), then by Equation 9.4 Δc_{wv}^{ta} for this part of the pathway is 0.05 mol m^{-3}. For the upper 1 m the resistance is 10 s m^{-1}, and so, for an average J_{wv} of 4 mmol m^{-2} s^{-1}, the decrease in water vapor

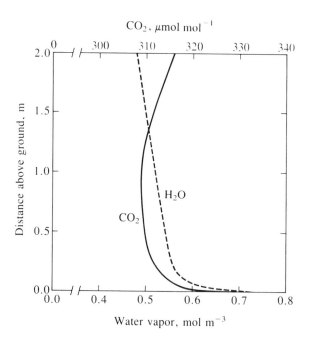

Figure 9.5

Possible variation of water vapor and CO_2 concentrations within a 2-m-high corn crop at noon on a sunny day. At the top of the canopy the wind speed is 2 m s^{-1}. In the turbulent air 30 m above the vegetation, c_{wv}^{ta} is 0.27 mol m^{-3}, and $c_{CO_2}^{ta}$ corresponds to 360 μmol mol^{-1} by volume (for actual values under field conditions, see Lemon et al., 1971). Such variations of atmospheric CO_2 and H_2O with height should be taken into account in canopy gas-exchange models that deal with leaves in various layers.

concentration would be 0.04 mol m^{-3} (Table 9.1). Thus, Δc_{wv}^{ta} is 0.15 mol m^{-3} over the 0.1 m just above the ground, 0.05 from 0.1 to 1.0 m, and 0.04 from 1.0 to 2.0 m, or 0.24 mol m^{-3} overall (see Table 9.1 and Fig. 9.5). If the turbulent air at the top of the canopy is at 20°C and 50% relative humidity, it would contain 0.48 mol water m^{-3}. The air near the soil would then contain approximately 0.48 + 0.24, or 0.72, mol H_2O m^{-3}, which corresponds to 75% relative humidity at 20°C ($c_{wv}^{*} = 0.96$ mol m^{-3} at 20°C; App. I). In summary, we note that (1) air close to the soil under a (fairly dense) plant canopy can have an appreciable relative humidity, (2) c_{wv}^{ta} continuously decreases as we move upward from the ground, (3) most of the overall drop in water vapor concentration occurs near the ground, and (4) most of the water vapor comes from the upper half of the corn crop in the present example.

Attenuation of PPFD

Before discussing J_{CO_2} within a plant community, we should consider how the amount of light varies down through the various layers of vegetation, because the PPFD (photosynthetic photon flux density, wavelengths of 400 to 700 nm) at each level helps determine the rate of photosynthesis there. The net rate of CO_2 fixation approaches light-saturation near a PPFD of 600 μmol m^{-2} s^{-1} for leaves of many C_3 plants, and it decreases to zero at light compensation. A comprehensive formulation—including effects of leaf angle, sun elevation in the sky, the finite width of the sun's disc, changes in spectral distribution of PPFD at various levels within the plant community, multiple reflections from leaves and other surfaces, as well as clumping versus uniform arrangement of leaves—leads to a nearly hopeless complication of the algebra. Instead, we will assume that the decrease in PPFD is due to absorption by the foliage in a manner analogous to Beer's law, $\ln(J_b/J_0) = -k_\lambda cb$ (Eq. 4.17). This approximation is particularly useful when the leaves are randomly distributed horizontally, as can occur in certain moderately dense plant communities (for measurements of leaf angle, accumulated leaf area at various levels in complex canopies, and other details, see Myneni et al., 1989; Nobel and Long, 1985; Norman and Campbell, 1989; Ross, 1981).

As we move downward into the vegetation, the PPFD decreases more or less exponentially with the amount of absorbing material encountered. For some canopies the greatest leaf area per interval of height occurs near the center (e.g., many grasses), and for others it occurs about three-quarters of the way up from the ground (e.g., many crops and trees). We will let F be the average cumulative total leaf area per unit ground area as we move down through the plant community. The dimensionless parameter F uses the area of only one side of a leaf, and thus it is expressed on the same basis as are our flux densities of H_2O and CO_2. F is zero at the top of the canopy and takes on its maximum value at ground level, a value generally referred to as the *leaf area index*. If the leaves in a particular plant community were all horizontal, the leaf area index would equal the average number of leaves above any point on the ground. In any case, the leaf area index equals the leaf area per plant divided by the ground area per plant.

We will represent the PPFD, J, incident on the leaf canopy by J_0. Primarily because of absorption by photosynthetic pigments, PPFD is attenuated down through the plant community. At any level in the vegetation, J is related to J_0 and F as follows:

$$\ln \frac{J_0}{J} = kF \tag{9.5}$$

where k is a dimensionless parameter describing the absorption properties of a particular type of foliage and is referred to as the *foliar absorption coefficient*. Because we are ignoring changes in spectral distribution at different levels in the vegetation, J_0 and J in Equation 9.5 can represent the flux density of photons from 400 to 700 nm or an energy flux density for these photons. Equation 9.5 was introduced into plant studies by Monsi and Saeki in 1953 (see T. Saeki in Evans, 1963; Šetlik, 1970; Zelitch, 1971).

Values of Foliar Absorption Coefficients

The foliar absorption coefficient k ranges from 0.3 to 1.3 for most plant canopies. Light penetrates the vertically oriented blades of grasses rather easily; in such cases, k can be near 0.4. At what cumulative leaf area per ground area would the incident PPFD be reduced by 95% for grasses with a foliar absorption coefficient of 0.4? By Equation 9.5, the accumulated leaf area per unit ground area in this case is

$$F = \frac{\ln \left(\dfrac{J_0}{0.05 J_0} \right)}{0.4} = 7.5$$

Thus, when the average leaf area index is 7.5 for such grasses, 5% of the PPFD reaches the soil surface. For 95% of the PPFD to be absorbed for a leaf area index of 3, k must be ln (1/0.05)/3 or 1.0 by Equation 9.5. Such a high foliar absorption coefficient applies to horizontal leaves with at least 0.5 g chlorophyll m^{-2}, which can occur for crops such as potato, soybean, sunflower, and white clover.

When the sun is overhead, vertical leaves absorb less sunlight and reflect more of it down into the vegetation per unit leaf area than do horizontal leaves. This accounts for the low values of k for grasses, because their leaves are generally rather erect. Moreover, leaves tend to be vertical near the top of certain plants, e.g., sugar beet, becoming on the average more horizontal toward the ground (Fig. 9.6; in other cases, leaves at the top of the canopy are more pendant, again becoming more horizontal toward the ground). This orientation reduces the foliar absorption coefficient of the upper leaves, and therefore more of the light incident on the plants is available for the lower leaves. In fact, optimal light utilization for photosynthesis generally occurs when the incident PPFD is distributed as uniformly as possible over the leaves, because the fraction of leaves exposed to PPFD levels above light saturation or below light compensation is then usually minimized. Indeed, canopies with leaf angles varying from vertical near the top to horizontal near the bottom (Fig. 9.6; see also McMillen and McClendon, 1979) tend to have higher productivities when the leaf area index exceeds about 3 (see Monsi *et al.*, 1973; Turitzin and Drake, 1981). Our arguments about the effect of leaf orientation on k presuppose that essentially all of the light is incident

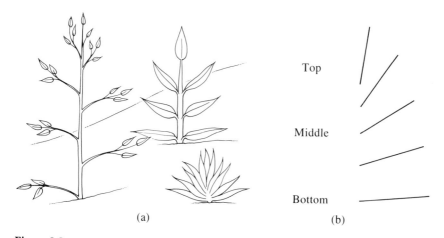

(a)   (b)

Figure 9.6

Variation in leaf angle and hence foliar absorption coefficient with distance above the ground for (a) various idealized plants and (b) sugar beet measured at various canopy positions (see Hodáňová, 1979). The greater erectness of the uppermost leaves leads to a lower k for them, and hence better penetration of PPFD down to the lower leaves.

on the top of the canopy. When much PPFD comes in from the sides, as for an isolated tree, foliar absorption coefficients determined for vertically incident light should not be used in Equation 9.5—indeed, k can be determined for other sun angles. Also, a foliar absorption coefficient can be determined for shortwave irradiation, instead of just for PPFD, as considered here (see Norman, 1980; Ross, 1981).

Light Compensation Point

Let us now consider the light compensation point for CO_2 fixation by leaves. As we mentioned in the last chapter (p. 447), light compensation generally occurs at a PPFD of about 8 μmol m^{-2} s^{-1} for a leaf temperature near 20°C and a CO_2 concentration of 350 μmol mol^{-1} ($c_{CO_2}^{ta}$ is somewhat below this level within the plant community, as we will see below). Suppose that a moderate PPFD of 400 μmol m^{-2} s^{-1} occurs on trees whose leaves have a foliar absorption coefficient of 0.8. At what cumulative area of leaves per unit ground area is a light compensation point of 8 μmol m^{-2} s^{-1} reached? By Equation 9.5, F is then ln (400/8)/0.8, or 4.9. Thus, only the upper five "layers" of leaves in a dense forest might be above light compensation for that part of the day when the PPFD on the canopy is 400 μmol m^{-2} s^{-1}. Of course, for a lower PPFD on the plant canopy, more leaves would be below the light compensation point. Occasional sunflecks of high PPFD reach the lower parts of the vegetation, complicating our analysis of where light compensation occurs.

Leaves that are below light compensation for most of the day do not contribute to the net photosynthesis of the plant. Such leaves generally lose 20% to 50% of their dry weight before dying and abscising. Following this loss of leaves on the lower branches of trees, the branches themselves die and eventually fall off or are blown off by the wind. Thus, tall trees in a dense forest often have few or no branches over the lower part of their trunks.

CO_2 Concentrations and Fluxes

In contrast to the concentration of water vapor, which continuously decreases with distance above the ground, on a sunny day the CO_2 concentration generally achieves a minimum somewhere within the plant community (Fig. 9.5). This occurs because both the turbulent air above the canopy and the soil can serve as sources of CO_2. During the day CO_2 diffuses toward lower concentrations from the soil up into the vegetation and from the overlying turbulent air down into the leaf canopy.

Respiration in root cells and in soil microorganisms can lead to a net upward CO_2 flux density from the ground of 1 to 3 μmol m^{-2} s^{-1} during the growing season (see Rosenberg et al., 1983). (An O_2 flux density of similar magnitude occurs in the opposite direction.) J_{CO_2} from the soil varies in phase with the soil temperature, which is higher during the daytime (see Ch. 7). We have already estimated that $r_{CO_2}^{ta}$, which is the same as r_{wv}^{ta}, might be 200 s m^{-1} for the first 0.01 m and 90 s m^{-1} for the next 0.09 m above the ground for a corn crop that is 2 m tall (Table 9.1). Using Equation 9.4 ($\Delta c_{CO_2}^{ta} = J_{CO_2} r_{CO_2}^{ta}$), we calculate for a moderate CO_2 flux density of 2 μmol m^{-2} s^{-1} emanating from the soil that the decrease in CO_2 concentration across the first 0.01 m above the ground is (2 $\times$ 10^{-3} mmol m^{-3})(200 s m^{-1}) or 0.40 mmol m^{-3}, which by the conversion factor in Table 8.2 represents a drop of 10 μmol CO_2 mol^{-1} at 20°C and 0.1 MPa air pressure. In the next 0.09 m, the CO_2 concentration might decrease by 0.18 mmol m^{-3}, which corresponds to about 4 μmol mol^{-1} (Table 9.1). Thus, the CO_2 level might decrease by 14 μmol mol^{-1} from 327 μmol mol^{-1} at the soil surface to 313 μmol mol^{-1} at 0.1 m above the ground (Fig. 9.5).

As we move further upward from the ground, the flux density of CO_2 initially increases as we encounter leaves that are below light compensation and thus have a net evolution of CO_2. For instance, J_{CO_2} directed upward may increase from 2 μmol m^{-2} s^{-1} at 0.1 m to 5 μmol m^{-2} s^{-1} at 0.5 m. The maximum upward J_{CO_2} occurs at the canopy level where light compensation occurs. As we move even higher and encounter leaves with net photosynthesis, the net flux density of CO_2 in the turbulent air decreases, and it may become zero at 1.0 m above the ground in a 2-m-tall corn crop with a high photosynthetic rate. Because J_{CO_2} equals $-K_{CO_2} \partial c_{CO_2}^{ta}/\partial z$ (Eq. 9.3), J_{CO_2} becomes zero (no net CO_2 flux upward or downward) when $\partial c_{CO_2}^{ta}/\partial z$ is zero, which corresponds to the local minimum in CO_2 concentration (Fig. 9.5). Thus J_{CO_2} may average 3 μmol m^{-2} s^{-1} from 0.1 to 1.0 m above the ground, an interval that has a resistance of 50 s m^{-1} (Table 9.1). This would lead to a $\Delta c_{CO_2}^{ta}$ of 0.15 mmol m^{-3}, which corresponds to a CO_2 decrease of 4 μmol mol^{-1}. Hence, the CO_2 concentration may reach its lowest value of 309 μmol mol^{-1} midway through the crop (see Fig. 9.5).

At noon on a sunny day, J_{CO_2} down into a cornfield might be 60 μmol m^{-2} s^{-1}. Essentially all of the net CO_2 flux from the turbulent air above the canopy is directed into the leaves in the upper half of the corn crop; e.g., J_{CO_2} may become -30 μmol m^{-2} s^{-1} at 1.5 m above the ground and zero at 1.0 m. Thus, the average CO_2 flux density in the upper half of the vegetation is about -30 μmol m^{-2} s^{-1}, and the resistance is 10 s m^{-1} (Table 9.1). Consequently,

$\Delta c_{CO_2}^{ta}$ for this upper portion of an actively photosynthesizing cornfield might be -0.30 mmol m^{-3}, which corresponds to a 7 μmol mol^{-1} decrease in CO_2 from the top of the canopy to 1.0 m below (Table 9.1, Fig. 9.5). We calculated on p. 482 that $c_{CO_2}^{ta}$ at the top of this canopy might be 316 μmol mol^{-1}. The CO_2 concentration at 1.0 above the ground would then be 309 μmol mol^{-1}, the same value that we estimated by working our way up from the ground (see Fig. 9.5). Two-thirds or more of the net photosynthesis generally occurs in the upper one-third of most canopies, as it does here for corn.

CO_2 concentrations in the air can vary over a wide range for different plant communities. For a cornfield exposed to a low wind speed (below 0.3 m s^{-1} at the top of the canopy), for a rapidly growing plant community, or for other dense vegetation where the eddy diffusion coefficient may be relatively small, the CO_2 concentration in the turbulent air within the plant stand can drop below 200 μmol mol^{-1} during a sunny day. On the other hand, for sparse desert vegetation, especially on windy or overcast days, $c_{CO_2}^{ta}$ generally does not decrease even 1 μmol mol^{-1} from the value at the top of the canopy.

CO_2 at Night

The CO_2 concentration at night is highest near the ground and continuously decreases as we go upward through the plants into the turbulent air above. J_{CO_2} emanating from the soil might be 1 μmol m^{-2} s^{-1}, and the respiratory flux density of CO_2 from the above-ground parts of plants can be 3 μmol m^{-2} s^{-1} at night. Respiration averaged over a 24-h period can be 20% of gross photosynthesis for a rapidly growing plant community and can increase to over 50% as the community matures. For certain climax communities, respiration can become nearly 100% of gross photosynthesis. When considered over a growing season, respiration for an entire plant is generally 30% to 50% of gross photosynthesis for agronomic crops (see Zelitch, 1971).

We will now estimate the decrease in CO_2 concentration that might occur in a 2-m-high corn crop at night. For purposes of calculation, we will assume that J_{CO_2} vertically upward increases from 1 μmol m^{-2} s^{-1} near ground level to 4 μmol m^{-2} s^{-1} at the top of the cornfield and that the lower wind speeds at night lead to somewhat higher resistances than the daytime values (see Fig. 9.7 and Table 9.1). Using Equation 9.4 ($\Delta c_{CO_2}^{ta} = J_{CO_2} r_{CO_2}^{ta}$), the drop in CO_2 concentration from the ground to the top of the canopy is 0.40 + 0.16 + 0.06, or 0.62 mmol m^{-3}, which corresponds to 15 μmol CO_2 mol^{-1} at 20°C (Fig. 9.7; conversion factor in Table 8.2). The resistance of the first 30 m of turbulent air above the plants might be 50 s m^{-1} at night (compare the value of 30 s m^{-1}

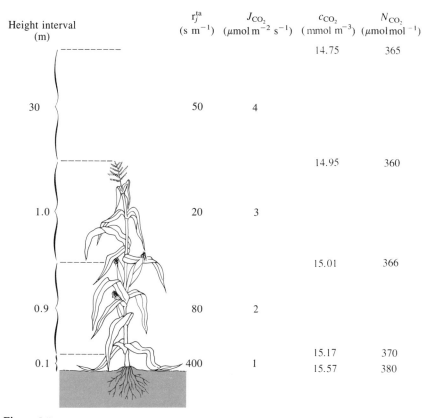

Figure 9.7
CO_2 resistances, flux densities, and concentrations within and above a cornfield at night.

during the daytime that we used earlier). If we let the total CO_2 flux density from the canopy be 4 μmol m⁻² s⁻¹, then $\Delta c_{CO_2}^{ta}$ for the first 30 m above the canopy would be 0.20 mmol m⁻³, which corresponds to 5 μmol CO_2 mol⁻¹. Therefore, assuming that $c_{CO_2}^{ta}$ is 360 μmol mol⁻¹ 30 m up in the turbulent air, the CO_2 concentration at night would be 365 μmol mol⁻¹ at the top of the canopy and 380 μmol mol⁻¹ at the soil surface (Fig. 9.7).

SOIL

Soils vary tremendously in their physical properties, such as the size of the individual particles. In a sandy soil many particles are over 1 mm in diameter, but in a clay most of the particles are less than 2 μm in diameter. Small soil particles have a much greater surface area per unit mass than do large particles. Sand, for example, can have under 1 m^2 of surface area/g, whereas most clays have 100 to 1 000 m^2 of surface area/g. Most soil minerals are aluminosilicates, with negatively charged surfaces that act as Donnan phases (see p. 138) with the mobile cations in the adjacent soil water. Because of their large surface areas per unit mass, clays dominate the ion exchange properties of many soils. The clay montmorillonite, for example, has 800 m^2 of surface area/g and can hold nearly 1 mmol of monovalent cations/g (see Hillel, 1982; Marshall and Holmes, 1988). Nutrient concentrations vary tremendously with soil type and water content; for moist agricultural soil, the phosphate level can be about 2 μM, with K^+ and NO_3^- being about 1 000-fold more concentrated (Ca^{2+}, Mg^{2+}, Na^+, and Cl^- can be even higher in concentration).

The irregularly shaped pores between the numerous particles in the soil contain both air and water (see Fig. 9.8). The soil pores, or voids, vary from somewhat under 40% to about 60% of the soil by volume. Thus, a soil whose pores are completely filled with water will contain 40% to 60% water by volume. In the vicinity of most roots, moist soil contains 8% to 30% water by volume, and the rest of the pore space is filled with air. The pores therefore provide many air-liquid interfaces where surface tension effects can lead to a negative hydrostatic pressure in the soil water. Such a negative P is generally the main contributor to the water potential in the soil, especially as the soil dries. Because the thermal

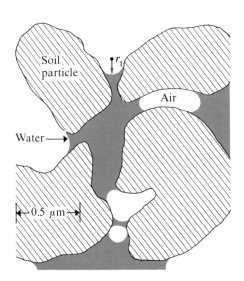

Figure 9.8

Schematic indication of the gas, liquid, and solid phases in a soil. The radii of curvature at the air-water interfaces help determine the negative hydrostatic pressures or tensions in the liquid phase.

properties of soil have already been discussed in Chapter 7, we will focus on the water relations here.

Soil Water Potential

As just indicated, the predominant influence on the soil water potential is usually the many air-liquid interfaces present in the soil (Fig. 9.8). Of course, the soil water contains dissolved solutes, which generally lead to an osmotic pressure (Π^{soil}) of between 0.01 and 0.2 MPa; the magnitude of Π^{soil} depends on the relative wetness of the soil, which varies greatly. Because of the relatively low osmotic pressures, we will often refer to the solution in the soil simply as water. In contrast to the generally small values for Π^{soil}, the many interfaces present in the rather small soil pores can lead to hydrostatic pressures of -2 MPa or even more negative. A few days after being saturated, a wet clay soil might retain 40% water by volume ("field capacity") and have a soil water potential (Ψ^{soil}) of -0.01 MPa, whereas permanent wilting from which many agricultural crops will not recover occurs when Ψ^{soil} is about -1.5 MPa and the volumetric water content of the clay is 20%. Field capacity and permanent wilting occur at about 30% and 10% water by volume, respectively, for loam, and 10% and 3% for sand (see Marshall and Holmes, 1988; Rendig and Taylor, 1989).

The surfaces of the air-water interfaces in the pores between the soil particles are usually concave when viewed from the air side, just as for capillaries (see Fig. 2.3). But the surfaces generally are not spherical or otherwise regularly shaped (the same restriction applies to the air-liquid interfaces in the pores of a cell wall). Nevertheless, we can define two principal radii of curvature for a surface.[5] Let us designate these radii, which occur in planes perpendicular to each other and to the liquid surface, by r_1 and r_2. The hydrostatic pressure in the soil water then is:

$$P = -\sigma\left(\frac{1}{r_1} + \frac{1}{r_2}\right) \tag{9.6}$$

5. If the arc formed by the intersection of a surface and a plane perpendicular to it is continued around to form a circle, the radius of the circle is the radius of curvature of the surface, r. A slightly curved surface has a large r, and r becomes infinite if the surface is flat in a particular direction. By convention, r is positive for a concave surface as viewed from the air side (see r_1 in Fig. 9.8) and negative for a convex surface.

where σ is the surface tension at an air-liquid interface.[6] The negative hydrostatic pressure or positive tension described by Equation 9.6 and resulting from the presence of air-water interfaces is often called the *soil matric potential*.

Instead of being concave, the water surface extending between adjacent soil particles may assume a semi-cylindrical shape, i.e., like a trough or channel. This means that one of the radii of curvature becomes infinite—e.g., $r_2 = \infty$; in such a case, the pressure is $-\sigma/r_1$ by Equation 9.6. If the air-liquid surface is convex when viewed from the air side, the radii as defined are negative; we would then have a positive hydrostatic pressure in the water. In the intermediate case—one radius positive and one negative (a so-called "saddle-shaped" surface)—whether the pressure is positive or negative depends on the relative sizes of the two radii of curvature.

Let us now estimate the hydrostatic pressure in the soil water within a wedge-shaped crevice between two adjacent soil particles, as is illustrated at the top of Figure 9.8. We will assume that the air-liquid surface is cylindrical, so r_2 equals ∞, and that r_1 is 0.1 μm. If we let σ be 0.0728 Pa m, the value for water at 20°C (App. I), then by Equation 9.6 the hydrostatic pressure is

$$P = \frac{-(0.0728 \text{ Pa m})}{(10^{-7} \text{ m})} = -7 \times 10^5 \text{ Pa}$$

$$= -0.7 \text{ MPa}$$

As the amount of soil water decreases, the air-water surface retreats into the crevice between the particles, the radius of curvature becomes less, and the pressure accordingly becomes more negative. Because Ψ^{soil} equals $P^{soil} - \Pi^{soil} + \rho_w g h$ (Eq. 2.13a), the soil water potential also becomes more negative as water is lost from such crevices.

Darcy's Law

Darcy in 1856 recognized that the flow of water through the soil is driven by a gradient in hydrostatic pressure (Darcy, 1856). We can represent this relation, known as Darcy's law, by the following expression:

---

6. To connect Equation 9.6 with Equation 2.22 ($P = -2\sigma \cos \alpha/r$, where r is the radius of the *capillary*), note that for a cylindrical capillary the two principal radii of the *surface* are the same ($r_1 = r_2$), and hence the factor $(1/r_1 + 1/r_2)$ is $2/r_1$, which equals $2 \cos \alpha/r$. Equation 9.6 was derived by Young and by Laplace in 1805 and is sometimes called the Young and Laplace equation (see Adamson, 1982).

$$J_V = -L^{\text{soil}} \frac{\partial P^{\text{soil}}}{\partial x} \qquad (9.7)$$

where J_V is the volume of solution crossing unit area in unit time and L^{soil} is the soil hydraulic conductivity coefficient.

Although Equation 9.7 is in a familiar form (flux density equals a proportionality coefficient times a force), we have used $-\partial P^{\text{soil}}/\partial x$ instead of the possibly more general force, $-\partial \Psi^{\text{soil}}/\partial x$ ($\Psi^{\text{soil}} = P^{\text{soil}} - \Pi^{\text{soil}} + \rho_w gh$, Eq. 2.13a). In Chapter 3 we derived an expression for J_V that incorporated a reflection coefficient: $J_V = L_P(\Delta P - \sigma \Delta\Pi)$ (Eq. 3.38). When σ is zero, as occurs for a porous barrier like soil, $\Delta\Pi$ does not lead to any volume flux density. Hence, we do not expect $\partial\Pi^{\text{soil}}/\partial x$ to influence the movement of water in the soil, and it is not included in Darcy's law. Because $\partial(\rho_w gh)/\partial x$ is not incorporated into Equation 9.7, the indicated form of Darcy's law applies only to horizontal flow in the soil ($\partial h/\partial x = 0$ when x is in a horizontal direction). Actually, $\rho_w g$ is only 0.01 MPa m^{-1} (App. I), so changes of $\rho_w gh$ in the vicinity of a root are relatively small. On the other hand, for percolation of water down appreciable distances into the soil, P^{soil} in Equation 9.7 should be replaced by $P^{\text{soil}} + \rho_w gh$. In fact, for drainage of very wet soil, the gravitational term can be the dominant factor in such a formulation of Darcy's law.

Soil Hydraulic Conductivity Coefficient

L^{soil} in Equation 9.7 has different units than L_P [$J_V = L_P(\Delta P - \sigma\Delta\pi)$, Eq. 3.38] or L_w [$J_{V_w} = L_w(\Psi^o - \Psi^i)$, Eq. 2.23]. Units for L_P and L_w are volume flux density/pressure—e.g., m s^{-1} Pa^{-1}. But L^{soil} has units of volume flux density/pressure gradient—e.g., (m s^{-1})/(Pa m^{-1}), or m^2 s^{-1} Pa^{-1}. We can use other self-consistent sets of units for Darcy's law; indeed, the soil hydraulic conductivity coefficient is expressed in many different ways in the literature.[7]

The soil hydraulic conductivity coefficient depends on the geometry of the pores in the soil. For geometrically similar pore shapes, and ignoring certain surface effects, the hydraulic conductivity coefficient is approximately propor-

---

7. Instead of using $-\partial P/\partial x$ [or $-\partial(P + \rho_w gh)/\partial x$], most soil literature expresses the force in Darcy's law as the negative gradient in the hydraulic head of water, the latter being the length of a vertical column of water yielding the same pressure, in which case the units for L^{soil} are the same as those for J_V. Because a 1-m height of water exerts a pressure of $\rho_w g \times 1$ m or 0.00979 MPa ($= 9.79 \times 10^3$ Pa) at sea level, 45° latitude, and 20°C (App. I), an L^{soil} of 10^{-10} m^2 s^{-1} Pa^{-1} corresponds to one of 9.79×10^{-7} m s^{-1}.

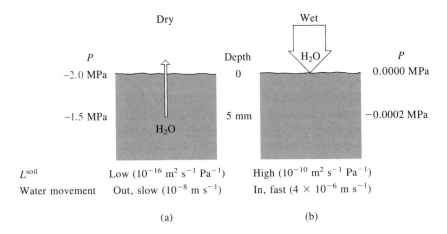

Figure 9.9

Summary of valve-like properties of upper layer of a loamy soil: (a) a fairly dry state and (b) while being wet by rain. Water fairly readily enters wet soil but is lost only gradually across a dry crust.

tional to the square of the pore size. However, the pores are so complex in shape that, in general, we cannot directly calculate L^{soil}. As soil dries, its water potential decreases and P^{soil} becomes less. When P^{soil} decreases below the minimum hydrostatic pressure that can occur in some fairly large pores, water flows out of them, but it will remain in pores that have smaller dimensions and can therefore have even more negative pressures—see Equation 9.6, $P = -\sigma(1/r_1 + 1/r_2)$. Not only is the higher conductivity of the larger pores thus lost, but also the remaining pathway for water flow becomes more tortuous, so L^{soil} decreases as the soil dries.

The soil hydraulic conductivity coefficient can be 10^{-17} m^2 s^{-1} Pa^{-1} or lower for a dry nonporous soil, and 10^{-13} m^2 s^{-1} Pa^{-1} or higher for a wet porous one. Specifically, for a porous clay whose pores are nearly filled with water, L^{soil} is usually 10^{-13} to 10^{-10} m^2 s^{-1} Pa^{-1}; and for a water saturated sandy soil, the soil hydraulic conductivity coefficient is generally 10^{-8} to 10^{-7} m^2 s^{-1} Pa^{-1} (see Hillel, 1982; Marshall and Holmes, 1988). (We will reconsider L^{soil} in terms of Poiseuille flow on p. 509.)

The ground is often covered by a rather dry crust where the soil hydraulic conductivity coefficient is quite low. Specifically, L^{soil} may average 10^{-16} m^2 s^{-1} Pa^{-1} in the upper 5 mm of the soil. If P^{soil} is -2.0 MPa near the surface and -1.5 MPa at 5 mm beneath it (Fig. 9.9), then by Darcy's law (Eq. 9.7) the volume flux density of water is

$$J_V = -(10^{-16} \text{ m}^2 \text{ s}^{-1} \text{ Pa}^{-1}) \left[\frac{(-2.0 \times 10^6 \text{ Pa}) - (-1.5 \times 10^6 \text{ Pa})}{(5 \times 10^{-3} \text{ m})} \right]$$

$$= 1 \times 10^{-8} \text{ m s}^{-1} \quad (0.9 \text{ mm day}^{-1})$$

This water flux density directed vertically upward at the soil surface equals (1 $\times$ 10^{-8} m^3 m^{-2} s^{-1})(1 mol/18 $\times$ 10^{-6} m^3), or 0.6 $\times$ 10^{-3} mol m^{-2} s^{-1}. When discussing water vapor movement in the previous section, we indicated that J_{wv} emanating from a shaded soil is generally 0.2 to 1 mmol m^{-2} s^{-1}, so our calculated flux density is consistent with the range of measured values. The calculation also indicates that a fairly large gradient in hydrostatic pressure can exist near the surface of a moist soil.

During a rainstorm the upper part of the soil can become nearly saturated with water. As a result, L^{soil} there might increase 10^6-fold from 10^{-16} m^2 s^{-1} Pa^{-1} up to 10^{-10} m^2 s^{-1} Pa^{-1} (Fig. 9.9). This facilitates the entry or infiltration of water into the soil, which initially can have a volume flux density of about 4 $\times$ 10^{-6} m s^{-1}. Such an infiltration rate, which corresponds to a 14 mm depth of water per hour, is maintained for only short times, as within minutes for a clay and after an hour or so for a sandy soil, the upper part of the soil becomes saturated with water and the infiltration rate decreases (see Hillel, 1982; Marshall and Holmes, 1988). In any case, we note that the uppermost layer or crust acts somewhat like a valve, retarding the outward movement of water when the soil is fairly dry (low L^{soil}), but promoting the infiltration of water upon moistening (high L^{soil}), as is indicated in Figure 9.9.

Besides moving as a liquid, water can also move as a vapor in the soil. Because water is continually evaporating from and condensing onto the many air-liquid interfaces in the soil, such movement can be relatively important, especially for dry soils where the liquid phases are discontinuous. The saturation value for water vapor partial pressure or concentration increases nearly exponentially with temperature (Fig. 8.5). Because the air between the soil particles generally is nearly saturated with water vapor (except near the surface), the amount of water vapor in the soil air increases rapidly with soil temperature. Consequently, the movement of water in the form of vapor tends to be greater at higher soil temperatures.

Fluxes for Cylindrical Symmetry

Although we have been mainly considering one-dimensional cases in Cartesian coordinates, the flow of soil water toward a root may be more appropriately described using cylindrical coordinates, because the length of a root is generally much greater than is its diameter. We will restrict our attention to the steady-state condition where the fluxes do not change with time. Also, we will consider the cylindrically symmetric case for which the fluxes and forces depend only on the radial distance r from the axis of the cylinder and not on any angle around it or location along its axis.

For a cylindrical surface of length l along the axis, the total volume of solution crossing per unit time is $J_V 2\pi r l$, where J_V is the volume flux density directed radially at a distance r from the axis of the cylinder. In the steady state, $J_V 2\pi r l$ is constant, so the magnitude of the flux density depends inversely on the radial distance. When L^{soil} is constant, we can represent the flux density for the cylindrically symmetric case by

$$J_V = \frac{1}{r} \frac{L^{\text{soil}}(P_a - P_b)}{\ln(r_a/r_b)} \tag{9.8}$$

where P_a is the hydrostatic pressure at a distance r_a from the axis of the cylinder, and P_b is the value at r_b. J_V is positive when the net flux density is directed into the root, as occurs when the hydrostatic pressure is higher ($P_a > P_b$) the further we are from the root ($r_a > r_b$). Equation 9.8 represents a general form for steady-state cases with cylindrical symmetry; e.g., Fick's first law (p. 11) then is $J_j = (1/r)D_j(c_j^a - c_j^b)/\ln(r_a/r_b)$, and Equation 7.12 gives the heat flux density for cylindrical symmetry.

The uptake of water by a young root that is 1 mm in diameter is usually 1×10^{-5} to 5×10^{-5} m³ day⁻¹ per m of root length (0.1 to 0.5 cm³ day⁻¹ per cm; see Kramer, 1983; Slatyer, 1967). This uptake occurs over a root surface area of $2\pi r l$, so the volume flux density of water at the root surface for a moderate uptake of 3×10^{-5} m³ day⁻¹ per m of root length is

$$J_V = \frac{(3 \times 10^{-5} \text{ m}^3 \text{ day}^{-1})}{(8.64 \times 10^4 \text{ s day}^{-1})[(2\pi)(0.5 \times 10^{-3} \text{ m})(1 \text{ m})]}$$

$$= 1.1 \times 10^{-7} \text{ m s}^{-1}$$

The influx of water is enhanced by the root hairs, which protrude from the epidermal cells (Fig. 1.4). They are often about 12 μm in diameter, up to 1 mm long, and generally vary in frequency from 0.5 to 50 per mm². J_V may be 10- to 100-fold less for older roots, because their outer surfaces generally become extensively cutinized and suberized.

For representative root spacing in a soil, water may move toward a root over a radial distance of about 10 mm. Let us next estimate the decrease in hydrostatic pressure that might occur over this interval. We will assume that Ψ^{soil} at an r_a of 10.5 mm is -0.3 MPa, made up of an osmotic pressure of 0.1 MPa and a P_a^{soil} of -0.2 MPa ($\Psi = P - \Pi + \rho_w gh$, Eq. 2.13a). We will let L^{soil} be 10^{-15} m^2 s^{-1} Pa^{-1}, as might apply to a loam of moderately low water content, and we will suppose that J_V at the surface of a root that is 1 mm in diameter (i.e., $r_b = 0.5$ mm) is 1.1×10^{-7} m s^{-1}. Using Equation 9.8, we can then calculate that the hydrostatic pressure near the root surface is

$$P_b^{soil} = \frac{r J_V \ln (r_a/r_b)}{L^{soil}} + P_a^{soil}$$

$$= -\frac{(0.5 \times 10^{-3} \text{ m})(1.1 \times 10^{-7} \text{ m s}^{-1}) \ln (10.5 \text{ mm}/0.5 \text{ mm})}{(10^{-15} \text{ m}^2 \text{ s}^{-1} \text{ Pa}^{-1})}$$

$$- 0.2 \text{ MPa}$$

$$= -0.4 \text{ MPa}$$

Thus, the hydrostatic pressure decreases 0.2 MPa across a distance of 10 mm in the soil next to the root. Assuming that the solute content of the soil water does not change appreciably over this interval, Π_b^{soil} is 0.1 MPa, and thus Ψ_b^{soil} adjacent to the root is -0.4 MPa $- 0.1$ MPa, or -0.5 MPa (see Table 9.2).

As the soil dries, L^{soil} decreases; therefore, the decreases in hydrostatic pressure and water potential must then be larger to maintain a given volume flux density toward a root. For example, L^{soil} is rather high for a wet soil with a Ψ^{soil} of -0.02 MPa; ΔP^{soil} over a radial distance of 10 mm from a root is generally 0.01 MPa or less in such a case. To maintain the same J_V for Ψ^{soil} equal to -1.5 MPa in the bulk soil, a decrease in P^{soil} of about 1.0 MPa might have to occur over the 10 mm adjacent to the root.

Fluxes for Spherical Symmetry

Sometimes the fluxes of water in the soil toward plant parts can approximate spherical symmetry. For instance, water movement can occur radially over a 10-mm interval toward a seed or a recently initiated root. For roots most water and nutrient uptake apparently takes place over the region containing the root hairs, which is proximal to the elongation region (see Fig. 1.4) and distal to where the periderm begins. The external cell layers of the periderm have suberin in the cell walls, which greatly limits water uptake. *Lenticels* occur in the suberized regions, consisting of a loose aggregation of cells facilitating gas exchange (lenticels also occur on stems). Lenticels and other interruptions of the suberized periderm can act as local sites toward which the flux of water converges. In such a case, the water movement from the surrounding soil toward the root can also be approximately spherically symmetric (see Caldwell, 1976).

For spherical symmetry we will use the same definitions as in the case of cylindrical symmetry (Eq. 9.8) but note that now $J_V 4\pi r^2$ is constant in the steady state. Thus, we obtain the following steady-state relation describing the volume flux density J_V at distance r from the center of a sphere when J_V varies only in the radial direction and L^{soil} is constant:

$$J_V = \frac{1}{r^2}\left(\frac{r_a r_b}{r_a - r_b}\right)L^{soil}(P_a - P_b) \qquad (9.9)$$

Equation 9.9 is similar to Equation 7.13 $[J_H^C = (r + \delta^{bl})K^{air}(T^{surf} - T^{ta})/(r\delta^{bl})]$ describing the heat flux density across an air boundary layer for spherical symmetry ($r_a = r + \delta^{bl}$ and $r_b = r$). A steady state is often not achieved in soils, so a "steady rate" is often used, where the rate of volumetric water depletion is constant, which leads to relations considerably more complicated than Equations 9.8 and 9.9 (see Caldwell, 1976).[8] We note that the equations describing water flow in a soil for the one-dimensional case (Eq. 9.7), for cylindrical symmetry (Eq. 9.8), and for spherical symmetry (Eq. 9.9) all indicate that the volume flux density of water is proportional to L^{soil} times a difference in hydrostatic pressure.

During germination the volume flux density of water into a seed is often limited by a seed coat (testa) of thickness δ^{sc}. The seed coat is often two to four cell layers thick, and thus is relatively thin compared with the radius of the seed (see Bewley and Black, 1978). For the volume flux density at the seed surface ($r = r_s$, $r_a = r_s$, and $r_b = r_s - \delta^{sc}$), Equation 9.9 becomes

---

8. In a steady state the water content of the soil does not change with time, whereas for a steady-rate situation the rate of change is set equal to a constant. In the general time-dependent case, a relation similar to $\dfrac{\partial c_j}{\partial t} = -\dfrac{\partial}{\partial x}\left(-D_j \dfrac{\partial c_j}{\partial x}\right)$ (Eq. 1.4), but in the proper co-ordinate system, must be satisfied to describe the water flow in the soil for cylindrical or spherical symmetry; D_j is replaced by a quantity analogous to L^{soil} that varies with water content and hence location in the soil.

$$J_V = \frac{1}{r_s^2} \frac{(r_s)(r_s - \delta^{sc})}{[r_s - (r_s - \delta^{sc})]} L^{sc} \Delta P^{sc}$$

$$= \frac{r_s - \delta^{sc}}{r_s} \frac{L^{sc}}{\delta^{sc}} \Delta P^{sc} \qquad (9.10)$$

$$\cong \frac{L^{sc}}{\delta^{sc}} \Delta P^{sc}$$

where L^{sc} is the hydraulic conductivity coefficient of a seed coat of thickness δ^{sc}, and the bottom line incorporates the supposition that $\delta^{sc} \ll r_s$. Equation 9.10 indicates that, when the region of interest is thin compared with the radial distance, the flow can be approximated by an equation appropriate for a one-dimensional movement across a seed coat of conductance L^{sc}/δ^{sc} (see Eq. 8.1b). Usually ΔP^{sc} in Equation 9.10 is replaced by $\Delta \Psi^{sc}$, because osmotic pressures can also influence water uptake by seeds.

As we have indicated for L^{soil}, L^{sc} also depends on water content, increasing more than 10^6-fold upon water uptake by the seed and subsequent rupture of the seed coat. L^{sc} then approaches values found for the L^{soil} of a moist soil (see Shaykewich and Williams, 1971). When dry, seeds can have a very negative water potential, e.g., -100 to -200 MPa. As the seeds imbibe water, the internal water potential rises; germination is initiated only when Ψ^{seed} and hence Ψ^{soil} are above about -1.0 MPa. Water uptake is generally not uniform over the whole seed surface, initially being higher near the micropyle (a small opening in the integuments of an ovule through which the pollen tube enters and which remains as a minute pore in the testa). Another crucial factor in the germination of seeds is the area of contact between the seed coat and the soil particles, which becomes progressively more important as the seed imbibes water and L^{sc} increases, leading to a higher J_V (see Eq. 9.10; during water uptake, L^{sc} increases more than $\Delta \Psi^{sc}$ decreases).

WATER MOVEMENT IN THE XYLEM AND THE PHLOEM

Under usual conditions, essentially all of the water entering a land plant comes from the soil by way of the root. The water is then conducted to the other parts of the plant mainly in the xylem. To reach the xylem in the mature part of a root, water must cross the root epidermis, the cortex, and then the endodermis. We will briefly describe the various cells encountered by water as it moves from the soil to the root xylem (see Fig. 1.4) and then discuss the dependence of water flow in the xylem on the gradient in hydrostatic pressure. Flow in the phloem likewise depends on $\partial P/\partial x$, although osmotic pressures are also important for interpreting water movement in this other "circulatory" system. The divisions and subdivisions of the vascular tissue in a leaf generally result in individual mesophyll cells being no more than three or four cells away from the xylem or the phloem. This proximity facilitates the movement of photosynthetic products into the phloem, such "loading" being necessary for the distribution of sugars like sucrose to "sinks" located in different parts of the plant.

Root Tissues

Longitudinal and cross-sectional views near the tip of a root are presented in Figure 1.4, indicating the various types of cells that can act as barriers to flow. Water may fairly easily traverse the single-cell layer of the root epidermis to reach the cortex. The root cortex often consists of 5 to 10 cell layers, the cytoplasm of adjacent cells being continuous because of the plasmodesmata. The collective protoplasm of the interconnected cells is referred to as the *symplasm* (see Ch. 1). In the symplasm, permeability barriers in the form of plasmalemmas and cell walls do not have to be surmounted for diffusion to occur from cell to cell, which facilitates the movement of water and solutes across the cortex. Much of the water flow across the root cortex actually occurs in an alternative pathway, the cell walls, which form the *apoplast* (see Passioura, 1988; Weatherley, 1982).

Roots of most woody and some herbaceous plants have mycorrhizal associations. The fungal hyphae invade the cortical region of the young roots and can extend over 10 mm away from the root surface. The relation appears to be mutualistic, because the fungus obtains organic matter from the plant and in return increases the effective area of contact of the plant with soil particles. The hyphae can remain active for older roots whose outer surface has become suberized and hence less conductive. The fungus thus increases the availability of certain nutrients like N and P to the host plant (see Kramer and Kozlowski, 1979).

Along that part of a root where cells have differentiated but where there is little root thickening due to secondary conducting tissue, an endodermis occurs inside the cortex (Fig. 1.4 and p. 8). At the endodermis, the cell wall pathway for water movement is blocked by the casparian strip, as realized by Caspary in 1865. This is a continuous hydrophobic band around the radial walls of the endodermal cells that blocks the apoplastic pathway along which water and solutes could have crossed the endodermal layer. The casparian strip is impregnated with suberin (a polymer of fatty acids) and lignin. It is generally considered part of the primary cell wall, but suberin and lignin are also deposited in the secondary wall (which often becomes quite thick) and all the way across the middle lamella. Water and ions therefore cannot flow across the endodermal cells in their cell walls but rather must move through their cytoplasm. Thus, the endodermis, with its casparian strip, serves to regulate the passage of solutes and water from the cortex to the xylem in the affected parts of the root. Inside the endodermis is a layer of parenchyma cells known as the *pericycle,* which surrounds the vascular tissue containing the conducting vessels of the xylem and the phloem (see Fig. 1.4). During water movement through a plant accompanying transpiration, the hydrostatic pressure in the root xylem is reduced and can become quite negative. This decreases the water potential ($\Psi = P - \Pi + \rho_w gh$, Eq. 2.13a) in the xylem and promotes water movement from the soil down a water potential gradient to the root xylem.

The Xylem

Before discussing the characteristics of flow in the xylem, we will briefly review some of its anatomical features. In general, the conducting xylem elements have thick, lignified secondary cell walls and contain no protoplasts; i.e., the xylem cells serve their special function of providing the plant with a low resistance conduit only when they are dead.

Two types of conducting cells are distinguished in the xylem: the vessel members (found in angiosperms) and the phylogenetically more primitive tracheids (in angiosperms, gymnosperms, and the lower vascular plants). Tracheids typically are tapered at the ends, whereas the generally shorter and broader vessel members often abut each other with rather blunt ends. The end cell walls of the vessel members are perforated, so the vessel members arranged end-to-end form a continuous tube called a *xylem vessel.* The end cell wall of a vessel member bearing the holes is referred to as a *perforation plate;* a simple perforation plate (Fig. 1.3) essentially eliminates the end walls between the individual members in a vessel. Although xylem elements vary considerably in their widths (from about 8 μm up to 500 μm), we will represent them by cylinders with radii of 20 μm for purposes of calculation. Conducting cells of the xylem generally range in length from 1 to 10 mm for tracheids and 0.2 to 3 mm for vessel members; vessels vary greatly in length, even within the same plant, ranging from about 0.01 to 10 m (see Carlquist, 1988; Zimmermann, 1983).

Besides vessel members and tracheids, other types of cells in the xylem are parenchyma cells and fibers (Fig. 1.3). Xylem fibers, which contribute to the structural support of a plant, are long thin cells having lignified cell walls; they are generally devoid of protoplasts at maturity but are nonconducting. The living parenchyma cells in the xylem are important for the storage of carbohydrates and for the lateral movement of water and solutes into and out of the conducting cells.

Poiseuille's Law

To describe fluid movement in the xylem quantitatively, we need to relate the flow to the particular driving force causing the motion. For cylindrical tubes, an appropriate relationship is the one determined experimentally by Hagen in 1839 and independently by Poiseuille in 1840. They found that the volume of fluid moving in unit time along a cylinder is proportional to the fourth power of its radius, and that the movement depends linearly on the drop in hydrostatic pressure. Wiedemann in 1856 showed that this rate of volume movement per tube could be represented as follows:

$$\text{Volume flow rate per tube} = -\frac{\pi r^4}{8\eta}\frac{\partial P}{\partial x} \qquad (9.11a)$$

where r is the radius of the cylinder, η is the viscosity of the solution, and $-\partial P/\partial x$ is the negative gradient of the hydrostatic pressure. Throughout this text we have been concerned with the volume flowing per unit time and $area$, J_V. For flow in a cylinder of radius r, and hence area πr^2, J_V is

$$J_V = -\frac{r^2}{8\eta}\frac{\partial P}{\partial x} \qquad (9.11b)$$

Because positive flow ($J_V > 0$) occurs in the direction of decreasing hydrostatic pressure ($\partial P/\partial x < 0$), the minus sign is necessary in Equation 9.11.

Equation 9.11 is generally referred to as $Poiseuille's\ law$ and sometimes as the Hagen–Poiseuille law. It assumes that the fluid in the cylinder moves in layers, or $laminae$, each layer gliding over the adjacent one (see Streeter and Wylie, 1985). Such laminar movement occurs only if flow is slow enough to meet a criterion deduced by Reynolds in 1883. Specifically, the dimensionless quantity $\rho J_V d/\eta$ (another form of the Reynolds number, Eq. 7.16) must be less than 2 000 (ρ is the solution density and d is the cylinder diameter).[9] Otherwise, a transition to turbulent flow occurs, and Equation 9.11 is no longer valid (we will shortly consider the Reynolds number for flow in the xylem). The fluid in Poiseuille (laminar) flow is stationary at the wall of the cylinder; the speed of solution flow increases in a parabolic fashion to a maximum value in the center of the tube, where it is twice the average speed, J_V (see Vogel, 1981). Thus, the flows in Equation 9.11 are actually the mean flows averaged over the entire cross section of cylinders of radius r. The dimensions for the viscosity η are pressure $\times$ time, e.g., Pa s (the cgs unit, dyne s cm^{-2}, is termed a *poise,* which equals 0.1 Pa s); the viscosity of water at 20°C is 1.002×10^{-3} Pa s (values of η_w at other temperatures are given in App. I).

Applications of Poiseuille's Law

As we have already indicated, the volume flux densities described by both Poiseuille's law [$J_V = -(r^2/8\eta)\, \partial P/\partial x$, Eq. 9.11b] and Darcy's law ($J_V = -L^{\text{soil}}\, \partial P^{\text{soil}}/\partial x$, Eq. 9.7) depend on the negative gradient of the hydrostatic pressure. Can we establish any correspondence between these two equations? If the soil pores were cylinders of radius r, all aligned in the same direction, we could have Poiseuille flow in the soil, in which case L^{soil} would equal $r^2/8\eta$— we are assuming for the moment that the pores occupy the entire soil volume. When r is 1 μm and η equals 1.002×10^{-3} Pa s, we obtain

$$\frac{r^2}{8\eta} = \frac{(1 \times 10^{-6}\ \text{m})^2}{(8)(1.002 \times 10^{-3}\ \text{Pa s})} \doteq 1.2 \times 10^{-10}\ \text{m}^2\ \text{s}^{-1}\ \text{Pa}^{-1}$$

which corresponds to the L^{soil} for a very wet soil (Fig. 9.9). Actually, L^{soil} for water-filled porous clay with average pore radii of 1 μm is about 10^{-11} m^2 s^{-1} Pa^{-1}, which is about 10-fold smaller than $r^2/8\eta$ calculated using the average pore radius. As a result, J_V in the soil is considerably less than for Poiseuille flow through pores of the same average radius. This is mainly because the soil pores are not in the shape of cylinders aligned in the direction of flow; also, the pores between soil particles occupy only about half of the soil volume.

---

9. Equation 7.16 indicates that Re is vd/v, where the kinematic viscosity v equals η/ρ (η is properly known as the dynamic viscosity). Footnote 14, Chapter 2, indicates that the volume flux density J_V equals the mean velocity of fluid movement v. Thus, Re can be represented by $J_V d/(\eta/\rho)$, or $\rho J_V d/\eta$.

Next, we will use Poiseuille's law to estimate the pressure gradient necessary to cause a specified volume flux density in the conducting cells of the xylem.[10] The speed of sap ascent in the xylem of a transpiring tree can be about 3.6 m h^{-1}, which equals 1 mm s^{-1} (see Briggs, 1967; Slatyer, 1967; Zimmermann, 1983). On p. 99 we indicated that v_w equals the volume flux density of water, J_{V_w}; the average speed of the *solution* equals J_V, the volume flux density of the solution (for a dilute aqueous solution such as occurs in the xylem, $J_V \cong J_{V_w}$; also see p. 170). Thus, J_V in the xylem of a transpiring tree might be 1 mm s^{-1}. For a viscosity of 1.0 mPa s and a xylem element with a lumen radius of 20 μm, the pressure gradient required to satisfy Equation 9.11b is

$$\frac{\partial P}{\partial x} = \frac{-8\eta J_V}{r^2} = \frac{-(8)(1.0 \times 10^{-3} \text{ Pa s})(1.0 \times 10^{-3} \text{ m s}^{-1})}{(20 \times 10^{-6} \text{ m})^2}$$

$$= -2 \times 10^4 \text{ Pa m}^{-1} = -0.02 \text{ MPa m}^{-1}$$

As expected, the pressure decreases along the direction of flow (Fig. 9.10).

The estimate of -0.02 MPa m^{-1} is consistent with experimental observations of the $\partial P/\partial x$ accompanying solution flow in *horizontal* xylem vessels. For the vertical vessels in a tree, however, a static hydrostatic pressure gradient of -0.01 MPa due to gravity exists even in the absence of flow.[11] For transpiring plants, the total $\partial P/\partial x$ in the vertical sections of the xylem is often about -0.03 MPa m^{-1} (Fig. 9.10). Based on the above calculations, this pressure gradient is sufficient to overcome gravity (-0.01 MPa m^{-1}) and to cause Poiseuille flow in the xylem vessels (-0.02 MPa m^{-1} needed in the above example where $r = 20$ μm; see Briggs, 1967). The above gradients refer to vessel members joined by simple perforation plates that offer little obstruction to flow (Fig. 1.3). Gradients tend to be greater for tracheids, because they are connected by numerous small pits containing locally thin regions of the cell wall. These pits can greatly

---

10. For application of Poiseuille's law to complex tissue such as the xylem, care must be taken to ensure that particular vessel elements or tracheids are conducting (e.g., not blocked by embolisms), the actual radii must be determined (note the r^4 dependence in Eq. 9.11a), and corrections may be necessary for lumen shape, tracheid taper, and cell wall characteristics including pits (see Calkin *et al.*, 1986; Schulte *et al.*, 1989a). For instance, if the lumen is elliptical with major and minor axes of a and b, respectively, then r^4 in Equation 9.11a should be replaced by $a^3b^3/(8a^2 + 8b^2)$.

11. We can appreciate this by considering a vertical column of pure water ($\Pi = 0$) at equilibrium, where the water potential Ψ is $P + \rho_w gh$; $\rho_w gh$ increases vertically upward, so P must decrease by the same amount for Ψ to remain unchanged, as it does at equilibrium. Because $\rho_w g$ equals 0.0098 MPa m^{-1} (App. I), the additional pressure gradient caused by gravity amounts to -0.01 MPa m^{-1} (see Fig. 9.10).

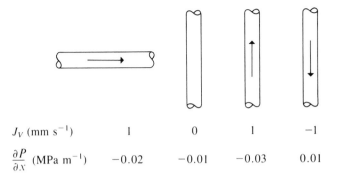

J_V (mm s^{-1})	1	0	1	−1
$\dfrac{\partial P}{\partial x}$ (MPa m^{-1})	−0.02	−0.01	−0.03	0.01

Figure 9.10

Flows and pressure gradients in cylindrical tubes. The tube radius is 20 μm and $\partial P/\partial x$ is calculated from Poiseuille's law (Eq. 9.11b) as modified by gravitational effects. For vertical tubes, x is considered positive upward. Arrows indicate the direction of flow.

restrict the flow, because some cell wall material must be traversed to move from one tracheid to another. For instance, for fern tracheids with small radii (<10 μm) the pressure drop is mainly along the lumen, reflecting the major influence of the lumen radius on Poiseuille flow, whereas for larger tracheids most is across the pits (see Calkin et al., 1986). In conifers half of the pressure drop along the xylem may occur across the pits (see Jarvis, 1975), and hence the overall $\partial P/\partial x$ for the same lumen diameter and flow rate would be about twice as large as that calculated here.

We have been approximating xylem vessels by cylinders that are 20 μm in radius, as might be appropriate for diffuse-porous (small-porous) trees. For ring-porous (large-porous) trees, the mean radii of xylem vessels are often about 100 μm (see Carlquist, 1988; Zimmermann, 1983). For a given pressure gradient, Poiseuille's law (Eq. 9.11b) indicates that a much larger speed of xylem sap movement occurs in the ring-porous trees, as is indeed observed. Using our above values and a water density of 1 000 kg m^{-3}, the Reynolds number for the diffuse-porous case is

$$\text{Re} = \frac{\rho J_V d}{\eta} = \frac{(1\ 000\ \text{kg m}^{-3})(1 \times 10^{-3}\ \text{m s}^{-1})(40 \times 10^{-6}\ \text{m})}{(1.0 \times 10^{-3}\ \text{Pa s})}$$

$$= 0.04\ \text{kg m}^{-1}\ \text{s}^{-2}\ \text{Pa}^{-1} = 0.04$$

Such a low value indicates that no turbulence is expected. Even if d were 5-fold larger and J_V were 10-fold higher, as can occur in a ring-porous tree, the

Reynolds number would still be far less than the value of 2 000 where turbulence generally sets in.

It is interesting to estimate the sort of pressure gradient necessary to promote a given flow along a cell wall. This use of Poiseuille's law will allow us to compare such a pathway with water movement in the lumens of xylem vessels. Because the interfibrillar spaces, or interstices, in a cell wall have diameters near 10 nm (100 Å), we will let r be 5 nm for purposes of calculation. (Complications due to the tortuosity of the aqueous channels through the interstices will be omitted here. We will also assume that the pores occupy the entire cell wall.) For J_V equal to 1 mm s^{-1}, Equation 9.11b indicates that the pressure gradient required for solution flow through the cell walls is $-(1 \times 10^{-3}$ m s$^{-1})$ $(8)(1.0 \times 10^{-3}$ Pa s)/$(5 \times 10^{-9}$ m)^2 or -3×10^5 MPa m^{-1}. In contrast, a $\partial P/\partial x$ of only -0.02 MPa m^{-1} is needed for the same J_V in the xylem element with a 20 μm radius. Thus, the $\partial P/\partial x$ for Poiseuille flow through the small interstices of a cell wall is over 10^7 times greater than for the same flux density through the lumen of the above xylem element. Because of the tremendous pressure gradients required to force water through the small interstices available for solution conduction in the cell wall, fluid could not flow rapidly enough up a tree in the cell walls—as has been suggested—to account for observed rates of water movement.

To help compare the relative effects on flow of a cell wall, a plasmalemma, and the lumen of a xylem vessel, we will calculate the hydrostatic pressure drops across each of them in a hypothetical case. Let us consider a xylem vessel member that is 1 mm long and 20 μm in radius. We have just calculated that the pressure gradient necessary for a Poiseuille flow of 1 mm s^{-1} in the lumen of this vessel member is -0.02 MPa m^{-1}, which amounts to $(-0.02$ MPa m$^{-1})(10^{-3}$ m), or -2×10^{-5} MPa for the 1 mm length of the cell. Let us suppose that a cell wall 1 μm thick is placed on both ends of the cell. In the previous paragraph we indicated that a pressure gradient of -3×10^5 MPa m^{-1} is necessary for a J_V of 1 mm s^{-1} through the interstices of a cell wall. To cross this cell wall (a total distance of 2 μm for both ends) would require a pressure change of $(-3 \times 10^5$ MPa m$^{-1})(2 \times 10^{-6}$ m), or -0.6 MPa. Finally, let us imagine that a plasmalemma with a typical hydraulic conductivity coefficient of 7×10^{-13} m s^{-1} Pa^{-1} is also placed around the cell. Using Equation 3.38 [$J_V = L_P(\Delta P - \sigma \Delta\Pi)$] with J_V equal to 1 mm s^{-1}, the difference in hydrostatic pressure required for such flow out the ends is $(1 \times 10^{-3}$ m s$^{-1})/(7 \times 10^{-13}$ m s^{-1} Pa^{-1}), or 1.4×10^9 Pa, which means a total pressure drop of 3×10^3 MPa to traverse the plasmalemma at each end of the cell. In summary, the pressure drops needed in this hypothetical case are 2×10^{-5} MPa to flow through the lumen, 0.6 MPa to cross the cell walls, and 3×10^3 MPa to cross the membranes.

In fact, membranes generally serve as the main barrier to water flow into or out of plant cells. The interstices of the cell walls provide a much easier pathway for such flow, and a hollow xylem vessel presents the least obstacle to flow (such as up a stem). Consequently, xylem provides a plant with tubes, or conduits, well suited to moving water over long distances. The region of a plant made up of cell walls and the hollow xylem vessels is often called the apoplast, as noted in Chapter 1. Water and the solutes that it contains can move fairly readily in the apoplast, but they must cross a membrane to enter the symplast (symplasm), the interconnected cytoplasm of the cells.

The Phloem

Water and solute movement in the phloem involves cooperative interactions among several types of cells. The conducting cells of the phloem generally have a high internal hydrostatic pressure, which often leads to artifacts during cytological investigations. In part because of the observational difficulties involved, there has been a lack of agreement on the actual internal structure of the cells and the mechanism for solute movement in the phloem. Our brief treatment will be primarily concerned with certain anatomical features of the phloem cells and an overall description of pressure-driven flow (for further details, see Cronshaw *et al.*, 1985; Kursanov, 1984; Zimmermann and Milburn, 1975).

The conducting cells of the phloem are the sieve cells in lower vascular plants and gymnosperms, and the generally shorter, wider, and less tapered sieve-tube members in angiosperms (Fig. 1.3). Both types of cells are collectively called *sieve elements*. Mature sieve elements almost invariably have lost their nuclei, and apparently the tonoplast has broken down, so no large central vacuole is present, although the plasmalemma remains intact. However, unlike the conducting elements of the xylem, the sieve elements contain cytoplasm and are living. The sieve elements in most plants range from 0.1 to 3 mm in length and tend to be longer in gymnosperms than in angiosperms. Typical radii of the lumens are 6 to 25 μm. Specialized cells are generally found adjacent to and in close association with the sieve elements, called *companion cells* in angiosperms (see Fig. 1.3) and *albuminous cells* in gymnosperms. These cells have nuclei and generally contain many mitochondria. Although their precise functions are still unclear, such cells are metabolically related to the conducting cells and may supply them with carbohydrates, ATP, and other materials. Moreover, companion and albuminous cells accumulate sugars and other solutes, which may either passively diffuse—e.g., through the plasmodesmata that are present—or be actively transported into the sieve elements.

Sieve-tube members usually abut end-to-end to form the *sieve tubes*. The pair of generally inclined end walls between two sequential sieve-tube members forms the *sieve plate* (Fig. 1.3; sieve cells are joined at less specialized sieve areas). Sieve plates have many pores, ranging in diameter from less than 1 μm up to about 5 μm, and ordinarily have strands of cytoplasmic material passing through them. The pores are lined, not crossed, by the plasmalemmas. Therefore, solution moving in the phloem does not have to cross any membranes as it flows from cell to cell along a sieve tube. A distinguishing feature of sieve-tube members in dicots and some monocots is the presence of *P protein* (for *phloem protein*), which often occurs in tubular as well as fibrillar form (see Cronshaw, 1981). Although the exact functions of this material are unclear, P protein can plug the sieve plate pores upon injury of the sieve tube. Specifically, when the phloem is opened by cutting a stem, the positive hydrostatic pressure in the phloem forces the contents of the conducting cells toward the incision, carrying the P protein into the sieve plate pores. As a further wound response, callose (a glucose polysaccharide) can also be deposited in the sieve plate pores, thereby closing them within minutes following injury.

The products of photosynthesis—*photosynthate*—are distributed throughout the plant by the phloem. Changes in phloem flow can affect the distribution of photosynthate, which in turn can affect the metabolism of mesophyll cells (see De Michele *et al.*, 1978). For high transpiration rates or a moderate moisture stress leading to reduced leaf water potentials, translocation of photosynthate can decrease more than photosynthesis in some plants, with the consequence that leaves and nearby parts of the stem accumulate more starch. In C_4 plants the profuse vascularization and close proximity of bundle sheath cells to the phloem (see Fig. 8.11) can lead to rapid removal of photosynthate from the leaves and little deposition of starch in the bundle sheath cells over the course of a day, whereas certain C_3 plants can have a large increase of starch in the mesophyll cells during the daytime. The number of sieve elements in a petiole can reflect the productivity of the leaf; e.g., petioles of sun leaves tend to have more sieve elements per unit leaf area than do petioles of shade leaves. The practice of "ringing" or "girdling" a branch (cutting away a band of phloem-containing bark all the way around) can stop the export of photosynthate, with the consequence that fruit on such a branch can become considerably larger than it otherwise would. The practice of girdling the trunk of a tree abolishes the export of photosynthate to the roots, ultimately leading to death of the tree.

Phloem Contents and Speed of Movement

An elegant way of studying the contents of certain sieve elements is by means of aphid stylets (see Zimmermann and Milburn, 1975). An aphid feeds on the phloem by inserting its stylet into an individual sieve element. After the aphid has been anesthetized and its body removed, the remaining mouth part forms a tube that leads the phloem solution from the sieve element to the outside. Solutes in the phloem solution extracted using this technique are over 90% carbohydrates, mainly sucrose and some other oligosaccharides. The concentration of sucrose is generally 0.2 to 0.7 M; values near 0.3 M are typical for small plants, whereas some tall trees have sucrose concentrations near the upper limit. The types and concentrations of the solutes exhibit daily and seasonal variations, and also depend on the tissues that the phloem solution is flowing toward or away from. For example, the solution in the phloem moving out of senescing leaves is low in sucrose but contains an appreciable concentration of amino acids and amides, sometimes as high as 0.5 M—an amino acid concentration near 0.05 M is representative of phloem solution in general. The initial movement of ions from the root to the rest of the plant occurs mainly in the xylem; subsequent recirculation can take place in the phloem, such as the movement of ions out of leaves just before their abscission.

Solutes can move rapidly in the phloem, flow being toward regions of lower osmotic pressure. Thus, photosynthetic products move from the leaves to storage tissues in the stem and the root, where they are generally converted to starch. At other times, sugars produced from the hydrolysis of such starch may move in the opposite direction, from the storage tissue to meristematic areas at the top of the plant. The speed of solute movement is ordinarily 0.2 to 2 m h^{-1}, the rate varying, among other things, with the plant species and vigor of growth (see Kursanov, 1984). Although a particular sugar generally moves in the phloem in the direction of a decrease in its concentration, diffusion is not the mechanism. First, the rate of solute movement far exceeds that of diffusion (see Ch. 1). Second, when a radioactive solute, a dye, or a heat pulse is introduced into conducting sieve elements, the "front" moves with a fairly constant speed (distance moved is proportional to time), whereas in a one-dimensional diffusional process, distance moved is proportional to the square root of time in such a case ($x_e^2 = 4D_j t_e$, Eq. 1.6).

Rather than the solute speed in the phloem, we are often more interested in how much dry matter is translocated. For example, if the sieve elements contain 0.5 M (500 mol m^{-3}) sucrose moving at an average speed of 0.6 m h^{-1}, what is the transfer rate of sucrose in kg m^{-2} h^{-1}? By Equation 3.6 ($J_j = \bar{v}_j c_j$), the flux density of sucrose is

$$J_{sucrose} = (0.6 \text{ m h}^{-1})(500 \text{ mol m}^{-3})$$

$$= 300 \text{ mol m}^{-2} \text{ h}^{-1}$$

Because sucrose has a mass of 0.342 kg mol^{-1}, this flux density corresponds to 100 kg m^{-2} h^{-1}. In the present example, the flow is per m^2 of sieve-tube *lumens*; the rate of flow per unit area of *phloem tissue* would be less by the ratio of the lumen cross-sectional area to the total phloem cross-sectional area, generally 0.2 to 0.5.

Mechanism of Phloem Flow

What causes the movement of solutes in the phloem? Primarily because of observational difficulties, this question has proved hard to answer. Another complication is that water may readily enter and leave the various types of cells in the phloem and surrounding tissue. The phloem therefore cannot be viewed as an isolated independent system. For example, when the water potential in the xylem decreases, as occurs during rapid transpiration, fluid in the phloem generally moves more slowly. Some water may move upward in the xylem and, later, downward in the phloem; but this is not the whole story, because movement in the phloem can be in either direction.

In 1930 Münch proposed that fluid movement in the phloem is caused by a gradient in hydrostatic pressure (see Münch, 1930). This would lead to flow analogous to that in the xylem as described by Poiseuille's law (Eq. 9.11b), $J_V = -(r^2/8\eta)(\partial P/\partial x)$. To examine this hypothesis, we will assume that the average speed of flow in the lumen of the conducting cells of the phloem, J_V, is 0.6 m h^{-1} (0.17 mm s^{-1}). Our sieve-tube members will be 12 μm in radius and 1 mm long, and the sieve plates will be 5 μm thick, with pores 1.2 μm in radius covering one-third of their surface area. Based on the relative areas available for conduction, J_V is three times as high in the sieve plates as in the lumen of the sieve tube—namely, 0.51 mm s^{-1} across the sieve plates. The viscosity of the solution in a sieve tube is greater than that of water; e.g., η^{phloem} may be about 1.7 mPa s at 20°C, the value for 0.5 M sucrose. The pressure change per cell expected from Poiseuille's law is then

$$\Delta P = \frac{-8\eta J_V}{r^2} \Delta x = \frac{-(8)(1.7 \times 10^{-3} \text{ Pa s})(0.17 \times 10^{-3} \text{ m s}^{-1})}{(12 \times 10^{-6} \text{ m})^2}(1 \times 10^{-3} \text{ m})$$

$$= -16 \text{ Pa}$$

in the lumen and -24 Pa across the sieve plate. For cells 1 mm long, the pressure gradient needed to cause Poiseuille flow is then -40×10^3 Pa m^{-1}, or -0.040 MPa m^{-1}.

In the above case, somewhat over half of the hydrostatic pressure gradient in the phloem is necessary to overcome the resistance of the sieve plate pores. When the end walls are steeply inclined to the axis of the sieve element, the pores of the sieve plate can occupy an area greater than that of the cross section of the sieve tube. This causes J_V in the pores to be less than in the lumen, which tends to reduce the resistance to flow in the phloem.

Values for Components of the Phloem Water Potential

We will now examine some of the consequences of a pressure-driven flow in the phloem. We will let the water potential in the xylem be -0.6 MPa at ground level and -0.8 MPa at 10 m above the ground (see Fig. 9.11 and Table 9.2). Because there is no evidence for active transport of water across membranes into the phloem, the water potential in the phloem 10 m above the ground, $\Psi^{\text{phloem}}_{10\,m}$, must be lower than -0.8 MPa if water is to enter passively—water entering the phloem in the leaves can come from the xylem. The water potential there might be -1.0 MPa, for example. By analogous reasoning, Ψ^{phloem} in the root must be higher than Ψ^{xylem} if water is to spontaneously leave the phloem in the region where the solutes, such as sucrose, are removed; an appropriate value for the water potential in the phloem at ground level, $\Psi^{\text{phloem}}_{0\,m}$, might be -0.4 MPa. Water may then flow down the phloem from a water potential of -1.0 MPa to one of -0.4 MPa. This appears to be an energetically uphill movement. However, as in the xylem, the term representing the force on solution moving within the phloem is ΔP, not $\Delta\Psi$.

The fact that the decrease in hydrostatic pressure, not the change in water potential, represents the driving force along the phloem deserves special emphasis. By Equation 3.38, J_V equals $L_P(\Delta P - \sigma \, \Delta\Pi)$, where the reflection coefficient σ within the phloem is zero because no membranes intervene between sequential members of a sieve tube. Differences in the osmotic pressure at various locations in the phloem, which help determine the overall change in water potential, therefore do not directly affect the movement of solution *along* the phloem. On the other hand, substances entering or leaving the phloem pass across a plasmalemma, which most likely has a mean reflection coefficient close to unity for solutes such as sucrose. J_V given by Equation 3.38 is then $L_P(\Delta P - \Delta\Pi)$, which equals $L_P\Delta\Psi$ by Equation 2.13a ($\Psi = P - \Pi + \rho_w gh$).

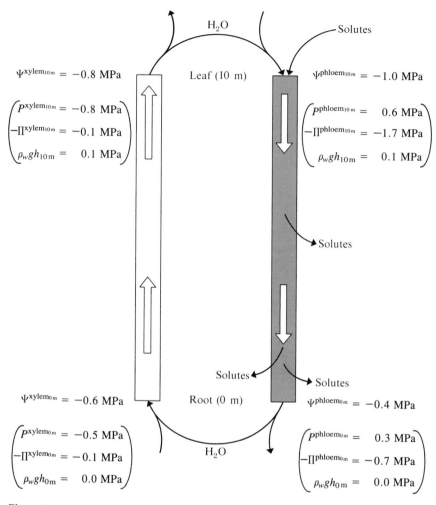

Figure 9.11

Idealized representation of xylem and phloem flow driven by gradients in hydrostatic pressure. Water enters and leaves the phloem by passively moving toward regions of lower water potential. Solutes either diffuse or are actively transported into and out of the sieve elements, leading to a decrease in the phloem osmotic pressure from 1.7 MPa in the leaf to 0.7 MPa in the root.

Consequently, the change in water potential probably does represent the driving force on water moving into or out of the phloem (see Fig. 9.11).

Next we will estimate the various components of the water potential at the upper end of the phloem tissue under consideration (Fig. 9.11). The osmotic pressure in a sieve tube in the phloem of a leaf 10 m above the ground,

$\Pi^{\text{phloem}}_{10\ \text{m}}$, might be due to the following solutes: 0.5 M sucrose, 0.1 M other sugars, 0.05 M amino acids, and 0.05 M inorganic ions, i.e., $\Sigma_j c_j = 0.5$ M + 0.1 M + 0.05 M + 0.05 M = 0.7 M, which is 700 mol m^{-3}. Using the Van't Hoff relation—$\Pi_s = RT \Sigma_j c_j$ (Eq. 2.10), where RT is 2.437 × 10^{-3} m^3 MPa mol^{-1} at 20°C (App. I)—we estimate that the osmotic pressure in the phloem 10 m above the ground is (2.437 × 10^{-3} m^3 MPa mol^{-1})(700 mol m^{-3}), or 1.7 MPa. Such a large osmotic pressure, caused by the high concentrations of sucrose and other solutes, suggests that active transport is necessary at some stage to move certain photosynthetic products from leaf mesophyll cells to the sieve elements of the phloem. From the definition of water potential, $\Psi = P - \Pi + \rho_w g h$ (Eq. 2.13a), we conclude that the hydrostatic pressure in the phloem of a leaf 10 m above the ground is

$$P^{\text{phloem}}_{10\ \text{m}} = \Psi^{\text{phloem}}_{10\ \text{m}} + \Pi^{\text{phloem}}_{10\ \text{m}} - \rho_w g h_{10\ \text{m}}$$

$$= -1.0\ \text{MPa} + 1.7\ \text{MPa} - 0.1\ \text{MPa}$$

$$= 0.6\ \text{MPa}$$

(see Fig. 9.11). This appreciable hydrostatic pressure causes sieve-tube members to exude when cut.

As we indicated above, the hydrostatic pressure gradient can be -0.04 MPa m^{-1} for Poiseuille flow in representative (horizontal) conducting cells of the phloem. P would therefore decrease 0.4 MPa for a distance of 10 m. Gravity also leads to a pressure gradient in a column of water (0.1 MPa increase in P when descending 10 m), so in the present case the ΔP needed to cause Poiseuille flow vertically *downward* is -0.3 MPa. Thus, the hydrostatic pressure in the phloem at ground level is 0.6 $-$ 0.3 or 0.3 MPa. The osmotic pressure there equals $P^{\text{phloem}}_{0\ \text{m}} + \rho_w g h_{0\ \text{m}} - \Psi^{\text{phloem}}_{0\ \text{m}}$ (Eq. 2.13a), which is 0.7 MPa. The values of Ψ^{phloem} and its components at $h = 0$ m and $h = 10$ m are summarized in Figure 9.11.

In our present example, the osmotic pressure of the phloem solution decreases from 1.7 MPa in the leaf to 0.7 MPa in the root (Fig. 9.11). Such a decrease in Π is consistent with the phloem's function of delivering photosynthetic products to different parts of the plant. Moreover, our calculations indicate that flow is in the direction of decreasing concentration, but that diffusion is not the mechanism. (Although the total concentration decreases in the direction of flow, the c_j of every solute does not necessarily do so.) Finally, we note the importance of removing solutes, either by active transport or by diffusion into the cells near the conducting cells of the phloem.

The known involvement of metabolism in translocation in the phloem could be due to active transport of solutes into the phloem of a leaf or other *source,* which is often referred to as *loading,* and/or to their removal, or *unloading,* in a root or other *sink,* such as a fruit. Indeed, loading appears to involve proton-sucrose cotransport via a carrier located in the plasmalemma of sieve elements, which moves this sugar from the apoplast into the conducting cells (movement of photosynthate among mesophyll cells and possibly to companion cells is mainly in the symplast). Low concentrations of translocated solutes in the cells outside the phloem in sink regions, which favor the diffusion of these solutes out of the sieve elements, could also be maintained by metabolic conversions of phloem solutes, such as by starch formation in the cells adjacent to the conducting cells of the phloem. In any case, active loading or unloading of the phloem sets up a gradient in osmotic pressure, which in turn generates the hydrostatic pressure gradient that can lead to Poiseuille flow along the phloem. Much remains to be learned about the details of such fluid movement in the phloem (see Gifford and Evans, 1981; Kursanov, 1984).

THE SOIL–PLANT–ATMOSPHERE CONTINUUM

Our final general topic is the movement of water from the soil, through a plant, out into the surrounding atmosphere. During a growing season, about 100 times more water is transpired by a plant than remains in it for growth. Thus, the amount of transpiration is a fairly accurate estimate of water uptake by the roots. Although the rate that water crosses each section of the pathway is essentially the same in the steady state, the resistances and the areas across which water flows differ markedly for the various components. Moreover, a steady state generally does not prevail, especially when environmental factors such as sunlight or air temperature are changing.

Values of Water Potential Components

Possible values for the water potential and its components in various parts of the soil–plant–atmosphere system are given in Table 9.2. The values do not apply to all plants, nor even to the same plant at all times. Rather, they serve to indicate representative contributions of P, Π, $\rho_w gh$, and relative humidity to Ψ in various parts of the soil–plant–atmosphere continuum.

First, we will consider the soil water potential, Ψ^{soil}. As we indicated earlier in this chapter, Ψ^{soil} is usually dominated by P^{soil}, which is negative because of surface tension effects at the numerous air–liquid interfaces in the soil. The

Table 9.2

Representative values for the various components of the water potential in the soil–plant–atmosphere continuum. Ψ equals $P - \Pi + \rho_w gh$ in the liquid phases (Eq. 2.13a) and $(RT/\overline{V}_w) \ln (\%$ relative humidity$/100) + \rho_w gh$ in the gas phases (Eq. 2.21), all at 25°C.

Location	Ψ (MPa)	P (MPa)	$-\Pi$ (MPa)	$\rho_w gh$ (MPa)	$\dfrac{RT}{\overline{V}_w} \ln \left(\dfrac{\% \text{ RH}}{100} \right)$ (MPa)
Soil 0.1 m below ground and 10 mm from root	-0.3	-0.2	-0.1	0.0	
Soil adjacent to root	-0.5	-0.4	-0.1	0.0	
Xylem of root near ground surface	-0.6	-0.5	-0.1	0.0	
Xylem in leaf at 10 m above ground	-0.8	-0.8	-0.1	0.1	
Vacuole of mesophyll cell in leaf at 10 m	-0.8	0.2	-1.1	0.1	
Cell wall of mesophyll cell at 10 m	-0.8	-0.4	-0.5	0.1	
Air in cell wall pores at 10 m (water vapor assumed to be in equilibrium with water in cell wall)	-0.8			0.1	-0.9
Air just inside stomata at 95% relative humidity	-6.9			0.1	-7.0
Air just outside stomata at 60% relative humidity	-70.0			0.1	-70.1
Air just across boundary layer at 50% relative humidity	-95.1			0.1	-95.2

magnitude of the water potential varies greatly with environmental conditions and the type of soil. After a rainfall or in freshly irrigated soil, Ψ^{soil} may be about -0.01 MPa, whereas permanent wilting of plants often occurs when it decreases below about -1.5 MPa.

The value of the soil water potential at which wilting of a plant occurs depends on the osmotic pressure in the vacuoles of its leaf cells, as the following argument will show (Π^{vacuole} is generally the same as Π^{cytosol}; see p. 79). Let us consider the case in Table 9.2 where the water potential in the vacuole of a leaf cell 10 m above the ground is initially -0.8 MPa. As the soil dries, Ψ^{soil} decreases and eventually becomes -1.0 MPa. When Ψ^{soil} becomes -1.0 MPa, Ψ^{leaf} must be less than this for water movement to continue from the soil to the leaf. Ψ^{leaf} could be -1.0 MPa when $P^{vacuole}_{10\ m}$ is 0.0 MPa, $\rho_w g h^{vacuole}_{10\ m}$ remains 0.1 MPa, and $\Pi^{vacuole}_{10\ m}$ remains 1.1 MPa (the latter two are values in Table 9.2; actually, as the hydrostatic pressure in a leaf cell decreases, the cell will shrink somewhat because of the elastic properties of the cell wall, so $\Pi^{vacuole}$ will increase slightly; see p. 87). Zero hydrostatic pressure in the vacuole means that the cell has lost turgor, and the leaf thus wilts in response to this low Ψ^{soil}. For certain xerophytes in arid areas, the osmotic pressure in the leaves can be 2.5 to 5.0 MPa under normal physiological conditions. The value of the soil water potential at which wilting occurs for such plants is considerably lower (i.e., more negative) than for the plant indicated in Table 9.2. A high osmotic pressure in the vacuoles of the leaf cells can therefore be viewed as an adaptation to the low soil water potentials in arid regions (see Slatyer, 1967).

Let us suppose that the soil dries even further from the level causing wilting. Because at least some transpiration still occurs because of the very low Ψ^{air}, Ψ^{leaf} will be less than Ψ^{soil}, and some cellular water will be lost from the leaf. The vacuolar contents then become more concentrated, and $\Pi^{vacuole}$ increases. For instance, if Ψ^{leaf} became -2.0 MPa for a Ψ^{soil} of -1.8 MPa, the osmotic pressure in the vacuole of a leaf cell 10 m above the ground would be 2.1 MPa, which represents a loss of nearly half of the cellular water ($\Pi^{vacuole}_{10\ m}$ was originally 1.1 MPa for this leaf cell; Table 9.2).

As we have indicated, the driving force for water movement in the xylem is the negative gradient in hydrostatic pressure, which can lead to a flow describable by Poiseuille's law (Eq. 9.11). In Table 9.2, P^{xylem} decreases by 0.3 MPa from the root to a leaf 10 m above the ground. The xylary sap, which contains chiefly water plus some minerals absorbed from the soil, usually does not have an osmotic pressure in excess of 0.2 MPa. The hydrostatic pressure, on the other hand, can have much larger absolute values and generally changes markedly during the day. When there is extremely rapid transpiration, large tensions (negative hydrostatic pressures) can develop in the xylem. These tensions are maintained by the cohesion of water molecules resulting from the intermolecular hydrogen bonding (Ch. 2). When transpiration essentially ceases, as it can at night or under conditions of very high relative humidity in the air surrounding the plant, the tension in the xylem becomes very small—in fact, the hydrostatic pressure can even become positive (reflecting water movement from the surrounding cells into the root xylem in response to Π^{xylem}). Such a positive hydrostatic pressure (termed *root pressure*) can cause guttation as xylem fluid is then exuded through specialized structures called hydathodes located near the ends of veins on the margins of leaves.

Water is conducted to and across the leaves in the xylem. It then moves to the individual leaf cells by flowing partly apoplastically in the cell walls and partly symplastically (only short distances are involved, because the xylem ramifies extensively in a leaf). The water potential is usually about the same in the vacuole, the cytosol, and the cell wall of a particular mesophyll cell (see values in Table 9.2). If this were not the case, water would redistribute by flowing energetically downhill toward lower water potentials. The water in the cell wall pores is in contact with air, where evaporation can take place, so a flow occurs along the interstices to replace lost water. This flow can be approximately described by Poiseuille's law (Eq. 9.11), which indicates that a (very small) hydrostatic pressure gradient exists across such cell walls.

As water continually evaporates from the cell walls of mesophyll cells, the accompanying solutes originally in the xylary sap are left behind and can accumulate in the cell wall water. Some solutes are of course needed for cell growth. For halophytes and xerophytes growing in soils of high salinity, excess inorganic ions can be actively excreted from a leaf by salt glands on its surface (see Lüttge, 1975). The periodic abscission of leaves is another way of "preventing" an excessive buildup of solutes in the cell wall water, as well as for returning mineral nutrients to the soil (see Epstein, 1972).

Resistances and Areas

We will now consider the resistances to water flow in those parts of the soil–plant–atmosphere continuum where water moves as a liquid. (We have already considered the gaseous parts of the pathway in Ch. 8.) If we let the flux density of water equal the drop in water potential across some component divided by its resistance, we would have only part of the story, because we should also consider the relative areas of each component as well as whether $\Delta\Psi$ represents the driving force. Moreover, $\Delta\Psi$ represents the relative energies of water only at constant T, and thus ideally we should compare water potentials only between locations at the same temperature. The root and adjacent soil are usually at the same temperature, as are the mesophyll cells and air in the intercellular air spaces of a leaf. But roots and leaves generally are *not* at the same temperature. Nevertheless, using the pressure gradient (Eq. 9.11) turns out to be sufficiently accurate for describing the flow in the xylem or the phloem, even when a temperature difference exists along the pathway (see Spanner, 1972).

Let us designate the average volume flux density of water across area A^j of component j by $J_{V_w}^j$. A^j could be the root surface area, the effective cross-sectional area of the xylem, or the area of one side of the leaves. In the steady state, the product $J_{V_w}^j A^j$ is essentially constant, because nearly all of the water taken up by the root is lost by transpiration, and thus the same volume of water moves across each component along the pathway per unit time. We will represent the drop in water potential across component j by $\Delta \Psi^j$, and define the resistance of component j (R^j) as follows:

$$J_{V_w}^j A^j = \frac{\Delta \Psi^j}{R^j} \cong \text{constant} \qquad (9.12)$$

where $J_{V_w}^j A^j$ is the volume of water crossing component j in unit time (e.g., m³ s⁻¹). A relation similar to Equation 9.12 was proposed by van den Honert in 1948 (van den Honert, 1948).

To illustrate the constancy of flow of water through a plant, let us reconsider its volumetric flux densities for various components of the pathway (see Table 9.3). In the last chapter we indicated that transpiration by an exposed leaf of a C_3 mesophyte might be 4.3 mmol m⁻² s⁻¹ (p. 423), which corresponds to a volume flux density of water of 0.77×10^{-7} m s⁻¹ (1 mol H_2O = 0.018 kg = 18×10^{-6} m³). In this chapter we calculated that J_{V_w} of a young root can be 1.1×10^{-7} m s⁻¹ (p. 502). For a plant having leaves and roots with such volume flux densities, equality of water flow across each component (Eq. 9.12) indicates that the area of one side of the leaves must be $(1.1 \times 10^{-7}$ m s⁻¹$)/(0.77 \times 10^{-7}$ m s⁻¹$)$, or 1.4 times larger than the surface area of the young roots (see Table 9.3). $J_{V_w}^{root}$ depends markedly on root age and hence varies along the length of a root; e.g., it is usually considerably lower for older roots (see Epstein, 1972;

Table 9.3
Values for the volume flux density of water and relative areas for its flow along a plant, illustrating the constancy of $J_{V_w}^j A^j$ (Eq. 9.12).

Component	$J_{V_w}^j$ (m s⁻¹)	Relative A^j
Young roots	1.1×10^{-7}	1.0
Stem xylem	1.0×10^{-3}	1.1×10^{-4}
Leaves	0.77×10^{-7}	1.4

Slatyer, 1967). Because the root system of many perennials can be quite extensive, A^{root} including the relatively nonconducting regions can be 20 to 100 times larger than the surface area of mature leaves. In the stem xylem J_V can be 1 mm s^{-1} (p. 510). Again using Equation 9.12, we conclude that the cross-sectional area of the conducting parts of the stem xylem is then $(1.1 \times 10^{-7}$ m s$^{-1})/(1.0 \times 10^{-3}$ m s$^{-1})$ or 1.1×10^{-4} times the surface area of the young roots (Table 9.3); i.e., a much smaller area is involved in plant water conduction than in water absorption or in water release. Relatively few measurements of A^j and $J_{V_w}^j$ have been made for the sequential components involved in water flow in plants.

Although Equation 9.12 can be used to describe certain overall characteristics of water flow in the soil–plant–atmosphere continuum, $\Delta\Psi^j$ does not always represent the driving force on water. For instance, a change in the osmotic pressure component of Ψ has no direct effect on the flow along the xylem or the phloem. Also, such a relation is not useful for a gas phase, because the value of the resistance R^j depends on the concentration of water vapor (see p. 409). When $J_{V_w}^j$ is in m s^{-1}, A^j in m^2, and $\Delta\Psi^j$ in MPa, Equation 9.12 indicates that the units of R^j are MPa s m^{-3}. For young sunflower and tomato plants, the resistance from the root surface to the leaf mesophyll cells, R^{plant}, is about 1.0×10^8 MPa s m^{-3} when the plants are approximately 0.3 m tall and $\Delta\Psi^{plant}$ is close to 0.2 MPa (see Kramer, 1983). Using Equation 9.12, we obtain

$$J_{V_w}^j A^j = \frac{(0.2 \text{ MPa})}{(1.0 \times 10^8 \text{ MPa s m}^{-3})} = 2 \times 10^{-9} \text{ m}^3 \text{ s}^{-1}$$

which gives the volume of water flowing across each component per unit time. For sunflower, bean, and tomato, $R^{root}:R^{stem}:R^{leaves}$ is about 2:1:1.5, whereas R^{root} is relatively higher for soybean and R^{leaves} is relatively higher for safflower (see Jarvis, 1975). For the above value for R^{plant}, R^{stem} is about $(1/4.5)(1.0 \times 10^8$ MPa s m$^{-3})$ or 2.2×10^7 MPa s m^{-3} for the young sunflower or tomato plants. Resistances can be considerably higher for other plants—e.g., R^{plant} can be 10^{10} MPa s m^{-3} for wheat and even higher for barley (see Jones, 1978).

The resistance for water movement along the stem can be separated into (1) a quantity expressing some inherent flow properties of the xylem and (2) the geometrical aspects of the conduits. By analogy with Ohm's law, where R equals $\rho \, \Delta x/A$ (pp. 121 and 397), we thus obtain

$$\rho^j = \frac{R^j A^j}{\Delta x^j} = \frac{1}{J_{V_w}^j} \frac{\Delta\Psi^j}{\Delta x^j} \tag{9.13}$$

where ρ^j is the hydraulic resistivity of the xylem tissue of length Δx^j and cross-sectional area A^j, and the second equality incorporates Equation 9.12 (we also note that $J_{V_w}^j = (1/\rho^j)(\Delta\Psi^j/\Delta x^j)$ by the definition of resistivity in Eq. 8.1c).

For many plants containing xylem vessels, ρ^{stem} equals 100 to 500 MPa s m^{-2}; hydraulic resistivities can be somewhat higher for plants with tracheids, such as conifers; e.g., it can be 1 600 MPa s m^{-2} for certain ferns (see Hunt et al., 1990; Milburn, 1979; Woodhouse and Nobel, 1982; Zimmermann, 1983). Just as for the gas fluxes discussed in Chapter 8, Equations 9.12 and 9.13 can be recast in terms of conductivities or conductances, which are the reciprocals of resistivities and resistances, respectively.[12] Also, the hydraulic conductance per unit length, K_h ($= \Delta x_j/R^j$; m^4 MPa^{-1} s^{-1}) is often determined for the xylem. By Equation 9.12, K_h equals $J_{V_w}^j A^j/(\Delta\Psi^j/\Delta x^j)$, which is the volume flow rate per unit water potential gradient. K_h is useful for comparing xylem characteristics among species.

Assuming a stem xylem cross-sectional area of 5 mm^2 for the young sunflower and tomato plants with a stem length of 0.3 m and an R^{stem} of 2.2 $\times$ 10^7 MPa s m^{-3}, Equation 9.13 indicates the following hydraulic resistivity for the stem xylem:

$$
\begin{aligned}
\rho^{stem\ xylem} &= \frac{R^{stem}\ A^{stem\ xylem}}{\Delta x^{stem}} \\[2mm]
&= \frac{(2.2 \times 10^7\ \text{MPa s m}^{-3})(5 \times 10^{-6}\ \text{m}^2)}{(0.3\ \text{m})} \\[2mm]
&= 370\ \text{MPa s m}^{-2}
\end{aligned}
$$

A tree 10 m tall might have 500 times more xylem area, which for the same hydraulic resistivity would lead to an R^{stem} of 1.5 $\times$ 10^6 MPa s m^{-3}. The smaller resistance results from the many more xylem vessels in the tree stem, even though the overall xylem length is much greater than for the tomato or sunflower plant. Likewise, the tree has a much higher hydraulic conductance per unit length, (10 m)/(1.5 $\times$ 10^6 MPa s m^{-3}) or 6.7 $\times$ 10^{-6} m^4 MPa^{-1} s^{-1}, than the tomato or sunflower, whose K_h (1.4 $\times$ 10^{-8} m^4 MPa^{-1} s^{-1}) is at the upper end of the range measured for a series of young stems and leaf petioles (see Schulte et al., 1989a).

---

12. Hydraulic conductivities ($1/\rho^j$) can be used for the phloem as well as the xylem; e.g., the phloem hydraulic conductivity can be about 1 $\times$ 10^{-4} m^2 MPa^{-1} s^{-1} for herbaceous species and 4 $\times$ 10^{-3} m^2 MPa^{-1} s^{-1} for trees; pressure gradients along the phloem are generally much higher in small herbaceous species than in tall trees (see J. A. Milburn in Zimmermann and Milburn, 1975).

We can relate the hydraulic resistivity of the xylem to flow characteristics predicted by Poiseuille's law [Eq. 9.11b, $J_v = -(r^2/8\eta) \, \partial P/\partial x$]. Specifically, by comparing Poiseuille's law with Equation 9.13 and identifying $-\partial P/\partial x$ with $\Delta\Psi^j/\Delta x^j$, we note that ρ^j equals $8\eta/r^2$. Using a representative value for $\rho^{\text{stem xylem}}$ and the viscosity of water at 20°C (App. I), we then have

$$r = \sqrt{\frac{8\eta}{\rho^{\text{stem xylem}}}} = \sqrt{\frac{(8)(1.0 \times 10^{-3} \text{ Pa s})}{(370 \times 10^6 \text{ Pa s m}^{-2})}}$$

$$= 5 \times 10^{-6} \text{ m}$$

which is a possible effective radius of xylem elements. Owing to the presence of cell walls and nonconducting cells, the lumen of the xylem vessels might correspond to only about one-fourth of the cross-sectional area of the xylem tissue in the stem, $A^{\text{stem xylem}}$. A lumen radius of 10 μm for one-fourth the area has the same J_v and $\partial P/\partial x$ as pores 5 μm in radius occupying the entire area (consider the r^2 factor in Eq. 9.11b). Other complications, such as the resistance of the perforation plates, cause the actual radius to be even greater than the effective radius. We should also note that, because of the inverse relationship between $\rho^{\text{stem xylem}}$ and r^2, the larger xylem vessels tend to conduct proportionally more than do the smaller ones in a given stem. Finally, recall that we used Poiseuille's law to equate the soil hydraulic conductivity coefficient L^{soil} to $r^2/8\eta$ (p. 509), which is analogous to our present consideration of a reciprocally related quantity, the xylem hydraulic resistivity. In fact, the xylem hydraulic conductivity could be equated to $r^2/8\eta$, where r is the effective radius of the xylem vessels averaged over the entire cross-sectional area of the xylem.

For the example in Table 9.2, the drop in water potential is 0.2 MPa across the soil part of the pathway, 0.1 MPa from the root surface to the root xylem, and 0.2 MPa along the xylem—values that suggest the relative magnitudes of the three resistances involved ($\Delta\Psi^j \cong \text{constant} \times R^j$, Eq. 9.12). As the soil dries, its hydraulic conductivity decreases, and the relative size of the water potential drop in the soil usually becomes larger (see p. 503). For soybean in pots R^{soil} becomes greater than R^{plant} below a Ψ^{soil} of -1.1 MPa (see Blizzard and Boyer, 1980). R^{plant} can also increase as Ψ^{leaf} decreases, perhaps because the entry of air breaks the water continuity (cavitation) in some of the xylem vessels, which thus become nonconducting. The rapid uptake of soil water by the root during those parts of the day when transpiration is particularly high can also lead to a large hydrostatic pressure gradient (consider Darcy's law, Eq. 9.7), and $\Delta\Psi^{\text{soil}}$ from the bulk soil to the root then increases. In fact R^{soil} is often the largest resistance for that part of the soil–plant–atmosphere continuum where water moves predominantly as a liquid (water can also move as a vapor in the soil; see p. 501). The resistance of the root epidermis, cortex, and endodermis is generally somewhat less than R^{soil}. We also note that, as the soil dries, a hydraulic resistance at the root-soil interface appears to develop, possibly representing a shrinking of the root away from the soil particles and hence less contact with the soil water (see Passioura, 1982; Weatherley, 1982). The resistance of the xylem is proportional to its length. In conifers and other plants, the conducting area of the stem xylem is also proportional to the leaf area (see Jarvis, 1975). If the transpiration rate per unit leaf area and the xylem element dimensions are the same, the pressure gradient for Poiseuille flow is then the same; i.e., a higher $J_V^{leaves}A^{leaves}$ is compensated for by a higher $A^{stem\ xylem}$ (see Eqs. 9.12 and 9.13), so $\Delta\Psi/\Delta x$ along the stem need not change in such cases.

Capacitance and Time Constants

The daily changes in hydrostatic pressure in the xylem can cause plants to have observable fluctuations in their stem diameters. When the transpiration rate is high, the large tension within the xylem vessels is transmitted to the water in the cell walls of the xylem vessels, then to communicating water in adjacent cells, and eventually all the way across the stem. The decrease in hydrostatic pressure can therefore cause a tree trunk to contract during the day. At night, the hydrostatic pressure in the xylem increases and may even become positive, and the trunk diameter increases, generally by 0.3% to 1%. Such changes in diameter, and therefore volume, represent net release of water during the day and net storage at night. In fact, the daily change in water content of a plant can be equal to the amount of water transpired in a few minutes to a few hours during the daytime. These changes in water content correspond to a capacitance effect that is superimposed on the resistance network for water flow (see Hunt *et al.*, 1990; Jarvis, 1975; Jarvis *et al.*, 1981; Kozlowski, 1972; Slatyer, 1967). The ability of various plant parts to store water and the time course for its release, both of which affect the water potential, are analogous to the effects on voltage of charge storage by electrical capacitors that we discussed in Chapter 3.

Let us begin by comparing the average daily evapotranspiration per unit ground area to the daily fluctuations in water content of a tree trunk. We will suppose that the trunks average 0.2 m in diameter, 10 m in height, 4 m apart in a rectangular grid, and transpire a depth of water of 4 mm day^{-1} (see p. 483), nearly all of which occurs during the daytime. If a 1% diurnal change in diameter reflects volumetric changes in water content, each trunk would daily vary (10 m)(π)[(0.100 m)^2 − (0.099 m)^2] or 0.006 m^3 in water volume; each tree transpires 0.06 m^3 water daily (4 m × 4 m × 0.004 m day^{-1}). Thus, the change in water content of the trunk could supply about 1 hour's worth of transpired water during the daytime. Absorption of water by the roots can thus lag behind the loss of water by leaf transpiration, a consequence of water coming from the storage capacity or *capacitance* of the trunk.

As transpiration increases following stomatal opening at dawn, the leaf water potential decreases. This lowers the hydrostatic pressure in the leaves, and concomitantly the water content of the leaves decreases, raising the average osmotic pressure and thereby further decreasing the leaf water potential ($\Psi = P - \Pi + \rho_w gh$, Eq. 2.13a). During this period, water uptake from the soil does not balance transpiration by the leaves, so the steady-state relation embodied in Equation 9.12 is not obeyed. The water content of the leaves may decrease 10% during the daytime. For a leaf area index of 5 and leaves 300 μm thick consisting of 70% water by volume (p. 347), the change in water content then corresponds to a depth of water of (5)(300 × 10^{-6} m)(0.70)(0.10), or 1.1 × 10^{-4} m (0.11 mm). This corresponds to 3% of the daily transpiration of 4 mm and thus could represent about 20 min of water loss during the daytime. Roots can also exhibit daily changes in water content. For some species roots have more biomass than do the leaves and trunk together; in such cases roots may store the equivalent of a few hours' worth of transpiration (see Jarvis, 1975).

Capacitance effects can also be seen on a longer time scale. For instance, the sapwood between the vascular cambium on the outside and the heartwood[13] on the inside can represent 20% to 40% of the radial dimension of a mature tree and even more for a young tree or sapling. The sapwood can store about 1 week's worth of water at moderate transpiration rates (even longer in drought periods with lowered rates of transpiration). A cactus stem (see Fig. 7.9) can store many months worth of transpired water during drought periods when stomatal opening is limited.

We will define the water capacitance C^j of plant part j as follows:

$$C^j = \frac{\text{change in water content of component } j}{\text{change in average water potential along component } j} \quad (9.14)$$

$$= \frac{\Delta V_w^j}{\Delta \overline{\Psi^j}}$$

---

13. The heartwood represents the central xylem, which is generally rather darkly pigmented. It has no living cells, conduction capacity, or food storage.

where V_w^j is the volume of water in component j. We estimated above that a tree trunk might change its water content by 0.006 m³ each day. This could be accompanied by a change in average xylem water potential from -0.3 MPa at dawn (the value of Ψ^{soil} 10 mm from root; Table 9.2) to -0.7 MPa (average of -0.6 MPa and -0.8 MPa, the values of Ψ at the two ends of the xylem indicated in Table 9.2), or -0.4 MPa overall. By Equation 9.14, C^{trunk} is then

$$C^{trunk} = \frac{(-0.006 \text{ m}^3)}{(-0.4 \text{ MPa})} = 1.5 \times 10^{-2} \text{ m}^3 \text{ (MPa)}^{-1}$$

The much smaller stem of a young tomato or sunflower would have a much lower water capacitance, e.g., its C^{stem} might be 1 000-fold less than the value calculated for the tree trunk.

Upon comparing Equation 9.14 with Equation 3.1 ($Q = C \Delta E$, where Q is the net charge accumulated that leads to an electrical potential change ΔE across a region of capacitance C), we note that $\Delta \overline{\Psi}^j$ in Equation 9.14 takes the place of ΔE in electrical circuits. In fact, we can again borrow from electrical circuit analysis to indicate how the initial average water potential along some component $\overline{\Psi}_0^j$ will approach a final average water potential $\overline{\Psi}_\infty^j$:

$$\overline{\Psi}^j - \overline{\Psi}_\infty^j = (\overline{\Psi}_0^j - \overline{\Psi}_\infty^j)e^{-t/\tau^j} \tag{9.15}$$

which is identical to Equation 7.23 except that $\overline{\Psi}^j$ replaces T^{surf}. Similar to the time constant for thermal changes (Eq. 7.24), we can identify a time constant τ^j for changes in the average water potential of component j:

$$\tau^j = R_s^j C^j \tag{9.16}$$

where C^j is defined by Equation 9.14 and R_s^j represents the effective resistance from the water storage region to the main transpiration pathway (Fig. 9.12). This τ^j indicates the time required for the water potential to change to within $1/e$ or 37% of its final value. As C^j decreases, the changes in Ψ^j occur faster (Eqs. 9.15 and 9.16). In the limit of no water storage in a plant (all C^j's equal to zero, see Eq. 9.14), the water potentials at various locations in the plant would instantaneously adjust to the new steady-state situation.

Because water can be stored all along the stem, C^{stem} equals the sum of the capacitances for individual parts of the pathway arranged in parallel ($C^j =$

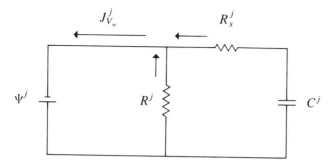

Figure 9.12

Electrical circuit portraying a plant component that can store water. The battery represents the drop in water potential along component j. The current represents the flux density of water, which can come across R^j as well as from storage in capacitor C^j and then across R_s^j.

$\Sigma_i\, C_i^j$ for capacitances in parallel).[14] The accompanying resistances of individual parts of the pathway are greater than the overall resistance ($1/R_s^j = \Sigma_i\, 1/R_{s_i}^j$ for resistances in parallel). The overall resistance for water movement from the water storage region to the xylem in the stem can be about 20% of the stem xylem resistance defined by Equation 9.12. Such movement can be via apoplastic or symplastic pathways, with the symplastic pathway being favored in certain cases (see Smith and Nobel, 1986).

The time constant $R_s^j\, C^j$ indicates how rapidly the average water potential along component j changes following changes in the water storage in C^j. Using the values introduced above, we obtain a time constant for the stem part of the pathway of

$$\tau^{\text{stem}} = (0.2)(2.2 \times 10^7 \text{ MPa s m}^{-3})[1.5 \times 10^{-5} \text{ m}^3 \text{ (MPa)}^{-1}] = 66 \text{ s}$$

for the young tomato or sunflower, indicating that water potentials in its stem respond fairly rapidly to environmentally induced changes in xylary water flow. For the 10-m-tall tree and again assuming that R_s^{stem} equals $0.2\, R^{\text{stem}}$, the time constant is 4.5×10^3 s (75 min). Thus, considerable time is required to move the water from the sapwood into the trunk xylem of a tree. In other words, because the trunk has a relatively high water capacitance, water released from

---

14. If $\Delta\Psi^j$ is the same all along the pathway, then we can simply add the component capacitances. Otherwise, we can let $C^j = \Sigma_i\, \Delta\overline{\Psi}_i^j C_i^j / \Delta\overline{\Psi}^j$, where $\Delta\overline{\Psi}_i^j$ is the average water potential change along component i that has capacitance C_i^j.

the trunk can affect the stem water potential for an extended period. The peak xylary sap flow at the base of the trunk can therefore lag several hours behind the peak transpiration of a tree. For the barrel cactus *Ferocactus acanthodes*, lags of 4 h can occur between the time of maximum transpiration and the time of maximum water uptake from the soil (see Schulte *et al.*, 1989b). The osmotic pressure increases (represented in the electrical circuit analogs by voltage sources) that accompany nocturnal CO_2 uptake by the stem of this CAM plant also affect its internal redistribution of water, leading to a tendency for the hydrostatic pressure in the outer, chlorophyll-containing tissue to increase during the night and to decrease during the daytime.

Daily Changes

We have already indicated daily changes that take place in the soil–plant–atmosphere continuum. For instance, the soil temperature changes, which affects the CO_2 evolution by respiration in soil microorganisms as well as in root cells. The hydrostatic pressure in the xylem is more negative when the rate of transpiration is high, causing plants to decrease slightly in diameter during the day. All of these changes are direct or indirect consequences of the daily variation in sunlight. Of course, the rate of photosynthesis also changes during the day. Photosynthesis is affected not only by the PPFD but also by the leaf temperature, which depends on the varying air temperature and net radiation balance for a leaf (Ch. 7).

Marked daily changes also occur in the water potentials in the soil–plant–atmosphere continuum. Let us consider a plant in a well-watered soil (Fig. 9.13). At night transpiration essentially ceases, because the stomata then close; the water potentials in the soil, root, and leaf may then all become nearly equal (and close to zero). At dawn, the stomata open. Transpiration then removes water from the leaf and Ψ^{leaf} decreases (Fig. 9.13). After a short lag, the length of which depends on the capacitances involved and the transpiration rate, Ψ^{root} begins to decrease, but only to about -0.3 MPa (Fig. 9.13) because plenty of water is initially available in the wet soil. At dusk, these changes in Ψ^{leaf} and Ψ^{root} are reversed. As soil moisture is lost, Ψ^{soil} becomes more negative day by day, so Ψ^{root} and Ψ^{leaf} also become more negative. (We are assuming that Ψ^{air} here remains essentially unchanged and is much lower than the other water potentials. Moreover, we are also assuming isothermal conditions, which though unrealistic will fortunately give us a useful approximation.) As the soil becomes drier, a steeper gradient in water potential is necessary to sustain the water flow up to the root, and therefore the difference between Ψ^{soil} and Ψ^{root} becomes

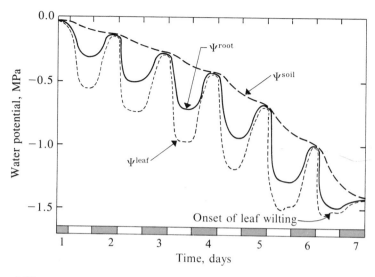

Figure 9.13

Schematic representation of daily changes in the water potentials in the soil, root, and leaf of a plant in an initially wet soil that dries over a one week period. Ψ^{soil} is the water potential in the bulk soil, Ψ^{root} is that in the root xylem, and Ψ^{leaf} is the value in a mesophyll cell. Shaded region indicates night. (Adapted from R. O. Slatyer, *Plant–Water Relationships*. Academic Press, New York, 1967. Used by permission.)

larger day by day (Fig. 9.13). On the other hand, $\Psi^{leaf} - \Psi^{root}$, to which the transpiration rate is proportional by Equation 9.12, is similar for the first four days. As Ψ^{leaf} becomes more negative on ensuing days, the leaf turgor pressure decreases, and eventually the hydrostatic pressure in the mesophyll cells of the leaf becomes zero some time during the day, e.g., when Ψ^{leaf} becomes -1.5 MPa in response to the decreasing Ψ^{soil}. The leaf thus wilts, but recovers at night. Permanent wilting and damage to the leaf may result when Ψ^{soil} becomes lower on subsequent days. This portrayal of successive daily changes of Ψ^{soil}, Ψ^{root}, and Ψ^{leaf} illustrates that even such nonequilibrium processes in the soil–plant–atmosphere continuum can be analyzed in terms of the water potential.

What new applications of physics and chemistry might we expect in plant physiology in the future? Our quantitative approach should be expanded to include the interdependence of forces and fluxes, such as for gas exchange. The permeability coefficient for CO_2 crossing cellular and organelle membranes needs to be determined. Nonisothermal conditions must be adequately handled. Of even greater potential impact is the application of a mathematical approach to the very complex field of plant growth and development, including hormone action, differentiation, photomorphogenesis, reproduction, and senescence. Many aspects of plant ecology are also ripe for physiochemical explanations, especially regarding root function. Mastery of the basic principles emphasized in this text as well as the adoption of a problem-solving approach will help us deal quantitatively with unresolved questions in the above areas and also allow us to predict plant responses to new environmental situations. Progress in any field requires some good fortune. But, in the words of Pasteur, "Chance favors only the prepared mind."

Problems

9.1 Suppose that J_{wv} above some canopy reaches a peak value equivalent to 1.0 mm of water h^{-1} during the daytime when the air temperature is 30°C and is 0.10 mm h^{-1} at night when T^{ta} is 20°C. Assume that during the daytime the relative humidity decreases by 20% across the first 30 m of turbulent air and that the eddy diffusion coefficient halves at night because of an ambient wind speed lower than during the daytime. (a) What are J_{wv} in mmol m^{-2} s^{-1} and Δc_{wv}^{ta} in mol m^{-3} during peak transpiration? (b) What is r_{wv}^{ta} (over the first 30 m) during peak transpiration and at night? (c) What is Δc_{wv}^{ta} at night in mol m^{-3}? To what drop in relative humidity does this correspond? (d) If J_{wv}/J_{CO_2} is -700 H_2O/CO_2 during peak transpiration, what are J_{CO_2} and $\Delta c_{CO_2}^{ta}$ then? (e) What is K_{CO_2} for the first 30 m of turbulent air at night? (f) How long would it take for water vapor to diffuse 1 m at night by eddy diffusion and by ordinary diffusion? Assume that Equation 1.6 ($x_e^2 = 4D_j t_e$) applies, where D_{wv} is 2.4×10^{-5} m^2 s^{-1} at 20°C, and K_{wv} has the value averaged over the first 30 m of turbulent air.

9.2 Suppose that the foliar absorption coefficient is 0.7 for trees with an average leaf area index of 8.0. (a) If a light compensation point for CO_2 fixation is at a PPFD of 8 μmol m^{-2} s^{-1}, what are the cumulative leaf areas for light compensation when 2 000, 200, 20, and 0 μmol m^{-2} s^{-1} are incident on the canopy? (b) Suppose that J_{wv}/J_{CO_2} for the soil is 200 H_2O/CO_2, that 90% of the water vapor passing out of the canopy comes from the leaves, and that the net photosynthetic rate for the forest (using CO_2 from above the canopy as well as from the soil) corresponds to 20 kg of carbohydrate hectare^{-1} h^{-1}. If J_{wv} from the soil is 0.6 mmol m^{-2} s^{-1}, what are the J_{CO_2}'s up from the soil and down into the canopy? (c) What is the absolute value of J_{wv}/J_{CO_2} above the canopy under the conditions of (b)? (d) When the leaves are randomly distributed with respect to distance above the ground and the trees are 16 m tall, at what level does the maximum upward flux of CO_2 occur when 200 μmol m^{-2} s^{-1} is incident on the canopy? (e) If the net rate of photosynthesis is proportional to PPFD, about where does $c_{CO_2}^{ta}$ achieve a minimum? Assume that the conditions are as in (b) and (d) and that the maximum upward J_{CO_2} is 5.0 μmol m^{-2} s^{-1}. Note that in the present case the PPFD halves every 2 m downward in the vegetation.

9.3 A horizontal xylem element has a cross-sectional area of 0.004 mm^2 and conducts water at 20°C at a rate of 20 mm^3 h^{-1}. (a) What is the mean speed of the fluid in the xylem element? (b) What pressure gradient is necessary to cause such a flow? (c) If the pressure gradient remained the same, but cell walls with interstices 20 nm across filled the xylem element, what would be the mean speed of fluid movement? Assume that the entire area of the cell walls is available for conduction.

9.4 A horizontal sieve tube of the phloem has an effective radius of 10 μm. Assume that $D_{sucrose}$ in the phloem solution, which has a viscosity of 1.5 mPa s, is 0.3 × 10^{-9} m^2 s^{-1} at 20°C. (a) If there is no flow in the sieve tube and a thin layer of ^{14}C-sucrose is inserted, how long would it take by diffusion for the radioactive label at 10 mm and at 1 m to be 37% of the value at the plane of insertion? (b) If there is a pressure gradient of -0.02 MPa m^{-1} that leads to Poiseuille flow in the phloem, what is J_V there? (c) For the conditions of (b), how long does it take on the average for sucrose to move 10 mm and 1 m in the sieve tube (ignore concomitant diffusion)? Compare your values with answers to (a). (d) For the pressure gradient in (b), what would J_V be in a vertical sieve tube?

9.5 Suppose that L^{soil} is 1 × 10^{-11} and 2 × 10^{-16} m^2 s^{-1} Pa^{-1} when Ψ^{soil} is -0.01 and -1.4 MPa, respectively. For a plant in the wet soil, J_V^{xylem} is 2 mm s^{-1} and Ψ^{leaf} is -0.2 MPa for a leaf 3 m above the ground. Assume that the roots are 3 mm in diameter and that their surface area is 5 times that of one side of the leaves and 10^5 times larger than the conducting area of the xylem. Assume that all temperatures are 20°C and ignore all osmotic pressures. (a) What are the radii of curvature at hemispherical air-liquid interfaces in the wet soil and in the dry soil? (b) If water moves a distance of 8 mm to reach the root, what is the drop in hydrostatic pressure over that interval in the wet soil? (c) Suppose that the root xylem is arranged concentrically in a ring 500 μm below the root surface. If the average conductivity of the epidermis, cortex, endodermis, and xylem cell walls in the root is like that of the dry soil, what is the drop in hydrostatic pressure across them? (d) If the water in the leaf xylem is in equilibrium with that in the vacuoles of mesophyll cells, what is the average hydrostatic pressure gradient in the xylem for the wet soil case (assume that A^{xylem} is constant throughout the plant)? (e) If A^{mes}/A is 20 and the cell wall pores are 10 nm in diameter and 1 μm long, what ΔP along them will account for the rate of transpiration for the plant in the wet soil? (f) If the relative humidity above the dry soil is increased to 99%, what is J_V^{xylem} then?

9.6 Consider a tree with a leaf area index of 6 and a crown diameter of 6 m. The trunk is 3 m tall, has a mean cross-sectional area of 0.10 m^2 of which 5% is xylem tissue, and varies from an average water potential along its length of -0.1 MPa at dawn to -0.5 MPa in the steady state during the daytime. (a) If the average water vapor flux density of the leaves is 1 mmol m^{-2} s^{-1}, what is the transpiration rate of the tree in m^3 s^{-1}? (b) How long could such a transpiration rate be supported by water from the leaves, if the volume of water in the leaves per unit ground area changes from an equivalent depth of 1.0 mm to 0.8 mm? (c) What is R^{trunk} if the water potential at the base is -0.3 MPa during the daytime? (d) What is the hydraulic resistivity and hydraulic conductivity of the trunk? (e) What is the capacitance and the time constant for water release from the trunk, if water equivalent to 0.8% of the trunk volume enters the transpiration stream? Assume that the resistance involved is the same as R^{trunk}.

References

Adamson, A. W. 1982. *Physical Chemistry of Surfaces,* 4th ed. Wiley, New York.

Bewley, J. D., and M. Black. 1978. *Physiology and Biochemistry of Seeds in Relation to Germination,* Vol. 1. Springer-Verlag, Berlin.

Blizzard, W. E., and J. S. Boyer. 1980. Comparative resistance of the soil and the plant to water transport. *Plant Physiology* 66:809–814.

Briggs, G. E. 1967. *Movement of Water in Plants.* Blackwell, Oxford, England.

Caldwell, M. M. 1976. Root extension and water absorption. In *Water and Plant Life: Problems and Modern Approaches,* O. L. Lange, L. Kappen, and E.-D. Schulze, eds. Ecological Studies 19. Springer-Verlag, Berlin. Pp. 63–85.

Calkin, H. W., A. C. Gibson, and P. S. Nobel. 1986. Biophysical model of xylem conductance in tracheids of the fern *Pteris vittata. Journal of Experimental Botany* 37:1054–1064.

Carlquist, S. 1988. *Comparative Wood Anatomy: Systematic, Ecological, and Evolutionary Aspects of Dicotyledon Wood.* Springer-Verlag, Berlin.

Cronshaw, J. 1981. Phloem structure and function. *Annual Review of Plant Physiology* 32:465–484.

Cronshaw, J., W. J. Lucas, and R. T. Giaquinta. 1985. *Phloem Transport.* Liss, New York.

Darcy, H. 1856. *Les Fontaines Publiques de la Ville de Dijon.* Dalmont, Paris.

De Michele, D. W., P. J. H. Sharpe, and J. D. Goeschl. 1978. Toward the engineering of photosynthetic productivity. *CRC Critical Reviews in Bioengineering 3,* Issue 1:23–91.

Epstein, E. 1972. *Mineral Nutrition of Plants: Principles and Perspectives.* Wiley, New York.

Evans, L. T. ed. 1963. *Environmental Control of Plant Growth.* Academic Press, New York.

Gifford, R. M., and L. T. Evans. 1981. Photosynthesis, carbon partitioning, and yield. *Annual Review of Plant Physiology* 32:485–509.

Grace, J. 1977. *Plant Responses to Wind.* Academic Press, London.

Hillel, D. 1982. *Introduction to Soil Physics.* Academic Press, New York.

Hodáňová, D. 1979. Sugar beet canopy photosynthesis as limited by leaf age and irradiance. Estimation by models. *Photosynthetica* 13:376–385.

Hunt, E. R., Jr., S. W. Running, and C. A. Federer. 1990. Resistances and capacitances in the liquid water pathway for soil–plant–atmosphere models at various temporal and spatial scales. *Agricultural and Forest Meteorology,* in press.

Hutchinson, B. A., and B. B. Hicks, eds. 1985. *The Forest-Atmosphere Interaction.* Reidel, Dordrecht, The Netherlands.

Jarvis, P. G. 1975. Water transfer in plants. In *Heat and Mass Transfer in the Biosphere, I, Transfer Processes in Plant Environment,* D. A. de Vries and N. H. Afgan, eds. Halsted Press, Wiley, New York. Pp. 369–394.

Jarvis, P. G., W. R. N. Edwards, and H. Talbot. 1981. Models of plant and crop water use. In *Mathematics and Plant Physiology,* D. A. Rose and D. A. Charles-Edwards, eds. Academic Press, London. Pp. 151–194.

Jones, H. G. 1978. Modelling diurnal trends in leaf water potential in transpiring wheat. *Journal of Applied Ecology 15*:613–626.

Jones, H. G. 1983. *Plants and Microclimate.* Cambridge University Press, Cambridge, England.

Kozlowski, T. T., ed. 1968. *Water Deficits and Plant Growth*, Vols. I and II. Academic Press, New York.

Kozlowski, T. T., ed. 1972. *Water Deficits and Plant Growth*, Vol. III. Academic Press, New York.

Kramer, P. J. 1983. *Water Relations of Plants.* Academic Press, San Diego, California.

Kramer, P. J., and T. T. Kozlowski. 1979. *Physiology of Woody Plants.* Academic Press, New York.

Kursanov, A. L. 1984. *Assimilate Transport in Plants.* Elsevier, Amsterdam.

Lemon, E., D. W. Stewart, and R. W. Shawcroft. 1971. The sun's work in a cornfield. *Science 174*:371–378.

Loomis, R. S., W. A. Williams, and A. E. Hall. 1971. Agricultural productivity. *Annual Review of Plant Physiology 22*:431–438.

Lüttge, U. 1975. Salt glands. In *Ion Transport in Plant Cells and Tissues*, D. A. Baker and J. L. Hall, eds. North Holland, Amsterdam. Pp. 335–376.

Marshall, T. J., and J. W. Holmes. 1988. *Soil Physics*, 2nd ed. Cambridge University Press, Cambridge, England.

McMillen, C. G., and J. H. McClendon. 1979. Leaf angle: An adaptive feature of sun and shade leaves. *Botanical Gazette 140*:437–442.

Milburn, J. A. 1979. *Water Flow in Plants.* Longman, London.

Millington, R. J., and D. B. Peters. 1969. Exchange (mass transfer) coefficients in crop canopies. *Agronomy Journal 61*:815–819.

Monsi, M., and T. Saeki. 1953. Über den Lichtfaktor in den Pflanzengesellschaften und seine Bedeutung für die Stoffproduktion. *Japanese Journal of Botany 14*:22–52.

Monsi, M., Z. Uchijima, and T. Oikawa. 1973. Structure of foliage canopies and photosynthesis. *Annual Review of Ecology and Systematics 4*:301–327.

Monteith, J. L. 1973. *Principles of Environmental Physics.* American Elsevier, New York.

Münch, E. 1930. *Die Stoffbewegungen in der Pflanze.* Fischer, Jena, Federal Republic of Germany.

Myneni, R. B., J. Ross, and G. Asrar. 1989. A review on the theory of photon transport in leaf canopies. *Agricultural and Forest Meteorology 45*:1–153.

Nobel, P. S., and S. P. Long. 1985. Canopy structure and light interception. In *Techniques in Bioproductivity and Photosynthesis*, 2nd ed., J. Coombs, D. O. Hall, S. P. Long, and J. M. O. Scurlock, eds. Pergamon, Oxford, England. Pp. 41–49.

Norman, J. M. 1980. Interfacing leaf and canopy light interception models. In *Predicting Photosynthesis for Ecosystem Models*, Vol. II, J. D. Hesketh and J. W. Jones, eds. CRC Press, Boca Raton, Florida. Pp. 49–67.

Norman, J. M., and G. S. Campbell. 1989. Canopy structure. In *Plant Physiological Ecology: Field Methods and Instrumentation*, R. W. Pearcy, J. Ehleringer, H. A. Mooney, and P. W. Rundel, eds. Chapman and Hall, London. Pp. 301–325.

Oke, T. R. 1987. *Boundary Layer Climates*, 2nd ed. Methuen, London.

Oliver, H. R. 1971. Wind profiles in and above a forest canopy. *Quarterly Journal of the Royal Meteorological Society 97*:548–553.

Passioura, J. B. 1982. Water in the soil–plant–atmosphere continuum. In *Physiological Plant Ecology*, O. L. Lange, P. S. Nobel, C. B. Osmond, and H. Ziegler, eds. *Encyclopedia of Plant Physiology, New Series*, Vol. 12B. Springer-Verlag, Berlin. Pp. 5–33.

Passioura, J. B. 1988. Water transport in and to roots. *Annual Review of Plant Physiology and Plant Molecular Biology 39*:245–265.

Rendig, V. V., and H. M. Taylor. 1989. *Principles of Soil–Plant Interrelationships*. McGraw-Hill, New York.

Rosenberg, N. J., B. L. Blad, and S. B. Verma. 1983. *Microclimate: The Biological Environment*, 2nd ed. Wiley, New York.

Ross, J. 1981. *The Radiation Regime and Architecture of Plant Stands*. W. Junk, The Hague.

Russell, G., B. Marshall, and P. G. Jarvis, eds. 1989. *Plant Canopies: Their Growth, Form and Function*. Cambridge University Press, Cambridge, England.

Schulte, P. J., A. C. Gibson, and P. S. Nobel. 1989a. Water flow in vessels with simple or compound perforation plates. *Annals of Botany 64*:171–178.

Schulte, P. J., J. A. C. Smith, and P. S. Nobel. 1989b. Water storage and osmotic pressure influences on the water relations of a dicotyledonous desert succulent. *Plant, Cell and Environment 12*:831–842.

Šetlik, I., ed. 1970. *Prediction and Measurement of Photosynthetic Productivity*. Pudoc, Centre for Agricultural Publishing and Documentation, Wageningen, The Netherlands.

Shaykewich, C. F., and J. Williams. 1971. Resistance to water absorption in germinating rapeseed (*Brassica napus* L.). *Journal of Experimental Botany 22*:19–24.

Slatyer, R. O. 1967. *Plant–Water Relationships*. Academic Press, New York.

Smith, J. A. C., and P. S. Nobel. 1986. Water movement and storage in a desert succulent: Anatomy and rehydration kinetics for leaves of *Agave deserti*. *Journal of Experimental Botany 37*:1044–1053.

Spanner, D. C. 1972. Plants, water, and some other topics. In *Psychrometry in Water Relations Research*, R. W. Brown and B. P. Van Haveren, eds. Utah Agricultural Experiment Station, Utah State University, Logan. Pp. 29–39.

Streeter, V. L., and E. B. Wylie. 1985. *Fluid Mechanics*, 8th ed. McGraw-Hill, New York.

Thom, A. S. 1975. Momentum, mass and heat exchange of plant communities. In *Vegetation and the Atmosphere*, Vol. 1, *Principles*, J. L. Monteith, ed. Academic Press, New York. Pp. 57–109.

Turitzin, S. N., and B. G. Drake. 1981. The effect of a seasonal change in canopy structure on the photosynthetic efficiency of a salt marsh. *Oecologia 48*:79–84.

van den Honert, T. H. 1948. Water transport as a catenary process. *Discussions of the Faraday Society 3*:146–153.

van Wijk, W. R., ed. 1966. *Physics of Plant Environment*, 2nd ed. North-Holland, Amsterdam.

Vogel, S. 1981. *Moving Fluids; the Physical Biology of Flow*. Willard Grant Press, Boston, Massachusetts.

Weatherley, P. E. 1982. Water uptake and flow in roots. In *Physiological Plant Ecology*, O. L. Lange, P. S. Nobel, C. B. Osmond, and H. Ziegler, eds. *Encyclopedia of Plant Physiology, New Series*, Vol. 12B. Springer-Verlag, Berlin. Pp. 79–109.

Woodhouse, R. M., and P. S. Nobel. 1982. Stipe anatomy, water potentials, and xylem conductances in seven species of ferns (Filicopsida). *American Journal of Botany* 69:135–140.

Zelitch, I. 1971. *Photosynthesis, Photorespiration, and Plant Productivity*. Academic Press, New York.

Zimmermann, M. H. 1983. *Xylem Structure and the Ascent of Sap*. Springer-Verlag, Berlin.

Zimmermann, M. H., and J. A. Milburn, eds. 1975. *Transport in Plants I, Phloem Transport. Encyclopedia of Plant Physiology, New Series*, Vol. 1. Springer-Verlag, Berlin.

APPENDICES

Numerical Values of Constants and Coefficients

Symbol	Description	Magnitude
c	speed of light in vacuum	2.998×10^8 m s^{-1}
c_{wv}^{*}	saturation concentration of water vapor (i.e., at 100% relative humidity)	See pp. 548–550 for values from $-30°C$ to $60°C$.
C_P^{water}	volumetric heat capacity of water at constant pressure (1 atmosphere, 0.1013 MPa)	4.217 MJ m^{-3} °C^{-1} at 0°C 4.175 MJ m^{-3} °C^{-1} at 20°C 4.146 MJ m^{-3} °C^{-1} at 40°C
C_P^{air}	volumetric heat capacity of dry air at constant pressure (1 atmosphere)	1.300 kJ m^{-3} °C^{-1} at 0°C 1.212 kJ m^{-3} °C^{-1} at 20°C 1.136 kJ m^{-3} °C^{-1} at 40°C
D_{CO_2}	diffusion coefficient of CO_2 in air (1 atmosphere, 0.1013 MPa)	1.33×10^{-5} m^2 s^{-1} at 0°C 1.42×10^{-5} m^2 s^{-1} at 10°C 1.51×10^{-5} m^2 s^{-1} at 20°C 1.60×10^{-5} m^2 s^{-1} at 30°C 1.70×10^{-5} m^2 s^{-1} at 40°C
D_{O_2}	diffusion coefficient of O_2 in air (1 atmosphere, 0.1013 MPa)	1.95×10^{-5} m^2 s^{-1} at 20°C
D_{wv}	diffusion coefficient of water vapor in air (1 atmosphere, 0.1013 MPa)	2.13×10^{-5} m^2 s^{-1} at 0°C 2.27×10^{-5} m^2 s^{-1} at 10°C 2.42×10^{-5} m^2 s^{-1} at 20°C 2.57×10^{-5} m^2 s^{-1} at 30°C 2.72×10^{-5} m^2 s^{-1} at 40°C
e	base for natural logarithm	2.71828 ($1/e = 0.368$)
	electronic charge	1.602×10^{-19} C
F	Faraday's constant	9.649×10^4 coulomb mol^{-1} 9.649×10^4 J mol^{-1} V^{-1} 2.306×10^4 cal mol^{-1} V^{-1} 23.06 kcal mol^{-1} V^{-1}

Symbol	Description	Magnitude
g	gravitational acceleration	9.780 m s^{-2} (sea level[1], 0° latitude)
		9.807 m s^{-2} (sea level[1], 45° latitude)
		9.832 m s^{-2} (sea level[1], 90° latitude)
		978.0 cm s^{-2} (sea level[1], 0° latitude)
		980.7 cm s^{-2} (sea level[1], 45° latitude)
		983.2 cm s^{-2} (sea level[1], 90° latitude)
h	Planck's constant	6.626×10^{-34} J s
		6.626×10^{-27} erg s
		0.4136×10^{-14} eV s
		1.584×10^{-37} kcal s
$h\dot{c}$		1.986×10^{-25} J m
		1 240 eV nm
H_{sub}	heat of sublimation of water	51.37 kJ mol^{-1} (2.847 MJ kg^{-1}) at -10°C
		51.17 kJ mol^{-1} (2.835 MJ kg^{-1}) at -5°C
		51.00 kJ mol^{-1} (2.826 MJ kg^{-1}) at 0°C
		12.27 kcal mol^{-1} (680 cal g^{-1}) at -10°C
		12.22 kcal mol^{-1} (677 cal g^{-1}) at -5°C
		12.18 kcal mol^{-1} (675 cal g^{-1}) at 0°C
H_{vap}	heat of vaporization of water	45.06 kJ mol^{-1} (2.501 MJ kg^{-1}) at 0°C
		44.63 kJ mol^{-1} (2.477 MJ kg^{-1}) at 10°C
		44.21 kJ mol^{-1} (2.454 MJ kg^{-1}) at 20°C
		44.00 kJ mol^{-1} (2.442 MJ kg^{-1}) at 25°C
		43.78 kJ mol^{-1} (2.430 MJ kg^{-1}) at 30°C
		43.35 kJ mol^{-1} (2.406 MJ kg^{-1}) at 40°C
		42.91 kJ mol^{-1} (2.382 MJ kg^{-1}) at 50°C
		40.68 kJ mol^{-1} (2.258 MJ kg^{-1}) at 100°C
k	Boltzmann's constant	1.381×10^{-23} J molecule^{-1} K^{-1}
		1.381×10^{-16} erg molecule^{-1} K^{-1}
		8.617×10^{-5} eV molecule^{-1} K^{-1}
kT		0.02354 eV molecule^{-1} at 0°C
		0.02526 eV molecule^{-1} at 20°C
		0.02569 eV molecule^{-1} at 25°C
		0.02699 eV molecule^{-1} at 40°C
K^{air}	thermal conductivity coefficient of dry air (1 atmosphere)[2]	0.0237 W m^{-1} °C^{-1} at -10°C
		0.0243 W m^{-1} °C^{-1} at 0°C
		0.0250 W m^{-1} °C^{-1} at 10°C
		0.0257 W m^{-1} °C^{-1} at 20°C
		0.0264 W m^{-1} °C^{-1} at 30°C
		0.0270 W m^{-1} °C^{-1} at 40°C
		0.0277 W m^{-1} °C^{-1} at 50°C

1. The correction for height above sea level is -3.09×10^{-6} m s^{-2} per m of altitude.

2. The pressure sensitivity is very slight, K^{air} increasing only about 0.0001 W m^{-1} °C^{-1} per atmosphere (0.1013 MPa) increase in pressure.

Symbol	Description	Magnitude
	thermal conductivity coefficient of moist air (100% relative humidity, 1 atmosphere)	0.0242 W m^{-1} °C^{-1} at 0°C 0.0255 W m^{-1} °C^{-1} at 20°C 0.0264 W m^{-1} °C^{-1} at 40°C
K^{water}	thermal conductivity coefficient of water	0.565 W m^{-1} °C^{-1} at 0°C 0.599 W m^{-1} °C^{-1} at 20°C 0.627 W m^{-1} °C^{-1} at 40°C
ln 2		0.6931
N	Avogadro's number	6.0220×10^{23} entities mol^{-1}
Nhc		0.1196 J mol^{-1} m 119 600 kJ mol^{-1} nm 28.60 kcal mol^{-1} µm 28 600 kcal mol^{-1} nm
N^*_{wv}	saturation mole fraction of water vapor (i.e., at 100% relative humidity) at 1 atmosphere (0.1013 MPa)	See pp. 548–550 for values from −30°C to 60°C.
P^*_{wv}	saturation vapor pressure of water	See pp. 548–500 for values from −30°C to 60°C.
	protonic charge	1.602×10^{-19} C
R	gas constant	8.314 J mol^{-1} K^{-1} 1.987 cal mol^{-1} K^{-1} 8.314 m^3 Pa mol^{-1} K^{-1} 8.314×10^{-6} m^3 MPa mol^{-1} K^{-1} 0.08205 litre atmosphere mol^{-1} K^{-1} 0.08314 litre bar mol^{-1} K^{-1} 83.14 cm^3 bar mol^{-1} K^{-1}
RT		2.271×10^3 J mol^{-1} (m^3 Pa mol^{-1}) at 0°C 2.437×10^3 J mol^{-1} (m^3 Pa mol^{-1}) at 20°C 2.479×10^3 J mol^{-1} (m^3 Pa mol^{-1}) at 25°C 2.271×10^{-3} m^3 MPa mol^{-1} at 0°C 2.437×10^{-3} m^3 MPa mol^{-1} at 20°C 2.479×10^{-3} m^3 MPa mol^{-1} at 25°C 542.4 cal mol^{-1} at 0°C 582.2 cal mol^{-1} at 20°C 2.271 litre MPa mol^{-1} at 0°C 2.437 litre MPa mol^{-1} at 20°C 22.71 litre bar mol^{-1} at 0°C 24.37 litre bar mol^{-1} at 20°C 22 710 cm^3 bar mol^{-1} at 0°C 24 370 cm^3 bar mol^{-1} at 20°C 22.41 litre atmosphere mol^{-1} at 0°C 24.05 litre atmosphere mol^{-1} at 20°C
2.303 RT		5.612 kJ mol^{-1} at 20°C 5.708 kJ mol^{-1} at 25°C 1.342 kcal mol^{-1} at 20°C 1.364 kcal mol^{-1} at 25°C 56 120 cm^3 bar mol^{-1} at 20°C

Symbol	Description	Magnitude
RT/F		25.3 mV at 20°C
		25.7 mV at 25 °C
2.303 RT/F		58.2 mV at 20°C
		59.2 mV at 25°C
		60.2 mV at 30°C
$RT/\overline{V}_w$		135.0 MPa at 20°C
		137.3 MPa at 25°C
		32.31 cal cm^{-3} at 20°C
		135.0 J cm^{-3} at 20°C
		1 350 bars at 20°C
		1 330 atmospheres at 20°C
2.303 $RT/\overline{V}_w$		310.9 MPa at 20°C
		316.2 MPa at 25°C
		3 063 atmospheres at 20°C
		3 109 bars at 20 °C
	solar constant	1 368 W m^{-2}
		1.960 cal cm^{-2} min^{-1}
		1.368 × 10^5 erg cm^{-2} s^{-1}
		0.1368 W cm^{-2}
	thermal capacity of water (mass basis)	4 218 J kg^{-1} °C^{-1} at 0°C
		4 182 J kg^{-1} °C^{-1} at 20°C
		4 179 J kg^{-1} °C^{-1} at 40°C
		1.0074 cal g^{-1} °C^{-1} at 0°C
		0.9988 cal g^{-1} °C^{-1} at 20°C
		0.9980 cal g^{-1} °C^{-1} at 40°C
	thermal capacity of water (mole basis)	75.99 J mol^{-1} °C^{-1} at 0°C
		75.34 J mol^{-1} °C^{-1} at 20°C
		75.28 J mol^{-1} °C^{-1} at 40°C
		18.14 cal mol^{-1} °C^{-1} at 0°C
		17.99 cal mol^{-1} °C^{-1} at 20°C
		17.98 cal mol^{-1} °C^{-1} at 40°C
$\overline{V}_w$	partial molal volume of water	1.805 × 10^{-5} m^3 mol^{-1} at 20°C
		18.05 cm^3 mol^{-1} at 20°C
ε_0	permittivity of a vacuum	8.854 × 10^{-12} coulomb^2 m^{-2} N^{-1}
		8.854 × 10^{-12} coulomb m^{-1} V^{-1}
η_{air}	viscosity of air	1.716 × 10^{-5} Pa s at 0°C
		1.813 × 10^{-5} Pa s at 20°C
		1.907 × 10^{-5} Pa s at 40°C
η_w	viscosity of water	1.787 × 10^{-3} Pa s at 0°C
		1.307 × 10^{-3} Pa s at 10°C
		1.002 × 10^{-3} Pa s at 20°C
		0.798 × 10^{-3} Pa s at 30°C
		0.653 × 10^{-3} Pa s at 40°C
		0.547 × 10^{-3} Pa s at 50°C
		0.01002 dyn s cm^{-2} at 20°C
		0.01002 poise at 20°C

Symbol	Description	Magnitude
ν_{air}	kinematic viscosity of air (dry, 1 atmosphere)	1.327×10^{-5} m^2 s^{-1} at 0°C 1.505×10^{-5} m^2 s^{-1} at 20°C 1.691×10^{-5} m^2 s^{-1} at 40°C
ν_w	kinematic viscosity of water	1.787×10^{-6} m^2 s^{-1} at 0°C 1.004×10^{-6} m^2 s^{-1} at 20°C 0.658×10^{-6} m^2 s^{-1} at 40°C
π	circumference/diameter of circle	3.14159
ρ_{air}	density of dry air (1 atmosphere, 0.1013 MPa)	1.293 kg m^{-3} at 0°C 1.205 kg m^{-3} at 20°C 1.128 kg m^{-3} at 40°C
	density of saturated air (1 atmosphere)[3]	1.290 kg m^{-3} at 0°C 1.194 kg m^{-3} at 20°C 1.097 kg m^{-3} at 40°C
ρ_w	density of water	999.8 kg m^{-3} (0.9998 g cm^{-3}) at 0°C 1 000.0 kg m^{-3} (1.0000 g cm^{-3}) at 4°C 999.7 kg m^{-3} (0.9997 g cm^{-3}) at 10°C 998.2 kg m^{-3} (0.9982 g cm^{-3}) at 20°C 995.6 kg m^{-3} (0.9956 g cm^{-3}) at 30°C 992.2 kg m^{-3} (0.9922 g cm^{-3}) at 40°C
$\rho_w g$		0.00979 MPa m^{-1} (20°C, sea level, 45° latitude) 0.0979 bar m^{-1} (20°C, sea level, 45° latitude) 979 dyn cm^{-3} (20°C, sea level, 45° latitude) 0.0966 atmosphere m^{-1} (20°C, sea level, 45° latitude)
σ	Stefan–Boltzmann constant	5.670×10^{-8} W m^{-2} K^{-4} 5.670×10^{-12} W cm^{-2} K^{-4} 8.130×10^{-11} cal cm^{-2} min^{-1} K^{-4} 5.670×10^{-5} erg cm^{-2} s^{-1} K^{-4}
σ_w	surface tension of water	0.0756 N m^{-1} (Pa m) at 0°C 0.0742 N m^{-1} (Pa m) at 10°C 0.0728 N m^{-1} (Pa m) at 20°C 0.0712 N m^{-1} (Pa m) at 30°C 0.0696 N m^{-1} (Pa m) at 40°C 7.28×10^{-8} MPa m at 20°C 72.8 dyn cm^{-1} at 20°C 7.18×10^{-5} atmosphere cm at 20°C 7.28×10^{-5} bar cm at 20°C

3. Moist air is less dense than dry air at the same temperature and pressure, because the molecular weight of water (18.0) is less than the average for air (29.0).

Temperature (°C)	c_{wv}^* (g m^{-3})	c_{wv}^* (mol m^{-3})	N_{wv}^* (at 1 atmosphere)	P_{wv}^* (kPa)
−30	0.34	0.019	0.00037	0.038
−25	0.55	0.031	0.00062	0.063
−20	0.88	0.049	0.00102	0.103
−19	0.97	0.054	0.00112	0.114
−18	1.06	0.059	0.00123	0.125
−17	1.16	0.064	0.00135	0.137
−16	1.27	0.070	0.00149	0.151
−15	1.39	0.077	0.00163	0.165
−14	1.52	0.084	0.00179	0.181
−13	1.65	0.092	0.00196	0.198
−12	1.80	0.100	0.00214	0.217
−11	1.96	0.109	0.00234	0.238
−10	2.14	0.119	0.00256	0.260
−9	2.33	0.129	0.00280	0.284
−8	2.53	0.141	0.00306	0.310
−7	2.75	0.153	0.00333	0.338
−6	2.99	0.166	0.00364	0.369
−5	3.25	0.180	0.00396	0.402
−4	3.52	0.195	0.00431	0.437
−3	3.82	0.212	0.00469	0.476
−2	4.14	0.230	0.00511	0.517
−1	4.48	0.249	0.00555	0.562
0	4.85	0.269	0.00604	0.611
1	5.20	0.288	0.00649	0.657
2	5.56	0.309	0.00697	0.706
3	5.95	0.330	0.00748	0.758
4	6.36	0.353	0.00803	0.816
5	6.80	0.378	0.00862	0.873
6	7.27	0.403	0.00923	0.935
7	7.76	0.431	0.00989	1.002
8	8.28	0.459	0.01059	1.073
9	8.82	0.490	0.01133	1.148
10	9.41	0.522	0.01212	1.228
11	10.02	0.556	0.01296	1.313
12	10.67	0.592	0.01384	1.403
13	11.35	0.630	0.01478	1.498

Temperature (°C)	c_{wv}^* (g m^{-3})	c_{wv}^* (mol m^{-3})	N_{wv}^* (at 1 atmosphere)	P_{wv}^* (kPa)
14	12.08	0.670	0.01578	1.599
15	12.84	0.713	0.01683	1.706
16	13.64	0.757	0.01795	1.819
17	14.49	0.804	0.01913	1.938
18	15.38	0.854	0.02037	2.064
19	16.32	0.906	0.02169	2.198
20	17.31	0.961	0.02308	2.339
21	18.35	1.018	0.02455	2.488
22	19.44	1.079	0.02610	2.645
23	20.59	1.143	0.02774	2.810
24	21.80	1.210	0.02946	2.985
25	23.07	1.280	0.03128	3.169
26	24.40	1.354	0.03319	3.363
27	25.79	1.432	0.03520	3.567
28	27.26	1.513	0.03723	3.782
29	28.79	1.598	0.03955	4.008
30	30.40	1.687	0.04190	4.246
31	32.08	1.781	0.04437	4.495
32	33.85	1.879	0.04696	4.758
33	35.70	1.981	0.04968	5.034
34	37.63	2.089	0.05253	5.323
35	39.65	2.201	0.05553	5.627
36	41.76	2.318	0.05868	5.945
37	43.97	2.441	0.06197	6.280
38	46.28	2.569	0.06543	6.630
39	48.69	2.703	0.06905	6.997
40	51.21	2.842	0.07285	7.381
41	53.83	2.988	0.07682	7.784
42	56.57	3.140	0.08098	8.205
43	59.43	3.299	0.08533	8.646
44	62.41	3.464	0.08989	9.108
45	65.52	3.637	0.09464	9.590
46	68.75	3.816	0.09962	10.09
47	72.12	4.003	0.1048	10.62
48	75.63	4.198	0.1102	11.17
49	79.28	4.401	0.1159	11.74

Temperature (°C)	c_{wv}^*		N_{wv}^* (at 1 atmosphere)	P_{wv}^* (kPa)
	(g m^{-3})	(mol m^{-3})		
50	83.08	4.611	0.1218	12.34
55	104.5	5.798	0.1555	15.75
60	130.3	7.217	0.1967	19.93

SOURCES: R. J. List, *Smithsonian Meterological Tables*, 6th rev. ed., Smithsonian Institution, Washington, D.C. (1968) for water vapor data below 0°C, which are values in equilibrium with ice, and L. Haar, J. S. Gallagher, and G. S. Kell, *NBS/NRC Steam Tables,* Hemisphere, Washington, D.C. (1984) at and above 0°C. Accuracy criteria for P_{wv}^* (see D. W. Scott and A. G. Osborn. 1979. Representation of vapor-pressure data. *Journal of Physical Chemistry 83*:2714–2723) indicate deviations from measurements at specific temperatures averaging 0.2% below 0°C and 0.02% at and above 0°C.

Conversion Factors and Definitions

Quantity	Equals	Quantity	Equals
acre	43 560 feet^2		760 mm Hg (at sea level,
	4 047 m^2		45° latitude)[1]
	0.4047 hectare		10.35 m water (at sea level,
	4.047 × 10^{-3} km^2		45° latitude, 20°C)[1]
ampere	1 coulomb s^{-1}		14.70 pound inch^{-2} (psi)
	1 V ohm^{-1}	bar	0.1MPa
Å	10^{-10} m		100 kPa
	0.1 nm		10^5 Pa
	10^{-8} cm		10^5 N m^{-2}
atmosphere	0.1013 MPa		10^5 J m^{-3}
	1.013 × 10^5 Pa		10^6 dyn cm^{-2}
	1.013 × 10^5 N m^{-2}		0.9869 atmosphere
	1.013 × 10^5 J m^{-3}		1.020 kg cm^{-2} (at sea level,
	1.013 bar		45° latitude)[1]
	1.013 × 10^6 dyn cm^{-2}		750 mm Hg (at sea level,
	1.033 × 10^4 kg m^{-2} (at sea level		45° latitude)[1]
	45° latitude)[1]		10.22 m water (at sea level,
	1.033 kg cm^{-2} (at sea level, 45°		45° latitude, 20°C)[1]
	latitude)[1]		14.50 pound inch^{-2} (psi)

1. Sometimes it proves convenient to express forces in units of mass. To see why this is possible, consider the force F exerted by gravity on a body of mass m. This force is equal to mg, g being the gravitational acceleration (see App. I). Thus F/g can be used to represent a force, but the units are those of mass. One atmosphere is quite often defined as 760 mm Hg (or 1.033 kg cm^{-2}), but the elevational and latitudinal effect on g should also be considered (see p. 544).

Quantity	Equals	Quantity	Equals
becquerel	1 disintegration s^{-1} 2.703 × 10^{-11} curie	CO_2 concentration	See Table 8.2.
British thermal unit (Btu)	1 055 J 252.0 cal	coulomb	1 J V^{-1} 1 ampere s
Btu h^{-1}	0.2931 W	coulomb V	1 J
cal	4.187 J 4.187 × 10^7 ergs	curie	3.7 × 10^{10} becquerels 3.7 × 10^{10} disintegrations s^{-1}
cal cm^{-2}	1 langley 4.187 × 10^4 J m^{-2}	dalton	1.661 × 10^{-24} g (1/12 mass of ^{12}C)
cal cm^{-2} min^{-1}	697.8 W m^{-2} 6.978 × 10^5 ergs cm^{-2} s^{-1} 1 langley min^{-1}	day	86 400 s 1 440 min
cal cm^{-1} °C^{-1} min^{-1}	6.978 W m^{-1} °C^{-1}	degree (angle)	0.01745 radian
cal cm^{-1} °C^{-1} s^{-1}	418.7 W m^{-1} °C^{-1}	dyn	1 erg cm^{-1} 1 g cm s^{-2} 10^{-5} N
cal cm^{-3}	41.87 bars	dyn cm	1 erg 10^{-7} J
cal g^{-1}	4 187 J kg^{-1}	dyn cm^{-1}	10^{-3} N m^{-1}
cal m^{-2} s^{-1}	4.187 W m^{-2}	dyn cm^{-2}	0.1 N m^{-2} 10^{-6} bar
cal min^{-1}	0.06978 J s^{-1} 0.06978 W	einstein	1 mol 6.022 × 10^{23} photons
cal s^{-1}	4.187 J s^{-1} 4.187 W	erg	1 dyn cm 1 g cm^2 s^{-2} 10^{-7} J 6.242 × 10^{11} eV 2.390 × 10^{-8} cal 2.390 × 10^{-11} kcal
candela	1 lumen steradian^{-1}		
°Celsius (°C)	(°C + 273.15) kelvin [(9/5)(°C) + 32] °F		
cm	0.3937 inch		
cm bar^{-1} s^{-1}	10^{-7} m Pa^{-1} s^{-1}		
cm^2 bar^{-1} s^{-1}	10^{-9} m^2 Pa^{-1} s^{-1} 10^{-3} m^2 MPa^{-1} s^{-1}	erg cm^{-2} s^{-1}	10^{-3} J m^{-2} s^{-1} 10^{-3} W m^{-2} 10^{-7} W cm^{-2} 1.433 × 10^{-6} cal cm^{-2} min^{-1}
cm s^{-1}	0.01 m s^{-1} 10 mm s^{-1} 0.03600 km h^{-1} 0.02237 mile h^{-1}		
cm^3	10^{-6} m^3 1 ml	eV	1.602 × 10^{-19} J 1.602 × 10^{-12} erg

Quantity	Equals	Quantity	Equals
eV	96.49 kJ mol^{-1}	J	1 N m
molecule^{-1}	23.06 kcal mol^{-1}		1 W s
F	1 coulomb V^{-1}		1 m^3 Pa
° Farenheit	(°F -32)(5/9) °C		1 kg m^2 s^{-2}
(°F)			1 coulomb V
			10^7 ergs
foot	30.48 cm		0.2388 cal
	0.3048 m		10 cm^3 bar
foot^2	9.290×10^{-2} m^2	J kg^{-1}	2.388×10^{-4} cal g^{-1}
foot^3	2.832×10^{-2} m^3	J mol^{-1}	0.2388 cal mol^{-1}
	28.32 litres	J s^{-1}	W
footcandle	1 lumen foot^{-2}		
	10.76 lux		
foot s^{-1}	1.097 km h^{-1}	kcal	4.187 kJ
	0.6818 mile h^{-1}		4.187×10^{10} ergs
		kcal h^{-1}	1.163 W
g	1 dyn s^2 cm^{-1}	kcal mol^{-1}	4.187 kJ mol^{-1}
g cm^{-2}	10 kg m^{-2}		0.04339 eV molecule^{-1}
g cm^{-3}	1 000 kg m^{-3}	kDa	1.661×10^{-21}g (1 000 ×
g dm^{-2} h^{-1}	27.78 mg m^{-2} s^{-1}		1/12 mass of ^{12}C)
gallon	4.546 litre	kelvin	(kelvin -273.15) °C
(British)	4.546×10^{-3} m^3	kg	2.2046 pound
gallon (U.S.,	3.785 litre		0.001 tonne (metric ton)
liquid)	3.785×10^{-3} m^3	kg hectare^{-1}	0.893 pound acre^{-1}
gallon (U.S.)	9.354 litre hectare^{-1}	kg m^{-2}	0.1 g cm^{-2}
acre^{-1}			9.807 Pa (at sea level, 45°
			latitude; see g, App. I)
hectare	10^4 m^2		9.807 N m^{-2} (at sea level,
	2.471 acres		45° latitude; see g, App.
hertz	1 cycle s^{-1}		I)
horsepower	745.7 W		
hour	3 600 s	kg m^{-3}	10^{-3} g cm^{-3}
		kJ mol^{-1}	0.01036 eV molecule^{-1}
inch	25.40 mm	km	0.6214 mile
inch^2	6.452×10^{-4} m^2	km^2	100 hectares
inch^3	1.639×10^{-5} m^3		247.1 acres
	0.01639 litre		0.3861 mile^2

Quantity	Equals	Quantity	Equals
km h^{-1}	0.2778 m s^{-1}	M	1 000 mol m^{-3}
	27.78 cm s^{-1}	mbar	0.1 kPa
	0.6214 mile h^{-1}		100 Pa
knot	1.852 km h^{-1}		10^{-3} bar
	1.151 mile h^{-1}		10^3 dyne cm^{-2}
kW	1 000 J s^{-1}		0.9869 × 10^{-3} atmosphere
	3.600 × 10^6 J h^{-1}	mho	1 ohm^{-1}
	859.8 kcal h^{-1}		1 siemens
kW h	3.600 × 10^6 J		1 ampere V^{-1}
	8.598 × 10^5 cal	mg cm^{-3}	1 kg m^{-3}
lambert	0.3183 candela cm^{-2}	mg dm^{-2} h^{-1}	2.778 × 10^{-2} mg
	3 183 candela m^{-2}		m^{-2} s^{-1}
langley	1 cal cm^{-2}	Mg hectare^{-1}	0.4461 ton (U.S.) hectare^{-1}
	4.187 × 10^4 J m^{-2}	m s^{-1}	100 cm s^{-1}
langley day^{-1}	0.4846 W m^{-2}		3.600 km h^{-1}
langley min^{-1}	697.8 W m^{-2}		2.237 miles h^{-1}
litre	0.001 m^3	micron	1 μm
	0.03531 foot^3	mile	1 609 m
	0.2200 gallon (British)		1.609 km
	0.2642 gallon (U.S.)		5280 feet
litre	24.20 cal	mile^2	640 acres
atmosphere	101.3 J		2.590 × 10^6 m^2
litre bar	100 J		2.590 km^2
	23.88 cal	mile h^{-1}	0.4470 m s^{-1}
litre hectare^{-1}	0.1069 gallon (U.S.) acre^{-1}		44.70 cm s^{-1}
ln	2.303 log		1.609 km h^{-1}
log	0.4343 ln	MJ	0.2778 kW h
lumen foot^{-2}	1 footcandle	ml	1 cm^3
	10.76 lux		10^{-6} m^3
lumen m^{-2}	1 lux	mM	1 mol m^{-3}
	0.09290 footcandle	month (mean)	2.630 × 10^6 s
lux	1 lumen m^{-2}	MPa	10^6 N m^{-2}
	0.09290 footcandle		10^6 J m^{-3}
m	3.280 foot		10 bars
	39.37 inch		1.020 × 10^5 kg m^{-2} (at sea

Quantity	Equals	Quantity	Equals
	level, 45° latitude; see g, App. I)	s m^{-1}	10^{-2} s cm^{-1}
mol m^{-3}	1 mM	siemens	1 mho (ohm^{-1})
	1 μmol cm^{-3}		1 ampere V^{-1}
nautical mile	1.151 mile	therm	1×10^5 Btu
ng cm^{-2} s^{-1}	10 μg m^{-2} s^{-1}		1.055×10^8 J
N	1 kg m s^{-1}	ton (U.S.)	2 000 pounds
	10^5 dyn		907.2 kg
			1 ton (short)
N m	1 J		0.8929 ton (long)
N m^{-1}	10^3 dyn cm^{-1}	ton (U.S.)	2.242 tonne hectare^{-1}
N m^{-2}	1 Pa	acre^{-1}	2.242 Mg hectare^{-1}
	10^{-6} MPa	tonne	1 000 kg
	10 dyn cm^{-2}	(metric ton)	1 Mg
	10^{-2} mbar		1.102 ton (U.S.)
ounce (avoirdupois)	28.35 g	tonne hectare^{-1}	0.4461 ton (U.S.) acre^{-1}
ounce (troy)	31.10 g	tonne m^{-3}	1 g cm^{-3}
Pa	1 N m^{-2}	torr	1 mm Hg
	1 J m^{-3}		133.3 Pa (at sea level, 45° latitude; see g, App. I)
	1 kg m^{-1} s^{-2}		
	10^{-5} bar		1.333 mbar (at sea level, 45° latitude; see g, App. I)
	9.869×10^{-6} atmosphere		
photosynthesis	See Table 8.2.		1 333 dyn cm^{-2} (at sea level, 45° latitude; see g, App. I)
poise	1 dyn s cm^{-2}		
	0.1 N s m^{-2}		
	0.1 Pa s	transpiration	See Table 8.2.
pound	0.4536 kg	V	1 ampere ohm
pound acre^{-1}	1.121 kg hectare^{-1}		1 J coulomb^{-1}
pound inch^{-2} (psi)	6.895 kPa	W	1 J s^{-1}
	0.06895 bar		1 kg m^2 s^{-3}
	0.06805 atmosphere		10^7 ergs s^{-1}
radian	57.30°		14.33 cal min^{-1}
revolution min^{-1} (rpm)	6° s^{-1}	W cm^{-2}	10^4 W m^{-2}
	0.1047 radian s^{-1}	W m^{-2}	1 J m^{-2} s^{-1}
s cm^{-1}	100 s m^{-1}		10^3 ergs cm^{-2} s^{-1}

Quantity	Equals	Quantity	Equals
	1.433×10^{-3} cal cm^{-2} min^{-1}	yard	0.9144 m
	1.433×10^{-3} langley min^{-1}	yard^2	0.8361 m^2
	0.2388 cal m^{-2} s^{-1}	yard^3	0.7646 m^3
	2.388×10^{-5} cal cm^{-2} s^{-1}	year (normal calendar)	3.154×10^7 s
W m^{-1} °C^{-1}	2.388×10^{-3} cal cm^{-1} °C^{-1} s^{-1}		5.256×10^5 min
			8 760 h
	0.1433 cal cm^{-1} °C^{-1} min^{-1}		365 days
W s	1 J	year (sidereal)	365.256 days
			3.156×10^7 s
week	6.048×10^5 s	μg cm^{-2} s^{-1}	10 mg m^{-2} s^{-1}

APPENDIX **III**

Mathematical Relations

Prefixes. Logarithms. Quadratic equation. Trigonometric functions. Differential equations.

Prefixes (for units of measure)

a	atto	10^{-18}	da[1]	deka	10
f	femto	10^{-15}	h[1]	hecto	10^2
p	pico	10^{-12}	k	kilo	10^3
n	nano	10^{-9}	M	mega	10^6
μ	micro	10^{-6}	G	giga	10^9
m	milli	10^{-3}	T	tera	10^{12}
c[1]	centi	10^{-2}	P	peta	10^{15}
d[1]	deci	10^{-1}	E	exa	10^{18}

Logarithms

The following relations are presented to facilitate the use of natural and common logarithms, their antilogarithms, and trigonometric functions. For those readers who are completely unfamiliar with such quantities, a text or handbook should be consulted first.

---

1. Not recommended by SI (Système International, the internationally accepted system for units).

$$\ln (xy) = \ln x + \ln y \qquad\qquad \ln x^a = a \ln x$$

$$\ln(x/y) = \ln x - \ln y \qquad\qquad \ln(1/x^a) = -a \ln x$$

$$\ln(1/x) = -\ln x \qquad\qquad \ln x = 2.303 \log x$$

$$\ln e = 1, \qquad \ln 10 = 2.303 \qquad \log 10 = 1, \qquad \log e = \frac{1}{2.303}$$

$$\ln e^y = y \qquad\qquad \log 10^y = y$$

$$e^{\ln y} = y \qquad\qquad 10^{\log y} = y$$

$$\ln(1 + x) = x - \frac{x^2}{2} + \frac{x^3}{3} - \frac{x^4}{4} + \cdots \qquad -1 < x \leq 1$$

$$e^x = 1 + x + \frac{x^2}{2!} + \frac{x^3}{3!} + \cdots$$

Quadratic Equation

The quadratic equation has the following form:

$$ax^2 + bx + c = 0 \qquad (a \neq 0)$$

Its two solutions (two roots) are

$$x = \frac{-b \pm \sqrt{b^2 - 4ac}}{2a}$$

Trignometric Functions

Consider a right triangle of hypotenuse r:

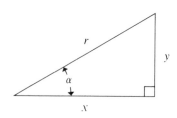

α in radians or $^\circ$
(2π radians = $360°$)

$$\sin \alpha = \frac{y}{r} \qquad -1 \leq \sin \alpha \leq 1$$

$$\cos \alpha = \frac{x}{r} \qquad -1 \leq \cos \alpha \leq 1$$

$$\tan \alpha = \frac{y}{x} \qquad -\infty \leq \tan \alpha \leq \infty$$

$$\sin \alpha = \alpha - \frac{\alpha^3}{3!} + \frac{\alpha^5}{5!} - \frac{\alpha^7}{7!} + \cdots \qquad \alpha \text{ in radians}$$

$$\cos \alpha = 1 - \frac{\alpha^2}{2!} + \frac{\alpha^4}{4!} - \frac{\alpha^6}{6!} + \cdots \qquad \alpha \text{ in radians}$$

Differential Equations

As a final topic in this appendix, we will consider the application of the integral calculus to the solution of differential equations. A differential equation expresses the relationship between derivatives (first-order as well as higher-order) and various variables or functions. The procedure in solving differential equations is first to put the relation into a form that can be integrated and then to carry out suitable integrations so that the derivatives are eliminated. To complete the solution of a differential equation, we must incorporate the known values of the functions at particular values of the variables, the so-called boundary conditions. We will illustrate the handling of differential equations by a simple, but extremely useful, example.

One of the most important differential equations in biology has the following general form:

$$\frac{dy}{dt} = -ky \qquad\qquad (III.1)$$

where k is a positive constant and t represents time. We encountered this equation in Chapter 4 (Eq. 4.9) in discussing the various competing pathways for the de-excitation of an excited singlet state. Equation III.1 also describes the process of radioactive decay, where y is the amount of radioisotope present at any time t. The equation indicates that the rate of change in time of the amount of radioactive substance (dy/dt) is linearly proportional to the amount present at that time (y). Because the radioisotope decays in time, dy/dt is negative, and hence there is a minus sign in Equation III.1. Any process that can be described by Equation III.1, such as a chemical reaction, is called a first-order rate process, and k is known as the first-order rate constant.

To put Equation III.1 into a form suitable for integration, we must separate the variables (y and t) so that each one appears on only one side of the equation:

$$\frac{dy}{y} = -k \, dt \tag{III.2}$$

which follows from Equation III.2 upon multiplying each side by dt/y. (Note that the same initial process of separation of variables applies to the integration of a more complicated example in Ch. 3, namely, Eq. 3.11.) When the variables are separated so that a possible integrand, e.g., $-k \, dt$, is expressed in terms of only one variable, we can integrate that integrand. On the other hand, the integrand $-ky \, dt$ cannot be integrated as it stands—i.e, we cannot perform $\int -ky \, dt$—because we do not known how y depends on t. In fact, the very purpose of solving Equation III.1 is to determine the functional relationship between y and t.

Next, we will insert integral signs into Equation III.2 and perform the integration:

$$\int \frac{dy}{y} = \ln y = -\int k \, dt = -kt + C \tag{III.3}$$

When we take exponentials of both sides of Equation III.3, we obtain the following expression for y:

$$e^{\ln y} = y(t) = e^{-kt+C} = e^C e^{-kt} \tag{III.4}$$

where e^C is a constant. In Equation III.4 we have replaced y with $y(t)$ to emphasize that y depends on the independent variable t. When t equals zero, Equation III.4 indicates that $y(0) = e^C e^{-k \times 0}$, which is simply e^C. Thus, the constant e^C is the value taken on by the dependent variable y when t is 0. This latter relationship, $e^C = y(0)$, is referred to as a boundary condition (or initial condition, because we are here dealing with time). Incorporating this boundary condition into Equation III.4, we obtain the following solution to the differential equation represented by Equation III.1:

$$y(t) = y(0)e^{-kt} \tag{III.5}$$

Because of the factor e^{-kt}, Equation III.5 indicates that y decays exponentially with time for a first-order rate process. Moreover, $y(t)$ decreases to $1/e$ of its initial value $[y(0)]$ when t satisfies the following relation:

$$y(\tau) = \frac{1}{e} y(0) = y(0)e^{-k\tau} \tag{III.6}$$

where the value of time, τ, that satisfies Equation III.6 is known as the *lifetime* of the process whose decay or disappearance is being considered. Equation III.6 indicates that e^{-1} equals e^{-kt}, so the first-order rate constant is equal to the reciprocal of the lifetime τ (see Eq. 4.13). Therefore, the solution (Eq. III.5) of the partial differential equation (Eq. III.1) describing a first-order rate process becomes

$$y(t) = y(0)e^{-t/\tau} \tag{III.7}$$

Gibbs Free Energy and Chemical Potential

Entropy and equilibrium. Gibbs free energy. Chemical potential. Pressure
dependence of μ_j. Concentration dependence of μ_j. References.

The concept of chemical potential is introduced in Chapter 2 and used throughout
the rest of the book. In order not to overburden the text with mathematical
details, certain points are stated without proof. Here we will justify the form of
the pressure term in the chemical potential and also provide insight into how the
expression for the Gibbs free energy arises.

Entropy and Equilibrium

A suitable point of departure is to reconsider the condition for equilibrium. The
most general statement we can make concerning the attainment of equilibrium
by a system is that it occurs when the entropy of the system plus its surroundings
is at a maximum. Unfortunately, entropy has proved to be a rather elusive concept
to master and a difficult quantity to measure. Moreover, reference to the surroundings—the "rest of the universe" in the somewhat grandiloquent language

of physics—is a nuisance. Consequently, thermodynamicists sought a function that would help describe equilibrium, but would depend only on readily measurable parameters of the system under consideration. As we will see, the Gibbs free energy is such a function for most applications in biology.

The concept of entropy is really part of our day-to-day observations. We know that an isolated system will spontaneously change in certain ways—a system proceeds toward a state that is more random or less ordered than the initial one. For instance, neutral solutes will diffuse toward regions where they are less concentrated. In so doing, the system lowers its capacity for further spontaneous change. For all such processes ΔS is positive, whereas ΔS becomes zero and S achieves a maximum at equilibrium. Equilibrium means that no more spontaneous changes will take place; entropy is therefore an *index for the capacity for spontaneous change*. It would be more convenient in some ways if entropy had been originally defined with the opposite sign. In fact, some authors introduce the quantity *negentropy*, which equals $-S$ and reaches a *minimum* at equilibrium. In any case, we must ultimately use a precise mathematical definition for entropy, such as $dS = dQ/T$, where dQ refers to the heat gain or loss in some reversible reaction taking place at temperature T.

We can represent the total entropy of the universe, S_u, as the entropy of the system under consideration, S_s, plus the entropy of the rest of the universe, S_r. We can express this in symbols as follows:

$$S_u = S_s + S_r \quad \text{or} \quad dS_u = dS_s + dS_r \quad \text{(IV.1)}$$

An increase in S_u accompanies all real processes—that is, incidentally, the most succinct way of stating the second law of thermodynamics. S_u is maximum at equilibrium.

The heat absorbed by a system during some process is equal to the heat given up by the rest of the universe. Let us represent the infinitesimal heat exchange of the system by dQ_s. For an isothermal reaction or change, dQ_s is simply $-dQ_r$, because the heat must come from the rest of the universe. From the definition of entropy,[1] $dS = dQ/T$, we can obtain the following relationship:

$$dS_r = \frac{dQ_r}{T} = -\frac{dQ_s}{T} = -\frac{dU_s + P\,dV_s}{T} \quad \text{(IV.2)}$$

---

1. This definition really applies only to reversible reactions, which we can in principle use to approximate a given change; otherwise, dQ is not uniquely related to dS.

The last step in Equation IV.2 derives from the principle of the conservation of energy for the case when the only form of work involved is mechanical—a common assumption in stating the first law of thermodynamics. It is thus possible to express dQ_s as the sum of the change in internal energy (dU_s) plus a work term ($P\,dV_s$). The internal energy (U_s) is a function of the state of a system; i.e., its magnitude depends on the characteristics of the system but is independent of how the system got to that state. PV_s is also a well-defined variable. However, heat (Q_s) is not a function of the state of a system.

As we indicated above, equilibrium occurs when the entropy of the universe is maximum. This means that dS_u then equals zero. By substituting Equation IV.2 into the differential form of Equation IV.1, we can express this equilibrium condition solely in terms of system parameters:

$$0 = dS_s + \left(- \frac{dU_s + P\,dV_s}{T} \right) \quad \text{or}$$

$$- T\,dS_s + dU_s + P\,dV_s = 0 \tag{IV.3}$$

Equation IV.3 suggests that there is some function of the *system* that has an extremum at equilibrium. In other words, we might be able to find some expression determined by the parameters describing the system and whose derivative is zero at equilibrium. If so, the rather abstract statement that the entropy of the universe is a maximum at equilibrium could then be replaced by a statement referring only to measurable attributes of the system—easily measurable ones, we hope.

In the 1870s J. W. Gibbs—perhaps the most brilliant thermodynamicist to date—chose a rather simple set of terms that turned out to have the very properties for which we are searching. This function is now referred to as the *Gibbs free energy*, and has the symbol G:

$$G = U + PV - TS \tag{IV.4a}$$

which, upon differentiating, yields

$$dG = dU + P\,dV + V\,dP - T\,dS - S\,dT \tag{IV.4b}$$

Equation IV.4b implies that, at constant temperature ($dT = 0$) and constant pressure ($dP = 0$), dG is

$$dG = dU + P\,dV - T\,dS \quad \text{at constant } T \text{ and } P \tag{IV.5}$$

By comparing Equation IV.5 with the equilibrium condition expressed by Equation IV.3, we see that dG for a system equals zero at equilibrium at constant temperature and pressure. Moreover, G depends only on U, P, V, T, and S of the *system*. It turns out that the extremum condition, $dG = 0$, actually occurs when G reaches a minimum at equilibrium. This useful attribute of the Gibbs free energy is strictly valid only when the overall system is at *constant temperature and pressure*, conditions that closely approximate those encountered in many biological situations. Thus, our criterion for equilibrium shifts from a maximum of the entropy of the universe to a minimum in the Gibbs free energy of the system.

Gibbs Free Energy

We will now consider how the internal energy, U, changes when material enters or leaves a system. This will help us derive an expression for the Gibbs free energy that is quite useful for biological applications.

The internal energy of a system changes when substances enter or leave it. For convenience, we will consider a system of fixed volume and at the same temperature as the surroundings so that there are no heat exchanges. If dn_j moles of species j enter such a system, U increases by $\mu_j dn_j$, where μ_j is an intensive variable representing the free energy contribution to the system per mole of species j entering or leaving. Work is often expressed as the product of an intensive quantity (such as μ_j, P, T, E, h) times an extensive one (dn_j, dV, dS, dQ, dm, respectively); i.e., the amount of any kind of work depends on both some thermodynamic parameter characterizing the internal state of the system and the extent or amount of change for the system. In our present example the extensive variable describing the amount of change is dn_j, and μ_j represents the contribution to the internal energy of the system per mole of species j. When more than one species crosses the boundary of our system, which is at constant volume and the same temperature as the surroundings, the term $\sum_j \mu_j \, dn_j$ is added to dU, where dn_j is positive if the species enters the system and negative if it leaves. In the general case, when we consider all the ways that the internal energy of a system can change, we can represent dU as follows:

$$dU = dQ - P\,dV + \sum_j \mu_j dn_j \qquad \text{(IV.6)}$$

Let us now return to the development of a useful relation for the Gibbs free energy of a system. When dU as expressed by Equation IV.6 is substituted into dG as given by Equation IV.5, we obtain

$$dG = T dS - P dV + \sum_j \mu_j dn_j + P dV - T dS$$

$$= \sum_j \mu_j dn_j \tag{IV.7}$$

where dQ has been replaced by $T dS$. Equation IV.7 indicates that the particular form chosen for the Gibbs free energy leads to a very simple expression for dG at constant T and P—namely, dG then depends only on μ_j and dn_j.

To obtain an expression for G, we must integrate Equation IV.7. To facilitate the integration we will define a new variable, α, such that dn_j equals $n_j d\alpha$, where n_j is the total number of moles of species j present in the final system; i.e., n_j is a constant describing the final system. The subsequent integration from $\alpha = 0$ to $\alpha = 1$ corresponds to building up the system by a simultaneous addition of all the components in the same proportions present in the final system. (The intensive variable μ_j is also held constant for this integration pathway; i.e., the chemical potential of species j does not depend on the size of the system.) Using Equation IV.7 and this rather easy integration pathway, we obtain the following expression for the Gibbs free energy:

$$G = \int dG = \int \sum_j \mu_j dn_j = \int_0^1 \sum_j \mu_j n_j \, d\alpha$$

$$= \sum_j \mu_j n_j \int_0^1 d\alpha = \sum_j \mu_j n_j \tag{IV.8}$$

The well-known relation between G and the μ_j's in Equation IV.8 can also be obtained by a method that is more elegant mathematically but somewhat involved.

In Chapter 6 we presented without proof an expression for the Gibbs free energy (Eq. 6.1 is essentially Eq. IV.8) and also noted some of the properties of G. For instance, at constant temperature and pressure, the direction for a spontaneous change is toward a lower Gibbs free energy; minimum G is achieved at equilibrium. Hence, ΔG is negative for such spontaneous processes. Spontaneous processes can in principle be harnessed to do useful work, where the maximum amount of work possible at constant temperature and pressure is equal to the absolute value of ΔG (some of the energy is dissipated by inevitable inefficiencies such as frictional losses, so $-\Delta G$ represents the *maximum* work possible). To drive a reaction in the direction opposite to that in which it proceeds spontaneously requires a free energy input of at least ΔG.

Chemical Potential

Let us now examine the properties of the intensive variable μ_j. Equation IV.8 ($G = \Sigma_j \mu_j n_j$) suggests a very useful way of defining μ_j. In particular, if we keep μ_i and n_i constant, we obtain the following expression:

$$\mu_j = \left(\frac{\partial G}{\partial n_j}\right)_{\mu_i, n_i} = \left(\frac{\partial G}{\partial n_j}\right)_{T,P,E,h,n_i} \tag{IV.9}$$

where n_i and μ_i refer to all species other than species j. Because μ_i can depend on T, P, E (the electrical potential), h (the height in a gravitational field), and n_i, the act of keeping μ_i constant during partial differentiation is the same as keeping T, P, E, h, and n_i constant, as indicated in Equation IV.9. Equation IV.9 indicates that the chemical potential of species j is the partial molal Gibbs free energy of a system with respect to that species, and that it is obtained when T, P, E, h, and the amount of all other species are held constant. Thus, μ_j corresponds to the intensive contribution of species j to the extensive quantity G, the Gibbs free energy of the system.

In Chapters 2 and 3 we argued that μ_j depends on T, $a_j (a_j = \gamma_j c_j$, Eq. 2.5), P, E, and h in a solution, and that the partial pressure of species j, P_j, is also involved for the chemical potential in a vapor phase. We can summarize the two relations as follows:

$$\mu_j^{\text{liquid}} = \mu_j^* + RT \ln a_j + \overline{V}_j P + z_j FE + m_j g h \tag{IV.10}$$

$$\mu_j^{\text{vapor}} = \mu_j^* + RT \ln \frac{P_j}{P_j^*} + m_j g h$$

The forms for the gravitational contribution ($m_j g h$) and the electrical one ($z_j FE$) can easily be understood. We showed in Chapter 3 that $RT \ln a_j$ is the correct form for the concentration term in μ_j. The reasons for the forms of the pressure terms in a liquid ($\overline{V}_j P$) and in a gas [$RT \ln (P_j/P_j^*)$] are not so obvious. Therefore, we will examine the pressure dependence of the chemical potential of species j in some detail.

Pressure Dependence of μ_j

To derive the pressure terms in the chemical potentials of solvents, solutes, and gases, we must rely on certain properties of partial derivatives as well as on commonly observed effects of pressure. To begin with, we will differentiate the chemical potential in Equation IV.9 with respect to P:

$$\left(\frac{\partial \mu_j}{\partial P}\right)_{T,E,h,n_i,n_j} = \left[\frac{\partial}{\partial P}\left(\frac{\partial G}{\partial n_j}\right)_{T,P,E,h,n_i}\right]_{T,E,h,n_i,n_j}$$

$$= \left[\frac{\partial}{\partial n_j}\left(\frac{\partial G}{\partial P}\right)_{T,E,h,n_i,n_j}\right]_{T,\ P,E,h,n_i}$$

(IV.11)

where we have reversed the order for partial differentiation with respect to P and n_j (this is permissible for functions such as G, which have well-defined and continuous first-order partial derivatives). The differential form of Equation IV.4 ($dG = dU + P\,dV + V\,dP - T\,dS - S\,dT$) gives us a suitable form for dG. If we substitute dU given by Equation IV.6 ($dU = T\,dS - P\,dV + \Sigma_j\,\mu_j dn_j$, where dQ is replaced by $T\,dS$) into this expression for the derivative of the Gibbs free energy, we can express dG in the following useful form:

$$dG = V\,dP - S\,dT + \sum_j \mu_j dn_j$$

(IV.12)

Using Equation IV.12 we can readily determine the pressure dependence of the Gibbs free energy as needed in the last bracket of Equation IV.11—namely, $(\partial G/\partial P)_{T,E,h,n_i,n_j}$ equals V by Equation IV.12. Next, we have to consider the partial derivative of this V with respect to n_j (see last equality of Eq. IV.11). But Equation 2.6 indicates that $(\partial V/\partial n_j)_{T,P,E,h,n_i}$ is $\overline{V}_j$, the partial molal volume of species j. Substituting these partial derivatives into Equation IV.11 leads us to the following useful expression:

$$\left(\frac{\partial \mu_j}{\partial P}\right)_{T,E,h,n_i,n_j} = \overline{V}_j$$

(IV.13)

Equation IV.13 is of pivotal importance in deriving the form of the pressure term in the chemical potentials of both liquid and vapor phases.

Let us first consider an integration of Equation IV.13 appropriate for a liquid. We will make use of the observation that the partial molal volume of a species in a solution does not depend on the pressure to any significant extent. For a solvent this means that the liquid generally is essentially incompressible. If we integrate Equation IV.13 with respect to P at constant T, E, h, n_i and n_j with $\overline{V}_j$ independent of P, we obtain the following relations:

$$\int \frac{\partial \mu_j}{\partial P} \, dP = \int_{\mu_j^*}^{\mu_j^{\text{liquid}}} d\mu_j = \mu_j^{\text{liquid}} - \mu_j^*$$
$$= \int \overline{V}_j \, dP = \overline{V}_j \int dP = \overline{V}_j P + \text{"constant"} \qquad (\text{IV}.14)$$

where the definite integral in the top line is taken from the chemical potential of species j in a standard state as the lower limit up to the general μ_j in a liquid as the upper limit. The integration of $\int \overline{V}_j dP$ leads to our pressure term $\overline{V}_j P$ plus a constant. Because the integration was performed while holding T, E, h, n_i, and n_j fixed, the "constant" can depend on all of these variables, but not on P.

Equation IV.14 indicates that the chemical potential of a liquid contains a pressure term of the form $\overline{V}_j P$. The other terms (in particular, μ_j^*, $RT \ln a_j$, $z_j FE$, and $m_j gh$) do not depend on pressure, a condition used throughout this text. The experimental observation that gives us this very useful form for μ_j is that $\overline{V}_j$ is generally not influenced very much by pressure; e.g., a liquid is often essentially incompressible. If this should prove invalid under certain situations, $\overline{V}_j P$ would then not be a suitable term in the chemical potential of species j for expressing the pressure dependence in a solution.

Next we turn to the form of the pressure term in the chemical potential of a gas, where the assumption of incompressibility that we used for a liquid is not valid. Our point of departure is the perfect or ideal gas law:

$$P_j V = n_j RT \qquad (\text{IV}.15)$$

where P_j is the partial pressure of species j and n_j is the number of moles of species j in volume V. Thus, we will assume that real gases behave like ideal gases, which appears to be justified for biological applications. Based on Equations IV.15 and 2.6 $\left[\overline{V}_j = \left(\dfrac{\partial V}{\partial n_j} \right)_{n_i, T, P, E, h} \right]$, the partial molal volume $\overline{V}_j$ for gaseous species j is RT/P_j. We also note that the total pressure P equals $\Sigma_j P_j$, where the summation is over all gases present (Dalton's law of partial pressures); hence, dP equals dP_j when n_i, T, and V are constant. When we integrate Equation IV.13, we thus find that the chemical potential of gaseous species j depends on the *logarithm* of its partial pressure:

$$\int \frac{\partial \mu_j}{\partial P} \, dP = \int_{\mu_j^*}^{\mu_j^{\text{vapor}}} d\mu_j = \mu_j^{\text{vapor}} - \mu_j^* = \int \overline{V}_j \, dP = \int \frac{RT}{P_j} \, dP_j$$
$$= RT \ln P_j + \text{"constant"} \qquad (\text{IV}.16)$$

where the "constant" can depend on T, E, h, and n_i, but not on n_j (or P_j). In particular, the "constant" equals $-RT \ln P_j^* + m_j gh$, where P_j^* is the saturation partial pressure for species j at atmospheric pressure and some particular temperature. Hence, the chemical potential for species j in the vapor phase (μ_j^{vapor}) is

$$\mu_j^{\text{vapor}} = \mu_j^* + RT \ln P_j - RT \ln P_j^* + m_j gh \qquad \text{(IV.17)}$$

$$= \mu_j^* + RT \ln \frac{P_j}{P_j^*} + m_j gh$$

which is essentially the same as Equations 2.20 and IV.10.

We have defined the standard state for gaseous species j, μ_j^*, as the chemical potential when the gas phase has a partial pressure for species j (P_j) equal to the saturation partial pressure (P_j^*), when we are at atmospheric pressure ($P = 0$) and the zero level for the gravitational term ($h = 0$), and for some specified temperature. Many physical chemistry texts ignore the gravitational term (we calculated that it has only a small effect for water vapor, Ch. 2) and define the standard state for the condition when P_j equals 1 atmosphere and species j is the only species present ($P = P_j$). The chemical potential of such a standard state equals $\mu_j^* - RT \ln P_j^*$ in our symbols.

The partial pressure of some species in a vapor phase in equilibrium with a liquid depends slightly on the total pressure in the system—loosely speaking, when the pressure on the liquid is increased, more molecules are squeezed out of it into the vapor phase. The exact relationship between the pressures involved, which is known as the Gibbs equation, is as follows for water:

$$\frac{\partial (\ln P_{wv})}{\partial P} = \frac{\bar{V}_w}{RT} \qquad \text{or} \qquad \frac{\partial P_{wv}}{\partial P} = \frac{\bar{V}_w}{\bar{V}_{wv}} \qquad \text{(IV.18)}$$

where the second equality follows from the derivative of a logarithm [$\partial \ln u/\partial x = (1/u)(\partial u/\partial x)$] and the ideal or perfect gas law ($P_{wv}V = n_{wv}RT$, Eq. IV.15, so $\partial V/\partial n_{wv} = \bar{V}_{wv} = RT/P_{wv}$). Because $\bar{V}_w$ is much less than $\bar{V}_{wv}$, the effect is quite small (e.g., at 20°C and atmospheric pressure, $\bar{V}_w = 1.8 \times 10^{-5}$ m³ mol⁻¹ and $\bar{V}_{wv} = 2.4 \times 10^{-2}$ m³ mol⁻¹). From the first equality in Equation IV.18, we see that $\bar{V}_w dP = RT\, d \ln P_{wv}$. Hence, if the chemical potential of the liquid phase (μ_w) increases by $\bar{V}_w dP$ as an infinitesimal pressure is applied, then an equal increase, $RT\, d \ln P_{wv}$, occurs in μ_{wv} (see Eq. IV.13 for a definition of μ_{wv}), and hence we will still be in equilibrium ($\mu_w = \mu_{wv}$). This relation can be integrated to give $RT \ln P_{wv} = \overline{V}_w P + \text{constant}$, where the constant is $RT \ln P_{wv}^0$ and P_{wv}^0 is the partial pressure of water vapor at atmospheric pressure ($P = 0$); hence, $RT \ln P_{wv}/P_{wv}^0$ equals $\overline{V}_w P$, a relation used in Chapter 2 (see footnote on p. 92). We note that effects of external pressure on P_{wv} can be of the same order of magnitude as deviations from the ideal gas law for water vapor, both of which are usually neglected in plant physiology.

Concentration Dependence of μ_j

We will complete our discussion of chemical potential by using Equation IV.17 to obtain the logarithmic term in concentration that is found for μ_j in a liquid phase. First, it should be pointed out that Equation IV.17 has no concentration term per se for the chemical potential of species j in a gas phase. However, the partial pressure of a species in a gas phase is really analogous to the concentration of a species in a liquid; e.g., $P_j V = n_j RT$ for gaseous species j (Eq. IV.15), and concentration means number/volume $= n_j/V = P_j/RT$.

Raoult's law states that at equilibrium the partial pressure of a particular gas above its volatile liquid is proportional to the mole fraction of that solvent in the liquid phase. A similar relation more appropriate for solutes is Henry's law, which states that P_j in the vapor phase is proportional to the N_j of that solute in the liquid phase. Although the proportionality coefficients in the two relations are different, they both indicate that P_j^{vapor} depends linearly on N_j^{solution}. For dilute solutions the concentration of species j, c_j, is proportional to its mole fraction, N_j (this is true for both solute and solvent). Thus, when P_j^{vapor} changes from one equilibrium condition to another, we expect a similar change in c_j^{solution}, because μ_j^{liquid} equals μ_j^{vapor} at equilibrium. In particular, Equation IV.17 indicates that μ_j^{vapor} depends on $RT \ln (P_j/P_j^*)$, and hence the chemical potential of a solvent or solute should contain a term of the form $RT \ln c_j$, as in fact it does (see Eqs. 2.4 and IV.10). As we discussed in Chapter 2, we should be concerned about the concentration that is thermodynamically active, a_j ($a_j = \gamma_j c_j$, Eq. 2.5), so the actual term in the chemical potential for a solute or solvent is $RT \ln a_j$, not $RT \ln c_j$. In Chapter 3, instead of the present argument based on Raoult's and Henry's laws, we used a comparison with Fick's first law to justify the $RT \ln a_j$ term. Moreover, the Boyle–Van't Hoff relation, which was derived *assuming* the $RT \ln a_j$ term, has been amply demonstrated experimentally. Consequently, the $RT \ln a_j$ term in the chemical potential for a solute or solvent can be justified or derived in a number of different ways, all of which depend on agreement with experimental observations.

General References

Castellan, G. W. 1983. *Physical Chemistry*, 3rd ed. Addison-Wesley, Reading, Massachusetts.

Denbigh, K. G. 1981. *The Principles of Chemical Equilibrium*, 4th ed. Cambridge University Press, Cambridge, England.

Katchalsky, A., and P. F. Curran. 1967. *Nonequilibrium Thermodynamics in Biophysics*. Harvard University Press, Cambridge, Massachusetts.

Morowitz, H. J. 1970. *Entropy for Biologists—An Introduction to Thermodynamics*. Academic Press, New York.

Noggle, J. H. 1985. *Physical Chemistry*. Little, Brown, Boston, Massachusetts.

Solutions to Problems

CHAPTER 1

1.1 (a) We note that c_j at $x = 3$ mm is 37% (i.e., $\sim e^{-1}$) of c_j at the origin ($x = 0$), so we can use Equation 1.6 to calculate D_j:

$$D_j = \frac{x_e^2}{4t_e} = \frac{(0.003 \text{ m})^2}{(4)(3\ 600 \text{ s})} = 0.63 \times 10^{-9} \text{ m}^2 \text{ s}^{-1}$$

(b) Similar to (a), but now we solve Equation 1.6 for t_e:

$$t_e = \frac{x_e^2}{4D_j} = \frac{(0.090 \text{ m})^2}{(4)(0.63 \times 10^{-9} \text{ m}^2 \text{ s}^{-1})} = 3.2 \times 10^6 \text{ s} . \quad (37 \text{ days})$$

(c) We will use Equation 1.5 and solve at the origin ($x = 0$, so $e^{-x^2/4D_j t} = 1$), leading to

$$M_j = 2c_j(\pi D_j t)^{1/2} = 2(100 \text{ mol m}^{-3})[(\pi)(0.63 \times 10^{-9} \text{ m}^2 \text{ s}^{-1})(3\ 600 \text{ s})]^{1/2}$$

$$= 0.53 \text{ mol m}^{-2}$$

(d) If t_e is the same and D_j is $1/100$ of the value in (b), Equation 1.6 indicates that x_e is then $1/10$ as large, or 9 mm.

1.2 (a) The mass of a mitochondrion is its volume times the density (mass/volume):

$$\text{mass of mito} = [0.30 \times (10^{-6} \text{ m})^3] (1\ 110 \text{ kg m}^{-3}) = 3.33 \times 10^{-16} \text{ kg}$$

Hence, the mass of Avogadro's number of mitochondria is:

mass/mol mito $= (3.33 \times 10^{-16}$ kg$)(6.022 \times 10^{23}$ mol$^{-1}) = 2.0 \times 10^8$ kg mol^{-1}

Therefore, the "molecular weight" of mitochondria is 2.0×10^{11}.

(b) If D_j is proportional to $1/(MW)^{1/3}$, then

$$D_{mito} = D_x \frac{(MW_x)^{1/3}}{(MW_{mito})^{1/3}} = (0.5 \times 10^{-9} \text{ m}^2 \text{ s}^{-1}) \frac{(200)^{1/3}}{(2 \times 10^{11})^{1/3}}$$

$$= 5 \times 10^{-13} \text{ m}^2 \text{ s}^{-1}$$

(c) Using Equation 1.6, $t_e = \dfrac{x_e^2}{4D_j} = \dfrac{(0.2 \times 10^{-6} \text{ m})^2}{(4)(5 \times 10^{-13} \text{ m}^2 \text{ s}^{-1})} = 0.62$ s

For x_e of 50 μm, $t_e = \dfrac{(50 \times 10^{-6} \text{ m})^2}{(4)(5 \times 10^{-13} \text{ m}^2 \text{ s}^{-1})} = 1\,250$ s   (21 min)

(d) Again using Equation 1.6, $t_e = \dfrac{(50 \times 10^{-6} \text{ m})^2}{(4)(0.3 \times 10^{-9} \text{ m}^2 \text{ s}^{-1})} = 2.1$ s

Because ATP diffuses much faster than do mitochondria across such cellular distances, it is much more expedient for ATP to diffuse to where it is to be used than for mitochondria to so diffuse.

1.3 (a) The times for CO_2 to diffuse across the two barriers can be compared using Equation 1.6:

$$\frac{t_e^{air}}{t_e^{cell \ wall}} = \frac{(1 \times 10^{-3} \text{ m})^2 \ (4) \ (D_{CO_2}^{cw})}{(2 \times 10^{-6} \text{ m})^2 \ (4)(10^6 \times D_{CO_2}^{cw})} = 0.25$$

(b) Again using Equation 1.6,

$$\frac{D_{CO_2}^{plasmalemma}}{D_{CO_2}^{cell \ wall}} = \frac{(8 \times 10^{-9} \text{ m})^2 \ (4)(t_e)}{(2 \times 10^{-6} \text{ m})^2 \ (4)(t_e)} = 1.6 \times 10^{-5}$$

(c) Using Equation 1.9 to define P_j, we obtain

$$\frac{P_{CO_2}^{cell \ wall}}{P_{CO_2}^{plasmalemma}} = \frac{(8 \times 10^{-9} \text{ m})(D_{CO_2}^{cell \ wall})(100 \times K_{CO_2}^{plasmalemma})}{(2 \times 10^{-6} \text{ m})(1.6 \times 10^{-5} \ D_{CO_2}^{cell \ wall})(K_{CO_2}^{plasmalemma})}$$

$$= 2.5 \times 10^4$$

1.4 (a) For the unstirred boundary layer, K_j is unity. Using Equation 1.9, P_j then is:

$$P_{H_2O}^{bl} = \frac{(2.6 \times 10^{-9} \text{ m s}^{-1})(1)}{(20 \times 10^{-6} \text{ m})} = 13 \times 10^{-5} \text{ m s}^{-1}$$

$$P_{methanol}^{bl} = \frac{(0.8 \times 10^{-9} \text{ m}^2 \text{ s}^{-1})(1)}{(20 \times 10^{-6} \text{ m})} = 4.0 \times 10^{-5} \text{ m s}^{-1}$$

$$P_{L\text{-}leucine}^{bl} = \frac{(0.2 \times 10^{-9} \text{ m}^2 \text{ s}^{-1})(1)}{(20 \times 10^{-6} \text{ m})} = 1.0 \times 10^{-5} \text{ m s}^{-1}$$

(b) For the two barriers in series,

$$\frac{1}{P_j^{\text{total}}} = \frac{1}{P_j^{\text{bl}}} + \frac{1}{P_j^{\text{memb}}}$$

and so

$$P_j^{\text{memb}} = \frac{P_j^{\text{bl}} \, P_j^{\text{total}}}{P_j^{\text{bl}} - P_j^{\text{total}}}$$

Using measured values of P_j^{total} and P_j^{bl} from (a), we have:

$$P_{\text{D}_2\text{O}}^{\text{memb}} = \frac{(1.3 \times 10^{-4} \text{ m s}^{-1})(1.0 \times 10^{-4} \text{ m s}^{-1})}{(1.3 \times 10^{-4} \text{ m s}^{-1}) - (1.0 \times 10^{-4} \text{ m s}^{-1})} = 4.3 \times 10^{-4} \text{ m s}^{-1}$$

$$P_{\text{methanol}}^{\text{memb}} = \frac{(4.0 \times 10^{-5} \text{ m s}^{-1})(2.0 \times 10^{-5} \text{ m s}^{-1})}{(4.0 \times 10^{-5} \text{ m s}^{-1}) - (2.0 \times 10^{-5} \text{ m s}^{-1})} = 4.0 \times 10^{-5} \text{ m s}^{-1}$$

$$P_{\text{L-leucine}}^{\text{memb}} = \frac{(1.0 \times 10^{-5} \text{ m s}^{-1})(3.0 \times 10^{-8} \text{ m s}^{-1})}{(1.0 \times 10^{-5} \text{ m s}^{-1}) - (3.0 \times 10^{-8} \text{ m s}^{-1})} = 3.0 \times 10^{-8} \text{ m s}^{-1}$$

(c) The unstirred layer is the bigger barrier for D_2O, because $P_{\text{D}_2\text{O}}^{\text{memb}} > P_{\text{D}_2\text{O}}^{\text{bl}}$; both barriers are of similar resistance for methanol, because $P_{\text{methanol}}^{\text{memb}} = P_{\text{methanol}}^{\text{bl}}$; the membrane is the bigger barrier for L-leucine, because $P_{\text{L-leucine}}^{\text{memb}} \ll P_{\text{L-leucine}}^{\text{bl}}$.

(d) If P_j^{memb} becomes large, then $1/P_j^{\text{memb}}$ approaches 0, and $1/P_j^{\text{total}} \cong 1/P_j^{\text{bl}}$. Thus the maximum value of P_j^{total} is P_j^{bl} [given in (a)].

1.5 (a) The volume to surface area ratio of a cylinder is $rl/[2(r + l)]$. As $c_j^o = 0$ and $c_j^i(t) = 0.1$ $c_j^i(0)$, Equation 1.12 becomes (after solving for t):

$$t = \frac{rl \, \ln 10}{2(r + l)P_j} = \frac{(0.5 \times 10^{-3} \text{ m})(100 \times 10^{-3} \text{ m})(2.303)}{2[(0.5 + 100) \times 10^{-3} \text{ m}](10^{-6} \text{ m s}^{-1})} = 573 \text{ s} \quad (9.6 \text{ min})$$

(b) Here, $\dfrac{V}{A}$ becomes $\pi r^2 l/(2\pi r^2)$ or $l/2$, and so

$$t = \frac{l \, \ln 10}{2 \, P_j} = \frac{(100 \times 10^{-3} \text{ m})(2.30)}{(2)(10^{-6} \text{ m s}^{-1})} = 1.2 \times 10^5 \text{ s} \quad (32 \text{ h})$$

(c) Here, $c_j^i(t)$ is $0.01 \, c_j^i(0)$, so the rearranged Equation 1.12 yields

$$\frac{t_{99\%}}{t_{90\%}} = \frac{\dfrac{V}{P_j A} \ln 100}{\dfrac{V}{P_j A} \ln 10} = \frac{\ln 10^2}{\ln 10} = 2$$

(d) Again using the rearranged Equation 1.12,

$$\frac{t_{new}}{t_{original}} = \frac{P_j^{original}}{P_j^{new}} = \frac{(10^{-6} \text{ m s}^{-1})}{(10^{-8} \text{ m s}^{-1})} = 100$$

1.6 (a) The surface area to volume ratio (A/V) for a sphere is $3/r$ and for a cylinder is $2(r + l)/rl$.

Thus, A/V for *Nitella* is $\dfrac{2 (0.5 \times 10^{-3} \text{ m} + 100 \times 10^{-3} \text{ m})}{(0.5 \times 10^{-3} \text{ m})(100 \times 10^{-3} \text{ m})}$ or $4\ 020 \text{ m}^{-1}$, for *Valonia* is

$\dfrac{3}{(5 \times 10^{-3} \text{ m})}$ or 600 m^{-1}, and for *Chlorella* is $\dfrac{3}{(2 \times 10^{-6} \text{ m})}$ or $1.5 \times 10^6 \text{ m}^{-1}$.

(b) *Chlorella* [see (a)]

(c) Because the concentration criterion and P_j are the same in all cases, Equation 1.12 leads to:

$$\frac{t_a}{t_b} = \frac{A_b/V_b}{A_a/V_a}$$

and so

$$t_{Nitella} = (1 \text{ s}) \frac{(1.5 \times 10^6 \text{ m}^{-1})}{(4\ 020 \text{ m}^{-1})} = 370 \text{ s} \qquad (6.3 \text{ min})$$

$$t_{Valonia} = (1 \text{ s}) \frac{(1.5 \times 10^6 \text{ m}^{-1})}{(600 \text{ m}^{-1})} = 2\ 500 \text{ s} \qquad (42 \text{ min})$$

(d) For a cylindrical cell like *Nitella*, Equation 1.16 gives a tangential stress of rP/t, which is twice as large as the longitudinal stress. For a spherical cell, drawing a section through the center and analyzing as for Figure 1.13 leads to

$$P\pi r^2 = 2\pi r t\sigma \qquad \text{so} \qquad \sigma = \frac{rP}{2t}$$

Here, P/t is the same in all cases, so the maximum stress is $(0.5 \times 10^{-3} \text{ m})$ P/t for *Nitella*, $(2.5 \times 10^{-3} \text{ m})$ P/t for *Valonia*, and $(1.0 \times 10^{-6} \text{ m})$ P/t for *Chlorella*. Therefore, *Valonia* will experience the highest cell wall stress.

CHAPTER 2

2.1 (a) Still 15 mm, because the tilt will not affect the vertical rise, although the length of the water column in the capillary will be greater when tilted.

(b) By Equation 2.2a $(h = 2\sigma \cos \alpha/rpg)$, the product hp is the same in each case, so

$$h_{sucrose} = h_{water} \frac{\rho_{water}}{\rho_{sucrose}} = (15 \text{ mm}) \frac{(998 \text{ kg m}^{-3})}{(1\ 200 \text{ kg m}^{-3})} = 12.5 \text{ mm}$$

(c) We can rearrange Equation 2.2a and solve:

$$\cos \alpha = \frac{hr\rho g}{2\,\sigma} = \frac{(7.5 \times 10^{-3}\ \text{m})(1 \times 10^{-3}\ \text{m})(998\ \text{kg m}^{-3})(9.8\ \text{m s}^{-2})}{(2)(0.0728\ \text{kg s}^{-2})}$$

$$= 0.50$$

$$\alpha = 60°$$

(d) Because all the parameters are the same as in (c), $h = 7.5$ mm.

(e) Because the height of the rise is inversely proportional to the radius (Eq. 2.2a or 2.2b), we can appropriately scale the rise given:

$$h = (15 \times 10^{-3}\ \text{m}) \frac{(1 \times 10^{-3}\ \text{m})}{(1 \times 10^{-6}\ \text{m})} = 15\ \text{m}$$

(f) The capillary wall is not fully wettable in (c) and (d), and surface tension acts over a much smaller circumference in (e). The weight of the fluid supported ($mg = \pi r^2 h \rho$) is the same and greatest in (a) and (b).

2.2 (a) Because the system is at equilibrium, the water potential is the same on the two sides of the barrier. From Equation 2.13a ($\Psi = P - \Pi + \rho_w g h$) and recognizing there is no change in vertical position, we obtain

$$P^A - \Pi^A = P^B - \Pi^B$$

and so

$$\Delta P = P^B - P^A = \Pi^B - \Pi^A$$

Using Equation 2.10 ($\Pi_s \cong RT \sum_j c_j$), we obtain

$$\Pi^A = (2.44 \times 10^3\ \text{m}^3\ \text{Pa mol}^{-1})(0.1 \times 10^3\ \text{mol m}^{-3})$$

$$= 0.244 \times 10^6\ \text{Pa} = 0.244\ \text{MPa}$$

and

$$\Pi^B = 10\ \Pi^A = 2.44\ \text{MPa}$$

$$\Delta P = 2.44\ \text{MPa} - 0.244\ \text{MPa} = 2.20\ \text{MPa, higher on 1 molal side}$$

(b) We cannot use Equation 2.13a, because water is not what moves across the barrier—species j does. We will use Equation 2.4 ($\mu_j = \mu_j^* + RT \ln a_j + \overline{V}_j P$, ignoring electrical and gravitational terms). At equilibrium, μ^A equals μ^B and so we obtain

$$\Delta P = P^B - P^A = \frac{RT}{\overline{V}_j}(\ln a_j^A - \ln a_j^A) = \frac{RT}{\overline{V}_j} \ln \frac{a_j^A}{a_j^B} = \frac{RT}{\overline{V}_j} \ln \frac{c_j^A}{c_j^B}$$

because activity coefficients are unity.

Now insert values:

$$\Delta P = \frac{(2.44 \times 10^3 \text{ m}^3 \text{ Pa mol}^{-1})}{(40 \times 10^{-6} \text{ m}^3 \text{ mol})} \ln \left(\frac{0.1 \text{ molal}}{1 \text{ molal}} \right)$$

$$= -140 \text{ MPa, higher on 0.1 molal side}$$

(c) We must change γ^B_{solute}:

$$P^A - P^B = \frac{RT}{\bar{V}_j} \ln \frac{c^A_j}{(0.5) \ c^B_j} = \frac{(2.44 \times 10^3 \text{ m}^3 \text{ Pa mol}^{-1})}{(40 \times 10^{-6} \text{ m}^3 \text{ mol}^{-1})} \ln \frac{(0.1 \text{ molal})}{(0.5)(1 \text{ molal})}$$

$$= 98 \text{ MPa, higher on 0.1 molal side}$$

(d) Side A would have the same composition as side B ($\Delta c_{\text{solute}} = 0$ molal), so $\Delta P = 0$ MPa.

(e) The value of μ_j is never known in absolute terms, because μ^*_j is arbitrary.

2.3 (a) For an ideal solution, γ_w equals 1. Therefore,

$$a_w = N_w = \frac{\dfrac{(1.00 \text{ kg water})}{(0.018 \text{ kg mol}^{-1} \text{ water})}}{\dfrac{(1.00 \text{ kg water})}{(0.018 \text{ kg mol}^{-1} \text{ water})} + \dfrac{(0.080 \text{ kg sorbitol})}{(0.182 \text{ kg mol}^{-1} \text{ sorbitol})}} = 0.992$$

(b) Rearrange Equation 2.7 and solve:

$$\Pi = -(135.0 \text{ MPa}) (\ln 0.992) = 1.08 \text{ MPa}$$

(c) Again using Equation 2.7, we find

$$a_w = e^{-(1.0 \text{ MPa}/135.0 \text{ MPa})} = 0.9926$$

which is an 0.74% reduction from the value for pure water (1.0000 . . .).

(d) We can rearrange Equation 2.10 and solve:

$$\sum_j c_j = \frac{(1.0 \text{ MPa})}{(2.437 \times 10^3 \text{ m}^3 \text{ Pa mol}^{-1})} = 410 \text{ mol m}^{-3}$$

(e) Equation 2.10 predicts that

$$\Pi_s = (2.437 \times 10^3 \text{ m}^3 \text{ Pa mol}^{-1})(0.25 \text{ mol m}^{-3}) = 6 \times 10^{-4} \text{ MPa}$$

Hence, the polymer must be influencing Π mainly by decreasing γ_w (see Eq. 2.11).

(f) Using Equations 2.7 and 2.13a (with the gravitational term omitted), the water potential of the cell is

$$\Psi = (0.8 \text{ MPa}) - (-135.0 \text{ MPa}) (\ln 0.98) = -1.9 \text{ MPa}$$

2.4 (a) As there are no barriers to the movement of water within the tank, the water potential will be the same at 0.1 m and at the bottom of the tank (-0.600 MPa).

(b) There are no barriers to diffusion, so Π is the same throughout the tank. Rearrange Equation 2.13a and solve for conditions at surface of tank:

$$\Pi = P + \rho_w gh - \Psi = 0 + (\rho_w g)(0) - (-0.600 \text{ MPa}) = 0.600 \text{ MPa}$$

(c) As Ψ equals $-\Pi$ [see (b)], P equals $-\rho_w gh$ and $\rho_w g$ is 0.0098 MPa m^{-1}: at $h = 0$ m, $\rho_w gh = 0$ MPa and $P = 0$ MPa; at $h = -0.1$ m, $\rho_w gh = -0.001$ MPa and $P = 0.001$ MPa; and at $h = -10$ m, $\rho_w gh = -0.098$ MPa and $P = 0.098$ MPa.

(d) At equilibrium Ψ_w equals Ψ_{wv}. Rearrange Equation 2.21 and solve:

$$\% \text{ relative humidity} = (100)(e^{(-0.600 \text{ MPa})/(135.0 \text{ MPa})})\% = 99.6\%$$

2.5 (a) From Equation 2.15, volume (V) is inversely related to external osmotic pressure (Π°). Plot volume versus $1/\Pi^\circ$ and determine that $V = 32$ μm^3 for $\Pi^\circ = 0.4$ MPa.

(b) Intercept in (a) on y axis is the nonaqueous volume per chloroplast (12 μm^3).

(c) Using values from (a) and (b), $(32 \text{ }\mu\text{m}^3 - 12 \text{ }\mu\text{m}^3)/(32 \text{ }\mu\text{m}^3) = 0.63$.

(d) Rearrange Equation 2.15 and solve for n in a chloroplast:

$$n = \frac{(32 \text{ }\mu\text{m}^3 - 12 \text{ }\mu\text{m}^3) (0.4 \text{ MPa})}{(2.437 \times 10^3 \text{ m}^3 \text{ Pa mol}^{-1})} = 3.3 \times 10^{-15} \text{ mol}$$

2.6 (a) Using Equations 2.23 and 2.13a,

$$J_{V_w} = L_w [\Psi^\circ - (P^i - \Pi^i)] = (10^{-12} \text{ m s}^{-1} \text{ Pa}^{-1})[0 - (0.6 \text{ MPa} - 1.0 \text{ MPa})]$$

$$= 4 \times 10^{-7} \text{ m s}^{-1} \quad \text{(flow is inward)}$$

(b) Using Equation 2.28 with $V/A = r/3$,

$$t_e = \frac{(0.5 \times 10^{-3} \text{ m})}{(3)(10^{-12} \text{ m s}^{-1} \text{ Pa}^{-1})(5 \text{ MPa} + 1 \text{ MPa})} = 28 \text{ s}$$

(c) At point of incipient plasmolysis Ψ° equals Ψ^i, so J_{V_w} equals 0 (Eq. 2.23).

(d) At equilibrium Ψ_{wv} equals Ψ°. Using Equation 2.21, $\Psi_{wv} = 135.0$ MPa ln (97/100) $= -4.1$ MPa. Hence, by Equation 2.23:

$$J_{V_w} = (10^{-12} \text{ m s}^{-1} \text{ Pa}^{-1})[-4.1 \text{ MPa} - (-0.4 \text{ MPa})]$$

$$= -3.7 \times 10^{-6} \text{ m s}^{-1} \quad \text{(flow is outward)}$$

(e) Because $\Pi^{\text{cell wall}}$ is 0, $P^{\text{cell wall}}$ equals Ψ^i or -0.4 MPa (Eq. 2.13a). Rearrange Equation 2.22 and solve:

$$\cos \alpha = \frac{-rP}{2\sigma} = -\frac{(10 \times 10^{-9} \text{ m})(-0.4 \text{ MPa})}{(2)(7.28 \times 10^{-8} \text{ MPa m})} = 0.027$$

Hence, $\alpha = 88°$.

CHAPTER 3

3.1 (a) The concentration of charge is Q/V, where the volume of such a thin layer (thickness t) is $4\pi r^2 \times t$, and so

$$\frac{\frac{4}{3}\pi(30 \times 10^{-6} \text{ m})^3 \, (10^{-3} \text{ mol m}^{-3})}{4\pi(30 \times 10^{-6} \text{ m})^2 \, (3 \times 10^{-9} \text{ m})} = 3.3 \text{ mol m}^{-3}$$

(b) Because a sulfate ion has a charge number of -2, the additional negative charge corresponds to

$$\frac{(2 \times 10^7)}{(6.022 \times 10^{23} \text{ mol}^{-1})} = 3.32 \times 10^{-17} \text{ mol}$$

The concentration of additional charge is

$$\frac{(3.32 \times 10^{-17} \text{ mol})}{\frac{4}{3}\pi \, (30 \times 10^{-6} \text{ m})^3} = 0.29 \times 10^{-3} \text{ mol m}^{-3}$$

Hence, C increases by 29%, so ΔE becomes -129 mV by Equation 3.2.

(c) The 10^7 sulfate ions carry charge (Q) equal to

$$Q = (-3.32 \times 10^{-17} \text{ mol}) (9.649 \times 10^4 \text{ C mol}^{-1}) = -3.2 \times 10^{-12} \text{ C}$$

The membrane potential changes from -100 mV to -129 mV during the time that the work is done, so the average difference in electrical potential across the membrane is -115 mV. Hence, the electrical work is about

$$Q \, \Delta E = (-3.2 \times 10^{-12} \text{ C}) (-115 \times 10^{-3} \text{ V}) = 3.7 \times 10^{-13} \text{ J}$$

3.2 (a) At equilibrium and with $\gamma = 1$, Equation 3.5 indicates that

$$E_M = E_{N_K} = \frac{RT}{z_K F} \ln \frac{a_K^o}{a_K^i} = \frac{(25.3 \text{ mV})}{(1)} \ln \frac{(1 \text{ mol m}^{-3})}{(160 \text{ mol m}^{-3})} = -128 \text{ mV}$$

(b) The ionic strength of the solution inside the cell is

$$\tfrac{1}{2}\sum_j c_j z_j^2 = \tfrac{1}{2}[(160 \text{ mol m}^{-3})(+1)^2 + (160 \text{ mol m}^{-3})(-1)^2] = 160 \text{ mol m}^{-3}$$

and that of the external solution is 1 mol m^{-3}. From Equation 3.3,

$$\ln \gamma_\pm^i = \frac{(1.17)(+1)(-1)\sqrt{160 \text{ mol m}^{-3}}}{32 + \sqrt{160 \text{ mol m}^{-3}}} = -0.331$$

and so $\gamma_\pm^i$ equals 0.718, and $\gamma_\pm^o$ is 0.965.

Using Equation 3.5a,

$$E_{N_K} = \frac{(25.3 \text{ mV})}{(1)} \ln \frac{(0.965)}{(0.718)} \frac{(1 \text{ mol m}^{-3})}{(160 \text{ mol m}^{-3})} = -121 \text{ mV}$$

(c) The concentration of ions inside the cell is as follows: $c_K = 160$ mol m^{-3} + 3 mol m^{-3} = 163 mol m^{-3}; $c_{Cl} = 160$ mol m^{-3}; and $c_{ATP} = 1$ mol m^{-3}. The ionic strength therefore is

$$\tfrac{1}{2}[(163 \text{ mol m}^{-3})(+1)^2 + (160 \text{ mol m}^{-3})(-1)^2 + (1 \text{ mol m}^{-3})(-3)^2] = 166 \text{ mol m}^{-3}$$

From Equation 3.3,

$$\ln \gamma_{K-ATP} = \frac{(1.17)(+1)(-3)\sqrt{166}}{32 + \sqrt{166}} = -1.008$$

and so γ_{K-ATP} equals 0.37, and a_{ATP} is $(0.37)(1 \text{ mol m}^{-3}) = 0.37 \text{ mol m}^{-3}$.

3.3 (a) $E_{N_j} = E_M$ when species j is in equilibrium. Using Equation 3.5b,

$$E_{N_K} = \frac{(25.7 \text{ mV})}{(+1)} \ln \frac{(1 \text{ mM})}{(100 \text{ mM})} = -118 \text{ mV} \qquad \text{K}^+ \text{ is in equilibrium}$$

$$E_{N_{Mg}} = \frac{(25.7 \text{ mV})}{(+2)} \ln \frac{(0.1 \text{ mM})}{(10 \text{ mM})} = -59 \text{ mV} \qquad \text{Mg}^{2+} \text{ is not in equilibrium}$$

(b) Note that $E_{N_{Na}} = E_{N_{Ca}} = E_M$ and rearrange Equation 3.5b:

$$a_{Na}^i = \frac{a_{Na}^o}{e^{-z_j E_M/25.7}} = \frac{(0.1 \text{ mM})}{(e^{(-118 \text{ mV})/(25.7 \text{ mV})})} = 10 \text{ mM} = c_{Na}^i \qquad (\gamma_{Na} = 1)$$

$$a_{Ca}^i = (1.0 \text{ mM})e^{(-118)(+2)/(25.7)} = 1.0 \times 10^{-4} \text{ mM} = 0.1 \text{ μM} = c_{Ca}^o \qquad (\gamma_{Ca} = 1)$$

(c) Here E_M equals $E_{N_{Cl}}$ − 177 mV, so $E_{N_{Cl}}$ equals 177 mV − 118 mV or 59 mV. Using Equation 3.5b ($\gamma_{Cl} = 1$, $c_{Cl}^o = 1.0$ mM + 0.1 mM + 0.2 mM = 1.3 mM), we obtain

$$c_{Cl}^i = \frac{(1.3 \text{ mM})}{(e^{(59 \text{ mV})(-1)/(25.7 \text{ mV})})} = 13 \text{ mM}$$

(d) Using Equation 3.24,

$$\frac{J_{Cl}^{in}}{J_{Cl}^{out}} = \frac{c_{Cl}^{o}}{c_{Cl}^{i} \, e^{(-1)(FE_M/RT)}} = \frac{(1.3 \text{ mM})}{(13 \text{ mM}) \, e^{-(-118 \text{ mV})/(25.7 \text{ mV})}} = 0.0010$$

(e) Using Equation 3.25,

$$\mu_{Cl}^{o} - \mu_{Cl}^{i} = (2.479 \times 10^3 \text{ J mol}^{-1})[\ln (0.0010)] = -17 \text{ kJ mol}^{-1}$$

From (a), $E_{N_{Mg}} = -59$ mV and so Equation 3.26a yields

$$\mu_{Mg}^{o} - \mu_{Mg}^{i} = 2F(E_{N_{Mg}} - E_M)$$

$$= (2)(9.65 \times 10^{-2} \text{ kJ mol}^{-1} \text{ mV}^{-1})(-59 \text{ mV} + 118 \text{ mV}) = 11 \text{ kJ mol}^{-1}$$

3.4 (a) Using Equation 3.10b ($u_- = 1.04 \, u_+$),

$$E^{old} - E^{new} = (59.2 \text{ mV}) \frac{(1.04 \, u_K - u_K)}{(1.04 \, u_K + u_K)} \log \left(\frac{1 \text{ mM}}{10 \text{ mM}} \right) = -1 \text{ mV}$$

(b) A Donnan potential that can be calculated using Equation 3.10b ($u_- = 0$):

$$E^{membrane} - E^{solution} = (59.2 \text{ mV}) \frac{-u_+}{+u_+} \log \left(\frac{200 \text{ mM}}{10 \text{ mM}} \right) = -77 \text{ mV}$$

(c) Using the Goldman equation (Eq. 3.19) with $c_{Cl}^{o} = 20$ mM, $c_K^{o} = c_{Na}^{o} = 10$ mM,

$c_{Cl}^{i} = c_K^{i} = 100$ mM, $c_{Na}^{i} = 10$ mM, $P_{Na} = 0.2 \, P_K$, and $P_{Cl} = 0.01 \, P_K$:

$$E_M = (25.7 \text{ mV}) \ln \left[\frac{P_K \, (10 \text{ mM}) + (0.2 \, P_K)(10 \text{ mM}) + (0.01 \, P_K)(100 \text{ mM})}{P_K \, (100 \text{ mM}) + (0.2 \, P_K)(10 \text{ mM}) + (0.01 \, P_K)(20 \text{ mM})} \right] =$$

$$-53 \text{ mV}$$

(d) If P_{Cl}/P_K equals 0, then using the Goldman equation we obtain:

$$E_M = (25.7 \text{ mV}) \ln \left[\frac{(10 \text{ mM}) + (0.2)(10 \text{ mM})}{(100 \text{ mM}) + (0.2)(10 \text{ mM})} \right] = -55 \text{ mV}$$

If $P_{Na}/P_K = P_{Cl}/P_K = 0$, then

$$E_M = (25.7 \text{ mV}) \ln \frac{(10 \text{ mM})}{(100 \text{ mM})} = -59 \text{ mV}$$

3.5 (a) For the steady state for the cell (with no change in anion j in the cellular compartments),

$$J_j^{out} = (10 \text{ nmol m}^{-2} \text{ s}^{-1}) \frac{(50)(4\pi)(2 \times 10^{-6} \text{ m})^2}{(4\pi)(20 \times 10^{-6} \text{m})^2} = 5 \text{ nmol m}^{-2} \text{ s}^{-1}$$

(b) Using Equation 3.25,

$$\mu_j^o - \mu_j^i = (2.479 \times 10^3 \text{ J mol}^{-1}) \ln \frac{(1 \text{ nmol m}^{-2} \text{ s}^{-1})}{(5 \text{ nmol m}^{-2} \text{ s}^{-1})} = -4.0 \text{ kJ mol}^{-1}$$

(c) If the concentration is the same on both sides of the membrane, then E_{N_j} equals 0. Using Equation 3.26b,

$$\mu_j^i - \mu_j^o = (-1)(9.65 \times 10^4 \text{ J mol}^{-1} \text{ V}^{-1})(-118 \text{ mV} - 0 \text{ mV}) = 11.4 \text{ kJ mol}^{-1}$$

Because the chemical potential of the anion is higher inside than outside the cell, the efflux is passive. The influx, however, requires energy of at least 11.4 kJ mol^{-1}.

(d) If one ATP is required per ion transported

$$\text{ATP consumed} = \frac{(0.1 \times 10^{-9} \text{ mol m}^{-2} \text{ s}^{-1})(1 \text{ ATP ion}^{-1})(4\pi)(20 \times 10^{-6} \text{ m})^2}{\dfrac{4\pi}{3}(20 \times 10^{-6} \text{ m})^3}$$

$$= 15 \text{ } \mu\text{mol m}^{-3} \text{ s}^{-1}$$

3.6 (a) Using Equation 3.22 with an energy requirement of 20 kJ mol^{-1} to break one mole of hydrogen bonds (U_{min}), we obtain

$$Q_{10} = \sqrt{\frac{(283 \text{ K} + 10 \text{ K})}{(283 \text{ K})}} \, e^{(10 \text{ K})(20 \text{ kJ mol}^{-1})/[(8.314 \text{ J mol}^{-1} \text{ K}^{-1})(283 \text{ K})(293 \text{ K})]}$$

(b) Here we rearrange Equation 3.22 to obtain

$$U_{min} = \frac{(8.3143 \text{ J mol}^{-1} \text{ K}^{-1})(283 \text{ K})(293 \text{ K})}{(10 \text{ K})} \ln \left(\frac{3.2}{\sqrt{\dfrac{293}{283}}} \right) = 79 \text{ kJ mol}^{-1}$$

Hence, the number of hydrogen bonds per molecule that must be broken to account for a U_{min} of 79 kJ mol^{-1} is approximately 4.

3.7 (a) Because $J_{K \, max}^{in} = J_{Na \, max}^{in}$ and $c_K^o = c_{Na}^o$, we can use Equation 3.27a to obtain

$$\frac{J_K^{in}}{J_{Na}^{in}} = \frac{K_{Na} + c_{Na}^o}{K_K + c_K^o} = \frac{(1.0 \text{ mol m}^{-3}) + (0.01 \text{ mol m}^{-3})}{(0.01 \text{ mol m}^{-3}) + (0.01 \text{ mol m}^{-3})} = 50.5$$

(b) Again using Equation 3.27a, we obtain

$$\frac{J_K^{in}}{J_{Na}^{in}} = \frac{(1.0 \text{ mol m}^{-3}) + (100 \text{ mol m}^{-3})}{(0.01 \text{ mol m}^{-3}) + (100 \text{ mol m}^{-3})} = 1.01$$

(c) Facilitated diffusion requires no energy so ATP is not involved. Using Equation 3.27a, we obtain

$$J_K^{in} = \frac{(10 \text{ nmol m}^{-2} \text{ s}^{-1})(0.1 \text{ mol m}^{-3})}{(0.01 \text{ mol m}^{-3}) + (0.1 \text{ mol m}^{-3})} = 9.1 \text{ nmol m}^{-2} \text{ s}^{-1}$$

3.8 (a) By Equation 3.40 we obtain

$$\sigma^o = \frac{\sum_j \sigma_j \Pi_j^o}{\Pi^o} = \frac{(1.00)(0.2 \text{ MPa}) + (0.30)(0.1 \text{ MPa}) + (0.80)(0.1 \text{ MPa})}{(0.2 \text{ MPa}) + (0.1 \text{ MPa}) + (0.1 \text{ MPa})} = 0.78$$

(b) Because J_V is zero, the stationary state applies, so Equation 3.40 with $\sum_j \sigma_j \Pi_j^i = \sigma_{glycerol} \Pi_{glycerol}^i + \sigma_{rest} \Pi_{rest}^i$ and $\tau^i = 0$ leads to

$$\sigma_{rest}^i = \frac{(0.78)(0.4 \text{ MPa}) - (0.80)(0.2 \text{ MPa}) + (0.5 \text{ MPa})}{(1.0 \text{ MPa})} = 0.65$$

(c) If the membrane is nonselective, then all σ_j's are zero. From Equation 3.39 with $\Delta P = P^o - P^i$, we obtain

$$J_V = L_P(\Delta P - \sum_j \sigma_j \Delta \Pi_j)$$
$$= (10^{-12} \text{ m s}^{-1} \text{ Pa}^{-1})(0.0 \text{ MPa} - 0.5 \text{ MPa}) = -5 \times 10^{-7} \text{ m s}^{-1}$$

(d) If the membrane is impermeable, then all σ_j's equal unity. Using Equation 3.39, we obtain

$$J_V = (10^{-12} \text{ m s}^{-1} \text{ Pa}^{-1})[(0.0 \text{ MPa} - 0.5 \text{ MPa}) - (0.4 \text{ MPa} - 1.2 \text{ MPa})]$$
$$= 3 \times 10^{-7} \text{ m s}^{-1}$$

3.9 (a) Because osmotic pressures are proportional to concentrations and sucrose is impermeant, we can use Equation 3.42:

$$\sigma_{glycine} = \frac{(0.3 \text{ molal})}{(0.4 \text{ molal})} = 0.75$$

(b) At the point of incipient plasmolysis, $\sigma^o \Pi^o$ equals $\sigma^i \Pi^i$. Using Equation 2.10 ($\Pi_s^o = RTc$) for sucrose ($\sigma = 1.00$), the effective osmotic pressure is

$$\sigma^o \Pi^o = (2.437 \times 10^3 \text{ m}^3 \text{ Pa mol}^{-1})(0.3 \times 10^3 \text{ mol m}^{-3}) = 0.73 \times 10^6 \text{ Pa}$$

Upon rearranging Equation 3.44, we note that V is proportional to $1/\Pi^o$. From Problem 2.5, V is 20 μm^3 at $1/\Pi^o$ of 1.0 MPa^{-1}, 28 μm^3 at 2.0 MPa^{-1}, and so at $1/(0.73 \text{ MPa}) = 1.37$ MPa^{-1}, V is 23 μm^3.

(c) The chloroplast volume did not change with the addition of glycine to the external solution, hence $\sigma_{glycine}$ is zero (see Eq. 3.42).

(d) Using Equation 3.42 and noting that osmotic pressures are proportional to concentrations, we obtain

$$\text{actual concentration} = \frac{\text{effective concentration}}{\sigma_{\text{glycine}}}$$

$$= \frac{(0.3 \text{ molal})}{(0.6)} = 0.5 \text{ molal}$$

Glycerol is permeant and so enters the chloroplasts, which leads to water entry. If there were no other solutes inside the chloroplast, the concentration of glycerol inside would eventually equal that outside. However, because there are at least some impermeant solutes inside chloroplasts, Ψ^i will be lower than Ψ^o even for equal concentrations of glycerol inside and outside. Consequently, water (and glycerol) will continue to move in until the chloroplasts burst.

CHAPTER 4

4.1 (a) Using Equation 4.2a ($E_\lambda = hc/\lambda_{\text{vacuum}}$ for a single photon), we obtain

$$E_{400} = \frac{(10^{20})(1.986 \times 10^{-25} \text{ J m})}{(400 \times 10^{-9} \text{ m})} = 50 \text{ J}$$

(b) Using Equation 4.2b, the energy absorbed is

$$E_{1\ 800} = \frac{(1 \text{ mol})(119\ 600 \text{ kJ mol}^{-1} \text{ nm})}{(1\ 800 \text{ nm})} = 66.4 \text{ kJ}$$

Using the appropriate volumetric heat capacity of water, the final temperature will be:

$$T + \Delta T = 0°C + \frac{(66.4 \times 10^3 \text{ J})}{(10^{-3} \text{ m}^3)(4.19 \times 10^6 \text{ J m}^{-3} \text{ °C}^{-1})} = 16°C$$

(c) The maximum photon flux density will occur when all photons have the lowest energy and hence the longest wavelength possible (600 nm). By Equation 4.2b, E_{600} is

$$E_{600} = \frac{(119\ 600 \text{ kJ mol}^{-1} \text{ nm})}{(600 \text{ nm})} = 199 \text{ kJ mol}^{-1}$$

Hence, an energy flux density of 1 W m^{-2} corresponds to

$$\frac{(1 \text{ J m}^{-2} \text{ s}^{-1})}{(199 \times 10^3 \text{ J mol}^{-1})} = 5.0 \times 10^{-6} \text{ mol m}^{-2} \text{ s}^{-1}$$

(d) Cannot be interconverted, unless know sensitivity of photometric device at 600 nm.

4.2 (a) From Equation 4.1 ($\lambda \nu = \upsilon$) or 4.2a, we note that $\lambda_{\text{vacuum}} = c/\nu$, and so

$$\lambda_{\text{vacuum}} = \frac{(3.0 \times 10^8 \text{ m s}^{-1})}{(0.9 \times 10^{15} \text{ s}^{-1})} = 3.33 \times 10^{-7} \text{ m} = 333 \text{ nm}$$

From page 195 and Equation 4.1, v_{air} is about $0.9997c$, so

$$\lambda_{air} = (0.9997)(333 \text{ nm}) = 333 \text{ nm} \qquad \text{(decrease is 0.1 nm)}$$

In the flint glass, we obtain

$$\lambda_{glass} = \frac{(2.0 \times 10^8 \text{ m s}^{-1})}{(0.9 \times 10^{15} \text{ s}^{-1})} = 2.22 \times 10^{-7} \text{ m} = 222 \text{ nm}$$

(b) To go from $S_{(\pi,\pi)}$ to $T_{(\pi,\pi^*)}$ for chlorophyll requires energy equivalent to the absorption of a photon with a wavelength a little longer than 680 nm, so a 333-nm photon would generally have enough energy. However, because a change in spin direction is also required, this transition is extremely improbable.

(c) A transition from π to π^* for a straight-chain hydrocarbon with six double bonds in conjugation can be caused by the absorption of a photon with a wavelength of about 350 nm (p. 234), so the 333 nm photon has enough energy.

(d) Wave number $= \dfrac{(0.9 \times 10^{15} \text{ s}^{-1})}{(3.0 \times 10^8 \text{ m s}^{-1})} = 3.0 \times 10^6 \text{ m}^{-1}$

4.3 (a) The quantum yield for ATP production is equal to the rate of ATP production divided by the rate at which photons are absorbed. If all incident 680 nm photons are absorbed,

$$\text{minimum photon flux density} = \frac{(0.2 \text{ mol m}^{-2} \text{ h}^{-1})}{(0.4)(3\ 600 \text{ s h}^{-1})} = 139 \ \mu\text{mol m}^{-2} \text{ s}^{-1}$$

(b) Using energy/mole from Equation 4.2b times the photon flux density, we obtain

$$\text{energy flux density} = \frac{(0.1196 \text{ J mol}^{-1} \text{ m})}{(680 \times 10^{-9} \text{ m})} (139 \times 10^{-6} \text{ mol m}^{-2} \text{ s}^{-1})$$

$$= 24 \text{ J m}^{-2} \text{ s}^{-1} = 24 \text{ W m}^{-2}$$

(c) ATP is synthesized using energy derived from $S^a_{(\pi,\pi^*)}$, which can be reached by all 680 nm photons but only 95% of the 430 nm photons, which also represent more energy. Hence, the ratio of energy conversion is

$$\text{ratio} = \frac{0.95/(hc/430 \text{ nm})}{1.00/(hc/680 \text{ nm})} = 0.60$$

(d) Using Equation 4.2b, we obtain

$$\lambda_{vacuum} = \frac{(119\ 600 \text{ kJ mol}^{-1} \text{ nm})}{(40 \text{ kJ mol}^{-1})} = 3\ 000 \text{ nm} = 3.0 \ \mu\text{m}$$

4.4 (a) Using Equation 4.13 $(1/\tau = \Sigma_j \ 1/\tau_j)$, we obtain

$$\tau = \frac{1}{\left(\dfrac{1}{10^{-8} \text{ s}} + \dfrac{1}{5 \times 10^{-9} \text{ s}} + \dfrac{1}{10^{-8} \text{ s}} \right)} = 2.5 \times 10^{-9} \text{ s}$$

(b) Electromagnetic radiation comes directly as fluorescence and also can come as phosphorescence from the triplet state. Using Equation 4.15, we obtain

$$\Phi_{em} = \frac{k_{fluor} + k_{phos}}{\sum\limits_{j} k_j} = \left(\frac{1}{\tau_{fluor}} + \frac{1}{\tau_{phos}}\right)\tau$$

$$= \left(\frac{1}{10^{-8}\ s} + \frac{1}{5 \times 10^{-9}\ s}\right)(2.5 \times 10^{-9}\ s) = 0.75$$

(c) By Equation 4.13 we obtain

$$\frac{1}{\tau} = k = k_{transfer} + k_{previous}$$

$$= (10^{12}\ s^{-1}) + \frac{1}{(2.5 \times 10^{-9}\ s)} \cong 10^{12}\ s^{-1}$$

hence $\tau \cong 10^{-12}$ s.

4.5 (a) Rearranging Beer's law (Eq. 4.18), we obtain

$$c_{cis} = \frac{\log J_0/J_b}{\varepsilon_\lambda b} = \frac{\log (1.00/0.35)}{(2.0 \times 10^3\ m^2\ mol^{-1})(10 \times 10^{-3}\ m)} = 23 \times 10^{-3}\ mol\ m^{-3}$$

(b) $A_{450}^{cis} = \log \left(\dfrac{1.00}{0.35}\right) = 0.46$

The absorbance is unaffected by changes in the incident photon flux density.

(c) Assuming that each photon absorbed excites a molecule and using Equation 4.15, we obtain

$$initial\ rate = (0.5)(0.65)(10^{17}\ molecules\ m^{-2}\ s^{-1})$$

$$= 3.3 \times 10^{16}\ molecules\ m^{-2}\ s^{-1}$$

For the cuvette pathlength, this corresponds to

$$\frac{(3.3 \times 10^{16}\ molecules\ m^{-2}\ s^{-1})}{(10 \times 10^{-3}\ m)} = 3.3 \times 10^{18}\ molecules\ m^{-3}\ s^{-1}$$

or

$$\frac{(3.3 \times 10^{18}\ molecules\ m^{-3}\ s^{-1})}{(6.02 \times 10^{23}\ molecules\ mol^{-1})} = 5.5 \times 10^{-6}\ mol\ m^{-3}\ s^{-1}$$

Using Equation 4.18, we obtain

$$\Delta A_{450} = \varepsilon_{450}^{cis}\ \Delta c^{cis} b = (2 \times 10^3\ m^2\ mol^{-1})(5.5 \times 10^{-6}\ mol\ m^{-3}\ s^{-1})(10^{-2}\ m)$$

$$= 1.1 \times 10^{-4}\ s^{-1}$$

(d) The quantum yields for the *cis*-to-*trans* and the *trans*-to-*cis* isomerizations are here the same (0.50), and hence after a long time (a "photostationary state") the relative amounts of the two forms depend inversely on the relative absorption coefficients:

$$cis/trans \text{ ratio} = \frac{(10^3 \text{ m}^2 \text{ mol}^{-1})}{(2.0 \times 10^3 \text{ m}^2 \text{ mol}^{-1})} = 0.5$$

The same conclusion can be deduced by noting that the same number of photons are absorbed by each isomer and using Equation 4.18.

4.6 (a) By Problem 4.2, wave number equals v/c or $1/\lambda_{vacuum}$. Here $1/\lambda$ equals $\dfrac{1}{(500 \text{ nm})} \pm 1.2 \times 10^5 \text{ m}^{-1}$, so

$$\lambda_{below} = \frac{1}{\left[\dfrac{1}{(500 \times 10^{-9} \text{ m})} - (1.2 \times 10^5 \text{ m}^{-1})\right]} = 5.32 \times 10^{-7} \text{ m} = 532 \text{ nm}$$

$$\lambda_{above} = \frac{1}{\left[\dfrac{1}{(500 \times 10^{-9} \text{ m})} + (1.2 \times 10^5 \text{ m}^{-1})\right]} = 4.72 \times 10^{-7} \text{ m} = 472 \text{ nm}$$

(b) The percentage absorbed depends on ε_λ, c, and b (Eq. 4.18). Here, $\varepsilon_{satellite} cb$ equals $\log (J_0/J_b) = \log (1.00/0.80)$, or 0.097. But ε_{main} equals $5\varepsilon_{satellite}$, so $\varepsilon_{main} cb$ equals 0.485, indicating that J_0/J_b equals 3.05, which means that $1 - (J_b/J_0)$ is 0.67, i.e., 67% is absorbed by the main band.

(c) Based on Equation 4.18, we obtain (maximum absorptance occurs for the maximum absorption coefficient):

$$c = \frac{A_\lambda}{\varepsilon_\lambda b} = \frac{(0.3)}{(5 \times 10^3 \text{ m}^2 \text{ mol}^{-1})(5 \times 10^{-3} \text{ m})} = 0.012 \text{ mol m}^{-3}$$

4.7 (a) The energy for each band is calculated using Equation 4.2b ($E_\lambda = Nhc/\lambda_{vacuum}$). The splitting between vibrational sublevels equals the energy difference between adjacent bands:

$$E_{431} - E_{450} = \frac{(119\ 600 \text{ kJ mol}^{-1} \text{ nm})}{(431 \text{ nm})} - \frac{(119\ 600 \text{ kJ mol}^{-1} \text{ nm})}{(450 \text{ nm})} = 12 \text{ kJ mol}^{-1}$$

(b) The temperature dependence of the minor band indicates a transition from an excited sublevel of the ground state. The Boltzmann factor, $e^{-E/RT}$, is 0.007 for an E of 12 kJ mol^{-1} at 20°C, so the transition is only 0.7% as probable as the 450-nm transition; compare the indicated ε_λ's. (Also, the population of the first excited vibrational sublevel of the ground state decreases as the temperature decreases.)

(c) By Equation 4.2b the energy difference between the 450-nm band and the 494-nm band is

$$E_{450} - E_{494} = \frac{(119\ 600 \text{ kJ mol}^{-1} \text{ nm})}{(450 \text{ nm})} - \frac{(119\ 600 \text{ kJ mol}^{-1} \text{ nm})}{(494 \text{ nm})}$$

$$= 24 \text{ kJ mol}^{-1}$$

which is the energy difference between two vibrational sublevels. As emission of fluorescence generally occurs from the lowest vibrational sublevel of the excited state, the 450-nm band apparently represents a transition from the first vibrational sublevel of the ground state (the most populated one according to the Boltzmann factor) to the third vibrational sublevel of the excited state.

(d) Originally the molecule had 11 double bonds in a single conjugated system. Therefore, using Equation 4.2b we obtain

$$E_{max}^{original} = \frac{(119\ 600\ kJ\ mol^{-1}\ nm)}{(450\ nm)} = 266\ kJ\ mol^{-1}$$

The new molecule has five double bonds in each of two conjugated systems. Therefore, Equation 4.2b indicates that

$$\lambda_{max}^{new} = \frac{Nhc}{E_{max}^{original} + (6)(25\ kJ\ mol^{-1})} = \frac{(119\ 600\ kJ\ mol^{-1}\ nm)}{(266\ kJ\ mol^{-1}) + (150\ kJ\ mol^{-1})} = 288\ nm$$

If ε_λ is proportional to the number of double bonds in a conjugated system and the new molecule has two conjugated systems per molecule, then

$$\varepsilon_{288} = \frac{(2)(5)(1.0\ \times\ 10^4\ m^2\ mol^{-1})}{(11)} = 0.91\ \times\ 10^4\ m^2\ mol^{-1}$$

CHAPTER 5

5.1 (a) The volume fraction of the cell occupied by chloroplasts is

$$\frac{(50)\left(\frac{4}{3}\ \pi\right)(2\ \times\ 10^{-6}\ m)^3}{\left(\frac{4}{3}\ \pi\right)(20\ \times\ 10^{-6}\ m)^3} = 0.05$$

(b) Cell volume times density equals the cell mass:

$$\frac{4}{3}\ \pi(20\ \times\ 10^{-6}\ m)^3\ \times\ 1000\ kg\ m^{-3} = 3.35\ \times\ 10^{-11}\ kg\ cell^{-1}$$

of which 10% or $3.35\ \times\ 10^{-12}\ kg\ cell^{-1}$ is dry weight. CO_2 fixation leads to the formation of carbohydrate (Fig. 5.1), which has a mass of 30 g per mol C. Hence, the rate of dry matter production is:

$$[0.1\ mol\ CO_2\ (g\ chlorophyll)^{-1}\ h^{-1}]\ (1\ g\ chlorophyll\ kg^{-1})\ \times$$
$$(3.35\ \times\ 10^{-11}\ kg\ cell^{-1})\ [30\ g\ (mol\ CO_2)^{-1}] = 10.05\ \times\ 10^{-11}\ g\ h^{-1}\ cell^{-1}$$

Hence, the time is $\dfrac{(3.35\ \times\ 10^{-12}\ kg\ cell^{-1})}{(10.05\ \times\ 10^{-14}\ kg\ h^{-1}\ cell^{-1})} = 33\ h$

(c) The molecular weight of Chl a is 893.5 (p. 248). For Chl b, a formyl group ($-$CHO) replaces a methyl group ($-$CH$_3$) of Chl a, so its molecular weight is 907.5. Using a weighted mean, the mean molecular weight of chlorophyll in the cell thus is

$$\frac{(3)(893.5) + (1)(907.5)}{(4)} = 897.0$$

(d) The concentration of chlorophyll is

(1 g chlorophyll kg^{-1}) [1 mol (897 g)$^{-1}$] (1 000 kg m^{-3}) = 1.11 mol chlorophyll m^{-3}

Maximum absorptance occurs along a cell diameter (40 μm), and so using Beer's law ($A_\lambda = \varepsilon_\lambda cb$, Eq. 4.18) we obtain:

$$A_{blue} = (1.21 \times 10^4 \text{ m}^2 \text{ mol}^{-1})(1.11 \text{ mol m}^{-3})(40 \times 10^{-6} \text{ m}) = 0.54$$

$$A_{red} = (0.90 \times 10^4 \text{ m}^2 \text{ mol}^{-1})(1.11 \text{ mol m}^{-3})(40 \times 10^{-6} \text{ m}) = 0.40$$

5.2 (a) The λ_{max} of fluorescence corresponds to a transition from the lowest sublevel of the excited state to some sublevel of the ground state, which is the lowest sublevel for the present case of the shortest λ_{max}. By Equation 4.2b ($E_\lambda = Nhc/\lambda_{vacuum}$), λ_{580} corresponds to an energy of (119 600 kJ mol^{-1} nm)/(580 nm), or 206 kJ mol^{-1}. For the first pigment, fluorescence is maximum at wavelengths corresponding to (3)(10 kJ mol^{-1}) less energy, so

$$\lambda_{max} = \frac{(119\ 600 \text{ kJ mol}^{-1} \text{ nm})}{(206 \text{ kJ mol}^{-1}) - (30 \text{ kJ mol}^{-1})} = 680 \text{ nm}$$

The transition for the pigment with 10 double bonds in conjugation occurs (2)(20 kJ mol^{-1}) lower in energy, so

$$\lambda_{max} = \frac{(119\ 600 \text{ kJ mol}^{-1} \text{ nm})}{(166 \text{ kJ mol}^{-1}) - (30 \text{ kJ mol}^{-1})} = 879 \text{ nm}$$

(b) To pass energy on by resonance transfer, the available energy in the donating molecule (indicated by its fluorescence emission spectrum) must match the energy that can be accepted by the receiving molecule (indicated by its absorption spectrum). Because λ_{max} for fluorescence by the pigment with 8 bonds in conjugation closely matches λ_{max} for absorption by the red band of Chl a, it can readily pass its excitation to Chl a whereas the other molecule cannot.

(c) Blue light excites the Soret band of Chl a, which rapidly becomes de-excited to the lower excited singlet (Fig. 4.5). Therefore, the energy available is given by the red fluorescence of Chl a (λ_{max} = 666 nm, see Fig. 5.3). Consequently, the first compound (λ_{max} = 580 nm) cannot be excited by Chl a. For the pigment with 10 double bonds in conjunction, Equation 4.15 indicates that the absorption maximum occurs at

$$\lambda_{max} = \frac{(119\ 600 \text{ kJ mol}^{-1} \text{ nm})}{(166 \text{ kJ mol}^{-1})} = 720 \text{ nm}$$

Thus, this pigment could become excited by resonance transfer from Chl a.

5.3 (a) Using Equation 4.18 [$A_\lambda = \log(J_0/J_b) = \varepsilon_\lambda cb$], we obtain

$$A_{675} = (0.6 \times 10^4 \text{ m}^2 \text{ mol}^{-1})(2 \times 10^{-6} \text{ m})(20 \text{ mol m}^{-3}) = 0.24$$

The fraction transmitted by the chloroplast (J_b/J_0) is $10^{-0.24}$ or 0.58, so the fraction absorbed is 0.42.

(b) The incident photon flux density is calculated by dividing the radiant energy flux density (40 W m^{-2} or 40 J s^{-1} m^{-2}) by the energy per mole of 675 nm photons, which by Equation 4.2b ($E_\lambda = Nhc/\lambda$) is 177 kJ mol^{-1}. If 0.42 of these photons are absorbed, then the rate at which photons are absorbed by a single chloroplast is

$$(0.42) \frac{(40 \text{ J s}^{-1} \text{ m}^{-2})}{(177 \times 10^3 \text{ J mol}^{-1})} = 95 \times 10^{-6} \text{ mol m}^{-2} \text{ s}^{-1} = 95 \text{ } \mu\text{mol m}^{-2} \text{ s}^{-1}$$

The number of chlorophyll molecules participating in this absorption for the single chloroplast is

$$(2 \times 10^{-6} \text{ m})(20 \text{ mol m}^{-3}) = 40 \times 10^{-6} \text{ mol m}^{-2} = 40 \text{ } \mu\text{mol m}^{-2}$$

(c) Each absorbed photon excites a chlorophyll, so the rate at which chlorophyll molecules are excited is

$$\frac{(95 \text{ } \mu\text{mol photon m}^{-2} \text{ s}^{-1})}{(40 \text{ } \mu\text{mol chlorophyll m}^{-2})} = 2.4 \text{ excitations chlorophyll}^{-1} \text{ s}^{-1}$$

The excitation frequency for a photosynthetic unit (PSU) is (2.4 excitations chlorophyll^{-1} s^{-1}) (250 chlorophylls PSU^{-1}), or 600 excitations s^{-1} PSU^{-1}. Because only 100 excitations s^{-1} PSU^{-1} can be processed, the processable fraction is (100/600), or 0.17.

(d) Because only 0.17 of excitations from the absorbed radiation are processable and 8 photons are required/O$_2$, the O$_2$ evolution rate is:

$$(0.17) \frac{(95 \text{ } \mu\text{mol photons m}^{-2} \text{ s}^{-1})}{(8 \text{ photons/O}_2)} = 2.0 \text{ } \mu\text{mol O}_2 \text{ m}^{-2} \text{ s}^{-1}$$

(e) Because the absorptance of three chloroplasts is three times that of the one chloroplast considered in (a), Equation 4.18 indicates that

$$J_{\text{transmitted}} = J^{\text{incident}} 10^{-0.72} = \frac{(40 \text{ J s}^{-1} \text{ m}^{-2})}{(177 \times 10^3 \text{ J mol}^{-1})} (0.191)$$

$$= 43 \times 10^{-6} \text{ mol photons m}^{-2} \text{ s}^{-1}$$

The fraction absorbed by the single chloroplast is again 0.42, so (0.42)(43 μmol m^{-2} s^{-1}), or 18 μmol photons are absorbed m^{-2} s^{-1}. Thus, the excitation frequency is

$$\frac{(18 \text{ } \mu\text{mol photons m}^{-2} \text{ s}^{-1})}{(40 \text{ } \mu\text{mol chlorophyll m}^{-2})} = 0.45 \text{ excitations chlorophyll}^{-1} \text{ s}^{-1}$$

which corresponds to 113 excitations $(250 \text{ chlorophylls})^{-1} \text{ s}^{-1}$, of which 100 excitations s^{-1} can be processed. Thus, $(100/113)$ or 0.88 (nearly all) of the excitations are processed in the shaded chloroplast, which is a much higher fraction than for the unshaded chloroplast in (c).

5.4 (a) The number of photons required per oxygen evolved is derived from the initial linear part of the curve describing the relationship between incident photon flux density and oxygen evolution:

$$\frac{(10 \times 10^{-6} \text{ mol photon absorbed m}^{-2} \text{ s}^{-1})}{(10^{-4} \text{ mol O}_2 \text{ m}^{-3})(10^{-2} \text{ m})} = 10 \text{ photons absorbed (O}_2 \text{ evolved)}^{-1}$$

(b) During an intense, brief flash of light almost all chlorophyll molecules become excited, although only one excitation can be processed per PSU. If 10 excitations need to be processed before a molecule of oxygen can be evolved [see (a)], then the concentration of PSU's will be equal to 10 times the concentration of oxygen evolved during a flash. Thus, the ratio of chlorophyll to PSU's is

$$\frac{(10 \times 10^{-3} \text{ mol chlorophyll m}^{-3})}{(10 \text{ PSU/O}_2)(5 \times 10^{-6} \text{ mol O}_2 \text{ m}^{-3})} = 200 \text{ chlorophylls PSU}^{-1}$$

(c) Here, 10^{-2} mol chlorophyll m^{-3} corresponds to 10^{-4} mol chlorophyll m^{-2}, which by (b) is 0.50×10^{-6} mol PSU m^{-2}. This processed 10 μmol photons $\text{m}^{-2} \text{ s}^{-1}$, so it could process 20 photons $\text{PSU}^{-1} \text{ s}^{-1}$ (0.05 s per photon).

(d) The uncoupling of ATP formation from electron flow is analogous to disengaging a clutch, which allows a motor to run faster; the O_2 evolution therefore initially speeds up. After a while, the lack of ATP formation causes no CO_2 to be fixed; the electron acceptors therefore stay reduced and electron flow to them is curtailed. Moreover, electron flow may eventually switch over to the pseudocyclic type, which involves no net O_2 evolution.

5.5 (a) The 710-nm light absorbed by Photosystem I leads to cyclic electron flow and accompanying ATP formation. In the idealized example given, 550-nm light by itself leads to excitation of Photosystem II only, and no photophosphorylation occurs.

(b) Accessory (or auxiliary) pigments, presumably carotenoids, which are isoprenoids.

(c) More photons are absorbed at 550 nm $[A_{550} = 1.0$ indicates that $J_b = J_0 \times 10^{-1.0} = 0.10J_0$ by Eq. 4.18 $(A_\lambda = \log J_0/J_b)$, so 90% are absorbed] than at 710 nm $(A_{710} = 0.1$ indicates that $J_b = J_0 \times 10^{-0.1} = 0.79J_0$, so 21% are absorbed). Thus, for every 90 photons absorbed at 550 nm, 21 will be absorbed at 710 nm. Because CO_2 fixation requires an equal number of photons to be absorbed by each of the two photosystems, all 21 of the 710 nm photons can potentially be used (maximum quantum efficiency of 1.00), whereas only 21 of the 90 or 0.23 of the photons at 550 nm can be used.

(d) The minimum photon flux density occurs if each absorbed photon bleaches one P_{700}. Again using Equation 4.18, we obtain

$$A_{700} = 10^{-5} = \varepsilon_{700}cb = (0.8 \times 10^4 \text{ m}^2 \text{ mol}^{-1})(cb)$$

Hence, $cb = \dfrac{A_{700}}{\varepsilon_{700}} = \dfrac{(10^{-5})}{(0.8 \times 10^4 \text{ m}^2 \text{ mol}^{-1})} = 1.3 \times 10^{-9} \text{ mol m}^{-2}$

CHAPTER 6

6.1 (a) By Equation 6.5, we have

$$\Delta G = -17.1 \text{ kJ mol}^{-1} + (2.479 \text{ kJ mol}^{-1}) \ln \frac{(1 \text{ molal})}{(1 \text{ molal})(1 \text{ molal})} = -17.1 \text{ kJ mol}^{-1}$$

Because ΔG for the reaction is negative, the reaction proceeds in the forward direction.

(b) Using Equation 6.5 as in (a), we calculate

$$\Delta G = -17.1 \text{ kJ mol}^{-1} + 2.479 \text{ kJ mol}^{-1} \ln \frac{(10^{-3} \text{ molal})}{(10^{-3} \text{ molal})(10^{-3} \text{ molal})}$$

$$= -17.1 \text{ kJ mol}^{-1} + 17.1 \text{ kJ mol}^{-1} = 0 \text{ kJ mol}^{-1}$$

Hence, the reactants and the product are in equilibrium, and thus the reaction does not proceed in either direction.

(c) As in (a) and (b) we use Equation 6.5 to calculate $\Delta G = 17.1 \text{ kJ mol}^{-1}$, indicating that the reaction proceeds in the backward direction.

(d) Equilibrium is attained when $\Delta G = 0$ [see (b)]. The equilibrium constant therefore is

$$\frac{(10^{-3} \text{ molal})}{(10^{-3} \text{ molal})(10^{-3} \text{ molal})} = 10^3 \text{ molal}^{-1}$$

6.2 (a) Rearranging Equation 6.9, we obtain

$$\ln \frac{(B)}{(B^+)} = \frac{E^*_{B^+ - B} - E_{B^+ - B}}{RT/F} = \frac{(0.118 \text{ V}) - (0.000 \text{ V})}{(0.0257 \text{ V})} = 4.59$$

Hence, $(B^+)/(B) = e^{-4.59} = 0.010$.

(b) Based on the ΔG^*'s, the first reaction has more of a tendency to go to the reduced form (A), meaning that it has a higher midpoint redox potential (e.g., consider beginning with unit activity of all species). Using Equation 6.7 ($\Delta G = -nF\Delta E$), we calculate that $E^{*,\text{H}}_A$ is higher than $E^{*,\text{H}}_B$ by

$$\frac{(8.37 \text{ kJ mol}^{-1}) - (2.93 \text{ kJ mol}^{-1})}{(96.49 \text{ kJ mol}^{-1} \text{ V}^{-1})} = 0.056 \text{ V}$$

and so $E^{*,\text{H}}_A = 0.056 \text{ V} + 0.118 \text{ V} = 0.174 \text{ V}$.

(c) When $(\text{oxidized})_j$ equals $(\text{reduced})_j$, E_j equals $E^{*,\text{H}}_j$. Thus, the difference in electrical potential initially is the difference between the two midpoint redox potentials:

$$\Delta E = E^{*,\text{H}}_A - E^{*,\text{H}}_B = 0.174 \text{ V} - 0.118 \text{ V} = 0.056 \text{ V}$$

Electrons spontaneously flow toward higher redox potentials, which is toward the A–A^+ half cell (couple).

(d) Consider the reaction $A + B^+ \rightleftharpoons A^+ + B$, which is obtained by subtracting the second reaction from the first. For this composite reaction, $\Delta G^* = (8.37 \text{ kJ mol}^{-1}) - (2.93 \text{ kJ mol}^{-1}) = 5.44 \text{ kJ mol}^{-1}$.

By Equation 6.6 ($\Delta G^* = -RT \ln K$),

$$K = e^{-(5.44 \text{ kJ mol}^{-1})/(2.479 \text{ kJ mol}^{-1})} = 0.111$$

At equilibrium, A^+ and B have the same concentration, as do A and B^+.

Thus, $K = \dfrac{(A^+)(B)}{(A)(B^+)} = \dfrac{(A^+)^2}{(A)^2}$, and so $(A^+) = \sqrt{K}(A) = \sqrt{0.111}(A) = 0.33(A)$.

Because all reactants and products were initially 1 molal, we have $(A) + (A^+) = 2$ molal $= 1.33(A)$, and so

$$(A) = (B^+) = 1.50 \text{ molal}$$
$$(A^+) = (B) = 0.50 \text{ molal}$$

(e) If instead of the chemical reaction in (d) we had two half-cell reactions, we could turn the difference in electron free energy into work instead of heat. However, the final concentrations at equilibrium would be the same in the two cases.

6.3 (a) Rearranging the expression for the equilibrium constant for ATP formation (Eq. 6.12), we obtain

$$[\text{ATP}] = (K_{\text{pH } 7})[\text{ADP}][\text{phosphate}] = (5.0 \times 10^{-6} \text{ M}^{-1})(2.0 \times 10^{-3} \text{ M})(5.0 \times 10^{-3} \text{ M})$$
$$= 5.0 \times 10^{-11} \text{ M}$$

(b) Because $\text{ADP} + \text{P}_i \rightleftharpoons \text{ATP} + \text{H}_2\text{O}$ (Eq. 6.11) and there was negligible ATP to begin with [see (a)], a decrease in ADP from 2 mM to 1 mM causes the ATP concentration to become 1 mM. Using Equation 6.13b for the ATP formation energy, we obtain

$$\Delta G = 30 + 5.71 \log \frac{(1.0 \times 10^{-3} \text{ M})}{(1.0 \times 10^{-3} \text{ M})(4.0 \times 10^{-3} \text{ M})} \text{ kJ (mol ATP)}^{-1}$$
$$= 44 \text{ kJ (mol ATP)}^{-1}$$

(c) Using Equation 6.15b, we can calculate the redox potential of the NADP^+–NADPH couple:

$$E_{\text{NADP}^+\text{–NADPH}} = -0.32 - 0.030 \log \left(\frac{0.03}{1.00}\right) \qquad V = -0.274 \text{ V}$$

Therefore, the difference in redox potential between the NADP^+–NADPH couple and ferredoxin is:

$$(-0.274 \text{ V}) - (0.58 \text{ V}) = 0.306 \text{ V}$$

(d) When two moles of electrons are transferred (NADPH can donate two electrons), Equation 6.7 indicates that

$$\Delta G = -(2)(96.5 \text{ kJ mol}^{-1} \text{ V}^{-1})(0.306 \text{ V}) = -59 \text{ kJ per 2 moles of electrons}$$

which is a large enough free energy decrease to form 1 mole of ATP [$\Delta G = 44$ kJ (mol ATP)$^{-1}$ in (b)].

(e) Assuming $E_M = 0$, Equation 6.16c indicates that

$$pH^o - pH^i = \frac{\mu_H^i - \mu_H^o}{5.71} = \frac{(44/2)}{(5.71)} = 3.9$$

when 2 H^+'s are involved and $(44/3)/(5.71) = 2.6$ when 3 H^+'s are involved.

6.4 (a) Using Equation 6.9, we obtain

$$E_b = (0.040 \text{ V}) - \frac{(0.0257 \text{ V})}{(1)} \ln\left(\frac{0.20}{1.00}\right) = 0.081 \text{ V}$$

(b) For Cyt c to be just able to pass electrons to Cyt b [see (a)], E_c has to be equal to E_b (or be slightly lower). Thus, by Equation 6.9 we obtain

$$E_c = 0.220 \text{ V} - \frac{(0.0257 \text{ V})}{(1)} \ln\frac{(1 \text{ mM})}{(Fe^{3+})} = 0.081 \text{ V}$$

Hence,

$$\ln\frac{(1 \text{ mM})}{(Fe^{3+})} = \frac{(0.220 \text{ V} - 0.081 \text{ V})}{(0.0257 \text{ V})} = 5.41$$

$$(Fe^{3+}) = \frac{(1 \text{ mM})}{(e^{5.41})} = \frac{(10^{-3} \text{ M})}{(224)} = 4.5 \times 10^{-6} \text{ M}$$

(c) The ΔE equivalent to a drop in free energy of 40 kJ mol^{-1} is calculated using Equation 6.7:

$$\Delta E = \frac{(-40 \text{ kJ mol}^{-1})}{-(1)(96.49 \text{ kJ mol}^{-1} \text{ V}^{-1})} = 0.415 \text{ V}$$

Thus E_c must be 0.415 V more positive than E_b, or $E_c = 0.081 \text{ V} + 0.415 \text{ V} = 0.496$ V.

(d) Because two H^+'s move per electron and the activity (or pH) term for H^+ is the same on the two sides of the membrane, the redox potential difference is twice the electrical potential difference, or 0.30 V.

(e) Assuming $pH^o = pH^i$, Equation 6.16c indicates that

$$E_M = \frac{\mu_H^i - \mu_H^o}{96.5} = \frac{(-40/2)}{(96.5)} = -0.21 \text{ V}$$

(f) When 3 H^+'s are required, the minimum E_M is $(-40/3)/(96.5) = -0.14$ V. When 30% of the energy is lost, E_M is $(-0.14 \text{ V})/(0.70) = -0.20$ V.

CHAPTER 7

7.1 (a) We know that absorbed IR is here equal to emitted IR for the leaf, and therefore we can set Equation 7.4 equal to Equation 7.5 and solve for T^{surr}:

$$a_{IR}\sigma\,[(T^{\text{surr}})^4 + (T^{\text{sky}})^4] = 2\,e_{IR}\sigma(T^{\text{leaf}})^4$$

$$T^{\text{surr}} = \left[\frac{2\,e_{IR}\sigma(T^{\text{leaf}})^4}{a_{IR}\sigma} - (T^{\text{sky}})^4\right]^{1/4}$$

$$= [2(T^{\text{leaf}})^4 - (T^{\text{sky}})^4]^{1/4}$$

because a_{IR} here equals e_{IR}. For the clear sky we therefore have

$$T^{\text{surr}} = [2(283 \text{ K})^4 - (233 \text{ K})^4]^{1/4} = 315 \text{ K} = 42°C$$

and for a cloudy sky, we obtain

$$T^{\text{surr}} = [2(283 \text{ K})^4 - (275 \text{ K})^4]^{1/4} = 290 \text{ K} = 17°C$$

(b) For emitted radiant energy, Wien's displacement law (Eq. 4.3) becomes $\lambda_{\max}T = 2.9 \times 10^6$ nm K. Therefore, for the leaf,

$$\lambda_{\max} = \frac{(2.9 \times 10^6 \text{ nm K})}{(283 \text{ K})} = 10\ 200 \text{ nm} = 10.2 \text{ } \mu m$$

Emitted radiant energy for the surroundings with the clear sky has

$$\lambda_{\max} = \frac{(2.9 \times 10^6 \text{ nm K})}{(315 \text{ K})} = 9\ 200 \text{ nm} = 9.2 \text{ } \mu m$$

Emitted radiant energy for the surroundings with the cloudy sky has

$$\lambda_{\max} = \frac{(2.9 \times 10^6 \text{ nm K})}{(290 \text{ K})} = 10\ 000 \text{ nm} = 10.0 \text{ } \mu m$$

(c) The amount of absorbed radiation is equal to the amount of absorbed solar radiation plus absorbed infrared radiation. Using Equations 7.3 [$a(1 + r)\,S$] and 7.4, total absorbed radiation equals:

absorbed shortwave and longwave $= (0.6)\,(1 + 0.10)\,(700 \text{ W m}^{-2})$

$$+ (0.96)(5.67 \times 10^{-8} \text{ W m}^{-2} \text{ K}^4) [(282 \text{ K})^4 + (233 \text{ K})^4]$$

$$= 462 \text{ W m}^{-2} + 505 \text{ W m}^{-2} = 967 \text{ W m}^{-2}$$

(d) The thermal (i.e., longwave or IR) radiation emitted is calculated using Equation 7.5:

$$\text{emitted IR} = (2)(0.96)(5.67 \times 10^{-8} \text{ W m}^{-2} \text{ K}^{-4})(283 \text{ K})^4 = 698 \text{ W m}^{-2}$$

Thus, 100 (698 W m^{-2})/(967 W m^{-2}) or 72% of the absorbed radiation calculated in (c) is emitted as thermal radiation.

(e) Using Equation 7.3, the solar (i.e., shortwave) radiation absorbed is

$$(0.60)(1 + 0.15)(250 \text{ W m}^{-2}) = 173 \text{ W m}^{-2}$$

and by Equation 7.4 the absorbed infrared radiation is

$$(0.96)(5.67 \times 10^{-8} \text{ W m}^{-2} \text{ K}^{-4}) [(282 \text{ K})^4 + (275 \text{ K})^4] = 656 \text{ W m}^{-2}$$

Therefore, the total absorbed radiation is 173 W m^{-2} + 656 W m^{-2}, or 829 W m^{-2}, and the outgoing radiation is 698 W m^{-2} from (d), so the net radiation is:

$$829 \text{ W m}^{-2} - 698 \text{ W m}^{-2} = 131 \text{ W m}^{-2}$$

7.2 (a) The average value of a function is $\dfrac{1}{b-a}\displaystyle\int_a^b f(x)\, dx$, i.e., the area under the function divided by the distance between the two endpoints of interest. For a circle, we can apply similar reasoning, thus

$$\bar{l} = \frac{A}{d} = \frac{\pi r^2}{2r} = \frac{\pi r}{2} = \frac{\pi (0.06 \text{ m})}{2} = 0.094 \text{ m}$$

(b) Using Equation 7.8, we have

$$\delta^{bl}_{(mm)} = 4.0 \sqrt{\frac{l_{(m)}}{v_{(m\ s^{-1})}}} = 4.0 \sqrt{\frac{(0.094 \text{ m})}{(0.80 \text{ m s}^{-1})}} = 1.37 \text{ mm}$$

Assuming that the mean distance across a leaf is the diameter, we have

$$\delta^{bl}_{(mm)} = 4.0 \sqrt{\frac{(0.12 \text{ m})}{(0.80 \text{ m s}^{-1})}} = 1.55 \text{ mm}$$

which is quite similar to the value determined using mean length.

(c) Using Equation 7.11, the heat flux density is:

$$J^C_H = \frac{2 K^{air} (T^{leaf} - T^{ta})}{\delta^{bl}} = \frac{(2)(0.0259 \text{ W m}^{-1} \ ^\circ\text{C}^{-1})(25^\circ\text{C} - 20^\circ\text{C})}{(1.37 \times 10^{-3} \text{ m})}$$

$$= 189 \text{ W m}^{-2}$$

(d) From (c) we know that conductive/convective heat losses are 189 W m^{-2} and we also know that the total heat load from net radiation is 300 W m^{-2}. Therefore, the amount of energy that must be dissipated as latent heat through transpiration is:

$$300 \text{ W m}^{-2} - 189 \text{ W m}^{-2} = 111 \text{ W m}^{-2}$$

Rearranging Equation 7.19 we thus find:

$$J_{wv} = \frac{J_H^T}{H_{vap}} = \frac{(111 \text{ J s}^{-1} \text{ m}^{-2})}{(44.0 \times 10^3 \text{ J mol}^{-1})} = 2.5 \times 10^{-3} \text{ mol m}^{-2} \text{ s}^{-1}$$

7.3 (a) Using the equation for net radiation balance in the soil (p. 380), we find that net radiation equals:

$$\begin{aligned} aS + a_{IR}\sigma(T^{surr})^4 - \varepsilon_{IR}\sigma(T^{soil})^4 &= 100 \text{ W m}^{-2} + (1)(5.67 \times 10^{-8} \text{ W m}^{-2} \text{ K}^{-4})(295 \text{ K})^4 \\ &\quad - (1)(5.67 \times 10^{-8} \text{ W m}^{-2} \text{ K}^4)(293 \text{ K})^4 \\ &= 100 \text{ W m}^{-2} + 429 \text{ W m}^{-2} - 418 \text{ W m}^{-2} \\ &= 111 \text{ W m}^{-2} \end{aligned}$$

(b) Modifying Equation 7.21, we find heat flux density down each stem to be

$$J_H^C = \frac{K^{water} (T^{veg} - T^{soil})}{\Delta z}$$

Because flux density is a rate per unit area, multiplying by the stem area gives the rate of heat conduction down a stem:

$$A^{stem} J_H^C = \frac{A^{stem} K^{water} (T^{veg} - T^{soil})}{\Delta z} = \frac{\pi(0.015 \text{ m})^2 (0.60 \text{ W m}^{-1}°\text{C}^{-1})(22°\text{C} - 20°\text{C})}{(0.8 \text{ m})}$$

$$= 0.00106 \text{ W} = 1.06 \text{ mW}$$

The average value of J_H^C per m^2 of ground is (1.06 mW plant^{-1}) (4 plants m^{-2} of ground) $= 4.2$ mW m^{-2} of ground.

(c) Modifying Equation 7.11 by removing the factor 2 because the soil conducts to the air only from one surface, we find

$$J_H^C = \frac{K^{air} (T^{soil} - T^{air})}{\delta^{bl}} = \frac{(0.0257 \text{ W m}^{-1} °\text{C}^{-1})(20°\text{C} - 21°\text{C})}{(4 \times 10^{-3} \text{ m})} = -6 \text{ W m}^{-2}$$

(d) Using Equation 7.19, latent heat loss is

$$J_H^T = J_{wv}H_{vap} = (0.3 \times 10^{-3} \text{ mol m}^{-2} \text{ s}^{-1})(44.2 \times 10^3 \text{ J mol}^{-1}) = 13 \text{ W m}^{-2}$$

The flux density of heat conducted into the soil can be calculated from the overall energy balance (see Eq. 7.2):

$$\begin{aligned} J_H^C &= \text{(net radiation)} - \text{(conduction/convection)} - \text{(latent heat)} \\ &= 111 \text{ W m}^{-2} - (-6 \text{ W m}^{-2}) - 13 \text{ W m}^{-2} = 104 \text{ W m}^{-2} \end{aligned}$$

(e) Rearranging Equation 7.21, we find

$$\frac{\partial T}{\partial z} = \frac{J_H^C}{-K^{soil}} = -\frac{(104 \text{ W m}^{-2})}{(0.60 \text{ W m}^{-1} \text{°C}^{-1})} = -173 \text{ °C m}^{-1}$$

7.4 (a) Using Equation 7.10 for a sphere, we obtain

$$\delta_{(mm)}^{bl} = 2.8 \sqrt{\frac{d_{(m)}}{v_{(m \text{ s}^{-1})}}} + \frac{0.25}{v_{(m \text{ s}^{-1})}} = 2.8 \sqrt{\frac{(0.2 \text{ m})}{(1.0 \text{ m s}^{-1})}} + \frac{(0.25)}{(1.0 \text{ m s}^{-1})} = 1.50 \text{ mm}$$

The amount of heat conducted across the boundary layer of a sphere can be calculated using Equation 7.13:

$$J_H^C = \frac{(r + \delta^{bl}) K^{air} (T^{surf} - T^{ta})}{r \delta^{bl}}$$

$$= \frac{(0.1 \text{ m} + 0.0015 \text{ m})(0.0259 \text{ W m}^{-1} \text{°C}^{-1})(25\text{°C} - 20\text{°C})}{(0.1 \text{ m})(0.0015 \text{ m})} = 88 \text{ W m}^{-2}$$

(b) Rearranging Equation 7.14, we find

$$h_c = \frac{J_H^C}{T^{surf} - T^{ta}} = \frac{(88 \text{ W m}^{-2})}{(25\text{°C} - 20\text{°C})} = 18 \text{ W m}^{-2} \text{°C}^{-1}$$

(c) Using Equation 7.9 for a cylinder, we obtain:

$$\delta_{(mm)}^{bl} = 5.8 \sqrt{\frac{d_{(m)}}{v_{(m \text{ s}^{-1})}}} = 5.8 \sqrt{\frac{(1.2 \times 10^{-3} \text{ m})}{(1.0 \text{ m s}^{-1})}} = 0.20 \text{ mm}$$

(d) Given that net radiation is due entirely to absorbed shortwave radiation and that heat loss does not occur through evaporation or conduction to the stem, Equation 7.2 indicates that absorbed solar irradiation must be balanced by convective heat loss (Eq. 7.12):

$$a\bar{S} = J_H^C = \frac{K^{air} (T^{spine} - T^{ta})}{r \ln\left(\dfrac{r + \delta^{bl}}{r}\right)}$$

The mean solar flux density on the spine, $\bar{S}$, is less than the maximum S:

$$\bar{S} = \frac{\text{projected area}}{\text{actual area}} S = \frac{dl}{\pi dl} S = \frac{S}{\pi} = \frac{100 \text{ W m}^{-2}}{\pi} = 32 \text{ W m}^{-2}$$

Therefore, we calculate

$$T^{spine} = \frac{(0.7)(32 \text{ W m}^{-2})(0.6 \times 10^{-3} \text{ m}) \ln\left(\dfrac{0.6 \text{ mm} + 0.2 \text{ mm}}{0.6 \text{ mm}}\right)}{(0.0257 \text{ W m}^{-1} \text{°C}^{-1})} + 20.0\text{°C} = 20.2\text{°C}$$

(e) In the absence of spines, IR is absorbed from one source (surroundings) and emitted from one source (cactus surface). Combining Equations 7.4 and 7.5, we obtain

net IR $= a_{IR}\sigma(T^{surr})^4 - e_{IR}\sigma(T^{surf})^4$

$= (0.97)(5.67 \times 10^{-8}$ W m^{-2} K$^{-4})[(253$ K$)^4 - (298$ K$)^4] = -208$ W m^{-2}

In the presence of spines, IR is absorbed from two sources (surroundings and spines) and emitted from one source (cactus surface). Therefore,

net IR $= a_{IR}\sigma[(T^{surr})^4 + (T^{spine})^4](\frac{1}{2}) - e_{IR}\ \sigma\ (T^{surf})^4$

$= (0.97)(5.67 \times 10^{-8}$ W m^{-2} K$^{-4})\ [\frac{1}{2}(253$ K$)^4 + \frac{1}{2}(293$ K$)^4 - (298$ K$)^4]$

$= -118$ W m^{-2}

(f) Because the plant is spherical, the total solar radiation absorbed equals its projected area (πr^2) times the solar flux density perpendicular to the solar beam (1 000 W m^{-2}) times the absorbance (0.30). To obtain the average solar radiation absorbed over the plant surface, we divide by the plant's surface area $(4\pi r^2)$. Thus,

$$a\bar{S} = \frac{\pi r^2(1\ 000\ \text{W m}^{-2})(0.30)}{4\pi r^2} = 75\ \text{W m}^{-2}$$

Using answers from (e) and (a), the average net energy balance is

absorbed solar + net IR − net heat convection
$= 75$ W m^{-2} − 118 W m^{-2} − 88 W m^{-2} = −131 W m^{-2}

Considering the whole plant surface area, the net energy balance equals $(-131$ W m$^{-2})(4\pi)(0.1$ m$)^2 = -16.5$ W. Given that the volumetric heat capacity is $(0.80)(4.175$ MJ m^{-3} °C$^{-1})$ $= 3.34$ MJ m^{-3} °C^{-1}, the change in tissue temperature can be calculated (Eq. 7.23):

$$\frac{\Delta T}{\Delta t} = \frac{\text{heat storage rate}}{C_P V} = \frac{(16.5\ \text{J s}^{-1})}{(3.34 \times 10^6\ \text{J m}^{-3}\ °\text{C}^{-1})(4/3)(\pi)(0.1\ \text{m})^3}$$

$= -0.00118$ °C s^{-1} = −4.2°C h^{-1}

CHAPTER 8

8.1 (a) Using Equation 8.3 we find

$$g_{wv}^{bl} = \frac{D_{wv}}{\delta^{bl}} = \frac{(2.42 \times 10^{-5}\ \text{m}^2\ \text{s}^{-1})}{(0.8 \times 10^{-3}\ \text{m})} = 0.030\ \text{m s}^{-1} = 30\ \text{mm s}^{-1}$$

If we rearrange Equation 8.3,

$$r_{wv}^{bl} = \frac{1}{g_{wv}^{bl}} = \frac{1}{(0.030 \text{ m s}^{-1})} = 33 \text{ s m}^{-1}$$

(b) The fraction of the leaf surface area occupied by stomatal pores is

$$na^{st} = (64 \text{ stomata mm}^{-2})(6 \times 10^{-3} \text{ mm})(20 \times 10^{-3} \text{ mm}) = 0.0077$$

The effective radius of a stomate (see p. 404) is

$$r^{st} = \left(\frac{a^{st}}{\pi}\right)^{1/2} = \left(\frac{(6 \text{ μm})(20 \text{ μm})}{\pi}\right)^{1/2} = 6 \text{ μm}$$

(c) The average flux density within an individual stomatal pore will be greater than that in the boundary layer by the ratio of the areas available for gas diffusion:

$$J_{wv}^{st} = \frac{A}{A^{st}} J_{wv}^{bl} = \frac{1}{(0.0077)} J_{wv}^{bl} = 130 J_{wv}^{bl}$$

(d) Using Equation 8.5 we find

$$g_{wv}^{st} = \frac{D_{wv} na^{st}}{\delta^{st} + r^{st}} = \frac{(2.42 \times 10^{-5} \text{ m}^2 \text{ s}^{-1})(0.0077)}{(25 \times 10^{-6} \text{ m}) + (6 \times 10^{-6} \text{ m})}$$

$$= 0.0060 \text{ m s}^{-1} = 6.0 \text{ mm s}^{-1}$$

Considering the discussion following Equation 8.8, we find

$$g_{wv}^{\prime st} = g_{wv}^{st} \frac{P}{RT} = (6.0 \times 10^{-3} \text{ m s}^{-1})(41.6 \text{ mol m}^{-3})$$

$$= 0.25 \text{ mol m}^{-2} \text{ s}^{-1} = 250 \text{ mmol m}^{-2} \text{ s}^{-1}$$

If the air pressure is reduced to 0.9 atmosphere, then D_{wv} increases by (1.0 atm/0.9 atm) (see Eq. 8.9); therefore g_{wv}^{st} increases by that amount (see Eq. 8.5):

$$g_{gv}^{st, \, 0.9 \text{ atm}} = \left(\frac{1 \text{ atm}}{0.9 \text{ atm}}\right)(6.0 \text{ mm s}^{-1}) = 6.7 \text{ mm s}^{-1}$$

Because $g_{wv}^{\prime st}$ is independent of pressure (p. 410), it remains at 250 mmol m^{-2} s^{-1}.

(e) Using Equation 8.6 we find

$$g_{wv}^{ias} = \frac{D_{wv}}{\delta^{ias}} = \frac{(2.42 \times 10^{-5} \text{ m}^2 \text{ s}^{-1})}{(0.5 \times 10^{-3} \text{ m})} = 0.048 \text{ m s}^{-1} = 48 \text{ mm s}^{-1}$$

For the other conductance, we have

$$g_{wv}^{\prime ias} = g_{wv}^{ias} \frac{P}{RT} = (48 \times 10^{-3} \text{ m s}^{-1})(41.6 \text{ mol m}^{-3}) = 2.0 \text{ mol m}^{-2} \text{ s}^{-1}$$

(f) Because the cavity is cylindrical, the effective radius equals the actual radius. Using Equation 8.5 (see also Eq. 8.14), we thus obtain

$$r_{wv}^{cavity} = \frac{\delta^{cav} + r^{cav}}{D_j na^{cav}} = \frac{(100 \times 10^{-6} \text{ m}) + (25 \times 10^{-6} \text{ m})}{(2.42 \times 10^{-5} \text{ m}^2 \text{ s}^{-1})(64 \times 10^{6} \text{ m}^{-2})(\pi)(25 \times 10^{-6} \text{ m})^2}$$

$$= 41 \text{ s m}^{-1}$$

8.2 (a) Using Equation 8.12, we can calculate the water vapor conductance of the leaf:

$$g_{wv}^{leaf} = \frac{g_{wv}^{ias} g_{wv}^{st}}{g_{wv}^{ias} + g_{wv}^{st}} + g_{wv}^{c} = \frac{(40 \text{ mm s}^{-1})(6 \text{ mm s}^{-1})}{(40 \text{ mm s}^{-1}) + (6 \text{ mm s}^{-1})} + (0.1 \text{ mm s}^{-1})$$

$$= 5.3 \text{ mm s}^{-1}$$

Then using Equation 8.13, we can incorporate the conductance through the boundary layer with that of the leaf to find the total water vapor conductance:

$$g_{wv}^{total} = \frac{(5.3 \text{ mm s}^{-1})(20 \text{ mm s}^{-1})}{(5.3 \text{ mm s}^{-1}) + (20 \text{ mm s}^{-1})} = 4.2 \text{ mm s}^{-1}$$

(b) If the cuticular pathway is ignored, we have three conductances in series:

$$g_{wv}^{total} = \frac{1}{\dfrac{1}{(40 \text{ mm s}^{-1})} + \dfrac{1}{(6 \text{ mm s}^{-1})} + \dfrac{1}{(20 \text{ mm s}^{-1})}} = 4.1 \text{ mm s}^{-1}$$

If the intercellular air spaces are ignored, stomatal and cuticular conductances are in parallel ($g_{wv}^{leaf} = g_{wv}^{st} + g_{wv}^{c}$) and in series with the boundary layer conductance:

$$g_{wv}^{total} = \frac{1}{\dfrac{1}{(6 \text{ mm s}^{-1}) + (0.1 \text{ mm s}^{-1})} + \dfrac{1}{(20 \text{ mm s}^{-1})}} = 4.7 \text{ mm s}^{-1}$$

If both the cuticle and the intercellular air spaces are ignored, then the stomatal and boundary layer conductances are in series:

$$g_{wv}^{total} = \frac{1}{\dfrac{1}{(6 \text{ mm s}^{-1})} + \dfrac{1}{(20 \text{ mm s}^{-1})}} = 4.6 \text{ mm s}^{-1}$$

(c) The conductance of the upper surface here equals the conductance of the lower surface [determined in (a)] and is in parallel with it:

$$g_{wv}^{total} = g_{wv}^{upper} + g_{wv}^{lower} = 2 \, g_{wv}^{lower} = (2)(4.2 \text{ mm s}^{-1}) = 8.4 \text{ mm s}^{-1}$$

(d) Here 0.72 of g_{wv}^{total} comes from the lower surface, whose conductance is determined in (a), so

$$g_{wv}^{total} = \frac{g_{wv}^l}{(0.72)} = \frac{(4.2 \text{ mm s}^{-1})}{(0.72)} = 5.8 \text{ mm s}^{-1}$$

(e) Using Equation 8.16 and $c_{wv}^{leaf} = (RH^{leaf}/100)c_{wv}^{*, \, leaf}$, we find

$$J_{wv}^{lower} = g_{wv}^{total} \, \Delta c_{wv}^{total} = g_{wv}^{total} \, (c_{wv}^{leaf} - c_{wv}^{ta})$$

$$= (4.2 \text{ mm s}^{-1} \, [(0.99)(30.40 \text{ g m}^{-3}) - (7.5 \text{ g m}^{-3})] = 95 \text{ mg m}^{-2} \text{ s}^{-1}$$

(f) Considering the discussion following Equation 8.8, we obtain:

$$g_{wv}^{\prime total} = g_{wv}^{total} \, \frac{P}{RT} = (4.2 \times 10^{-3} \text{ m s}^{-1}) \frac{(1 \text{ atm})}{(0.08205 \text{ litre atm mol}^{-1} \text{ K}^{-1})(303 \text{ K})} \left(\frac{1\,000 \text{ litre}}{\text{m}^3} \right)$$

$$= 0.169 \text{ mol m}^{-2} \text{ s}^{-1} = 169 \text{ mmol m}^{-2} \text{ s}^{-1}$$

Equation 8.16 was modified as described in the text (p. 417), so that

$$J_{wv} = g_{wv}^{\prime total} \, \Delta N_{wv}^{total} = g_{wv}^{\prime total} (N_{wv}^{leaf} - N_{wv}^{ta})$$

$$= (169 \text{ mmol m}^{-2} \text{ s}^{-1})[(0.99)(0.04190) - (0.0103)] = 5.3 \text{ mmol m}^{-2} \text{ s}^{-1}$$

(g) Rearranging Equation 8.16, the drop in water vapor concentration is

$$\Delta c_{wv}^{st} = \frac{J_{wv}}{g_{wv}^{st}} = \frac{(0.095 \text{ g m}^{-2} \text{ s}^{-1})}{(6.0 \times 10^{-3} \text{ m s}^{-1})} = 15.8 \text{ g m}^{-3}$$

Similarly, rearranging Equation 8.16 for the mole fraction case we obtain:

$$\Delta N_{wv}^{st} = \frac{J_{wv}}{g_{wv}^{\prime st}} = \frac{J_{wv}}{g_{wv}^{st} \dfrac{P}{RT}} = \frac{(5.3 \times 10^{-3} \text{ mol m}^{-2} \text{ s}^{-1})}{(6.0 \times 10^{-3} \text{ m s}^{-1})(40.2 \text{ mol m}^{-3})} = 0.022$$

8.3 (a) We will calculate the total surface area of the palisade mesophyll cells (rectangular parallelepipeds) and spongy mesophyll cells (spheres) and express it per unit area of the leaf surface:

$$\frac{A^{mes}}{A} = \frac{[(4)(100 \text{ μm})(30 \text{ μm}) + (2)(30 \text{ μm})(30 \text{ μm})] + (2)(4\pi)(15 \text{ μm})^2}{(30 \text{ μm})(30 \text{ μm})} = 22$$

(b) The value for the sun leaf is

$$\frac{A^{mes}}{A} = \frac{(2)[(4)(100 \text{ μm})(30 \text{ μm}) + (2)(30 \text{ μm})(30 \text{ μm})] + (4\pi)(15 \text{ μm})^2}{(30 \text{ μm})(30 \text{ μm})} = 34$$

(c) Using Equation 8.19, the maximum $r_{CO_2}^{cw}$ will be achieved when $K_{CO_2}^{cw}$ is minimal, namely 0.91 (i.e., no HCO_3^- or H_2CO_3; see Table 8.3):

$$r_{CO_2}^{cw} = \frac{A \, \Delta x^{cw}}{A^{mes} \, D_{CO_2}^{cw} \, K_{CO_2}^{cw}} = \frac{(1)(0.2 \times 10^{-6} \text{ m})}{(22)(5.0 \times 10^{-10} \text{ m}^2 \text{ s}^{-1})(0.91)} = 20 \text{ s m}^{-1}$$

(d) Using Equation 8.19, we find

$$r_{CO_2}^{pl} = r_{CO_2}^{clm} = \frac{A}{A^{mes} \, P_{CO_2}} = \frac{1}{(22)(1.0 \times 10^{-3} \text{ m s}^{-1})} = 45 \text{ s m}^{-1}$$

(e) Using Equation 8.19, we find

$$r_{CO_2}^{cyt} = \frac{A \, \Delta x^{cyt}}{A^{mes} \, D_{CO_2}^{cyt} K_{CO_2}^{cyt}} = \frac{(1)(0.1 \times 10^{-6} \text{ m})}{(22)(1.0 \times 10^{-9} \text{ m}^2 \text{ s}^{-1})(1)} = 5 \text{ s m}^{-1}$$

$$r_{CO_2}^{stroma} = \frac{A \, \Delta x^{stroma}}{A^{mes} \, D_{CO_2}^{stroma} K_{CO_2}^{stroma}} = \frac{(1)(0.2 \times 10^{-6} \text{ m})}{(22)(1.0 \times 10^{-9} \text{ m}^2 \text{ s}^{-1})(1)} = 23 \text{ s m}^{-1}$$

(f) Using Equations 8.18a and 8.18b, the total liquid-phase resistance to CO_2 diffusion for the shade leaf is:

$$r_{CO_2}^{liquid} = r_{CO_2}^{cw} + r_{CO_2}^{pl} + r_{CO_2}^{cyt} + r_{CO_2}^{clm} + r_{CO_2}^{stroma}$$

$$= 20 \text{ s m}^{-1} + 45 \text{ s m}^{-1} + 5 \text{ s m}^{-1} + 45 \text{ s m}^{-1} + 23 \text{ s m}^{-1}$$

$$= 138 \text{ s m}^{-1}$$

The total liquid-phase resistance to CO_2 diffusion for the sun leaf is smaller than that of the shade leaf because the sun leaf has a higher A^{mes}/A:

$$r_{CO_2}^{liquid} = \left(\frac{22}{34}\right)(138 \text{ s m}^{-1}) = 89 \text{ s m}^{-1}$$

8.4 (a) Rearranging Equation 8.22, we find

$$V_{max} = \frac{v_{CO_2} \, (K_{CO_2} + c_{CO_2}^{chl})}{c_{CO_2}^{chl}} = \frac{(4 \text{ mol m}^{-3} \text{ s}^{-1})(5 \text{ }\mu M + 9 \text{ }\mu M)}{(9 \text{ }\mu M)} = 6.2 \text{ mol m}^{-3} \text{ s}^{-1}$$

(b) Again using Equation 8.22, we have

$$0.9 \, V_{max} = \frac{V_{max} c_{CO_2}^{chl}}{5 \text{ }\mu M + c_{CO_2}^{chl}} \qquad \text{or} \qquad 4.5 \text{ }\mu M + 0.9 \, c_{CO_2}^{chl} = c_{CO_2}^{chl}$$

$$c_{CO_2}^{chl} = \frac{4.5 \text{ }\mu M}{0.1} = 45 \text{ }\mu M$$

(c) Utilizing Equation 8.26 with the flux densities expressed as on p. 449, we find that the total resistance to CO_2 diffusion is:

$$r_{CO_2}^{total} = r_{CO_2}^{bl_l} + r_{CO_2}^{leaf_l} + r_{CO_2}^{mes} + \left(\dfrac{1}{1 - \dfrac{J_{CO_2}^{r+pr}}{J_{CO_2}^{ps}}}\right) r^{chl}$$

$$= 60 \text{ s m}^{-1} + 250 \text{ s m}^{-1} + 150 \text{ s m}^{-1} + \left(\dfrac{1}{1 - 0.45}\right)(100 \text{ s m}^{-1})$$

$$= 642 \text{ s m}^{-1}$$

Using Equation 8.26, we find

$$J_{CO_2} = \dfrac{c_{CO_2}^{ta} - c_{CO_2}^{chl}}{r_{CO_2}^{total}} = \dfrac{(13 \times 10^{-3} \text{ mol m}^{-3}) - (9 \times 10^{-3} \text{ mol m}^{-3})}{(642 \text{ s m}^{-1})}$$

$$= 6.2 \times 10^{-6} \text{ mol m}^{-2} \text{ s}^{-1}$$

(d) We use the same equations as in (c) and find

$$r_{CO_2}^{total} = 60 \text{ s m}^{-1} + 250 \text{ s m}^{-1} + 150 \text{ s m}^{-1} + \left(\dfrac{1}{1 - 0.05}\right)(100 \text{ s m}^{-1}) = 565 \text{ s m}^{-1}$$

$$J_{CO_2} = \dfrac{(13 \times 10^{-3} \text{ mol m}^{-3}) - (7 \times 10^{-3} \text{ mol m}^{-3})}{(565 \text{ s m}^{-1})} = 10.6 \times 10^{-6} \text{ mol m}^{-2} \text{ s}^{-1}$$

(e) We need to determine $J_{CO_2}^{r+pr}$, e.g., using Equation 8.23 and J_{CO_2} from (d) and knowing that $J_{CO_2}^{r+pr} = 0.05 J_{CO_2}^{ps}$:

$$J_{CO_2} = J_{CO_2}^{ps} - J_{CO_2}^{r+pr} = 20 J_{CO_2}^{r+pr} - J_{CO_2}^{r+pr} = 19 J_{CO_2}^{r+pr}$$

$$J_{CO_2}^{r+pr} = J_{CO_2}/19 = (10.6 \times 10^{-6} \text{ mol m}^{-2} \text{ s}^{-1})/(19) = 0.56 \times 10^{-6} \text{ mol m}^{-2} \text{ s}^{-1}$$

We also note that 10 μmol mol^{-1} CO_2 is about 0.41 mmol m^{-3} (Table 8.2). Thus, rearranging Equation 8.25 and noting that in the steady state for a confined bag J_{CO_2} is zero, we find

$$c_{CO_2}^{chl} = c_{CO_2}^{ta} - (J_{CO_2}^{r+pr})(r_{CO_2}^{chl})$$

$$= (0.41 \text{ mmol m}^{-3}) - (0.56 \times 10^{-3} \text{ mmol m}^{-2} \text{ s}^{-1})(100 \text{ s m}^{-1})$$

$$= 0.35 \text{ mmol m}^{-3} = 0.35 \text{ } \mu\text{M}$$

(f) By considering Figure 8.12 and the discussion on p. 448, we can obtain a relation for the CO_2 efflux at night:

$$c_{CO_2}^{mito} - c_{CO_2}^{ta} = J_{CO_2}^{i} (r_{CO_2}^{i} + r_{CO_2}^{mes} + r_{CO_2}^{leaf_l} + r_{CO_2}^{bl_l})$$

$$c_{CO_2}^{mito} = (0.56 \times 10^{-3} \text{ mmol m}^{-2} \text{ s}^{-1})(500 \text{ s m}^{-1} + 150 \text{ s m}^{-1} + 250 \text{ s m}^{-1}$$

$$+ 60 \text{ s m}^{-1}) + 13 \text{ mmol m}^{-3} = 14 \text{ mmol m}^{-3} = 14 \text{ } \mu\text{M}$$

If stomatal closure causes $r_{CO_2}^{leaf}$ to become 5 000 s m^{-1}, then

$$c_{CO_2}^{mito} = (0.56 \times 10^{-3} \text{ mmol m}^{-2} \text{ s}^{-1})(500 \text{ s m}^{-1} + 150 \text{ s m}^{-1} + 5\,000 \text{ s m}^{-1}$$

$$+ 60 \text{ s m}^{-1}) + 13 \text{ mmol m}^{-3} = 16 \text{ mmol m}^{-3} = 16 \text{ } \mu M$$

8.5 **(a)** We can use Equation 8.16 to find g_{wv}^{total}:

$$g_{wv}^{total} = \frac{J_{wv}}{c_{wv}^{leaf} - c_{wv}^{air}} = \frac{(5.0 \times 10^{-3} \text{ mol m}^{-2} \text{ s}^{-1})}{(2.20 \text{ mol m}^{-3}) - (0.32)(1.69 \text{ mol m}^{-3})}$$

$$= 3.0 \times 10^{-3} \text{ m s}^{-1} = 3.0 \text{ mm s}^{-1}$$

For conductances in series, the reciprocal of the total conductance ($1/g^{total}$) is the sum of the reciprocals of the individual conductances (see p. 413), so

$$\frac{1}{g_{wv}^{st}} = \frac{1}{g_{wv}^{total}} - \frac{1}{g_{wv}^{bl}} - \frac{1}{g_{wv}^{ias}}$$

$$= \frac{1}{(3.0 \text{ mm s}^{-1})} - \frac{1}{(15 \text{ mm s}^{-1})} - \frac{1}{(30 \text{ mm s}^{-1})} = 0.23 \text{ s mm}^{-1}$$

$$g_{wv}^{st} = \frac{1}{(0.23 \text{ s mm}^{-1})} = 4.3 \text{ mm s}^{-1}$$

(b) Stomatal conductance will decrease fourfold, so total conductance is

$$g_{wv}^{total} = \frac{1}{\dfrac{1}{(15 \text{ mm s}^{-1})} + \dfrac{1}{(0.25)(4.3 \text{ mm s}^{-1})} + \dfrac{1}{(30 \text{ mm s}^{-1})}} = 1.0 \text{ mm s}^{-1}$$

Because g_{wv}^{total} is 1/3 of that in (a), transpiration will be reduced threefold (to 1.7 mmol m^{-2} s^{-1}).

(c) Leaf temperature will increase because of the decrease in latent heat loss accompanying transpiration.

(d) First we find the original gas-phase CO$_2$ conductance (p. 450):

$$g_{CO_2}^{gas} = \frac{g_{wv}^{total}}{1.60} = \frac{(3.0 \text{ mm s}^{-1})}{(1.60)} = 1.88 \text{ mm s}^{-1}$$

The new gas-phase conductance is

$$g_{CO_2}^{gas} = \frac{g_{wv}^{total}}{1.60} = \frac{(1.0 \text{ mm s}^{-1})}{(1.60)} = 0.63 \text{ mm s}^{-1}$$

If we use the original gas-phase conductance, we can find the liquid-phase conductance to CO$_2$, which is unchanged:

$$g_{CO_2}^{liquid} = \cfrac{1}{\cfrac{1}{g_{CO_2}^{total}} - \cfrac{1}{g_{CO_2}^{gas}}} = \cfrac{1}{\cfrac{1}{(0.70 \text{ mm s}^{-1})} - \cfrac{1}{(1.88 \text{ mm s}^{-1})}} = 1.12 \text{ mm s}^{-1}$$

$$\frac{g_{CO_2}^{total}{}_{new}}{g_{CO_2}^{total}{}_{original}} = \cfrac{\cfrac{g_{CO_2}^{gas}{}_{new} \, g_{CO_2}^{liq}}{g_{CO_2}^{gas}{}_{new} + g_{CO_2}^{liq}}}{0.70 \text{ mm s}^{-1}} = \cfrac{\cfrac{(0.63 \text{ mm s}^{-1})(1.12 \text{ mm s}^{-1})}{(0.63 \text{ mm s}^{-1}) + (1.12 \text{ mm s}^{-1})}}{(0.70 \text{ mm s}^{-1})}$$

$$= \frac{(0.40 \text{ mm s}^{-1})}{(0.70 \text{ mm s}^{-1})} = 0.57$$

and hence photosynthesis is decreased 43%.

By Equation 8.29, WUE is proportional to photosynthesis/transpiration, so

$$\frac{WUE^{new}}{WUE^{original}} = \frac{(0.57)/(0.33)}{(1.00)/(1.00)} = 1.73$$

and WUE is increased 73%.

(e) Given that $\delta^{bl} = 4.0 \sqrt{\dfrac{l}{v}}$ (Eq. 7.8), a fourfold increase in windspeed will change δ^{bl} by a factor of $\sqrt{\frac{1}{4}} = \frac{1}{2}$. Because g_{wv}^{bl} equals $\dfrac{D_{wv}}{\delta^{bl}}$ (Eq. 8.3), g_{wv}^{bl} will be doubled and so becomes (15 mm s^{-1})(2) or 30 mm s^{-1}. Therefore, total water vapor conductance becomes

$$g_{wv}^{total} = \cfrac{1}{\cfrac{1}{g_{wv}^{bl}} + \cfrac{1}{g_{wv}^{st}} + \cfrac{1}{g_{wv}^{ias}}} = \cfrac{1}{\cfrac{1}{(30 \text{ mm s}^{-1})} + \cfrac{1}{(4.3 \text{ mm s}^{-1})} + \cfrac{1}{(30 \text{ mm s}^{-1})}}$$

$$= 3.3 \text{ mm s}^{-1}$$

Because g_{wv}^{total} was originally 3.0 mm s^{-1}, J_{wv} will also increase by 10%.

(f) Because δ^{bl} is halved, J_H^C initially doubles; T^{leaf} becomes closer to T^{ta}. If T^{leaf} is above T^{ta}, as generally occurs during the day, the decrease in T^{leaf} in response to a higher wind will reduce c_{wv}^e (which is nearly at the saturation value). If the accompanying fractional reduction in $c_{wv}^e - c_{wv}^{ta}$ is more than the fractional increase in g_{wv}^{total}, the increase in wind speed will lead to a *decrease* in transpiration, because $J_{wv} = g_{wv}^{total} (c_{wv}^e - c_{wv}^{ta})$ by Equations 8.15 and 8.16.

CHAPTER 9

9.1 (a) We must convert the transpiration value from mm H$_2$O h^{-1} to the more common units of J_{wv}, mmol m^{-2} s^{-1}, noting that 1 m^3 H$_2$O corresponds to 996 kg at 30°C and 18.0 g corresponds to 1 mol:

$$1 \text{ mm h}^{-1} = \frac{10^{-3} \text{ m}^3 \text{ m}^{-2}}{3\,600 \text{ s}} = \frac{996 \text{ g m}^{-2}}{3\,600 \text{ s}} = \frac{(996/18.0) \text{ mol m}^{-2}}{3\,600 \text{ s}} = 0.0154 \text{ mol m}^{-2} \text{ s}^{-1}$$

The change in water vapor concentration during peak transpiration is

$$\Delta c_{wv}^{ta} = (c_{wv}^*)\left(\frac{\Delta RH}{100}\right) = (1.69 \text{ mol m}^{-3})\left(\frac{20}{100}\right) = 0.34 \text{ mol m}^{-3}$$

(b) Using Equation 9.4, r_{wv}^{ta} during peak transpiration is

$$r_{wv}^{ta} = \frac{\Delta c_{wv}^{ta}}{J_{wv}} = \frac{(0.34 \text{ mol m}^{-3})}{(15.4 \times 10^{-3} \text{ mol m}^{-2} \text{ s}^{-1})} = 22 \text{ s m}^{-1}$$

Because the value of K_{wv} at night is half the daytime value, by Equation 9.4 r_{wv}^{ta} at night must be twice the daytime r_{wv}^{ta}:

$$r_{wv}^{ta} = \frac{\Delta z}{K_{wv}} = (22 \text{ s m}^{-1})(2) = 44 \text{ s m}^{-1}$$

(c) At night, J_{wv} is only 10% of the peak value and thus is $(15.4 \text{ mmol m}^{-2} \text{ s}^{-1})$ (0.10), or 1.54 mmol m^{-2} s^{-1}. Therefore, using Equation 9.4 we find

$$\Delta c_{wv}^{ta} = J_{wv} r_{wv}^{ta} = (1.54 \text{ mmol m}^{-2} \text{ s}^{-1})(44 \text{ s m}^{-1}) = 68 \text{ mmol m}^{-3}$$

At an air temperature of 20 °C, the drop in relative humidity is

$$100 \frac{\Delta c_{wv}^{ta}}{c_{wv}^*} = (100) \frac{(0.068 \text{ mol m}^{-3})}{(0.96 \text{ mol m}^{-3})} = 7\%$$

(d) Because we know J_{wv} from (a), we can calculate J_{CO_2}:

$$J_{CO_2} = \frac{J_{wv}}{-700} = \frac{(15.4 \text{ mmol m}^{-2} \text{ s}^{-1})}{(-700)} = -0.022 \text{ mmol m}^{-2} \text{ s}^{-1} = -22 \text{ }\mu\text{mol m}^{-2} \text{ s}^{-1}$$

Using Equation 9.4, we find

$$\Delta c_{CO_2}^{ta} = J_{CO_2} r_{CO_2}^{ta} = (-0.022 \text{ mmol m}^{-2} \text{ s}^{-1})(22 \text{ s m}^{-1}) = -0.48 \text{ mmol m}^{-3}$$

(e) Rearranging Equation 9.4, we find

$$K_{CO_2} = \frac{\Delta z}{r_{CO_2}^{ta}} = \frac{(30 \text{ m})}{(44 \text{ s m}^{-1})} = 0.68 \text{ m}^2 \text{ s}^{-1}$$

(f) Using Equation 1.6, we can calculate the time for eddy diffusion at night:

$$t_e = \frac{x_e^2}{4 K_{wv}} = \frac{(1 \text{ m})^2}{(4)(0.68 \text{ m}^2 \text{ s}^{-1})} = 0.37 \text{ s}$$

Also using Equation 1.6, we can calculate the time for ordinary diffusion:

$$t_e = \frac{x_e^2}{4D_{wv}} = \frac{(1\text{ m})^2}{(4)(2.4 \times 10^{-5}\text{ m}^2\text{ s}^{-1})} = 10\ 400\text{ s} = 2.9\text{ h}$$

9.2 (a) We use Equation 9.5 to calculate the cumulative leaf areas at different incident light levels, where $F = (\ln J_0/J)/K$:

For $J_0 = 2\ 000\ \mu\text{mol m}^{-2}\text{ s}^{-1}$, $F = \dfrac{\ln(2\ 000/8)}{(0.7)} = 7.9$ (i.e., essentially none of the leaves are then below light compensation).

For $J_0 = 200\ \mu\text{mol m}^{-2}\text{ s}^{-1}$, $F = \dfrac{\ln(200/8)}{(0.7)} = 4.6$.

For $J_0 = 20\ \mu\text{mol m}^{-2}\text{ s}^{-1}$, $F = \dfrac{\ln(20/8)}{(0.7)} = 1.3$.

For $J_0 = 0\ \mu\text{mol m}^{-2}\text{ s}^{-1}$, all of the leaves are below light compensation (i.e., $F = 0.0$).

(b) We can calculate $J_{CO_2}^{soil}$ coming up from the soil as follows:

$$J_{CO_2}^{soil} = \frac{J_{wv}^{soil}}{200} = \frac{(0.6\text{ mmol m}^{-2}\text{ s}^{-1})}{(200)} = 0.0030\text{ mmol m}^{-2}\text{ s}^{-1} = 3.0\ \mu\text{mol m}^{-2}\text{ s}^{-1}$$

We can calculate $J_{CO_2}^{total}$ (which includes CO_2 from above the canopy and from the soil) by conversion of the net photosynthetic rate (see Table 8.2):

$$J_{CO_2}^{total} = (20\text{ kg ha}^{-1}\text{ h}^{-1})\left(0.92\ \frac{\mu\text{mol m}^{-2}\text{ s}^{-1}}{\text{kg ha}^{-1}\text{ h}^{-1}}\right) = 18.4\ \mu\text{mol m}^{-2}\text{ s}^{-1}$$

The flux from above the canopy is then $18.4\ \mu\text{mol m}^{-2}\text{ s}^{-1} - 3.0\ \mu\text{mol m}^{-2}\text{ s}^{-1}$, or 15.4 $\mu\text{mol m}^{-2}\text{ s}^{-1}$.

(c) We are given that 10% of the water comes from the soil ($J_{wv}^{soil} = 0.6\text{ mmol m}^{-2}\text{ s}^{-1}$), so J_{wv} above the canopy is $6.0\text{ mmol m}^{-2}\text{ s}^{-1}$. From (b) we calculated J_{CO_2} above the canopy to be $-15.4\ \mu\text{mol m}^{-2}\text{ s}^{-1}$, so

$$\frac{J_{wv}}{J_{CO_2}} = \left|\frac{(6\ 000\ \mu\text{mol m}^{-2}\text{ s}^{-1})}{(-15.4\ \mu\text{mol m}^{-2}\text{ s}^{-1})}\right| = 390\ H_2O/CO_2$$

(d) The maximum upward flux of CO_2 will occur for the level at the light compensation point (leaves below this level have a net evolution of CO_2). At $200\ \mu\text{mol m}^{-2}\text{ s}^{-1}$ incident on the canopy, light compensation occurs at $F = 4.6$ [calculated in (a)]. Because the LAI is 8.0, F increases by 1.0 every 2.0 m (16 m/8.0). Therefore, an F of 4.6 occurs at 9.2 m from the top of the canopy, or 6.8 m above the ground.

(e) The minimum in $c_{CO_2}^{ta}$ occurs at the level corresponding to the cumulative uptake of all the CO_2 coming down into the canopy ($15.4\ \mu\text{mol m}^{-2}\text{ s}^{-1}$); net uptake of CO_2 occurs only between 6.8 m and 16 m in the canopy, and the magnitude of this uptake is 20.4 μmol m^{-2} s^{-1} (i.e., 15.4 + 5.0, the latter from below 6.8 m). We will divide the canopy into 2-m thick layers across which F increases by 1.0, and we will let x equal that part of the total flux taken up by the first layer of leaves. Because PPFD is halved every 2 m [by Eq. 9.5, ln (J_0/J) = kF = (0.7)(1.0) = 0.7, or J_0/J = 2.0] and because J_{CO_2} is proportional to PPFD by supposition, CO_2 uptake is also halved every 2 m. Therefore, we obtain

$$\text{total uptake} = 20.4 \; \mu\text{mol m}^{-2} \text{ s}^{-1} = x + \tfrac{1}{2}x + (\tfrac{1}{2})^2x + (\tfrac{1}{2})^3x + \left(\frac{8.0 - 6.8 \text{ m}}{2.0 \text{ m}}\right)(\tfrac{1}{2})^4x$$

$$= 1.91x$$

$$x = 10.7 \; \mu\text{mol m}^{-2} \text{ s}^{-1}$$

The first 2-m layer thus takes up 10.7 μmol m^{-2} s^{-1}, and the second layer one-half of this, or 5.3 μmol m^{-2} s^{-1}. Thus, at 4 m from the top of the canopy, the total uptake is 16.0 μmol m^{-2} s^{-1}, which is slightly more than the flux of CO_2 from the air above the canopy. Therefore, $c_{CO_2}^{ta}$ will be minimal at slightly above 12 m from the ground.

9.3 (a) The mean speed of the fluid in the xylem element is

$$J_V = \frac{\text{volume flow rate per tube}}{\text{tube area}} = \frac{(20 \text{ mm}^3 \text{ h}^{-1})}{(0.004 \text{ m}^2)(3\ 600 \text{ s h}^{-1})} = 1.4 \text{ mm s}^{-1}$$

(b) Rearranging Poiseuille's law (Eq. 9.11b), we get

$$\frac{\Delta P}{\Delta x} = \frac{-J_V 8\eta}{r^2} = \frac{-J_V 8\eta}{A/\pi} = \frac{-(1.4 \text{ mm s}^{-1})(8)(1.002 \times 10^{-3} \text{ Pa s})}{(0.004 \text{ mm}^2/\pi)}$$

$$= -8.8 \text{ Pa mm}^{-1} = -0.0088 \text{ MPa m}^{-1}$$

(c) Again using Poiseuille's law, we find

$$J_V = -\frac{r^2}{8\eta}\frac{\Delta P}{\Delta x} = \frac{-(10 \times 10^{-9} \text{ m})^2}{(8)(1.002 \times 10^{-3} \text{ Pa s})}(-0.0088 \times 10^6 \text{ Pa m}^{-1})$$

$$= 1.1 \times 10^{-10} \text{ m s}^{-1}$$

9.4 (a) We can rearrange Equation 1.6 to find the time required for diffusion:

$$\text{For 10 mm: } t = \frac{x_e^2}{4D_{\text{sucrose}}} = \frac{(1 \times 10^{-2} \text{ m})^2}{(4)(0.3 \times 10^{-9} \text{ m}^2 \text{ s}^{-1})} = 8.3 \times 10^4 \text{ s} = 23.1 \text{ h}$$

$$\text{For 1 m: } \quad t = \frac{x_e^2}{4D_{\text{sucrose}}} = \frac{(1 \text{ m})^2}{(4)(0.3 \times 10^{-9} \text{ m}^2 \text{ s}^{-1})} = 8.3 \times 10^8 \text{ s} = 26 \text{ years}$$

(b) We use Poiseuille's law (Equation 9.11b) to find

$$J_v = -\frac{r^2}{8\eta}\frac{\Delta P}{\Delta x} = -\frac{(10 \times 10^{-6} \text{ m})^2}{(8)(1.5 \times 10^{-3} \text{ Pa s})}(-0.02 \times 10^6 \text{ Pa m}^{-1})$$

$$= 1.7 \times 10^{-4} \text{ m s}^{-1} = 0.17 \text{ mm s}^{-1} \qquad (0.6 \text{ m h}^{-1})$$

(c) Because J_v is the mean velocity, to travel 10 mm takes [10 mm/(0.17 mm s^{-1})] or 59 s and to travel 1 m takes [1 m/(0.17 $\times$ 10^{-3} m s^{-1})] or 5 900 s, which is 1.6 h, considerably less than that calculated for diffusion in (a). Also note that the distance traveled in the phloem vessel is proportional to time rather than to the square root of time, as it is for diffusion (see Eq. 1.6).

(d) A vertical sieve tube has a static gradient in the absence of flow and due to gravity of -0.01 MPa m^{-1}. The gradient leading to flow would be either -0.01 MPa m^{-1} (upward) or -0.03 MPa m^{-1} (downward). Thus, J_v is either half that in (b), namely 0.3 m h^{-1} upward, or 0.9 m h^{-1} downward.

9.5 (a) If we ignore Π^{soil}, then Ψ^{soil} equals P^{soil}, which equals $-\sigma\left(\dfrac{1}{r_1} + \dfrac{1}{r_2}\right)$ by Equation 9.5. For hemispherical interfaces r_1 equals r_2, so Ψ^{soil} equals $\dfrac{-2\sigma}{r}$. Therefore, for the wet soil we obtain

$$r = \frac{-2\sigma}{\Psi^{\text{soil}}} = \frac{-(2)(0.0728 \text{ Pa m})}{(-0.01 \times 10^6 \text{ Pa})} = 1.5 \times 10^{-6} \text{ m} = 15 \text{ }\mu\text{m}$$

For the dry soil, we obtain

$$r = \frac{-2\sigma}{\Psi^{\text{soil}}} = \frac{-(2)(0.0728 \text{ Pa m})}{(-1.4 \times 10^6 \text{ Pa})} = 1.0 \times 10^{-7} \text{ m} = 0.10 \text{ }\mu\text{m}$$

(b) First, we must find J_v of the root from the following relationship based on Equation 9.12:

$$J_v^{\text{root}} = J_v^{\text{xylem}}\frac{A^{\text{xylem}}}{A^{\text{root}}} = (2 \times 10^{-3} \text{ m s}^{-1})\left(\frac{1}{10^5}\right) = 2 \times 10^{-8} \text{ m s}^{-1}$$

Then assuming cylindrical symmetry we rearrange Equation 9.8 to find the drop in hydrostatic pressure in the wet soil:

$$\Delta P = P_a - P_b = \frac{J_v r \ln (r_a/r_b)}{L^{\text{soil}}}$$

$$= \frac{(2 \times 10^{-8} \text{ m s}^{-1})(1.5 \times 10^{-3} \text{ m}) \ln \left(\dfrac{8 \text{ mm} + 1.5 \text{ mm}}{1.5 \text{ mm}}\right)}{(1 \times 10^{-11} \text{ m}^2 \text{ s}^{-1} \text{ Pa}^{-1})} = 5.5 \text{ Pa}$$

(c) Again, by rearranging Equation 9.8 we find the drop in hydrostatic pressure to be

$$\Delta P = \frac{J_V r \ln (r_a/r_b)}{L^{\text{soil}}} = \frac{(2 \times 10^{-8} \text{ m s}^{-1})(1.5 \times 10^{-3} \text{ m}) \ln \left(\dfrac{1.5 \text{ mm}}{1.5 \text{ mm} - 0.5 \text{ mm}} \right)}{(2 \times 10^{-16} \text{ m}^2 \text{ s}^{-1} \text{ Pa}^{-1})}$$

$$= 6.1 \times 10^4 \text{ Pa} = 0.061 \text{ MPa}$$

(d) First, we should determine the water potential in the root xylem:

$$\Psi^{\text{root xylem}} = \Psi^{\text{soil}} - \Delta \Psi^{\text{soil}} - \Delta \Psi^{\text{root}}$$

$$= (-0.01 \text{ MPa}) - (5.5 \times 10^{-6} \text{ MPa}) - (0.061 \text{ MPa}) = -0.071 \text{ MPa}$$

$$= P^{\text{root xylem}} - \Pi^{\text{root xylem}} + \rho_w g h = P^{\text{root xylem}}$$

In the leaf 3 m above the ground, Ψ^{leaf} (-0.20 MPa) is $P^{\text{leaf}} - \Pi^{\text{leaf}} + \rho_w g h$, which is $P^{\text{leaf}} - 0 + (0.01 \text{ MPa m}^{-1})(3 \text{ m})$ or $P^{\text{leaf}} + 0.03$ MPa, so P^{leaf} is -0.20 MPa $- 0.03$ MPa, or -0.23 MPa. Thus, the gradient is

$$\frac{\Delta P}{\Delta x} = \frac{P^{\text{leaf}} - P^{\text{root xylem}}}{\Delta x} = \frac{(-0.23 \text{ MPa}) - (0.07 \text{ MPa})}{(3 \text{ m})} = -0.053 \text{ MPa m}^{-1}$$

(e) We must first find J_V of the leaf based on Equation 9.12:

$$J_V^{\text{leaf}} = J_V^{\text{xylem}} \frac{A^{\text{xylem}}}{A^{\text{root}}} \frac{A^{\text{root}}}{A^{\text{leaf}}} = (2 \times 10^{-3} \text{ m s}^{-1})(10^{-5})(5) = 1.0 \times 10^{-7} \text{ m s}^{-1}$$

We must then determine J_V of the mesophyll:

$$J_V^{\text{mes}} = J_V^{\text{leaf}} \frac{A^{\text{leaf}}}{A^{\text{mes}}} = (1.0 \times 10^{-7} \text{ m s}^{-1}) \left(\frac{1}{20} \right) = 5 \times 10^{-9} \text{ m s}^{-1}$$

Now we apply Poiseuille's law (Eq. 9.11b) to find the drop in hydrostatic pressure that could account for the rate of transpiration for the plant in the wet soil:

$$\Delta P = \frac{-J_V 8\eta \, \Delta x}{r^2} = \frac{-(5 \times 10^{-9} \text{ m s}^{-1})(8)(1.002 \times 10^{-3} \text{ Pa s})(1 \times 10^{-6} \text{ m})}{(5 \times 10^{-9} \text{ m})^2}$$

$$= -1.6 \text{ Pa, which is very small}$$

(f) If we use Equation 2.21, we find that at 99% relative humidity

$$\Psi^{\text{ta}} = \frac{RT}{V_w} \ln \frac{RH}{100} = (135 \text{ MPa}) \ln \left(\frac{99}{100} \right) = -1.4 \text{ MPa}$$

Because Ψ^{ta} then equals Ψ^{soil}, water will not move from the dry soil to the air through the plant, so J_V^{xylem} is zero.

9.6 (a) The total leaf area is equal to the leaf area index times the projected area of one leaf layer, which is $(6)(\pi)(3 \text{ m})^2$ or 170 m^2. Noting that 1 mol of H_2O is 18 g, which occupies $18 \times$

10^{-6} m^3, the transpiration rate is:

$$\text{transpiration rate} = J_{wv} \, (18 \times 10^{-6} \text{ m}^3 \text{ mol}^{-1}) A^{\text{leaf}}$$

$$= (1 \times 10^{-3} \text{ mol m}^{-2} \text{ s}^{-1})(18 \times 10^{-6} \text{ m}^3 \text{ mol}^{-1})(170 \text{ m}^2)$$

$$= 3.1 \times 10^{-6} \text{ m}^3 \text{ s}^{-1}$$

(b) We must first determine the amount of water available for transpiration:

$$(1.0 \times 10^{-3} \text{ m} - 0.8 \times 10^{-3} \text{ m})(\pi)(3 \text{ m})^2 = 5.7 \times 10^{-3} \text{ m}^3$$

We can now calculate how long this will last:

$$\frac{(5.7 \times 10^{-3} \text{ m}^3)}{(3.1 \times 10^{-6} \text{ m}^3 \text{ s}^{-1})} = 1\ 840 \text{ s} = 31 \text{ min}$$

(c) We use Equation 9.12 to find R^{trunk} using Ψ^{trunk} of -0.3 MPa at the base and -0.5 MPa at midheight:

$$R^{\text{trunk}} = \frac{\Delta \Psi}{J_{wv} A^{\text{leaf}}} = \frac{2(\Psi^{\text{base}} - \Psi^{\text{mid}})}{(J_{wv} A^{\text{leaf}})} = \frac{2[(-0.3 \text{ MPa}) - (-0.5 \text{ MPa})]}{(3.1 \times 10^{-6} \text{ m}^3 \text{ s}^{-1})}$$

$$= 1.3 \times 10^5 \text{ MPa s m}^{-3}$$

(d) Using Equation 9.13, we obtain

$$\rho = \frac{R^{\text{trunk}} A^{\text{trunk}}}{\Delta x} = \frac{(1.3 \times 10^5 \text{ MPa s m}^{-3})(0.10 \text{ m}^2)(0.05)}{(3 \text{ m})} = 220 \text{ MPa s m}^{-2}$$

Hydraulic conductivity is the inverse of hydraulic resistivity:

$$\frac{1}{\rho} = \frac{1}{(220 \text{ MPa s m}^{-2})} = 4.6 \times 10^{-3} \text{ m}^2 \text{ s}^{-1} \text{ MPa}^{-1}$$

(e) We use Equation 9.14 to find the capacitance:

$$C^{\text{trunk}} = \frac{\Delta V^{\text{trunk}}}{\Delta \Psi^{\text{day}}} = \frac{(0.10 \text{ m}^2)(3 \text{ m})(0.008)}{(-0.1 \text{ MPa}) - (-0.5 \text{ MPa})} = 6.0 \times 10^{-3} \text{ m}^3 \text{ MPa}^{-1}$$

We use Equation 9.15 to calculate the time constant for water release from the trunk:

$$t = R^{\text{trunk}} C^{\text{trunk}} = (1.3 \times 10^5 \text{ MPa s m}^{-3})(6 \times 10^{-3} \text{ m}^3 \text{ MPa}^{-1}) = 780 \text{ s} = 13 \text{ min}$$

Index

Page numbers in **boldface** refer to figures or structural formulas.

617